BRAIN CORTICOSTEROID RECEPTORS

STUDIES ON THE MECHANISM, FUNCTION, AND NEUROTOXICITY OF CORTICOSTEROID ACTION

ANNALS OF THE NEW YORK ACADEMY OF SCIENCES

Volume 746

BRAIN CORTICOSTEROID RECEPTORS

STUDIES ON THE MECHANISM, FUNCTION, AND NEUROTOXICITY OF CORTICOSTEROID ACTION

Edited by E. Ronald de Kloet, Efrain C. Azmitia, and Philip W. Landfield

The New York Academy of Sciences
New York, New York
1994

Cover: *Glucocorticoid receptor (red) and mineralocorticoid receptor (green) in cell nuclei of rat hippocampal CA1 neurons, visualized by dual labeling immunocytochemistry and confocal laser microscopy. Yellow indicates co-localization of both receptor types. Methodology is described in: "Confocal Scanning Laser Microscopy of Steroid Receptors in Brain,"* Methods in Neurosciences, *Vol. 22: 162–174 (1994). Courtesy of Bas van Steensel, Erica P. Van Binnendijk, and Roel van Driel, University of Amsterdam, E.C. Slater Institute, The Netherlands.*

Library of Congress Cataloging-in-Publication Data

Brain corticosteroid receptors : studies on the mechanism, function, and neurotoxicity of corticosteroid action / edited by E. Ronald de Kloet, Efrain C. Azmitia, and Philip W. Landfield.
p. cm. — (Annals of the New York Academy of Sciences, ISSN 0077-8923 : v. 746)
Includes bibliographical references and index.
ISBN 0-89766-907-X (cloth : alk. paper). — ISBN 0-89766-908-8 (pbk. : alk. paper)
1. Adrenocortical hormones—Receptors—Congresses. 2. Hypothalamus-pituitary-adrenal axis—Congresses. 3. Neurohormones—Receptors—Congresses. 4. Neurotoxicology—Congresses. 5. Transcription factors—Congresses. I. de Kloet, E. R. II. Azmitia, Efrain C. III. Landfield, Philip W. IV. New York Academy of Sciences. V. Series.
[DNLM: 1. Receptors, Steroid—physiology—congresses. 2. Brain—physiology—congresses. W1 AN626YL v. 746 1994]
Q11.N5 vol. 746
[QP572.A3]
500 s—dc20
[612.8]
DNLM/DLC
for Library of Congress 94-38769
CIP

&/PCP
Printed in the United States of America
ISBN 0-89766-907-X (cloth)
ISBN 0-89766-908-8 (paper)
ISSN 0077-8923

ANNALS OF THE NEW YORK ACADEMY OF SCIENCES

Volume 746
November 30, 1994

BRAIN CORTICOSTEROID RECEPTORS

STUDIES ON THE MECHANISM, FUNCTION, AND NEUROTOXICITY OF CORTICOSTEROID ACTION[a]

Editors and Conference Chairs

E. RONALD DE KLOET, EFRAIN C. AZMITIA, AND PHILIP W. LANDFIELD

CONTENTS

[a]This volume contains the papers from a conference entitled *Brain Corticosteroid Receptors: Studies on the Mechanism, Function, and Neurotoxicity of Corticosteroid Action,* which was held by the New York Academy of Sciences in Arlington, Virginia on March 2–5, 1994.

Part III. Molecular and Cellular Actions of Corticosteroids

Part IV. Corticosteroids in Development, Aging, and Neurodegeneration

Part V. Neurological and Psychiatric Disorders

Poster Papers

Financial assistance was received from:

Major funder

- NATIONAL INSTITUTE ON AGING/NIH

Supporter

- ELI-LILLY NEDERLAND B.V.
- THE HEALTH FOUNDATION

Contributors

- GLAXO RESEARCH INSTITUTE
- HOECHST-ROUSSEL PHARMACEUTICALS INC.
- MILES INC.
- ORGANON INTERNATIONAL B.V.
- THE UPJOHN COMPANY
- R. W. JOHNSON PHARMACEUTICAL RESEARCH INSTITUTE
- SOLVAY-DUPHAR B.V.

Preface

Late in the day of November 13, 1991 at the Neuroscience meeting in New Orleans, the editors concluded that there were at least three good reasons to hold an International Conference on Brain Corticosteroid Receptors.

First, the field has experienced substantial development. Since the discovery about 25 years ago by Bruce McEwen and colleagues that the brain contains corticosteroid receptors, research in this area has continued to expand. The field has now reached, we believe, a critical mass needed to sustain an international conference integrating multiple levels of basic and clinical research.

Second, the degree of conceptual understanding of underlying mechanisms has reached a sufficiently advanced level such that new insights and synthesis might be possible on the regulation, function and pathological consequences of dysregulation of brain corticosteroid receptors.

Third, the experimental approaches used to study brain corticosteroid receptor function at the molecular, cellular and organismic levels of biological organization have become highly sophisticated. This would ensure a lively multidisciplinary conference.

The eventual result was an exciting Conference on Brain Corticosteroid Receptors, which was held on March 2–5, 1994 in Arlington, Virginia, sponsored by the New York Academy of Sciences. We thank the participants for sharing an unforgettable experience.

This volume provides a comprehensive analysis of brain corticosteroid receptor function. It contains chapters written by speakers and chairpersons and includes, after each contribution, one of the highlights of the conference: the discussions. We wish the reader much enjoyment and inspiration.

E. Ronald de Kloet
Efrain C. Azmitia
Philip W. Landfield

Corticosteroid Hormones and Signal Specificity

JOHN W. FUNDER[a]

Baker Medical Research Institute
Melbourne, Australia

When the human mineralocorticoid receptor was first cloned by Arriza *et al.*,[1] it was shown to be expressed at high levels in the hippocampus, and to have indistinguishably high affinity for aldosterone, corticosterone and cortisol. That renal mineralocorticoid receptors and 'corticosterone-prefering sites' in the rat hippocampus have identical affinity for a range of steroids had been demonstrated some years before;[2] taken together, these studies posed two questions. The first of these—the function and roles of 'mineralocorticoid receptors' in nonepithelial tissues such as hippocampus—is the principal subject of this paper. The second, given the much higher circulating levels of cortisol/corticosterone, is how aldosterone can occupy and activate mineralocorticoid receptors in Na^+ transporting epithelia.

The second of these questions has been answered, at least in part, by the demonstration of the operation of the enzyme 11β-hydroxysteroid dehydrogenase in tissues such as kidney and colon.[3,4] Subsequent studies exploring the physiological roles of the enzyme in the kidney showed that mineralocorticoid specificity resides neither with the receptor nor the hormone response element. In these studies adrenalectomized rats were given carbonexolone, to block the action of 11β-hydroxysteroid dehydrogenase, and the urinary electrolyte effects of corticosterone, or the selective glucocorticoid agonist RU28362, determined.[5]

Under these conditions both RU28362 and corticosterone produced an effect on urinary electrolytes indistinguishable from that of aldosterone. In addition, the effect of corticosterone was not blocked by either RU486, the glucocorticoid receptor antagonist, or RU28318, the selective mineralocorticoid receptor antagonist, when either were given alone; given together, however, they completely reversed the urinary electrolyte effects of corticosterone.

We have interpreted these data as follows. First, in epithelial tissues mineralocorticoid receptors are 'protected' by 11β-hydroxysteroid dehydrogenase. Second, corticosterone is a mineralocorticoid receptor agonist, in that it can product Na^+ retention when glucocorticoid receptors are blocked. Third, glucocorticoid receptors also appear able to produce a benchmark mineralocorticoid effect on urinary electrolytes, occupied either by RU28362 or corticosterone in the presence of RU28318. Finally, given the apparently indistinguishable effects of mineralocorticoid and glucocorticoid receptor activation on urinary electrolyte ratios, it would seem most probable that their effects are mediated via relatively simple hormone response elements. Such response elements, of the kind represented by the consensus sequence GGAACAnnnTGTTCT, are classically unable to discriminate between activated glucocorticoid receptors, mineralocorticoid receptors, progesterone receptors and androgen receptors.

This is clearly not the case in nonepithelial tissues. There are a number of studies which might be chosen to illustrate this difference, for instance on synapsin[6] or serotonin 5HT la receptors;[7,8] for this presentation the example chosen is that of the

[a]Correspondence: Professor John W. Funder, Baker Medical Research Institute, P.O. Box 348, Prahran, Victoria, Australia 3181

central effects of steroids on blood pressure. When aldosterone is infused intracerebroventricularly into rats sensitized by uninephrectomy and salt loading, the blood pressure rises.[9] Infusion of the potent, glucocorticoid receptor selective agonist RU26988 neither mimics nor blocks the response to aldosterone. Infusion of corticosterone alone is without effect on blood pressure, but co-infused with aldosterone corticosterone blocks the effect of aldosterone, at doses similar to those of aldosterone. Taken together, these data suggest that in the central nervous system 'mineralocorticoid receptors' are largely unprotected; that the physiological occupant of these receptors is corticosterone, in the rat, and presumably cortisol in most species; that where aldosterone appears as agonist, corticosterone is antagonist, and vice versa; that activated glucocorticoid receptors do not mimic the effects of activated mineralocorticoid receptors; and finally, that these effects may thus be mediated via composite response elements rather than simple nonselective response elements.

One of the consequences of these high affinity 'mineralocorticoid receptors' in the central nervous system being unprotected is that they are much more likely to be occupied at relatively low glucocorticoid levels than are classical glucocorticoid receptors. Previously it had been suggested that such receptors might be ≥90% occupied,[10] although these calculations have been challenged on the basis of differences in estimating total hippocampal 'mineralocorticoid receptors' concentrations.[11] On the other hand, recent studies from our laboratory[12] indirectly bu cogently argue for a physiological 'always-occupied' role for such receptors in the guinea-pig, and by extrapolation in other species.

The guinea-pig appears to have a series of radical differences in various components of the pituitary-adrenal axis compared with other species. It has long been known that guinea-pigs have very high circulating cortisol levels, and low-affinity, low-capacity transcortin, so that free cortisol levels are relatively even higher. The cause of such high cortisol levels appears to be a C for G substitution in the sequence for ACTH,[13] leading to a Ala-Pro substitution at amino acid 24 in the erstwhile 'invariant' steroidogenic domain, and a superagonist ACTH molecule.[14] Similarly consistent with the high circulating cortisol levels is that guinea-pigs have very low-affinity glucocorticoid receptors, with approximately one-twentieth the affinity for dexamethasone as those in the mouse;[15] this is largely, but not entirely, due to a substitution encoding Trp rather than Cys in the ligand binding domain.[16]

Given the very high circulating cortisol levels in the guinea-pig, but plasma aldosterone levels similar to those in other species,[17] it seemed possible that in the guinea-pig mineralocorticoid receptors may have intrinsically higher affinity for aldosterone than for cortisol, in contrast with man[1] and the rat.[2] In fact, this is not the case; in cytosols from hippocampus, colon, kidney and heart, both guinea-pig and rat mineralocorticoid receptors have the same high affinity for aldosterone (K_d 1–2 nM), and a hierarchy of steroid binding (corticosterone ≥ aldosterone = cortisol) which is identical in the two species examined, and indistinguishable from that reported for recombinant human mineralocorticoid receptors.[1]

There is no way of knowing whether a mutation in one particular component of the guinea-pig pituitary-adrenal axis represents in an evolutionary sense the primordial difference, with the other changes representing subsequent evolutionary events to accommodate to/lessen the impact of the initial change. For the sake of illustration, however, if the primordial change were a mutation in ACTH to produce a superagonist molecule, then the other known changes can be envisaged as a cascade of events caused by this change (high cortisol) or as the result of selection pressure to accommodate to the high cortisol (low-affinity glucocorticoid receptors, high clearance rates, low transcortin affinity and capacity to allow major first pass hepatic clearance). Despite the very high cortisol levels, it appears that there has been no evolutionary drive for the 'mineralocorticoid receptors' in the hippocampus, for example, to have lower affinity than they do in species with much lower circulating glucocorticoid levels. The simplest interpretation of this is that such unprotected receptors may operate not as signal-detecting mechanisms which may be either more or less occupied by agonist, and correspondingly less or more unoccupied, but as receptors which are operationally always occupied.

There would appear to be at least three ways in which such an always-occupied receptor may function as a contributor to signal transduction. It was previously suggested[10] that in such systems the bulk of the variation in concentration of steroid-receptor complexes may reflect variation in receptor levels, rather than those of steroid. If this is the case, then such steroid-receptor complexes may act as negative regulators of glucocorticoid receptor action via composite response elements. Composite response elements are so-called because they contain both a binding site for a transcription factor such as AP-1, and a site which binds both activated glucocorticoid receptors and activated mineralocorticoid receptors; activated glucocorticoid receptors abrogate AP-1-induced transcription, whereas activated mineralocorticoid receptors do not,[18] consistent with a role for always-occupied in the central nervous system as tonic inhibitory dampeners of glucocorticoid action via classical glucocorticoid receptors.[19]

A second possible way in which always-occupied 'mineralocorticoid receptors' in the central nervous system may operate to modulate glucocorticoid action in the central nervous system is by the formation of 'mineralocorticoid receptor'-glucocorticoid receptor heterodimers. Recently the formation of such dimers was suggested by cotransfection studies in a human neuroblastoma cell line, in which at low (<2 nM) cortisol concentrations reporter gene expression with both receptors present was very much higher than the sum of the individual responses, and directly demonstrated by supershift studies.[10] In these studies a 27-mer MMTV-LTR containing a canonical consensus sequence bound activated mineralocorticoid receptors, which were supershifted by antibodies to mineralocorticoid receptors but not to glucocorticoid receptors, and activated glucocorticoid receptors, with the complex supershifted by glucocorticoid receptor antibodies but not by those to mineralocorticoid receptors. When mineralocorticoid and glucocorticoid receptors were added together, the entire labeled complex could be supershifted by either antibody. Finally, in studies in which varying ratios of mineralocorticoid and glucocorticoid receptors were coincubated with the labeled MMTV-LTR fragment, the putative heterodimers appear to have significantly higher affinity for the response element than homodimers of either receptor.

Though both such mechanisms (binding to composite response elements, formation of heterodimers) may play a role in glucocorticoid actions via 'mineralocorticoid receptors' in the central nervous system they are in a sense undimensional, in that they cannot sense fluctuations in glucocorticoid signal. A third possibility is that although such receptors are of such high affinity that they are always occupied, such occupancy under physiological conditions in not uniquely by cortisol. From recent studies on the affinity of progesterone for mineralocorticoid receptors, it appears very likely that in the luteal phase and throughout pregnancy such receptors may be significantly occupied by progesterone, putatively opposing the effects of cortisol occupancy, excitatory or inhibitory.

In the original specificity studies on the cloned human mineralocorticoid receptor progesterone was reported to have 10–15% the affinity for the receptor as aldosterone or cortisol.[1] More recently, a human mineralocorticoid receptor expression system has been used to explore the affinity of progesterone for mineralocorticoid receptors.[21] With 5 nM [^{3}H]aldosterone as tracer, nonradioactive progesterone at 10^{-10} M was reported to lower specific binding of [^{3}H]aldosterone to less than 20%, and to even lower levels at higher concentrations. On the basis of this study, the authors claimed an affinity of less than 10^{-11} M for progesterone binding to mineralocorticoid receptors. When, however, they used 5 nM [^{3}H]progesterone as tracer to label mineralocorticoid receptors, they reported displacement of tracer to less than 70% with 10^{-10} M nonradioactive progesterone, and to less than 30% with 10^{-9} M. Even if 5 nM tracer were to completely saturate the receptors, 10^{-10} M nonradioactive progesterone (*i.e.*, one fiftieth of the tracer concentration) could not be expected to lower binding to less than 98%, and 10^{-9} M to below 86%. It thus appears that the value of $<10^{-11}$ M is not correct for the affinity of progesterone for mineralocorticoid receptors under equilibrium conditions, and that it would thus be appropriate to attempt to define this value. In addition, given the propensity of progesterone to bind nonspecifically to glass and plastic surfaces, which is often reflected in inadvertently low free concentrations, we have used siliconized glassware throughout the study.

In these experiments (Funder and Myles, manuscript in preparation) we have used cytosol extracts from rat and guinea-pig. Both epithelial (colonic) and nonepithelial (hippocampus) sources of mineralocorticoid receptors were used, and both [^{3}H]progesterone and [^{3}H]aldosterone were used as tracer, to monitor any possible nonspecific sequestration of progesterone (either tracer or competitor). Finally, all studies were performed both in the presence of 1 μm (500-fold excess) of RU38486, to block tracer binding to glucocorticoid and any classical progesterone receptors, and in its absence. Under these conditions, in both guinea-pig tissues with either tracer, progesterone and aldosterone have equivalent affinity for mineralocorticoid receptors; in the rat, progesterone appears to have slightly (~3-fold) higher affinity.

These studies thus raise several questions of physiology. First of these is the possible mechanism(s) operating to exclude progesterone from physiological mineralocorticoid receptors in the kidney and other epithelial tissues. Under certain circumstances progesterone can be 21-hydroxylated to deoxycorticosterone in the kidney, which would convert an antagonist into an agonist; although progesterone binds to the species of 11β-hydroxysteroid dehydrogenase that has been cloned and sequenced, it does not appear to be a substrate for the enzyme; neither of these mechanisms would thus appear able to satisfactorily exclude progesterone from renal and colonic receptors, on the presumption that it is so excluded. Secondly, if progesterone does bind to 'mineralocorticoid receptors' in the central nervous system, the extent to which such occupancy is involved in the cardiovascular changes of pregnancy, or the changes in affect that may occur postpartum, remains to be determined.

In summary, it seems most appropriate to explore the roles of central nervous system 'mineralocorticoid receptors' by apposing them to the roles of mineralocorticoid receptors in epithelial tissues. In epithelial tissues mineralocorticoid receptors can be occupied by aldosterone in response to sodium restriction or depletion because they are protected from glucocorticoid receptor occupancy by the enzyme 11β-hydroxysteroid dehydrogenase. In addition, they are equivalently activated by aldosterone and cortisol/corticosterone, and appear to act via simple hormone response elements, in that very similar effects can be procured by activated glucocorticoid receptors. In the central nervous system, in contrast, there appear to be clearly distinct effects of 'mineralocorticoid' and glucocorticoid receptor occupancy; in many studies, corticosterone is agonist and aldosterone antagonist, or vice versa. In the guinea-pig mineralocorticoid receptors appear to be the only limb of the pituitary-adrenal system examined to date which is not substantially different from that in other species; the lack of evolutionary drive towards a low-affinity mineralocorticoid receptor (as is the case for the glucocorticoid receptor) we interpret as evidence that such receptors in the central nervous system are always essentially fully occupied. Such an 'always-occupied' receptor may modulate glucocorticoid receptor activity at composite response elements, or by forming heterodimers with activated glucocorticoid receptors. It is also consistent with being always occupied by a balance of cortisol (antagonist) and progesterone (agonist), a possibility given our recent demonstration that progesterone has similar affinity for mineralocorticoid receptors as aldosterone/corticosterone/cortisol, and consistent with the roles imputed to progesterone in the cardiovascular and behavioral syndromes of the luteal phase, pregnancy and the puerperium.

REFERENCES

1. Arriza, J. L., C. Weinberger, T. Glasser, B. L. Handelin, D. E. Housman & R. M. Evans. 1987. Cloning of human mineralocorticoid receptor complementary DNA; structural and functional kinship with the glucocorticoid receptor. Science **237:** 268–275.
2. Krozowski, Z. S. & J. W. Funder. 1983. Renal mineralocorticoid receptors and hippocampal corticosterone binding species have identical intrinsic steroid specificity. Proc. Natl. Acad. Sci. USA **80:** 6056–6060.
3. Funder, J. W., P. T. Pearce, R. Smith & A. I. Smith. 1988. Mineralocorticoid action: target-tissue specificity is enzyme, not receptor-mediated. Science **242:** 583–585.
4. Edwards, C. R. W., P. M. Stewart, D. Burt, M. A. McIntyre, E. R. de Kloet, L. Brett, W. S. Sutanto & C. Monder. 1988. Localization of 11β-hydroxysteroid dehydrogenase-tissue specific protector of the mineralocorticoid receptor. Lancet **ii:** 986–989.
5. Funder, J. W., P. Pearce, K. Myles & L. P. Roy. 1990. Apparent mineralocorticoid excess, pseudohypoaldosteronism and urinary electrolyte excretion: towards a redefinition of "mineralocorticoid" action. FASEB J. **4:** 3234–3238.
6. Nestler, E. J., T. C. Rainbow, B. S. McEwen & P. Greengard. 1981. Corticosterone increases the level of protein 1, a neuron-specific protein in rat hippocampus. Science **212:** 1162–1164.
7. de Kloet, E. R., D. H. G. Versteeg & G. L. Kovacs. 1983. Aldosterone blocks the response to corticosterone in the raphe-hippocampal serotonin system. Brain Res. **264:** 323–327.
8. de Kloet, E. R., H. Sybesma & J. M. H. M. Reul. 1986. Selective control by corticosterone of serotonin$_1$ receptor capacity in raphe-hippocampal system. Neuroendocrinology **42:** 513–521.
9. Gomez-Sanchez, E. P., M. T. Venkataraman, D. Thwaites & C. Fort. 1990. Intracerebroventricular infusion of corticosterone antagonized ICV-aldosterone hypertension. Am. J. Physiol. **258:** E649–E653.
10. Reul, J. M. H. M. & E. R. de Kloet. 1985. Two receptor systems for corticosterone in rat brain: microdistribution and differential occupation. Endocrinology **117:** 2505–2511.
11. Stephenson, G., M. Hammet, G. Hadaway & J. W. Funder. 1984. Ontogeny of renal mineralocorticoid receptors and urinary electrolyte responses in the rat. Am. J. Physiol. **247:** F665–F671.
12. Myles, K. & J. W. Funder. 1994. Type I (mineralocorticoid) receptors in the guinea-pig. Am. J. Physiol. In press.
13. Keightley, M-C., J. W. Funder & P. J. Fuller. 1991. Molecular cloning and sequencing of a guinea-pig pro-opiomelanocortin cDNA. Mol. Cell. Endocrinol. **82:** 89–98.
14. Smith, A. I., C. A. Wallace, R. L. Moritz, R. J. Simpson, L. B. Schmauk-White,

E. A. WOODCOCK & J. W. FUNDER. 1987. Isolation, amino-acid sequence and superagonist steroidogenic activity of guinea-pig ACTH. J. Endocrinol. **115:** 5–8.
15. KRAFT, N., A. J. HODGSON & J. W. FUNDER. 1979. Glucocorticoid receptor and effector mechanisms: a comparison of the corticosensitive mouse with the corticoresistant guinea-pig. Endocrinology **104:** 344–349.
16. KEIGHTLEY, M-C. & P. J. FULLER. 1994. Unique sequences in the guinea-pig glucocorticoid receptor induce constitutive transactivation and decrease steroid sensitivity. Mol. Endocrinol. In press.
17. WHIPP, G. T., E. M. WINTOUR, J. P. COGHLAN & B. A. SCOGGINS. 1976. Regulation of aldosterone in the guinea-pig; effect on oestrous cycle, pregnancy and sodium status. Aust. J. Exp. Biol. Med. **54:** 1.
18. PEARCE, D. & K. R. YAMAMOTO. 1993. Mineralocorticoid and glucocorticoid receptor activities distinguished by nonreceptor factors at a composite response element. Science **259:** 1161–1165.
19. FUNDER, J. W. 1993. Mineralocorticoids, glucocorticoids, receptors, and response elements. Science **259:** 1132–1133.
20. TRAPP, T., R. RUPPRECHT & F. HOLSBOER. 1994. Coexpression of the mineralocorticoid receptor and glucocorticoid receptor: characterization of transactivation and DNA-binding properties. J. Cell. Biochem. **18B:** 349.
21. RUPPRECHT, R., J. M. H. M. REUL, B. VAN STEENSEL, D. SPENGLER, M. SODER, B. BERNING, F. HOLSBOER & K. DAMM. 1993. Pharmacological and functional characterization of human mineralocorticoid and glucocorticoid receptor ligands. Eur. J. Pharmacol. **247:** 145–154.

DISCUSSION OF THE PAPER

M. F. DALLMAN (*University of California, San Francisco*): In the guinea pig, if adrenalectomized, is there a change in ACTH?

FUNDER: We have not looked at this. These animals are adrenalectomized overnight and used the next day. So they have at least resting normal levels of ACTH. But I cannot answer your question.

DALLMAN: If the MR are always occupied in the guinea pig, then would you not expect any change in the ACTH with adrenalectomy?

FUNDER: We have not fathomed the extent of derangement of the hippocampal hypothalamic-pituitary-adrenal axis in the guinea pig. After fifteen years of picking bits off the problem, these findings with the MR came as a sort of nasty shock. But I think they are actually telling us something also.

A. MUNCK (*Dartmouth Medical School*): With regard to the affinities derived from the Scatchard plots, were the binding experiments you did on the various tissues done at 37°C?

FUNDER: They were done overnight at 4°C.

MUNCK: Do you have any idea how those affinities might change with temperature?

FUNDER: No, not in these studies. And certainly not in any comparison of rat and guinea pig.

MUNCK: Is it possible that the prominently occupied mineralocorticoid receptors could convey permissive effects of glucocorticoids or for that matter mineralocorticoids? Many years ago we speculated that if there were very high affinity receptors, moment-to-moment regulation would not make much sense physiologically. But they could conceivably convey permissive effects. These are things that have to be there in order for everything else to work.

FUNDER: Operationally that could be the case. However, I am uncertain what the evolutionary pressure is to cope with a system with absolutely no variation of steroid input signal as opposed to one with a circadian rhythm.

I. S. EDELMAN (*Columbia University*): The Yamamoto model is interesting, but it may have limited applicability in a variety of organ systems. It was evolved with a bias toward the glucocorticoid response elements. In tissues that are highly dependent on aldosterone for regulation of salt transport, the response elements have never been defined. There are some proteins known to be induced by aldosterone. But no clear-cut definition has been made of the events beyond the immediate regulation of transcription. So I suspect that that model will undergo substantial modification before a final view of the situation is obtained.

FUNDER: I very much agree. Ascribing it to a simple response element is based on the activated glucocorticoid receptor and mineralocorticoid receptor triggering the same response. Nobody has looked at androgen or progesterone in this connection.

EDELMAN: Bill Rosner recently raised an intriguing issue by providing evidence that the binding globulins may be active components in the steroid regulatory system. He found that sex steroid binding globulins have specific receptors on the surface of target cells and that the binding event is dependent on occupancy of the globulin by the steroid. If the nature of the signal transduction event can be identified, one has to add that as another level of complexity. Has anybody looked at corticosteroid-binding globulin (CBG) in this respect?

FUNDER: There are some Russian studies on CBG, with which I am not familiar. They predate Bill's studies, and he acknowledges them.

E. R. DE KLOET (*Sylvius Laboratories*): With regard to the question of Dr. Munck about permissive effects, we have evidence obtained by examining a functional response of the hippocampus that some effects mediated by glucocorticoid receptors can only be seen when mineralocorticoid receptors are also occupied. That would support a permissive effect of the mineralocorticoid receptor. But with respect to the high affinity of progesterone for these receptors, do you think that especially in females there may be a situation wherein part of the mineralocorticoid receptors are not available for their biological activity due to the fact that they are occupied by progesterone?

FUNDER: Certainly. In the luteal phase progesterone levels are 20 to 90 pg/ml. During pregnancy they are over 100 pg/ml, which is up into the cortisol range. We are now asking what keeps progesterone out of renal mineralocorticoid receptors and whether there is some balance of renal conversion of progesterone into deoxycorticosterone (DOC). A question remains about the availability of the free steroids in the hippocampus, in the spongy high nonspecific binding brain tissue. For instance, if I inject animals with the same dose of DOC and aldosterone, I get four times as much free DOC in the brain as I do free aldosterone. This is one of the issues that needs to be resolved before we can really determine what progesterone is doing in these sites.

DE KLOET: Does this mean that there may exist a physiological situation where the MRs are fully occupied with progesterone? They are not biologically active then.

FUNDER: Indeed, there may be a significant occupancy of those receptors by progesterone depending on whether one is pregnant or in a luteal phase. The presumption is that it will be an antagonist occupancy, clearly.

Brain Mineralocorticoid Receptor Function[a]

E. RONALD de KLOET, NYNKE Y. ROTS, DÉSIRÉE T. W. M. VAN DEN BERG, AND MELLY S. OITZL

Division of Medical Pharmacology
Leiden/Amsterdam Center for Drug Research
Leiden University
P.O. Box 9503
2300 RA Leiden, The Netherlands

Corticosterone (B), the principal glucocorticoid secreted by the rat adrenal gland, controls energy metabolism and the response to stress. These actions are aimed at maintaining homeostasis by control of the sensitivity or threshold of the stress response system and at restoring homeostasis after a stress-induced disturbance (see paper by Munck). B has a pleiotropic action and induces multiple changes in peripheral responses that indirectly influence brain function. This has been shown for responses of the immune system, the cardiovascular system and after B-induced changes in carbohydrate and mineral metabolism. However, the corticosteroid hormone also exerts a potent and direct action on specific neuronal circuits. This direct action of B is mediated by high-affinity mineralocorticoid receptors (MR) and lower-affinity glucocorticoid receptors (GR) which are colocalized in brain cells. MR and GR expression is particularly abundant in hippocampal neurons.[1]

In this contribution we will first briefly summarize the present information on the properties and function of central MR and GR. Next we will present experiments that challenge the view that an MR, overwhelmingly occupied with endogenous hormone, would be excluded as a dynamic regulatory component of the stress response system even under conditions of high B. Finally, we will show an experiment designed to relate MR properties in hippocampus to individual-specific behavioral and neuroendocrine responses to stress.

Differential Occupancy of MR and GR

We will focus on the function of B mediated by the apparent 'nonselective' brain MR occurring in nonepithelial nerve cells, *i.e.*, the hippocampal neurons. The 'nonselectivity' refers to the high affinity of aldosterone (ALDO) and B binding to these MR. Half maximal occupation of the 'nonselective' brain MR is already at low B concentrations. It is, therefore, reasonable to assume that under most circumstances this receptor is extensively occupied. Some time ago we postulated that such MR would mediate a 'tonic' action of B involving maintenance of homeostasis, as opposed to the 'phasic' actions via GR contributing to restoration of homeostasis.[1,2]

In vitro in cytosol B binds to MR with an affinity constant in the order of 0.2–0.5 nM.[3–6] Similar high affinity in MR is observed for ALDO, progesterone (PROG) and deoxycorticosterone (DOC). B binds with a tenfold lower affinity to GR, while the affinity of ALDO, DOC and PROG to GR is even an order of magnitude lower.[7]

[a]These studies were supported by the Netherlands Organization for Scientific Research (NWO) Grants 546-092 and 564-025, and by the Dutch Heart Foundation Grant D 88002.

Circulating B (and ALDO) readily enter the brain, and are retained extremely well by hippocampal neurons.[8] *In vivo* MR and GR in hippocampus and other brain regions are differentially occupied by the endogenous ligands. Half maximal occupancy is already attained at very low levels of circulating free B at the circadian trough, when receptor occupancy is assessed with a post-hoc *in vitro* radioligand binding assay. B or ALDO administered in tracer amounts (about 1 nmol ^{3}H B) to the adrenalectomized (ADX) male rat results in an about 50% MR occupancy.[9] In spite of their high MR affinity, PROG and DOC are poorly retained in cell nuclei *in vivo*, but only the latter appears an effective competitor for B uptake.[10–12]

In vivo MR in epithelial cells in periventricular and anterior hypothalamic region retain ALDO, but not B.[13] It is assumed, but not yet proved that these cells contain the high-affinity isoform of the enzyme 11β-hydroxysteroid dehydrogenase (11β-OHSD), which inactivates B, but not ALDO. This enzyme presumably is identical to the 11β-OHSD proposed to confer aldosterone selectivity in kidney.[14] Its chemical identity still needs to be assessed (see paper by Funder).

MR and GR are unevenly distributed in brain. While most glia cells and neurons express both receptor types, their density is highest in neurons of the hippocampus (see papers by Fuxe and Bohn). Confocal laser scanning microscopy of receptor antigens using immunocytochemistry and double staining with different chromophores showed colocalization in hippocampal neurons (Van Steensel, de Kloet & Van Driel, see front cover). We found significant activation of the immunoreactive GR signals with low amounts of circulating B.[15] A post-hoc *in vitro* binding assay revealed that GR is occupied progressively with rising levels of B following stress and towards the circadian peak (see FIG. 1).

In situ hybridization and immunocytochemistry showed that MR is evenly expressed in all hippocampal neurons from early life into adulthood. Subsequently, the MR mRNA, the immunoreactive MR protein and particularly the MR binding sites slowly decrease in capacity during the aging process.[15–18] In contrast, in the 2-day-old rat GR is poorly expressed in dentate gyrus neurons, while it is abundant in CA3; at two weeks of age GR expression in CA3 has diminished, while that in the dentate gyrus has rapidly increased.[19–22] GR is downregulated by high corticosteroid levels. MR shows a profound increase after glucocorticoid administration. These findings clearly demonstrate differential regulation of MR and GR in hippocampal neurons.[23,24]

In conclusion, the physiological B signal produces predominant MR activation already at very low basal B levels and binding studies, mainly performed with hippocampal tissue, suggest that MR has already become saturated at low concentrations of B. This concentration of B is far below the maximal physiological level attained by this steroid and raises the question to what extent MR contributes to the effect of high B. After stress and at the circadian peak the high circulating B levels occupy GR in particular, in addition to MR.

MR and GR Function in Brain

Our strategy to examine B action on centrally regulated functions is based on removal of adrenals and dose-dependent B replacement for gradual and sequential activation of both receptor populations. A second approach employs administration of receptor antagonists in minute amounts in the brain. The experimental designs also take into account that the putative genomic B effects require at least several minutes to develop and will last for several hours.

Using this strategy we have reported findings on the cellular, physiological and

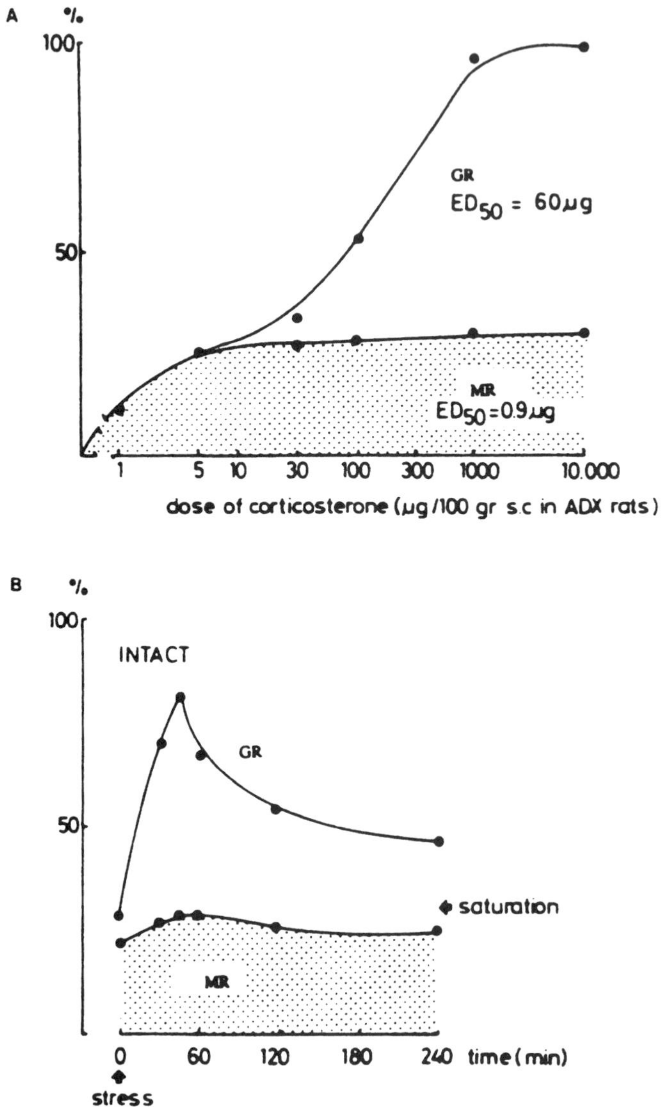

FIGURE 1. Cumulative occupancy of soluble MRs and GRs in hippocampus. Occupancy is calculated from labelling available MRs and GRs with receptor-specific radioligands after treatment. This amount of receptor is subtracted from total number of sites (MR + GR) measured in untreated (24-h ADX) rats. Binding capacity based on scatchard analysis. (**A**) 1 h after subcutaneous administration of increasing doses of B. (**B**) After exposure of rats to a novel environment. (Data from Reul *et al.*[23])

behavioral level of analysis which demonstrate that MR- and GR-mediated actions exerted in hippocampus are distinct in time domain and mechanism, and proceed in a coordinate fashion.[25,26] Dose-response curves of B often display a U-shape. Low doses of B predominantly activate MR and activate GR to a lesser extent. The MR-mediated effects can be blocked by MR antagonists. These include synthetic compounds such as RU 28318 and spironolactone, but presumably also naturally occurring PROG and 11β-OH-PROG (see poster paper by Carey & De Kloet and paper by Damm *et al.*). High doses of B activate both MR and GR; the latter effect exerted by B can often be blocked by the synthetic glucocorticoid antagonist RU486. Interestingly, blockade of the high B effect by the MR antagonist also appears to be effective (see below). This observation is counterintuitive, since the binding studies suggest saturation of MR at low B concentrations. Apparently, the MR-antagonist also blocks interaction of MR and GR.

TABLE 1 summarizes our data on the function of MR and GR in centrally regulated processes.[27] Low B, resulting in a predominant MR activation maintains (stabilizes) neuronal activity (see paper by Joëls), is of relevance for the threshold or sensitivity of the neuroendocrine stress response system (see paper by Dallman) and contributes to control of behavioral reactivity and response selection (see below). This 'stabilizing' function predominantly operating via MR by low B contrasts with the effect of high B in 'recovery' of disturbed homeostasis requiring MR + GR occupancy. Thus, high B operating via a coordinate MR + GR activation restores excitability transiently raised by excitatory stimuli, facilitates termination of the stress-induced neuroendocrine response and promotes storage of information on stress-provoking events.

Based on these observations two models for coordinate MR- and GR-mediated action in hippocampus have been developed. FIGURE 2A shows progressive MR and GR occupancy by increasing B concentrations. The scheme is drawn in a U-shape to accommodate the increasing body of evidence that in long-term absence of B as well as during excess of B similar responses are measured in hippocampal electrophysiology and hippocampal-associated behavior. FIGURE 2B represents the MR/GR balance hypothesis, which is based on the effects observed after administration of selective MR or GR antagonists. The MR/GR balance representing opposite MR- and GR-mediated effects may explain the U-shaped dose-response curve obtained with the mixed agonist B.

In previous reports we stated that the extensive occupancy of hippocampal MR already at low levels of circulating B suggests this receptor mediates a tonic influence on hippocampus function. In addition, the occupancy data obtained with radioligand

TABLE 1. Functions of Brain Corticosteroid Receptors

Receptor	Localization	Function
MR (aldo-selective)	periventricular anterior hypothalamus	salt homeostasis cardiovascular control
MR (cort-selective)	hippocampus septum, amygdala	stabilization excitability sensitivity HPA response behavior reactivity
GR (+ MR)	widely distributed, stress centers	restoration excitability suppression HPA response facilitation memory processes

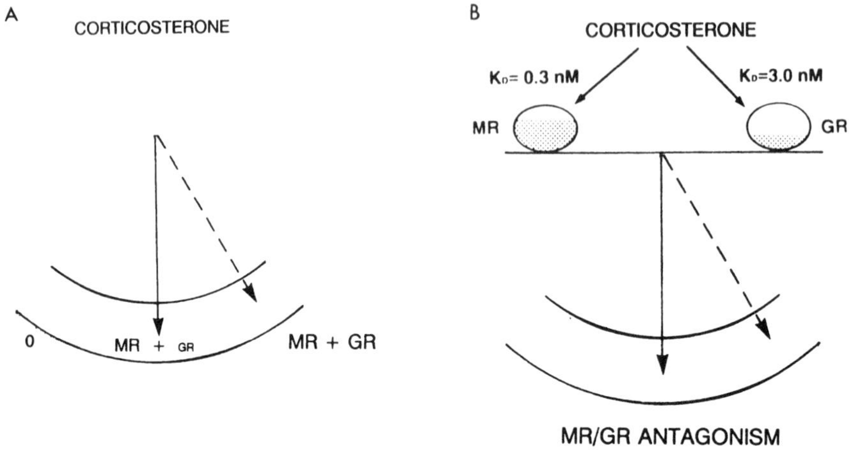

FIGURE 2. The MR/GR balance hypothesis.

binding suggest that MR operates only in the low range of B concentrations. This hypothesis will be tested by three experiments, which are described in the next sections. Experiment 1 examines hippocampal MR function at high levels of exogenous B. Experiment 2 is designed to study MR function during conditions of hyperthermia and stress, which supposedly enhance the action of endogenous hormones via MR. Finally, experiment 3 exploits individual differences in MR capacity in search for changes in MR-related functional correlates.

Experiment 1. Central MR Function after High Exogenous B: Behavioral Response

The hippocampus is critically involved in learning and memory processes, in particular with respect to spatial orientation.[28] This can be assessed in the Morris maze test, which includes rats learning the location of a spatially fixed underwater-platform with the help of distant cues during several training sessions. The time and swimming distance required for the rats to find the platform serve as criteria for acquisition, consolidation and retrieval of spatial information. Melly S. Oitzl recently observed corticosteroid modulation of behavioral performance of male Wistar rats in the Morris maze.[29] The effects of the corticosteroid hormones on behavioral performance in this test could be distinguished in distinct MR- and GR-mediated effects. Intracerebral administration of the GR antagonist in amounts as low as 100 ng before and after the first training session resulted in impaired performance 24 h later. Thus, blockage of GR, and not of MR, interfered with the learning process by disrupting consolidation of spatial information. Since administration prior to the first and second training session did not alter performance during these sessions, neither acquisition nor retrieval of learned information was affected.[29]

Another parameter, namely, the strategy rats use to locate the platform is analyzed in the absence of the platform (the so-called free swim trial). Blockade of central MR, prior to the test on the second day after removal of the platform, results in an altered behavioral search pattern. This finding suggests implication of MR function in the choice of behavioral responses. Performance of spatial navigation of ADX rats showed deficiency both in MR and GR function: when testing started three days after ADX, these rats showed impaired spatial learning as well as altered strategies.[29]

The hippocampus also plays a key role in the response to novelty.[30] Disturbances of hippocampal functioning reduced the animals' tendency to flexibly explore its environment resulting in preservation of acquired behavior. This is not a matter of stimulus detection, but of a different reactivity to stimuli. Since MR is involved in the process of response selection as demonstrated by the altered search strategy in the water maze, we proposed a particular involvement of MRs in the structure of the behavioral response to a novel environment. Therefore, a task was designed allowing the animal to distribute its exploratory activity between an object in the center and the periphery of an open field.[30]

We found a U-shaped relation between the exploration pattern and the circulating level of B and consequent receptor occupancy (FIG. 3). Tested one day after ADX, rats explored significantly more the center area and its object compared to sham-operated and intact rats. Treatment of ADX rats with a low dose of B, *i.e.*, preferential occupancy of MR one hour before the test reduced exploration of the center area and resulted in an exploration pattern similar to sham-operated and intact rats. ADX and intact rats injected with a high dose of B, *i.e.*, complete occupancy of MR and GR, showed more exploration of the center. Blockade of central MR by 100 ng

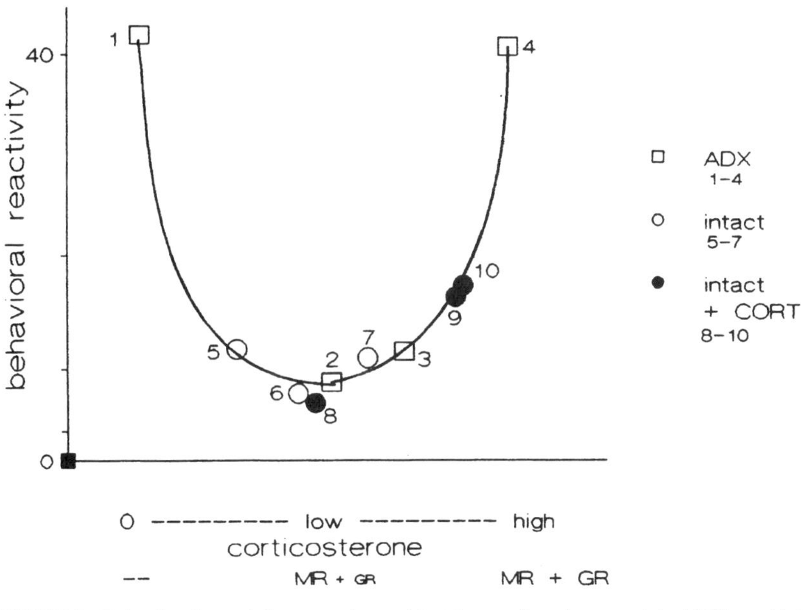

FIGURE 3. Behavioral reactivity towards an object in novel environment. 1: ADX + vehicle; 2: ADX + low B (50 μg/kg sc); 3: Sham + vehicle; 4: ADX + high B (1000 μg/kg sc); 5: MR antagonist icv (100 ng/μl); 6: Vehicle icv (1 μl); 7: GR antagonist icv (100 ng/μl); 8: MR antagonist + high B; 9: GR antagonist + high B; 10: Vehicle + high B. (Data from Oitzl & de Kloet.[30])

of the antagonist RU 28318 prevented the high B-induced increase in exploration of the center area, while administration of the GR antagonist had no effect. The total locomotor activity did not differ between the groups.

In conclusion, B controls the behavioral response of an animal to novelty in a U-shaped manner over a concentration range from below 25 nM up to stress levels of 1000 nM B. Minimal reactivity towards the object in the center is observed at about 100 nM of total B, which is consistent with predominant MR and little GR occupancy. At any concentration of B, its effect can be blocked by pretreatment with MR antagonist given icv. This finding suggests a critical role for hippocampal MRs in control of the animals' response to novelty even under conditions of high B.

Experiment 2. Central MR Function, Hyperthermia and Stress: Autonomous and Neuroendocrine Response

The pioneering studies of Brody[31] have demonstrated that mineralocorticoid (ALDO or DOC)-induced changes in blood pressure are coordinated by centrally regulated functions. These functions include central control of sodium homeostasis and of sympathetic nervous activity, which are influenced by the mineralocorticoids and by the neuropeptides, angiotensin II and natriuretic peptide, acting in coordination in the anteroventral part of the periventricular hypothalamus (AV3V).

Désirée Van den Berg[32] showed that in conscious, normotensive Wistar rats a single icv injection of a few ng's of ALDO induced an increase in systolic blood pressure at 8 hrs after administration, whereas icv injection of the MR antagonist RU28318 caused a long-lasting decrease.[33] Other researchers showed that icv infusion of RU28318 abolished the development of DOCA-salt hypertension[34,35] and blocked the centrally evoked aldosterone-induced hypertension.[36]

Our observations on the central effect of the mineralocorticoids were made with the indirect tail cuff method to measure blood pressure. This procedure involves—for reasons of habituation to the experimental procedure—a training period of 14 days to the conditions of the blood pressure measurement. Animals are exposed for 30 min daily to a warm lamp resulting in an elevated ambient temperature (32°C) and elevated core temperature (38.5°C). Next the animal is placed into a small cage for 1 min to allow application of the tail cuff (*i.e.*, restraint).

When arterial pressure was directly measured in freely moving rats the icv administered MR agonists and antagonists were ineffective. Also pressure of the rats to the condition of chronic discontinuous hyperthermia only (no restraint) did not alter the arterial pressure. However, the subsequent brief restraint stress evoked a pressor response. Pretreatment with the MR antagonists icv suppressed this stress-induced pressor response, but only if the rats had been trained for two weeks to the condition of chronic discontinuous hyperthermia and restraint stress. The effect of the MR antagonist developed slowly and was significant after 7 hrs (FIG. 4).

We postulate that if rats are conditioned to chronic discontinuous exposure to hyperthermia and restraint, they activate central and humoral mechanisms to facilitate hyperthermia loss and to inhibit sodium loss in anticipation to this condition. The mechanism would include enhanced MR-mediated action of ALDO, enhanced action of angiotensin II and a reduced action of natriuretic peptides aimed to adjust fluid and sodium homeostasis as well as sympathetic outflow. In fact, it seems as if the pressor response has become dependent upon ALDO, since the rise in arterial pressure is suppressed by MR antagonist pretreatment.[36]

Plasma levels of B, ALDO and ACTH were significantly increased after a 30-min exposure to 32°C. Moreover, B levels were increased even further after exposure to

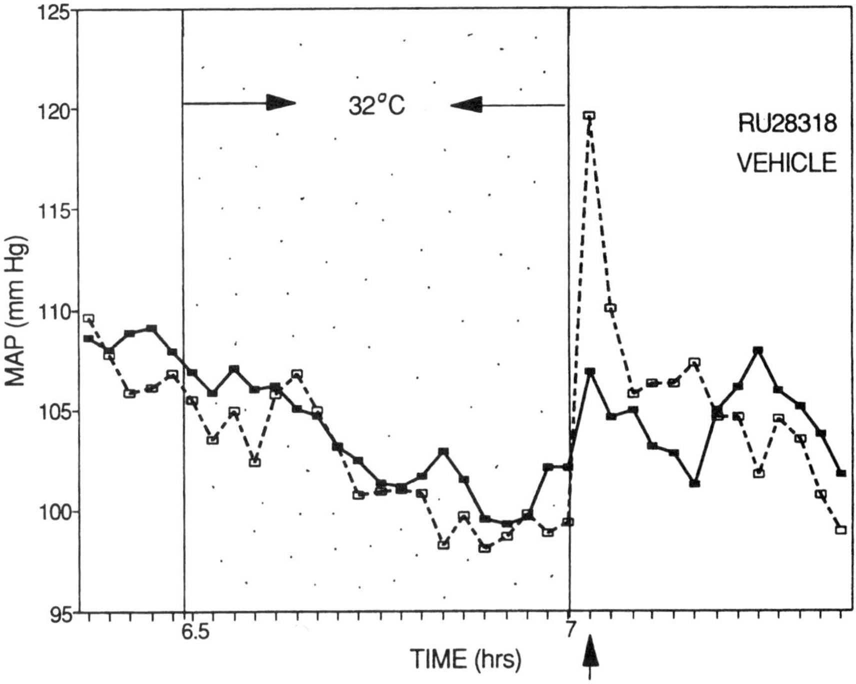

FIGURE 4. Effect of mineralocorticoid antagonist on mean arterial pressure of rats exposed to an ambient temperature of 32°C for 30 min during 14 days. *Dotted area* indicates the warmest period. *Arrow* refers to the time point of 1.5 min restraint stress; (. . . .. . .) vehicle-treated rats; (solid line) RU 28318-treated rats. (Data from van den Berg.[32])

the training procedure than after acute exposure. The MR antagonist suppressed (−25%) this elevation of B at 2 hrs after icv administration to the rats that were subjected to chronic discontinuous hyperthermia.[32,35] Previously we observed that the MR antagonist given icv to Wistar rats not subjected to hyperthermia actually increased the basal level of circulating B and enhanced (+25%) the stress-induced elevation of B.[37]

In conclusion, chronic discontinuous exposure to hyperthermia shifts pressor and neuroendocrine responses to stress to a condition of enhanced MR dependency. Accordingly, blockade of the central ALDO-selective MR (presumably located in the AV3V) by the MR antagonist icv then results in attenuation of stress-induced, autonomic and neuroendocrine responses, where they were originally enhanced by this antagonist.

Experiment 3. Hippocampal MR Capacity: Correlation with Stress Responsiveness?

As was pointed out in the previous two sections, the effect of pharmacological blockade of the central MR by an antagonist seemed to be dependent upon the extent of MR occupancy, achieved either by administration of B or by activation of an endogenous agonist of these receptors. If, as experiment 1 suggests, B effects involving MR in the CNS can best be described by a U-shaped curve (see also FIGS. 2A and 3) it can be predicted that the extent of occupancy (= receptor activation) determines the direction (suppression or activation) of the effect exerted by the MR antagonist.

Occupancy is a function of ligand concentration as well as of the capacity and binding affinity of MR. In experiment 3, by Nynke Rots, the attention is focused on the number of MRs in hippocampus, as having a predictive value for stress responsiveness of the animal. The presumption is that if MRs are extensively occupied the number of MR may be somehow correlated with the magnitude of the B effect on a hippocampus-associated function such as the reactivity of the HPA system, in response to a 'cognitive' stressor.

In order to approach this problem experimentally we compared two pharmacogenetically selected rat lines with respect to their stress responsiveness and B receptor properties in brain. Two rat lines were selected that differed in their responsiveness to the dopamine agonist apomorphine. This drug induces stereotypic gnawing behavior by stimulating dopamine neurons in the neostriatum.[38] The apomorphine susceptible rats (apo-sus) display at least a twentyfold higher gnawing score, in a 45-min observation interval, than the apomorphine unsusceptible (apo-unsus) rats. A breeding programme with these selected lines was initiated a few years ago by Alex R. Cools from the University of Nijmegen[38] to explore the genetic background of individual differences in behavioral and stress responsiveness of rats. We performed experiments with the 12th up to the 14th generation.

In behavioral studies apo-sus rats showed an increased novelty-induced locomotor activity and a fleeing response upon defeat when placed into the home cage of a resident, while apo-unsus rats reveal a freezing response in the same situation. It has been proposed that apo-sus rats perform fear-motivated behavioral responses elicited and directed by external stimuli, while apo-unsus rats are marked by a high degree of intrinsic organization (self-organization).[38] Interestingly, results of a recent study[39] suggest that apo-sus rats are less susceptible to the induction of experimental allergic encephalomyelitis (EAE), an autoimmune disease, than the apo-unsus rats. The different dopamine responsiveness was recently substantiated by clear-cut differences in tyrosine hydroxylase mRNA in the A_9 dopaminergic cell body area and in striatal dopamine D_2-receptor binding (Ref. 40; unpublished observation).

Thus, apo-sus rats selected for high responsiveness of the nigrostriatal dopamine system showed increased behavioral reactivity to environmental challenges and increased resistance to induction of autoimmunity. Under basal a.m. condition apo-sus rats showed a similar plasma ACTH and total (CBG-bound + free) B level, as compared to the apo-unsus rats. Their p.m. ACTH and total B levels were increased, however. In apo-sus rats exposure to a novel environment or a conditioned emotional response resulted in an enhanced and more prolonged plasma ACTH response, while B was not much different from apo-unsus rats. However, basal free B level was lower in apo-sus rats, while the stress-induced circulating free B level was enhanced and prolonged (FIG. 5).

In vitro studies with tissue of apo-sus rats showed increased capacity of hippocampal and pituitary MR, but not of GR, while binding affinities were not different. In frontal cortex GR capacity was increased. Other brain regions were not different in corticosteroid receptor properties.[41] In line with these *in vitro* observations a higher retention of ^{3}H-B was observed *in vivo* in hippocampal neurons of apo-sus animals. These binding data suggest increased hippocampal MR activity in the

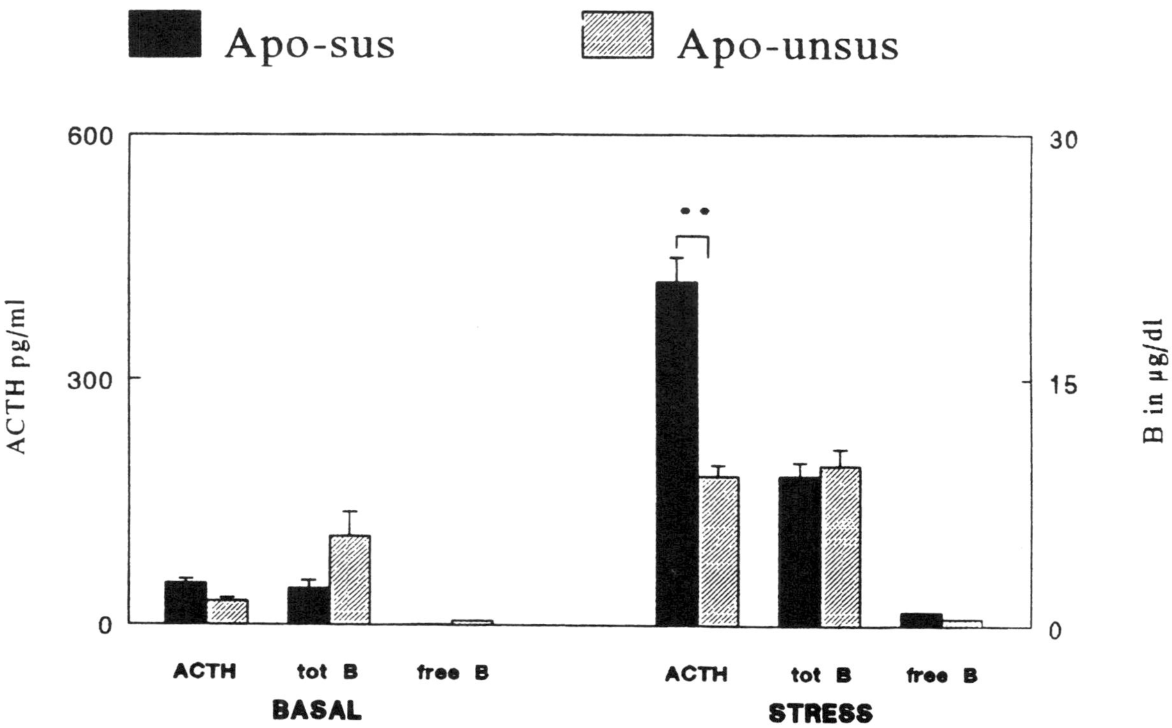

FIGURE 5. Basal and stress-induced levels of ACTH, total and free B in plasma of male apo-sus and apo-unses rats. Basal = blood sampling at 9.00 a.m. Stress = conditioned emotional response; blood samples are taken at 15 min after stress. Free B levels are basal: apo-sus 10 ng/dl; apo-unsus 240 ng/dl. 15 min after stress; apo-sus 820 ng/dl; apo-unsus 430 ng/dl.

apo-sus rat line characterized by reduced basal free B. This very same rat line, however, also shows an enhanced pituitary ACTH response to emotional stress, suggesting a reduced inhibitory corticosteroid feedback signal (Ref. 40; Rots *et al.*, unpublished observation). This observation raises a question about the role of hippocampal MR in neuroendocrine regulation: Is the rise in MR in the more "reactive" apo-sus line an adaptive response aimed to maintain homeostasis or in other words, aimed to counteract this enhanced intrinsic reactivity by an MR-directed suppressive action on the sensitivity of the stress response system?

In conclusion, the two rat lines selected for high and low nigrostriatal dopamine activity are considered to represent the extremes in behavioral and neuroendocrine responsiveness to stress that coexist in a normal population of rats. We find here that MR capacity is rather related to basal than to stress-induced free B levels.

CONCLUDING REMARK

The three experiments have provided novel information on MR functioning in rat brain. Experiment 1 demonstrated that central MRs are responsive to high B, even though the binding studies suggest that these MRs are already saturated at much lower concentrations. The same experiment revealed a MR-mediated U-shaped dose response curve; behavioral reactivity towards a novel object was highest in the absence of B or after administration of high doses of B to adrenally-intact rats.

Other researchers have shown a U-shaped ALDO effect on salt appetite mediated by central MR. Low doses of ALDO suppress the high salt intake, which develops after ADX, whereas high doses of exogenous ALDO stimulate salt intake of ADX rats and of adrenally-intact animals.[42] Behavioral reactivity and salt appetite were both affected by central mineralocorticoid antagonist administration. The direction of the antagonist effect (suppression or facilitation) depended on the extent of MR activation by the exogenous agonist.

Experiment 2 showed that conditioning to hyperthermia increased MR dependency of the autonomic and neuroendocrine response to stress. The enhanced "endogenous" MR activation could be attenuated by the mineralocorticoid antagonist, which now suppressed the stress response. In previous work with rats that were not conditioned to hyperthermia, we found that the MR antagonist enhanced the adrenocortical response. We propose, therefore, that also in response to endogenous ligands brain MRs operate in a U-shaped mode.

Synaptic or transmitter stimulation of hippocampal CA1 neurons also displayed U-shaped characteristics in response to corticosteroid hormones. Minimal responses were observed with predominant MR activation and these responses were very stable even with persistent synaptic stimulation. Serotonin-1A and muscarinic cholinergic receptor-mediated responses, low threshold calcium conductance and the Ca-dependent K after-hyperpolarization, all were maximal in the absence of hormone as well as in the presence of high amounts of B. Interestingly, for some of the responses to high B the MR activation was obligatory (Refs. 25,43; Joëls, this volume).

According to Munck and Náray-Fejes-Tóth (Ref. 44; see Munck, this volume) a U-shaped dose response may be representative for permissive and suppressive effects of corticosteroids. It is thought such a U-shaped response is the result of the interplay between corticosteroid responsive mediators and subsequent adaptive changes in receptor properties. Such a relationship can easily be demonstrated by studies on effects of cytokines during various corticosteroid regimes.[44,45] The present experiments suggest that such U-shaped B effects may involve MR, or GR or a coordinate MR and GR interaction. A mechanism for the latter possibility was recently proposed by Pearce and Yamamoto.[46] MRs would facilitate AP_1 (cfos and cjun dimers) activation after binding to composite response elements which then would be suppressed after subsequent competition by GRs at higher concentration of B.

ACKNOWLEDGMENT

The editorial assistance of Ellen M. Heidema is greatly appreciated.

REFERENCES

1. DE KLOET, E. R. 1991. Front. Neuroendocrinol. **12:** 95–164.
2. DE KLOET, E. R. & J. M. H. M. REUL. 1987. Psychoneuroendocrinology **12:** 83–105.
3. MOGUILEWSKI, M. & J. P. RAYNAUD. 1980. J. Steroid Biochem. **12:** 309–314.
4. VELDHUIS, H. D. & E. R. DE KLOET. 1982. Endocrinology **110:** 153–159.
5. KROZOWSKI, Z. K. & J. W. FUNDER. 1983. Proc. Natl. Acad. Sci. USA **80:** 6056–6060.
6. REUL, J. M. H. M. & E. R. DE KLOET. 1985. Endocrinology **117:** 2505–2512.
7. SUTANTO, W. & E. R. DE KLOET. 1987. Endocrinology **121:** 1405–1411.
8. MCEWEN, B. S., J. M. WEISS & L. S. SCHWARTZ. 1968. Nature **220:** 911–912.
9. REUL, J. M. H. M., J. R. VAN DEN BOSCH & E. R. DE KLOET. 1987. J. Endocrinol. **115:** 459–467.
10. DE KLOET, E. R., J. VAN DER VIES & D. DE WIED. 1974. Endocrinology **94:** 61–73.
11. DE KLOET, E. R., G. WALLACH & B. S. MCEWEN. 1975. Endocrinology **96:** 598–609.
12. MCEWEN, B. S., E. R. DE KLOET & G. WALLACH. 1976. Brain Res. **105:** 129–136.
13. SUTANTO, W. & E. R. DE KLOET. 1991. Med. Res. Rev. **11:** 716–639.
14. NARAY-FEYES-TÓTH, A., E. RUSVAI, D. DENAULT, D. L. ST. GERMAN & G. FEYES-TÓTH. 1993. Am. J. Physiol. **265:** F896–F900.
15. VAN EEKELEN, J. A. M., J. Z. KISS, H. M. WESTPHAL & E. R. DE KLOET. 1987. Brain Res. **436:** 120–128
16. VAN EEKELEN, J. A. M., P. ROSENFELD, S. LEVINE, H. M. WESTPHAL & E. R. DE KLOET. 1987. Neurosci. Res. Commun. **1:** 129–133.
17. VAN EEKELEN, J. A. M., W. JIANG, E. R. DE KLOET & M. C. BOHN. 1988. J. Neurosci. Res. **21:** 88–94.
18. VAN EEKELEN, J. A. M., M. C. BOHN & E. R. DE KLOET. 1991. Dev. Brain Res. **61:** 33–43.
19. ROSENFELD, P., W. SUTANTO, S. LEVINE & E. R. DE KLOET. 1988. Dev. Brain Res. **42:** 113–118.
20. ROSENFELD, P., J. A. M. VAN EEKELEN, H. M. WESTPHAL, S. LEVINE & E. R. DE KLOET. 1988. Dev. Brain. Res. **42:** 119–127.
21. ROSENFELD, P., W. SUTANTO, S. LEVINE & E. R. DE KLOET. 1990. Dev. Brain Res. **52:** 57–62.
22. VAN EEKELEN, J. A. M., N. Y. ROTS, W. SUTANTO & E. R. DE KLOET. 1991. Neurobiol. Aging **13:** 159–170.
23. REUL, J. M. H. M., F. R. VAN DEN BOSCH & E. R. DE KLOET. 1987. J. Endocrinol. **115:** 459–467.
24. REUL, J. M. H. M., P. T. PEARCE, J. W. FUNDER & Z. S. KROZOWSKI. 1989. Mol. Endocrinol. **3:** 1674–1680.
25. JOËLS, M. & E. R. DE KLOET. 1992. TINS **15:** 25–30.
26. DE KLOET, E. R., M. S. OITZL & M. JOELS. 1993. Cell. Mol. Neurobiol. **13:** 433–455.
27. DE KLOET, E. R. 1992. Ann. N.Y. Acad. Sci. **663:** 357–371.
28. CHAN PALAY, V. & C. KOHLER. 1989. The Hippocampus: New Vista. Allan R. Liss. New York.

29. Oitzl, M. S. & E. R. de Kloet. 1992. Behav. Neurosci. **106:** 62–71.
30. Oitzl, M. S. & E. R. de Kloet. 1994. Eur. J. Neurosci. **6:** 1072–1079.
31. Brody, M. J. & A. K. Johnson. 1980. Neuroendocrinology **6:** 249–292.
32. van den Berg, D. T. W. M. 1993. Stress, cardiovascular regulation and brain corticosteroid receptors. Ph.D. thesis. University of Leiden, The Netherlands.
33. van den Berg, D. T. W. M., E. R. de Kloet, H. H. van Dijken & W. de Jong. 1990. Endocrinology **126:** 118–124.
34. Janiak, P. & M. J. Brody. 1988. Am. J. Physiol. **255:** R166–R173.
35. van den Berg, D. T. W. M., W. de Jong & E. R. de Kloet. Am. J. Physiol. In press.
36. Gómez-Sánchez, E. P. 1991. Am. J. Hypertens. **4:** 374–381.
37. Ratka, A., W. Sutanto, M. Bloemers & E. R. de Kloet. 1989. Neuroendocrinology **50:** 117–123.
38. Cools, A. R., R. Brachten, D. Heeren, A. Willemsen & B. Ellenbroek. 1990. Brain Res. Bull. **74:** 49–69.
39. Cools, A. R., N. Y. Rots, B. Ellenbroek & E. R. de Kloet. 1993. Neuropsychobiology **28:** 100–105.
40. Rots, N. Y., M. S. Oitzl, A. Berod, W. Rostene, A. R. Cools & E. R. de Kloet. 1993. Soc. Neurosci. Abstr. no. 387.6.
41. Sutanto, W., E. R. de Kloet, F. de Bree & A. R. Cools. 1989. Neurosci. Res. Commun. **5:** 19–26.
42. Fregly, M. J. & I. W. Waters. 1966. Physiol. Behav. **1:** 65–74.
43. Joëls, M. & E. R. de Kloet. 1993. J. Neurosci. **13:** 4082–4090.
44. Munck, A. & A. Náray-Fejes-Tóth. 1992. Mol. Cell. Endocrinol. **90:** C1–C4.
45. Schöbitz, B., F. Holsboer, W. Sutanto, G. Gross, E. Schönbaum & E. R. de Kloet. 1994. Neuroendocrinology **59:** 387–395.
46. Pearce, S. & K. Yamamoto. 1993. Science **259:** 1161–1165.

DISCUSSION OF THE PAPER

Question: Does the high level of mineralocorticoid receptors in the apomorphine-susceptible rats explain the phenomenon that you see in those rats?

de Kloet: We know that antimineralocorticoids disinhibit pituitary-adrenal activity, and this suggests that hippocampal MR are involved in control of ACTH. Accordingly, if the affinity or capacity of the MR are decreased, one would predict a rise in ACTH. In the apomorphine-susceptible rats compared to the non-susceptible rats we observed an increased MR capacity and an increased ACTH response to stress. This is at variance with the predicted enhanced inhibition. On the other hand, free basal corticosterone levels are lower in the apomorphine-susceptible rats, which corresponds to increased MR. Thus, at present we ask questions like: how are MRs regulated and what is the relationship between MR properties and pituitary-adrenal activity?

H. Akil (*University of Michigan*): How much occupancy of MR is there and how can MR function be modulated to yield the kinds of findings you are reporting? If things are maximally occupied at all time, it seems that you should not get that nice dose response as a function of MR.

de Kloet: We did not expect we would be able to modify MR function with a high level of corticosteroid, but yet we saw the dose-dependent MR-mediated effects. Now there are two possibilities. Either what we measure as occupancy in radioligand binding studies is an artifact, in other words, there are more MR sites available than we measure; or the mineralocorticoid receptors and glucocorticoid receptors cooperate in some unknown manner.

AKIL: Do you see any circadian differences in this MR function?

DE KLOET: All that can be said is that we saw the effect of stress on repeatedly warmed rats, the second experiment in the evening. So the next experiment should be to compare the responses in the morning and evening.

M. F. DALLMAN (*University of California, San Francisco*): Have you studied the effect of adrenalectomy in rats selected for apomorphine susceptibility? The vast difference in MR concentration in the brain may or may not be related to the difference in behavior, susceptibility to encephalitis, corticosteroid-binding globulin levels. If you adrenalectomize the animals, do they show the same differences in the various behaviors?

DE KLOET: That is a question I cannot answer, since we did not do the experiment.

S. LEVINE (*Stanford University Medical School*): While it is quite easy to find a function for the MR or a function for the GR, and clearly the roles of the receptors are different for different behaviors, I still have a great deal of difficulty thinking of the MR as implicated in regulating behavior.

DE KLOET: I agree that neither MR nor a steroid regulates behavior. We believe that steroid action on a neural circuit changes the probability that a certain response in a certain condition will occur. We have learned a lot, however, about the mechanism of corticosteroid action in the hippocampus from elecrophysiology and neurochemistry.

E. M. STERNBERG (*National Institute of Mental Health*): With your apomorphine-susceptible rats, it looks almost as if you have made a Fischer rat. Your apomorphine-susceptible rats are resistant to encephalitis. It is a stress hyperresponsive rat that has higher MR in the hippocampus. I shall show the same data correlated with high MR in the hippocampus of the Fischer rat.

DE KLOET: However, this was not a Fischer rat.

STERNBERG: But I am saying that the phenotype is the same.

DE KLOET: With apomorhine we have selected rats out of a normal Wistar population. You might also find a subpopulation of Fischer rats which look like the apomorphine-susceptible rats. That hippocampal MR is increased in the rats with high MR was not predicted by us, and presents an interesting new twist in the story, I agree.

STERNBERG: The point that is fascinating is the high MR. I do not know what it means, but we have found the same thing.

DE KLOET: Thank you for your comment.

Corticosteroids and the Control of Function in the Hypothalamo-Pituitary-Adrenal (HPA) Axis[a]

MARY F. DALLMAN, SUSAN F. AKANA, NANCY LEVIN, C-D. WALKER, MARGARET J. BRADBURY, S. SUEMARU, AND KAREN S. SCRIBNER

Department of Physiology
University of California San Francisco
San Francisco, California 94143-0444

The central nervous system (CNS) through its control of secretion of releasing and inhibiting factors from the neuroendocrine hypothalamus into the hypothalamo-hypophysial portal system regulates hormonal synthesis and secretion from the anterior pituitary, and, via its control of the pituitary hormones, regulates activity of peripheral target endocrine glands. In order to control activity of the peripheral endocrine glands, the CNS must receive information about their function. This information is conveyed by neurons in the CNS that contain receptors for a specific target hormone. The degree to which hormone receptors are occupied by the appropriate target hormone determines input to, and/or the activity of the hypothalamic neuroendocrine cells. Occupancy of the receptors is determined primarily by the concentration of free, unbound hormone in the circulation which is free to diffuse or be transported across the blood-brain barrier to receptor sites in the CNS.

The characteristics of function in the HPA axis pose severe demands if there is to be tight feedback regulation by adrenal corticosteroids. There is a challenging range of circulating corticosteroid levels that is associated with the normal physiology of the adrenocortical system. The circadian rhythm in basal activity of the HPA axis results in total circulating corticosteroid concentrations that may be <10 nM at the nadir of the rhythm and about 700 nM at the peak of the rhythm in both man and rats. Of the total corticosteroid concentration, 99–95% is tightly bound in the circulation to transcortin and is unavailable for diffusion to brain sites. The amount of steroid bound depends on the concentrations of both steroid and transcortin in the circulation. Furthermore, at all times of day, the system can be stimulated by stressors to cause corticosteroid values that may exceed 1 μM. Nonetheless, the secretion of corticotropin-releasing factor (CRF) and vasopressin (AVP), the major hypothalamic neuropeptides that regulate ACTH synthesis in and secretion from the corticotropes of the anterior pituitary, appears to be under tight corticosteroid feedback control during the basal circadian trough and peak as well as during CRF/AVP responses to stressors. Although ACTH-secreting cells in the anterior pituitary respond to corticosteroids *in vitro*, with inhibition of ACTH synthesis and secretion, it seems likely that this effect is primarily a coarse gain feedback that is involved during high level corticosteroid secretion.[1–3]

The tight feedback control exerted on activity of the HPA system by the CNS appears to occur because of two corticosteroid receptors of different affinities that

[a]Supported in part by a grant from the National Institutes of Health (DK 28172).

provide information to the hypothalamic CRF/AVP neurons; there is also a marked and complex interaction between occupancy of both receptor types. The high affinity receptor (the type I corticosteroid or mineralocorticoid receptor, MR) has a K_D of ~0.5 nM or less for both corticosterone (B) and cortisol[4–7] and is singularly suited, when occupied, to maintain adrenal corticosteroid secretion at the low levels observed during the circadian trough in activity of the HPA system. The lower affinity receptor (the type II corticosteroid or glucocorticoid receptor, GR) has a K_D of 2.5–20 nM[6–9] and has an affinity appropriate to detect and regulate circadian peak or stressor-induced corticosteroid secretion.

Below, we will discuss our views of how the occupancy of these two receptor types results in the very tight regulation of function in the adrenocortical system that occurs under the many levels of activity imposed on the system. Subsequently, we will examine how, given the regulatory corticosteroid feedback, hyperactivity in the adrenocortical system can occur. Our conclusions have arisen from several decades of experimental results from many laboratories, and have involved study of both adrenalectomized and intact animals, primarily rats which secrete corticosterone (B).

Feedback Role(s) of the Type I Corticosteroid or Mineralocorticoid Receptors (MR)

Normally, in man and rats there is little, if any hypothalamic drive to ACTH and corticosteroid secretion during the diurnal minimum in basal activity. Immunoneutralization of CRF does not alter ACTH or B secretion during the trough of the circadian rhythm in HPA activity, suggesting that any activity during the basal trough is a consequence of constitutive pituitary secretion (reviewed in Ref. 10). After bilateral adrenalectomy and complete removal of steroid feedback, increased CRF and AVP mRNA, peptide concentrations and secretion occur.[11–13] The mRNA for the precursor of ACTH increases as does ACTH synthesis and secretion.[1,11,13] Adrenalectomized rats still exhibit circadian rhythms in basal activities in neurosecretogogs and ACTH,[3,11] and also respond to stress with increases in hypothalamic secretogogs and ACTH.[14] However, this activity is increased to a new, higher level than that observed in the rat with adrenals. When ADX rats are treated with B in amounts that produce plasma B concentrations that can occupy primarily type I corticosteroid receptors in the CNS, the ADX-induced increase in activity of hypothalamic neuroendocrine neurons that synthesize CRF and AVP, of corticotropes in the anterior pituitary and of ACTH secretion is restored during the trough of the circadian rhythm to the level observed in sham-adrenalectomized (adrenally intact) controls.[11,13,15,16] However, systemic infusions of dexamethasone, a poor MR agonist but good GR agonist *in vivo*,[17] do not restore AVP mRNA levels to normal in ADX rats during either trough or peak times in basal activity.[15] Furthermore, in ADX rats provided with a low level of B that did restore AVP mRNA and peptide and ACTH levels to normal in the a.m., treatment with an MR antagonist, spironolactone, increased ACTH secretion.[15] Because spironolactone had no effect on ACTH secretion in ADX rats in the absence of B, the results suggested strongly that activation of AVP (and probably CRF) mRNA after ADX is primarily a consequence of loss of type 1 corticosteroid receptor activation.

In summary, when MRs in neurons associated with feedback regulation of the HPA axis are chronically empty (after ADX), tonic increases of CRF and AVP synthesis in parvocellular neurons of the paraventricular nuclei (PVN) occur, and ACTH synthesis and secretion from the pituitary is tonically driven even during the circadian trough in basal activity. These events occur either by release of the neuroendocrine cells from tonic inhibition by MR-containing neurons or by tonic stimulation of the neuroendocrine cells by neurons that no longer are regulated by the steroid-receptor complex.

For several reasons, we believe from studies of ADX rats that feedback inhibition of CRF/AVP is driven transsynaptically, rather than exerted directly at the CRF/AVP cell bodies. First, the pioneering studies by McEwen *et al.*,[18] confirmed and expanded by others,[19] showed uptake and retention in several limbic and midbrain brain sites, but not in hypothalamus, of very small doses of ^{3}H-corticosterone (0.7 ug) given ip to ADX rats. The concentrations of steroid in plasma achieved after these injections would have been sufficient to occupy MR, but probably not to occupy the GR that are found in abundance in hypothalamus.[6,20–24] Second, although present, the mRNA and protein for MR are not abundant in the CRF/AVP cells of the paraventricular nuclei, unlike the mRNA and protein for GR. Although very low levels of MR binding have been reported in punched sections of the PVN,[6] and hypothalamus,[20] the quantities are small. Third, lesions which do not damage the PVN but which do block input to the nuclei have been shown in ADX, B-treated rats or intact rats to raise basal trough CRF mRNA and/or ACTH levels,[2,26,27] suggesting that the lesions have disrupted steroid-induced negative feedback input to the PVN. Finally, a more flexible system results from corticosteroid feedback exerted remotely, rather than directly at the CRF/AVP neurons. Rather than directly inhibiting the capacity of the neuroendocrine neurons to synthesize and secrete their neurohormones, a remote feedback system allows immediate override of the steroid-induced inhibition when other, direct (*e.g.*, stimulus-induced) inputs to the neuroendocrine cells occur.

Ratka *et al.*[28] show, using a synthetic MR antagonist in intact rats, that both trough, basal and stress-induced responses in the HPA axis are augmented by MR blockade, suggesting that occupancy of MR mediates ACTH feedback. Similarly, basal trough cortisol levels are increased after treatment with an MR antagonist in man.[29] It is also clear that persistent occupancy of the MR, provided by very small, passive elevations of plasma B in intact rats, causes compensatory inhibitory adjustments in regulation of both the circadian peak and stress-induced ACTH secretion.[30] Normally, trough plasma B levels in unstressed rats are indistinguishable from those in bilaterally ADX animals. When those values are increased by exogenously administered B to levels of 30 to 150 nM total B, or ~0.6–6 nM freely diffusible steroid, both the circadian basal peak B levels and the ACTH responses to restraint applied during the circadian trough are reduced in a dose-related fashion by the endogenous system. Although there is as yet no proof that the exogenously supplied B acted through occupancy of the MR, it seems highly probable, since the lowest dose used had no effect on any of the glucocorticoid endpoints measured (*e.g.*, thymus weight), although it did consistently inhibit endpoints associated with the activity of the HPA axis (*e.g.*, adrenal weight). The overall effect of the increased feedback signal provided by exogenous B during the circadian trough was to maintain mean daily total plasma B levels at the normal value of ~145 nM, thus minimizing the effects of elevated steroid on target tissues.[30]

Feedback Role(s) of the Type II Corticosteroid or Glucocorticoid Receptors (GR)

Occupancy of the lower affinity GR appears, not unexpectedly, to be required in ADX rats for restraining circadian peak and stressor-induced ACTH secretion.[16,28] Concentrations of constant B which normalize hypothalamic peptides and ACTH secretion in ADX rats during the circadian trough are inadequate during the circadian peak, and neuropeptide mRNA levels and ACTH are elevated above those in intact rats.[10,11,31,32] Similarly, ADX rats that have been treated with constant low doses of B, but not exposed to a normal daily peak in plasma B, have higher peak ACTH responses to acute stressors applied during the time of the circadian trough.[32] In elegant experiments, Jacobson and Sapolsky have shown that regulation of the ACTH response to hypoxia requires acute occupancy of GR as well as MR.[33] Higher tonic, circulating concentrations of B do restore ACTH secretion in ADX rats to normal or subnormal during the circadian peak. However, constant elevations in B of ~350 nM also completely block ACTH responses to stressors applied during the trough of the circadian rhythm in both ADX and intact rats.[30,34] These results suggest strongly that tonic elevations of B in the range in which both MR and GR are occupied completely inhibit endogenous activity in the HPA system.

A clear interaction between occupancy of MR and partial occupancy of GR on circadian peak ACTH concentrations was demonstrated in experiments in which low doses of constant B and dexamethasone, singly or together, were infused into ADX rats.[15] The effects of the steroids on peak basal ACTH and on thymus weight were evaluated. The thymus gland contains only GR, so that inhibition of thymus weight by corticosteroids must be mediated by association of the steroids only with GR. This makes thymus weight a very convenient endpoint to use in assessing the biological effect of steroids in association with type II corticosteroid receptors. Neither a very low dose of B nor of dexamethasone alone decreased peak ACTH; when given together, the two steroids interacted to markedly inhibit circadian peak basal ACTH concentrations. By contrast, the steroids had a strictly additive effect on thymus weight as would be expected of a single receptor system that could be occupied by both steroids. Thus, it appears over a long time-domain that control of the amplitude of the circadian peak in basal activity and ACTH responses to stimuli both require occupancy of GR as well as MR.[15]

CNS Site(s) of B Feedback

Both receptor types have been reported to occur in the same cells, and it is possible that the interaction between the occupied receptors on control of ACTH is mediated within single cells.[7] Implantation of dexamethasone directly over the PVN has been shown to markedly decrease CRF and AVP content and ACTH secretion in adrenalectomized rats.[35,36] However, bilateral implants of crystalline corticosterone over the PVN did not inhibit basal a.m. ACTH to normal in adrenalectomized rats,[37] suggesting that the decided inhibitory effects exerted by dexamethasone were not physiologically relevant. Further, the fact referred to above, that lesions which isolate the PVN from other brain sites may stimulate CRF and AVP mRNA levels in PVN and also ACTH secretion suggests that the inhibitory effect on HPA axis function induced by corticosterone is exerted at (a) site(s) remove from the PVN.

It is somewhat discouraging that there is as yet little clear evidence for CNS site(s) that normalize(s) activity in the HPA axis under either basal or stimulated conditions. It seems highly likely that feedback sites are stressor-specific and widely distributed throughout the CNS. A recent report by Diorio *et al.*[38] shows excellent evidence for a feedback site that modulates the ACTH and B responses to restraint in the prefrontal cortex. Lesions of the site augmented, and implantation of B diminished HPA responses to restraint. The manipulations in prefrontal cortex had effects that were specific to restraint stress; they did not affect either the ACTH and B responses to ether or the basal circadian rhythm.[38]

The hippocampal formation, because of its plenitude of both receptor types, has frequently been examined as a site of corticosteroid feedback and results are about equally divided, for and against.[39,40] Bradbury *et al.* recently reported that bilaterally complete sections of the fimbria-fornix, the major hippocampal efferent pathway, do not alter sensitivity of ACTH secretion to corticosterone in the adrenalectomized rats.[41] However, in intact rats with lateral lesions of the fornix, Herman *et al.*[42] reported that CRF and AVP mRNAs in parvocellular cells of the PVN were increased as was plasma ACTH; a similar result was reported for the effects of lateral fornix lesions on plasma B by Fischette *et al.*[43] These results, together with the inhibitory effect of corticosteroid implants in ventral hippocampus reported by Slusher,[44] suggest that there may be a feedback site in the ventral hippocampal-subiculum area.

Distinguishing How Chronic Increases in HPA Activity Occur

A manipulation that causes increased CRF/AVP expression and secretion and ACTH secretion may be a consequence of altering sensitivity of the HPA axis to B feedback. However, activity in the HPA axis may be increased either by increased drive from stimuli (facilitation),[10,45,46] by a decrease in input from a B feedback site,[2,26,47] or by bypassing B feedback.[48] Under all of these conditions, manifestations of hypercorticoidism occur: adrenal weight is increased and thymic atrophy is present.

Using the information gained from study of normal and adrenalectomized rats implanted with exogenous B, detailed above, we have examined how corticosteroid hypersecretion occurs in three experimental conditions: chronic stress, lactation and genetic obesity; we have also tested the effects of two lesions in the CNS on feedback sensitivity of ACTH to B. From the earlier experiments reported above, it is apparent that measurement in adrenalectomized rats of the inhibition of ACTH by experimentally fixed B levels should reveal facilitated acute drive to CRF/AVP secretion, or decreased efficacy in type I or type II receptor-mediated feedback.

Because inhibition of ACTH secretion by corticosteroids is proportional to stimulus strength, *i.e.*, a given dose of steroid acts to inhibit a percentage of the 0 steroid ACTH secretion,[3] then increased drive to ACTH secretion would be observed if the experimental ACTH inhibition curve were elevated above the control. However, the IC_{50} for steroid inhibition should not be shifted to the right. In contrast, if the efficacy of corticosteroid feedback inhibition of ACTH were diminished, but drive to ACTH secretion were not affected, then the experimental ACTH inhibition curve would be shifted to the right, and if there were no additional stimulation, the curve would overlie the control at 0 steroid levels.

Chronic stimuli to the HPA axis in rats, such as streptozotocin-induced diabetes or cold, with time result in quite normal basal activity in ACTH and corticosterone. However, characteristic of the animal experiencing a chronic stressor, HPA responses to novel stimuli are greater than normal, and sensitivity of ACTH secretion after novel stressors to inhibition by exogenously administered corticosteroids is reduced (*e.g.*, Ref. 10). We recently reported[46] that sensitivity of ACTH to B feedback in diabetic adrenalectomized rats replaced with exogenous B is normal under basal conditions both in the a.m. and p.m. However, acute stress-induced ACTH secretion was resistant to the fixed B signal in these rats, and B inhibition under acute stress conditions was less effective. Because the sensitivity of ACTH to B feedback was normal both in the a.m. (primarily mediated by occupancy of type I corticosteroid receptors) and in the p.m. (requiring occupancy of both type I and type II corticosteroid receptors), we concluded that the decreased B feedback in these rats observed only after acute stimulation was a consequence of facilitated stress-induced drive to the CRF/AVP cells, and that corticosteroid feedback sensitivity was normal.[46]

Lactating rats also exhibit elevated basal trough B levels but unlike chronically stressed rats, they have decreased basal peak B levels and stress responses.[48,49] When lactating and virgin rats were adrenalectomized and replaced with constant B, sensitivity of basal ACTH to B feedback was normal in both the a.m. and p.m. in the lactating rats (*i.e.*, the inhibition curve was not shifted to the right as would be anticipated as a consequence of reduced B feedback efficacy).[48] These results suggest strongly that corticosteroid feedback sites act normally in lactating rats and that the hypersection of corticosteroids during the morning is a consequence of what was subsequently clearly shown to be the CRF-releasing stimulus of suckling.[48,49] In the intact lactating rat, however, elevated trough B may in part serve to reduce both the p.m. rise in basal activity and stress-induced ACTH responses.

Genetically obese rodents[50] and women with abdominal obesity[51,52] have slightly elevated basal corticosteroid secretion during the diurnal trough. In obese rats, peak basal B levels are, if anything, decreased compared to lean littermates.[50] When adrenalectomized and replaced with exogenous B, there was a tendency for a right-shift of the basal a.m. ACTH inhibition curve in the obese rats. The differences were particularly noticeable over the low range of 2–4 ug B/dl, the range in which MR are occupied. Additionally, there was no shift to the right of the calculated IC_{50} for ACTH with stress in obese rats although the expected right shift occurred with stress in the lean littermates. The results of these experiments showed clearly that sensitivity of obese rats to type II receptor, GR, occupancy was normal and suggested that overactivity in the HPA axis was a consequence of decreased feedback inhibition exerted by occupancy of MR.[50]

Another experimental model of obesity is produced by lesions of the ventromedial hypothalamus in rats. Such rats hypersecrete B during the trough of the circadian rhythm in basal activity, and like genetically obese rats, peak B and stress-induced activity in the HPA may be normal or even reduced (*e.g.*, Ref. 47). The sensitivity of basal trough ACTH in adrenalectomized rats with VMH lesions to exogenously supplied B feedback has been tested in two studies. King *et al.*[26] showed significant inhibition of ACTH by B in sham-lesioned rats but no significant inhibitory relationship of ACTH to B in rats with VMH lesions. These results suggested that there was a deficit in feedback efficacy of B on ACTH in female rats with lesions of the VMH. Another study, performed on male rats with lesions of the VMH produced similar results.[47] After adrenalectomy and low levels of B replacement, there was a significant shift to the right in the inhibition of ACTH by B under basal trough conditions.[47] Thus, obesity induced by genetic abnormalities or by CNS lesions appears to reduce the feedback efficacy provided by occupancy of MR during the trough of the basal circadian rhythm, accounting for the observed adrenal hypertrophy and thymic atrophy.

Finally, to test whether disruption of the major hippocampal outflow through the fornix fimbria altered feedback sensitivity of ACTH to B, lesioned rats were adrenalectomized, replaced with exogenous B and sampled during the trough and peak of the circadian rhythm as well as during and after stimulation by hypoxia or ether.[41] No evidence was found for altered B efficacy on ACTH under any of these conditions, leading us to conclude that lesioning this hippocampal outflow did not tonically disrupt a feedback site which projects to the CRF/AVP cells in the PVN.[41]

The results of the experiments detailed above suggest that in the various conditions examined, corticosteroid feedback efficacy was altered under basal conditions only in genetically obese rodents and in rats with lesions of the VMH. Increased drive to CRF/AVP secretion better explains the hyperactivity of the HPA system observed in chronically stressed and in lactating mother rats. A right-shift in feedback sensitivity of ACTH to B under basal conditions is an initial criterion that must be met in order to study further central B feedback sites in situations in which altered feedback efficacy is suspected. Although this criterion is necessary, it is not sufficient. Further studies must show, at a minimum, that: when chemically lesioned so that cell bodies but not fibers are damaged, the site reduces B feedback efficacy; the site has cells which contain MR and/or GR; local infusions of B into the site reduce CRF/AVP in PVN and reduce ACTH secretion under either basal conditions or after specific stimuli in adrenalectomized rats; infusions of MR or GR antagonists into a suspected feedback site, or neutralization of endogenous MR or GR in the site, should cause a shift in sensitivity of specific HPA axis responses to B to the right. If all of these criteria are met for a suspected feedback site, then the presumptive evidence is good that it is, indeed, a feedback site. Subsequent experiments would then be required to determine the mechanism(s) and neural pathway(s) by which the inhibitory effect of B on activity in the HPA axis is achieved.

CONCLUSIONS

Considerable progress in understanding the physiology of corticosteroid feedback regulation in the HPA system has been made during the last decade. The understanding that feedback is mediated by occupancy of two corticosteroid receptors of differing affinities is important and has altered experimental designs so that the role of each receptor type in modulation of activity in the HPA axis can be determined. Further, it is clear that the HPA axis is physiologically regulated by corticosteroid feedback under most conditions throughout the 24-hour day. When hyperactivity of the HPA axis is observed, it is important to distinguish the mechanism by which this occurs: is the system hyperactive because of increased drive to the CRF/AVP neurons from chronic or intermittent stimuli or lesions, or because alteration of input from a corticosteroid feedback site has decreased feedback efficacy? Finally, as yet no CNS site has been unequivocally shown to exert inhibition of activity in the HPA axis through occupancy of specified corticosteroid receptors. We hope that this goal will be achieved in the decade to come.

REFERENCES

1. Dallman, F. M., G.B. Makara, J. L. Roberts, N. Levin & M. Blum. 1985. Corticotrope response to removal of releasing factors and corticosteroids. Endocrinology **117:** 2190–2197.
2. Levin, N., J. Shinsako & M. F. Dallman. 1988. Corticosterone acts on the brain to inhibit adrenalectomy-induced adrenocorticotropin secretion. Endocrinology **122:** 694–701.
3. Dallman, M. F., S. F. Akana, C. S. Cascio, D. N. Darlington, L. Jacobson & N. Levin. 1987. Inhibition of ACTH secretion: variations on a theme of B. Rec. Prog. Horm. Res. **43:** 113–173.
4. Beaumont, K. & D. D. Fanestil. 1983. Characterization of a rat brain aldosterone receptors reveals high affinity for corticosterone. Endocrinology **113:** 2043–2051.
5. Krozowski, Z. S. & J.W. Funder. 1983. Renal mineralocorticoid receptors and hippocampal corticosterone binding species have identical intrinsic steroid specificity. Proc. Natl. Acad. Sci. USA **80:** 6056–6060.

6. Reul, J. M. H. M. & E. R. de Kloet. 1985. Two receptor systems for corticosterone in rat brain: microdistribution and differential occupation. Endocrinology **117:** 2505–2511.
7. Arriza, J. L., R. B. Simerly, L. W. Swanson & R. M. Evans. 1988. The neuronal mineralocorticoid receptor as a mediator of glucocorticoid response. Neuron **1:** 887–900.
9. Rousseau, G. G., J. D. Baxter & G. M. Tomkins. 1972. Glucocorticoid receptors: relations between steroid binding and biological effects. J. Mol. Biol. **67:** 99–115.
9. McEwen, B. S. & G. Wallach. 1973. Corticosterone binding to hippocampus: nuclear and cytosol binding *in vitro.* Brain Res. **57:** 373–386.
10. Dallman, M. F., S. F. Akana, K. A. Scribner, M. J. Bradbury, C.-D. Walker, A. M. Strack & C. S. Cascio. 1992. Stress, feedback and facilitation in the hypothalamo-pituitary-adrenal axis. J. Neuroendocrinol. **5:** 517–526.
11. Kwak, S., M. I. Morano, E. A. Young, S. J. Watson, & H. Akil. 1993. The diurnal CRF mRNA rhythm in the hypothalamus: decreased expression in the evening is not dependent on endogenous glucocorticoids. Neuroendocrinology **57:** 96–105.
12. Sawchenko, P. E. 1987. Adrenalectomy-induced enhancement of CRF and vasopressin immunoreactivity in parvocellular neurosecretory neurons: anatomic, peptide and steroid specificity. J. Neurosci. **7:** 1093–1106.
13. Byer, H. S., S. G. Matta & B. M. Sharp. 1988. Regulation of the messenger ribonucleic acid for corticotropin-releasing factor in the paraventricular nuclei and other brain sites in the rat. Endocrinology **123:** 2117–2123.
14. Bradbury, M. J., C. S. Cascio, K. A. Scribner & M. F. Dallman. 1991. Stress-induced adrenocorticotropin secretion: diurnal responses and decreases during stress in the PM are not dependent on corticosterone. Endocrinology **128:** 680–688.
15. Bradbury, M. J., S. F. Akana & M. F. Dallman. 1994. Roles of type I and II corticosteroid receptors in regulation of basal activity in the HPA axis during the diurnal trough and the peak: evidence for a non-additive effect of combined receptor occupation. Endocrinology. In press.
16. Dallman, M. F., N. Levin, C. S. Cascio, S. F. Akana, L. Jacobson & R. W. Kuhn. 1989. Pharmacological evidence that the inhibition of diurnal ACTH secretion by corticosteroids in mediated via type I corticosterone-preferring receptors. Endocrinology **124:** 2844–2850.
17. Spencer, R. L., A. H. Miller, M. Stein & B. S. McEwen. 1991. Corticosterone regulation of type I and type II adrenal steroid receptors in brain, pituitary and immune tissue. Brain Res. **549:** 236–246.
18. McEwen, B. S., J. M. Weiss & L. S. Schwartz. 1969. Uptake of corticosterone by rat brain and its concentration by certain limbic structures. Brain Res. **16:** 227–241.
19. Birmingham, M. K., M. Sar & W. E. Stumpf. 1984. Localization of aldosterone and corticosterone in the central nervous system, assessed by quantitative autoradiography. Neurochem. Res. **9:** 333–350.
20. Liposits, Zs., R. M. Uht, R. W. Harrison, F. P. Gibbs & M. C. Bohn. 1987. Ultrastructural localization of glucocorticoid receptor (GR) in hypothalamic paraventricular neurons synthesizing corticotropin-releasing factor (CRF). Histochemistry **87:** 407–412.
21. Uht, R. M., J. F. McKelvy, R. W. Harrison & M. C. Bohn. 1988. Demonstration of glucocorticoid receptor-like immunoreactivity in glucocorticoid-sensitive vasopressin and corticotropin-releasing factor neurons in the hypothalamic paraventricular nucleus. J. Neurosci. Res. **19:** 405–411.
22. Chao, H., P. Choo & B. McEwen. 1989. Glucocorticoid and mineralocorticoid receptor mRNA expression in rat brain. Neuroendocrinology **50:** 365–371.
23. Agnati, L. F., K. Fuxe, Z. Y. Uy, A. Harfstrand, S. Okret, A. C. Wikstrom, M. Goldstein, M. Zoli, W. Vale & J. A. Gustafsson. 1985. Morphometrical analysis of the distribution of corticotrophin releasing factor, glucocorticoid receptor and phenylethanolamine-N-methyltransferase immunoreactive structures in the paraventricular hypothalamic nucleus of the rat. Neurosci. Lett. **54:** 147–152.
24. Aronsson, M., K. Fuxe, Y. Dong, L. Agnati, S. Okret & J.-A. Gustaffson. 1988.

Localization of glucocorticoid receptor mRNA in the male rat brain by *in situ* hybridization. Proc. Natl. Acad. Sci. USA **85:** 9331–9335.
25. Spencer, R. L., A. H. Miller, H. Moday, M. Stein & B. S. McEwein. 1993. Diurnal differences in basal and acute stress levels of type I and type II adrenal steroid receptor activation in neural and immune tissues. Endocrinology **133:** 1941–1950.
26. King, B. M., C. A. Zansler, A. A. Tatford, K. L. Neville, H. Sam, J. M. Kass, & M. F. Dallman. 1993. Level of corticosterone replacement determines body weight gain in adrenalectomized rats with VMH lesions. Physiol. Behav. **54:** 1187–1190.
27. Herman, J. P., S. J. Wiegand & S. J. Watson. 1990. Regulation of basal corticotropin-releasing hormone and arginine vasopressin messenger ribonucleic acid expression in the paraventricular nucleus: effects of selective hypothalamic deafferentiations. Endocrinology **127:** 2408–2417.
28. Ratka, A., W. Sutano, M. Bloemers & E. R. de Kloet. 1989. On the role of mineralocorticoid (type I) and glucocorticoid (type II) receptors in neuroendocrine regulation. Neuroendocrinology **50:** 117–123.
29. Dodt, C., W. Kern, H. L. Fehm & J. Born. 1993. Antimineralocorticoid canrenoate enhances secretory activity of the hypothalamus-pituitary-adrenocortical (HPA) axis in humans. Neuroendocrinology **58:** 570–574.
30. Akana, S. F., K. A. Scribner, M. J. Bradbury, A. M. Strack, D.-D. Walker & M. F. Dallman. 1992. Feedback sensitivity of the rat hypothalamo-pituitary-adrenal axis and its capacity to adjust to exogenous corticosterone. Endocrinology **131:** 585–594.
31. Swanson, L. W. & D. M. Simmons. 1989. Differential steroid hormone and neural influences on peptide mRNA levels in CRH cells of the paraventricular nucleus: a hybridization histochemical study in the rat. J. Comp. Neurol. **285:** 413–435.
32. Jacobson, L., S. F. Akana, C. S. Cascio, J.Shinsako & M. F. Dallman. 1988. Circadian variations in plasma corticosterone permit normal termination of adrenocorticotropin responses to stress. Endocrinology **122:** 1343–1348.
33. Jacobson, L. & R. M. Sapolsky. 1993. Augmented ACTH responses to stress in adrenalectomized rats replaced with constant physiological levels of corticosterone are partially normalized by acute increases in corticosterone. Neuroendocrinology **58:** 420–429.
34. Akana, S. F., C. S. Cascio, J. Shinsako & M. F. Dallman. 1985. Corticosterone: narrow range required for normal body and thymus weight and ACTH. Am. J. Physiol. **249:** R527–R532.
35. Kovacs, K. J., Z. J. Kiss & G. B. Makara. 1986. Glucocorticoid implants around the hypothalamic paraventricular nucleus prevent the increase of corticotropin-releasing factor and arginine vasopressin immunostaining induced by adrenalectomy. Neuroendocrinology **44:** 229–234.
36. Sawchenko, P. E. 1987. Evidence for a local site of actions for glucocorticoids in inhibiting CRF and vasopressin expression in the paraventricular nucleus. Brain Res. **403:** 213–224.
37. Kovacs, K. J. & G. B. Makara. 1988. Corticosterone and dexamethasone act at different brain sites to inhibit adrenalectomy-induced adrenocorticotropin hypersecretion. Brain Res. **474:** 205–210.
38. Diorio, D., V. Viau & M. J. Meaney. 1993. The role of the medial prefrontal cortex (cingulate gyrus) in the regulation of hypothalamic-pituitary-adrenal responses to stress. J. Neurosci. **13:** 3839–3847.
39. Jacobson, L. & R. M. Sapolsky. 1991. The role of the hippocampus in feedback regulation of the hypothalamic-pituitary-adrenal axis. Endocr. Rev. **12:** 118–134.
40. Canny, B. J. 1990. Hippocampal glucocorticoid receptors and the regulation of ACTH secretion. Mol. Cell. Endocrinol. **71:** C35–C38.
41. Bradbury, M. J., A. M. Strack & M. F. Dallman. 1993. Lesions of the hippocampal efferent pathway (fimbria-fornix) do not alter sensitivity of adrenocorticotropin to feedback inhibition by corticosterone in rats. Neuroendocrinology **57:** 396–407.
42. Herman, J. P., M. K.-H. Schafer, E. A. Young, R. Thompson, J. Douglass, H. Akil & S. J. Watson. 1989. Evidence for hippocampal regulation of neuroendocrine neurons of the hypothalamo-pituitary-adrenocortical axis. J. Neurosci. **9:** 3072–3082.

43. FISCHETTE, C. T., B.R. KOMISARUK, H. M. EDINGER, H. H. FEDER, & A.SIEGEL. 1980. Differential fornix ablations and the circadian rhythmicity of adrenal corticosteroid secretion. Brain Res. **195:** 373–387.
44. SLUSHER, M. A. 1966. Effects of cortisol implants in the brainstem and ventral hippocampus on diurnal corticosteroid levels. Exp. Brain Res. **1:** 184–194.
45. AKANA, S. F. & M. F. DALLMAN. 1992. Feedback and facilitation in the adrenocortical system: unmasking facilitation by partial inhibition of the glucocorticoid response to prior stress. Endocrinology **131:** 57–68.
46. SCRIBNER, K. A., S. F. AKANA, C.-D. WALKER, & M. F. DALLMAN. 1993. Streptozotocin-diabetic rats exhibit facilitated adrenocorticotropin responses to acute stress but normal sensitivity to feedback by corticosteroids. Endocrinology **133:** 2667–2674.
47. SUEMARU, S., M. F. DALLMAN, S. F. AKANA, D. N. DARLINGTON & J. SHINSAKO. 1988. Ventromedial hypothalamic lesions: a neural model for Cushing's disease. Abstract 8th Int. Congress of Endocrinology. Kyoto, Japan.
48. WALKER, C.-D., S. L. LIGHTMAN, M. K. STEELE & M. F. DALLMAN. 1992. Suckling is a persistent stimulus to the adrenocortical system of the rat. Endocrinology **130:** 115–125.
49. WALKER, C.-D., M. F. DALLMAN, A. A. PALMER & M. K. STEELE. 1993. Involvement of central corticotropin-releasing factor (CRF) in suckling-induced inhibition of luteinizing hormone secretion in lactating rats. J. Neuroendocrinol. **5:** 451–459.
50. WALKER, C.-D., K. A. SCRIBNER, J. S. STERN & M. F. DALLMAN. 1992. Obese Zucker (fa/fa) rats exhibit normal target sensitivity to corticosterone and increased drive to adrenocorticotropin during the diurnal trough. Endocrinology **131:** 2629–2637.
51. WILLIAMS, D. P., T. W. BOYDEN, R. W. PAMENTER, T. G. LOHMAN & S. B. GOING. 1993. Relationship of body fat percentage and fat distribution with dehydroepiandrosterone sulate in premenopausal females. J. Clin. Endocrinol. Metab. **77:** 80–85.
52. PASQUALI, R., S. CANTOBELLI, F. CASIMIRRI, M. CAPELLI, L. BORTOLUZZI, R. FLAMIA, A. M. M. LABATE & L. BARBERA. 1993. The hypothalamo-pituitary-adrenal axis in obese women with different patterns of body fat distribution. J. Clin. Endocrinol. Met. **77:** 341–346.

DISCUSSION OF THE PAPER

P. C. WHITE (*Cornell University Medical Center*): When you gave intact rats pellets and looked at stress in the a.m. and concluded that mineralocorticoid recptors are involved in determining the size of the ACTH response, why exactly does that experiment tell you that mineralocorticoid receptors are important for the size of the stress response?

DALLMAN: It is an assumption and is certainly not proved. But if circulating corticosteroid concentrations are elevated tonically to 1.2 micrograms per deciliter, which is what happened, that should provide very little, if any, occupancy of the glucocorticoid receptor. And it is because of that, that we assume that the effects we saw due to that very small elevation were mediated through MR.

QUESTION: Those are really lovely experiments, particularly the ones where you show small changes in the a.m. affecting the p.m. How do you envision this mechanism that keeps track of or averages the levels of steroids across the rhythm?

DALLMAN: The two receptor types talk to each other, but what mechanism it is that does the 24-hour averaging I am unable to say.

QUESTION: Do you think the hippocampus is involved?

DALLMAN: No. When Margaret Bradbury made fornix lesions (and this is a third case in which we have adrenalectomized rats and then exposed them to fixed corticosterone concentrations in the circulation), there was absolutely no shift in the sensitivity of ACTH to corticosterone inhibition in the morning or in the evening in the hippocampal lesioned rats. Therefore, we do not believe that the sensitivity of ACTH to occupancy of MR is mediated by at least dorsal hippocampal structures.

QUESTION: The data for the involvement of MR in ACTH control is terrific. But in the spironolactone experiment, the dose of spironolactone is enormous.

DALLMAN: We gave industrial doses of spironolactone, but it had no effect on thymus in either the adrenalectomized rat with no steroid or the adrenalectomized rat with GR occupying levels of steroid. It was not acting on thymus, at least.

QUESTION: But was it acting to block the steroid on the thymus?

DALLMAN: No, the thymus is clean. Spironolactone does not affect it under either condition.

M. J. MEANEY (*McGill University*): With regard to the ventromedial hypothalamine (VHM) lesion, do you know if the steroid effect involves the lesion site per se or the subsequent obesity? Would not the VMH region be an interesting steroid target in control of ACTH?

DALLMAN: It seems that the VMH lesion is blocking input from elsewhere, certainly not from the VMH region per se. It is not secondary to the obesity, because it happens before the obesity occurs.

I. S. EDELMAN (*Columbia University*): There are four unaccounted for actors in the drama of the adrenalectomized animal, namely, epinephrine and norepinephrine of adrenal origin and MSH and the endorphins, which are co-regulated with ACTH because it is encoded in the precursor protein. Those four actors may play trivial roles or maybe no role at all, but one or more of them may play a significant role.

Corticosteroid Action and Neuroendocrine-Immune Interactions

ROEL DERIJK AND ESTHER M. STERNBERG[a]

Unit on Neuroendocrine Immunology and Behavior
Clinical Neuroendocrinology Branch
National Institute of Mental Health
National Institutes of Health
Building 10, Room 3S231
9000 Rockville Pike
Bethesda, Maryland 20892

INTRODUCTION

Many recent studies have shown that the immune system–central nervous system (CNS) counter-regulatory loop plays an important role in susceptibility to inflammatory/autoimmune disease.[14,42,46] One component of this interaction, the hypothalamic-pituitary-adrenal axis (HPAA), is a regulatory factor of immune responses through the effects of the glucocorticoids. Thus, in relation to inflammatory disease, premorbid reactivity as well as postinflammatory alterations in HPAA activity play an important role in the susceptibility to as well as patterns of immune responses and course and severity of illness. Glucocorticoids are traditionally considered antiinflammatory, and their effects can be divided into three classes. First, glucocorticoids have profound effects on the distribution of immune cells.[20] Generally, neutrophilic leukocytosis together with eosinopenia, monocytopenia and lymphocytopenia are seen together with diminished infiltration of neutrophils and monocytes at the inflammatory site. Second, glucocorticoids can inhibit the production and release of many proinflammatory substances.[33] These include cytokines, complement components, arachidonic acid-derived products, histamine and bradykinins. Third, glucocorticoids enhance several aspects of the immune response seemingly in the context of their antiinflammatory actions. For instance, glucocorticoids inhibit the production of C3, the key component of the complement system, while expression of factor H, an intrinsic inhibitor of the system, is induced.[30] Recently, it was proposed that glucocorticoids regulate a delicate balance between cytokine and the corresponding cytokine receptor generating an optimized response.[15,34] Moreover, glucocorticoids are additional factors in the maturation of T cells from the Th0 to the Th2 stage, producing primarily B cell maturation supporting cytokines including IL-4, IL-5, IL-6, IL-10, and IL-13, while inhibiting the induction the so-called Th1 stage, producing a primarily cellular immunity pattern of lymphokine production: IL-2, IFN, and TNF.[13,36]

Although the initial evidence was obtained from pharmacological experimental data,[24,33] the best evidence that disruption of this immune system–central nervous system communication is not simply an epiphenomenon but has physiologic relevance for the severity of and susceptibility to inflammation, is derived from animal models of inflammatory or infectious disease in which either genetic, surgical, pharmacological or toxic disruptions of the loop are associated with enhanced susceptibility to or severity of inflammatory disease (reviewed in Ref. 42). Such

[a]To whom requests for reprints should be addressed.

interruptions can occur at the level of the hypothalamus, pituitary or adrenal glands, either on a genetic basis, by surgical manipulation, or pharmacological treatment with drugs which interrupt or reconstitute the axis at various anatomical, biochemical or molecular levels.

Lewis and Fischer Rats as a Model for Genetic Disruption of the HPAA-Immune Interaction

Lewis (LEW/N) rats are a strain of rats that are highly susceptible to inflammatory illness compared to other strains. Compared to histocompatible F344/N rats, LEW/N rats exhibit a profound blunting of corticosterone, adrenocorticotropin hormone (ACTH) and corticotropin releasing hormone (CRH) responses to a variety of inflammatory and other stressful stimuli.[42–44] While the pattern of inflammatory disease that develops in LEW/N rats is related to the antigen or proinflammatory stimulus to which they are exposed, their overall susceptibility to inflammatory disease is related to their profoundly blunted HPAA responses to these stimuli.[43,44] Although morning basal corticosterone levels do not differ in LEW/N and F344/N rats, LEW/N rats show a blunting of circadian rhythm of corticosterone levels compared to F344/N rats, with maximal differences occurring between 2 p.m. and 10 p.m.[16,40] In addition, expression of enkephalin mRNA is also blunted in the paraventricular nucleus of LEW/N rats in response to proinflammatory stimuli such as intraperitoneal streptococcal cell walls, suggesting that regulatory mechanisms controlling expression of both CRH and enkephalin may be impaired in these rats.[44]

Factors Affecting CRH Release in Lewis and Fischer Rats

Glucocorticoid and Mineralocorticoid Receptor Levels in Lewis and Fischer Rats. The hippocampus, containing high levels of both the mineralocorticoid (MR, type I) and glucocorticoid receptor (GR, type II), has been implicated in the regulation of the HPAA. In addition, recent evidence indicates a direct regulatory effect of the MR and GR on the regulation of CRH and AVP mRNA under nonstress conditions.[2]

Several differences have been noted between MR and GR numbers in LEW/N and F344/N rats, depending on the tissue measured, and the assay used, while the affinity of these receptors is not different in the two strains.[16,41] MR number, as determined by ^{3}H-aldosterone binding, is significantly higher in hippocampus and pituitary of adrenalectomized female F344/N compared to female LEW/N rats, but does not differ significantly in hypothalamus or spleen. These differences are not apparent in male F344/N and LEW/N rats, adrenalectomized 16 hours prior to performance of binding assays.[16] Although the sex difference in these two studies might explain such binding number discrepancies, procedural differences, notably time after adrenalectomy are a more likely explanation. Thus, these differences in receptor numbers determined by radioligand binding methods probably reflect differences in the reequilibration capacity of receptors postadrenalectomy, rather than actual static differences between strains, and highlight inherent difficulties in comparing and interpreting corticosteroid binding studies, which are performed in adrenalectomized animals. In support of this contention is the finding, in *in situ* hybridization studies of the hippocampus of LEW/N and F344/N rats pre- and postadrenalectomy, that intact LEW/N rats express higher GR mRNA levels compared to F344/N rats, but do not show the postadrenalectomy increase in GR mRNA seen in F344/N rats.[41] Intact LEW/N and F344/N rats show no difference in hippocampal MR mRNA expression, and LEW/N rats also do not show an increase in MR mRNA postadrenalectomy. MR mRNA is significantly lower in adrenalectomized LEW/N compared to adrenalectomized F344/N rats. Such lack of upregulation of GR and MR mRNA postadrenalectomy in LEW/N rats may reflect their low integrated preadrenalectomy corticosteroid levels, with the resultant lack of effect of removal of corticosteroids by adrenalectomy. Alternatively, it could reflect an intrinsic inability of this strain to upregulate receptors postadrenalectomy. Regardless of the cause, the effect of such lack of upregulation would be to amplify strain differences compared to strains that do upregulate, when receptor numbers are determined by binding studies performed in adrenalectomized animals.

In immune tissue, GR number, as determined by ^{3}H-dexamethasone binding, is significantly higher in thymus of adrenalectomized LEW/N compared to F344/N rats.[41] These differences in GR binding number in the thymus of LEW/N rats, also noted in studies by Dhabnar *et al.*,[16] could be secondary to differential HPAA responses of the two strains, since the integrated levels of corticosterone over time would be expected to be higher in F344/N rats, perhaps predisposing to steroid-induced thymocyte cell death in this strain. Corticotropin binding globulin (CBG) levels are also markedly higher in F344/N thymus compared to LEW/N, and less, but significantly higher in F344/N plasma and spleen.[16] Although such CBG levels might be expected to buffer F344/N tissues from the effects of their higher corticosterone levels, such buffering, if it occurs, is probably not complete, as evidenced by the smaller thymus and larger adrenal weights in F344/N compared to LEW/N rats.[43]

CRH Neurotransmitter Regulatory Pathways. Several neurotransmitter systems known to regulate hypothalamic CRH differ basally in these strains, including the central serotonin 5-HT1A and benzodiazepine/γ-aminobutyric acid (GABA) systems. LEW/N rats express significantly fewer hippocampal and frontal cortical 5-HT1A binding sites and less 5-HT1A mRNA than do F344/N or outbred HSD rats, while levels of hippocampal 5-HT are significantly higher in F344/N and Harlan Sprague-Dawley (HSD) rats compared to LEW/N rats. These differences are consistent with the relative HPAA responsiveness of these strains. GABA/benzodiazepine receptor number is higher in LEW/N compared to F344/N hypothalamus, again consistent with the relative HPAA activity of the two strains.[40] CRH, ACTH and corticosterone responses to adrenergic, serotonergic and cholinergic agonists are blunted in LEW/N compared to F344/N rats, although hypothalamic muscarinic, α1 and α2 adrenergic and corticol 5-HT2 receptor numbers do not differ in the two strains.[5] Thus, Lewis rats exhibit lower endogenous agonist levels and higher endogenous inhibitor levels in response to stimulation, consistent with an overall diminished HPAA activity.

HPGA Differences in LEW/N and F344/N Rats. There is general consensus that the hypothalamic-pituitary-gonadal axis (HPGA) has a strong impact on the immune system which can best be illustrated by the increased susceptibility of females of several species to autoimmune diseases including rheumatoid arthritis. Elevations of both estrogens (E) and progesterone (P) result in increased activity of both cellular and humoral immunity.[23,39] Comparison of female LEW/N and F344/N rats also reveals differences in pituitary-gonadal axis hormones with LEW/N rats having greater P and estradiol levels, and differences in estrous cycle pattern compared to F344/N rats (Smith *et al.*, Neuroimmunomodulation (1994), in press). Interestingly, LEW/N rats show increased *in vitro* IL-1 responses compared to arthritis-resistant rats.[3] Although the effect of E and P on the immune system can be direct, it has also been shown that these hormones affect the regulation of the HPAA, probably by altering central GR and/or MR expression.[37,45] Therefore, the effect E and P on immune regulation can also be mediated by an indirect effect on glucocorticoid feedback of the HPAA.

Adrenomedullary and Adrenocortical Responses in Lewis and Fischer Rats. In contrast to the marked differences in LEW/N and F344/N HPAA responses to most stressors, both strains show equally high ACTH responses to insulin-induced hypoglycemia, and minimal to moderate differences in epinephrine response to this stressor.[21] Early norepinephrine (NE), dihydroxyphenyl-glycol (DHPG), dihydroxyphenylalanine (DOPA) and dihydroxyphenylacetic acid (DOPAC) responses are also similar, but late (>30 minutes after insulin challenge) NE, DHPG, DOPA and DOPAC responses are markedly and significantly less in LEW/N compared to F344/N rats. This pattern suggests that the adrenomedullary responses of LEW/N rats are relatively intact compared to their sympathoneuronal responses. LEW/N rats' adrenocortical responses are also significantly different compared to F344/N rats, with significantly lower adrenal weights compared to F344/N rats, and low plasma corticosterone responses to exogenous CRH, probably secondary to chronic understimulation of the adrenal glands by chronically low ACTH and CRH.

Pharmacologic and Surgical Manipulations of the HPA Axis

Effect of Adrenalectomy on Basal Neurotransmitters in LEW/N and F344/N Rats. Adrenalectomy increases GABA/benzodiazepine binding in F344/N rats to approach levels in nonadrenalectomized LEW/N rats. Dexamethasone treatment of adrenalectomized F344/N rats normalizes GABA/benzodiazepine binding to pre-adrenalectomy levels. Thus these neurotransmitter differences are likely to be secondary to the strains' differential corticosterone responses.[40] In contrast, while adrenalectomy is associated with increases in hypothalamic 5-HIAA in both strains as well as in outbred HSD rats, the strain differences in 5-HIAA levels are maintained postadrenalectomy.

Effects on Inflammatory and Infectious Susceptibility. Interrupting the CNS-immune system interaction with drugs, such as the glucocorticoid receptor antagonist RU 486, renders previously inflammatory-resistant F344/N rats highly susceptible to the lethal proinflammatory effects of bacterial products such as streptococcal cell walls.[42] Also in another model of autoimmunity (experimental allergic encephalomyelitis, EAE) resembling human multiple sclerosis (MS), injection of myelin basic protein (MPB) induces EAE in susceptible LEW/N rats. High-dose corticosteroid replacement prevents development of EAE in these animals.[32] Adrenalectomy without glucocorticoid replacement renders otherwise resistant PVG rats highly susceptible to the lethal effects of myelin basic protein, while adrenalectomy with low-dose corticosteroid replacement renders these rats susceptible to development of EAE in response to this antigen.

Conversely, transplantation of hypothalami from inflammatory-resistant F344/N rats into the third ventricle of LEW/N rats is associated with greater than 85% reduction in the carrageenan-induced peripheral inflammatory response. This reduction of inflammatory response is less in LEW/N rats transplanted with LEW/N hypothalami, and was not apparent in LEW/N rats transplanted with F344/N spinal cord, or receiving sham surgery (Poltorak and Sternberg, submitted).

These results strongly support the hypothesis that endogenous corticosteroids and intact HPAA responses are involved in the restriction of inflammatory processes.

HPAA and Infections: Evidence that the HPAA Plays a Role in Susceptibility to Infectious Diseases

Since stress has profound effects on the central CRH-system regulating circulating glucocorticoid concentrations,[8] the stress paradigm has often been used to study the role of glucocorticoids in infections. In a number of human and animal studies it has been shown that stress can alter susceptibility to and the course of infectious diseases, including viral infections such as rhinovirus,[11] influenza,[17,25] tuberculosis (TB),[4] and bacterial infection.[18]

A series of studies in mice infected with influenza virus[17,25] have shown that restraint stress, which induces a profound HPAA response, is associated with greater severity of infection than in mice not exposed to such stressful stimuli. Simultaneous treatment of restrained animals with the glucocorticoid receptor antagonist RU 486 reconstitutes the viral resistance. Similar studies in mice infected with TB also show that RU 486 reverses the effects of restraint stress on both increased corticosterone responses and progression of mycobacterium infection.[4]

In a recent study on stress and the common cold in humans, a correlation between psychological stress and severity of infection was shown,[11] although the effects were relatively small and the mechanism involved remains obscure. Studies of HPAA function in HIV-positive humans suggest that ACTH and cortisol responses are blunted compared to HIV-negative individuals.[29] Studies in patients with the advanced stage of HIV infection that precedes the development of AIDS, show varying ACTH and cortisol responses to infusions of ovine CRH.

Variables which complicate interpretations of results in human studies may be related to varying stages of the disease and concurrent drug therapy in patients. Moreover, one cannot discriminate if the change in HPAA activity is primary or secondary to the disease process. Nevertheless, a correlation is observed between pathological stages and changed HPAA activity.

SUMMARY AND CONCLUSIONS

Glucocorticoid feedback during inflammation depends on several factors, including effective corticosteroid concentrations, the temporal coordination between the immune system and the HPAA, and sensitivity of the target organ to corticosteroids.

In the case of LEW/N rats, several lines of evidence indicate that CRH is aberrantly regulated, and as a consequence, the circulating corticosterone concentrations are blunted. In addition, basal ACTH and corticosterone responses differ compared to SD and F344/N rats.[16,40] Therefore glucocorticoid feedback during inflammation is impaired leading to uncontrolled inflammation as observed in various models. We found that in the LEW/N rat, the regulation of the GR and MR, at least in the hippocampus, is different from F344/N rats.[41] A similar impaired upregulation of the GR has been described in aged rats, and this has been postulated to be the cause of the altered HPAA reactivity after various stimuli in these rats.[9,19,38] Moreover, adult LEW/N rats exhibit HPAA responses which resemble those observed during the stress hyporesponsive period of immature SD or F344/N rats.[1] These latter data suggest that in the LEW/N rat the ontogeny and/or regulation of the

HPAA is disturbed. Taken together, the impaired regulation of HPAA reactivity observed in LEW/N rats could be due to an impaired regulation of the central MR and/or GR. Although no data of GR dynamics in the context of local inflammation are available yet, this could be an additional factor determining tissue sensitivity and immune responses to corticosteroids.

In humans, determination of the role of the HPAA in inflammation is more complex due to the much higher intrinsic variability between subjects. Furthermore, difference in plasma cortisol concentrations between healthy controls and patients with immunological diseases such as rheumatoid arthritis are small; in the past few years, several groups have tested basal corticosterone levels and the responsivity of the HPAA in patients with autoimmune diseases, including rheumatoid arthritis,[6,7,35] fibromyalgia,[22] and multiple sclerosis (Michelson, J. Clin. Endocrinol. Metab. (1994), in press). In general, only small differences were seen in basal corticosteroid levels, while stimulation by CRH-infusion, exercise or hypoglycemia induced an ACTH response while the corticosteroid response was either unchanged or only slightly different. Thus, it is unlikely that the plasma cortisol concentration alone is a factor directly leading to pathology.

However additional corticosteroid-related factors may be involved. First, the timing of the cortisol response in relation to immune activation is crucial. For example, during the first two hours following T cell stimulation, several tightly regulated sequential events in the T cell take place,[12] while 10 min after stimulation T cells become refractory to inhibition by corticosteroids.[10] Therefore, a subtle change in the timing of HPAA reactivity following immune challenge could lead to altered immunoreactivity. Second, the sensitivity of the target tissue determines to a large degree the efficacy of cortisol. Several factors, including CBG, hormone-inactivating enzymes, heat shock proteins and receptor expression are involved. Also cytokines, released during inflammation, can further alter both Kd and Bmax of the GR in different systems.[26,27] Interestingly, the sensitivity of the HPAA to dexamethasone suppression, as used as a diagnostic tool in depressive disorders and Cushing's syndrome, was found to correlate with *in vitro* sensitivity of T cells to the inhibitory action of dexamethasone.[31] In these studies it has been proposed that the regulation of the GR, possibly by corticosteroids itself, determined corticosteroid sensitivity or resistance. Moreover, mutations in the GR itself have been shown to be associated with the chronic fatigue syndrome in a special case.[28]

Taken together, these findings indicate that the regulation of the GR is an important factor in determining corticosteroid sensitivity of target tissue. Thus the dynamics and the regulation of the GR, in addition to HPAA reactivity with corticosteroids as effector hormones, may play a role in the control of inflammation.

REFERENCES

1. Aksentijevich, S., H. J. Whitfield, W. Scott Young III, R. L. Wilder, G. P. Chrousos, P. W. Gold & E. M. Sternberg. 1992. Arthritis-susceptible Lewis rats fail to emerge from the stress hyporesponsive period. Dev. Brain Res. **65:** 115–118.
2. Bradbury, M. J., S. F. Akana & M. F. Dallman. 1994. Roles of type I and II corticosteroid receptors in regulation of basal activity in the hypothalamo-pituitary-adrenal axis during the diurnal trough and the peak: evidence for a nonadditive effect of combined receptor occupation. Endocrinology **134:** 1286–1296.
3. Bristol, L. A., S. K. Durum & S. P. Eisenberg. 1993. Differential regulation of group A streptococcal peptidoglycan polysaccharide (PG-APS) stimulated macrophage production of IL-1 by rat strains susceptible and resistant to PG-APS induced arthritis. Cell Immunol. **149:** 130–143.

4. Brown, D. H., J. F. Sheridan, D. Pearl & B. S. Zwillig. 1993. Regulation of mycobacterial growth by the hypothalamus-pituitary-adrenal axis: differential responses of mycobacterium bovis BCG-resistant and -susceptible mice. Infect. Immun. **61:** 4793–4800.
5. Calogero, A. E., E. M. Sternberg, G. Bagdy, C. Smith, R. Bernardini, S. Aksetijevich, R. L. Wilder, P. W. Gold & G. P. Chrousos. 1992. Neurotransmitter-induced hypothalamic-pituitary-adrenal axis responsiveness is defective in inflammatory disease-susceptible Lewis rats: *in vivo* and *in vitro* studies suggesting globally defective hypothalamic secretion of corticotropin-releasing hormone. Neuroendocrinology **55:** 600–608.
6. Cash, J. M., L. J. Crofford, W. T. Gallucci, E. M. Sternberg, P. W. Gold, G. P. Chrousos & R. L. Wilder. 1992. Pituitary-adrenal axis responsiveness to ovine corticotropin releasing hormone in patients with rheumatoid arthritis treated with low dose prednisone. J. Rheumatol. **19:** 1692–1696.
7. Chikanza, I. C., P. Petoru, G. Kingsley, G. Chrousos & G. S. Panayi. 1992. Defective hypothalamic response to immune and inflammatory stimuli in patients with rheumatoid arthritis. Arthritis Rheum. **35:** 1281–1288.
8. Chrousos, G. P. & P. W. Gold. 1992. The concepts of stress and stress system disorders: overview of physical and behavioral homeostasis. JAMA **267:** 1244–1252.
9. Cizza, G., A. E. Calogero, L. S. Brady, G. Bagdy, E. Bergamini, M. R. Blackman, G. P. Chrousos & P. W. Gold. 1994. Male Fischer 344/N rats show a progressive central impairment of the hypothalamic pituitary adrenal axis with advancing age. Endocrinology **134:** 1611–1620.
10. Claman, H. N. 1972. Corticosteroids and lymphoid cells. N. Engl. J. Med. **287:** 388–397.
11. Cohen, S., D. A. J. Tyrell & A. P. Smith. 1991. Psychological stress and susceptibility to the common cold. N. Engl. J. Med. **325:** 606–612.
12. Crabtree, G. R. 1989. Contingent genetic regulatory events in T lymphocyte activation. Science **243:** 355–361.
13. Daynes, R. A., B. A. Araneo, T. A. Dowell, K. Huang & D. Dudley. 1990. Regulation of murine lymphokine production *in vivo.* J. Exp. Med. **171:** 979–996.
14. DeRijk, R. H. & F. Berkenbosch. 1991. The immune-hypothalamo-pituitary adrenal axis and autoimmunity. Int. J. Neurosci. **59:** 91–100.
15. DeRijk, R. H. & F. Berkenbosch. 1994. Suppressive and permissive actions of glucocorticoids; a way to control innate immunity and to facilitate specificity of adaptive immunity? *In* Bilateral Communication between the Endocrine and Immune Systems. C. J. Grossman, Ed. 73–95. Springer-Verlag. New York.
16. Dhabhar, F. S., B. S. McEwen & R. L. Spencer. 1993. Stress response, adrenal steroid receptor levels and corticosteroid-binding globulin levels—a comparison between Sprague-Dawley, Fischer 344 and Lewis rats. Brain Res. **616:** 89–98.
17. Dobbs, C. M., M. Vasquez, R. Glaser & J. F. Sheridan. 1993. Mechanisms of stress-induced modulation of viral pathogenesis and immunity. J. Neuroimmunol. **48:** 151–160.
18. Edwards III, C. K., L. M. Yunger, R. M. Lorence, R. Dantzer & K. W. Kelley. 1991. The pituitary gland is required for protection against lethal effects of *Salmonella typhimurium.* Proc. Natl. Acad. Sci. **88:** 2274–2277.
19. Eldridge, J. C., D. G. Fleenor, D. S. Kerr & P. W. Landfield. 1989. Impaired upregulation of type II corticosteroid receptors in hippocampus of aged rats. Brain Res. **478:** 248–256.
20. Fauci, A. S. 1976. Glucocorticoid therapy: mechanisms of action and clinical considerations. Ann. Inter. Med. **84:** 304–315.
21. Goldstein, D. S., M. Garty, G. Bagdy, K. Szemerdi, E. M. Sternberg, S. Listwak, K. Pacek, A. Deka-Starosta, P. C. Chang, R. Stull, P. W. Gold & I. J. Kopin. 1993. Role of CRH in glucopenia-induced adrenomedullary activation in rats. J. Neuroendocrinol. **5:** 475–486.
22. Griep, E. N., J. W. Boersma & E. R. de Kloet. 1993. Altered reactivity of the hypothalamic-pituitary-adrenal axis in the primary fibromyalgia syndrome. J. Rheumatol. **20:** 469–474.

23. GROSSMAN, C. J., G. A. ROSELLE & C. L. MENDENHA. 1991. Sex steroid regulation of autoimmunity. J. Steroid Biochem. **40:** 649–659.
24. HENCH, P. S., E. C. KENDALL, C. H. SLOCUMB & H. F. POLLEY. 1949. The effect of a hormone of the adrenal cortex (17-hydroxy-11-dehydrocorticosterone: compound E) and of pituitary adrenocorticotropic hormone on rheumatoid arthritis. Proc. Staff Meet. Mayo Clin., Rochester **24:** 181–197.
25. HERMANN, G., C. A. TOVAR, F. M. BECK, C. ALLEN & J. F. SHERIDAN. 1993. Restraint stress differentially affects the pathogenesis of an experimental influenza viral infection in three inbred strains of mice. J. Neuroimmunol. **47:** 83–94.
26. HILL, M. R., R. D. STITH & R. E. MCCALLUM. 1988. Human recombinant IL-1 alters glucocorticoid receptor function in reuber hepatoma cells. J. Immunol. **141:** 1522–1528.
27. KAM, J. C., S. J. SZEFLER, W. SURS, W. SHER & D. Y. M. LEUNG. 1993. Combination IL-2 and IL-4 reduces glucocorticoid receptor-binding affinity and T cells response to glucocorticoids. J. Immunol. **151:** 3460–3466.
28. KARL, M., S. W. J. LAMBERTS, S. D. DETERA-WADLEIGH, I. ENCIO, C. STRATAKIS, D. M. HURLEY, D. ACCILI & G. CHROUSOS. 1993. Familial glucocorticoid resistance caused by a splice site deletion in the human glucocorticoid receptor gene. J. Clin. Endocrinol. Metab. **76:** 683–689.
29. KUMAR, M., A. M. KUMAR, R. MORGAN, J. SZAPOCZNIK & C. EISDORFER. 1993. Abnormal pituitary-adrenocortical response in early HIV-1 infection. J. AIDS **6:** 61–65.
30. LEMERCIER, C., N. JULEN, M. COULPIER, H. DAUCHEL, D. OZANNE, M. FONTAINE & J. RIPOCHE. 1992. Differential modulation by glucocorticoids of alternative complement protein secretion in cells of the monocyte/macrophage lineage. Eur. J. Immunol. **22:** 909–915.
31. LOWY, M. T., A. T. REDER, J. P. ANTEL & H. Y. MELTZER. 1984. Glucocorticoid resistance in depression: the dexamethasone suppression test and lymphocyte sensitivity to dexamethasone. Am. J. Psychiatry **141:** 1365–1370.
32. MASON, D. W., I. MACPHEE & F. ANTONI. 1990. The role of the neuroendocrine system in determining genetic susceptibility to experimental allergic encephalomyelitis in the rat. Immunology **70:** 1–5.
33. MUNCK, A., P. M. GUYER & N. Y. HOLBROOK. 1984. Physiological functions of glucocorticoids in stress and their relation to pharmacological actions. Endocr. Rev. **5:** 25–44.
34. MUNCK, A. & A. NARAY-FEJES-TOTH. 1992. The ups and downs of glucocorticoid physiology: permissive and suppressive effects revisited. Mol. Cell. Endocrinol. **90:** C1–C4.
35. NEECK, G., K. FEDERLIN, V. GRAEF, D. RUSCH & K. L. SCHMIDT. 1990. Adrenal secretion of cortisol in patients with rheumatoid arthritis. J. Rheumatol. **17:** 24–29.
36. PAUL, W. E. & R. A. SEDER. 1994. Lymphocyte responses and cytokines. Cell **76:** 241–251.
37. PEIFFER, A., B. LAPOINTE & N. BARDEN. 1991. Hormonal regulation of type II glucocorticoid receptor messenger ribonucleic acid in rat brain. Endocrinology **129:** 2166–2174.
38. SAPOLSKY, R. S., L. C. KREY & B. S. MCEWEN. 1983. Corticosterone receptors decline in a site-specific manner in the aged rat brain. Brain Res. **289:** 235–240.
39. SCHUURS, A. H. M. 1990. Effects of gender and sex steroids on the immune response. J.Steroid Biochem. **35:** 157–172.
40. SMITH, C. C., E. HAUSER, N. K. RENAUD, A. LEFF, S. AKSETIJEVICH, G. P. CHROUSOS, R. L. WILDER, P. W. GOLD & E. M. STERNBERG. 1992. Increased hypothalamic [^{3}H]flunitrazepam binding in hypothalamic-pituitary-adrenal axis hyporesponsive Lewis rats. Brain Res. **569:** 295–299.
41. SMITH, C. C., R. J. OMELJANIUK, H. J. WHITFIELD, S. AKSENTIJEVICH, M. W. FELLOWS, E. ZELAZOWSKI, P. W. GOLD & E. M. STERNBERG. 1994. Differential mineralocorticoid (type I) and glucocorticoid (type II) receptor expression in Lewis and Fischer rats. Neuroimmunomodulation **1:** 66–73.
42. STERNBERG, E. M. 1992. The stress response and the regulation of inflammatory disease. Ann. Inter. Med. **117:** 854–866.
43. STERNBERG, E. M., J. M. HILL, G. P. CHROUSOS, T. KAMILARIS, S. J. LISTWAK, P. W.

GOLD & R. L. WILDER. 1989. Inflammatory mediator-induced hypothalamic-pituitary-adrenal axis activation is defective in streptococcal cell wall arthritis susceptible Lewis rats. Proc. Natl. Acad. Sci. USA **86:** 2374–2378.
44. STERNBERG, E. M., W. S. YOUNG III, R. BERNARDINI, A. E. CALOGERO, G. P. CHROUSOS, P. W. GOLD & R. L. WILDER. 1989. A central nervous system defect in biosynthesis of corticotropin-releasing hormone is associated with susceptibility to streptococcal cell wall arthritis in Lewis rats. Proc. Natl. Acad. Sci. USA **86:** 4771–4775.
45. VIAU, V. & M. J. MEANEY. 1991. Variations in the hypothalamic-pituitary-adrenal response to stress during the estrous cycle in the rat. Endocrinology **129:** 2503–2511.
46. WICK, G., Y. HU, S. SCHWARZ & G. KROEMER. 1993. Immunoendocrine communication via the hypothalamo-pituitary-adrenal axis in autoimmune diseases. Endocr. Rev. **14:** 539–563.

DISCUSSION OF THE PAPER

B. S. MCEWEN (*The Rockefeller University*): I wonder if you could add some comments with regard to other CRH systems in the brain. In addition there is now information about the local proinflammatory effects of CRH perhaps produced by macrophages. Have you or has anyone else had a chance to look at this in Lewis and Fischer rats?

STERNBERG: We are studying CRH in the amygdala in these strains, but we do not have the results yet. Drs. George Chrousos and Ronald Wilder with our group have studied CRH in the periphery. Lewis rats do make significantly more CRH in the periphery at sites of inflammation, compared to Fischer rats.

QUESTION: You showed that there was no CRH response in the Lewis rats despite the fact that they did have significant corticosterone effects. What happens with arginine vasopressin hormone (AVP) in Lewis rats?

STERNBERG: Lewis rats have a greater AVP response than Fischer. However, one wonders why Lewis rats do not die when treated with streptococcal cell wall peptidoglycan polysaccaride. They just go on to develop arthritis, whereas Fischer rats given RU 486 plus SCW die. And it is the same in the EAE model. If you adrenalectomize and give them the antigen, they die. So Lewis rats clearly have evolved a compensation for this life-threatening condition and possibly the mechanism is through AVP.

J. P. HERMAN (*University of Kentucky*): One of the things that hypothalamic transplants do is crank up growth factors. Did you see any increase in endogenous CRH in those animals?

STERNBERG: I did not show those experiments because they are still in progress. But, yes, we saw an increase in endogenous CRH in the rats that received the Fischer hypothalami, the Lewis hypothalami, and even the spinal cords, but not in the sham-operated animals. Thus, it is probably not only hypothalamic CRH which suppresses peripheral inflammation, because the only rats that had a suppression of inflammation were the rats that received the Fischer hypothalami, despite the fact that the endogenous CRH was cranked up in all cases. Central CRH is therefore a necessary but probably not a sufficient condition for suppressing or predisposing resistance to inflammation.

QUESTION: What is known about the genetics of the Lewis rat? Is this a single gene disease?

STERNBERG: Those experiments are currently in progress; perhaps I shall tell you the results next year.

Glial and Neuronal Glucocorticoid Receptor Immunoreactive Cell Populations in Developing, Adult, and Aging Brain[a]

A. CINTRA,[b] M. BHATNAGAR,[b], G. CHADI,[b] B. TINNER,[b]
J. LINDBERG,[b] J-Å. GUSTAFSSON,[c] L. F. AGNATI,[d]
AND K. FUXE[b]

[b]Department of Neuroscience
Division of Cellular and Molecular Neurochemistry
Karolinska Institutet
171 77 Stockholm, Sweden

[c]Department of Medical Nutrition
Huddinge University Hospital
Huddinge, Sweden

[d]Department of Human Physiology
University of Modena
Modena, Italy

INTRODUCTION

Since 1985[1] we have charted glucocorticoid receptor (GR) immunoreactive (IR) nerve and glial cell populations in the rat brain using a mouse monoclonal IgG 2a anti-GR antibody.[1–5] By means of double immunolabeling procedures it has also been possible to demonstrate that especially the monoamine and peptidergic neurons of various classes represent important targets for glucocorticoid action.[1–4] Recently for the first time we were able to objectively and extensively characterize GR immunoreactive nerve cell and glial cell populations in the adult brain.[7] This work will now be summarized in the present article together with the GR mapping in the developmental and aged brain, giving new aspects on the role of the GR in neuronal plasticity.

Regional and Cellular Localization of the Glucocorticoid Receptor in the Adult Rat CNS

Mapping Based on Subjective Evaluations

GR immunoreactive charts were obtained by our group[5–7] and by Ahima and Harlan (1990).[8] These maps are based on subjective evaluations of GR IR. A

[a]This work was supported by grants from the Swedish Medical Research Council (04X-715), from the Knut and Alice Wallenberg Foundation. GC was supported by a grant (91/2989-9) from FAPESP, São Paulo, Brazil.

comparison of these subjective evaluations with the application of Spearman's nonparametric correlation coefficient was made by evaluating the differences between each pair of scores. It was found that these two subjective evaluations only are in agreement with regard to the telencephalon, although also in this region a large variability among the two sets of scores is found. It must be considered that the subjective evaluations have a tendency to become uncertain for various reasons, such as the variability in the scales of scores, the wideness of each class and how each group considers the density and intensity of a staining in their evaluation. Finally, with different backgrounds it becomes difficult to obtain a subjective evaluation of the intensity. It must be considered, however, that the discrepancies found between the two available GR immunoreactive charts of the rat CNS in certain regions, such as the CA3 and CA4 areas of the hippocampal formation, and the Purkinje cell layer and the arcuate nucleus represent true differences. Thus, the two monoclonal antibodies used may recognize different epitopes of the GR and therefore differential posttranslational modification of the GR may lead to the regional differences in GR staining observed. Also the two monoclonal antibodies may recognize different conformational states of the GR. It should be pointed out that in areas with discrepancies in GR immunocytochemistry, GR mRNA levels can still be observed, suggesting that monoclonal GR antibodies do not recognize epitopes of other proteins.[9,10] It is questionable, however, if significant GR IR exists within the supraoptic nucleus, since very low GR mRNA levels have been observed in this region. In our laboratory we also failed to observe any significant GR IR within the supraoptic nucleus. Therefore, the findings of Kiss *et al.* (1988)[11] of GR IR in the supraoptic nucleus must be regarded with caution. There is also an agreement that the vast majority of the magnocellular neurons of the paraventricular hypothalamic nucleus lack GR IR and GR mRNA levels.[6,9,10]

Overall GR recognized by immunocytochemistry[8,12] shows substantial similarities with the distribution patterns of the GR obtained with *in situ* hybridization, biochemical binding techniques and receptor autoradiography.[9,10,13–15]

Quantitative Mapping

We performed a microdensitometric and morphometric analysis of GR IR at various rostrocaudal levels of the rat brain and the spinal cord[5] in order to obtain for the first time an objective mapping of GR immunoreactive glial and neuronal profiles in the CNS. TABLE 1 presents brain areas with number of GR immunoreactive profiles per 10^4 μm^2 and intensity of staining (mean grey value). High densities of GR immunoreactive neuronal profiles can be noticed within the anterior olfactory nuclei, most layers of the cingulate cortex, within the outer layers of the frontal and parietal cortex, the retrosplenial agranular cortex as well as within the nucleus caudate putamen, the nucleus accumbens, the tuberculum olfactorium and the ventromedial thalamic nucleus. In contrast, the highest intensity of GR IR is found in restricted nerve cell nuclei of the hypothalamus, the hippocampal formation and the lower brain stem, such as the locus coeruleus and the other noradrenaline (NA)-rich cell body regions and adrenaline-rich areas, as well as in the raphe regions rich in serotonin (5-HT)-containing neurons.[1,6,16]

By means of GR immunocytochemistry in combination with a morphometric method it was also possible to describe nerve cell clusters in the dorsal striatum and in the nucleus accumbens of the male rat,[7] giving evidence for a new basic cytoarchitectonic organization of these regions in the rat. Islands, often containing 3 to 10 GR immunoreactive nerve cells appear to represent a major building block

TABLE 1. Density and Mean Gray Value (MGV) of Glucocorticoid Receptor Immunoreactive Profiles in Some Areas of the Rat Brain as Analyzed at Various Rostrocaudal Levels in Coronal Sections[a]

Bregma Level	Brain Area	Sampled Area 10^4 μm²	n Profiles per 10^4 μm²	MGV	Bregma Level	Brain Area	Samples Area 10^4 μm²	n Profiles per 10^4 μm²	MGV
6.2	IGr	12	6.1		−2.3	VM	23.0	8.1	47
5.2	AOD	12	11.3		−2.3	LH	15.0	10.9	
5.2	AOL	14	10.8	88	−3.3	Arc	5.0	24.8	99
5.2	AOV	12	11.7		−3.3	VMH	13.0	28.0	
5.2	AOM	14	14.3		−3.3	DMD	4.0	33.5	104
4.2	Cg (2-3)	11	9.7		−3.3	CA1 (alv)	1.6	21.6	68
4.2	Fr	13	9.7	81	−3.3	CA1 (or)	6.0	7.2	79
2.2	Acb	29	12.2	54	−3.3	CA1 (Py)	2.9	50.0	108
2.2	Pir II	7	13.3		−3.3	CA1 (Rad)	13.0	3.8	74
2.2	Tu II	16	19.2	58	−3.3	CA1 (LMol)	2.5	10.7	61
1.0	CPu dm	29	16.5		−3.3	CA 2 (Py)	2.9	45.3	111
1.0	CPu vl	29	14.6		−3.3	CA 3 (Py)	3.9	17.3	
1.0	Par II	16	11.6	56	−3.3	DG (Mol)	8.0	5.6	44
1.0	Par III-IV	10	16.3		−3.3	DG (Gr)	3.3	76.0	64
1.0	Par V-VI	16	12.9		−4.3	LG	25.0	12.9	90
0.2	LSI	21	12.9		−5.8	MGB	18.0	12.0	
0.2	BST	17	18.0	45	−5.8	SNC	4.8	8.2	
−0.3	MPA	28	14.5	55	−5.8	SNR	19.0	5.4	
−0.3	HDB	29	3.1		−5.8	VTA	7.5	11.0	81
−0.3	GP	29	3.0	34	−5.8	SC	29.0	9.0	52
−0.9	SFO	16	17.7		−5.8	CG	17.0	14.1	68

−0.9	LPO	23	11.1		−7.8	DR	18.0	11.5	89
−0.9	MPO	17	16.0		−9.3	RPn	6.0	6.6	80
−1.3	DEn	20	11.0		−9.3	LC	6.0	7.6	124
−1.3	AHA	27	15.4		−9.3	LPB	17.0	13.4	60
−1.3	SCh	5.5	4.5		− 10.3	Sp 5	20.0	11.3	
−1.3	AV	11	16.4		−10.3	Cb (Mol)	19.2	9.4	
−1.8	HL II)	10	19.0	84	−10.3	Cb (P)	1.7	33.5	
−1.8	HL III	10	14.2	78	−10.3	Cb (Gr)	14.8	56.0	51
−1.8	HL IV	13	9.0	64	−10.3	Cb (wm)	5.1	10.9	35
−1.8	HL V	13	12.6	74	−10.3	RMg	22.0	8.8	
−1.8	HL VI	15	12.6	81	−12.7	C3	11.0	4.0	
−1.8	Cg	8	30.3	62	−12.7	C1	6.6	7.0	
−1.8	PAFP	10	18.4	107	−13.3	Sol	8.2	4.0	93
−2.3	Ce	23	10.6	72	−13.3	dmX	0.6	17.0	
−2.3	LD	27	9.7	51	−13.3	Sp 5	3.5	4.7	64
−2.3	Me	21	15.0	61		Sp Cervical	1.9	25.1	
−2.3	VMH	12.0	12.2	67		IML	0.9	10.0	

[a]Four male rats were used. The sections from three rat brains were stained for GR IR without nickel enhancement. Out of these three series of brain sections the rat showing most frequently the median values was chosen to be shown in this table. The value obtained from each rat was the mean value of 2–8 determinations. The values obtained in this way should only be regarded as a way to obtain a better rank among regions with regard to density of GR IR profiles.

in the organization of the dorsal striatum and the nucleus accumbens. The previously demonstrated dopamine (DA), enkephalin (ENK) and DARPP-32 immunoreactive islands and striae of the dorsal striatum are substantially larger and less uniformly distributed than GR IR islands. Thus, it must be emphasized that GR IR is uniformly distributed in the striatum and will not determine the formation of chemical-anatomical compartments. Instead, GR input to the striatal cells may increase the formation of possible integrative compartments as well as the complexity of the pattern in their temporal activation to reach a final integrative compartment.[7]

Density histograms and correlations between densities of GR immunoreactive neurons and GR immunoreactive glial cells have revealed substantial differences between these GR immunoreactive cell populations.[5] Thus, while GR immunoreactive neurons are present in practically an equal number of regions with high and low density without a high modal value, GR immunoreactive glial cells are present in a large number of brain regions with a low density of GR immunoreactive profiles. Thus, under physiological conditions it appears likely that the glucocorticoid regulation of glial cells in the brain takes place only in a restricted population of such cells. Of special interest is the observed inverse correlation between the density of GR immunoreactive neurons and glial cells in large numbers of brain regions, with the glial cells appearing in larger amounts in the midbrain, pons, medulla oblongata and spinal cord. It should be noticed, however, that the glial nuclear profiles dominate in some regions of the tel- and diencephalon, such as the molecular layers of the cerebral cortex, the CA3 and CA4 area of the hippocampal formation, the nucleus of the horizontal limb of the diagonal band, the globus pallidus and the zona reticulata of the substantia nigra. Thus, it seems clear that in some brain regions glucocorticoids mainly affect metabolism via direct actions on glial cells, while in most brain areas the major target is represented by nerve cell populations.

By means of the quantitative analysis of GR immunoreactive neuronal and glial profiles it has become possible to perform a taxonomy of GR immunoreactive brain regions.[5] Through the use of a similarity index of GR-related parameters it becomes possible to show plots demonstrating links among regions showing the highest similarity. In this analysis eight forebrain regions were found to be connected at a 95% similarity level although belonging to different neuronal circuitries. The major parameter linking these regions together appeared to be a high density of GR immunoreactive neurons. The forebrain regions involved were the frontal cortex, the dorsal and ventral striatum, the limbic cortex and the anterior olfactory nuclei. It was therefore concluded that glucocorticoids affect various circuitries of the sensory-motor cortex and of the limbic system in order to optimize their performances and to reach an appropriate tuning in these networks to obtain optimal integration.

Cellular Localization

A large number of papers[6,8,17] have clearly demonstrated that GR IR is predominantly found in the nuclei of nerve and glial cells, using two types of anti-GR monoclonal mouse antibodies.[18,19] However, it must be underlined that in large nerve cells, such as the pyramidal nerve cells of the neocortex and of the hippocampal formation, a cytoplasmatic GR IR has been demonstrated, and it has been suggested that cytoplasmic GR IR may be preferentially sensitive to the detergent Triton X-100 used in the present immunocytochemical studies.[20] Furthermore, the available evidence also suggests that the monoclonal antibody mainly recognizes the activated form of the GR, probably due to the conformational changes taking place leading to exposure of antigenic determinants in the GR. It is probably the lack of binding of the monoclonal antibody which may be the major explanation for the disappearance of nuclear GR IR in glial and neuronal cells seen after adrenalectomy.[6,8] These results are of special importance when interpreting the results obtained in mapping GR IR in the CNS. Thus, the mapping of GR IR will be dependent not only upon the density of the GR but also upon how many of the GRs that are activated by corticosterone, so that the GR distribution pattern in nerve and glial cell populations will be dependent upon the endocrine state of the animal.

Cytological studies on the GR in nonneuronal cell cultures, however, have more clearly demonstrated the presence of the GR both within the nucleus as well as in the cytoplasm.[21,22] It is of substantial interest that during mitosis the GR was found in the centriolar region and in the mitotic spindle apparatus accompanied by a diffuse staining of the surrounding cytoplasm. Also the glucocorticoids did not appear to produce a compartment shift but seemed to increase the affinity of the GR for its nuclear and cytoplasmic docking sites. It was also speculated that the structural association of the GR with microtubules and centrioles may in fact represent sites of GR action. However, all the results obtained in the CNS with a predominant nuclear localization of the GR strongly implies that the major pool of GR are involved in nuclear gene transcriptional control.

GR immunoreactive nuclear neuronal profiles dominated over nuclear glial profiles in the vast majority but not in all of the areas analyzed in the grey matter of the CNS. In the white matter GR immunoreactive glial profiles represented both oligodendroglia cells mainly found within the telencephalon, and fibrillary astroglia mainly found in the lower brain stem and the spinal cord. The distribution pattern within the grey matter and double labelling experiments with glial fibrillary acidic protein (GFAP) antibodies (FIG. 1) suggested the existence of GR IR both within protoplasmatic astroglia as well as in perineuronal oligodendroglia.[5] Thus, although in the past we emphasized the neuronal actions of glucocorticoids, it must be kept in mind that the various glial populations of the grey and white matter of the CNS also represent major targets for the actions of glucocorticoids[6,23] as seen, *e.g.*, from the ability of dexamethasone to increase glial angiotensinogen mRNA levels in discrete brain areas.[24]

GR IR in Transmitter-Identified Neurons of the Adult Rat CNS

Neurons Containing Classical Transmitters

A large number of studies have demonstrated that many classes of monoamine and peptidergic neurons represent targets for glucocorticoid action.[1–4] Thus, the vast majority of the NA and 5-HT cell groups contain strong nuclear GR IR, which is true also for the adrenaline cell groups of the medulla oblongata. Furthermore, 40%–75% of the total population of GR immunoreactive midbrain DA cells contain moderate GR IR. Various DA cell groups of the hypothalamus with the exception of the A11 DA cell group in the posterior hypothalamus showed a moderate GR IR.[1] Also large numbers of glutamate neurons contain strong GR IR, such as the pyramidal nerve cells of the CA1 and CA2 area, and the small pyramidal cells of the layer 3 of the cerebral cortex. Thus, it seems probable that the glucocorticoids can strongly modulate glutamate transmission and metabolic events also in these types of neurons. This is probably true also for large numbers of GABAergic neurons, since, *e.g.*, the vast majority of striatal nerve cells represent GABAergic efferent neurons to the

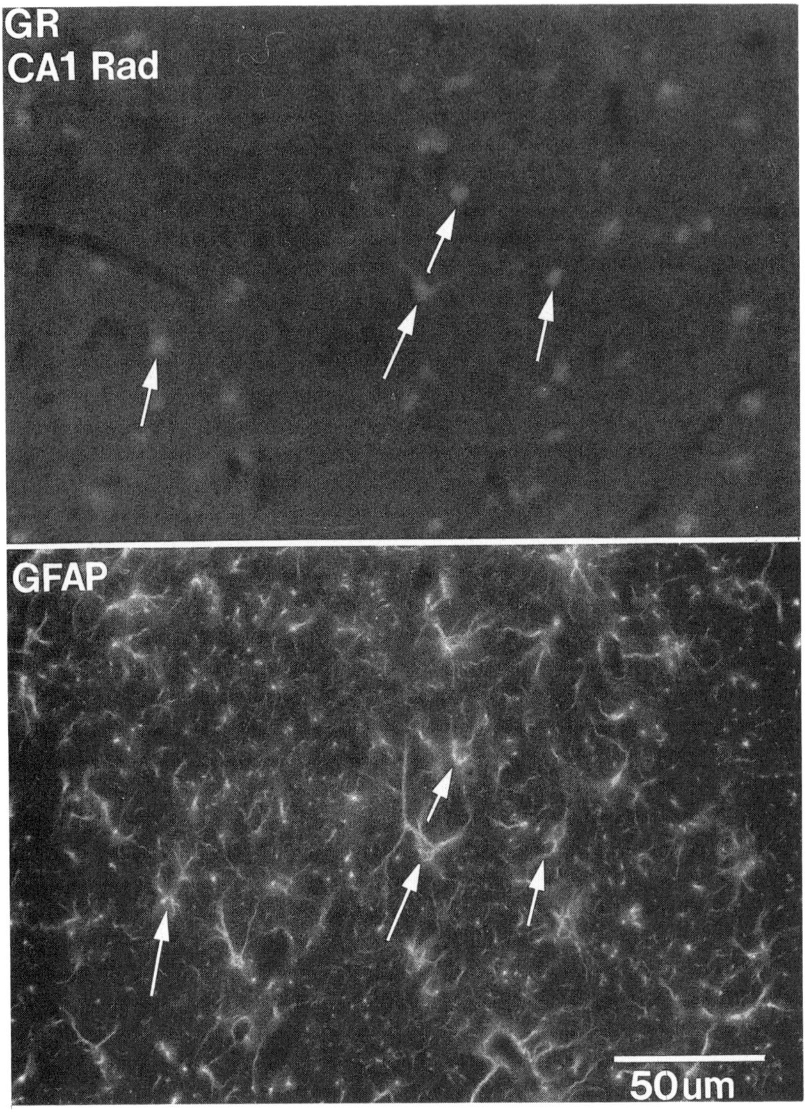

FIGURE 1. Double immunolabelling of a coronal section of the dorsal hippocampus (CA1 stratum radiatum) in adult rat brain using a rabbit polyclonal antiserum against GFAP (1:2000) and a mouse monoclonal GR antiserum (1:200). The GR IR was visualized by FITC fluorescence and the GFAP IR by biotin streptavidin Texas-Red fluorescence. *Arrows* point to examples of glial cells containing both markers (nuclear GR IR and GFAP IR processes).

globus pallidus and the substantia nigra. As previously described,[7] the vast majority of middle sized striatal neurons contain GR IR. However, cholinergic neurons within the nuclei of the diagonal band and the medial septal nucleus, giving rise to cholinergic pathways to the cerebral cortex including the hippocampal formation, do not show any GR IR. This is also true for the large cholinergic interneurons of the striatum.[7] The cholinergic motor neurons of the cranial nerve nuclei, such as the facial moyor nucleus and the trigeminal motor nucleus, contain weak if any GR nuclear IR.

Neurons Containing Neuropeptides

The vast majority of corticotropin releasing factor (CRF) immunoreactive neurons of the hypothalamus and the preoptic area and of the amygdaloid complex contain strong nuclear GR IR, which is true also for growth hormone releasing factor immunoreactive nerve cells of the arcuate nucleus.[2,25] A number of thyrotropin release hormone immunoreactive neurons in various preoptic and hypothalamic nuclei of the male rat also contain strong GR IR. With regard to somatostatin immunoreactive neurons projecting into the median eminence, two thirds of them show GR IR of strong intensity.[3,25] GR IR has also been demonstrated in luteinizing hormone release hormone immunoreactive neurons of the female and male rat.[26] Thus, the vast majority of hypophysiotrophic hormone containing neurons of the hypothalamus and the preoptic area show moderate-to-strong GR IR, indicating a major modulatory role of glucocorticoids in the control of the neuroendocrine system.

It must be emphasized, however, that most families of neuropeptide neurons in the brain can demonstrate nuclear GR IR but that peptidergic neurons reveal marked heterogeneities among brain areas. Thus, the amount of GR IR in the various types of peptide neurons of the brain appears to be determined by the function of the region in which the peptidergic cell bodies are located.[3] Thus, within the arcuate nucleus practically all the peptidergic neurons show moderate-to-strong GR IR. Other peptidergic neurons which have been shown in some regions to be targets for glucocorticoid action are neuropeptide Y (NPY), galanin (GAL), neurotensin (NT), ENK and β-endorphin (β-END) immunoreactive neurons.[3,4,25]

Functional Implications

The results clearly indicate that large numbers of monoamine and peptidergic neurons are modulated by glucocorticoids via genomic actions, which *inter alia* may lead to changes in peptide and monoamine synthesis and break-down. It must be emphasized, however, that many regions innervated by monoamine and peptidergic nerve terminals also contain a high density of GR immunoreactive neurons. Thus, it seems likely that glucocorticoids via both pre- and postsynaptic effects at the genomic level can exert concerted effect on monoamine and peptide neurotransmission, leading to an increase or decrease of certain transmission lines. In view of the fact that the vast majority of the central synapses operate via multiple transmission lines it also seems likely that the actions of glucocorticoids on the monoamine and peptide systems can involve plastic changes in transmission so that some transmission lines are favored over others.[27,28]

Evidence has also recently been obtained that the peptidergic neurons of the spinal ganglia can contain nuclear GR IR. Thus, double immunofluorescence techniques demonstrate that one third of the CGRP and substance P containing neurons of the spinal and trigeminal ganglia possess substantial amounts of nuclear GR IR.[29] Thus, the possibility exists that glucocorticoids can control peptide synthesis also in substance P and CGRP neurons of the spinal ganglia and the trigeminal ganglia which may lead to alterations in pain perception. In agreement with this view, adrenalectomy has been found to significantly increase the number of cells containing substance P, suggesting that glucocorticoids exert an inhibitory action on substance P synthesis in these primary sensory neurons. Such effects on substance P synthesis may be of relevance of the antiflammatory and analgesic effects of glucocorticoids.[30]

The majority of monoaminergic and peptidergic neurons including hippocampal glutamate containing nerve cells exist within regions traditionally involved with stress actions. Therefore, it seems likely that glucocorticoids via these global effects on stress-related circuitries can reset the entire circuitries involved in the control of adrenocorticotropin secretion and also coordinate stress-related behaviors with the activity of the CRF neurons of the paraventricular hypothalamic nucleus. Also the presence of GR IR in a large number of nerve cells in the limbic regions as well as in pathways projecting into the limbic parts of the forebrain indicate that mood can be controlled by glucocorticoids at multiple sites.

The mapping of GR immunoreactive neurons in the adult rat has also demonstrated their presence in areas not traditionally involved in the control of stress actions. The map indicates, *e.g.*, that it is the afferent parts of the basal ganglia and not the efferent parts which are under a strong glucocorticoid control giving new aspects on how glucocorticoids released in stress could modulate sensory motor integration in the basal ganglia. This mapping has also indicated that the antinociceptive systems from the lower brain stem into the spinal cord, mainly substantia gelatinosa, as exemplified by 5-HT and NA descending neurons are under a strong glucocorticoid control. Thus, glucocorticoids may reduce pain responses not only by reducing pain transmission, *e.g.*, by effects on primary substance P immunoreactive sensory neurons, but also by activating antinociceptive systems. The mapping has also indicated that the somatosensory systems as well as the visual and acoustic pathways can be regulated by glucocorticoids at the thalamic and cortical level.[5]

Regional and Cellular Localization of the GR During Development and Aging

Prenatal Period

The recent mapping of prenatal GR mRNA levels and GR IR in the neuroepithelia of the medulla oblongata to the telencephalon gives clear evidence that GR in late embryonic life importantly controls proliferation within the neuroepithelia of the brain.[31] The mapping of the prenatal development of GR mRNA and IR also gives support for the view that glucocorticoids participate in the trophic regulation of the endocrine system in embryonic life as well as in the regulation of the dendritic branching and synaptogenesis of 5-HT and NA neurons.[32] A major GR mRNA signal is also observed in the arcuate nucleus and in the paraventricular hypothalamic nucleus, suggesting that the glucocorticoids may importantly take part also in the development and maturation of the neuroendocrine neurons representing key regions for endocrine control.

Diaz *et al.*[33] (and personal communication) have in line with this morphological analysis also demonstrated marked alterations in the pre- and postsynaptic properties of the nigrostriatal DA pathways following implantation of corticosterone pellets (release rate: 2.4 mg/day for 7 days) to pregnant dames during the last trimester of gestation. Increases in tyrosine hydroxylase mRNA levels in the ventral midbrain and increased striatal DA metabolism were associated with a reduced ability of apomorphine, a DA receptor agonist, to activate locomotion as evaluated in the prepubertal rat.[33] These results open up the possibility that stress in pregnancy can result via *inter alia* secretion of glucocorticoids to long-term changes in neuronal circuitries as exemplified in the nigrostriatal DA pathway.

Neonatal Period

In the neonatal rat brain strongly GR immunoreactive neurons have been demonstrated within arcuate and paraventricular neurons as well as in 5-HT and NA neurons, while a gradual development of GR IR takes place in cortical regions including hippocampal formation.[34] Postnatal corticosterone treatment has in fact also been shown to result in a long-term reduction of GR IR within CA1 hippocampal neurons but not within granular neurons of the dentate gyrus.[35] Thus, the treatment with corticosterone could result in a delay in the maturation processes of these neurons and/or in a long-term downregulation of GR IR in these cells. Mild stress has been shown to produce similar changes in GR IR.[36] The mild stress was induced by handling of the rat pups. In contrast, strong stress induced by maternal deprivation failed to produce such a long-term downregulation of GR IR within the pyramidal cells. One explanation for the difference between mild and severe stress may be that protooncogenes, such as c-fos, which become activated by severe stress can antagonize the activation by corticosterone of GR via antagonistic protein-protein interactions and/or via interference at the transcriptional level of the GR gene.[37,38]

It is well known that glucocorticoids in the neonatal period produce inhibitory actions on nerve cell proliferation. Nevertheless, it has been possible to demonstrate that neonatal treatment with corticosterone (10 mg/kg/day on postnatal day 2, 4, 6 and 8) can counteract the degeneration of tuberoinfundibular DA neurons and of arcuate growth hormone releasing hormone immunoreactive neurons induced by monosodium glutamate treatment.[39] Thus, in contrast to the case in hippocampal formation, where glucocorticoids appear to enhance the excitotoxic activity of glutamate,[40] corticosterone in the neonatal period in the arcuate nucleus appears capable of counteracting the activity at glutamate receptors. However, it must also be considered that corticosterone treatment also activates mineralocorticoid receptors (MR) and that the ratio of GR to MR in each individual nerve cell population may play a crucial role in determining the effects of corticosterone on the trophic state of the nerve cell population in question. GR and MR IR can be shown to be colocalized in the hippocampal neurons of the CA1 area as well as in the granular cells of the dentate gyrus. It should be mentioned that aldosterone treatment antagonizes the cell death induced in the dentate gyrus in the adult rat by adrenalectomy.[41–44] Based on these studies it seems clear that the de- and regenerative actions of corticosterone may be determined by the age of the animal, by the area studied and by the ratio of GR to MR in each individual nerve cell population. In addition, one may consider the presence of GR immunoreactive astrocytes among the neuronal cell populations analyzed as well as how the GR and MR activation altered the patterns of synthesis of neurotrophic factors in the individual nerve cell population.

GR and Neuronal Plasticity

In the Adult Brain

The GR mapping in the CNS summarized above has clearly indicated that glucocorticoids released in stress exert a major modulatory influence in information processing within large parts of the brain. However, the existence of GR IR in large numbers of glial cell populations together with its presence also in brain areas not directly involved with stress responses clearly indicate that glucocorticoids via activation of GR can also modulate neuronal and glial metabolic functions including the regulation of neuronal plasticity. Also GR IR has been found to be present in basic fibroblast growth factor (bFGF) immunoreactive astroglial cells (FIGS. 2 and 3) and can thus directly control astroglial trophism.[45] In line with this view, corticosterone in doses which mimic the serum levels of corticosterone found in stress can increase bFGF IR in nigral astrocytes.[46] Barbany and Persson[47,48] have also demonstrated that glucocorticoids can control neurotrophin mRNA expression within the hippocampal formation of the adult rat giving evidence that neuronal trophism can also be controlled by glucocorticoids. The available evidence suggests that GR activation may exert both a negative and a positive control of trophic factors.[45]

Recently we analyzed how tooth pulp deafferentation by removal of the coronal tooth pulp in maxillary molars unilaterally can lead to alterations in astroglial GR IR and astroglial bFGF IR within the nucleus tractus spinalis nervi trigemini, pars caudalis (FIG. 4). Thus, it was postulated based on previous work that degenerative and regenerative changes occurring in the central branches of the trigeminal somatosensory system[49] can lead to changes in astroglial trophism within this nucleus. Degenerative changes in the primary afferents of the trigeminal system have in fact been demonstrated.[50–52] The findings give clear indications that 1 and 7 days after the tooth pulp deafferentation there were substantial increases in the number and intensity of GR immunoreactive astroglial cells in the nucleus tractus spinalis nervi trigemini, pars caudalis associated also with an increase in the number of bFGF immunoreactive astroglial cells (Bhatnagar *et al.*, 1994, personal communication). These results indicate that long lasting increases in astroglial bFGF and GR IR probably related to increases in their gene expression can take place upon induction of degenerative events occurring in the peripheral branches of the trigeminal pulp neurons. The results open up the possibility that increased expression of GR in astroglia contribute to the increased expression of bFGF in the same astroglial population. Thus, glucocorticoids may also increase astroglial trophism within the nucleus tractus spinalis nervi trigemini, pars caudalis as previously indicated in the zona reticulata of the substantia nigra (see above).

In the Aged Brain

Previously, in the 24-month-old male Sprague-Dawley rat a selective reduction in GR IR was demonstrated in the CA1 pyramidal cells associated with the loss of nerve cell bodies.[53] The reduction of GR IR was also demonstrated in the central amygdaloid nucleus of the 24-month-old Sprague-Dawley rat but in no other area. In this case, however, there was no loss of nerve cell bodies. Thus, in the Sprague-Dawley rat the loss of GR immunoreactive neurons in the aging process may only take place in the CA1 area. It may be that it is the combined activation of MR and

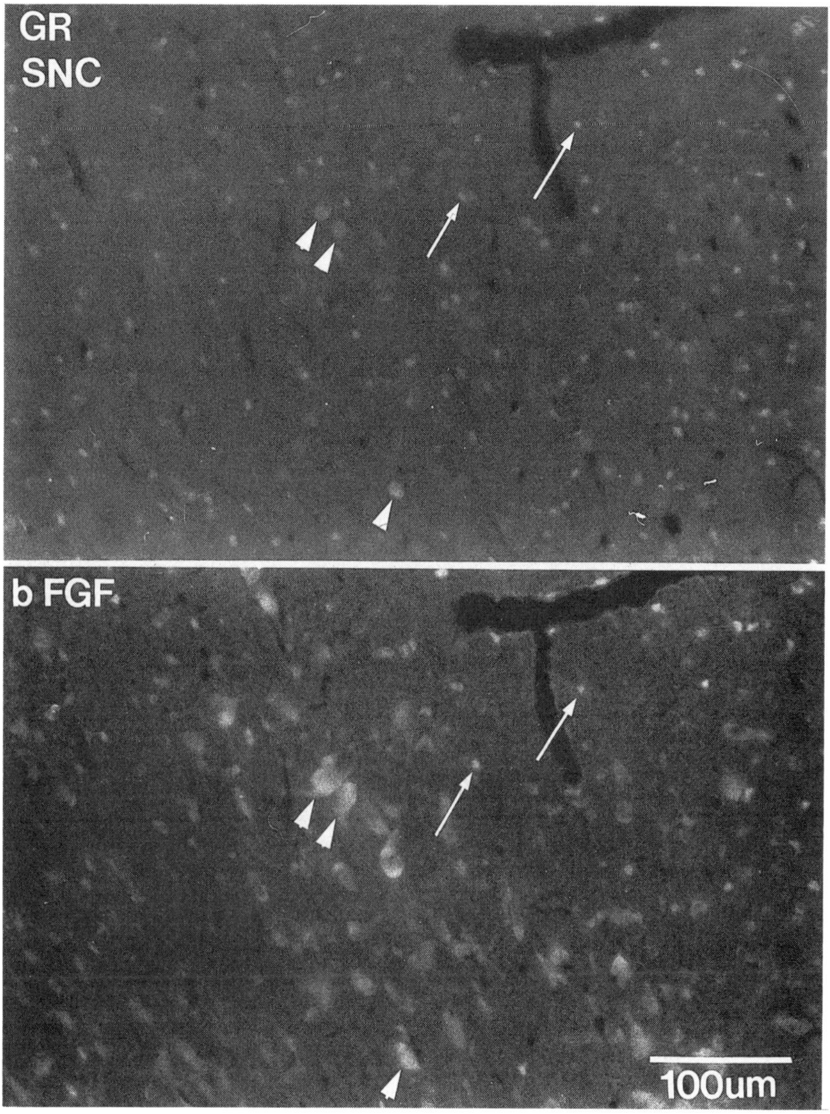

FIGURE 2. Immunocytochemical localization of GR IR and bFGF IR in the substantia nigra pars compacta (SNC) using double immunolabelling procedures. The GR IR was visualized by FITC technique and the bFGF was visualized using the biotin streptavidin Texas-Red complex as a marker. *Arrows* point to examples of glial cells containing both nuclear GR and bFGF IR and *arrowheads* to examples of nerve cells containing both nuclear GR IR and cytoplasmic bFGF IR.

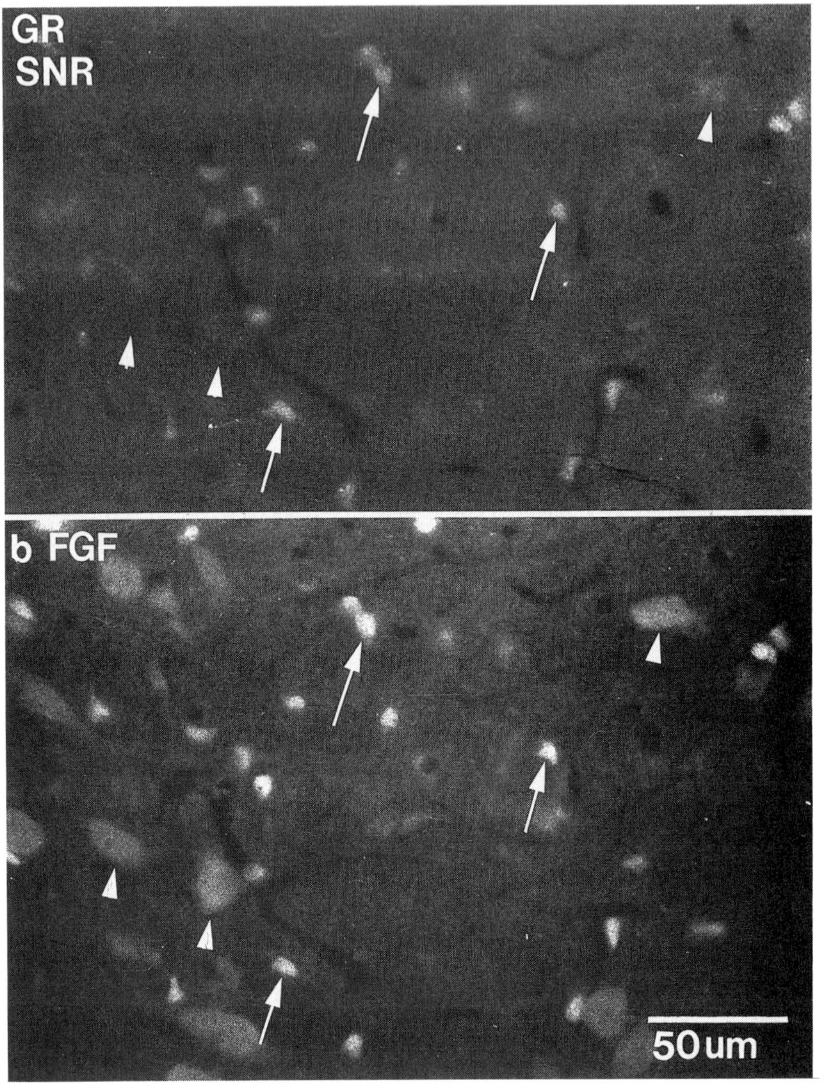

FIGURE 3. Immunocytochemical localization of GR IR and bFGF IR in the substantia nigra pars reticulata (SNR) using double immunolabelling procedures. The GR IR was visualized using the FITC fluorescence technique and the bFGF was visualized using the biotin streptavidin Texas-Red complex as a marker. *Arrows* point to examples of glial cells containing both markers (nuclear GR and bFGF IR) and *arrowheads* to examples of bFGF IR nerve cells lacking GR.

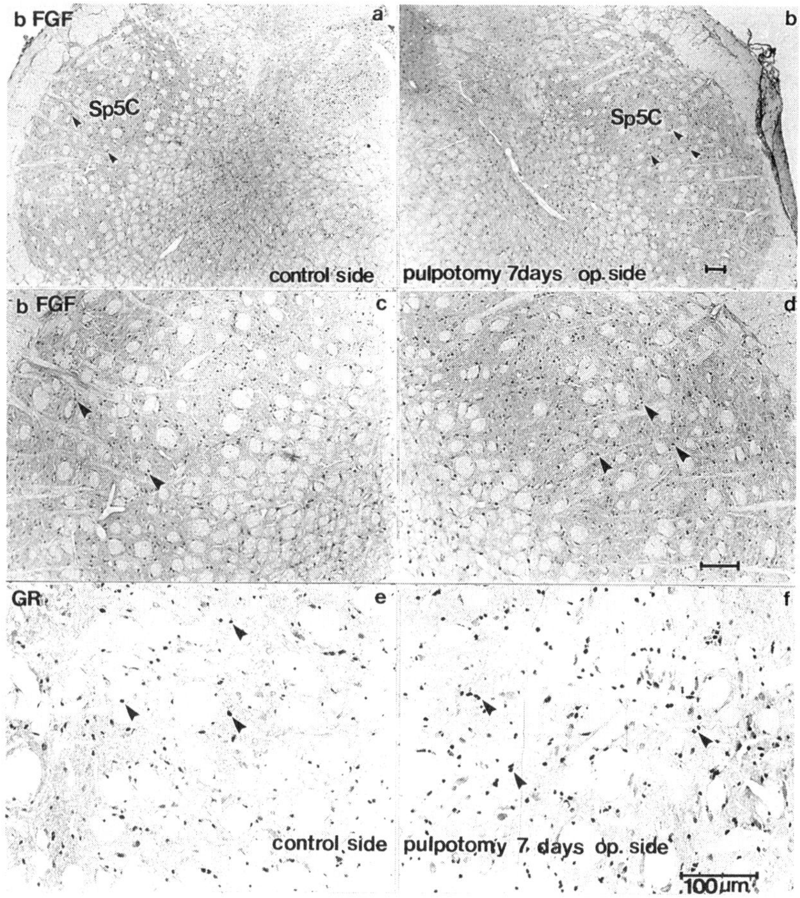

FIGURE 4. Immunohistochemical localization of bFGF and GR IR in the caudal spinal trigeminal nucleus (Sp 5C) after a 7-day pulpotomy using the ABC technique. Both antisera were diluted 1:1000. Panels **(a)** and **(c)** show nuclear glial bFGF IR on the *control side* in a low and high magnification and **(b)** and **(d)** show nuclear glial bFGF IR on the *operated side*. Substantial increases in bFGF IR are found on the operted site. *Arrowheads* indicate labelled glia cells. Panel **(e)** shows nuclear glial GR IR on the *control side* in high magnification, and panel **(f)** shows nuclear glial GR IR on the *operated site. Arrowheads* indicate labelled glia cells. Substantial increases of GR IR are found on the operated side.

GR in the CA1 pyramidal cells which may lead to the disappearance of nerve cells in this region. However, CA3 neurons degenerate in the aged rat.[40] These neurons contain high amounts of MR IR but only a weak GR IR and low GR mRNA levels.[6,9,10] In the male brown Norway rat, however, GR immunocytochemistry and measurements of the number of immunoreactive hippocampal cells using stereological techniques demonstrated that the total estimated number of GR immunoreactive cells in the CA1 and CA2 area was similar in the 3- and 36-month-old rat.[54]

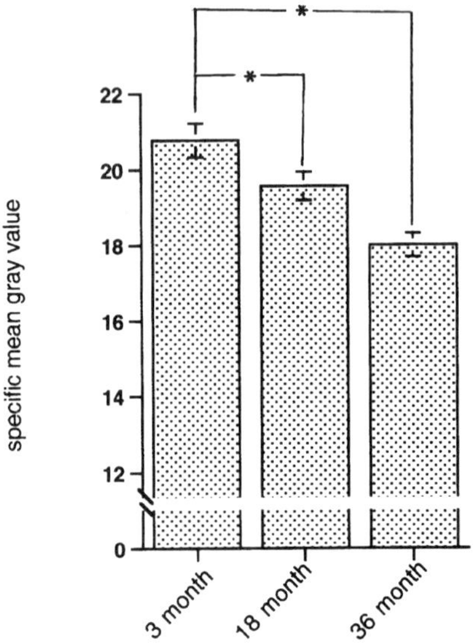

FIGURE 5. Effects of aging on the GR IR in the pyramidal cell layer of the CA1 area of the dorsal hippocampus of the brown Norway rat. The specific mean grey values (MGV) are shown as means ± SEM (n = 4–5). A significant reduction of MGV was found both in 18- and 36-month-old rats. One way ANOVA with Fisher's PLSD test. *: p <0.5. The area of the sample field is 14805.90 ± 164.30 μm^2.

However, the intensity of the GR IR was gradually decreased in the aged animals (FIG. 5). In view of reports[55] of reductions in GR density in the brown Norway rat in aging our findings suggest a possible reduced ability of GR in aging to undergo conformational changes upon activation by glucocorticoids necessary for translocation and/or activation of GR response elements and leading to reduced binding of the antibody.[56] It is of substantial interest that in this study a significant reduction was observed in the number of astroglial cells, demonstrating bFGF IR. Thus, astroglial trophism may become reduced in the aged hippocampus of the brown Norway rat but hypertrophy of the hippocampal astrocytes still exists. Thus, a dysfunction of gene expression in the astroglia[57] may take place in aging which may be of relevance for the demonstrated reductions in this strain of rats for GR but not MR IR within the CA1 pyramidal cells (FIG. 6). Thus, according to these findings the MR signal may dominate in the aged hippocampus and contribute to the acceleration of the aging processes of the pyramidal nerve cells.

In line with the results obtained in the aging hippocampus of the brown Norway rat, endothelin-1 (ET-1)-induced ischemic injury in the neostriatum was not enhanced by treatment with corticosterone as it was observed in the adult rat.[58] Thus, activation of GR within the neostriatum may more easily be obtained in the adult rat, leading to a possible further reduction in energy metabolism as suggested by Sapolsky.[40]

In conclusion, the present analysis together with other studies clearly indicate an important role of glucocorticoids in the control of information handling and of neuronal plasticity in the prenatal, postnatal, adult and aging rat. Thus, stress via release of adrenocortical hormones can markedly modulate neuronal plasticity especially in the pre- and postnatal rat and appropriate tune networks of different areas to optimize integrative processes. Such modulation of networks is essential for the survival of the animal under life-threatening conditions.

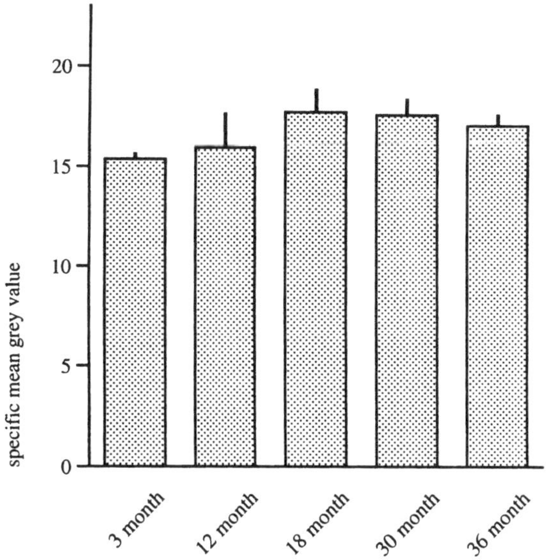

FIGURE 6. Effects of aging on the mineralocorticoid receptor (MR) IR in the pyramidal cell layer of the CA1 area of the dorsal hippocampus of the brown Norway rat. The specific mean grey (MGV) values are given as means ± SEM (n = 4–5). No changes in MGV for the MR IR were observed in the different age groups. The area of the sample field in CA1 is 56962.22 ± 638.30 μm^2.

SUMMARY

A detailed mapping of glucocorticoid receptor (GR) immunoreactivity (IR) in rat CNS was performed employing a mouse monoclonal antibody against rat liver GR. Subjective comparisons were made between the present results and the available data in the literature. A semiquantitation of GR immunostaining was found necessary and was obtained by microdensitometric and morphometric techniques, which enabled the distinction of neuronal and glial cell populations containing GR IR in various CNS regions.

GR IR in the CNS was mainly found in the nuclear compartment. The GR was present in neuronal populations with classical neurotransmitters, especially monoamines and glutamate and with various neuropeptides. The degree of colocalization varied according to the function of the brain area. Functional implications were made in relation to stress sensitivity, mood and nociception/antinociception. The global control of networks by glucocorticoids may allow an optimal integration of different types of circuits.

The GR is found already in the fetal rat and the development of GR mRNA and receptor protein was followed during the pre- and postnatal periods. The GR appears to be a major factor in brain maturation and in modulation of stress responses.

In aged Brown Norway rat brain GR IR but not mineralocorticoid receptor (MR) IR is reduced in the hippocampal nerve cells. The intensity of GR IR but not the number of nerve cells is altered, indicating a reduced activation of the GR in aging in this rat strain.

Overall GR participates in neuronal plasticity from fetal and postnatal life to adult life and aging.

ACKNOWLEDGMENTS

We are grateful to Ms. Anne Edgren for excellent secretarial assistance.

REFERENCES

1. Härfstrand, A., K. Fuxe, A. Cintra, L. Agnati, I. Zini, A.-C. Wikström, S. Okret, Z.-U. Yu, M. Goldstein, H. Steinbusch, A. Verhofstad & J.-Å. Gustafsson. 1986. Demonstration of glucocorticoid receptor immunoreactivity in monoamine neurons of the rat brain. Proc. Natl. Acad. Sci. USA **83:** 9779.
2. Cintra, A., K. Fuxe, A. Härfstrand, L. Agnati, A.-C. Wikström, S. Okret, W. Vale & J.-Å. Gustafsson. 1987. Presence of glucocorticoid receptor immunoreactivity in corticotropin releasing factor immunoreactive neurons of the rat di- and telencephalon. Neurosci. Lett. **77:** 25–30.
3. Cintra, A., K. Fuxe, V. Solfrini, L. Agnati, B. Tinner, A.-C. Wikström, W. Staines, S. Okret & J.-Å. Gustafsson. 1991. Central peptidergic neurons as targets for glucocorticoid action. Evidence for the presence of glucocorticoid receptor immunoreactivity in various types of classes of peptidergic neurons. J. Steroid Biochem. Mol. Biol. **40:** 93–103.
4. Cintra, A. & F. Bortolotti. 1992. Presence of strong glucocorticoid receptor immunoreactivity within hypothalamic and hypophyseal cells containing proopiomelanocortic peptides. Brain Res. **577:** 127–133.
5. Cintra, A., M. Zoli, L. Rosén, L. Agnati, S. Okret, A.-C. Wikström, J.-Å. Gustafsson & K. Fuxe. 1994. Mapping and computer assisted morphometry and microdensitometry of glucocorticoid receptor immunoreactive neurons and glial cells in the rat central nervous system. Neuroscience. In press.
6. Fuxe, K., A. C. Wikström, S. Okret, L. F. Agnati, A. Härfstrand, Z.-Y. Yu, L. Granholm, M. Zoli, W. Vale & J. Å. Gustafsson. 1985. Mapping of glucocorticoid receptor immunoreactive neurons in the rat tel-and diencephalon using a monoclonal antibody against rat liver glucocorticoid receptor. Endocrinology **177(5):** 1803–1812.
7. Zoli, M., A. Cintra, I. Zini, L. Hersh, J.-Å. Gustafsson, K. Fuxe & L. Agnati. 1990. Nerve cell clusters in dorsal striatum and nucleus accumbens of the male rat demonstrated by glucocorticoid receptor immunoreactivity. J. Chem. Neuranat. **3:** 355–366.
8. Ahima, R. & R. Harlan. 1990. Charting of type II glucocorticoid receptor-like immunoreactivity in the rat central nervous system. Neuroscience **39:** 579–604.
9. Aronsson, M., K. Fuxe, Y. Dong, L. Agnati, S. Okret & J.-Å. Gustafsson. 1988. Localization of glucocorticoid receptor mRNA in the male rat brain by *in situ* hybridization. Proc. Natl. Acad. Sci. USA **85:** 9331–9335.
10. Sousa, R. J., N. H. Tannery & E. M. Lafer. 1989. *In situ* hybridization mapping of the glucocorticoid receptor messenger ribonucleic acid in rat brain. Mol. Endocrinol. **3:** 481–494.
11. Kiss, J., J. van Eekelen, J. Reul, H. Westphal & E. de Kloet. 1988. Glucocorticoid receptor in magnocellular neurosecretory cells. Endocrinology **122:** 444–449.
12. Fuxe, K., A. Cintra, A. Härfstrand, L. F. Agnati, M. Kalia, M. Zoli, A.-C. Wikström, S. Okret, M. Aronsson & J.-Å. Gustafsson. 1987. Central glucocorticoid receptor immunoreactive neurons: new insights into the endocrine regulation of the brain. Ann. N.Y. Acad. Sci. **512:** 362–393.

13. REUL, J. & E. DE KLOET. 1985. Two receptor systems for corticosterone in rat brain: microdistribution and differential occupation. Endocrinology **117:** 2505–2511.
14. REUL, J. M. & E. R. DE KLOET. 1986. Anatomical resolution of two types of corticosterone receptor sites in the rat brain with *in vitro* autoradiography and computerized image analysis. J. Steroid Biochem. Mol. Biol. **24:** 269–272.
15. VAN EKELEN, J., J. KISS, H. WESTPHAL & E. DE KLOET. 1987. Immunocytochemical study on the intracellular localization of the type 2 glucocorticoid receptor in the rat brain. Brain Res. **436:** 120–128.
16. FUXE, K., A. HÄRFSTRAND, L. F. AGNATI, Z. YU, A. CINTRA, A.-C. WIKSTRÖM, S. OKRET, E. CANTONI & J. Å. GUSTAFSSON. 1985. Immunocytochemical study on the localization of glucocorticoid receptor immunoreactive nerve cells in the lower brain stem and spinal cord of the male rat using a monoclonal antibody against rat liver glucocorticoid receptor. Neurosci. Lett. **60:** 1–6.
17. AHIMA, R. & R. HARLAN. 1991. Differential corticosteroid regulation of type II glucocorticoid receptor-like immunoreactivity in the rat central nervous system: topography and implications. Endocrinology **129:** 226–236.
18. ANTAKLY, T. & H. EISEN. 1984. Immunocytochemical localization of glucocorticoid receptor in target cells. Endocrinology **115:** 1984–1989.
19. OKRET, S., A.-C. WIKSTRÖM, Ö. WRANGE, B. ANDERSSON & J. Å. GUSTAFSSON. 1984. Monoclonal antibodies against the rat liver glucocorticoid receptor. Proc. Natl. Acad. Sci. USA **81:** 1609–1613.
20. WIKSTRÖM, A.-C., O. BAKKE, S. OKRET, M. BRÖNNEGÅRD & J.-Å. GUSTAFSSON. 1987. Intracellular localization of the glucocorticoid receptor: evidence for cytoplasmatic and nuclear localization. Endocrinology **120:** 1232–1242.
21. AKNER, G., K. G. SUNDQVIST, M. DENIS, A. WIKSTRÖM & J. GUSTAFSSON. 1990. Immunocytochemical localization of glucocorticoid receptor in human gingival fibroblasts and evidence for a colocalization of glucocorticoid receptor with cytoplasmic microtubules. Eur. J. Cell Biol. **53:** 390–401.
22. AKNER, G., K. MOSSBERG, A. WIKSTRÖM, K. SUNDQVIST & J. GUSTAFSSON. 1991. Evidence for colocalization of glucocorticoid receptor with cytoplasmic microtubules in human gingival fibroblasts using two different monoclonal anti-GR antibodies, confocal laser scanning microscopy and image analysis. J. Steroid Biochem. Mol. Biol. **39:** 419–432.
23. VIELKIND, U., A. WALENCEWICZ, J. LEVINE & M. BOHN. 1990. Type II glucocorticoid receptors are expressed in oligodendrocytes and astrocytes. J. Neurosci. Res. **27:** 360–373.
24. BUNNEMANN, B., A. LIPPOLDT, J. AGUIRRE, A. CINTRA & R. METZGER. 1993. Glucocorticoid regulation of angiotensinogen gene expression in discrete areas of the male rat brain. Neuroendocrinology **57:** 856–862.
25. CECCATELLI, S., A. CINTRA, T. HÖKFELT, K. FUXE, A.-C. WIKSTRÖM & J.-Å. GUSTAFSSON. 1989. Coexistence of glucocorticoid receptor-like immunoreactivity with neuropeptides in the hypothalamic paraventricular nucleus. Exp. Brain Res. **78:** 33–42.
26. AHIMA, R. & R. HARLAN. 1992. Glucocorticoid receptors in LHRH neurons. Neuroendocrinology **56:** 845–850.
27. SWANSON, L., P. SAWCHENKO & R. LIND. 1986. Regulation of multiple peptides in CRF parvocellular neurosecretory neurons: implications for the stress responses. *In* Progress in Brain Research. T. Hökfelt, K. Fuxe & B. Pernow, Eds.: 169–190. Elsevier. Amsterdam.
28. FUXE, K., A. CINTRA, S. TANGANELLI, G. VON EULER, U.-B. FINNMAN, K. LUNDGREN, J. AGUIRRE, B. BJELKE, M. ZOLI, P. ENEROTH & L. AGNATI. 1989. Integration of humoral and neuronal signals in neuroendocrine regulation. Focus on interaction of glucocorticoids with central neuropeptide and monoamine neurotransmission. *In* Recent Advances in Basic and Clinical Neuroendocrinology. F. Casanueva & D. Dieguez, Eds. 109–126. Elsevier Sci. Publ. Amsterdam.
29. DELEÓN, M., R. COVEÑAS, G. CHADI, J. NARVÁEZ, K. FUXE & A. CINTRA. 1994. Subpopulations of primary sensory neurons show coexistence of neuropeptides and glucocorticoid receptors in the rat spinal and trigeminal ganglia. Brain Res. **636:** 338–342.

30. Coveñas, R., M. De León, G. Chadi, A. Cintra & K. Fuxe. 1994. Adrenalectomy increases the number of substance P and somatostatin immunoreactive nerve cells in the rat lumbar dorsal root ganglia. Brain Res. **640:** 352–356.
31. Cintra, A., V. Solfrini, B. Bunnemann, S. Okret, F. Bortolotti, J.-Å. Gustafsson & K. Fuxe. 1993. Prenatal development of glucocorticoid receptor gene expression and immunoreactivity in the rat brain and pituitary gland: a combined *in situ* hybridization and immunocytochemical analysis. Neuroendocrinology **57:** 1133–1147.
32. Slotkin, T. A., S. E. Lappi, E. C. McCook, M. I. Tayyeb, J. P. Eylers & F. J. Seidler. 1992. Glucocorticoids and the development of neuronal function: Effects of prenatal dexamethasone exposure on central noradrenergic activity. Biol. Neonate **61:** 326–336.
33. Diaz, R., S.-O. Ögren, M. Blum & K. Fuxe. 1993. Prenatal corticosterone differentially affects locomotor activity of postnatal male and female offspring. *In* Neuroscience annual meeting. Eds. Society for Neuroscience. Washington, DC.
34. Cintra, A., V. Solfrini, L. Agnati, J.-Å. Gustafsson & K. Fuxe. 1991. Strongly glucocorticoid receptor immunoreactive neurons in the neonatal rat brain. NeuroReport **2:** 85–88.
35. Zoli, M., L. Agnati, K. Fuxe, F. Ferraguti, G. Biagini, A. Cintra & J.-Å. Gustafsson. 1990. Long-lasting reduction of glucocorticoid receptor immunoreactivity in the hippocampal field CA1 but not in the dentate gyrus after neonatal treatment with corticosterone in the rat. Acta Physiol. Scand. **138:** 577–579.
36. Agnati, L., A. Cintra, G. Biagini, M. Zoli, P. Marama, C. Carani, J.-Å. Gustafsson & K. Fuxe. 1992. Central glucocorticoid receptors, dopamine and their involvement in stress responses in postnatal rats. *In* Stress: Neuroendocrine and Molecular Approaches. R. Kvetnansky, R. McCarty & J. Axelrod, Eds. 751–763. Gordon and Breach Sci. Publ. New York.
37. Yang-Yen, H.-F., J.-C. Chambard, Y.-L. Sun, T. Smeal, T. J. Schmidt, J. Drouin & M. Karin. 1990. Transcriptional interference between c-jun and the glucocorticoid receptor: Mutual inhibition of DNA binding due to direct protein-protein interaction. Cell **62:** 1205–1215.
38. Coveñas, R., M. deLeón, A. Cintra, B. Bjelke, J.-Å. Gustafsson & K. Fuxe. 1993. Coexistence of c-Fos and glucocorticoid receptor immunoreactivities in the CRF immunoreactive neurons of the paraventricular hypothalamic nucleus of the rat after acute immobilization stress. Neurosci. Lett. **149:** 149–152.
39. Zoli, M., F. Ferraguti, G. Biagini, A. Cintra, K. Fuxe & L. Agnati. 1991. Corticosterone treatment counteracts lesions induced by neonatal treatment with monosodium glutamate in the mediobasal hypothalamus of the male rate. Neurosci. Lett. **132:** 225–228.
40. Sapolsky, R. M., L. C. Krey & B. S. McEwen. 1986. The neuroendocrinology of stress and aging: the glucocorticoid cascade hypothesis. Endocr. Rev. **7:** 284–301.
41. Armstrong, J. N., D. C. McIntyre, S. Neubort & R. S. Sloviter. 1993. Learning and memory after adrenalectomy-induced hippocampal dentate granule cell degeneration in the rat. Hippocampus. **3:** 359–371.
42. Sloviter, R., E. Dean & S. Neubort. 1993. An electron microscopic analysis of adrenalectomy-induced hippocampal granule cell degeneration in the rat; apoptosis in the adult central nervous system. J. Comp. Neurol. **330:** 337–351.
43. Sloviter, R. S., A. L. Sollas, E. Dean & S. Neubort. 1994. Adrenalectomy-induced granule cell degeneration in the rat hippocampal dentate gyrus; characterization of an *in vivo* model of controlled neuronal death. J. Comp. Neurol. **330:** 324–336.
44. Woolley, C. S., E. Gould, R. R. Sakai, R. L. Spencer & B. S. McEwen. 1991. Effects of aldosterone or RU28362 treatment on adrenalectomy-induced cell death in the dentate gyrus of the adult rat. Brain Res. **554:** 312–315.
45. Fuxe, K., A. Cintra, R. Diaz, G. Chadi, J.-Å. Gustafsson & L. F. Agnati. 1994. Central glucocorticoid receptors and neuronal plasticity. *In* Methods in Neuroscience, Neurobiology of Steroids. E. R. de Kloet & W. Sutanto, Eds. Academic Press. Orlando, FL. In press.
46. Chadi, G., L. Rosén, A. Cintra, B. Tinnèr, M. Zoli, R. Pettersson & K. Fuxe.

1993. Corticosterone increases FGF-2 (bFGF) immunoreactivity in the substantia nigra of the rat. NeuroReport **4:** 783–786.
47. Barbany, G. & H. Persson. 1992. Regulation of neurotrophin mRNA expression in the rat brain by glucocorticoids. Eur. J. Neurosci. **4:** 396–403.
48. Barbany, G. & H. Persson. 1993. Adrenalectomy attenuates kainic acid-elicited increases of messenger RNAs for neurotrophins and their receptors in the rat brain. Neuroscience **54:** 909–922.
49. Arvidsson, J. & K Johansson. 1988. Changes in the central projection pattern of vibrissae innervating primary sensory neurons after nerve injury in the rat. Neurosci. Lett. **84:** 120–124.
50. Johnson, L., L. Westrum, M. Henry & R. Canfield. 1987. Transganglionic degeneration following dental lesions. *In* Effects of Injury on Trigeminal and Spinal Somatosensory Systems. L. Pubols & B. Sessle, Eds. 151–158. Alan R. Liss. New York.
51. Gobel, S. 1988. An electron microscopic analysis of the transsynaptic effects of peripheral nerve injury subsequent to tooth pulp extirpations on neurons on laminae I and II of the medullary dorsal horn. J. Neurosci. **4:** 2281–2290.
52. Gobel, S. & J. Binck. 1977. Degenerative changes in primary trigeminal axons and in neurons in nucleus caudalis following tooth pulp extirpations in the cat. Brain Res. **132:** 347–354.
53. Zoli, M., F. Ferraguti, J.-Å. Gustafsson, G. Toffano, K. Fuxe & L. F. Agnati. 1991. Selective reduction of glucocorticoid receptor immunoreactivity in the hippocampal formation and central amygdaloid nucleus of the aged rat. Brain Res. **545:** 199–207.
54. Cintra, A., G. Akner, A.-C. Wikström, L. Agnati, J.-Å. Gustafsson & K. Fuxe. 1994. Immunocytochemical studies on the glucocorticoid receptor. Methods Neurosci. In press.
55. van Eekelen, J. A., N. Y. Rots, W. Sutanto & E. R. de Kloet. 1992. The effect of aging on stress responsiveness and central corticosteroid receptors in the Brown Norway rat. Neurobiol. Aging **13:** 159–170.
56. Cintra, A., J. Lindberg, G. Chadi, B. Tinner, A. Möller, J.-Å. Gustafsson, E. de Kloet, M. Oitzl, L. Agnati & K. Fuxe. 1994. Basic fibroblast growth factor and steroid receptors in the aging hippocampus of the brown Norway rat: immunocytochemical analysis in combination with stereology. Neurochem. Int. **25:** 39–45.
57. Wagner, A. P., G. Reck & D. Platt. 1993. Evidence that V+fibronectin, GFAP and S100β mRNAs are increased in the hippocampus of aged rats. Exp. Gerontol. **28:** 135–143.
58. Agnati, L., F. Benfenati, V. Solfrini, G. Biagini, K. Fuxe, D. Giodolin, C. Carani & I. Zini. 1992. Brain aging and neuronal plasticity. *In* Physiological Processes of Aging. Towards a Multicausal Interpretation. Ann. N.Y. Acad. Sci. **673:** 180–186.

DISCUSSION OF THE PAPER

Question: Could you comment on glucocorticoid receptor co-location in CRH neurons in the olive nucleus?

Fuxe: We have not done any double labelling studies in the inferior olive region. My guess would be that you would find a co-location as you found in other areas.

Question: Do you have any thoughts about mineralocorticoid receptor localization in glia?

Fuxe: I was hoping to bring some slides with me on that aspect to the meeting, but I decided at the last minute that I was not quite convinced about the specificity of the MRs, so I left it out.

Question: How quickly does the GR increase in the glia after a lesion?

FUXE: So far we have done one day and seven days. In both cases we observed a substantial increase in the GR activity and of the basic FGF activity in the astroglia.

QUESTION: Do the radial glia show GR expression during development?

FUXE: I have not seen anything in development in the glia. It may be an insensitivity of our technique, but the dominant picture is rather an exclusive neuronal localization of GR in development.

QUESTION: Have you tried to quantitate upregulation or downregulation of the receptor by the hormone?

FUXE: Yes, we did that once. We wanted to see if, during the stress hyporesponsive period—the injection of stress-related doses of corticosterone repeatedly over this period—one could see GR changes in prepubertal and adult life. We ended up losing our adult animals, but the prepubertal rats came through. In the latter we found a highly significant reduction of GR activity selectively in the CA1-CA2 area, but not in the dentate gyrus granule layer. Either there was a delay in the maturation of these pyramidal cells or there was a true downregulation of the GR mRNA message, leading to that long-lasting deficit in GR activity.

E. R. DE KLOET (*Sylvius Laboratories*): You observed that some neurons like the cholinergic neurons did not contain glucocorticoid receptor immunoreactivity. That in itself is a remarkable observation that there are cells which lack these receptors. Do glucocorticoids modify cholinergic transmission and if so, are the glucocorticoid receptors then postsynaptically localized?

FUXE: Yes, indeed the cholinergic neurons are unique in that there is no detectable presynaptic GR; the effect is selectively on the postsynaptic level.

R. BAKER (*National Institute of Mental Health*: It seems that most of the cells in the arcuate nucleus are GR positive. To what extent is GR in glia cells in the arcuate?

FUXE: The dominant effect in the arcuate is the neuronal population. So there is a rather low frequency density of astrocytes which are positive for GR in the arcuate.

P. W. LANDFIELD (*University of Kentucky*): We have also used John Cidlowski's antibodies. He has two antibodies termed #57 and #53, I believe, and I forget which is which. But one stains glia tremendously and one shows the very same pattern of neuronal staining that you have seen and that Jim Herman and Stan Watson have also seen. So it is very important which antibody is being used. I do not have an explanation for that; perhaps you do. In all of these very systematic and elegant studies on colocalization, is there some sort of functional theme that you see emerging on what processes the glucocorticoid receptor is activating? What sorts of systems? Is there some underlying unifying theme?

FUXE: No, there is no unifying theme. GR are found both in stress-related regions and in non-stress-related regions. In nerve cells with strong GR IR like CRF and locus coeruleus neurons GR can have a special role in regulating a large number of genes. In the high density areas that are not traditionally involved with stress, there is the tuning of the networks for survival and optimum integration. In terms of actions at the cellular level, there could be substantial differences from region to region in the brain. Whatever actions occur with glucocorticoids will be dependent on a number of factors like the age and the stress state of the animal. Corticosterone given in postnatal life protects against glutamate. Given later on, it enhances the cell loss, according to Sapolsky. The importance of the region also has to be analyzed. The ratio of MR to GR could be critical as well as the overall synaptic input to the region which will modulate the impact of the glucocorticoids on the formation of neurotrophic factors. The involvement of basic FGF means that the impact of the glucocorticoids will vary very much with all these factors. It is difficult to find any simple way. Those steroids are there, and they have a special role to play in time and space in the brain. We just have to go on defining that role in detail.

LANDFIELD: We shall see later on, from Marian Joël's work and our own work, that at least in the hippocampus the GR has an inhibitory effect on the neurons. It reduces excitability by increasing calcium influx and the afterhyperpolarization, so where those receptors are found may be neurons that are shut down during stress.

FUXE: It would be wonderful if you could do that also in early life because we are fascinated about these differential effects at different times of the life span of the animal. We are doing age-dependent studies, not early in life but in late life. With aging there seems to be a greater impact of the glucocorticoids on the afterhyperpolarization in hippocampal neurons.

I. S. EDELMAN (*Columbia University*): Monoclonal antibodies are bound to give discrepancies in intensities because the epitopes can either be exposed or hidden depending on the physiological circumstances in which they are tried. What would be more mysterious is if the polyclonal antobodies gave discrepancies. That would not be expected, because the multiplicity of epitopes would insure that one would be apt to see all the species.

General Discussion

B. S. McEwen (*The Rockefeller University*): Dr. Funder emphasized that in the brain the MR seems to be largely or fully occupied at all times. I know of examples that exogenous mineralocorticoids have an effect if given to animals with intact adrenals. For example, aldosterone affects blood pressure and induces salt appetite by central actions. Are these particular effects some kind of pharmacologic artifact? Are there somewhere in the brain unoccupied mineralocorticoid receptors that we do not know about, or are those receptors that we think are more or less fully occupied really not that fully occupied? Or thirdly, is there some other mechanism of action that might be in addition to the genomic effect? One example from work that Dr. Randall Sakai has been doing in Philadelphia is that certain A-ring reduced metabolites of both aldosterone and deoxycorticosterone are able to rapidly turn on salt appetite in adrenally intact animals. This is an effect that seems to be mimicked by flunitrazipam, raising the possibility that a nongenomic membrane action perhaps through a GABA-A receptor might be involved. These are issues of cellular mechanism. Also do we really know for sure that all the brain mineralocorticoid receptors are always fully occupied, or is there room for some alternative steroid signaling mechanism?

I. S. Edelman (*Columbia University*): The best example of the lack of correlation between full occupancy in fact is in the case of glucocorticoid receptors. The characteristic of a limited number of binding sites is saturation. With glucocorticoids administered either to animals or humans you never saturate the response. You never fail to get some additional effects beyond anything that can be attributed to occupancy. Thus the loss of muscle, the skin effects, the brain effects. None of these saturates in the intact animal. There are other subtleties that have to be taken into account. Full occupancy does not imply a lack of regulation, because depending on which steroids are occupying which sites, all the receptors are degenerate with respect to their recognition specificities. It will probably make a difference whether the mineralocorticoid receptor is occupied for 30 percent by corticosterone, for 70 percent by aldosterone or for 20 percent by progesterone, etc. Thus, by modulating which steroid is involved in the occupancy of the site you can modulate the effects. In addition, the state of differentiation and the phenotype of the cell may regulate the consequences of occupancy by interaction with other protein components and transcription factors. Then effects can be expressed which otherwise would not be expressed even though the receptor is fully occupied. So full occupancy is not the end of this physiological story. There is room to maneuver.

J. W. Funder (*Baker Institute*): To the extent that they're genomic (leave Dr. Sakai's work to the side for a moment), I think that is very interesting. Studies from your laboratory showed that the effects of aldosterone on salt appetite could neither be blocked nor mimicked by corticosterone whereas they could be blocked by spironolactone. They had all the characteristics of a specific aldosterone-recognizing system. On the other hand, data from Elise Gómez-Sánchez's laboratory show that corticosterone blocks the aldosterone effect on blood pressure. That still begs the question you asked which is: what is the physiology of the mineralocorticoid receptor in brain?

M. F. Dallman (*University of California, San Francisco*): If you had a fully occupied receptor but were able to up- or downregulate it by some mechanism independent of steroid, you could generate a signal that could increase or decrease even though at every stage those receptors might be fully occupied. I do not know of any examples of this. But there may be a regulation of MR and GR that is not through steroids.

EDELMAN: I would solicit Dr. Funder's or anybody's response to whether they think the elevation in corticosterone in the morning in intact rats by one microgram percent is working through GR or whether it is overoccupying some quantity of unoccupied MR in the rat to cause its effects.

E. R. DE KLOET (*Sylvius Laboratories*): I think this is part of the question which Dr. McEwen brought up first. If in an intact animal the MR binds progesterone, deoxycorticosterone, corticosterone and aldosterone, which are all present in varying amounts, the steroids have the potential to differentially regulate the activity of this MR as agonist or as antagonist. This can be tested experimentally in an adrenalectomized animal which is given corticosterone or aldosterone in a dose-response study. The experiments I showed were done with adrenalectomized animals given high doses of corticosterone. There was a U-shaped dose response curve with maximal responses in the absence and with excess of corticosterone. Is this U-shape solely due to effects mediated by MR, or also in coordination with GR? By including selective antagonists we have shown that U-shaped dose response curves solely occur via MR, or GR, or MR + GR. Finally, I would like to remind everbody that steroid effects are time-dependent phenomena, which sometimes take hours to develop. Thus, it really depends on the specific time period one is looking for the steroid response.

QUESTION: If a student asked me what I thought about this whole issue of occupancy, I would say that our binding techniques are so fraught with difficulties that it is near impossible to truly answer this question without some kind of a leap of faith. First, we have to adrenalectomize the animal to measure binding. Second, sometimes the ligands interact, so you have to suppress one while you are measuring the other, and different laboratories come up with different numbers. So, all we have to assume is that some time during the course of the day there is not full occupancy. If we can assume that and that the system has some means of keeping track of an average and looks at timing not just at occupancy, then we do not have to invoke anything more complicated than just the receptor itself. On top of that, I am very sure that there are interactions between MR and GR and that there may also be membrane effects and all these wonderful things that we have to worry about. I hope that I am not being unfair to the binding data by saying this.

FUNDER: I would like to take issue with Dr. de Kloet's description of aldosterone as agonist, because I think that that is a hangover from thinking of the mineralocorticoid receptor in the kidney. When your own experiments with adrenalectomized animals show that serotonin turnover drops in hippocampus, restoring them to values in intact animals with corticosterone, antagonizing that effect aldosterone, I would make a plea for not calling aldosterone an agonist. It depends what we are looking at.

DE KLOET: It is accepted. No problem. The studies you are citing deal with complex transmitter circuits that may be subject to various steroid regulations at sites and mechanisms distal of the endpoint of serotonin turnover. In other words, the steroid effect on a complex process such as serotonin turnover involves a number of individual steps, each of which may be subject to receptor-mediated control, which on this level leads to assessment of corticosterone as agonist and aldosterone as antagonist. In the electrophysiological studies using single cells and single membrane events as endpoint, aldosterone and corticosterone both appear as agonists in hippocampus.

QUESTION: For those of us that work in cortisol-secreting animals and perhaps corticol-secreting animals that have no circadian rhythms for corticol secretion, could someone comment on how well we can extrapolate rat data to corticol-secreting animals?

DALLMAN: It depends on what corticol secretor you are talking about. Sheep do have a diurnal rhythm. They have very low CBG levels and thus high free corticol levels circulating in a diurnal rhythm. But there is no diurnal rhythm in dogs. The HPA regulation in dogs is very different from either rat, man or sheep.

C. WOOD (*University of Florida*): If you feed a sheep the way most veterinarians like to feed sheep in animal care facilities, which is give them a bucket of food and let them empty it, they have a beautiful diurnal rhythm. However, if you let them eat food like they do when they're grazing, they have absolutely no diurnal rhythm if you maintain humidity and temperature constant.

D. B. DEFRANCO (*University of Pittsburgh*): How do you change sensitivity of glucocorticoid receptors throughout the cell cycle? Dr. Munck and I have both found that during certain phases of the cell cycle and particularly G2 we have a fully loaded GR that is completely unable to transactivate. It looks like the mechanism responsible for that may be hyperphosphorylation of the receptor. So there you have a mechanism whereby you can have a fully loaded receptor that is inactive and the regulation could be via phosphorylation/dephosphorylation.

E. C. AZMITIA (*New York University*): I want to respond to the issue that Dr. McEwen brought up, and how I interpreted that. Coming from the field of serotonin—about ten years ago, we only had two serotonin receptors—a high affinity and a low affinity binding, and now we are up to something like 14 different receptors. How sure are we that there are only glucocorticoid receptors and mineralocorticoid receptors in light of the fact that we really do not have the drugs to tease apart the two types of receptors? Based just on binding, it would be very hard to look at subsets of receptors. Like the serotonin field with the develpment of new drugs and new antogonists and new agonists, the possibility of there being more than two receptors may become real.

EDELMAN: That is a real issue. We now know that less than 20 percent of the genes in the major mammals have been defined. In humans that means that there are about 80,000 genes that nobody has ever defined anything about at all at any level. And that is true essentially of all mammals. We also know that more and more gene families and super gene families are being identified. When Arriza and Evans did their original studies, they identified the mineralocorticoid receptor on the basis of its ability to activate a system that is responsive to progesterone and glucocorticoids. In other words, they did not have a unique mineralocorticoid response element to identify this previously unknown protein as being a unique mineralocorticoid receptor. This was an issue that was discussed at a meeting a couple of years ago in France. The problem is with the technology for looking at variance. You can have two types of variance. First, you can have multiple genes which can have homologies up to 95 percent, but the divergence in sequence could be critical. Second, you can have either posttranscriptional alternative splicing or posttranslational modifications, which give you a multiplicity of receptors. Therefore, I do not believe that we know how many corticosteroid receptors there are. Nor do I believe that we know the subtypes.

AZMITIA: A further caveat is that even when you have identified a single receptor—let us say with the serotonin system, the 1A receptor—the type of cell that this receptor contains is going to influence its response by a second messenger system. So that the 1A responding in a glia cell is going to respond differently than the 1A in a pyramidal neuron which is going to respnd differently than the 1A in a raphe neuron. So even if you have a single receptor, the cell that receptor is in and its state of maturation is going to influence its response.

EDELMAN: That is absolutely correct.

J. A. CIDLOWSKI (*University of North Carolina, Chapel Hill*): Dr. Sternberg, it seems to me that a lot of your data are consistent with a dysfunctional glucocorticoid receptor. Have you looked at the GR gene in those animals specifically to see if there is any mutation? Also have you tried to switch immune systems by doing a bone marrow transplant between these two animals, and does that cure the response?

E. M. STERNBERG (*National Institute of Mental Health*): We have not tried to switch immune systems. I am not sure that I agree with you that it looks like a dysfunctional GR. The affinity is the same. On what basis are you saying that?

CIDLOWSKI: Primarily on the basis of the RU486 data—the fact that when you block GR function in the Fischer rats with RU486, you get a symptom similar to that in the Lewis rats.

STERNBERG: I do not think that that proves that it is a dysfunctional GR.

CIDLOWSKI: No, it does not prove that it is a dysfunctional GR.

QUESTION: Just a follow-up on Dr. Cidlowski's point. If I remember correctly, you actually cured the Lewis rat by giving it steroid—by giving it glucocorticoid.

STERNBERG: That is right.

QUESTION: So, it is responsive.

STERNBERG: It is responsive, yes, and highly responsive, just as in rheumatoid patients who are exquisitely sensitive to low doses of steroids. In pharmacological doses corticosteroids obviously suppress, but already very low doses are active in Lewis rats. It is interesting because in human diseases like rheumatoid arthritis versus lupus, the rheumatoids are notoriously sensitive to both the antiinflammatory effects of steroids and the Cushingoid side effects of these steroids. Lupus patients who also have inflammatory disease are notoriously resistant to the effects of glucocorticoid. So it may be that lupus is a disease of the glucocorticoid receptor and rheumatoid arthritis is a disease somewhere else in that inflammatory pathway.

The Functional Relevance of the Heteromeric Structure of Corticosteroid Receptors[a]

CLAUDIO A. CAAMAÑO,[b,d] M. INÉS MORANO,[b]
STANLEY J. WATSON, JR.,[b] FRIEDRICH C. DALMAN,[c]
WILLIAM B. PRATT,[c] AND HUDA AKIL[b]

[b]*Mental Health Research Institute*
and
[c]*Department of Pharmacology*
The University of Michigan Medical School
Ann Arbor, Michigan 48109

Steroid receptors are transacting factors that must move through the cytoplasm to the nucleus to interact with the DNA. They are members of a superfamily with a well-known topology. The carboxy terminal region represents the ligand-binding domain and subsumes the specific recognition site for ligands. In the middle of the protein is the DNA-binding domain, which contains so-called cysteine-rich zinc fingers and is critical for DNA interactions. The amino terminal domain is thought to be important for transcriptional modulation. These domains appear to be functionally separable, and numerous chimeras have been constructed which retain appropriate binding selectivities and DNA recognition properties.[1,2] In this paper, we shall focus on two of the steroid receptors, the glucocorticoid receptor (GR) and the mineralocorticoid receptor (MR), which bind the stress hormones cortisol/corticosterone and their analogs. They play a key role in the negative feedback mechanisms of the hypothalamic-pituitary-adrenal axis, and in the control of its circadian rhythm.

It is now generally accepted that the association of heat shock proteins (hsps) with the steroid receptors is important for the stabilization of the steroid-binding protein in a non-DNA-binding and transcriptional inactive form. This "repressor" function has been typically considered to be the main, if not the only, role of the associated proteins on the receptor function.[3] According to this view, GR would appear to be a unique steroid receptor, in that it additionally requires the association of hsps for adopting a high steroid-binding affinity state.[4–7] We recently found similar requirements for the MR.[8]

From a broader perspective, heat shock or stress proteins are considered to be part of the family of molecular chaperones. The members of this family were recently defined by Hendrick & Hartl[9] as "proteins that bind to and stabilize an otherwise unstable conformer of another protein, and by controlled binding and release of the substrate protein, facilitate its correct folding *in vivo*: be it folding, oligomeric assembly, transport to a particular subcellular compartment, or controlled switching between active/inactive conformations." Within this context, the interaction between steroid-binding proteins and chaperones seems to illustrate only a particular example of the more general importance of chaperones in cell physiology. An interesting

[a]This work was supported by grants NIDA R01-DA02265, NIMH PO1-MH42251-06, and NIADDKD P30-AM34933.

[d]Address correspondence to Dr. Claudio A. Caamaño, Mental Health Research Institute, The University of Michigan Medical School, Ann Arbor, MI 48109-0720.

model supported by Pratt and his collaborators[10] basically conceives the steroid-free receptor complex as one of the products generated upon reaction of the steroid binding unit with a rather large chaperone complex. This heteromeric complex would contain hsp90, hsp70 and hsp56, and would be present in cytosol, independent of steroid receptors.

In systems of *in vitro* translation using reticulocyte lysate, and likely in situations *in vivo* as well, the association of the GR with hsps occurs as soon as or perhaps before the receptor is fully translated.[5] Subsequently, the potential of the lysate preparations in receptor assembly was successfully used to reconstitute previously transformed, hsp-free, receptors.[11,12] The feasibility of what can be regarded as a "reversal transformation" moved us to attempt the use of lysate for the *de novo* assembly of recombinant receptors generated in bacteria, a host that proved to be ineffective for the assembly of other steroid receptors.[8,13–17] We found that the bacterially expressed corticosteroid receptors MR and GR both reacted with lysate in a way similar to the *in vitro* transformed receptors, forming complexes including the 90-kDa heat shock protein, hsp90.[8,18] The resulting heteromeric complexes, unlike the uncomplexed receptors, closely resembled the native receptors in terms of their functional characteristics. However, the lysate untreated receptors evidenced some characteristics that differ from those expected for a typical transformed receptor. This result is difficult to fit within a model in which the only effect of the assembly is to prevent an intrinsically active receptor from being permanently active, without affecting its structure, and leads us to reconsider the potential role of hsps in the functioning of steroid receptors.

By combining the expression of recombinant corticosteroid receptors in bacteria and in eukaryotic cells, we recently learned more about the molecular and functional changes caused by receptor assembly. In this paper, we shall summarize our findings and try to interpret them within the context of other reports which, as a whole, point to a more general importance of the assembly process.

Expression of Corticosteroid Receptors in Bacteria

The DNA- and the steroid-binding domains of the rat MR and GR were expressed in *E. coli* as C-terminal fusion proteins of glutathione-S-transferase (GST) as previously reported for the rMR.[8] The subsequent biochemical characterization of both recombinant proteins was performed using two different bacterial preparations, referred to as crude and pure preparations, respectively. The crude preparations consisted of the supernatants resulting from the bacterial lysis and centrifugation. The pure preparations were obtained by subsequent affinity purification of the crude preparations in immobilized glutathione. For receptor assembly, the pure preparations were incubated with rabbit reticulocyte lysate,[8] resulting in the pure-treated preparations.

Characterization of MR and GR in Crude Preparations

As can be seen in TABLE 1, the crude preparations of MR and GR exhibited steroid-binding specificities that were consistent with those reported for the native receptors. For the MR, the potency of different steroids in [^{3}H]aldosterone competition-binding assays followed the order corticosterone ≥ deoxycorticosterone ≥ aldosterone > cortisol > spironolactone > progesterone > dexamethasone >>> RU26988/38486, testosterone or estradiol. For the GR, using [^{3}H]corticosterone, the

TABLE 1. Apparent K_i Values (nM) for Different Steroids[a]

Steroid	GR Crude	GR Pure-Treated	MR Crude	MR Pure-Treated
RU 28362	49 ± 1	1.9 ± 0.2	>1,000	>100
Dexamethasone	82 ± 7	2.2 ± 0.2	680 ± 106	20.2 ± 2.2
Corticosterone	122 ± 12	3.3 ± 0.1	20 ± 2	0.2 ± 0.1
Cortisol	208 ± 64	5.4 ± 0.2	95 ± 12	1.6 ± 0.2
Progesterone	453 ± 93	11.7 ± 1.2	500 ± 43	17.8 ± 1.8
Aldosterone	583 ± 99	17.4 ± 0.8	40 ± 5	0.6 ± 0.1
Deoxycorticosterone	600 ± 75	19.0 ± 1.0	23 ± 3	0.4 ± 0.1
Spironolactone	>1,000	>100	340 ± 19	6.5 ± 0.6
Testosterone	>1,000	>100	>1,000	>100
Estradiol	>1,000	>100	>1,000	>100

[a]At least three separate binding assays were performed with serial dilutions of the different steroid competitors (0.1–5,000 nM). The radioligand used was either [^{3}H]corticosterone (100 nM for crude GR and 5 nM for pure-treated GR preparations) or [^{3}H]aldosterone (50 nM for crude MR and 0.5 nM for pure-treated MR preparations).

rank order was RU28362 > dexamethasone ≥ corticosterone > cortisol > progesterone > aldosterone ≥ deoxycorticosterone >>>> spironolactone, testosterone or estradiol.

The reported apparent equilibrium dissociation constant (K_D, 4°C) values for corticosterone of native, cytosolic MR and GR approximately average 0.5 nM and 2.5 nM, respectively. However, the K_D values for the recombinant MR and GR in the crude preparations were significantly higher, implying a 78-fold and a 60-fold reduction in the steroid-binding affinity, respectively (see TABLE 2). The hydrodynamic characteristics of the receptors in crude preparations also differed from those observed in the native receptors. Sedimentation velocity analysis in sucrose density gradients of both recombinant MR and GR revealed that the steroid-bound

TABLE 2. Characteristics of the Bacterial Preparations of GR and MR[a]

Preparation	B_{max} (fmoles/mg Protein)	K_D (nM)	Sedimentation Coefficient (S)	Association with hsp90
GR				
Crude	443 ± 79	95 ± 28	≃ 4	–
Pure-treated	69,883 ± 9,884	3.8 ± 0.3	≃ 9	+
MR				
Crude	793 ± 158	47 ± 5	≃ 4	–
Pure-treated	21,700 ± 3,500	0.6 ± 0.1	≃ 9	+

[a]The binding capacity (B_{max}) and K_D values were determined by at least three separate Scatchard analyses using serial dilutions of [^{3}H]corticosterone for GR, and [^{3}H]aldosterone for MR. The sedimentation coefficient of the steroid-bound receptors was determined in linear sucrose gradients in the presence of molybdate. The presence of hsp90 in the receptor complexes was determined by the shift of the 9 S forms towards the bottom of the gradients when samples were incubated, prior to sedimentation, with an anti-hsp90 monoclonal antibody.

complexes migrated as 3.7–4.0 S species, as opposed to the typical 8.5–9.0 S species, hsp90-associated, observed in native receptor preparations. The presence in the gradients of molybdate or tungstate, which stabilize steroid receptors in the 9 S heteromeric forms, did not affect the sedimentation velocity of the recombinant receptors. This result strongly suggests that the steroid-binding protein does not interact with bacterial proteins or factors, at least in a manner stable enough to withstand the analytical procedure. Our results are in agreement with previous bacterial expression studies of glucocorticoid,[17] progesterone,[13–14] and androgen[15] receptors, which did not evidence the formation of definite heteromeric complexes upon sedimentation analysis. Interestingly, in these expression studies, the only receptor evidencing an abnormally low steroid-binding affinity was GR.[17] In addition, a similar low affinity was obtained for liver cytosolic GR preparations that had been previously activated to a DNA-binding, 4 S form.[16] On the basis of this similarity, it was postulated that the component(s) associated with the nonactivated GR, possibly the hsp90, would play an important role in stabilizing the receptor in a high-affinity state for steroids.[16] However, given that *in vitro* assembly of 9 S receptor complexes had proved unsuccessful at that time, the role of the associated components on receptor function could not be established.

In order to determine the DNA-binding characteristics of the steroid-receptor complexes generated in crude preparations, DNA-cellulose binding at 0°C was performed using the GR preparations. Interestingly, less than 10% of the radioactive complexes bound to the DNA-matrix. When samples were heated at 30°C for 1 h, or treated with 0.4 M KCI, conditions that typically promote receptor transformation, the percentage of the complexes bound to DNA did not significantly change.

Characterization of MR and GR in Pure Preparations

The incubation of affinity-purified preparations of MR and GR with concentrations up to 60 nM of different radiolabelled steroids, including [^{3}H]aldosterone, [^{3}H]corticosterone and [^{3}H]dexamethasone, resulted in negligible specific binding. However, after pretreatment of the preparations with rabbit reticulocyte lysate at 30°C to obtain "pure-treated" receptors, steroid-binding activity was detectable. This activating effect was gradual, and at 30°C the maximum binding was reached at 60 min and 30 min of treatment, for MR and GR, respectively. Incubation of the pure preparations alone at 30°C or in the presence of lysate at 0°C resulted in negligible steroid-binding activity. Besides the temperature dependence, the reaction with lysate also evidenced a strict Mg^{2+}/ATP dependence and best results were obtained when an ATP-regenerating system was used.[8] This time, in contrast with the crude receptor preparations, the observed K_D values (see TABLE 2) were in agreement with those reported for cytosolic receptors in rat brain[19–22] or for recombinant receptors expressed in eukaryotic cells.[23–26] The binding specificity pattern, evaluated as the potency rank order of different steroids in competition-binding assays, was the same as that observed with the crude preparations. However, the apparent inhibition equilibrium constant (K_i) values were at least one order of magnitude lower, reflecting the higher steroid-binding affinity of the receptors in the pure lysate-treated preparations. Interestingly, the sedimentation analysis of both lysate-treated preparations revealed that the steroid-receptor complexes migrate as approximately 9 S species in sucrose gradients containing 20 mM molybdate (TABLE 2). When molybdate was replaced by 0.4 M KCI, the sedimentation coefficient shifted to approximately 4 S, as occurs with native cytosolic receptors.

The presence of hsp90 in the steroid-receptor complexes was assessed by sedimentation-shift analysis[8] using the hsp90-reactive, 8D3 monoclonal antibody.[27] Only the 9S, and not the 4 S, complexes were shifted to 12–14 S species by preincubation of the samples with the antibody, indicating that hsp90 is a component of the 9.0 S forms of the lysate-treated recombinant MR and GR (TABLE 2).

DNA-cellulose binding revealed that both pure-treated complexes, MR and GR, can be activated to a DNA-binding form. Treatment of GR with 0.4 M KCI at 4°C for 30 min, or heating at 30°C for 1 h, significantly increased (~ 300%), over the control values, the fraction of receptor complexes bound to DNA. Treatment of MR with 100 mM thiocyanate, a chaotropic anion known to be effective for MR activation,[28] for 30 min at 30°C, or with 0.4 M KCI for 30 min at 4°C, also significantly increased the DNA retention values (79% and 37%, respectively) over the control values. In all the cases, the activating effect of the salt or temperature treatments was completely prevented when complexes were preformed in the presence of 20 mM molybdate.

Taken together, our results indicate that the reticulocyte lysate treatment of the pure receptor preparations promotes the *de novo* formation of heteromeric complexes comprising at least the steroid-binding protein and hsp90. Several lines of evidence support this conclusion:

a. The hydrodynamic behavior, the steroid-binding affinity, and the DNA-binding characteristics of the receptor complexes present in the pure lysate-treated preparations, which completely mimic the native receptors.

b. The reaction of the receptor complexes with the 8D3 monoclonal antibody against hsp90.

c. The temperature and energy dependence of the reaction between the pure preparations and the retyculocite lysate, which has also been observed in the reassociation of immunopurified progesterone and glucocorticoid receptors with heat shock proteins.[11–12,29]

Interestingly, the reassociation with heat shock proteins has only been observed when purified receptor preparations are used, the results being consistent with the presence of unknown inhibitory factors in crude cytosolic preparations. Our results are in agreement with this observation. Lysate treatment of crude bacterial preparations does not generate detectable amounts of hsp90-receptor complexes, although the nature of this inhibition is still unknown.

In addition to hsp90, other proteins, including hsp70 and hsp56, have been described to immunoprecipitate with the steroid-binding proteins of the different steroid receptors (for review, see Refs. 30–31). We are presently conducting experiments to determine the probable association of these proteins with the recombinant receptors during the lysate-mediated assembly.

The comparison between the hydrodynamic behavior, and the steroid- and DNA-binding characteristics of the receptors present in crude versus pure lysate-treated bacterial preparations strongly supports the important role of the assembly in receptor function. Our data show that the assembly process is necessary for the generation of corticosteroid receptor complexes with high affinity for the steroid. In agreement, rat epididymal sperm, which appears to lack hsp90, contains GR that fails to bind ligand.[32] In addition, our data show that the assembly process is required for the formation of receptor forms capable of undergoing transformation to a DNA-binding state. This last requirement may explain why the glucocorticoid receptor, when expressed in a strain of yeast with limited expression of hsp, is ineffective in promoting transcriptional activation from a glucocorticoid-responsive reporter plasmid.[33] The results in yeast constituted the first *in vivo* suggestion of the implication of hsp90 in the process of receptor activation. However, given that hsp90 is essential for yeast survival, the expression of hsp90 was not completely abolished and only a decrease in the ratio of hsp90 to receptor levels could be achieved. As opposed to yeast hsp90, the bacterial hsp90 homologue C62.5[34] does not appear to form stable complexes with corticosteroid receptors.[8,13–17] This difference could reflect the fact that the predicted amino acid sequence of C62.5 is only 42% identical to human hsp90, and also lacks a fifty amino acid stretch which is believed to be important for interaction with steroid receptors.[35]

In agreement with the *in vivo* evidence, our results with the crude preparations suggest a key role of hsps in the appropriate folding and transformation of the receptors and is at odds with a model in which hsp90 is seen primarily as a mechanism for preventing the steroid-binding protein from interacting with target DNA as proposed by Baulieu *et al.*[3]

Expression of the Glucocorticoid Receptor in Mammalian Cells

In order to further understand the structural requirements for the *in vivo* GR function, mutant GRs were expressed as the rat full-length receptors in mammalian Cos-1 cells. For comparison, the same mutants, lacking the N-terminal region, were also expressed in bacteria as fusion proteins and treated with reticulocyte lysate as described above. Two mutants were designed with substitutions in the hsp90-binding region of GR[36] (see FIG. 1). In the GRO mutant, Pro 643, located in a conserved hydrophobic stretch of the hsp90-binding region, was replaced by Ala. In the GR8, two substitutions were made in the conserved N-terminal end of the hsp90-binding region, which consisted of the replacement of Lys 597 by Ile, and Pro 600 by Leu.

In both expression systems tested, the GRO mutant exhibited a significant 4-fold reduction in steroid-binding affinity, in comparison with the wild type control. Interestingly, the binding specificity of GRO for different steroids did not change in either of the assayed systems. Since the steroid-binding activity of the GR8 mutant was substantially reduced in both systems, an accurate binding characterization could not be performed with this mutant.

Differences between the GRO mutant and the wild type GR were also detected in the sucrose-gradient sedimentation characteristics. In the presence of molybdate, under conditions in which the steroid-bound complexes of GR migrated exclusively as 9 S forms, more than 50% of the GRO complexes migrated as smaller, 4 S forms. This result was identical in both expression systems (mammalian and bacterial) and strongly suggests that the stability of the hsp90-containing 9 S forms is reduced in the GRO mutant. Moreover, under activating conditions, both receptors, GR and GRO, migrated exclusively as 4 S forms.

In order to evaluate the possible functional implications of the structural changes in the GRO mutant, transcriptional activation assays were conducted. Interestingly, when a glucocorticoid-responsive reporter plasmid was used, the GRO mutant showed only 30% of the hormonal response of the wild type GR. This decrease was observed even at maximal concentrations of ligand, eliminating the differences due to affinity changes; furthermore, the decreased transcriptional activation was seen after correcting for levels of expression and binding (B_{max}), indicating that the loss cannot be simply explained on the basis of diminished receptor protein, decreased ligand availability or decreased binding. In summary, in comparison with the wild GR, GRO exhibited: (i) normal, full *in vitro* dissociation from hsp90 under activating conditions, (ii) reduced but not suppressed steroid- and hsp90-binding, and (iii) reduced transcriptional activation. These facts lead us to suggest that proper receptor-hsp90 interaction, and/or stability of the heteromeric form of the receptor as a whole, are required for the transcriptional activation process. The chaperone role of

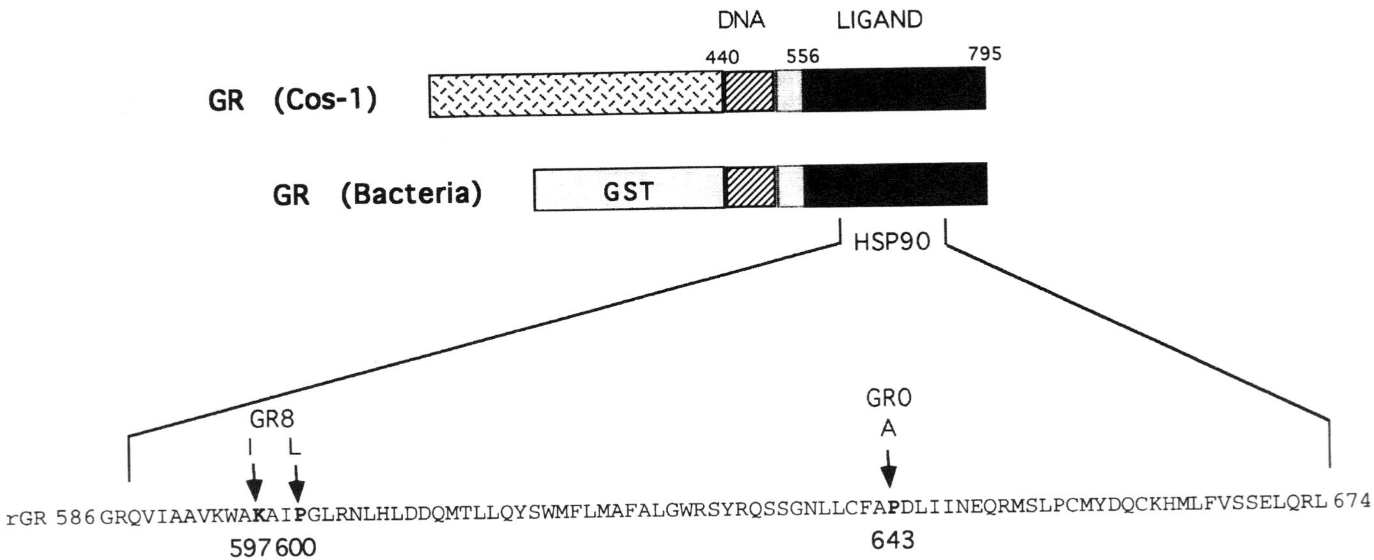

FIGURE 1. Schematic representation of the GR mutants generated in Cos-1 cells and in bacteria. GST, glutathione-S-transferase; DNA-, ligand-, and hsp90-binding domains of the receptor. The numbers below the amino acid sequence correspond to the rat GR sequence.

hsp90[37,38] supports the conclusion that the mutation might alter the hormonally controlled switching between the transcriptionally active/inactive conformations of the receptor complex. Consequently, the conserved hydrophobic region in GR, involving the Pro 643, might be a key mediator in receptor transformation.

It has been shown that mutant steroid receptors that do not associate with hsp90 are constitutive activators of transcription.[39–41] On this basis, an "inactivation" function was postulated for hsp90. In mutational analysis, however, the complete steroid-binding domain had to be deleted in order to completely abolish the hsp90 association. Importantly, the same receptor domain has been reported to influence the intracellular solubility of overexpressed receptors.[42]

Our findings are in accordance with these latter observations, and provide experimental support to an emerging model in which hsp90 association, or the assembly in general, are required to permit, and not to impede, function of the full-length, wild type receptor.

REFERENCES

1. GODOWSKI, P. J., D. PICARD & K. R. YAMAMOTO. 1988. Signal transduction and transcriptional regulation by glucocorticoid receptor-LexA fusion proteins. Science **241:** 812–816.
2. SCHERRER, L. C., D. PICARD, E. MASSA, J. M. HARMON, S. S. SIMONS, K. R. YAMAMOTO & W. B. PRATT. 1993. Evidence that the hormone binding domain of steroid receptors confers hormonal control on chimeric proteins by determining their hormone-regulated binding to heat-shock protein 90. Biochemistry **32:** 5381–5386.
3. PRATT, W. B., D. J. JOLLY, D. V. PRATT, S. M. HOLLENBERG, V. CIGUERE, F. M. CADEPOND, G. SCHWEIZER-GROYER, M. G. CATELLI, R. M. EVANS & E. E. BAULIEU. 1988. A region in the steroid binding domain determines formation of the non-DNA-binding, 9 S glucocorticoid receptor complex. J. Biol. Chem. **263:** 267–273.
4. BRESNICK, E. H., F. C. DALMAN, E. R. SANCHEZ & W. B. PRATT. 1989. Evidence that the 90-kDa heat shock protein is necessary for the steroid binding conformation of the L cell glucocorticoid receptor. J. Biol. Chem. **264:** 4992–4997.
5. DALMAN, F. C., E. H. BRESNICK, P. D. PATEL, G. H. PARDEW, S. J. WATSON & W. B. PRATT. 1989. Direct evidence that the glucocorticoid receptor binds to hsp90 at or near the termination of receptor translation *in vitro.* J. Biol. Chem. **264:** 19815–19821.
6. MESHINCHI, S., E. R. SANCHEZ, K. J. MARTELL & W. B. PRATT. 1990. Elimination & reconstitution of the requirement for hormone in promoting temperature-dependent transformation of cytosolic glucocorticoid receptors to the DNA-binding state. J. Biol. Chem. **265:** 4863–4870.
7. NEMOTO, T., Y. OHARA-NEMOTO, M. DENIS & J. A. GUSTAFSSON. 1990. The transformed glucocorticoid receptor has a lower steroid-binding affinity than the non-transformed receptor. Biochemistry **29:** 1880–1886.
8. CAAMAÑO, C. A., M. I. MORANO, P. D. PATEL, S. J. WATSON & H. AKIL. 1993. A bacterially expressed mineralocorticoid receptor is associated *in vitro* with the 90-kilodalton heat shock protein and shows typical hormone- and DNA-binding characteristics. Biochemistry **32:** 8589–8595.
9. HENDRICK, J. P. & F.-U. HARTL. 1993. Molecular chaperone functions of heat shock proteins. Annu. Rev. Biochem. **62:** 349–384.
10. PRATT, W. B. 1990. Interaction of hsp90 with steroid receptors: organizing some diverse observations and presenting the newest concepts. Mol. Cell. Endocrinol. **74:** C69–C76.
11. SMITH, D. F., D. B. SCHOWALTER, S. L. KOST & D. O. TOFT. 1990. Reconstitution of progesterone receptor with heat shock proteins. Mol. Endocrinol. **4:** 1704–1711.
12. SCHERRER, L. C., F. C. DALMAN, E. MASSA, S. MESCHINI & W. B. PRATT. 1990. Structural and functional reconstitution of the glucocorticoid receptor-hsp90 complex. J. Biol. Chem. **265:** 21397–21400.

13. EUL, J., M. E. MEYER, L. TORA, M. T. BOCQUEL, C. QUIRIN-STRICKER, P. CHAMBON & H. GRONEMEYER. 1989. Expression of active hormone and DNA-binding domains of the chicken progesterone receptor in *E. coli.* EMBO J. **8:** 83–90.
14. POWER, R. F., O. M. CONNEELY, D. P. MCDONNELL, J. H. CLARK, T. R. BUTT, W. T. SCHRADER & B. W. O'MALLEY. 1990. High level expression of a truncated chicken progesterone receptor in *Escherichia coli.* J. Biol. Chem. **265:** 1419–1424.
15. YOUNG, C. Y. F., S. QIU, J. L. PRESCOTT & D. J. TINDALL. 1990. Overexpression of a partial human androgen receptor in *E. coli*: characterization of steroid binding, DNA binding, and immunological properties. Mol. Endocrinol. **4:** 1841–1849.
16. NEMOTO, T., Y. OHARA-NEMOTO, M. DENIS & J. A. GUSTAFSSON. 1990. The transformed glucocorticoid receptor has a lower steroid-binding affinity than the non-transformed receptor. Biochemistry **29:** 1880–1886.
17. OHARA-NEMOTO, Y., P. E. STROMSTEDT, K. DAHLMAN-WRIGHT, T. NEMOTO, J. A. GUSTAFSSON & J. CARLSTEDT-DUKE. 1990. The steroid-binding properties of recombinant glucocorticoid receptor: a putative role for heat shock protein. J. Steroid Biochem. Mol. Biol. **37:** 481–490.
18. CAAMAÑO, C. A., M. I. MORANO & H. AKIL. 1992. A recombinant mineralocorticoid receptor expressed in bacteria adopts a high steroid-binding affinity state. J. Cell. Biochem. **16C:** 49.
19. MOGUILEWSKY, M. & J. P. RAYNAUD. 1980. Evidence for a specific mineralocorticoid receptor in rat pituitary and brain. J. Steroid Biochem. **12:** 309–314.
20. MCEWEN, B. S., E. R. DE KLOET & W. ROSTENE. 1986. Adrenal steroid receptors and actions in the nervous system. Physiol. Rev. **66:** 1121–1188.
21. WRANGE, O. & Z. Y. YU. 1983. Mineralocorticoid receptor in rat kidney and hippocampus: characterization and quantitation by isoelectric focusing. Endocrinology **113:** 243–250.
22. REUL, J. M. H. M. & E. R. DE KLOET. 1985. Two receptor systems for corticosterone in rat brain: microdistribution and differential occupation. Endocrinology **117:** 2505–2511.
23. ALNEMRI, E. S., A. B. MAKSYMOWYCH, N. M. ROBERTSON & G. LITWACK. 1991. Characterization and purification of a functional rat glucocorticoid receptor overexpressed in a baculovirus system. J. Biol. Chem. **266:** 3925–3936.
24. ALNEMRI, E. S., A. B. MAKSYMOWYCH, N. M. ROBERTSON & G. LITWACK. 1991. Overexpression and characterization of the human mineralocorticoid receptor. J. Biol. Chem. **266:** 18072–18081.
25. ARRIZA, J. L., C. WEINBERGER, G. CERELLI, T. M. GLASER, B. L. HANDELIN, D. E. HOUSMAN & R. M. EVANS. 1987. Cloning of human mineralocorticoid receptor complementary DNA: structural and functional kinship with the glucocorticoid receptor. Science **237:** 268–275.
26. BINART, N., M. LOMBES, M. E.RAFESTIN-OBLIN & E. E. BAULIEU. 1991. Characterization of human mineralocorticoid receptor expressed in the baculovirus system. Proc. Natl. Acad. Sci. USA **88:** 10681–10685.
27. PERDEW, G. H. 1988. Associationof the Ah receptor with the 90-kDa heat shock protein. J. Biol. Chem. **263:**13802–13805.
28. LUTTGE, W. G. & S. M. EMADIAN. 1988. Transformation of the [^{3}H]aldosterone-type I receptor complex to a DNA-binding species. Life Sci. **43:** 837–843.
29. SMITH, D. F., B. A. STENSGARD, W. J. WELCH & D. O. TOFT. 1990. Assembly of progesterone receptor with heat shock proteins and receptor activation are ATP-mediated events. J. Biol. Chem. **267:** 1350–1356.
30. PRATT, W. B. 1990. Interaction of hsp90 with steroid receptors: organizing some diverse observations and presenting the newest concepts. Mol. Cell. Endocrinol. **74:** C69–C76.
31. SMITH, D. F. & D. O. TOFT. 1993. Steroid receptors and their associated proteins. Mol. Endocrinol. **7:** 4–11.
32. KAUFMANN, S. H., W. W. WRIGHT, S. OKRET, A. C. WIKSTROM, J. A. GUSTAFSSON, N. L. SHAPER & J. H. SHAPER. 1992. Evidence that rodent epididymal sperm contain the Mr approximately 94,000 glucocorticoid receptor but lack the Mr approximately 90,000 heat shock protein. Endocrinology **130:** 3074–3084.

33. PICARD, D., B. KHURSHEED, M. J. GARABEDIAN, M. G. FORTIN, S. LINDQUIST & K. R. YAMAMOTO. 1990. Reduced levels of hsp90 compromise steroid receptor action *in vivo.* Nature (London) **348:** 166–168.
34. BARDWELL, J. C. A. & E. A. CRAIG. 1987. Eukaryotic Mr 83,000 heat shock protein has a homologue in *Escherichia coli.* Proc. Natl. Acad. Sci. USA **84:** 5177–5181.
35. BAULIEU, E. E., N. BINART, F. CADEPOND, M. G. CATELLI, B. CHAMBRAUD, J. GARNIER, J. M. GASC, G. GROYER-SCHWEIZER, M. E. OBLIN, C. RADANYI, G. REDEUILH, J. M. RENOIR & M. SABBAH. 1989. Do receptor-associated nuclear proteins explain the earliest steps of steroid hormone function? *In* The Steroid/Thyroid Hormone Receptor Family and Gene Regulation. J. Carlstedt-Duke, H. Erikson & J.-Å Gustafsson, Eds.: 301–318. Birkhäuser Verlag. Basel.
36. CAAMAÑO, C. A., M. I. MORANO, S. J. WATSON, F. DALMAN, W. B. PRATT & H. AKIL. 1994. Point mutations in the hsp90-binding region of the rat glucocorticoid receptor affect the steroid-binding characteristics of the receptor. J. Cell. Biochem. **18B:** 340.
37. WIECH, H., J. BUCHNER, R. ZIMMERMANN & U. JAKOB. 1992. Hsp90 chaperones protein folding *in vitro.* Nature **358:** 169–170.
38. MIYATA, Y. & I. YAHARA. 1992. The 90-kDa heat shock protein, hsp90, binds and protects casein kinase II from self-aggregation and enhances its kinase activity. J. Biol. Chem. **267:** 7042–7047.
39. PRATT, W. B., D. J. JOLLY, D. V. PRATT, S. M. HOLLENBERG, V. GIGUERE, F. M. CADEPOND, G. SCHWEIZER-GROYER, M. G. CATELLI, R. M. EVANS & E. E. BAULIEU. 1988. A region in the steroid binding domain determines formation of the non-DNA-binding, 9 *S* glucocorticoid receptor complex. J. Biol. Chem. **263:** 267–273.
40. SCHOWALTER, D. B., W. P. SULLIVAN, N. J. MAIHLE, A. B. DOBSON, O. M. CONNEELY, B. W. O'MALLEY & D. O. TOFT. 1991. Characterization of progesterone receptor binding to the 90-and 70-kDa heat shock proteins. J. Biol. Chem. **266:** 21165–21173.
41. CADEPOND, F., G. SCHWEIZER-GROYER, I. SEGARD-MAUREL, N. JIBARD, S. M. HOLLENBERG, V. GIGUERE, R. M. EVANS & E. E. BAULIEU. 1991. Heat shock protein 90 as a critical factor in maintaining glucocorticosteroid receptor in a nonfunctional state. J. Biol. Chem. **266:** 5834–5841.
42. ALNEMRI, E. S. & G. LITWACK. 1993. The steroid binding domain influences the intracellular solubility of the baculovirus overexpressed glucocorticoid and mineralocorticoid receptors. Biochemistry **32:** 5387–5393.

DISCUSSION OF THE PAPER

D. B. DEFRANCO (*University of Pittsburgh*): Is the assumption you are making that hsp-90 has no nuclear function?

AKIL: It is not something that I have any direct opinion about. It is the model, and the model may be incorrect. I absolutely agree with you that the conformation change may not lead to the shedding of hsp-90. What do you think?

DEFRANCO: I think what your data suggest is that there might, in fact, be some role for hsp-90 in the nucleus.

AKIL: The main reason I am showing the data in this rather physiological meeting is that there is a lot more to the molecular interactions. But whether the receptor acts through heat shock proteins or through folding and interaction with other proteins or directly, I do not know.

Note: The following discussion focused on some data showed by Dr. Akil regarding the existence of multiple MR isoforms, with different 5′untranslated regions, called α, β, γ, which have differential expression in brain regions and across development. This aspect is not covered in her paper.

DEFRANCO: Do you have any data on MR protein during development, particularly where you see differences in isoforms?

AKIL: We know that the pattern of MR binding more or less parallels the overall pattern of MR expression.

QUESTION: Do you see the profile with the different messenger isoforms in terms of protein?

AKIL: I cannot yet link these isoform data to a change in protein.

DEFRANCO: If you transfect one isoform versus another, do you see any difference?

AKIL: At least with the MR beta form we know it is very hard to express, while the alpha form expressed perfect MR activity. Something in β-MR mRNA may be inhibiting translation. Gamma has not been tried in our laboratory.

QUESTION: There are also some really old data from Liao, suggesting that the androgen receptor messenger might have some function itself interacting with the receptor protein. Do you have any thoughts on that?

AKIL: I have thoughts but no data.

QUESTION: When you did the localization of the different isoforms in the brain, did you find a coincidence of the expression of the kidney form with the localization of the aldosterone-specific mineralocorticoid receptors? It would make good sense because in the hippocampus the MR also binds corticosterone. But we should not forget that there is a true face for mineralocorticoid in the receptor which mediates aldosterone action on salt appetite. It might be that even if that is the exact same protein as the other one, that the angulation is different. In fact, you would expect them to be different. Do you see co-localization with that type of mineralocorticoid receptor in the brain with the kidney isoform?

AKIL: The kidney isoform and the other sort of more brain-specific isoform in all likelihood show up in the same cells except for some differences in magnitude of expression, as in the CA2 are, for example, or indusium griseum where one is expressed and not the other. Our guess is that all these forms are going to be co-expressed either in the same, very same cell, or at least in very similar cells in the neighborhood of each other.

QUESTION: Are you saying that in the brain the different isoforms are differently expressed?

AKIL: They are differently expressed in different subregions, and they are differentially regulated by adrenalectomy. Developmentally they are differentially expressed and regulated as well. But there is no region that is full of alpha and has zero of beta.

F. ANTONI (*University of Edinburgh*): In the 3′ mutants, you obtained mutants with respect to steroid binding. Are these also mutants with respect to hsp binding?

AKIL: Yes. they had a two- to threefold decrease in affinity towards all the major GR ligands. At the same time, they seem to have a more rapid rate of dissociation from hsp. They associated with hsp-90 at the same rate as far as we could tell from time experiments.

QUESTION: Did you test the mutant for its ability to bind to DNA? Was it constitutively active?

AKIL: No, that has not been done yet. We obviously need to do it. That was also Dr. DeFranco's question. The only evidence we have is indirect ability to activate in a CAT assay. And in that case it did not do nearly so well as the wild type. We need to do the more direct binding kinetics and see if it is directly on the DNA or through hsp interactions or dimer formation.

The Binary Corticosteroid Response: Transcriptional Regulation by Mineralo- and Glucocorticoid Receptors

KLAUS DAMM,[a] RAINER RUPPRECHT, AND FLORIAN HOLSBOER

Department of Neuroendocrinology
Max-Planck-Institute of Psychiatry
Kraepelinstrasse 2-10
80804 Munich, Germany

INTRODUCTION

The cloning of the human glucocorticoid receptor (hGR) provided the first complete structure of a steroid receptor and revealed a homology to the oncogene v-erbA in what was postulated to be the DNA binding domain of the receptor.[1] The cellular homologue of v-erbA was subsequently identified as a thyroid hormone receptor providing the first direct evidence that steroid and thyroid hormones function through receptors with mechanistically similar properties. The isolation of sequences encoding nuclear hormone receptors has since proceeded by many avenues and led to the isolation of cDNAs encoding all the known steroid, thyroid hormone, retinoic acid and vitamin D receptors as well as a number of "orphan" receptors with unknown ligand and function.[2] Through the availability of the cloned hormone receptor genes, molecular techniques can be employed to characterize the contribution of individual receptor systems to the integrated and complex biological responses inherent in development and homeostasis. The structure and functions of the glucocorticoid receptor have been extensively studied and have revealed separate receptor domains that determine DNA binding, ligand binding and transcriptional regulatory properties.[2,3] The central region, responsible for specific DNA sequence recognition, is very highly conserved within the nuclear receptor gene family and is characterized by nine invariant cysteine residues forming zinc-binding finger structures. The primary function of this domain is to appropriately position the receptor on target gene sequences in chromatin. The carboxyl terminal region of the hGR contains the hormone-binding function and is consequently less well conserved with other members of the steroid receptor family. In the absence of ligand this domain inhibits the DNA binding and activation functions.[3–6] The amino terminal portions of the hormone receptors vary in size and show no sequence relatedness. This variability may facilitate functionally distinct tissue- and gene-specific properties and is particularly evident in a comparison of the hGR with the human mineralocorticoid (hMR) and progesterone receptors (hPR).[7–10] Because of the near identity of the DNA binding domain, hGR, hMR and hPR can interact with the same or closely related target DNA sequences and as a result of extensive sequence conservation in the ligand binding domain, hGR and

[a]Present address: Klaus Damm, Ph.D., Dr. Karl Thomae GmbH, Dept. of Pharmacological Research, Birkendorfer Str. 65, 88397 Biberach, Germany.

hMR in particular are responsive to many of the same ligands.[2,11] The unique characteristics of the amino terminal region of each receptor may significantly influence the efficiency with which a receptor can regulate the expression of a particular gene. The existence of multiple receptor systems for glucocorticoid hormones raises important issues for the coordinate responses to this stimulus and the use of recombinant nuclear hormone receptors together with the reconstitution of hormone-dependent gene regulation in heterologous cell lines enables an accurate assessment of the functional consequences of ligand-receptor interaction on gene expression. The ability of steroid hormones to induce transcription from the hormone-responsive mouse mammary tumor virus (MTV) promoter provides an excellent *in vitro* system in which the efficacy of various steroids can be determined.[3,10,12,13] hMR or hGR expression vectors were contransfected with the MTV-LUC receptor gene in the human neuroblastoma cell line SK-N-MC. In this system, the agonistic activity of steroid ligands is proportional to the induction of the MTV promoter as measured by the extent of luciferase production in the cells. Utilizing this assay the transcriptional responses of hGR and hMR to various steroids can be compared at fixed hormone concentrations. The hGR, for example, exhibits a strong response to the synthetic glucocorticoids RU28362 and dexamethasone as well as to cortisol and corticosterone, the natural glucocorticoids.[3,10] Only a weak response to the mineralocorticoid aldosterone is found. The hMR, in contrast, is activated by both mineralocorticoids and glucocorticoids.[7,8,10] There are, however, several caveats to this type of assay. First, by performing these analyses at one fixed steroid concentration, the differential binding affinities of the hMR and hGR are ignored. Thus, while the hMR response to cortisol may be optimal at 10 nM, the hGR response is not at a maximum. Therefore, hormone response profiles provide a more accurate comparison of hormone-dependent gene regulation.[3,8,10] Second, the MTV promoter has several receptor binding sites and the hGR in particular has been shown to act cooperatively at several HREs. To distinguish between intrinsic transactivation properties of the receptors and the cooperative activity resulting from the binding of multiple receptor molecules, reporter genes containing single or multiple binding sites upstream of heterologous promoters, such as the herpes simplex virus thymidine kinase or the simian SV40 early gene promoter can be used.[12,14]

Functional Properties of hMR and hGR Ligands

The agonist properties of steroid ligands were determined by assessing the potency of the steroids to induce transcription from the hormone-responsive MTV-luciferase reporter gene. Dose-response curves for a series of natural and synthetic steroid hormones were obtained by cotransfecting the reporter gene together with hormone receptor expression plasmids in the human neuroblastoma cell line SK-N-MC. The transfected cells were incubated with increasing concentrations of steroid hormones. In this system, the agonistic activity of steroids is proportional to the induction of gene expression as measured by the extent of luciferase production in the cells (FIG. 1). Cortisol and the synthetic compounds dexamethasone, RU28362 and 9α-fluorocortisol induce a strong dose-dependent activation of the hGR with maximal induction levels between 700- and 1000-fold and an ED_{50} of 10.0, 2.0, 1.0 and 4.0 nM, respectively. The mineralocorticoids aldosterone and deoxycorticosterone, progesterone and the antiprogestin RU486 display only a weak agonistic activity and the antimineralocorticoid spironolactone induces no stimulation of the MTV promoter via the hGR (TABLE 1). The hMR was found to be very sensitive to the mineralocorticoids (aldosterone, deoxycorticosterone) as well as 9a-fluorocor-

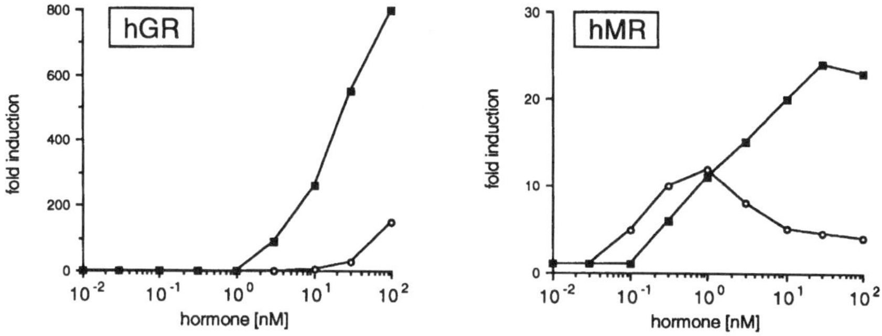

FIGURE 1. Transactivation properties of hMR and hGR. SK-N-MC cells cotransfected with MTV-LUC and either hMR or hGR expression plasmid were incubated with the indicated concentrations of aldosterone (○) and corticosterone (■). The induction of the MTV promoter was calculated from the luciferase activity measured in the individual cell extracts. Results are presented as the average of three to six individual experiments.

tisol, with ED_{50} values below 1 nM (TABLE 1). Glucocorticoids (cortisol, dexamethasone) also show a clear agonistic activity via the hMR with maximal transactivation potencies in the range of aldosterone and deoxycorticosterone and ED_{50} values of 0.4 and 1.1 nM, respectively. RU28362, RU486 and spironolactone are inactive as inducers of transcription, whereas progesterone displays a very weak partial agonistic activity (TABLE 1). The A-ring reduced pregnane derivatives of progesterone and deoxycorticosterone, allopregnanolone (THP) and tetrahydro-deoxycorticosterone (THDOC), which have been shown to activate the PR,[15] do not bind to the corticosteroid receptors and are only very weak activators of hGR and hMR (TABLE 1). To verify weak agonistic effects of hMR ligands, which are hardly detectable by the wild-type hMR due to its low transactivation potential, the chimeric

TABLE 1. Activation, Functional Sensitivity and Binding Affinity of hGR and hMR for Natural and Synthetic Steroid Ligands

	hGR			hMR		
	Fold Induction	ED_{50}	k_i	Fold Induction	ED_{50}	k_i
Aldosterone	160	50.0	63	12	0.1	0.16
Cortisol	750	10.0	15	24	2.0	0.37
Deoxycorticosterone	250	40.0	40	17	0.4	<0.01
Dexamethasone	900	2.0	1.1	13	2.0	1.1
9-α-Fluorocortisol	1000	4.0	96	42	0.2	0.35
RU28362	700	1.0	0.4	1	n.m.	>450
RU486	22	3.0	2.2	1	n.m.	>450
Spironolactone	1	n.m.	79	1	n.m.	5.7
Progesterone	36	50	16	2	30	<0.005
THDOC	5	>100	n.m.	3	100	>150
THP	6	>100	n.m.	1	n.m.	n.m.

receptor GMM provides a more sensitive assay system.[10,12] GMM resembles the hMR in its relative ligand-binding specificity but acts as a much stronger inducer of gene expression due to the presence of the hGR-derived transactivation domain $\tau 1$ in the amino terminus (FIG. 2). These assays resulted in a GMM-mediated 1000–1200-fold induction by aldosterone, deoxycorticosterone and cortisol and revealed a 240-fold induction with THDOC and a 56-fold induction using progesterone. Similar to the data observed with hMR, spironolactone, RU28362 and RU486 did not show any agonist activity.

Antagonist Activity of Steroid Ligands

A comparison of the ED_{50} values obtained in the cotransfection assay and binding affinities measured in *in vitro* binding experiments revealed a good agreement between both parameters when the various compounds were assayed with the hGR. The ED_{50} of the antiglucocorticoid RU486 is in the range of hGR agonists, and progesterone also has an ED_{50} value comparable to aldosterone and deoxycorticosterone. However, these two compounds induce only a low transcriptional response suggesting that despite their partial agonistic activity they tend to function as hGR antagonists *in vivo.* Spironolactone fails to induce detectable transcriptional activity and may function as a pure antagonist. The binding affinity and the ED_{50} values required for biological activity of several ligands are significantly different with regard to the hMR. For example, aldosterone shows a higher functional sensitivity for the hMR than deoxycorticosterone or cortisol (TABLE 1). The extremely high binding affinity of the hMR for progesterone (k_i <0.005 nM) in combination with the very low agonistic activity indicates that progesterone may act as a potent hMR antagonist. To demonstrate the antagonistic activity of these compounds at the gene expression level, the cotransfection assay described above was adopted. Activation of the hGR with 1 nM RU28362 results in an approximately 400-fold activation. One thousandfold excesses of RU486, progesterone and spironolactone decrease this response to approximately 10%, 45% and 60%, respectively, indicating that only RU486 has potent antiglucocorticoid activity (data not shown). Similar experiments with the hMR, activated with 1 nM aldosterone, revealed a dose-dependent antag-

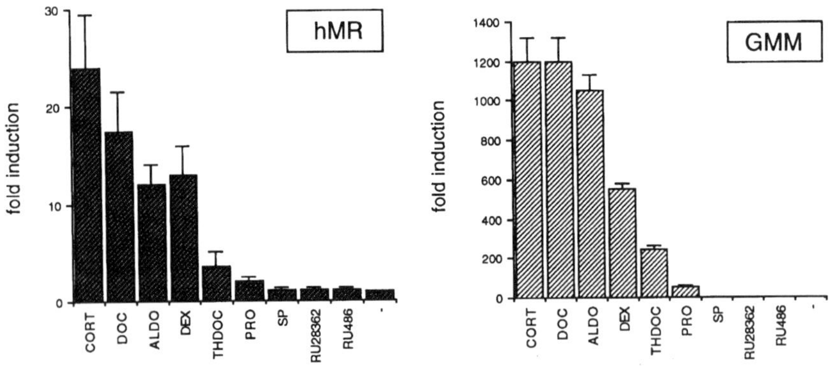

FIGURE 2. Maximal transcriptional activation by hMR and hGR. Dose-response curves for the indicated steroid hormones were obtained as described in FIGURE 1 and the values corresponding to the maximal transcriptional induction are shown.

onistic activity of spironolactone. However, increasing doses of progesterone nearly completely abolished the aldosterone-induced hMR response, demonstrating that progesterone can act as an antimineralocorticoid that is even more potent that spironolactone. Because of the low transactivation potential of the hMR the antagonistic properties of spironolactone and progesterone were assayed also on the chimeric receptor GMM. Spironolactone exhibits a 50% inhibition of the aldosterone response at 100 nM (IC50); however, similar to the results obtained with the wild-type hMR, progesterone is even more effective as an aldosterone antagonist with an IC50 value of 2–3 nM (FIG. 3).

Transactivation Properties of hMR, hGR and Chimeric Receptors

The molecular basis of the disparity in gene activation by the hMR and hGR was determined by comparing the cortisol responses of hybrid molecules carrying all possible combinations of amino terminal, DNA-binding, and ligand-binding domains for these two receptors (FIG. 4). An exchange of the DNA binding domains of both receptors produced only small effects on the ability of these hybrids to regulate the activity of the MTV promoter, suggesting that the DNA binding domains are functionally equivalent. The functional importance of the amino terminus in stimulating transcription was demonstrated in hybrid constructions in which this region was deleted (ΔGG and ΔMM) or exchanged (GMM and MGG). Removing the entire amino terminus of the hGR reduces the activity of ΔGG to 2% of the activity of the parent GRnx while ΔMM retained 25% of the activity of the MRnx. Transferring the hMR amino terminus to ΔGG, thereby creating hybrid MGG, did

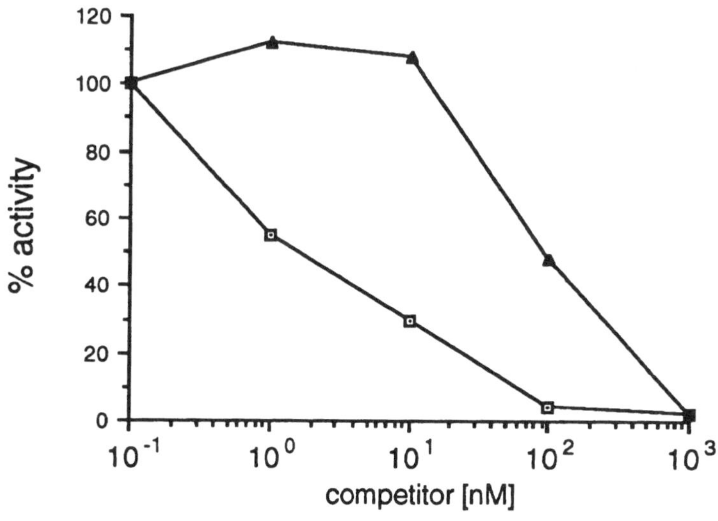

FIGURE 3. Competition of mineralocorticoid action by spironolactone and progesterone using the GMM expression plasmid. Results are expressed as % activity following co-incubation of aldosterone (1 nM) and increasing concentrations of spironolactone (▲) or progesterone (⊡) as indicated and represent the mean of three independent experiments.

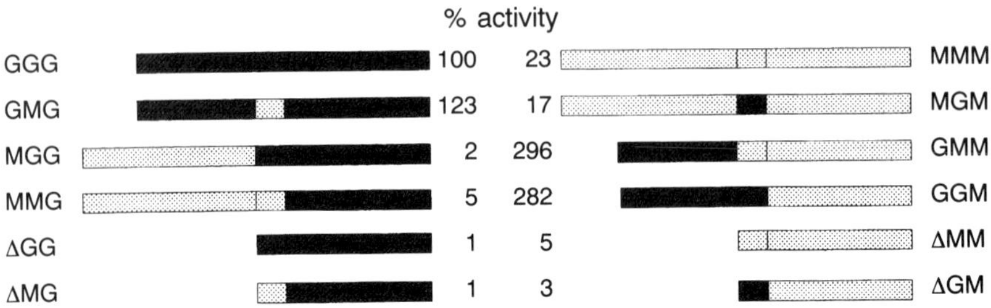

FIGURE 4. Structure and activity of GR/MR chimeric molecules. *Hatched bars* indicate fragments derived from hGR (G); *dotted bars* indicate hMR domains (M). Lettering refers to the origin of the domains. All constructs were assayed for induction of the MTV promoter in the absence or presence of 100 nM cortisol. Results are expressed as the percent activity related to hGRnx activity and represent th emean of four to six independent experiments.

not significantly increase activity (3%), whereas the reciprocal hybrid GMM exhibited activation properties 58-fold higher than ΔMM and 13-fold higher than MRnx (231%). Thus, the hGR amino terminus, which contains the t1 activation regions, strongly potentiates transcriptional response while the hMR amino terminus provides no apparent contribution to promoter activation. The properties of wild-type and chimeric receptors were similar in CV1 cells and also when the specific ligands RU28362 (hGR) and aldosterone (hMR) were used to activate the receptors (data not shown).

Cooperative Transcriptional Activation by hMR and hGR

To distinguish between intrinsic transactivation properties of the receptors and the cooperative activity resulting from the binding of multiple receptor molecules, we analyzed full-length and chimeric receptors on reporter constructs harboring one or two perfect palindromic GRE sequences (TABLE 2). When assayed on a single GRE (ΔM-GRE-LUC) GRnx exhibited an approximately 100-fold activation in response to 100 nM cortisol and transcriptional activity increased to 400-fold when an additional binding site was introduced (ΔM-GRE2-LUC). The activation ratio (2GRE/1GRE) of 3.8 is indicative of a synergistic (*i.e.*, more than additive) induction on the two adjacent response elements. Deletion of the hGR amino terminus in the construct ΔGG reduced transcriptional activation to a 13-fold and 92-fold induction

TABLE 2. Transactivation and Synergistic Activity of GR/MR Chimeric Receptors

	1 GRE	2 GRE
GRnx	107 ± 19	404 ± 91
ΔGG	13 ± 2	92 ± 13
MGG	9 ± 4	10 ± 4
MRnx	10 ± 2	13 ± 3
ΔMM	19 ± 3	115 ± 38
GMM	173 ± 57	552 ± 142

on ΔM-GRE-LUC and ΔM-GRE(2)-LUC. However, synergistic activity was maintained or even enhanced. The hMRnx, in contrast, displayed only approximately 10% of the hGRnx transcriptional activity when assayed on a single GRE and showed no synergistic and not even an additive increase on two GREs. Deletion of the hMR amino terminus in the construct ΔMM resulted in no reduction in transactivation on a single GRE but enabled cooperativity of the truncated molecule on two GREs (115-fold induction). Transfer of the hGR amino terminus (GMM) increased the transactivation potency (173-fold induction) without impairing synergistic activity on two GREs. The hybrid MGG is devoid of the cooperativity observed with the truncated molecule ΔGG indicating that the hMR amino terminus interferes with synergistic activity in this system. Similar results were obtained when the two GREs were separated by insertions of 35- or 70-bp linker sequences and by insertion of the GREs upstream of the heterologous thymidine kinase (tk) promoter. A single GRE was sufficient to render expression driven by the tk-promoter hormone inducible and the cooperativity of GRnx, GMM, ΔGG and ΔMM at two adjacent GREs was even more pronounced than in the context of the deleted MTV promoter (data not shown).

The Binary Corticosteroid Response System

The cotransfection of the human corticosteroid receptors in a human neuronally derived cell line provides a sensitive *in vitro* system suitable for the analysis of the binary corticosteroid response system in the brain. Our experiments reveal important differences in the hormonal response mediated by the two corticosteroid receptors.[10,12] Several compounds display a differential ability in transactivation potency and radioligand displacement. For example, the ED_{50} and k_i values of the antiglucocorticoid RU486 are in the range of hGR agonists and the equivalent values for progesterone are in the range of mineralocorticoids. However, the maximal transcriptional activation in response to these compounds is only a fraction of that obtained with dexamethasone or RU28362, suggesting that binding of RU486 or progesterone does not exposure the full transactivation functions of the hGR. Similar variations are observed with the hMR. Aldosterone shows a lower ED_{50} than deoxycorticosterone or cortisol, which display a 10-fold lower or a similar k_i value, respectively, thereby corroborating the previously reported functional preference of the hMR for aldosterone.[7,8] However, the preferential ligand of the hMR appears to be the sex steroid progesterone which binds the hMR with an affinity higher than aldosterone. In the cotransfection system, progesterone displays only a weak partial agonistic activity. Instead, increasing doses of progesterone almost completely abolish the aldosterone-induced hMR-mediated response of the MTV promoter. The transcriptional activity of RU28362-stimulated hGR is only partially antagonized by progesterone, indicating that its antiglucocorticoid activity is less pronounced. These studies as well as clinical and pharmacological evidence suggest a physiological role of progesterone in the regulation of the hypothalamic-pituitary-adrenocortical system via hMR, hGR and presumably also hPR. The regulation of gene expression by steroid receptors requires a cascade of various steps such as hormone binding, dissociation from the hsp90 complex, dimerization, DNA binding and interaction with the transcription machinery, which in combination results in the final transcriptional response. The partial agonistic activity of RU486-hGR, progesterone-hGR and progesterone-hMR complexes requires an interaction with the cognate response elements and argues against a role of these two "antihormones" as stabilizers of cytoplasmic receptor-hsp90 complexes. Rather they appear to interfere with transcriptional activation by agonists at later steps, possibly by inducing a different structure of the ligand binding domain.[16]

The differential ligand affinities and activation responses detected in the assays described above may reflect important mechanisms of physiological control. Additional complexity but also flexibility may be provided by the existence of multiple nuclear receptors responsive to the same hormone. Furthermore, the ability of hGR, hMR and also hPR to interact with the same target sites on the DNA may modify transcriptional control by increasing the possibilities for synergistic or competitive interactions. Contributing to functionally distinct gene-specific and tissue-specific properties and also to the efficiency with which a receptor can regulate the activity of a particular promoter are the unique amino termini of the steroid receptors. Our experiments revealed that the hGR amino terminus, which contains an acidic transactivation domain, is essential for full transcriptional activity, whereas the hMR amino terminus does not provide this strong transactivation function. When the receptors were assayed on the tandem GREs, amino terminal deletion mutants or the presence of the hGR amino terminus enabled synergistic activity. hMR and hybrid molecules containing the hMR amino terminus did not show cooperativity, demonstrating that the hMR amino terminus interferes with the transcriptional response. These data suggest: 1) that synergism can be separated from intrinsic transactivation; 2) that synergism is mediated mainly via the DNA- and ligand-binding domains of the receptors; and 3) that amino terminal sequences exert a differential control of this activity. The mechanism behind this difference is not clear. The amino terminus of the hMR is large (602 amino acids) and might block receptor interaction by a steric hindrance mechanism. Alternatively, the differential activity of hMR and hGR amino termini might be dependent on the specific spatial arrangement of the response elements and binding sites for other transcription factors, suggesting that despite the apparently identical DNA-binding properties of the two receptors, the different amino termini influence promoter specificity by the formation of different types of protein-protein interactions with other parts of the receptor molecule or potential target transcription cofactors. Amino terminal variations have been described for several other members of the steroid/thyroid receptor gene family and the amino terminus of estrogen and progesterone receptors has been implicated in specific regulation of some target genes. Since the promoters assayed to this point may all be classified as glucocorticoid-responsive *in vivo*, it is not known whether the hMR is a weak activator of gene transcription in general or has increased activity on a set of genes not yet identified. This question is particularly relevant to cell types in which the MR and GR may be coexpressed, such as in neuronal tissues.[17] The hybrid receptors and the hormonal response studies described in our studies will provide a basis for future investigations of MR-and GR-mediated regulation of neuronal target genes.

ACKNOWLEDGMENTS

The authors would like to thank Drs. J. Arriza, R. M. Evans, J. M. H. M. Reul and B. v. Steensel for critical contributions and discussions to the described experiments and B. Berning and M. Söder for excellent technical assistance.

REFERENCES

1. Hollenberg, S. M., C. Weinberger, E. S. Ong, G. Cerelli & A. Oro, R. Lebo, E. B. Thompson, M. G. Rosenfeld & R. M. Evans. 1985. Nature **318:** 635–641.

2. Evans, R. M. 1988. Science **240:** 889.
3. Giguere, V., S. M. Hollenberg, M. G. Rosenfeld & R. M. Evans. 1986. Cell **46:** 645.
4. Hollenberg, S. M. & R. M. Evans. 1988. Cell **55:** 899–906.
5. Picard, D. & K. R. Yamamoto. 1987. EMBO J. **6:** 3333–3340.
6. Danielsen, M., J. P. Northrop, J. Jonklaas & G. M. Ringold. 1987. Mol. Endocrinol. **1:** 816–822.
7. Arriza, J. L., C. Weinberger, G. Cerelli, T. M. Glaser, B. L. Handelin, D. E. Housman & R. M. Evans. 1987. Science **237:** 268–275.
8. Arriza, J. L., R. B. Simerly, L. W. Swanson & R. M. Evans. 1988. Neuron **1:** 887.
9. Bagchi, M. K., J. F. Elliston, S. Y. Tsai, D. P. Edwards, M. J. Tsai & B. W. O Malley. 1988. Mol. Endocrinol. **2:** 1221–1229.
10. Rupprecht, R., J. L. Arriza, D. Spengler, J. M. H. M. Reul, R. M. Evans, F. Holsboer & K. Damm. 1993. Mol. Endocrinol. **7:** 597.
11. Beato, M. 1989. Cell **56:** 335–344.
12. Rupprecht, R., J. M. H. M. Reul, B. Van Steensel, D. Spengler, M. Söder, B. Berning, F. Holsboer & K. Damm. 1993. Eur. J. Pharmacol. Mol. Pharm. Sect. **247:** 145–154.
13. Cordingley, M. G., A. Riegel & G. L. Hager. 1987. Cell **48:** 261–270.
14. Strähle, U., W. Schmid & G. Schütz. 1988. EMBO J. **7:** 3389–3395.
15. Rupprecht, R., J. M. H. M. Reul, T. Trapp, B. Van Steensel, C. Wetzel, K. Damm, W. Zieglgänsberger & F. Holsboer. 1993. Neuron **11:** 523–530.
16. Vegeto, E., G. F. Allan, W. T. Schrader, M. J. Tsai, D. P. McDonnell & B. W. O Malley. 1992. Cell **69:** 703–713.
17. de Kloet, E. R. 1991. Front. Neuroendocrinol. **12:** 95.

DISCUSSION OF THE PAPER

Question: If you overexpress just the amino terminal region of the MR in the same cell that you have GR with those constructs where you have two GREs, do you block the synergism? Is it some kind of squelching phenomenon or not?

Damm: Just the amino termini? We have not done this experiment, so I have no answer.

Question: It is striking that in the case where you have the two GREs and you look at MR function, not only is it not synergistic but if I am not mistaken, it was not additive either. Does that mean that binding of one MR precludes binding of another?

Damm: We do not think that it is steric hindrance, because we have spaced the two GREs by 70 nucleotides. That should give room enough for the two molecules to bind. Also, the GR is almost as large as the MR. The progesterone receptor is even larger, and both can bind to tandem GREs.

Question: It seems by testing one would eliminate whether there was some kind of squelching or synergism. One might squelch synergism but not squelch transactivation.

Damm: It would certainly be an interesting experiment.

Question: Is the MR also phosphorylated in the amino terminus?

Damm: I am not aware of phosphorylation studies using the MR. Maybe somebody in the audience has an answer to that.

E. R. de Kloet (*Sylvius Laboratories*): You mentioned the theroretical possibility that there may be heterodimers of mineralocorticoid and glucocorticoid receptors. I would just like to remind you of this morning's discussion where we went over a number of possible modulations of mineralocorticoid receptor-mediated action. Are there facts showing these heterodimers in situ? Somehow it should be possible to demonstrate such complexes. Are you aware of that?

DAMM: As John Funder mentioned, T. Trapp (Max-Plank-Institute of Psychiatry) presented a poster at the Keystone meeting three weeks ago, with data demonstrating heterodimers of GR and MR.

DE KLOET: Are these heterodimers made only in the test tube, and does compartmentalization of MR and GR prevent heterodimerization in the cell nucleus? I am saying this because our confocal laser microscopy studies showed that MR and GR receptors have different spatial distribution in the nucleus. The receptors also responded differently in transport into the nucleus if stimulated with a ligand. So does heterodimerization occur *in vivo*?

DAMM: We tested it by functional cotransfection studies and by *in vitro* DNA binding experiments using different amounts of protein. Otherwise, we have no *in vivo* data for that.

F. HOLSBOER (*Max-Plank-Institute of Psychiatry*): Has anybody done some single-cell PCR to prove that both these receptors are in the same cell? Are the transfection data really telling us that they are cooperative in regulating *in vivo* function? We really cannot answer the functional specificity of this finding.

DE KLOET: Immunocytochemistry demonstrated colocalization of MR and GR in the same cell.

J. W. FUNDER (*Baker Institute*): Regarding the straight line for affinity versus EC50 for GR but not MR, the three outliers in terms of MR were progesterone (PR), deoxycorticosterone (DOC), and spironolactone. Otherwise everything else appeared to line up as a straight line. The 10^{-11} molar affinities you quote for PR and DOC are difficult to accept, though I do not know the data. Was the spironolactone an agonist or an antagonist? Was the EC50 its antagonist effect or its agonist effect?

DAMM: I think it is a complete antagonist. It was out of the EC50 range, and was just put at the top.

FUNDER: I see. It is not really adding to whether or not that line is straight. But when you lined up the two difficult ones, DOC and PR, you gave them an affinity of 10^{-11} molar each, which is an incredibly high affinity. Don't you think that is wrong?

DAMM: The data are from Hans Reul, who has done binding studies for many years. What should be kept in mind here is that we are analyzing the human receptors. So you really cannot compare them to the rat or guinea pig receptors.

FUNDER: Whatever the receptor is, one cannot suspend mass action. My point this morning with progesterone was that one cannot drop binding to 30 percent by adding one-fifth as much cold, unless there is a 50-fold difference in the binding between hot and cold, which there is no evidence for at all. So the progesterone data are really not credible. Were the DOC data generated under the same circumstances?

DAMM: It was all done with cell extracts and in a parallel study.

Corticosteroid Receptor Modulation in Transgenic Mice[a]

N. BARDEN[b]

Molecular Psychogenetics Laboratory
Laval University Hospital Centre
and
Department of Physiology
Laval University
Québec, Canada

INTRODUCTION

Antisense RNA has now been shown to prevent the normal expression of several different gene products in both cell cultures and intact mice.[1–4] When expressed in the nucleus, antisense RNA is believed to form a double stranded complex with the endogenous mRNA precursor, and once this hybrid molecule is formed it is rapidly degraded by nucleases specific for double stranded regions. By use of an appropriate promoter, antisense RNA formation can be directed to specific cell types, and this approach can affect the expression of a specific gene in a single tissue and at a certain time in development. Introduction into the germ line of mice of transgenes that generate antisense RNA molecules thus offers the possibility of generating precise animal models for investigation of the normal physiological role of defined proteins as well as for human disease states.

Corticosteroid hormones govern a large range of processes affecting both development and physiological homeostasis. They regulate cell growth and differentiation, modify neuronal metabolism and may influence feeding and behavioral patterns.[5] However, in order to exert these biological actions, corticosteroids must first bind to a cellular receptor. Corticosterone, the major adrenal corticosteroid found in rodents, can bind to two different types of receptor. While the type II, or glucocorticoid receptor, is widespread throughout the brain and also found in the pituitary gland, the type I, or mineralocorticoid receptor has a much more restricted distribution and, in the central nervous system, is localized to the hippocampus, septum, amygdala and brain stem nuclei. Both types of receptor have been purified,[6,7] their functional domains characterized[8] and the cDNAs for rat, mouse and human receptors cloned.[9–12] Although adrenal steroids have well-known negative feedback effects on the hypothalamic-pituitary-adrenal (HPA) axis which lead to suppression of corticosterone secretion, their precise site of action remains controversial.[13] One means of studying the physiological role of corticosteroid receptors would be the partial gene expression knock-out induced by formation of corticosteroid receptor antisense RNA in brain tissues. We previously constructed an antisense RNA, complementary to the 3′ non-coding region of the glucocorticoid receptor mRNA, and showed its efficacy in decreasing functional glucocorticoid receptor levels in

[a]This work, supported in part by grants from the Medical Research Council of Canada and The Stanley Foundation, was carried out in the author's laboratory in Québec and at the Max-Planck Institute for Psychiatry, Munich during a sabbatical leave of absence.

[b]Mailing address: Dr. N. Barden, Molecular Psychogenetics, CHUL, 2705 Blvd. Laurier, Ste-Foy, Québec, G1V 4G2 Canada.

cells maintained in culture.[14] Extension of this work to the mineralocorticoid receptor and incorporation of these antisense RNA-generating constructs in transgenic animals offers possibilities for elucidating the roles of these two receptor types in HPA axis regulation.

Antisense RNA Generation

Corticosteroid receptor antisense RNAs were transcribed either from 1815 bp of the 3′ non-coding region of the rat glucocorticoid receptor cDNA inverted downstream from a human neurofilament (NF-L) gene promoter element or from a 1185-bp fragment of the human mineralocorticoid receptor cDNA also under control of the NF-L promoter (FIG. 1). In order to avoid cross inhibition of other receptor activity, antisense constructs directed against the 3′ non-coding region of the glucocorticoid receptor mRNA or the 5′ end of the mineralocorticoid receptor cDNA were used since these sequences contain the least homology with other members of the steroid hormone superfamily of genes.

Specificity of Corticosteroid Receptor Antisense RNA Action

When cotransfected with a glucocorticoid responsive reporter plasmid (MMTV-CAT) into Ltk$^-$ cells the glucocorticoid receptor antisense RNA construct (pNFLAsGR) efficiently reduced functional glucocorticoid receptor levels (FIG. 2). Stable transfectants bearing the same glucocorticoid receptor antisense gene fragment construction had a 50–70% decrease in glucocorticoid receptors as evidenced by a ligand binding assay with the type II glucocorticoid receptor-specific ligand ^{3}[H]-RU 28362.[14] The specificity of antisense RNA action can be seen by the capacity of the mineralocorticoid antisense construct (pNFLAsMR), but not the glucocorticoid antisense RNA (pNFLAsGR), to inhibit mineralocorticoid receptor-mediated activation of MMTV-CAT in the presence of aldosterone (FIG. 3). The corticosteroid receptor antisense RNAs used here effectively diminish their respective corticosteroid receptor production when transfected into cell lines. Their mechanism of action is not known with certainty, but the very low level of transgene transcripts in tissue extracts[14] together with the markedly reduced corticosteroid receptor mRNA levels suggests that double stranded sense:antisense RNA hybrids are formed and are rapidly degraded by nucleases.

Corticosteroid Receptor Antisense RNA Transgenic Mouse

We obtained transgenic mice following microinjection of the glucocorticoid receptor antisense gene into the pronucleus of fertilized oocytes obtained from superovulated prepubertal females. The presence of the antisense construct in founder mice and their offspring was established by Southern blot analysis of DNA purified from the tail. Mice which were positive for integration of the chimeric NFL AsGR served as founder animals for the development of individual transgenic lines. Of the five founder mice which initially screened positive for the presence of the transgene,[15] three were bred to homozygosity.

All three permanent transgenic lines established display a characteristic phenotype, the most remarkable aspect of which is a late onset development of marked obesity (FIG. 4). Transgenic animals have a greatly increased fat deposition and, at

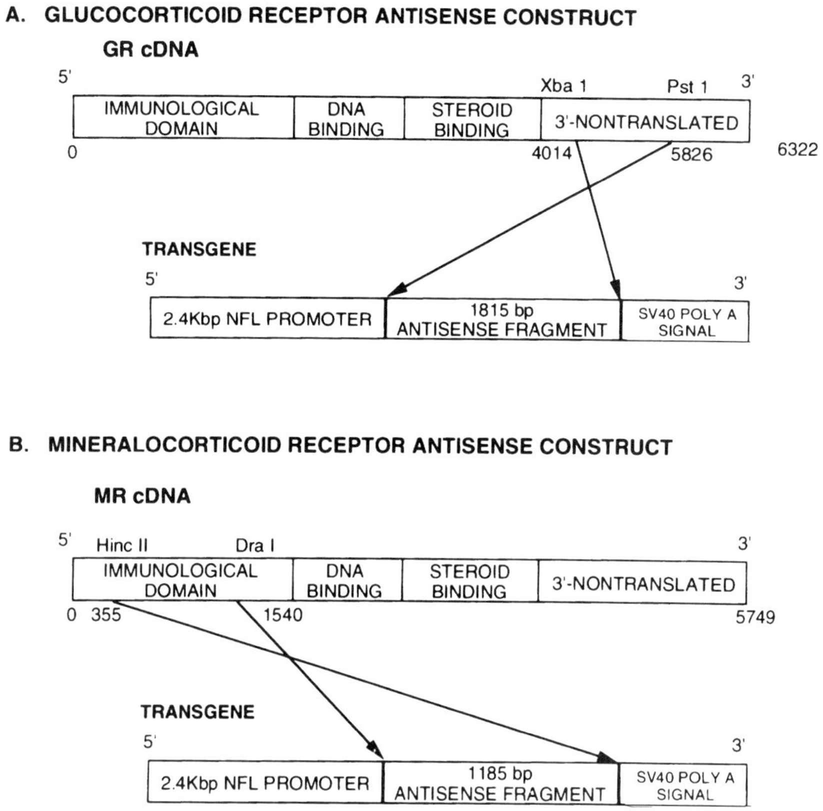

FIGURE 1. Schematic representation of the structure and construction of the NFL-AsGR and NFL-AsMR constructs used to direct the transcription of antisense RNA with a sequence complementary to that of regions of the mouse glucocorticoid receptor or mineralocorticoid receptor mRNA molecules. Plasmid NFL-AsGR was constructed by cloning a 1815-bp fragment of the 3′ non-coding region of the rat glucocorticoid receptor cDNA in the reverse orientation downstream from the human neurofilament gene promoter. The plasmid containing the human neurofilament L gene promoter (pNF-L, gift of Dr. J. P. Julien) was linearized by digestion with Hind III, which cleaves at a sequence at the 3′ end of the promoter. Extruding sequences were filled in with the Klenow fragment of DNA polymerase and a blunted Xba 1-Pst 1 fragment of the pSG-1 plasmid (gift of Dr. R. Miesfeld: this cDNA clone is a derivative of pRM16[9]) was inserted into pNF-L in a reverse orientation. Finally, we added a Pst 1-Bam H1 fragment of the VP1 gene of SV40 which contains a polyadenylation signal. pNFL-AsMR was constructed in a similar manner. A 1185-bp Hinc II-Dra 1 fragment of the human mineralocorticoid receptor cDNA (phMR 3750; gift from Dr. J. L. Arriza) was inserted in the reverse orientation between the human NF-L gene promoter and the SV40 polyadenylation signal. cDNA sequences containing the least degree of homology with other steroid receptors (3′ non-translated region for glucocorticoid receptor, 5′-immunologic domain for mineralocorticoid receptor) were used to generate antisense RNA and thus avoid cross inhibition of other steroid receptor activity.

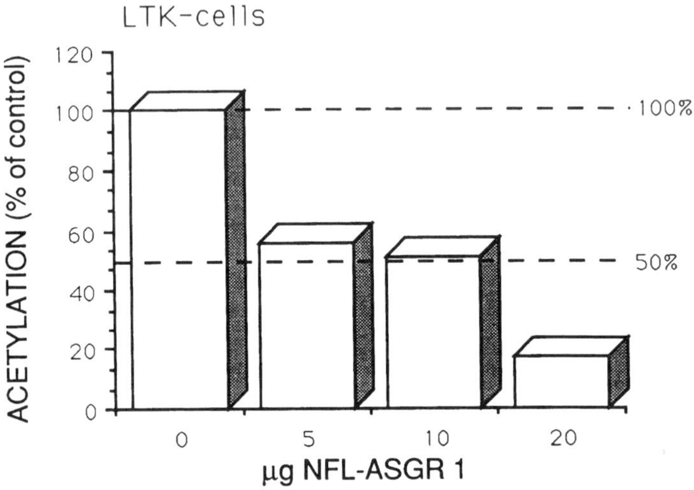

FIGURE 2. Inhibition of glucocorticoid receptor activity by antisense RNA. The promoter activity of the reporter plasmid (pMTVCAT) when transiently cotransfected with NFL-AsGR in Ltk⁻ cells was determined. The recombinant DNA construct (NFL-AsGR, 1–20 μg) was precipitated with pRSV Lac Z and the plasmid reporter pMTVCAT and introduced for 5 hours into 2×10^6 Ltk⁻ cells. Dexamethasone (0.1 μM) was added 48 hours after transfection and the cells were harvested 24 hours later and assayed for CAT activity (at constant β-galactosidase activity for each CAT assay).

around 6 months of age, are as much as twice the weight of normal mice. This accumulation of fat in different adipose tissue regions is observed despite the fact that transgenic mice eat 15% less chow than do normal mice.[16] Metabolic changes accompanying the development of obesity include a marked increase in serum triglyceride concentrations and a decrease in energy expenditure, suggested by a decreased oxygen consumption, during the first hours of the dark phase of the diurnal cycle. A role for glucocorticoids in the development of obesity caused by different etiologies has often been noted.[17–19] Characteristics similar to what we see in the transgenic animals (lower B_{max} for glucocorticoid binding with elevated serum corticosterone and ACTH concentrations) have also been reported,[20] although not universally acclaimed,[21] in the obese fa/fa Zucker rat. The animals we have developed may thus be important, not only for investigation of the role of corticosteroid receptors in the regulation of HPA axis activity, but also for studies on the role of corticosteroids in the etiology of obesity.

HPA Axis Regulation in Transgenic Mice

Effects of the antisense RNA on the level of glucocorticoid receptor mRNA in various tissue was measured by Northern blot analysis. Endogenous glucocorticoid receptor mRNA strands were detected in all tissue analyzed in transgenic mice. In different brain regions of the three different transgenic lines analyzed, (hypothalamus, frontal cortex, hippocampus) we observed a maximal 50–70% decrease of the glucocorticoid receptor mRNA compared to controls. While the transgene is not

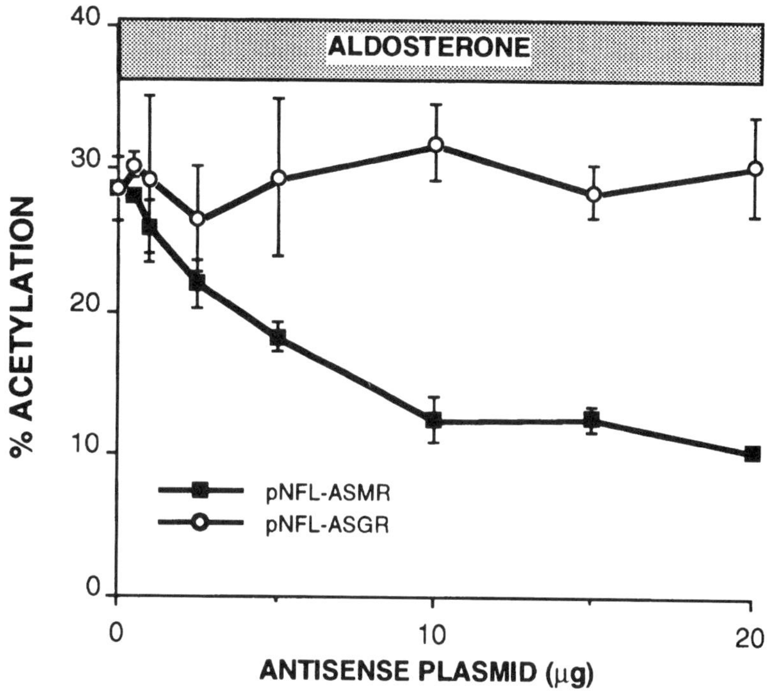

FIGURE 3. Specificity of mineralocorticoid receptor antisense RNA action. The recombinant DNA antisense-generating construct (pNFL-AsMR, 0.5–20 μg) was precipitated with pRShMR (1 μg), pRSVLacZ (2.5 μg) and the plasmid reporter pMTVCAT (10 μg) and introduced for 14–16 hours into 1 × 10^6 CV1 cells. Aldosterone (1 μM) was added 24 hours after transfection and the cells were harvested 24 hours later and assayed for CAT activity (at constant β-galactosidase activity for each CAT assay).

expressed in all tissues, some expression can be detected outside the central nervous system including the pituitary gland.[15] The 2.4-Kb neurofilament promoter region used in these animals is thus not sufficient to generate complete tissue-specific expression, as has been noted in other transgenic mice incorporating this element.[22]

The decreased glucocorticoid receptor mRNA concentrations were reflected by decreases in functional glucocorticoid binding activity. Scatchard analysis of glucocorticoid binding to cerebral cortex extracts showed 30–50% fewer glucocorticoid receptor binding sites when compared to control mice (FIG. 5).

The transgene incorporated into the genome was designed to disrupt normal HPA axis function. This was found to be the case as evidenced by both elevated early morning ACTH and corticosterone levels.[15] However, when measured in the afternoon or early evening, corticosterone and ACTH concentrations did not differ between control and transgenic animals, and the typical pattern of diurnal secretion of these hormones was flattened in the transgenics. In addition to causing alterations in the baseline levels of ACTH and corticosterone secretion patterns, the presence of the antisense RNA also rendered the mouse HPA axis less sensitive to glucocorticoid feedback inhibition.[23] While, in normal mice, injection of 1 μg/100 g body

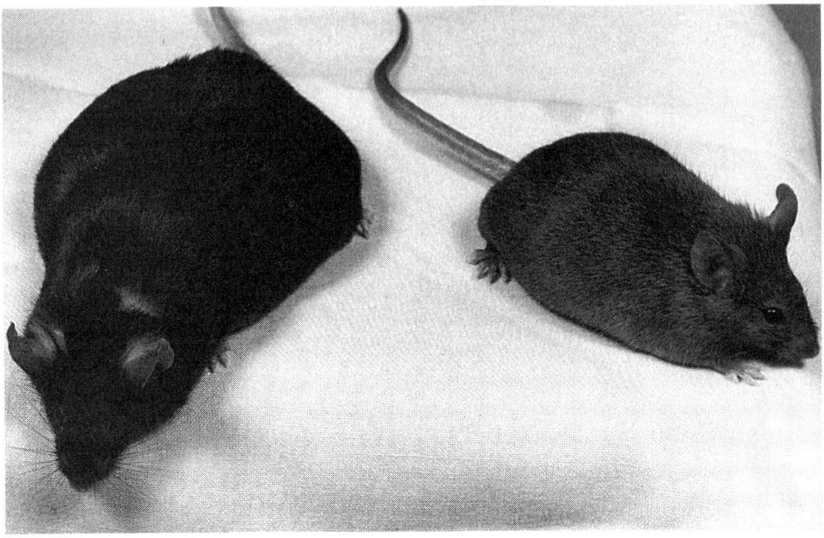

FIGURE 4. Transgenic mice expressing glucocorticoid receptor antisense RNA. Glucocorticoid receptor antisense transgenic mice (transgenic mouse, line 1.3, *left*; normal mouse, *right*) were produced by microinjection of an ~4.9-kb Nde1-Ase1 fragment excised from pNFL-AsGR plasmid (see FIG. 1) into the male pronucleus of fertilized mouse oocytes obtained from superovulated prepubertal females (C3H X c57BL/6F2) 20 hours after treatment with human chorionic gonadotropin.[30] The DNA fragment had been previously gel purified by filtration through an Elutip-D column (Schleicher and Schuel, Inc.) and its concentration adjusted to 3 μg/ml in 10 mM Tris and 0.1 mM EDTA. Injected embryos were reimplanted into pseudopregnant outbred CD1 females and offspring were screened for transgene integration by tail blotting with ^{32}P-labelled probe. DNA was extracted from 1–2 cm tail pieces by incubating in 0.7 ml of 50 mM Tris hydrochloride-100 mM EDTA, pH 8.0, 0.5% sodium dodecyl sulfate and 350 μg of proteinase K for 16 hours at 55°C with agitation. DNA was phenol-chloroform extracted, ethanol precipitated, and resuspended in 10 mM Tris hydrochloride–1 mM EDTA, pH 8.0. For Southern blot analysis,[31] 15 μg of total genomic DNA of each mouse was digested with Bam H1, electrophoresed on 1.0% agarose gels and transferred to nylon filters (Hybond N^+, Amersham). Filters were prehybridized for 2 hours at 42°C (in 50% formamide, 5 X SSPE (1 X SSPE is 0.15 M NaCl, 0.01 M NaH_2PO_4, 0.001 M EDTA) 6 X Denhardt's, 0.5% SDS and 200 μg/ml denatured salmon sperm DNA) and hybridized overnight at 42°C to a ^{32}P-labelled Bam H1-Bam H1 fragment containing a part of both the NF-L promoter and the antisense glucocorticoid receptor cDNA fragment. After hybridization, filters were washed twice for 15 min each in 2 X SSC with 0.5% SDS (sodium dodecyl sulfate) at room temperature, once for 30 min in 2 X SSC (1 X SSC is 0.15 M NaCl, 0.015 M sodium citrate), 0.5% SDS at 42°C and once for 30 min in 0.1 X SSC, 0.5% SDS at 65°C. Founder animals positive for transgene integration were bred as lines by mating first with (C3H X c57BL/6)F1 wild-type mice and subsequently as homozygotes. Homozygotic lines were derived from 3 of the founder transgenic mice.

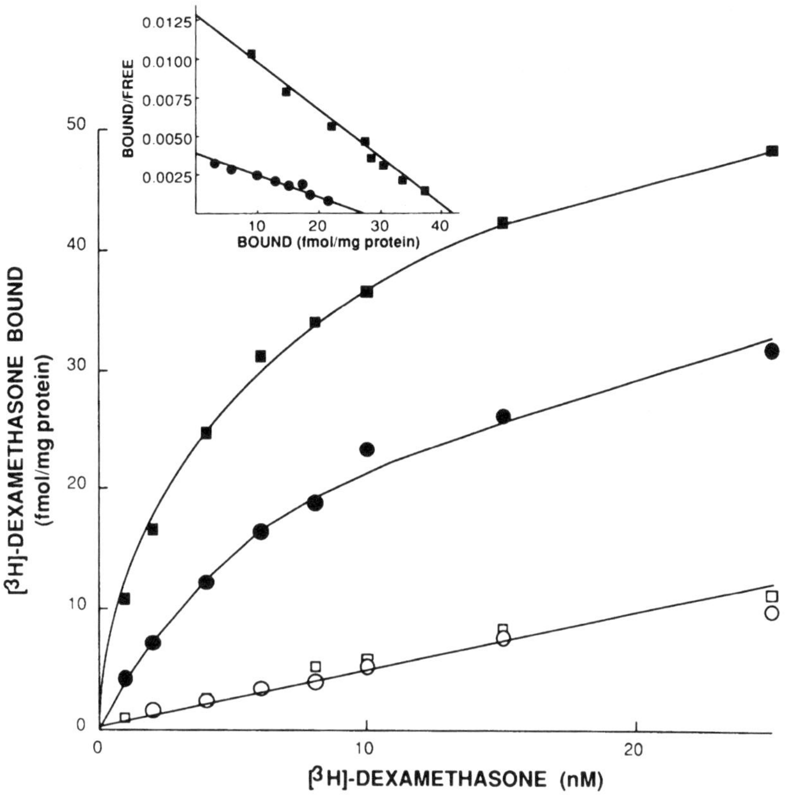

FIGURE 5. ^{3}[H]-dexamethasone binding activity of cerebral cortex in normal and transgenic mice. Scatchard plots (*inset*) of saturation binding date of normal (*squares; open symbols* nonspecific binding, *full symbols* total binding) and transgenic line 1.3 (*circles; open symbols* nonspecific binding, *full symbols* total binding) were performed with [^{3}H]-dexamethasone binding data. Tissue was homogenized in 30 mM Tris, 1 mM EDTA, 10 mM molybdate, 10% (v/v) glycerol and 1 mM dithiothreitol (TEDGM, pH 7.4). After centrifugation at 105,000 g for 45 min at 0°C, an aliquot of the cytosol was incubated with [^{3}H]-dexamethasone (Spec. act. 44.7 Ci/mmol, Amersham). Concentration range 1–25 nM for Scatchard analysis[32] for 20–24 hours at 4°C. The amount of nonspecific binding was determined in parallel incubations of the labelled steroid in the presence of a 500-fold excess of unlabeled RU 28362. Sephadex LH20 (Pharmacia) columns equilibrated with TEDGM buffer were used to separate bound from unbound steroid. Following incubation, 100 μl aliquots of the incubates were loaded into the columns, washed with 100 μl of TEDGM and eluted with 400 μl of TEDGM into minivials. The vials were then filled with 5 ml of aqueous counting cocktail Formula A-963 (New England Nuclear, Boston, MA) and counted in a LKB scintillation counter at 40% efficiency. Protein count was determined by the method of Bradford.[33]

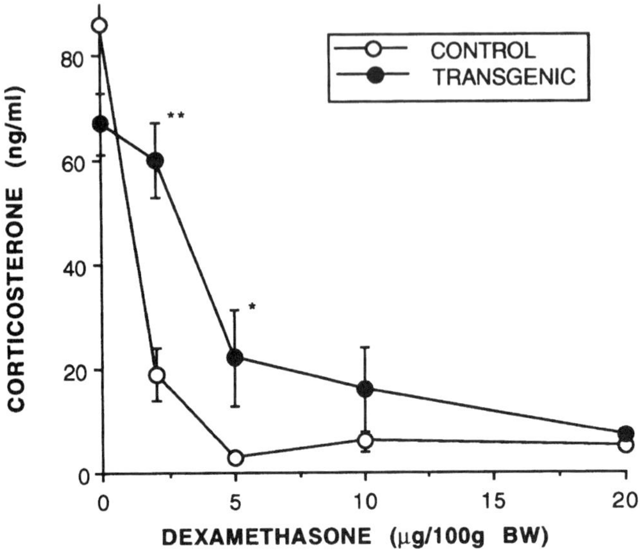

FIGURE 6. Resistance of transgenic mouse corticosterone to suppression by dexamethasone. Dexamethasone, dissolved in ethanol, was diluted with 0.9% saline and at 8:00 a.m. injected i.p. at dosages varying from 0 to 20 μg/100 g body weight. Mice were killed six hours later and trunk blood collected in EDTA-coated tubes. ACTH and corticosterone concentrations were measured by radioimmunoassay using commercially available kits (ICN Biomedical Inc., USA). Results shown are the mean ± SEM. *, $p < 0.05$; **, $p < 0.01$, control vs transgenic.

weight of dexamethasone six hours prior to blood sampling was sufficient to completely suppress corticosterone levels, in transgenic mice a tenfold higher concentration was required to achieve the same effect (FIG. 6).

DISCUSSION

Although introduction, into eukaryotic cells, of either antisense RNA or vectors expressing antisense RNA has been widely used to decrease the level of the final product of the complementary sense mRNA, only a limited success has been achieved with these same approaches in transgenic animals.[1,24] While the reasons for this discrepancy are not clear, antisense RNA technology has been successfully used here to decrease the glucocorticoid receptor level in mouse tissues.

The increased plasma ACTH concentration of transgenic animals could reflect a failure of glucocorticoids to inhibit HPA axis activity at either, or both, central and pituitary sites. Disruption of the normal diurnal rhythm of secretion argues in favor of at least one site of action in the central nervous system. However, since glucocorticoid receptor concentrations are also decreased in the pituitary gland,[15] it is possible that this represents an additional site at which glucocorticoid feedback inhibition of ACTH secretion could be affected. Morning corticosterone concentrations are elevated in these transgenic mice, but not to the same extent as is ACTH, suggesting that the adrenal gland may become refractory to continuous stimulation. Glucocorticoids are known to decrease the number of glucocorticoid receptors in both hippocampus and pituitary gland.[25,26] Thus the elevated corticosterone levels of the transgenic mice, which are secondary to decreased glucocorticoid receptor levels, could contribute to even further glucocorticoid receptor downregulation.

Glucocorticoid feedback inhibition is the principal event that could explain the pathophysiology of the HPA axis dysfunction in patients with major depression[27] and the dexamethasone suppression test is one of the most widely used tool for identifying HPA axis abnormalities in psychiatry. In view of the known action of antidepressant drugs to increase corticosteroid receptor concentrations,[28,29] it is possible that the dexamethasone resistance seen in depression results from defective corticosteroid receptor gene regulation in a similar way that we seen in the transgenic mice.

REFERENCES

1. Katsuki, M., M. Sato, M. Kimura, M. Yofohama, K. Kobayashi & T. Nomura. 1988. Conversion of normal behavior to shiverer by myeline basic protein antisense cDNA in transgenic mice. Science **241:** 593–595.
2. Izant, J. G. & H. Weintraub. 1985. Constitutive and conditional suppression of exogenous genes by anti-sense RNA. Science **229:** 345–351.
3. Kim, S. K. & B. J. Wold. 1985. Stable reduction of thymidine kinase activity in cells expressing high levels of anti-sense RNA. Cell **42:** 129–138.
4. McGarry, T. J. & S. Lindquist. 1986. Inhibition of heat shock protein synthesis by heat-inducible antisense RNA. Proc. Natl. Acad. Sci. USA **83:** 399–403.
5. Funder, J. W. & K. Sheppard. 1987. Adrenocortical steroids and the brain. Ann. Rev. Physiol. **49:** 397–411.
6. Wrange, O., S. Okret, M. Radojéiè, C. Carlstedt-Duke, & J.-A. Gustafsson. 1984. Characterization of the purified activated glucocorticoid receptor from rat liver cytosol. J. Biol. Chem. **259:** 4534–4541.
7. Wrange, O., C. Carlstedt-Duke & J.-A. Gustafsson. 1979. Purification of the glucocorticoid receptor from rat liver cytosol. J. Biol. Chem. **254:** 9284–9290.
8. Gustafsson, J.-A., J. Carlstedt-Duke, L. Poellinger, A. Okret, A.-C. Wikström, M. Brönnegård, N. M. Gillner, Y. Dong, K. Fuxe, A. Cintra, A. Harfstrand & L. F. Agnati. 1987. Biochemistry, molecular biology, and physiology of the glucocorticoid receptor. Endocr. Rev. **8:** 185–234.
9. Miesfeld, R., S. Okret, A.-C. Wikström, O. Wrange, J.-A. Gustafsson & K. R. Yamamoto. 1984. Characterization of a steroid hormone receptor gene and mRNA in wild-type and mutant cells. Nature **312:** 779–781.
10. Miesfeld, R., S. Rusconi, P. J. Godowski, B. A. Maler, S. Okret, A.-C. Wikström, J.-A. Gustafsson & K. R. Yamamoto. 1986. Genetic complementation of a glucocorticoid receptor deficiency by expression of cloned receptor cDNA. Cell **46:** 389–399.
11. Hollenberg, S. M., C. Weinberger, E. S. Ong, G. Cerelli, A. Oro, R. Lebo, E. B. Thompsen, M. G. Rosenfeld & R. M. Evans. 1985. Primary structure and expression of a functional human glucocorticoid receptor cDNA (abstract). Nature **318:** 635–641.
12. Arriza, J. L., R. B. Simerly, L. W. Swanson & R. M. Evans. 1988. The neuronal mineralocorticoid receptor as a mediator of glucocorticoid response. Neuron **1:** 887–900.
13. Bradbury, M. J., A. M. Strack & M. F. Dallman. 1993. Lesions of the hippocampal efferent pathway (fimbria-fornix) do not alter sensitivity of adrenocorticotropin to feedback inhibition by corticosterone in rats. Neuroendocrinology **58:** 396–407.
14. Pepin, M.-C. & N. Barden. 1991. Decreased glucocorticoid receptor activity following glucocorticoid receptor antisense RNA gene fragment transfection. Mol. Cell. Biol. **11:** 1647–1653.

15. PEPIN, M.-C., F. POTHIER & N. BARDEN. 1992. Impaired type II glucocorticoid receptor function in mice bearing antisense RNA transgene. Nature **355:** 725–728.
16. RICHARD, D., S. CHAPDELAINE, Y. DESHAIES, M.-C. PEPIN & N. BARDEN. 1993. Energy balance and lipid metabolism in transgenic mice bearing an antisense GCR gene construct. Am. J. Physiol. **265:** R146–R150.
17. YURKIMURA, Y., G. A. BRAY & A. R. WOLFSEN. 1978. Some effects of adrenalectomy in the fatty rat. Endocrinology **103:** 1924–1928.
18. FLETCHER, J. M. 1986. Effects of adrenalectomy before weaning in the genetically obese Zucker rat (fa/fa). Br. J. Nutrition **56:** 141–151.
19. FREEDMAN, M. R., B. A. HORWITZ & J. S. STERN. 1986. Effects of adrenalectomy and glucocorticoid replacement on development of obesity. Am. J. Physiol. **250:** R595–R607.
20. DOUGLAS WHITE, B. & R. J. MARTIN. 1990. Alterations in the binding characteristics of glucocorticoid receptors from obese Zucker rats. J. Steroid Biochem. **36:** 681–686.
21. LANGLEY, S. C. & D. A. YORK. 1990. Increased type II glucocorticoid-receptor numbers and glucocorticoid-sensitive enzyme activities in the brain of the obese Zucker rat. Brain Res. **533:** 268–274.
22. JULIEN, J.-P., L. BEAUDET, I. TRETJAKOFF & A. PETERSON. 1990. Neurofilament gene expression in transgenic mice. J. Physiol. Paris **84:** 50–52.
23. STEC, I., N. BARDEN, J. M. H. REUL & F. HOLSBOER. 1994. Dexamethasone non-suppression in transgenic mice expressing antisense RNA to the glucocorticoid receptor. J. Psychiatr. Res. In press.
24. INDREES MUNIR, M., B. J. F. ROSSITER & C. T. CASKEY. 1990. Antisense RNA production in transgenic mice. Somatic Cell Mol. Genet. **16:** 383–394.
25. SVEC, F. & M. RUDIS. 1981. Glucocorticoids regulate the glucocorticoid receptors in the AtT-20 cell. J. Biol. Chem. **256:** 5984–5987.
26. TORNELLO, S., E. ORTI, A. F. DENICOLA, T. C. RAINBOW & B. S. MCEWEN. 1982. Regulation of glucocorticoid receptor in brain by corticosterone treatment of adrenalectomized rats. Neuroendocrinology **35:** 411–417.
27. HOLSBOER, F., D. SPENGLER & I. HEUSER. 1992. The role of corticotropin-releasing hormone in the pathogenesis of Cushing's disease, anorexia nervosa, alcoholism, affective disorders and dementia. Prog. Brain Res. **93:** 385–417.
28. PEPIN, M.-C., S. BEAULIEU & N. BARDEN. 1989. Antidepressants regulate glucocorticoid receptor messenger RNA concentrations in primary neural cultures. Mol. Brain Res. **6:** 77–83.
29. PEIFFER, A., S. VEILLEUX & N. BARDEN. 1991. Antidepressant and other centrally-acting drugs regulate glucocorticoid receptor messenger RNA levels in rat brain. Psychoneuroendocrinology **16:** 505–515.
30. HOGAN, B., F. COSTANTINI & E. LACY. 1986. Manipulating the Mouse Embryo. A Laboratory Manual. Cold Spring Harbor Laboratory Press. Cold Spring Harbor, NY.
31. SOUTHERN, E. M. 1975. Detection of specific sequences among DNA fragments separated by gel electrophoresis. J. Mol. Biol.**98:** 503–517.
32. SCATCHARD, G. (1949). The attractions of proteins for small molecules and ions. Ann. N.Y. Acad. Sci. **51:** 660.
33. BRADFORD, M. M. 1976. A rapid method for quantification of microgram quantities of protein utilizing the principle of protein-dye binding. Anal. Biochem. **72:** 248–254.

DISCUSSION OF THE PAPER

QUESTION: What happened to MR in the transgenic mice?

BARDEN: We do get a decrease in MR as well. This is not a direct effect of the antisense, but it may be expected as a secondary effect of the increase in HPA activity.

QUESTION: Have you looked at vasopressin in these transgenics?

BARDEN: No, we haven't yet. This is obviously on the books to do.

QUESTION: Given your commentary about eating less and the oxygen utilitzation during a feeding period and the fact that it is not published yet, still homozygous GR knockout mice from the Schültz's laboratory developmentally have no adrenal medulla. Have you examined the adrenal medulla in your mice, because with the promotor you are using, one would expect expression in those cells?

BARDEN: No, we haven't. Data on oxygen utilization and feeding are published.[16]

QUESTION: Do you have any data on catecholamines?

BARDEN: Only a little in brown adipose tissue and heart, where there are decreased levels of catecholamines.

E. R. DE KLOET (*Sylvius Laboratories*): You proposed your transgenic mouse as a model for depression and you certainly showed some arguments. For instance, there is a flattened circadian rhythm of corticosterone, which is what you would like to see.

BARDEN: I should be very careful. I do not like saying it is a model for depression. I should say it is a model for neuroendocrinological changes that we see in depression, because I do not know whether the mice are depressed or not.

DE KLOET: What you show is a reduction in glucocorticoid receptor capacity and also that the magnitude of the feedback response is reduced, which I find an interesting observation. The feedback signal would also be weaker if the affinity of these receptors is reduced. What do you think about this?

BARDEN: We do not see any significant changes in the affinity of the receptors, so the feedback mechanism is presumably operating through concentrations.

QUESTION: Concerning the range of antidepressants used for the effect on GR, did you use the pure serotinergic drugs of those with additional catecholaminergic activity? How specific is this effect?

BARDEN: We did not do an exhaustive screening of all antidepressants. We used antidepressants that are more specific either for noradrenergic or for serotonergic systems. We also used some atypical antidepressants. We found stimulation of corticosteroid receptor mRNA concentrations with all of these types. As I say, it appears to be completely unrelated to action on the catecholamine systems.

QUESTION: So, do you think that an action on GR is the common factor for antidepressants?

BARDEN: It is too early to definitely say that this is the common site of action, but this possibility deserves to be kept in mind.

QUESTION: Did you also have control agents that are related? For instance, did you test neuroleptic drugs that are not effective antidepressants?

BARDEN: Yes, we did.

QUESTION: And did they also have activity on GR?

BARDEN: No, they did not.

L. S. BRADY (*National Institute of Mental Health*): This is consistent with some data from our group, where a number of structurally unrelated antidepressants all share the common property that they reduce CRH mRNA in the brain. So that would certainly be consistent with your findings.

QUESTION: Did I understand you correctly? In the cell model, it seemed that upregulation of the glucocorticoid receptor occurred after acute treatment and in the animals you suggested that it was after chronic treatment?

BARDEN: It does. In cells we can get effects in 24 to 48 hours. But in the animals it appears to take longer.

QUESTION: And it does not occur acutely?

BARDEN: No.

QUESTION: Would you like to speculate on possible mechanisms?

BARDEN: What is interesting about this is that the time course of action of the antidepressants, at least *in vivo*, coincides with their time course of action in restoring normality in depressive patients. But why it is so fast in the cells and then takes longer *in vivo*, I do not really see.

Mechanistic and Functional Studies of Rapid Corticosteroid Actions[a]

MILES ORCHINIK,[b] FRANK L. MOORE,[c]
AND JAMES D. ROSE[d]

[b]*Laboratory of Neuroendocrinology*
Rockefeller University
New York, New York 10021

[c]*Department of Zoology*
Oregon State University
Corvallis, Oregon 97331–2914

[d]*Departments of Psychology and Zoology-Physiology*
University of Wyoming
Laramie, Wyoming 82071

INTRODUCTION

Studies using nonmammalian vertebrates have provided insights into mechanisms and functions of corticosteroid actions that mammalian studies have not afforded. In part, this may reflect a tendency for corticosteroids to exert more robust suppressive actions on reproductive behaviors in amphibians, reptiles and birds than in mammals.[1–4] By studying the potent behavioral actions of corticosterone in an amphibian species, *Taricha granulosa,* we have discovered receptor mechanisms for rapid steroid actions that are mediated through neuronal membranes. In addition, we are beginning to understand how corticosteroids modulate the neural substrates mediating reproductive behavior.

Behavioral Actions of Corticosterone

Stress-Induced Suppression of Courtship Clasping

In *Taricha granulosa,* the roughskin newt, male courtship behavior consists of a prolonged amplectic clasp that serves to facilitate receptivity in the female.[5] This courtship behavior is stimulated by an interaction between androgens and neuropeptides, particularly arginine vasotocin, such that seasonal increases in androgen secretion enhance the responsiveness of males to the stimulatory actions of vasotocin.[6] Courtship behavior is also under inhibitory neuroendocrine control. Of particular interest is the finding that brief exposure to confinement stress during the breeding season profoundly suppresses the clasping behavior.[7] Several lines of evidence indicate that endogenous corticosterone is a causal factor in this behavioral response to stress: 1) The confinement stress that suppresses reproductive behavior also induces a significant increase in plasma levels of corticosterone. 2) Intraperitoneal

[a]This work was supported by NSF Grants BNS8901500 to F.L.M. and M.O., BNS8909173 to F.L.M., and NIH Grant NS13748, including a Javits Neuroscience Investigator award, to J.D.R.

(ip) injection of corticosterone rapidly inhibits male clasping in a dose-dependent manner. 3) Pharmacological blockade of corticosterone synthesis using metyrapone blocks both stress- and CRF-induced suppression of behavior.[7]

Neurophysiological Correlates

The conclusion that corticosterone mediates behavioral responses to stress is supported by neurophysiological studies on *Taricha* hindbrain neurons.[8] Descending hindbrain neurons play a critical role in regulating spinal motor systems that control limb movements, such as those involved in courtship clasping. In addition to reticulospinal neurons, a population of nonreticulospinal medullary neurons is also implicated in the neural system regulating clasping. These neurons respond to various forms of sensory stimulation, including pressure applied to the face, trunk and cloaca, with an increase in firing rate. Cloacal pressure is of interest because this stimulus elicits reflexive clasping in male *Taricha.* Extracellular recordings from single cells in the medulla demonstrated that ip injection of corticosterone produced multiple electrophysiological effects in both reticulospinal neurons and in nonreticulospinal, sensory-responsive neurons.[8] In reticulospinal neurons, there was a rapid decrease in the amplitude of antidromically activated action potentials following corticosterone injection. Sensory-responsive medullary neurons showed a decrease in spontaneous firing following corticosterone injection and a selective loss of responsiveness to tactile stimulation of the cloaca, but not the face or trunk (FIG. 1).

These behavioral and neurophysiological studies suggest that the stress-induced release of corticosterone suppresses courtship behavior, at least in part, by altering the neuronal processing of behaviorally relevant sensory information.

Rapid Actions of Corticosterone

One of the hallmarks of the stress response is the speed with which the hypothalamic-pituitary-adrenal axis is activated. Many vertebrate neurons possess the machinery to mount rapid responses to these dynamic fluctuations in corticosterone concentrations.[9] This is well illustrated in the neuronal and behavioral reorientation induced by stress in *Taricha.* Changes in the electrophysiological excitability and sensory responsiveness of hindbrain neurons are apparent within 3–8 minutes following ip injection of corticosterone and are maximal within 20–30 minutes.[8] Consistent with the latency of the neurophysiological responses, courtship behavior is significantly suppressed, relative to saline-injected controls, within 8 minutes following ip injection of physiological doses of corticosterone.[10] These rapid responses are not consistent with the model of intracellular receptor-directed changes in protein synthesis, but appear to be mediated through receptors in neuronal membranes.

Characterization of Membrane Receptors

Pharmacological Criteria

When characterizing membrane receptor systems using radioligand binding methods, a number of standard pharmacological criteria are applied to determine if the radioligand/receptor interaction follows the law of mass action, and to distinguish physiologically relevant receptors from artifactual binding sites.[11] These criteria generally include the following: 1) Equilibrium binding should be specific and saturable, with appropriate affinity (expressed as the equilibrium dissociation constant; K_d). 2) The rate of association should be consistent with physiological responses and dissociation studies should reveal reversible radioligand binding. The kinetically derived K_d should agree with the K_d derived from equilibrium saturation studies. 3) In control experiments, specific binding should be eliminated under conditions that denature or degrade proteins. In the case of steroid receptors, one should demonstrate that plasma steroid binding globulins are not interfering with the binding assay. 4) Receptors should exhibit pharmacological specificity which parallels the specificity of compounds in eliciting functional responses. 5) Receptors should be enriched in appropriate subcellular fractions derived from physiologically responsive tissues, but not unresponsive tissues. 6) Activation of the receptor should result in a functional response. 7) One should ultimately perform receptor purification and reconstitution studies, or receptor cloning, sequencing and functional expression.

We have addressed all but the last of these criteria, described below, and conclude that high-affinity receptors for corticosterone in neuronal membranes mediate rapid behavioral responses to stress.

Radioligand Binding Studies

We examined the binding of [^{3}H]corticosterone to a well washed P2 membrane fraction prepared from whole *Taricha* brains, using vacuum filtration over glass-fiber filters to separate bound from free radioligand.[10,11] Equilibrium saturation analysis demonstrated that the binding of [^{3}H]corticosterone to membranes was specific, saturable, and of high affinity. The data were fit with a one-site model yielding a K_d = 0.14 nM and a modest density of binding sites, B_{max} = 183 fmol/mg protein (FIG. 2). The K_d is similar to the affinity of corticosterone for the intracellular mineralocorticoid receptor, but represents much higher affinity binding than that described for rat brain membranes.[12,13] At concentrations near the K_d, approximately 75–80% of total binding was specific.

The association of 0.5 nM [^{3}H]corticosterone with neuronal membrane recognition sites was relatively rapid, and the specific binding was completely reversible. Both association and dissociation were described by monophasic rate equations (dissociation rate constant of 0.013 ± 0.001 min^{-1}; k_{obs} = 0.054 ± 0.005 min^{-1}, calculated k_{+1} = 0.079 min^{-1} nM^{-1}), and the kinetically derived K_d (0.2 nM) was similar to the K_d derived under equilibrium conditions.[10] This concordance supports the validity of the binding methodology.

As expected of a proteinaceous binding site, specific binding was eliminated by heat denaturation of membranes and protease treatment. Saline-perfusion of brains prior to decapitation did not alter [^{3}H]corticosterone binding, suggesting that plasma steroid binding globulins do not play a significant role in the binding of corticosterone to neuronal membranes.[10]

The recognition site displayed a distinct pharmacological specificity, differing from the intracellular corticoid receptors in being highly specific for corticosterone and cortisol, but having low affinity for both aldosterone and dexamethasone (TABLE 1). In subcellular fractionation studies, the specific binding of [^{3}H]corticosterone was most enriched in the synaptosomal membrane fraction relative to other fractions

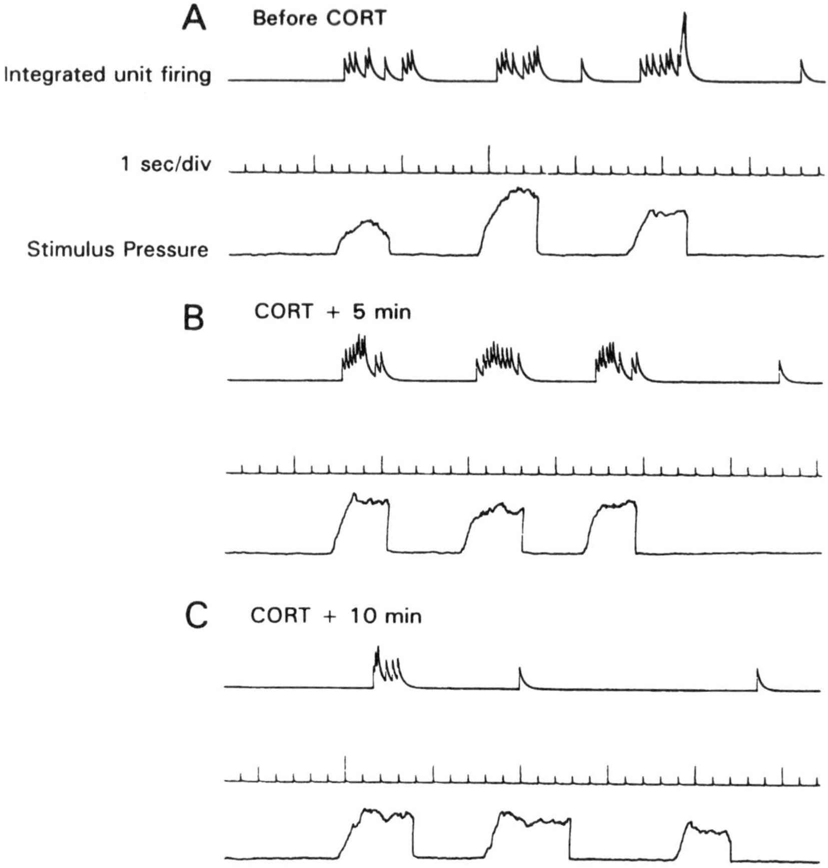

FIGURE 1. Response of a nonreticulospinal neuron to cloacal stimulation and elimination of that response by corticosterone. In each frame **(A-E)** the *top trace* shows integrated pulses from a discriminator that was triggered by spikes from the medullary neuron, the *middle trace* shows a 1-second time mark and the *bottom trace* shows the output of a pressure transducer signalling the application of cloacal pressure. **(A)** Before corticosterone injection, each application of cloacal pressure elicited trains of spikes from the neuron. **(B)** 5 min after ip injection of 25 μg corticosterone, strong responses are still elicited by cloacal pressure. **(C)** 10

prepared by discontinuous sucrose gradient centrifugation.[10] Using the rapid filtration assay, [^{3}H]corticosterone binding to the cytosolic and nuclear fractions was negligible, making it unlikely that the binding in the synaptosomal fraction could be due to contamination by intracellular corticoid receptors. *In vitro* receptor autoradiographic studies also supported the conclusion that binding sites were found in the regions of synaptic terminals. In the presence of unlabeled dexamethasone and ZK91587 added to occupy intracellular receptors, [^{3}H]corticosterone specific binding sites were localized not over regions of cell bodies, but in the neuropil. This

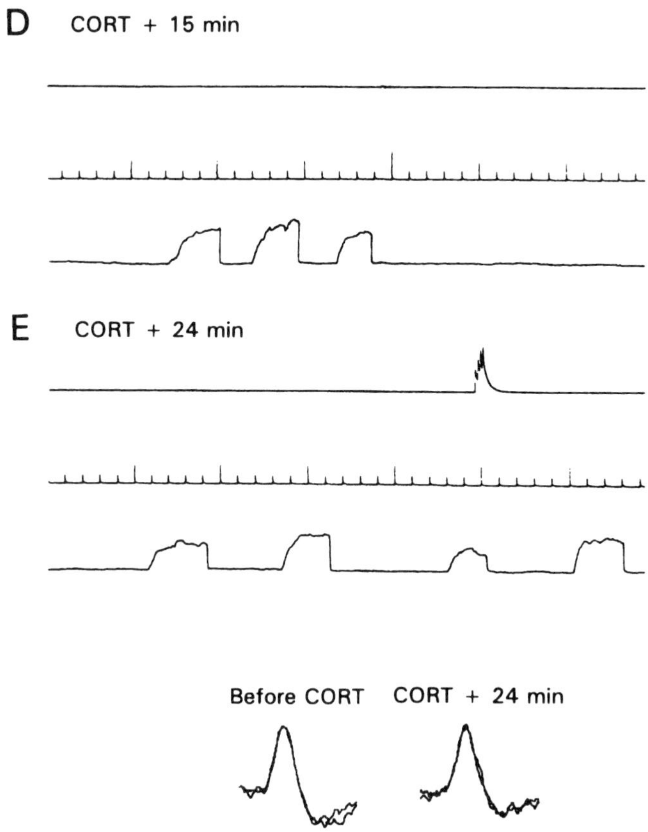

min after corticosterone, the neuron's response is weaker and less reliable. **(D)** 15 min after corticosterone, the response to cloacal pressure is gone, although responses to light pressure on the fact (not shown) were still present. **(E)** 24 min after corticosterone, a weak response was evident. The tracings at the *lower right* are superimposed spike waveforms (sweep duration = 2.5 ms) recorded from the neuron just before and 24 min after corticosterone injection. This spike did not show a pronounced change in amplitude or waveform after the corticosterone injection, in contrast to antidromically activated neurons. (From Rose *et al.*[8] Reprinted by permission from *Neuroendocrinology*.)

included the neuropil surrounding preoptic area neurons, a region known to regulate male sexual behavior in vertebrates.

Together, these studies meet a number of pharmacological criteria used to define genuine receptors in neuronal membranes; equilibrium and kinetic studies were consistent with a ligand/receptor interaction that follows the principle of mass action. The studies suggest that corticosterone binds to a high-affinity recognition site that has a unique pharmacological specificity and is enriched in synaptic membrane fractions.

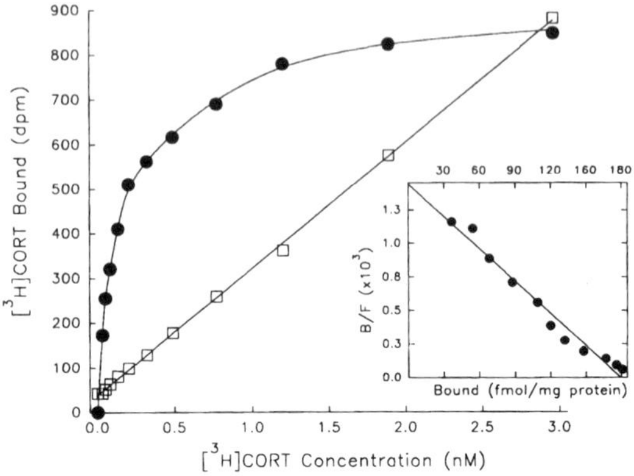

FIGURE 2. Equilibrium saturation binding of [^{3}H]corticosterone to crude synaptosomal membranes. Crude synaptosomal neuronal membranes from whole *Taricha* brains were prepared by homogenization in 0.32 M sucrose and then subjected to repeated washes in cold buffer. Approximately 75 μg of membrane protein was incubated with the concentrations of [^{3}H]corticosterone shown for 2 hr at 30°C in buffer containing 10 mM $MgCl_2$. The reaction was terminated, and bound radioligand separated from unbound by rapid filtration over glass-fiber filters and 9 ml rinse with ice cold buffer. Specific binding (*circles*) equals total binding (not shown) minus nonspecific binding (*squares*). Data were best fit with a one-site model with K_d = 0.14 ± 0.01 nM and B_{max} = 183 ± 4 fmol/mg protein. *Inset*: Scatchard replot of specific binding data was linear; Hill coefficient = 1.08. (Modified from Orchinik *et al.*[20])

TABLE 1. Potency of Steroids as Inhibitors of [^{3}H]Corticosterone Binding.[a,d]

Compound	Ki (nM)	Maximal Inhibition (%)[b]	Slope
Steroids			
Corticosterone	0.11 ± .006	100	0.96 ± 0.03
Cortisol	3.75 ± 0.56	100	0.92 ± 0.08
Aldosterone	293 ± 13	69	1.06 ± 0.06
5α-THDOC[c]	297 ± 14	70	0.96 ± 0.05
RU 28362	569 ± 40	36	1.14 ± 0.24
Progesterone	759 ± 113	37	0.99 ± 0.10
Testosterone	1,138 ± 63	32	0.96 ± 0.03
ZK 91587	>5,000	13	
Dexamethasone	>5,000	11	
RU 38486	>5,000	8	
3α-OH-DHP[c]	>5,000	5	

[a]Membranes were incubated 0.5 nM [^{3}H]corticosterone and a range of unlabeled steroids. Data are mean ± SEM values. K_i values obtained from nonlinear regression analysis.

[b]Maximal inhibition reported for 1 μM competitor.

[c]Potent modulators of $GABA_A$ receptors.

[d]From Orchinik *et al.*[10] Reprinted by permission from *Science*.

Transduction Mechanisms

Ligand-Gated Ion Channel or G Protein-Coupled Receptor?

A functional receptor not only binds ligand, but also elicits cellular responses. There appear to be two classes of transduction mechanisms used by steroid receptors in neuronal membranes. The best characterized mechanisms involve steroid recognition sites on $GABA_A$ receptors.[14,15] $GABA_A$ receptors are ligand-gated ion channels that bind γ-aminobutyric acid and regulate a hyperpolarizing Cl^- conductance. Select steroids, in particular ring A-reduced metabolites of progesterone and deoxycorticosterone, are potent allosteric modulators of GABA-stimulated Cl^- flux. However, structure-activity studies using both mammalian[14,15] and *Taricha* brains[16] have indicated that corticosterone does not meet the stringent structural requirements for modulation of $GABA_A$ receptor function.

Less well understood are the rapid steroid effects on effector mechanisms that are regulated by guanyl nucleotide regulatory binding proteins (G proteins). One possibility is that corticosterone binds to a G protein-coupled receptor, since well over a hundred hormone and neurotransmitter receptors belong to a superfamily of transmembrane receptors that couple to G proteins.[17] A model of cyclic receptor-G protein-effector interactions appears to describe the regulation of cellular responses mediated by many of these receptors.[18,19] In this model, the binding of hormone induces a conformational change in the receptor that favors the formation of a ternary complex of hormone-receptor-G protein that displays high-affinity agonist binding. This step, and the exchange of GTP for GDP that activates the G protein, is facilitated by Mg^{2+}. The binding of GTP induces an isomerization step that favors the dissociation of the ternary complex; dissociated G proteins regulate effector mechanisms while the receptor, free from G protein, displays low affinity for hormone binding. Dissociation of ligand from the receptor permits the receptor to recycle and thereby activate multiple G protein molecules. In this model for G protein-coupled receptors, agonist binding is subject to heterotropic negative modulation by guanine nucleotides.[18,19] Therefore, as described below, we performed studies to determine if [^{3}H]corticosterone binding in *Taricha* brain membranes is negatively modulated by guanyl nucleotides.

Radioligand Binding Studies

As with many G protein-coupled receptors, the binding of [^{3}H]corticosterone was enhanced up to 50% in a concentration-dependent manner by $MgCl_2$ (ED_{50} = 0.5 mM), whereas NaCl and $CaCl_2$ had no effect.[20] This increase in binding may reflect the ability of Mg^{2+} to promote the formation and stabilization of a high-affinity ternary complex. In titration experiments, nonhydrolyzable GTP analogs, particularly GTP-γS, inhibited [^{3}H]corticosterone binding in a dose-dependent manner; GTP-γS inhibited close to 90% of CORT binding (FIG. 3). The negative modulation of [^{3}H]corticosterone binding by GTP-γS may reflect a nucleotide-induced dissociation of receptor from G protein that yields an uncoupled receptor with low affinity for agonist binding. We have not been able to directly demonstrate two affinity states of the receptor in equilibrium saturation binding studies, because we do not have a receptor antagonist that would allow us to quantify agonist inhibition of radiolabeled antagonist binding. However, in a number of receptor systems, the guanyl nucleotide-induced low-affinity state of the receptor reflects an increase in the rate of hormone dissociation from receptors. Consistent with those findings, GTP-γS in-

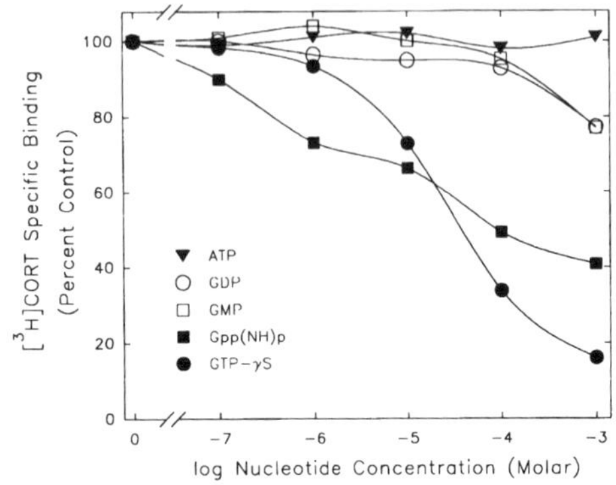

FIGURE 3. Effect of guanyl nucleotides on equilibrium binding of [^{3}H]corticosterone. *Taricha* brain membranes were incubated with 0.75 nM [^{3}H]CORT for 2 hr at 30°C and the concentrations of nucleotide shown. Nonhydrolyzable guanyl nucleotides inhibited specific binding with IC_{50} estimates of 43.2 ± 11 μM for GTP-γS and 104 ± 65 μM for Gpp(NH)p. (From Orchinik *et al.*[20] Reprinted by permission from the National Academy of Sciences.)

duced a rapid phase of [^{3}H]corticosterone dissociation from membrane binding sites that was not induced by unlabeled corticosterone alone (FIG. 4). Therefore, GTP-γS appears to induce a rapid dissociation of the ternary complex with resultant loss of a fraction of high-affinity agonist binding sites. The slower terminal dissociation phase presumably reflects dissociation of [^{3}H]corticosterone from receptors coupled to G protein.

These results are consistent with the formation of a ternary complex of steroid, receptor, and G protein that is subject to negative allosteric modulation by guanyl nucleotides. It is therefore likely that the rapid actions of corticosterone on *Taricha* neuronal activity and behavior involve signal transduction mechanisms using G protein-coupled receptors in neuronal membranes.

Functional Significance

Behavioral Studies

To determine if this putative receptor mediates rapid behavioral responses to corticosterone, we examined the short-latency effects of five steroids, having a wide range of affinities for the [^{3}H]corticosterone binding site, on courtship clasping. During the breeding season, male *Taricha* were injected ip with one of six doses of steroid to produce dose responses for each steroid. Five minutes later behavior testing was initiated, and clasping behavior was monitored for 20 minutes. Male clasping was significantly inhibited by injection of as little as 1 μg corticosterone, consistent with high-affinity binding of corticosterone to membrane receptors. The potencies of the five steroids to rapidly inhibit male sex behavior were linearly related to their potencies to inhibit [^{3}H]corticosterone binding to neuronal membranes (FIG. 5). The strong correlation between the half-maximal effective doses to produce behavioral responses and compete for corticosterone binding sites suggests that the receptor is behaviorally relevant.

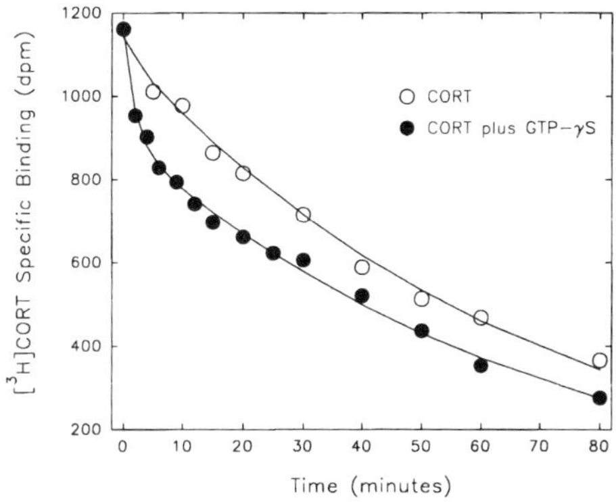

FIGURE 4. Effect of GTP-γS on dissociation of [3H]corticosterone. *Taricha* neuronal membranes were equilibrated with 0.75 nM [3H]CORT in buffer with 10 mM $MgCl_2$ for 2 hr at 30°C, at which time dissociation was initiated by the addition of 25 μl of unlabelled CORT or CORT plus 1 mM GTP-γS. For CORT alone, dissociation was monophasic, with k_{-a} = 0.014 ± 0.0006 min^{-1}. The addition of GTP-γS plus CORT resulted in biphasic dissociation; the initial dissociation was described by k_{-1} = 0.515 ± 0.135 min^{-1}, the terminal dissociation described by k_{-1} = 0.015 ± 0.0004 min^{-1}. (From Orchinik *et al.*[20] Reprinted by permission from the National Academy of Sciences.)

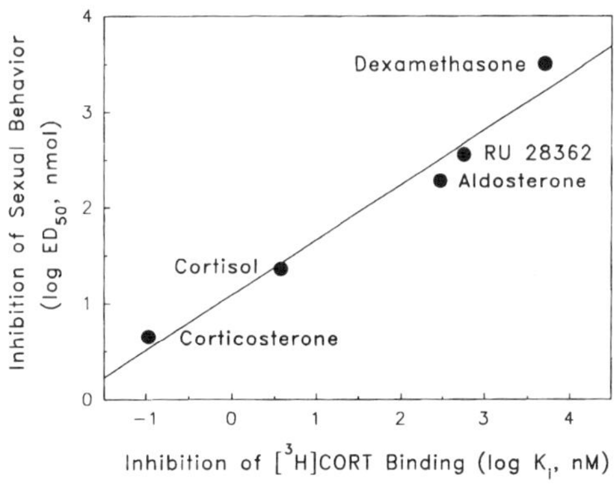

FIGURE 5. Linear relationship between potency of steroids to inhibit [3H]corticosterone binding and potency to inhibit sexual behavior. Males were injected with one of 5–7 doses of steroid, or of vehicle (n = 24 for each dose of each steroid; except RU 28362, n = 14). Behavioral testing was initiated 5 minutes post-injection and data were recorded as the number of claspers in 20-minute tests. ED_{50} values to inhibit male clasping behavior versus controls were obtained by probit analysis. (From Orchinik *et al.*[10] Reprinted by permission from *Science*.)

We have observed that clasping performance was suppressed in corticosterone-injected males even though these males tend to approach and position themselves above female newts in a manner similar to the saline-injected males. This suggests that corticosterone is rapidly and selectively altering aspects of sensory processing or motor output.

Neurophysiological Studies

Reticulospinal neurons that may regulate motor output involved in the clasping reflex as well nonreticulospinal, sensory-responsive medullary neurons in the neural system regulating clasping display rapid electrophysiological changes following corticosterone injection.[8] The various changes in firing rate, sensory responsiveness, and spike-generating responses of reticulospinal neurons emerge within 3–8 minutes of corticosterone injection and are consistent with a membrane site of action. Further, the pharmacological specificity of these neurophysiological responses was in accord with the specificity of the membrane-bound corticosterone receptor. A behaviorally-effective dose of corticosterone (11 μg ip) elicited pronounced, selective changes in these rostral medullary neurons. In contrast, dexamethasone (25 μg) had little or no rapid neurophysiological effect and little effect on clasping, consistent with mediation of responses by the membrane receptor, which has high affinity for corticosterone but very low affinity for dexamethasone.

The functional interaction between the inhibitory actions of corticosterone and the facilitatory actions of vasotocin on the clasping reflex may also be reflected in electrophysiological studies. A number of the neurophysiological actions of vasotocin on medullary neurons are opposite to the actions of corticosterone. For example, application of vasotocin enhances the spontaneous firing rate of rostral medullary neurons and potentiates the responsiveness of these neurons to cloacal stimulation.[21] If corticosterone is injected prior to vasotocin application, these excitatory effects of vasotocin are blocked or reversed. It appears that vasotocin enhances the responsiveness of medullary neurons to clasp-eliciting stimuli, and this may contribute to the enhanced expression of courtship clasping. One mechanism by which corticosterone suppresses courtship behavior may be to inhibit these medullary neurons and to selectively antagonize the vasotocin-enhanced neuronal responsiveness to behaviorally relevant sensory information.[6]

CONCLUSIONS

The rapid stress-induced suppression of courtship behavior in *Taricha* appears to be mediated by corticosteroid receptors in neuronal membranes. Radioligand binding studies indicated that there are high-affinity recognition sites for [^{3}H]corticosterone in brain membranes that are specific for naturally occurring corticosteroids, having low affinity for aldosterone and dexamethasone. Subcellular fractionation studies and *in vitro* receptor autoradiographic studies both indicated that the specific binding sites are most enriched in synaptic membranes. The negative modulation of [^{3}H]corticosterone binding by guanyl nucleotides is consistent with a receptor-G protein interaction, suggesting that signal transduction through these receptors involves G proteins in neuronal membranes. Altogether, the binding of [^{3}H]corticosterone to membranes appears to follow the law of mass action, and is consistent with the presence of genuine steroid receptors in neuronal membranes. The pharmacology, kinetics, and site of action of corticosterone in both behavioral and neurophysiological studies support the conclusion that these are functional receptors that mediate rapid behavioral responses to stress or corticosterone. Furthermore, corticosterone appears to suppress courtship clasping, in part, through rapid and selective effects on the neuronal processing of sensory information.

Vertebrate neurons appear to possess receptors in neuronal membranes that can rapidly respond to sudden fluctuations in circulating corticosteroids, as well as intracellular receptors that can exert more lasting effects on neuronal activity through the regulation of gene expression. Therefore, it is likely that corticosteroids modulate brain function through the integration of multiple receptor mechanisms.

Unfortunately, the receptors mediating the membrane actions of corticosteroids have not been extensively characterized in the mammalian brain. This may be due, in large part, to the formidable technical problems that arise from the combination of low-affinity binding sites and high nonspecific binding of steroids to membrane lipids. However, low-affinity corticosterone binding sites, such as those described in rat brain membranes,[12,13] would be ideally suited for responding rapidly to stress levels of corticosterone in the rat. A fuller characterization of membrane receptors, at the mechanistic and functional levels, will contribute to a more integrated understanding of corticosteroid actions in the brain.

ACKNOWLEDGMENTS

The invaluable assistance of Thomas F. Murray is acknowledged with gratitude.

REFERENCES

1. Moore, F. L. & M. Orchinik. 1991. Multiple molecular actions for steroids in the regulation of reproductive behaviors. Sem. Neurosci. **3:** 489–496.
2. Tokarz, R. R. 1987. Effects of corticosterone treatment on male aggressive behavior in a lizard (*Anolis sagrei*). Horm. Behav. **21:** 358–370.
3. DeNardo, D. F. & P. Licht. 1993. Effects of corticosterone on social behavior of male lizards. Horm. Behav. **27:** 184–199.
4. Wingfield, J. C. & B. Silverin. 1986. Effects of corticosterone on territorial behavior of free-living male song sparrows *Melospiza melodia*. Horm. Behav. **20:** 405–417.
5. Propper, C. R. 1991. Courtship in the rough-skinned newt *Taricha granulosa*. Anim. Behav. **41:** 547–554.
6. Moore, F. L., Lowry, C. A. & J. D. Rose. 1994. Steroid-neuropeptide interactions that control reproductive behaviors in an amphibian. Psychoneuroendocrinology. In press.
7. Moore, F. L. & L. J. Miller. 1984. Stress-induced inhibition of sexual behavior: corticosterone inhibits courtship behaviors of a male amphibian (*Taricha granulosa*). Horm. Behav. **18:** 400–410.
8. Rose, J. D., Moore, F. L. & M. Orchinik. 1993. Rapid neurophysiological effects of corticosterone on medullary neurons: relationship to stress-induced suppression of courtship clasping in an amphibian. Neuroendocrinology **57:** 815–824.
9. Orchinik, M. & B. S. McEwen. 1994.Rapid steroid actions in the brain: a critique of genomic and nongenomic mechanisms. *In* Genomic and Nongenomic Actions of Aldosterone. M. Wehling, Ed. CRC Series in Toxicology and Pharmacology. In press.
10. Orchinik, M., T. F. Murray & F. L. Moore. 1991. A corticosteroid receptor in neuronal membranes. Science **252:** 1848–1851.
11. Orchinik, M. & T. F. Murray. 1994. Steroid hormone binding to membrane receptors. *In* Neurobiology of Steroids, Methods in Neuroscience. E. R. de Kloet & W. Sutanto, Eds. Academic Press. In press.

12. Towle, A. C. & P. Y. Sze. 1983. Steroid binding to synaptic plasma membrane: differential binding of glucocorticoid and gonadal steroids. J. Steroid Biochem. **18:** 135–143.
13. Chen, Y.-Z., H. Fu & Z. Guo. 1993. Membrane receptors for glucocorticoids in mammalian neurons. *In* Receptors: Model Systems and Specific Receptors, Methods in Neurosciences. P. M. Conn, Ed. Vol. 11: Ch. 2. Academic Press. New York.
14. Lan, N. C., M. B. Bolger & K. W. Gee. 1991. Identification and characterization of a pregnane steroid recognition site that is functionally coupled to an expressed $GABA_A$ receptor. Neurochem. Res. **16:** 347–356.
15. Paul, S. M. & R. H. Purdy. 1992. Neuroactive steroids. FASEB J. **6:** 2311–2322.
16. Orchinik, M., T. F. Murray & F. L. Moore. 1994. Steroid modulation of $GABA_A$ receptors in an amphibian brain. Brain Res. **646:** 258–266.
17. Simon, M. I., M. P. Strathmann & N. Gautam. 1991. Diversity of G proteins in signal transduction. Science **252:** 802–808.
18. Birnbaumer, L., J. Abramowitz & A. M. Brown. 1990. Receptor-effector coupling by G proteins. Biochim. Biophys. Acta **1031:** 163–224.
19. Gilman, A. G. 1987. G Proteins: transducers of receptor-generated signals. Annu. Rev. Biochem. **56:** 615–649.
20. Orchinik, M., T. F. Murray, P. H. Franklin & F. L. Moore. 1992. Guanyl nucleotides modulate binding to steroid receptors in neuronal membranes. Proc. Natl. Acad. Sci. USA **89:** 3830–3834.
21. Rose, J. D., J. R. Kinnaird & F. L. Moore. 1993. Interactive effects of vasotocin and corticosterone on sensory responses of medullary neurons: implications for stress effects on neuropeptide control of reproductive behavior in an amphibian. Soc. Neurosci. Abstr. **19:** 585.

DISCUSSION OF THE PAPER

Question: One of the issues in this session was that the mineralocorticoid receptor in rat has an affinity for corticosterone comparable to the one that you show here. This implies 70 percent receptor occupancy due to the circulating levels of steroids. How is that in these newts? What are the levels of steroids that circulate? Is there any circadian variation?

Moore: Amphibians have circadian variation in corticosterone. The levels are lower than in the rat. The resting level for corticosterone would be about one or two nanograms per ml. A stressed animal would be up to 20 to 40 maybe 50 nanograms for corticosterone. Aldosterone levels are in the same range. We found a specific binding site for aldosterone in neuronal membranes and, although we have not characterized it completely, it is certainly different from the corticosterone membrane binding site.

Question: One of the reasons I am asking the question is because you saw the behavioral changes with low levels of corticosterone. You inferred that the corticosterone receptor has a behavioral importance. Do the levels of corticosterone during stress get high enough for the male to drop the female, since that is basically what the hormone is doing?

Moore: Yes, I did not go into that. That was work we did before the binding studies. We showed that stress increases the plasma levels of corticosterone and that those increases in corticosterone mediate inhibition of sex behavior.

M. Joëls (*University of Amsterdam*): You give evidence that a G protein is involved in the corticosterone effect. However, do you know whether it is a G protein that is coupled to a second messenger system, or is it a G protein that directly couples to an ion channel?

MOORE: No work has been done on the second messengers involved in signal transduction mediated by this receptor.

QUESTION: To follow up Dr. Joël's question, a resting level of one or two nanograms per ml is in fact three to six nanomolar, which is presumably largely bound. To what proportion is transcortin involved in these amphibians?

MOORE: Amphibians do have plasma steroid binding globulins, but we don't know how much CORT is bound to a plasma-binding globulin. Related to this question, all the binding studies were done with brains that were removed and homogenized. In control studies, brains were perfused with saline to remove blood but binding was not eliminated, suggesting that plasma binding globulins do not play a major role in corticosterone binding to membranes.

QUESTION: Have you tried administering the corticosterone directly onto these neurons and seeing how rapid the response is, rather than using the injection route?

MOORE: Dr. James Rose has conducted neurophysiological studies with newts and found that the administration of corticosterone directly onto medullary neurons can produce very rapid neuronal responses. But these reponses to direct application of corticosterone have not yet been fully characterized.

QUESTION: How long does the behavioral effect persist?

MOORE: The behavioral studies described in my talk only lasted for 20 minutes, because we wanted to determine whether the potency of a steroid to inhibit reproductive behavior rapidly was related to its binding affinity to the corticosterone receptor in neuronal membranes. Other behavioral studies indicate that an injection of corticosterone can inhibit male behaviors for many hours, exactly how long this effect lasts seems to depend on the amount of corticosterone, the season of the year, and how much the males have been handled or stressed.

QUESTION: I mean, it is very lovely that you let them go. But are they good for anything?

MOORE: Yes, they probably are happier being let go. I say this because most of our studies produce physiological changes that are similar to those associated with moderate stress responses. An important and related question is how much of the long-lasting effect of corticosterone is mediated through the nongenomic mechanism, and how much is mediated through genomic mechanisms. To answer this question, we will need to conduct behavioral studies using ligands that specifically target the membrane associated receptors versus the intracellular receptors. But we have not done these experiments yet.

QUESTION: Can you rule out the possibility that you are looking at a binding site of the steroid on a transmitter receptor, given the fact that you do not seem to see direct effects of steroids alone but that you have to stimulate the system and then see your steroid responses? You seem to have to do something to the male or the female in order to see your steroid effect. That could mean that you are actually looking at a site on one of the neurotransmitter receptors.

MOORE: Yes, it is possible that the corticosterone receptor is a binding site on one of the neurotransmitter receptors, other than the $GABA_A$ receptor. But so far we do not have conclusive evidence supporting or rejecting this hypothesis.

QUESTION: Could the corticosterone receptor be a site on transcortin?

MOORE: Because, as mentioned above, control studies found that specific binding was not eliminated when brains were perfused to remove blood from the brains, it does not seem likely. However, we cannot completely rule out the possibility. If the corticosterone receptor were a site on a binding globulin, then that would mean that this binding globulin is involved in controlling a rapid behavior response, and that would be very interesting.

QUESTION: Do these animals show any breeding cycles, and if so, does it correlate with changes in the density of these receptors?

MOORE: Good question. They have very specific breeding cycles which start in February and go until June, a prolonged breeding season. We do not know whether corticosterone receptor density changes seasonally, although it appears that newts are more sensitive to stress towards the end of their breeding season.

QUESTION: With respect to what kind of measures? Stress and suppression of the clasping reflex?

MOORE: Yes, suppression of clasping when handled by experimenters or injected with corticosterone.

Glucocorticoids and Stress: Permissive and Suppressive Actions

ALLAN MUNCK AND ANIKÓ NÁRAY-FEJES-TÓTH

Department of Physiology
Dartmouth Medical School
Lebanon, New Hampshire 03756-0001

INTRODUCTION

Stress and corticosteroids have been linked in the minds of endocrinologists since the 1930s, when Hans Selye began to develop his theories of the alarm reaction, the general adaptation syndrome, and diseases of adaptation.[1,2] Although mineralocorticoids were once assigned major roles in some of these phenomena, glucocorticoids were eventually identified as the main corticosteroids that protect the organism in stress. Mineralocorticoids have recently reentered the stress arena through their receptors, which bind glucocorticoids with high affinity unless protected by 11β-hydroxysteroid dehydrogenase.

As described in a decade-old review,[3] the question of how glucocorticoids at high concentrations protect against stress fell into abeyance with the discovery in the late 1940s of the antiinflammatory effects, which were incompatible with the prevailing notion that glucocorticoids protect by enhancing defense mechanisms. Physiologists responded to the antiinflammatory actions by relegating them to pharmacology—so effectively that in important later reviews on the role of the adrenal cortex in homeostasis they are not even mentioned.[4,5]

Our first goal in this article is to update evidence bearing on the alternative hypothesis proposed in the 1984 review,[3] namely, that antiinflammatory and other suppressive actions of glucocorticoids manifest a central physiological function of these hormones in stress to prevent stress-activated defense mechanisms from overshooting and damaging the organism. That hypothesis, as we learned later, had been suggested in germinal form long before by Marius Tausk,[6] but had never entered the regular endocrine literature. Numerous results since 1984 support the hypothesis, as will be discussed below, and no results that we are aware of invalidate it.

Our second goal is to present a speculative view of glucocorticoid physiology that integrates the suppressive actions with the classically recognized permissive actions.[7] Permissive actions, identified and elucidated by Dwight Ingle,[8] are those by which glucocorticoids at basal levels permit or normalize responses to stress and various agents, including other hormones, by priming some of the body's homeostatic defense mechanisms.

FIGURE 1 anticipates the latter goal with a graph which, for now, we shall interpret qualitatively, but shall later invest with more concrete meaning. It displays the relation observed between plasma glucocorticoid concentrations and the activity to which certain defense mechanisms can be stimulated by a challenge. At low levels, glucocorticoids are required to permissively enhance that activity; at high levels they suppress it.

Missing in this figure is the dimension of time. Different glucocorticoid effects have different time-courses of onset and decay. Suppressive effects often become more pronounced with time of hormone exposure, so that hormone levels that are

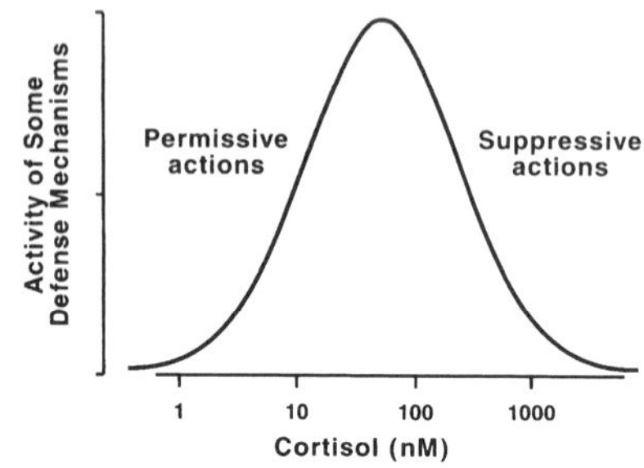

FIGURE 1. Bell-shaped curve, derived from the quantitative model described for FIGURE 2, that shows qualitatively how the responses of some homeostatic defense mechanisms to stress-inducing challenges are permissively enhanced by basal levels of glucocorticoids and suppressed by stress-induced levels.

beneficial in the short run may be harmful in the long run. Perhaps that is why basal glucocorticoid concentrations vary through the day. Levels of free cortisol in humans[9] and of free corticosterone in rats[10] oscillate diurnally over the lower range of the curve, up to about the concentration for peak activity. Stress-induced levels can rise well into the suppressive range.[9] During therapy concentrations may be held in that range for extended periods, simultaneously bestowing therapeutic benefit and physiological damage.

In the next sections we summarize some significant results of the last decade on how suppressive actions of glucocorticoids protect in stress. At the end we return to the relation between permissive and suppressive actions.

Antiinflammatory and Immunosuppressive Actions

Several reviews have recently appeared on topics related to this subject.[11–15]

The hypothesis that antiinflammatory actions of glucocorticoids are physiological[3] implies that in the absence of glucocorticoids inflammatory responses should be exacerbated. That implication—which also assumes that mechanisms of inflammation do not require permissive activation by glucocorticoids—was tested directly by Flower and colleagues,[16] who showed that in rats the inflammatory response to carrageenin is enhanced by adrenalectomy. Laue *et al.*[17] obtained a similar result, and in addition found that the inflammatory response in normal rats is enhanced by administration of the glucocorticoid antagonist RU486. Derijk and Berkenbosch[18] furthermore demonstrated control by endogenous glucocorticoids of autoimmune reactions.

A striking example of the role of endogenous glucocorticoids has come from the studies of Sternberg *et al.*[12,19,20] with Lewis and Fischer rats. Lewis rats, which develop much more severe arthritis after a challenge with streptococcal cell wall polysaccharide (SCW) than Fischer rats, were found to have a defect in biosynthesis of CRH that curtails their glucocorticoid responses. They can be protected from the reaction to SCW by injection with dexamethasone, whereas Fischer rats become more sensitive to SCW if pretreated with RU486, pointing to glucocorticoids as the main antiinflammatory and immunosuppressive agents of the hypothalmo-pituitary-adrenal (HPA) axis.

Lewis rats are also susceptible to developing experimental allergic encephalomyelitis (EAE) after injection with guinea pig myelin basic protein.[21] Other factors besides glucocorticoids may well be important here, since in certain rat strains no genetic linkage is found between resistance to EAE and high corticosterone levels.[22] Lewis rats, for example, have high plasma levels of arginine vasopressin (AVP), which may contribute to their exaggerated inflammatory responses.[23]

It should be noted that although glucocorticoids are probably the prime immunoregulators secreted during stress, they are not the only ones.[13] That first became evident when experiments showed that stress of adrenalectomized or hypophysectomized rats causes inhibition of mitogen-stimulated lymphocyte proliferation.[24] CRH is one among several peptides with immunoregulatory activity.[13,25] Whether such alternative immunoregulators operate under normal physiological conditions, or serve in back-up capacities, is unclear.

Glucocorticoids and Shock

Results from two studies lend strong support to the tentative suggestion[3] that glucocorticoids protect against shock by suppressing defense reactions activated by the stress that causes shock.

First is the discovery that the well-known protective effect of glucocorticoids against endotoxic shock is probably accounted for by suppression of secretion of tumor necrosis factor (TNF), which is activated by endotoxin and is the main cause of endotoxic shock.[26–28]

Second are the observations by Darlington *et al.*[29] on survival after hemorrhage of normal and adrenalectomized rats, treated or not with corticosterone. Those rats that died within 5 hours after hemorrhage had much higher plasma levels of vasopressin (VP), renin, and norepinephrine than those that survived. Untreated adrenalctomized rats, none of which survived, had the highest levels of vasoactive agents, and normal (sham operated) rats, all of which survived, had the lowest levels. The authors suggest that glucocorticoids protect from hemorrhagic shock by restraining the outpouring of vasoactive hormones that occurs in response to hemorrhage.[29]

Actions on the Central Nervous System

Understanding of the mechanisms underlying actions of corticosteroids on the nervous system was greatly enhanced by the recognition[30–32] that the corticosterone binding sites found in the brain by McEwen, Weiss and Schwartz[33] corresponded to the previously characterized mineralocorticoid[34–36] and glucocorticoid[37–39] receptors. Furthermore, the natural glucocorticoids, cortisol and corticosterone, have turned out to bind to mineralocorticoid receptors with affinities comparable to aldosterone,[30–32,40–42] behaving as mineralocorticoid agonists. Since glucocorticoids are normally present at concentrations far above their dissociation constants for mineralocorticoid receptors, aldosterone has access to those receptors only in cells where glucocorticoids are rapidly inactivated by 11β-hydroxysteroid dehydrogenase.[41,42] Which mineralocorticoid receptors in the brain are protected from glucocorticoids in this way is still unclear.

Noradrenergic, serotonergic, and possibly other neurotransmitter systems that are activated in response to stress *in vivo* are subsequently restrained by glucocorticoids.[43] Using isolated hippocampal tissue from adrenalectomized animals, Joëls and de Kloet[44,45] showed that pretreatment of the tissue *in vitro* with corticosterone or the pure glucocorticoid agonist RU28362 at concentrations (100–1000 nM) that saturate mineralocorticoid receptors and occupy most glucocorticoid receptors, enhances afterhyperpolarization of neuronal membranes and reduces norepinephrine-evoked neuronal excitability. The effects were blocked by the glucocorticoid antagonist

RU38486. Kerr *et al.*[46] obtained similar results with corticosterone administered either *in vitro* or *in vivo*, and found that the effects increase with age.

Low concentrations of corticosterone or aldosterone (1 nM), however, which occupy mainly mineralocorticoid receptors, have the opposite effect, decreasing afterhyperpolarization and enhancing neuronal excitability.[45,47] When a moderately high concentration of corticosterone is used, an initial enhancing effect is gradually overridden by the suppressive effect.[47] Thus in this system, natural glucocorticoids at low concentrations permissively enhance excitability through mineralocorticoid receptors, and at high concentrations suppress excitability through glucocorticoid receptors.

Control of Cytokines and Other Mediators

Prominently featured in the 1984 review[3] was the subject of lymphokines and monokines, the "hormones of the immune system" now referred to as cytokines. Interleukin 1 (IL-1), IL-2, colony stimulating factor (CSF) and gamma interferon (IFN-γ), almost the only cytokines identified at the time, were known to be suppressed by glucocorticoids, accounting in part for antiinflammatory and immunosuppressive effects. Many more cytokines have since been discovered[48,49]—interleukins alone number into the double digits—and several have been found to play important roles in classic endocrine reactions such as control and responses of the HPA axis.[12,13,50–52]

TABLE 1 lists the cytokines and other mediators that are suppressed by glucocorticoids (only references not cited in the earlier review[3] are included; sources on mediators for which no references are given are in that review). Categories to which the mediators are assigned are somewhat arbitrary. Many mediators could be listed under two headings, and some, like IL-1, under all three. Like the mediators originally discussed, those in this greatly expanded list are induced by some form of stress and are important or essential components of the organism's responses to stress. Their suppression by glucocorticoids supports the idea that inhibition of stress-induced mediators is a general mechanism by which glucocorticoids control defense reactions.[3]

Not all mediators are suppressed by glucocorticoids. The catecholamines are suppressed in some circumstances but not others,[83] and cardiac epinephrine synthesis

TABLE 1. Mediators Suppressed by Glucocorticoids

Hormones and Neurotransmitters	Cytokines	Inflammatory Mediators
CRH[53]	IFN-γ[54]	Eicosanoids[55,56]
VP[29,53,57,58]	TNF[27,28,59–62]	Bradykinin
ACTH[63–65]	GM-CSF[66–70]	Serotonin
β-Endorphin	IL-1β[70–74]	Histamine
Insulin[75,76]	IL-2[54,77–80]	Plasminogen activator
IGF-1[81]	IL-3[82]	Collagenase
Epinephrine[60,83]	IL-5[84]	Elastase
Norepinephrine[29,85,86]	IL-6[60,61,68,87,88]	
Nitric oxide[89]	IL-8[61,68,90–92]	
Substance P[93]		

may be increased.[94] Among the cytokines that have been tested, IL-4,[95] monocyte-macrophage colony-stimulating factor (M-CSF),[68] and transforming growth factor beta (TGF-β)[49,74,96] are not suppressed.

Molecular mechanisms by which glucocorticoids suppress the protein and peptide mediators vary. Plasminogen activator is blocked by a glucocorticoid-induced inhibitor.[97,98] ACTH is blocked at the levels of transcription and secretion.[99,100] IL-1 transcription, translation, and secretion are suppressed, and its mRNA is destabilized.[71–73,101] IL-2,[54,78,80] IL-3,[82] and IL-8[91] are blocked at transcriptional levels. Synthesis of TNF[59] and granulocyte-macrophage colony-stimulating factor (GM-CSF)[67,102,103] is reduced by destabilization of their mRNAs.

The effects on eicosanoids require comment. At one time it was thought the glucocorticoids inhibit eicosanoid synthesis by inducing a protein, lipocortin-1, that blocks activity of the phospholipase A_2 that releases arachidonic acid from phospholipids. Many questions have arisen about that mechanism, and about whether glucocorticoids inhibit synthesis of eicosanoids at all.[11,104,105] Current work favors the idea that in certain tissues glucocorticoids reduce eicosanoid synthesis by inhibiting expression of an inducible form of cyclooxygenase,[55,56,106–108] a key enzyme in eicosanoid synthesis. There is also new evidence for inhibition of phospholipases A_2[55,109] and C,[110] for induction of lipocortin-1,[111] and for a role of lipocortin-1 in the regulation of immune reactions.[112]

Almost all the initial observations on glucocorticoid suppression of cytokines were made with cultured cells.[3] Many of those results have since been extended to whole organisms. Corticosteroid treatment has been shown to suppress levels of TNF,[60,61] IL-2 mRNA,[77] IL-6,[60,61] and IL-8[61] in humans, and of IL-6 and IL-6 mRNA in rats.[88] Time-course studies[60] revealed that in human subjects given endotoxin to increase plasma levels of TNF and IL-6, cortisol administered together with, or 6 hours before, endotoxin suppresses or delays the increases, but cortisol administered 12, 36, 72 or 144 hours before endotoxin enhances the increases. Explanations for these results may lie in the permissive effects of glucocorticoids that are discussed below.

Although glucocorticoid suppression of cytokines and other mediators listed in TABLE 1 can be viewed as preventing damage caused by overshooting of defense mechanisms activated by those mediators, many of the mediators can be toxic in their own right.[3,11,49] TNF, as mentioned, is responsible for endotoxic shock.[28,113] IFN-γ,[114,115] GM-CSF,[116] IL-1,[117–119] IL-2[120,121] and IL-6[122] are toxic at high levels, as are most of the inflammatory mediators.[3] Glucocorticoid suppression therefore also protects the organism from toxicity of the mediators.

Model of Permissive and Suppressive Actions of Glucocorticoids

The main goal of the 1984 review was to deal with the controversial questions of whether the suppressive actions of glucocorticoids are physiologically significant, and if so, what is their role in stress. Permissive actions got little attention. They are observed on many functions associated with the body's defense mechanisms, including pressor, lipolytic, and glycogenolytic actions of catecholamines,[123] gluconeogenesis by glucagon,[124] and stimulation of humoral immune responses.[125] They may account for growth-promoting influences of glucocorticoids on cells in culture.[126] How they could sometimes be related to the suppressive actions has emerged from the paradoxical observation that glucocorticoids both suppress certain mediators and increase their receptor number, effects that would seem to cancel one another.

Glucocorticoids have been found to increase receptor and/or receptor mRNA levels of IFN-γ in human monocytes,[127] of IL-1 in human peripheral blood lymphocytes[125] and a glioblastoma cell line,[128] of IL-6 in human epithelial cells,[129] of VP[130,131] and of CRH[132] in the rat pituitary, of serotonin in the rat cerebral cortex but not hippocampus,[133] of insulin in IM-9 lymphocytes,[75] in rat skeletal muscle,[134] and in HepG2 hepatoma cells[135] (only the B isoform in HepG2 cells). They induce mRNA and/or expression of β_2-adrenergic receptors in hamster DDT_1 smooth muscle cells[136–139] and adipocytes[140,141] (but suppress them in brain cortex[133]), and induce α1B-adrenergic receptors in DDT_1 cells.[139] Increased receptor number often increases sensitivity to the mediator, but sometimes, as in the case of insulin receptors in skeletal muscle,[134,142] sensitivity may be modulated by other changes and remains unaltered. The significance of induction of IL-1 receptors is obscured by the finding that a dexamethasone-induced molecule, IL-R II, that binds but does not transmit signals from IL-1, actually reduces IL-1 activity.[143] As with glucocorticoid suppression of mediators, induction of receptors appears to be cell and tissue specific.

Induction of mediator receptors has been suggested to account for permissive enhancement by glucocorticoids of activities of various mediators.[123,125,126] Here this idea is used in a simple mathematical model[7] to explore the interactions of suppressive and permissive actions. We determine how the activity of a mediator M will vary if its concentration, [M], is lowered by glucocorticoids, and the concentration of its receptors, [R], is simultaneously increased. As shown in the upper part of FIGURE 2, which is plotted as a function of cortisol concentration (on a logarithmic scale), [M] and [R] are assumed to have similar dose-response curves with opposite signs. The curves are proportional to the amount of cortisol bound to glucocorticoid receptors[144] with a dissociation constant of 30 nM.

The lower, bell-shaped curve represents [MR], the equilibrium concentration of mediator-receptor complex formed at each cortisol concentration through the reversible reaction M + R $\leftrightarrow$ MR. Mediator activity, which is assumed to be proportional to [MR], follows the same curve. With cortisol concentration increasing from low levels, mediator receptors increase so that mediator activity is enhanced "permissively." Activity reaches a peak value, and then is suppressed as mediator concentration is lowered by higher cortisol levels. As discussed for FIGURE 1, which shows the same bell-shaped curve in a more general context, basal levels of free glucocorticoids vary over the range to the left of the peak, and stress-induced levels can increase well beyond the peak.

Because permissive effects are most evident at low hormone concentrations it has been suggested that they might be exerted through high-affinity receptors, and that suppressive actions such as the antiinflammatory effects, which clinically sometimes require high doses, might be exerted through low-affinity receptors.[144] The model shows, however, that a single kind of glucocorticoid receptor can account reasonably for both actions. Nonetheless, the results of Joëls and de Kloet described above demonstrate clearly that permissive effects of glucocorticoids in isolated hippocampal tissue are indeed mediated through high affinity receptors, the mineralocorticoid receptors. How general that phenomenon is remains to be determined.

Receptors with such high affinity for glucocorticoids that they would normally be permanently saturated *in vivo* and unable to mediate regulatory functions, could still be responsible for permissive actions.[144] That may be the condition *in vivo* of mineralocorticoid receptors unprotected by 11β-hydroxysteroid dehydrogenase. In FIGURE 2, if the dose-response curve for [R] results from glucocorticoid binding to mineralocorticoid receptors, with a dissociation constant of 0.5 nM,[32] the peak of the bell-shaped curve will shift substantially to the left and basal cortisol concentrations

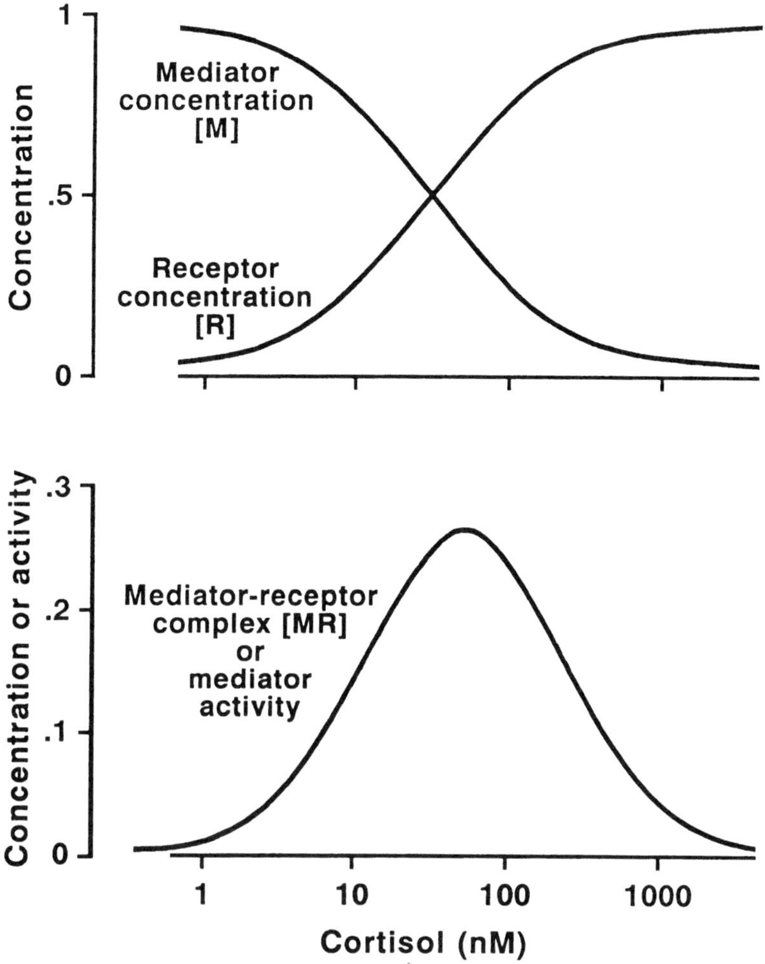

FIGURE 2. Mathematical model of glucocorticoid regulation of the activity of a hypothetical mediator system through permissive induction of mediator receptors [R], and suppression of mediator concentration, [M]. *Upper section*: dependence of [M] and [R] on plasma cortisol concentrations assuming that both are proportional to the amount of cortisol bound to glucocorticoid receptors with a dissociation constant of 30 nM. *Lower section*: bell-shaped curve for [MR], the equilibrium concentration of mediator-receptor complex formed at each cortisol concentration through the reversible reaction M + R $\leftrightarrow$ MR. Mediator activity is assumed to be proportional to [MR]. (From Munck & Náray-Fejes-Tóth.[7] Reprinted by permission from *Molecular and Cellular Endocrinology*.)

will exert almost maximum permissive effects over their whole range. Thus the quantitative relation between permissive and suppressive actions of glucocorticoids will vary considerably depending on whether the permissive actions are mediated through glucocorticoid or mineralocorticoid receptors. Qualitatively, however, as already implied by the discussion of FIGURE 1, most of the conclusions regarding glucocorticoid physiology and pathophysiology derived from the bell-shaped curve are unaffected by quantitative changes and by assumptions about the origins of the permissive and suppressive effects.[7]

SUMMARY

Protection against stress by glucocorticoids is discussed in relation to their permissive and suppressive actions. Evidence from the last decade is summarized regarding the physiological nature of the suppressive actions, and the hypothesis that they prevent stress-activated defense mechanisms from overshooting and damaging the organism. Support for this hypothesis has come from observations on how endogenous or administered glucocorticoids control inflammatory and immune responses, protect in endotoxic and hemorrhagic shock, regulate central nervous system responses to stimuli, and moderate many defense reactions through suppression of cytokines and other mediators. Studies showing that glucocorticoids permissively induce receptors for several mediators that they suppress have led to a model in which stimulated activity of a mediator system is increased permissively through induction of mediator receptors and decreased through suppression of mediator production.

REFERENCES

1. SELYE, H. 1936. Thymus and adrenals in the response of the organism to injuries and intoxications. Br. J. Exp. Pathol. **17:** 234–248.
2. SELYE, H. 1946. The general adaptation syndrome and the diseases of adaptation. J. Clin. Endocrinol. Metab. **6:** 117–230.
3. MUNCK, A., P. M. GUYRE & N. J. HOLBROOK. 1984. Physiological functions of glucocorticoids in stress and their relation to pharmacological actions. Endocr. Rev. **5:** 25–44.
4. INGLE, D. J. 1952. The role of the adrenal cortex in homeostasis. J. Endocrinol. **8:** xxiii xxxvii.
5. HOFFMAN, F. G. 1971. Role of the adrenal cortex in homeostasis and growth. *In* The Human Adrenal Cortex. N. P. Christy, Ed. 303–316. Harper & Row. New York.
6. TAUSK, M. 1951. Hat die Nebenniere tatsächlich eine Verteidigungsfunktion? Das Hormon (Organon, Holland) **3:** 1–24.
7. MUNCK, A. & A. NÁRAY-FEJES-TÓTH. 1992. The ups and downs of glucocorticoid physiology. Permissive and suppressive effects revisited. Mol. Cell. Endocrinol. **90:** C1–C4.
8. INGLE, D. J. 1954. Permissibility of hormone action. A review. Acta Endocrinol. **17:** 172–186.
9. BAXTER, J. D. & J. B. TYRRELL. 1987. The adrenal cortex. *In* Endocrinology and Metabolism. 2nd Edit. P. Felig, J. D. Baxter, A. E. Broadus & L. A. Frohman, Eds. 511–560. McGraw-Hill Book Company. New York.
10. JACOBSON, L., S. F. AKANA, C. S. CASCIO, J. SHINSAKO & M. F. DALLMAN. 1988. Circadian variations in plasma corticosterone permit normal termination of adrenocorticotropin responses to stress. Endocrinology **122:** 1343–1348.
11. MUNCK, A. & P. M. GUYRE. 1991. Glucocorticoids and immune function. *In* Psycho-

neuroimmunology. 2nd Edit. R. Ader, D. L. Felten & N. Cohen, Eds. 447–474. Academic Press. New York.

12. Sternberg, E. M., G. P. Chrousos, R. L. Wilder & P. W. Gold. 1992. The stress response and the regulation of inflammatory disease. Ann. Intern. Med. **117:** 854–866.
13. Reichlin, S. 1993. Neuroendocrine-immune interactions. N. Engl. J. Med. **329:** 1246–1253.
14. Wick, G., S. Schwarz & G. Kroemer. 1993. Immunoendocrine communication via the hypothalamo-pituitary-adrenal axis in autoimmune diseases. Endocr. Rev. **14:** 539–563.
15. Boumpas, D. T., G. P. Chrousos, R. L. Wilder, T. R. Cupps & J.E. Balow. 1993. Glucocorticoid therapy for immune-mediated diseases: basic and clinical correlates. Ann. Intern. Med. **119:** 1198–1208.
16. Flower, R. J., L. Parente, P. Persico & J. A. Salmon. 1986. A comparison of the acute inflammatory response in adrenalectomised and sham-operated rats. Br. J. Pharmacol. **87:** 57–62.
17. Laue, L., S. Kawai, D. D. Brandon, D. Brightwell, K. Barnes, R. A. Knazek, D. L. Loriaux & G. P. Chrousos. 1988. Receptor-mediated effects of glucocorticoids on inflammation: enhancement of the inflammatory response with a glucocorticoid antagonist. J. Steroid Biochem. **29:** 591–598.
18. DeRijk, R. & F. Berkenbosch. 1991. The immune-hypothalamo-pituitary-adrenal axis and autoimmunity. Int. J. Neurosci. **59:** 91–100.
19. Sternberg, E. M., J. M. Hill, G. P. Chrousos, T. Kamilaris, S. J. Listwak, P. W. Gold & R. L. Wilder. 1989. Inflammatory mediator-induced hypothalamic-pituitary-adrenal axis activation is defective in streptococcal cell wall arthritis-susceptible Lewis rats. Proc. Natl. Acad. Sci. USA **86:** 2374–2378.
20. Sternberg, E. M., S. W. Young, III, R. Bernardini, A. E. Calogero, G. P. Chrousos, P. W. Gold & R. L. Wilder. 1989. A central nervous system defect in biosynthesis of corticotropin-releasing hormone is associated with susceptibility to streptococcal cell wall-induced arthritis in Lewis rats. Proc. Natl. Acad. Sci. USA **86:** 4771–4775.
21. Mason, D. 1991. Genetic variation in the stress response: susceptibility to experimental allergic encephalomyelitis and implications for human inflammatory disease. Immunol. Today **12:** 57–60.
22. Villas, P. A., M. J. Dronsfield & E. P. Blankenhorn. 1991. Experimental allergic encephalomyelitis and corticosterone studies in resistant and susceptible rat strains. Clin. Immunol. Immunopathol. **61:** 29–40.
23. Patchev, V. K., K. T. Kalogeras, P. Zelazowski, R. L. Wilder & G. P. Chrousos. 1992. Increased plasma concentrations, hypothalamic content, and *in vitro* release of arginine vasopressin in inflammatory disease-prone, hypothalamic corticotropin-releasing hormone-deficient Lewis rats. Endocrinology **131:** 1453–1457.
24. Keller, S. E., S. J. Schleifer & M. K. Demetrikopoulos. 1991. Stress-induced changes in immune function in animals: hypothalamo-pituitary-adrenal influences. *In* Psychoneuroimmunology. 2nd Edit. R. Ader, D. L. Felten & N. Cohen, Eds. 771–787. Academic Press, Inc. New York.
25. Jain, R., D. Zwickler, C. S. Hollander, H. Brand, A. Saperstein, B. Hutchinson, C. Brown & T. Audhya. 1991. Corticotropin-releasing factor modulates the immune response to stress in the rat. Endocrinology **128:** 1329–1336.
26. Beutler, B., N. Krochin, I. W. Milsark, C. Luedke & A. Cerami. 1986. Control of cachetin (tumor necrosis factor) synthesis: mechanisms of endotoxin resistance. Science **232:** 977–980.
27. Beutler, B. & A. Cerami. 1987. Cachectin: more than a tumor necrosis factor. N. Engl. J. Med. **316:** 379–385.
28. Bertini, R., M. Bianchi & P. Ghezzi. 1988. Adrenalectomy sensitizes mice to the lethal effects of interleukin 1 and tumor necrosis factor. J. Exp. Med. **167:** 1708–1712.
29. Darlington, D. N., G. Chew, T. Ha, L. C. Keil & M. F. Dallman. 1990. Corticosterone, but not glucose, treatment enables fasted adrenalectomized rats to survive moderate hemorrhage. Endocrinology **127:** 766–772.

30. KROZOWSKI, Z. S. & J. W. FUNDER. 1983. Renal mineralocorticoid receptors and hippocampal corticosterone-binding species have identical steroid specificity. Proc. Natl. Acad. Sci. USA **80:** 6056–6060.
31. BEAUMONT, K. & D. D. FANESTIL. 1983. Characterization of rat brain aldosterone receptors reveals high affinity for corticosterone. Endocrinology **113:** 2043–2051.
32. REUL, J. M. H. M. & E. R. DE KLOET. 1985. Two receptor systems for corticosterone in rat brain: microdistribution and differential occupation. Endocrinology **117:** 2505–2511.
33. MCEWEN, B. S., J. M. WEISS & L. S. SCHWARTZ. 1968. Selective retention of corticosterone by limbic structures in rat brain. Nature **220:** 911–912.
34. EDELMAN, I. S., R. BOGOROSH & G. A. PORTER. 1963. On the mechanism of action of aldosterone on sodium transport: the role of protein synthesis. Proc. Natl. Acad. Sci. USA **50:** 1169–1177.
35. SHARP, G. W. G., C. L. KOMACK & A. LEAF. 1966. Studies on the binding of aldosterone in the toad bladder. J. Clin. Invest. **45:** 450–459.
36. FANESTIL, D. D. & I. S. EDELMAN. 1966. Characteristics of the renal nuclear receptors for aldosterone. Proc. Natl. Acad. Sci. USA **56:** 872–879.
37. MUNCK, A. & T. BRINCK-JOHNSEN. 1967. Specific metabolic and physicochemical interactions of glucocorticoids *in vivo* and *in vitro* with rat adipose tissue and thymus cells. Excerpta Med. Int. Cong. Ser. **132:** 472–481.
38. MUNCK, A. & T. BRINCK-JOHNSEN. 1968. Specific and nonspecific physicochemical interactions of glucocorticoids and related steroids with rat thymus cells *in vitro.* J. Biol. Chem. **243:** 5556–5565.
39. SCHAUMBURG, B. P. & E. BOJESEN. 1968. Specificity and thermodynamic properties of the corticosteroid binding to a receptor of rat thymocytes *in vitro.* Biochim. Biophys. Acta **170:** 172–188.
40. ARRIZA, J. L., C. WEINBERGER, G. CERELLI, T. M. GLASER, B. L. HANDELIN, D. E. HOUSMAN & R. M. EVANS. 1987. Cloning of human mineralocorticoid receptor complementary DNA: structural and functional kinship with the glucocorticoid receptor. Science **237:** 268–275.
41. STEWART, P. M., A. M. WALLACE, R. VALENTINO, D. BURT, C. H. L. SHACKLETON & C. R. W. EDWARDS. 1987. Mineralocorticoid activity of licorice: 11-beta-hydroxysteroid dehydrogenase deficiency comes of age. Lancet **ii:** 821–824.
42. FUNDER, J. W., P. T. PEARCE, R. SMITH & A. I. SMITH. 1988. Mineralocorticoid action: target tissue specificity is enzyme, not receptor, mediated. Science **242:** 583–585.
43. MCEWEN, B. S., R. R. SAKAI & R. L. SPENCER. 1993. Adrenal steroid effects on the brain: versatile hormones with good and bad effects. *In* Hormonally Induced Changes in Mind and Brain. J. Schulkin, Ed. 157–189. Academic Press. New York.
44. JOËLS, M. & E. R. DE KLOET. 1989. Effects of glucocorticoids and norepinephrine on the excitability in the hippocampus. Science **245:** 1502–1505.
45. JOËLS, M. & E. R. DE KLOET. 1992. Control of neuronal excitability by corticosteroid hormones. Trends Neurosci. **15:** 25–30.
46. KERR, D. S., L. W. CAMPBELL, S.-Y. HAO & P. W. LANDFIELD. 1989. Corticosteroid modulation of hippocampal potentials: increased effect with aging. Science **245:** 1505–1509.
47. JOËLS, M. & E. R. DE KLOET. 1990. Mineralocorticoid receptor-mediated changes in membrane properties of rat CA1 pyramidal neurons *in vitro.* Proc. Natl. Acad. Sci. USA **87:** 4495–4498.
48. ARAI, K.-I., F. LEE, A. MIYAJIMA, S. MIYATAKE, N. ARAI & T. YOKOTA. 1990. Cytokines: coordinators of immune and inflammatory responses. Annu. Rev. Biochem. **59:** 783–836.
49. LOWRY, S. F. 1993. Cytokine mediators of immunity and inflammation. Arch. Surg. **128:** 1235–1241.
50. RIVIER, C., R. CHIZZONITE & W. VALE. 1989. In the mouse, the activation of the hypothalamic-pituitary-adrenal axis by a lipopolysaccharide (endotoxin) is mediated through interleukin-1. Endocrinology **125:** 2800–2805.
51. BESEDOVSKY, H. O., A. DEL REY, I. KLUSMAN, H. FURUKAWA, G. M. ARDITI & A.

KABIERSCH. 1991. Cytokines as modulators of the hypothalamus-pituitary-adrenal axis. J. Steroid Biochem. Mol. Biol. **40:** 613–618.
52. ZHOU, D., A. W. KUSNEKOV, M. R. SHURIN, M. DEPAOLI & B. S. RABIN. 1993. Exposure to physical and psychological stressors elevates plasma interleukin 6: relationship to the activation of the hypothalamic-pituitary-adrenal axis. Endocrinology **133:** 2523–2530.
53. CANNY, B. J., J. W. FUNDER & I. J. CLARKE. 1989. Glucocorticoids regulate ovine hypophysial portal levels of corticotropin-releasing factor and arginine vasopressin in a stress-specific manner. Endocrinology **125:** 2532–2539.
54. ARYA, S. K., F. WONG-STAAL & R. C. GALLO. 1984. Dexamethasone-mediated inhibition of human T cell growth factor and γ-interferon messenger RNA. J. Immunol. **133:** 273–276.
55. COYNE, D. W., M. NICKOLS, W. BERTRAND & A. R. MORRISON. 1992. Regulation of mesangial cell cyclooxygenase synthesis by cytokines and glucocorticoids. Am. J. Physiol. **263:** F97–F102.
56. MASFERRER, J. L., K. SEIBERT, B. ZWEIFEL & P. NEEDLEMAN. 1992. Endogenous glucocorticoids regulate an inducible cyclooxygenase enzyme. Proc. Natl. Acad. Sci. USA **89:** 3917–3921.
57. DAVIES, L. G., R. ARENTZEN, J. M. REID, R. W. MANNING, B. WOLFSON, K. H. LAWRENCE & F. BALDINO, JR. 1986. Glucocorticoid sensitivity of vasopressin mRNA levels in the paraventricular nucleus of the rat. Proc. Natl. Acad. Sci. USA **83:** 1145–1149.
58. RAFF, H. 1987. Glucocorticoid inhibition of neurohypophysial vasopressin secretion. Am. J. Physiol. **252:** R635–R644.
59. BEUTLER, B. & A. CERAMI. 1988. Cachectin (tumor necrosis factor): a macrophage hormone governing cellular metabolism and inflammatory response. Endocr. Rev. **9:** 57–66.
60. BARBER, A. E., S. M. COYLE, M. A. MARANO, E. FISCHER, S. E. CALVANO, Y. FONG, L. L. MOLDAWER & S. F. LOWRY. 1993. Glucocorticoid therapy alters hormonal and cytokine responses to endotoxin in man. J. Immunol. **150:** 1999–2006.
61. SANTOS, A. A., M. R. SCHELTINGA, E. LYNCH, E. F. BROWN, P. LAWTON, E. CHAMBERS, J. BROWNING, C. A. DINARELLO, S. M. WOLFF & D. W. WILMORE. 1993. Elaboration of interleukin 1-receptor antagonist is not attenuated by glucocorticoids after endotoxemia. Arch. Surg. **128:** 138–144.
62. BRENNER, T., A. YAMIN, O. ABRAMSKY & R. GALLILY. 1993. Stimulation of tumor necrosis factor-α production by mycoplasmas and inhibition by dexamethasone in cultured astrocytes. Brain Res. **608:** 273–279.
63. KELLER-WOOD, M. E. & M. F. DALLMAN. 1984. Corticosteroid inhibition of ACTH secretion. Endocr. Rev. **5:** 1–24.
64. DALLMAN, M. F., S. F. AKANA, C. S. CASCIO, D. N. DARLINGTON, L. JACOBSON & N. LEVIN. 1987. Regulation of ACTH secretion: variations on a theme of B. Recent Prog. Horm. Res. **43:** 113–173.
65. DAYANITHI, G. & F. A. ANTONI. 1989. Rapid as well as delayed inhibitory effects of glucocorticoid hormones on pituitary adrenocorticotropic hormone release are mediated by type II glucocorticoid receptors and require newly synthesized messenger ribonucleic acid as well as protein. Endocrinology **125:** 308–313.
66. KELSO, A. & A. MUNCK. 1984. Glucocorticoid inhibition of lymphokine secretion by alloreactive T lymphocyte clones. J. Immunol. **133:** 784–791.
67. SHAW, G. & R. KAMEN. 1986. A conserved AU sequence from the 3′ untranslated region of GM-CSF mRNA mediates selective mRNA degradation. Cell **46:** 659–667.
68. TOBLER, A., R. MEIER, M. SEITZ, B. DEWALD, M. BAGGIOLINI & M. F. FEY. 1992. Glucocorticoids downregulate gene expression of GM-SCF, NAP-1/IL-8, and IL-6, but not M-CSF in human fibroblasts. Blood **79:** 45–51.
69. HER, E., J. FRAZER, K. F. AUSTEN & W. F. OWEN, JR. 1991. Eosinophil hematopoietins antagonize the programmed cell death of eosinophils. Cytokine and glucocorticoid effects on eosinophils maintained by endothelial cell-conditioned medium. J. Clin. Invest. **88:** 1982–1987.

70. LANE, S. J., J. R. W. WILKINSON, G. M. COCHRANE, T. H. LEE & J. P. ARM. 1993. Differential *in vitro* regulation by glucocorticoids of monocyte-derived cytokine generation in glucocorticoid-resistant bronchial asthma. Am. Rev. Resp. Dis. **147:** 690–696.
71. KNUDSEN, P. J., C. A. DINARELLO & T. B. STROM. 1987. Glucocorticoids inhibit transcriptional and post-transcriptional expression of interleukin 1 in U937 cells. J. Immunol. **139:** 4129–4134.
72. KERN, J. A., R. J. LAMB, J. C. REED, R. P. DANIELE & P. C. NOWELL. 1988. Dexamethasone inhibition of interleukin 1 beta production by human monocytes: posttranscriptional mechanisms. J. Clin. Invest. **81:** 237–244.
73. LEE, S. W., A. P. TSOU, H. CHAN, J. THOMAS, K. PETRIE, E. M. EUGUI & A. C. ALLISON. 1988. Glucocorticoids selectively inhibit the transcription of the interleukin-1β gene and decrease the stability of interleukin-1β mRNA. Proc. Natl. Acad. Sci. USA **85:** 1204–1208.
74. KHALIL, N., C. WHITMAN, L. ZUO, D. DANIELPOUR & A. GREENBERG. 1993. Regulation of alveolar macrophage transforming growth factor-β secretion by corticosteroids in bleomycin-induced pulmonary inflammation in the rat. J. Clin. Invest. **92:** 1812–1818.
75. MCDONALD, A. R. & I. D. GOLDFINE. 1988. Glucocorticoid regulation of insulin receptor gene transcription in IM-9 cultured lymphocytes. J. Clin. Invest. **81:** 499–504.
76. PHILIPPE, J. & M. MISSOTTEN. 1990. Dexamethasone inhibits insulin biosynthesis by destabilizing insulin messenger ribonucleic acid in hamster insulinoma cells. Endocrinology **127:** 1640–1645.
77. PINKSTON, P., C. SALTINI, J. MÜLLER-QUERNHEIM & R. G. CRYSTAL. 1987. Corticosteroid therapy suppresses spontaneous interleukin 2 release and spontaneous proliferation of lung T lymphocytes of patients with active pulmonary sarcoidosis. J. Immunol. **139:** 755–760.
78. VACCA, A., S. MARTINOTTI, I. SCREPANTI, M. MARODER, M. P. FELLI, A. R. FARINA, A. GISMONDI, A. SANTONI, L. FRATI & A. GULINO. 1990. Transcriptional regulation of the interleukin 2 gene by glucocorticoid hormones. J. Biol. Chem. **265:** 8075–8080.
79. ALWAMI, W. Y., M. L. LIPMAN, A. C. STEVENS, B. ZANKER, E. T. HADRO & T. B. STROM. 1991. Abrogation of glucocorticosteroid-mediated inhibition of T cell proliferation by the synergistic action of IL-1, IL-6, and IFN-γ. J. Immunol. **146:** 3523–3527.
80. PALIOGIANNI, F., A. RAPTIS, S. S. AHUJA, S. M. NAJJAR & D.T. BOUMPAS. 1993. Negative transcriptional regulation of human interleukin 2 (IL-2) gene by glucocorticoids through interference with nuclear transcription factors AP-1 and NT-AT. J. Clin. Invest. **91:** 1481–1489.
81. UNTERMAN, T. G., J. J. JENTEL, D. T. OEHLER, R. G. LACSON & J. F. HOFERT. 1993. Effects of glucocorticoids on circulating levels and hepatic expression of insulin-like growth factor (IGF)-binding proteins and IGF-I in the adrenalectomized streptozotocin-diabetic rat. Endocrinology **133:** 2531–2539.
82. CULPEPPER, J. A. & F. LEE. 1985. Regulation of IL 3 expression by glucocorticoids in cloned murine T lymphocytes. J. Immunol. **135:** 3191–3197.
83. ROMANOFF, P. A. & J. W. FUNDER. 1994. Differential glucocorticoid effects on catecholamine responses to stress. Am. J. Physiol. **266:** E118–E128.
84. WANG, Y., H. D. CAMPBELL & I. G. YOUNG. 1993. Sex hormones and dexamethasone modulate interleukin-5 gene expression in T lymphocytes. J. Steroid Biochem. Mol. Biol. **44:** 203–210.
85. KVETNANSKY, R., K. FUKUHARA, K. PACÁK, G. CIZZA, D. S. GOLDSTEIN & I. J. KOPIN. 1993. Endogenous glucocorticoids restrain catecholamine synthesis and release at rest and during immobilization stress in rats. Endocrinology **133:** 1411–1419.
86. PACAK, K., R. KVETNANSKY, M. PALKOVITS, K. FUKUHARA, G. YADID, I. J. KOPIN & D. S. GOLDSTEIN. 1993. Adrenalectomy augments *in vivo* release of norepinephrine in the paraventricular nucleus during immobilization stress. Endocrinology **133:** 1404–1410.

87. GUERNE, P.-A., D. A. CARSON & M. LOTZ. 1990. IL-6 production by human articular chondrocytes. Modulation of its synthesis by cytokines, growth factors, and hormones *in vitro*. J. Immunol. **144:** 499–505.
88. SCHÖBITZ, B., M. VAN DEN DOBBELSTEEN, F. HOLSBOER, W.SUTANTO & R. DE KLOET. 1993. Regulation of interleukin 6 gene expression in rat. Endocrinology **132:** 1569–1576.
89. CETKOVIC-CVRLJE, M., S. SANDLER & D. L. EIZIRIK. 1993. Nicotinamide and dexamethasone inhibit interleukin-1-induced nitric oxide production by RINm5F cells without decreasing messenger RNA expression for nitric oxide synthase. Endocrinology **133:** 1739–1743.
90. STANDIFORD, T. J., S. L. KUNKEL, M. W. ROLFE, H. L. EVANOFF, R. M. ALLEN & R. M. STRIETER. 1992. Regulation of human alveolar macrophage- and blood monocyte-derived interleukin-8 by prostaglandin E_2 and dexamethasone. Am. J. Respir. Cell Mol. Biol. **6:** 75–81.
91. MUKAIDA, N., G. L. GUSELLA, T. KASAHARA, Y. KO, C. O. ZACHARIAE, T. KAWAI & K. MATSUSHIMA. 1992. Molecular analysis of the inhibition of interleukin-8 production by dexamethasone in a human fibrosarcoma cell line. Immunology **75:** 674–679.
92. WERTHEIM, W. A., S. L. KUNKEL, T. J. STANDIFORD, M. D. BURDICK, F. S. BECKER, C. A. WILKE, A. R. GILBERT & R. M. STRIETER. 1993. Regulation of neutrophil-derived IL-8: the role of prostaglandin E_2, dexamethasone, and IL-4. J. Immunol. **151:** 2166–2175.
93. FREIDIN, M. & J. A. KESSLER. 1991. Cytokine regulation of substance P expression in sympathetic neurons. Proc. Natl. Acad. Sci. USA **88:** 3200–3203.
94. KENNEDY, B. & M. G. ZIEGLER. 1991. Cardiac epinephrine synthesis. Regulation by a glucocorticoid. Circulation **84:** 891–895.
95. BERTOGLIO, J. H. & E. LEROUX. 1988. Differential effects of glucocorticoids on the proliferation of a murine helper and a cytolytic T cell clone in response to IL-2 and IL-4. J. Immunol. **141:** 1191–1196.
96. OURSLER, M. J., B. L. RIGGS & T. C.SPELSBERG. 1993. Glucocorticoid-induced activation of latent transforming growth factor-β by normal human osteoblast-like cells. Endocrinology **133:** 2187–2196.
97. CWIKEL, B. J., P. A. BAROUSKI-MILLER, P. L. COLEMAN & T. D. GELEHRTER. 1984. Dexamethasone induction of an inhibitor of plasminogen activator on HTC hepatoma cells. J. Biol. Chem. **259:** 6847–6851.
98. COLEMAN, P. L., P. D. PATEL, B. J. CWIKEL, U. M. RAFFERTY, R. SZNYCER-LASZUK & T. D. GELEHRTER. 1986. Characterization of the dexamethasone-induced inhibitor of plasminogen activator in HTC hepatoma cells. J. Biol. Chem. **261:** 4352–4357.
99. LUNDBLAD, J. R. & J. L. ROBERTS. 1988. Regulation of proopiomelanocortin gene expression in pituitary. Endocr. Rev. **9:** 135–158.
100. DROUIN, J., Y. L. SUN & M. NEMER. 1989. Glucocorticoid repression of pro-opiomelanocortin gene transcription. J. Steroid Biochem. **34:** 63–69.
101. LEW, W., J. J. OPPENHEIM & K. MATSUSHIMA. 1988. Analysis of the suppression of IL-1α and IL-1β production in human peripheral blood mononuclear adherent cells by a glucocorticoid hormone. J. Immunol. **140:** 1895–1902.
102. CAPUT, D., B. BEUTLER, K. HARTOG, R. THAYER, S. BROWN-SHIMER & A. CERAMI. 1986. Identification of a common nucleotide sequence in the 3′ untranslated region of mRNA molecules specifying inflammatory mediators. Proc. Natl. Acad. Sci. USA **83:** 1670–1674.
103. THORENS, B., J.-J. MERMOD & P. VASSALLI. 1987. Phagocytosis and inflammatory stimuli induce GM-CSF mRNA in macrophages through posttranscriptional regulation. Cell **48:** 671–679.
104. NÁRAY-FEJES-TÓTH, A., B. ROSENKRANZ, J. C. FRÖLICH & G. FEJES-TÓTH. 1988. Glucocorticoid effect on arachidonic acid metabolism *in vivo*. J. Steroid Biochem. **30:** 155–159.
105. GOULDING, N. J. & P. M. GUYRE. 1992. Regulation of inflammation by lipocortin 1. Immunol. Today **13:** 295–297.

106. O'BANION, M. K., H. B. SADOWSKI, V. WINN & D. A. YOUNG. 1991. A serum- and glucocorticoid-regulated 4-kilobase mRNA encodes a cyclooxygenase-related protein. J. Biol. Chem. **266:** 23261–23267.
107. O'BANION, M. K., V. D. WINN & D. A. YOUNG. 1992. cDNA cloning and functional activity of a glucocorticoid-regulated inflammatory cyclooxygenase. Proc. Natl. Acad. Sci. USA **89:** 4888–4892.
108. WINN, V. D., M. K. O'BANION & D. A. YOUNG. 1992. Anti-inflammatory glucocorticoid action: inhibition of griPGHS, a new cyclooxygenase. J. Lipid Mediators. In press.
109. VISHWANATH, B. S., F. J. FREY, M. J. BRADBURY, M. F. DALLMAN & B. M. FREY. 1993. Glucocorticoid deficiency increases phospholipase A_2 activity in rats. J. Clin. Invest. **92:** 1974–1980.
110. AKOMPONG, T., R. L. SPENCER & B. S. MCEWEN. 1993. Glucocorticoids inhibit soluble phospholipase C activity and cytosolic guanine nucleotide regulatory protein-αi immunoreactivity in spleen. Endocrinology **133:** 1963–1970.
111. VISHHWANATH, B. S., F. J. FREY, M. BRADBURY, M. F. DALLMAN & B. M. FREY. 1992. Adrenalectomy decreases lipocortin-I messenger ribonucleic acid and tissue protein content in rats. Endocrinology **130:** 585–591.
112. PERRETTI, M. & R. J. FLOWER. 1993. Modulation of IL-1-induced neutrophil migration by dexamethasone and lipocortin 1. J. Immunol. **150:** 992–999.
113. TRACEY, K. J., B. BEUTLER, S. F. LOWRY, J. MERRYWEATHER, S. WOLPE, I. W. MILSARK, R. J. HARIRI, T. J. FAHEY, III, A. ZENTELLA, J. D. ALBERT, G. T. SHIRES & A. CERAMI. 1986. Shock and tissue injury induced by recombinant human cachectin. Science **234:** 470–474.
114. SHERWIN, S. A., K. A. FOON, P. G. ABRAMS, M. R. HEYMAN, J. J. OCHS, T. WATSON, A. MALUISH & R. K. OLDMAN. 1984. A preliminary phase I trial of partially purified interferon-γ on patients with cancer. J. Biol. Response Modifiers **3:** 599–607.
115. AULT, B. H., F. B. STAPLETON, L. GABER, A. MARTIN, S. ROY, III & S. B. MURPHY. 1988. Acute renal failure during therapy with recombinant human gamma interferon. N. Engl. J. Med. **319:** 1397–1400.
116. LANG, R. A., D. METCALF, R. A. CUTHBERTSON, I. LYONS, E. STANLEY, A. KELSO, G. KANNOURAKIS, D. J. WILLIAMSON, G. K. KLINTWORTH, T. J. GONDA & A. R. DUNN. 1987. Transgenic mice expressing a hemopoietic growth factor gene (GM-CSF) develop accumulations of macrophages, blindness, and a fatal syndrome of tissue damage. Cell **51:** 675–686.
117. BENDTZEN, K., T. MANDRUP-POULSEN, J. NERUP, J. H. NIELSEN, C. A. DINARELLO & M. SVENSON. 1986. Cytotoxicity of human pI 7 interleukin-1 for pancreatic islets of langerhans. Science **232:** 1545–1547.
118. SANDLER, S., A. ANDERSON & C. HELLERSTRÖM. 1987. Inhibitory effects of interleukin 1 on insulin secretion, insulin biosynthesis, and oxidative metabolism of isolated rat pancreatic islets. Endocrinology **121:** 1424–1431.
19. BOSWELL, J. M., M. A. YUI, D. W. BURT & V. E. KELLEY. 1988. Increased tumor necrosis factor and IL-1 β gene expression in the kidneysof mice with lupus nephritis. J. Immunol. **141:** 3050–3054.
120. ROSENBERG, S. A., M. T. LOTZE, L. M. MUUL, S. LEITMAN, A. E. CHANG, S. E. ETTINGHAUSEN, Y. L. MATORY, J. M. SKIBBER, E. SHILONI, J.T. VETTO, C. A. SEIPP, C. SIMPSON & C. M. REICHERT. 1985. Observations on the systemic administration of autologous lymphokine-activated killer cells and recombinant interleukin-2 to patients with metastatic cancer. N. Engl. J. Med. **313:** 1485–1492.
121. WALDMANN, T. A. 1993. The IL-2/IL-2 receptor system: a target for rational immune intervention. Immunol. Today **14:** 264–270.
122. CAMPBELL, I. L., C. R. ABRAHAM, E. MASLIAH, P. KEMPER, J. D. INGLIS, M. B. A. OLDSTONE & L. MUCKE. 1993. Neurologic disease induced in transgenic mice by cerebral overeexpression of interleukin 6. Proc. Natl. Acad. Sci. USA **90:** 10061–10065.
123. DAVIES, A. O. & R. J. LEFKOWITZ. 1984. Regulation of β-adrenergic receptors by steroid hormones. Annu. Rev. Physiol. **46:** 119–130.

124. Friedmann, N., J. H. Exton & C. R. Park. 1967. Interaction of adrenal steroids and glucagon on gluconeogenesis in perfused rat liver. Biochem. Biophys. Res. Comm. **29:** 113–119.
125. Akahoshi, T., J. J. Oppenheim & K. Matsushima. 1988. Induction of high-affinity interleukin 1 receptor on human peripheral blood lymphocytes by glucocorticoid hormones. J. Exp. Med. **167:** 924–936.
126. Baker, J. B., G. S. Barsh, D. H. Carney & D. D. Cunningham. 1978. Dexamethasone modulates binding and action of epidermal growth factor in serum-free culture. Proc. Natl. Acad. Sci. USA **75:** 1882–1886.
127. Strickland, R. W., L. M. Wahl & D. S. Finbloom. 1986. Corticosteroids enhance the binding of recombinant interferon-γ to cultured human monocytes. J. Immunol. **137:** 1577–1580.
128. Gottschall, P. E., K. Koves, K. Mizuno, I. Tatsuno & A. Arimura. 1991. Glucocorticoid upregulation of interleukin 1 receptor expression in a glioblastoma cell line. Am. J. Physiol. **261:** E362–E368.
129. Snyers, L., L. De Wit & J. Content. 1990. Glucocorticoid up-regulation of high-affinity interleukin-6 receptors on human epithelial cells. Proc. Natl. Acad. Sci. USA **87:** 2838–2842.
130. Antoni, F. A., M. C. Holmes & J. Z. Kiss. 1985. Pituitary binding of vasopressin is altered by experimental manipulations of the hypothalamo-pituitary-adrenocortical axis in normal as well as homozygous (di/di) Brattleboro rats. Endocrinology **117:** 1293–1299.
131. Lutz-Bucher, B., K. Kovacs, G. Makara, E. Stark & B. Koch. 1986. Central nervous system control of pituitary vasopressin receptors: evidence for involvement of multiple factors. Neuroendocrinology **43:** 618–624.
132. Wynn, P. C., J. P. Harwood, K. J. Catt & G. Aguilera. 1985. Regulation of corticotropin-releasing-factor (CRF) receptors in the rat pituitary gland: effects of adrenalectomy on CRF receptors and corticotroph responses. Endocrinology **116:** 1653–1659.
133. Kuroda, Y., M. Mikuni, N. Nomura & K. Takahashi. 1993. Differential effect of subchronic dexamethasone treatment on serotonin-2 and beta-adrenergic receptors in the rat cerebral cortex and hippocampus. Neurosci. Lett. **155:** 195–198.
134. Giorgino, F., A. Almahfouz, L. J. Goodyear & R. J. Smith. 1993. Glucocorticoid regulation of insulin receptor and substrate IRS-1 tyrosine phosphorylation in rat skeletal muscle *in vivo.* J. Clin. Invest. **91:** 2020–2030.
135. Kosaki, A. & N. J. G. Webster. 1993. Effect of dexamethasone on the alternative splicing of the insulin receptor mRNA and insulin action in HepG2 hepatoma cells. J. Biol. Chem. **268:** 21990–21996.
136. Collins, S., M. G. Caron & R. J. Lefkowitz. 1988. β2-adrenergic receptors in hamster smooth muscle cells are transcriptionally regulated by glucocorticoids. J. Biol. Chem. **263:** 9067–9070.
137. Hadcock, J. R. & C. C. Malbon. 1988. Regulation of β-adrenergic receptors by "permissive" hormones: glucocorticoids increase steady-state levels of receptor mRNA. Proc. Natl. Acad. Sci. USA **85:** 8415–8419.
138. Hadcock, J. R., H.-Y. Wang & C. C. Malbon. 1989. Agonist-induced destabilization of β-adrenergic receptor mRNA. Attenuation of glucocorticoid-induced up-regulation of β-adrenergic receptors. J. Biol. Chem. **264:** 19928–19933.
139. Sakaue, M. & B. B. Hoffman. 1991. Glucocorticoids induce transcription and expression of the α1B adrenergic receptor gene in DTT1 MF-2 smooth muscle cells. J. Clin. Invest. **88:** 385–389.
140. Ros, M., J. K. Northup & C. C. Malbon. 1989. Adipocyte G-proteins and adenylate cyclase. Effect of adrenalectomy. Biochem. J. **257:** 737–744.
141. Fève, B., L. J. Emorine, M.-M. Briend-Sutren, F. Lasnier, A. D. Strosberg & J. Pairault. 1990. Differential regulation of β_1- and β_2-adrenergic receptor protein and mRNA levels by glucocorticoids during 3T3-F442A adipose differentiation. J. Biol. Chem. **265:** 16343–16349.
142. Saad, M. J., F. Folli, J. A. Kahn & C. R. Kahn. 1993. Modulation of insulin receptor,

insulin receptor substrate-1, and phosphatidyl 3-kinase in liver and muscle of dexamethasone-treated rats. J. Clin. Invest. **92:** 2065–2072.

143. Colotta, F., F. Re, M. Muzio, R. Bertini, N. Polentarutti, M. Sironi, J. G. Giri, S. K. Dower, J. E. Sims & A. Mantovani. 1993. Interleukin-1 type II receptor: a decoy target for IL-1 that is regulated by IL-4. Science **261:** 472–475.

144. Munck, A. 1976. General aspects of steroid hormone-receptor interactions. *In* Receptors and Mechanism of Action of Steroid Hormones, Part I. J. R. Pasqualini, Ed. 1–40. Marcel Dekker, Inc. New York.

General Discussion

A. MUNCK (*Dartmouth Medical School*): Are there questions about the last presentation or any other subject raised in this session?

E. M. STERNBERG (*National Institute of Mental Health*): Are you saying that there is evidence that as corticol levels go up or corticosterone levels go up, the mediator receptors also go up?

MUNCK: Yes, the model assumes this. There is a lot of experimental evidence for it with various receptors, such as that the interleukin 6 and the interleukin 1 receptors are induced by cortisol.

STERNBERG: This model seems to apply within certain given systems. But how would it apply within the HPA axis, the talmudic system?

MUNCK: There are many other kinds of mechanisms suggested for the permissive effects. One of these is to activate the G protein-coupled transmitter systems. So there are other ways to get the up-slope on the curve. I cannot apply this concretely to the HPA axis. To the immune system, it is a little easier. The gamma-interferon receptor is also induced by glucocorticoids, and there are data from the gamma-interferon system. It depends on which receptors are sensitive to glucocorticoids, and whether there is a receptor that is sensitive and a mediator that is suppressed. I do not happen to know whether glucocorticoids are necessary to induce ACTH receptors, CRH receptors, or any of those. But since glucocorticoids suppress those mediators if they were to induce their receptors, you would get exactly the same phenomenon that you are seeing here.

H. AKIL (*University of Michigan*): Within the brain what is happening would be made nonlinear by the fact that there are so many circuits that have the ability to both amplify and suppress. There are a lot of receptors which may or may not directly modulate anything, but may still complex repercussions on transmitters which in turn modulate CRH. This may put a monkey wrench into your beautiful mathematical model.

MUNCK: The point here was not to try to demonstrate the full complexity of any system; it is just that a very stripped-down system like this can already display fairly complex behavior.

QUESTION: I was pleased that you say the IL-6 receptor is induced, but I am not sure about IL-1. There are two IL-1 receptors, type 1 and type 2. Type 2 seems to be without a single transduction mechanism and is to my knowledge induced by dexamethasone. So it could be that IL-1 type 2 receptor could act like a functional antagonist to the IL-1 receptor, which has a single transduction mechanism and in that sense is in fact acting against the effects of IL-1. I would like to have your comment on that.

MUNCK: IL-1 is induced as well, yes.

QUESTION: But is the IL-1 type 1 or type 2? It could make a very big difference.

MUNCK: That I cannot tell. Sorry.

L. BRADY (*National Institute of Mental Health*): I have some comments on Dr. Barden's work on the neuroendocrine model of depression. I have been looking at the effects of different antidepressant drugs in naive animals over a longer time course. Over an approximately eight-week period I have seen an upregulation of mRNA expression in the hippocampus of the mineralocorticoid receptor with imipramine or fluoxetine. With α-andrenergic antagonists acting via a different pharmacologic mechanism we see an upregulation of GR. Over the eight weeks of administration we see a downregulation of CRH mRNA expression in the PVN. So

I think it does implicate these mechanisms in regulating the output of the HPA axis and they may be very important in mediating the therapeutic effects of antidepressant agents. The transgenic mouse model of Dr. Barden is very interesting.

F. A. ANTONI (*University of Edinburgh*): After all the criticism Dr. Munck has received, actually I would like to support him. There is reasonably good evidence in the HPA axis that the pituitary vasopressin receptor is upregulated by glucocorticoids. Two independent studies about eight years ago showed that vasopressin is reduced, though it is still undecided whether it is mediated by the mineralocorticoid or glucocorticoid receptor. Certainly that particular system in the HPA axis, the regulation of the pituitary vasopressin receptor versus the regulation of vasopressin release from the PVN parvocellular system, would support your mathematical model.

MUNCK: I am very glad to hear that. I did not know the vasopressin receptor story.

QUESTION: We had two excellent presentations today on the distribution of the glucocorticoid receptor and the localization in the nucleus. I know in the audience there are people who have published that it is in the nucleus all the time and does not reside in the cytoplasm. How specific is this property, whether it is within various tissues or cell types, that the receptor sits in the nucleus versus the cytoplasm? What sort of kinases are implicated if it is in the nucleus or in the cytoplasm, and how do we look at the shuttling?

MUNCK: Very briefly. The phosphorylation of the glucocorticoid receptor, which is probably the one in which this has been identified the best by my colleague Dr. Jack Bodwell, is entirely in the N-terminal domain of the receptor, the domain that was discussed before in switching between the glucocorticoid and the mineralocorticoid receptor. We do not know what happens in the mineralocorticoid receptor. There are two interesting phenomena. First, the kinases. Almost all the sites in the glucocorticoid receptor as well as the ones that are being identified in both the progesterone and the estrogen receptors are in consensus sequences for cell cycle-dependent kinases. That observation, and other studies particularly by Donald DeFranco and by John Cidlowski, prompted us to look at phosphorylation of the glucocorticoid receptor through the cell cycle. We have only looked at that carefully in one particular cell in which there are overexpressed glucocorticoid receptors, since we need overexpression in order to be able to measure phosphorylation adequately. The second phenomenon is that in the phase of the cell cycle where the cell is sensitive to glucocorticoids, late G1 and S phase, there is a low level of basal phosphorylation which substantially increases when hormone is added. In G2, which is where it has been shown that the cell is insensitive to glucocorticoids in their ability to induce enzymes, EGF receptors, and other proteins, there is a very high level of basal phosporylation. When hormone is added, it does not get any higher. So there is a big difference in phosphorylation between the sensitive and the insensitive phase of the cell cycle. That is the whole story so far.

QUESTION: Does this correlate with the ability of the receptor to translocate in the nucleus or not?

MUNCK: The question of nuclear translocation is still unanswered. Some have looked at translocation to the nucleus, DNA binding and so on and have found weakened or no translocation in the G2 or insensitive phase of the cycle. Both John Cidlowski and Donald DeFranco could say more about this than I can.

QUESTION: Dr. Antoni, have the studies on AVP receptor upregulation by glucocorticoids been validated in Brattleboro rats? You wondered if it was a direct or indirect effect.

ANTONI: I cannot remember whether Bernard Koch has done it. I do not think so.

The way these studies were done is that the hypothalamic input to the median eminence was cut with lesions either by complete de-afferentiation or by elimination of the paraventricular nucleus. That produced a marked reduction in vasopressin receptor number. They then treated the animals with dexamethasone (I think in the case of Bernard Koch's work with corticosterone). It could restore to normal the pituitary vasopressin receptor as measured by bind-and-grind assay.

COMMENT: Given Dr. Akil's cautionary tale about applying this to the brain, I am reminded of a review paper that Bruce McEwen published in *Biochemical Pharmacology*. If I am not mistaken, there may be some relationship here with serotonin in the hippocampus, because there are complimentary changes between glucocorticoids and 5HT1A receptors in the hippocampus and potentially their effects on serotonin synthesis and release. So there may be a neural example, and with regard to Dr. Akil's comment, it may be relevant to the brain as well.

Corticosteroids and Hippocampal Plasticity

BRUCE S. McEWEN

Laboratory of Neuroendocrinology
Rockefeller University
1230 York Avenue
New York, New York 10021

INTRODUCTION

Since the pioneering work of Selye,[1] adrenal steroids have been irrevocably linked to the concept of "stress" and its physiological and pathophysiological consequences. And yet, adrenal steroids do much more than mediate the effects of stress. They are part of a complex signalling system between the external environment, the brain, and the rest of the body; this system links waking and sleeping states, food-seeking and cognitive behaviors with metabolism of food and movement of cells of the immune system throughout the body.[2,3] Because of this role and the fact that the secretion of adrenal steroids is acutely sensitive to emotional and physical states that "stress" the body by pulling it away from its basal state, it is perhaps not so surprising that the hippocampal formation—part of the limbic system—is one of the principal target areas for adrenal steroids in the brain.

The hippocampus is a brain structure that is associated with aspects of learning and memory, especially those aspects related to spatial memory and "working" memory.[4] One way that this association manifests itself in relation to the link between waking activity and metabolism is that the anticipation of food normally increases adrenal steroid secretion, while the failure to receive food at the expected time increases secretion further,[5] unless the hippocampus is damaged.[6] One way of generalizing this phenomenon is to say that the hippocampus is activated when an event defies the *expectation* of what would occur.[7] Events are often stressful because they violate expectations,[8] and the stress-induced secretion of adrenal steroids is regulated, at least in part, through the activity of the hippocampal formation.[9]

Thus adrenal steroids are very relevant to the function of the hippocampus, even though, in 1968, when adrenal steroid uptake and retention was discovered in the hippocampus,[10] this was not known or even suspected. Since 1968, much has been learned about the hippocampus and about the receptor sites in it and in other brain regions, and many aspects of this unfolding story are being told in this volume. One aspect of adrenal steroid action in hippocampus is the structural and functional plasticity that is manifested in various ways in Ammon's horn and in the dentate gyrus. Such plasticity will be the subject of this brief review, along with a consideration of the functions of the two types of adrenal steroid receptors, type I and type II, that were discovered through the work of Funder and colleagues[11] and Reul and de Kloet.[12]

Adrenal Steroids and Hippocampal Neuronal Atrophy

Aus der Muhlen and Ockenfels[13] reported that ACTH or cortisone administration to mature guinea pigs caused neurons in hippocampus and other forebrain regions to stain darkly and appear necrotic, as if undergoing atrophy. While the interpretation of these findings may be questioned in light of more recent findings regarding artifacts that can develop in fixing brains for histological analysis,[14] they were instrumental in stimulating the work of Landfield and colleagues (Ref. 15 and this volume) which showed that aging in the rat results in some pyramidal neuron loss in hippocampus and that this loss was retarded by adrenalectomy in mid-life.[15] Robert Sapolsky then demonstrated that daily corticosterone (CORT) injections into young adult rats over 12 weeks mimicked the pyramidal neuron loss seen in aging.[16] He went on to demonstrate that excitatory amino acids play an important role in the cell loss by showing, first, that CORT exacerbates kainic acid-induced damage to hippocampus, as well as ischemic damage; and, second, that glucocorticoids potentiate excitatory amino acid-killing of hippocampal neurons in culture.[17]

In order to examine what actually happens to hippocampal neurons as a result of high levels of glucocorticoids, we used the single section Golgi technique to demonstrate that, after 21 d of daily CORT exposure, the apical dendritic tree of CA3 pyramidal neurons underwent atrophy.[18] Moreover, this atrophy was prevented by the anti-epileptic drug, phenytoin, given prior to CORT each day, thus implicating the release and actions of excitatory amino acids.[19] See FIGURE 1.

If CORT injections cause atrophy, would repeated stress do the same? Application of restraint stress (6 h/day for 21 d) produced the same pattern of atrophy as corticosterone treatment, namely, atrophy of apical, but not basal, dendrites of CA3c pyramidal neurons.[20] Such atrophy was also blocked by daily phenytoin administration, confirming that excitatory amino acid (EAA) release is involved.[19] Recent work showed that NMDA receptor blockade is also effective in preventing stress-induced dendritic atrophy (Magarinos and McEwen, unpublished). Thus EAA release is involved in both CORT-and stress-induced atrophy of CA3 pyramidal neurons; yet this does not necessarily mean that CORT is involved in stress-induced atrophy. CORT is released in response to restraint stress,[19] but in much lower amounts than that which is given exogenously to cause atrophy.[18]

In order to explore the role of endogenous CORT release in dendritic atrophy evoked by stress, we have used the steroid synthesis inhibitor, cyanoketone, to reduce the magnitude of CORT secretion in response to restraint stress.[21] Cyanoketone attenuates the adrenal steroid stress response but allows basal CORT secretion to occur.[22] We found that cyanoketone treatment blocked the atrophy of dendrites induced by daily restraint stress.[21] This result is consistent with a role for CORT secretion, but the story is not quite so simple because repeated restraint stress for 21 d does not consistently cause adrenal hypertrophy; moreover, the CORT response to repeated restraint stress habituates over time, showing a progressively earlier shut-off during the 6-h restraint session.[20] Thus the magnitude of CORT released in response to restraint stress is considerably smaller than that given in the exogenous CORT paradigm, and CORT probably plays a much smaller role in the atrophy produced by restraint stress. Nevertheless, it appears to play some role, and a possible link between CORT and EAA release is provided by the recent demonstration that adrenalectomy markedly reduces the magnitude of the EAA release evoked by restraint stress.[23] Future studies need to examine the role of cyanoketone in blocking EAA release during restraint stress, as further evidence for this link; and further work is needed to clarify the role of type I or type II adrenal steroid receptors in stress-induced atrophy, using antagonists or agonists applied systemically during stress to rats with adrenocortical insufficiency.

Another important issue is the relationship between the atrophy of CA3c neurons induced by repeated CORT treatment or by repeated restraint stress and the loss of pyramidal neurons that has been reported after both 12 weeks of CORT treatment

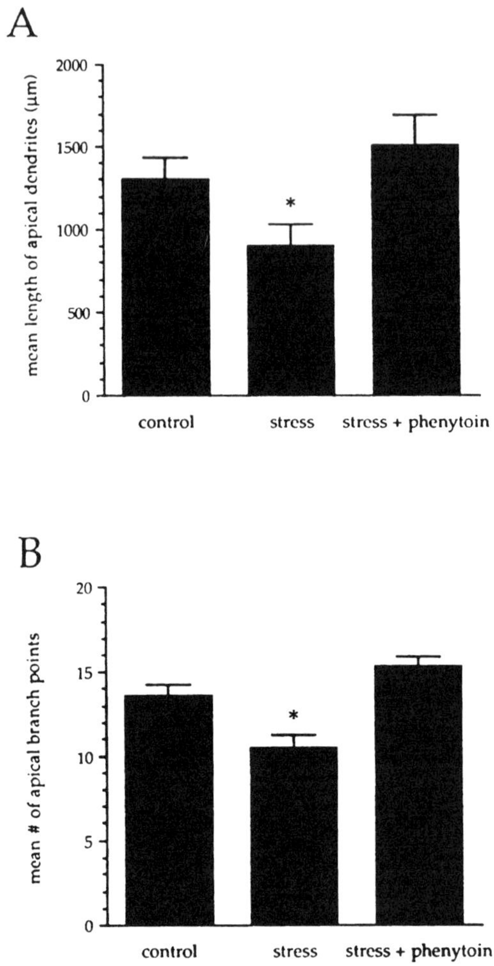

FIGURE 1. Effects of stress and phenytoin given prior to daily stress on **(A)** dendritic length and **(B)** dendritic branch points in CA3c pyramidal neurons of rat hippocampus (mean ± SEM). Significant difference from control, $p < 0.05$. (From Watanabe *et al.*[19] Reprinted by permission from *Hippocampus.*)

and severe social stress.[16,17] Indeed, we have seen that chronic social stress in tree shrews causes atrophy of CA3c pyramidal neurons, much as restraint stress does in rats (Magarinos, Fuxe, Flugge and McEwen, unpublished). However, the atrophy produced by stress in rats is reversible, within 7–14 d after the termination of stress (Magarinos and McEwen, unpublished). A possible link may be the fate of inhibitory interneurons that receive intense innervation from mossy fibers of dentate gyrus granule neurons and which are especially vulnerable to a variety of insults.[24] If some of these neurons were to die as a result of repeated restraint stress, then there might be a cumulative effect over time, in which repeated bouts of stress might progressively deplete the dentate gyrus of the buffering action that these inhibitory neurons appear to provide.

Does repeated stress impair the performance of behaviors that depend on the hippocampus? A recent study at Hunter College in New York demonstrated an impairment of performance on an 8-arm radial maze in rats that received 21 d of restraint stress prior to maze training.[25] This impairment was in the same direction, but not as great as impairment found in aging rats; moreover, stress effects were prevented by prior treatment of rats with phenytoin under the same conditions in which phenytoin was able to prevent the stress-induced atrophy of CA3 pyramidal neurons.[25] There are reports from MRI studies on human subjects that elevated cortisol levels, as in Cushing's syndrome, are associated with mild cognitive impairment and atrophy of hippocampal volume.[26,27] Therefore, future studies should include more detailed examination of the effects of elevated adrenal steroids and prolonged periods of stress on hippocampal volume and cognitive function in human populations as well as in experimental animals.

Adrenal Steroids and Neuronal Turnover in Dentate Gyrus

In contrast to pyramidal neurons of Ammon's horn, granule neurons of the adult dentate gyrus depend on adrenal steroids for their survival. Moreover, continuous replacement of granule neurons occurs in the adult dentate gyrus; and both neurogenesis and cell death increase following adrenalectomy (ADX) suggesting that it may be the death of neurons that signals the formation of replacement neurons.[28] The first clue as to the magnitude of the ADX effect on neuronal death in dentate gyrus was the finding that, three months after ADX of adult rats, some rats showed almost total loss of dentate gyrus granule neurons.[29] This finding attracted considerable attention because it conflicted with the prevalent view of stress- and glucocorticoid-induced neuronal death in Ammon's horn (see above); and it was also very puzzling because only some ADX rats showed the loss of the entire dentate gyrus.

We now have more information that allows us to better understand these events, based on the unique nature of adrenal steroid actions in dentate gyrus. ADX has been shown to induce apoptotic death of dentate granule neurons within 3–7 d in virtually every rat from which all adrenal tissue has been removed.[30] This phenomenon appears to be an acceleration of neuronal turnover in the dentate gyrus, which arises postnatally and continues to generate new neurons in adult life.[29] The loss of the entire dentate gyrus only in some rats may well be explained by the absence of accessory adrenal tissue in those rats; when not removed at the time of ADX, this tissue can supply enough adrenal steroids to prevent neuronal loss. Very low levels of adrenal steroids, sufficient to occupy type I adrenal steroid receptors, completely block dentate gyrus neuronal loss.[31] See FIGURE 2. The role of type I and type II adrenal steroid receptors will be discussed below.

Adrenalectomy is not a natural event, and thus one must consider the physiological role of this type of plasticity. One explanation for why the dentate gyrus can make new cells in adult life, as well as get rid of them, is that the hippocampus may change during adult life in relation to its function in spatial orientation and memory.[32] Birds that use space around them to hide and locate food, and voles as well as deer mice that use their environment to find mates have larger hippocampal volumes than species which do not; moreover, there are indications that hippocampal volume may change under some circumstances, *e.g.*, with season of the year.[32,33] It remains to be seen whether it is the dentate gyrus which changes under these changing conditions.

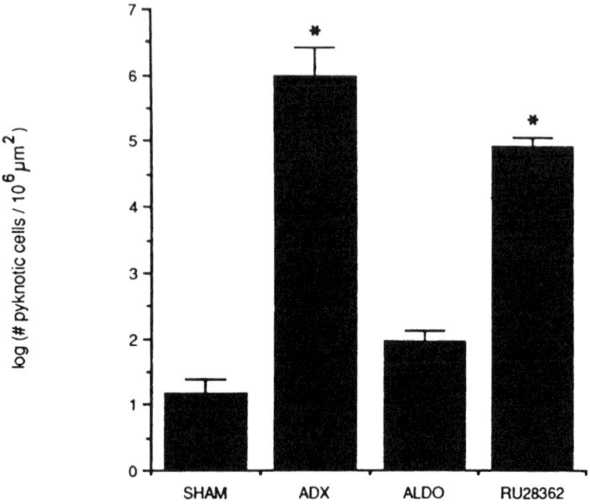

FIGURE 2. Pyknotic cells in dentate gyrus granule cell layer (log # pyknotic cells/10^6 μm^2 ± SEM) of adult rats that were either sham operated, ADX, or ADX and given either ALDO or RU28362. Significant difference from Sham and ALDO, $p < 0.001$. (From Woolley *et al.*[31] Reprinted by permission from *Brain Research.*)

Another important question concerns the impairment of hippocampal function that may accompany adrenocortical insufficiency. Several reports have indicated that long-term ADX rats that have damage to the dentate gyrus like that described by Sloviter *et al.*[29] show modest deficits in spatial memory.[34,35] Recent work in New York has revealed that deficits in performance of an 8-arm radial maze task is evident in rats ADX for ca 70 d, without there being evidence for ongoing neuronal loss or degeneration in the dentate gyrus.[36] One possibility, currently being tested (P. Vaher, personal communication) is that there are acute actions of adrenal steroids on brain function that are not manifested in terms of dentate neuronal turnover. These actions are potentially relevant to the acute actions of adrenal steroids on long-term potentiation that will now be described.

Adrenal Steroids and Long-Term Potentiation

A single burst of high frequency stimulation can immediately alter the responsiveness of neurons to further stimuli, an effect lasting over a time course of many hours to days. This type of plasticity is called long-term potentiation (LTP), and a related phenomenon is called primed-burst potentiation (PBP). Recent studies in a number of laboratories indicate that LTP in hippocampus is regulated by adrenal steroids in a biphasic manner. A number of *in vitro* studies have demonstrated in the hippocampal CA1 field that acute stress produces an impairment in LTP or in PBP.[37,38] In these studies, a negative correlation was also seen between the degree of induced LTP or PBP and plasma CORT levels. More recently, an inverted U-shaped curve was found with regard to induction of PBP in CA1 in relation to circulating CORT levels.[37] That is, both adrenocortical insufficiency as well as high levels of CORT had a detrimental effect on PBP, whereas optimal potentiation was obtained within an intermediate range (10–20 ug/dl) of circulating CORT.

Within the dentate gyrus, an acute administration of CORT has also been shown to produce a decrement in LTP (see Ref. 38). More recent data, however, indicates that LTP can be modulated biphasically by adrenal steroids acting, respectively, via type I and type II receptors. In the dentate gyrus of ADX rats, LTP was found to be enhanced within an hour by adrenal steroids acting via type I receptors, whereas type II receptor stimulation in ADX rats was found to rapidly suppress LTP.[39] These biphasic effects may help to explain the biphasic dose-response curve with respect to induction of PBP in the CA1 region of Ammon's horn, noted above, in which low levels of CORT at the nadir and rising phase of daily adrenocortical activity facilitate PBP, whereas stress levels of corticosterone inhibit PBP.[37]

The CA1 responds in a similar manner as the dentate gyrus to corticosteroids. This should not be surprising, for at least two reasons. First, it is very likely that type I and type II receptors coexist in pyramidal neurons and in dentate gyrus granule neurons.[40] Second, in CA1 pyramidal neurons, corticosteroids biphasically modulate excitability: *i.e.*, type I receptor stimulation facilitates excitability by disinhibiting a 5-HTIA receptor input; whereas type II receptor stimulation inhibits excitability by suppressing a beta receptor-mediated noradrenergic input.[41] It remains to be seen whether biphasic actions of adrenal steroids on LTP and PBP may be a general feature of the hippocampal formation.

Future studies will explore the structural and neurochemical changes that accompany the steroid enhancement of LTP and PBP, including the induction of immediate early genes and possible changes in structural protein and neurotrophin gene expression. New data from studies of LTP in the dentate gyrus of awake, freely-behaving rats, indicates that several forms of the immediate early gene, JUN, are induced by LTP in both dentate gyrus and Ammon's horn, whereas FOS is not induced; moreover, enhancement of LTP by ALDO administration lasts for at least 24 h, at which time it is still markedly higher than ADX rats given only vehicle treatment before LTP induction (Pavlides, unpublished).

Role of Adrenal Steroid Receptor Subtypes in Structural and Neurochemical Plasticity

It is evident from the distribution of mRNA and binding for type I and type II adrenal steroid receptor subtypes that neurons in dentate gyrus and Ammon's horn most likely contain both types of receptors (*e.g.*, Ref. 40). Very recently we began to learn some of the functions that these two receptors mediate, such as the enhancement of LTP produced via type I receptor activation and the suppression of LTP produced via type II receptor activation; and the containment of the dentate gyrus neuronal turnover mediated by type I receptors.

Ongoing work in our laboratory has revealed other specific effects mediated by adrenal steroid receptor types in the hippocampus and other brain regions. In these studies we applied steroids via subcutaneously-implanted Alzet minipumps. The type II agonist RU28362 (362) was administered in a dose (10 ug/h) that suppresses almost to zero the CRHmRNA signal in the paraventricular nuclei (PVN); the type I agonist aldosterone (ALDO) was given at a dose of 10 ug/h and it had no effect on CRHmRNA levels in PVN.[42] In the hippocampus of the same animals, ALDO prevented the upregulation of 5-HTIA receptors that was produced by ADX, whereas 362 was without any effects.[43] Other work in progress has revealed that ADX-induced decreases in dynorphin mRNA in dentate gyrus and in kainate receptors in the stratum lucidum of CA3 (representing the mossy fiber zone of terminals from dentate granule neurons) are reversed by ALDO treatment but not by 362. Thus most of the known actions of adrenal steroids in the hippocampus appear to be mediated by type I receptors. Exceptions are the type II-mediated inhibition of LTP, noted above, and the involvement of type II receptors in the excitotoxin-induced damage to hippocampal neurons in culture.[44]

SUMMARY

The unexpected discovery in 1968 that the hippocampus takes up and retains adrenal steroids has led to equally unanticipated findings regarding the actions of hormones on the brain and the ways in which the brain is capable of changing in response to the hormonal milieu. First were indications that adrenal steroids adversely affect pyramidal neurons of the hippocampus and cause damage during aging and as a result of severe and prolonged stress. Atrophy of dendrites, particularly in the CA3 region, appears to be an early index of these effects. Second was evidence that the dentate gyrus undergoes atrophy and granule neuron death after adrenalectomy; perhaps as a result of this neuronal death, neurogenesis is stimulated in dentate gyrus of adult rats. Third are recent indications that excitability of hippocampal neurons, including the ability to generate long-term potentiation (LTP), is regulated biphasically by adrenal steroids. One important goal of current research is to understand the role of type I and type II receptors for adrenal steroids in hormonally-induced hippocampal plasticity. Type I receptors appear to play a role in containing programmed-cell death and the rate of neurogenesis; they also regulate key neurochemical features of dentate gyrus and Ammon's horn; and they facilitate LTP. Present information indicates that type II receptors inhibit LTP and may play a role in the degenerative changes in Ammon's horn.

REFERENCES

1. Selye, H. 1941. The anesthetic effect of steroid hormones. Proc. Soc. Exp. Biol. Med. **46:** 116–121.
2. McEwen, B. S., R. L. Spencer & R. R. Sakai. 1993. Adrenal steroid actions upon the brain: versatile hormones with good and bad effects. *In* Hormonally Induced Changes in Mind and Brain, J. Schulkin, Ed. Academic Press. New York.
3. Ottaway, C. & A. J. Husband. 1992. Central nervous system influences on lymphocyte migration. Brain. Behav. Immun. **6:** 97–116.
4. Eichenbaum, H. & T. Otto. 1992. The hippocampus—what does it do? Behav. Neural Biol. **57:** 2–36.
5. Levine, S., J. Weinberg & L. P. Brett. 1979. Inhibition of pituitary-adrenal activity as a consequence of consummatory behavior. Psychoneuroendocrinology **4:** 275–286.
6. Osborne, B. 1986. Behavioral and adrenal responses and meal expectancy in rats with fornix transection. Physiol. Behav. **37:** 499–502.
7. Gray, J. 1982. Precis of the neuropsychology of anxiety: an enquiry into the functions of the septo-hippocampal system. Behav. Brain Sci. **5:** 469–534.
8. McEwen, B. S. & E. Stellar. 1993. Stress and the individual: mechanisms leading to disease. Arch. Internal Med. **153:** 2093–2101.
9. Jacobson, L. & R. Sapolsky. 1991. The role of the hippocampus in feedback regulation of the hypothalamic-pituitary-adrenocortical axis. Endocr. Rev. **12:** 118–134.
10. McEwen, B. S., J. M. Weiss & L. Schwartz. 1968. Selective retention of corticosterone by limbic structures in rat brain. Nature **220:** 911–912.

11. FUNDER, J. W., D. FELDMAN & I. S. EDELMAN. 1973. The roles of plasma binding and receptor specificity in the mineralocorticoid action of aldosterone. Endocrinology **92:** 994–1004.
12. REUL, J. M. & E. R. DE KLOET. 1985. Two receptor systems for corticosterone in rat brain: microdistribution and differential occupation. Endocrinology **117:** 2505–2511.
13. AUS DER MUHLEN, K. & H. OCKENFELS. 1969. Morphologische veranderungen im diencephalon und telencephalon: storungen des regelkreises adenohypophysenebennierenrinde. Z. Zellforsch. Mikrosck. Anat. **93:** 126–141.
14. CAMMERMEYER, J. 1978. Is the solitary dark neuron a manifestation of postmortem trauma to the brain inadequately fixed by perfusion? Histochemistry **56:** 97–115.
15. LANDfiELD, P. 1987. Modulation of brain aging correlates by long-term alterations of adrenal steroids and neurally-active peptides. Prog. Brain Res. **72:** 279–300.
16. SAPOLSKY, R., L. KREY & B. S. McEWEN. 1985. Prolonged glucocorticoid exposure reduces hippocampal neuron number: implications for aging. J. Neurosci. **5:** 1222–1227
17. SAPOLSKY, R. 1992. Stress, the Aging Brain and the Mechanisms of Neuron Death. The MIT Press. Cambridge, MA. 423 pp.
18. WOOLLEY, C., E. GOULD & B. S. McEWEN. 1990. Exposure to excess glucocorticoids alters dendritic morphology of adult hippocampal pyramidal neurons. Brain Res. **531:** 225–231.
19. WATANABE, Y., E. GOULD, H. CAMERON, D. DANIELS & B. S. McEWEN. 1992. Phenytoin prevents stress- and corticosterone-induced atrophy of CA3 pyramidal neurons. Hippocampus **2:** 431–436.
20. WATANABE, Y., E. GOULD & B. S. McEWEN. 1992. Stress induces atrophy of apical dendrites of hippocampal CA3 pyramidal neurons. Brain Res. **588:** 341–345.
21. MAGARINOS, A. & B. S. McEWEN. 1993. Blockade of corticosterone synthesis by cyanoketone prevents chronic stress-induced dendritic atrophy of hippocampal neurons in the rat. Abstr. Soc. Neurosci. **17:** # 71.2, p. 168.
22. AKANA, S., J. SHINSAKI & M. DALLMAN. 1983. Drug-induced adrenal hypertrophy provides evidence for reset in the adrenocortical system. Endocrinology **113:** 2232–2237.
23. LOWY, M. T., L. GAULT & B. K. YAMAMOTO. 1993. Adrenalectomy attenuates stress-induced elevations in extracellular glutamate concentrations in the hippocampus. J. Neurochem. **61:** 1957–1960.
24. HSU, M. & G. BUZSAKI. 1993. Vulnerability of mossy fiber targets in the rat hippocampus to forebrain ischemia. J. Neurosci. **13:** 3964–3979.
25. VILLEGAS, M., C. MARTINEZ, V. N. LUINE & B. S. McEWEN. 1993. Chronic stress impairs spatial memory in rats. Behav. Brain Res. In press.
26. STARKMAN, M., S. GEBARSKI, S. BERENT & D. SCHTEINGART. 1992. Hippocampal formation volume, memory dysfunction, and cortisol levels in patients with Cushing's syndrome. Biol. Psychiatry **32:** 756–765.
27. AXELSON, D., P. DORAISWAMY, W. McDONALD, O. BOYKO, L. TYPLER, L. PATTERSON, C. B. NEMEROFF, E. H. ELLINWOOD, JR. & K. R. R. KRISHAN. 1993. Hypercortisolemia and hippocampal changes in depression. Psychiat. Res. **47:** 163–173.
28. GOULD, E. & B. S. McEWEN. 1993. Neuronal birth and death. Curr. Opin. Neurobiol. **3:** 676–682.
29. SLOVITER, R. S., G. VALIQUETTE, G. M. ABRAMS, C. RONK, A. I. SOLLAS, L. A. PAUL & S. A. NEUBORT. 1989. Selective loss of hippocampal granule cells in the mature rat brain after adrenalectomy. Science **243:** 535–538.
30. GOULD, E., C. S. WOOLLEY & B. S. McEWEN. 1990. Short-term glucocorticoid manipulations affect neuronal morphology and survival in the adult dentate gyrus. Neuroscience **37:** 367–375.
31. WOOLLEY, C., E. GOULD, R. SAKAI, R. L. SPENCER & B. S. McEWEN. 1991. Effects of aldosterone or RU28362 treatment on adrenalectomy-induced cell death in the dentate gyrus of the adult rat. Brain Res. **554:** 312–315.
32. SHERRY, D. F., L. F. JACOBS & S. J. GAULIN. 1992. Spatial memory and adaptive specialization of the hippocampus. Trends Neurosci. **15:** 298–303.
33. GALEA, L. A. M., M. KAVALIERS, K.-P. OSSENKOPP, D. INNES & E. L. HARGREAVES.

1994. Sexually dimorphic spatial learning varies seasonally in two populations of deer mice. Brain Res. **635:** 18–16.
34. ARMSTRONG, J. D., D. C. MCINTYRE, S. NEUBORT & R. S. SLOVITER. 1993. Learning and memory after adrenalectomy-induced hippocampal dentate granule cell degeneration in the rat. Hippocampus **3:** 359–371.
35. CONRAD, C. D. & E. J. ROY. 1993. Selective loss of hippocampal granule cells following adrenalectomy: implications for spatial memory. J. Neurosci. **13:** 2582–2590.
36. VAHER, P., V. N. LUINE, E. GOULD & B. S. MCEWEN. 1993. Spatial memory deficit after adrenalectomy in rats. Abstracts Soc. Neurosci. **17:** in press.
37. DIAMOND, D. M., M. C. BENNET, M. FLESHNER & G. M. ROSE. 1992. Inverted-U relationship between the level of peripheral corticosterone and the magnitude of hippocampal primed burst potentiation. Hippocampus **2:** 421–430.
38. PAVLIDES, C., Y. WATANABE & B. S. MCEWEN. 1993. Effects of glucocorticoids on hippocampal long-term potentiation. Hippocampus **3:** 183–192.
39. PAVLIDES, C., Y. WATANABE, A. M. MAGARINOS & B. S. MCEWEN. 1992. Effects of glucocorticoid type I and type II agonists on hippocampal long-term potentiation. Abstracts, Soc. Neurosci. **18:** # 148.2, p. 343.
40. HERMAN, J., P. PATEL, H. AKIL & S. J. WATSON. 1989. Localization and regulation of glucocorticoid and mineralocorticoid receptor mRNAs in the hippocampal formation of the rat. Mol. Endocrinol. **3:** 1886–1894.
41. JOELS, M. & E. R. DE KLOET. 1992. Control of neuronal excitability by corticosteroid hormones. Trends Neurosci. **15:** 25–30.
42. ALBECK, D. S., N. B. HASTINGS & B. S. MCEWEN. 1994. Effects of adrenalectomy and type I or type II glucocorticoid receptor activation on AVP and CRH mRNA in the rat hypothalamus. *Mol Brain Res* in press.
43. KURODA, Y., Y. WATANABE, D. S. ALBECK, N. B. HASTINGS & B. S. MCEWEN. 1994. Effects of adrenalectomy and type I and type II adrenal steroid receptor activation on 5-HTIA and 5-HT2 receptor binding and 5-HT transporter mRNA expression in rat brain. Brain Res, **648:** 157–161.
44. PACKAN, D. R. & R. M. SAPOLSKY. 1990. Glucocorticoid endangerment of the hippocampus: tissue, steroid and receptor specificity. Neuroendocrinology **51:** 613–618.

DISCUSSION OF THE PAPER

QUESTION: Is apoptosis involved in the brain neurons?

MCEWEN: All we know is that the appearance of the neurons, the grape-like clusters of condensed chromatin, has some resemblance to what has been described in, say, a mature thymus lymphocyte after glucocorticoid exposure. And everything would suggest some kind of cell death gene that is perhaps inhibited by the presence of activated mineralocorticoid receptors. But what that gene is, we do not know. Our present view is that it is cell death that triggers cell division. There is no evidence that mineralo- and glucocorticoid receptors are pressed before the early thymidine-labelled cells express neuron-specific markers. So probably those cells do not respond directly to steroids but to some other signal from the dying neuron or from some adjacent glial cells, maybe from the radial glia.

M. JOËLS (*University of Amsterdam*): You propose that CA3 cells lose distal dendrites due to an increased excitatory amino acid transmission from dentate gyrus. If that is true, one would expect that all neurons suffer if they receive amino acid transmission and have GRs. There are quite a lot of neurons which fall into that category. But apparently that is not the case, because it seems to be quite specific for CA3 cells. Therefore, in this respect, is it not very important that these CA3 cells are quite special because they have only MRs? Also, you assumed that high CORT levels would give an increased amount of amino acids or an increased response to amino acid input. Dr. Landfield mentioned earlier, and I totally agree with him, that high CORT depresses the response to excitatory amino acid input. But the CA3 cells do not have glucocorticoid receptors to give this depression. So the loss of distal dendrites may be due to the lack of GRs in these cells and thus lack of depression of the amino acid response.

McEWEN: Those are two very good comments, because certainly the mossy fiber input is anatomically and neurochemically very different from any other excitatory input within the hippocampus, and I have been thinking more in terms of how the presynaptic input could be so powerful. I appreciate your point about the reduced amount of glucocoticoid receptor in the CA3 neuron, which would then put some of the emphasis at least on how the CA3 neuron is handling the situation.

JOËLS: I also wondered to what extent other things like calcium influx through voltage-gated channels are important in this respect, because phenytoin that was protective, is also a calcium blocker.

McEWEN: Good point, too. Thank you.

H. AKIL (*University of Michigan*): I have always thought about hippocampus as almost a straight-through system from dentate gyrus via CA3 to CA1. While there is a lot of collateralization and interaction, it is not an opposing system, and the individual components are not that independent of each other. Yet we constantly see, and you report also, results where one area will get remodeled and another will not. Or where mRNA will change in one place and not in another. These are very distinctive profiles of activity and responsiveness, which speak against a homogeneous patterning of response by a glucocorticoid. What do you think all this is doing as a net effect on some function that you think the hippocampus is performing?

McEWEN: I wish I knew. However, referring only to the three-cell circuit is a vast oversimplification of hippocampal anatomy—the interactions between different cell groups, the fact that there is a direct perforant path input to the CA3, and so on. There are many aspects that we are not even considering so far in all of this. Also, a whole group of studies suggests that there are important differences functionally between dorsal and ventral hippocampus. We are fortunate to have something that we can study given the greater level of complexity that we shall eventually have to address.

QUESTION: Other than at the extremes when the animal is bathing in steroids or not at all when cells are dying, but from moment to moment, do you really think that this all translates into something functionally very important?

McEWEN: I like the view of hippocampal function as posed by people like Jeffrey Gray saying the hippocampus comes on line when something happens in an animal's experience that defies what it expects to happen. If nothing defies expectation things go on more or less normally. This is true of working memory. I am not sure how we can fit spatial memory into it, but certainly that could be a separate function. I think the hippocampus may be particularly important when expectation is mismatched with what actually is going on.

P. EPSTEIN (*Chicago*): The idea of dentate gyrus changes having something to do with seasonal changes is interesting. Are there some specific observations or work that has been done which would corroborate that?

McEWEN: Polygamous voles show, at least in the males, an enlargement of hippocampal volume around the mating season. No one knows if this is the dentate gyrus that is changing or simply pyramidal neurons that are getting larger.

S. LEVINE (*Stanford University Medical School*): It seems to me a glorious paradox that there is a certain time in development when clearly the whole machinery of the elevation of CORT is actively being inhibited, at which point there is a very high amount of cell death. And at some point later in life, glucocorticoids have the function of causing cell death. I find this difficult to put together.

McEWEN: If you are referring to the adrenalectomy-induced cell death, that probably has a very special role like the seasonal change, which is still very hypothetical. The loss of pyramidal neurons happens under very severe conditions over long periods of time as in aging. The reason it does not happen more readily is that a series of checks and balances are built in. I even suggested that perhaps the atrophy is part of some kind of weird protective mechanism. These extreme conditions are not going on all the time. The turnover of neurons in the dentate gyrus serves a very specific purpose, which we have not identified yet. The main changes in the hippocampus with increasing levels of glucocorticoids are going to be more at the electrophysiological level implying short-term changes and perhaps depression of cognitive function, which are entirely reversible.

LEVINE: But there are periods of time when certain types of receptors are probably highest at some point in development and are then never seen again. We know that GR appear in some areas early in development and never appear again. This is all going on during a period of time when the HPA axis is markedly inhibited.

McEWEN: In terms of the development and the stress-hypo-responsive period, one certainly realizes that it is not just the absence of adrenal steroids but also the containment through excitatory amino acid and the GABA receptors that adds another dimension to this, because electrical activity can also then, presumably, regulate neurogenesis and neural destruction. And the regulation of the expression of the NMDA receptor is another important event.

E. C. AZMITIA (*New York University*): I am troubled over equating the presence of pyknotic cells with cell death. What is the evidence you are basing this on, since the cell shows pyknotic bodies two or three days or a week after ADX? Do those neurons actually die?

McEWEN: If you use a so-called degeneration stain to look at neurons you see evidence of degeneration granule neurons very clearly in a three- to seven-day adrenalectomized animal and perhaps at longer intervals. These are very clearly the older neurons in the outer part of the granule cell layer, whereas the new neurons are in the inner part. The degeneration stain is something that many people would expect as further evidence for actual degeneration and death of the cells.

AZMITIA: We too are going to present data in our paper showing that if you wait long periods of time, it appears that these cells are "dead." If you then treat with steroids, you can bring them back, almost to their normal levels. The amount of neurogenesis that occurs is really minimal.

McEWEN: In terms of that particular phenomenon, but it is an ongoing process. Perhaps in those animals it proceeds at a lower rate if there has been some accessory adrenal tissue. That may be something we can talk about.

B. BOHUS (*University of California, Irvine*): About the short-term changes in LTP and LTD, your data remind me of one of our experiments about ten years ago with Ronald de Kloet and George Cottrel. We published data on adrenalectomy and kindling of the hippocampus, another model of neuronal plasticity in the brain. It was surprising that we got a complex but very definitive effect of adrenalectomy which disappeared after three days. If we gave corticosterone, the post-ADX changes were normalized. So the time period of the ADX effect of LTP almost fits the kindling data. However, I am puzzled with the possible mechanism.

McEWEN: I would refer you to Marian Joëls. She goes much more into the mechanistic details, whereas I was describing an overall phenomenon.

Gene Products of Corticosteroid Action in Hippocampus[a]

NANCY R. NICHOLS[b] AND CALEB E. FINCH

Andrus Gerontology Center
and
Department of Biological Sciences
University of Southern California
Los Angeles, California 90089-0191

INTRODUCTION

Glucocorticoids mediate cellular adaptation to stress by a variety of mechanisms including receptor-mediated changes in gene expression.[1–4] Many studies on corticosteroids and stress in the brain have focused on factors that regulate the hypothalamic-pituitary-adrenal (HPA) axis and hence the output of glucocorticoids, or on negative feedback regulation of the HPA axis by glucocorticoids.[4–12] Few studies have addressed the role of changes in gene expression in glucocorticoid target cells in the brain following activation of the HPA axis and elevation of circulating glucocorticoids. However, the basis of physiological changes mediating restoration of homeostasis after stress or injury to the brain must reside with the integration of neural, endocrine, cellular and molecular signals. Additional mechanisms for adaptation to stress might be revealed by examining patterns of gene expression in brain in response to glucocorticoid manipulation.

The hippocampus is both a glucocorticoid target tissue with a high concentration of receptors and a site of negative feedback regulation of the HPA axis.[4,9,13,14] Glucocorticoid secretion is necessary for effects on neurotransmitter synthesis,[4,14] normal development of the dentate gyrus,[15] and control of neuronal excitability[16] in the rat hippocampus. These cellular effects of glucocorticoids mediate behavioral adaptation and can oppose the effects of stress on neuronal activity.[17] However, chronic high levels of glucocorticoid and chronic stress have been linked to neurodegenerative changes in pyramidal cell layers of the rat hippocampus similar to those that occur during aging.[18,19] In contrast, the removal of corticosteroids by adrenalectomy (ADX) induces neurodegeneration and apoptosis in the adult dentate gyrus.[20–22] Therefore, the responses of glucocorticoid-regulated mRNAs following brain lesioning and during aging were investigated to identify potential markers for studies on target responses to glucocorticoids during age-related neurodegeneration.

Isolation and Identification of Glucocorticoid-Regulated mRNAs from Adult Rat Hippocampus

Messenger RNAs that are responsive to corticosterone (CORT) in rat hippocampus were identified using a combination of two methods. For these experiments,

[a]This work was supported by the Office of Naval Research Grant No. N00014-85-K-0770, The John D. and Catherine T. MacArthur Foundation Network in Successful Aging, PHS Grant AG 07909, and The Brookdale Foundation.

[b]Address correspondence to: Nancy R. Nichols, Ph.D., Andrus Gerontology Center, University of Southern California, Los Angeles, CA 90089-0191.

hippocampal RNA was isolated from rats that were ADX for three days and injected subcutaneously with 1 ml oil as vehicle or 40 mg/kg CORT in oil for three days. Initially, two-dimensional (2-d) gel analysis of *in vitro* translation products[23] was used to describe select changes in four poly(A)-containing RNAs, coding for 50-, 35-, 33-, and 20-kd polypeptides, in response to CORT.[24] The results of the 2-d gel analysis were used to evaluate different animal experiments for reproducible responses to CORT against a background of similar nonchanging *in vitro* translation products. Two sets of samples were then pooled to make a cDNA library in λgt10 from which glucocorticoid-regulated mRNAs were cloned by differential hybridization with probes made from ADX or CORT-treated rats.[25] Primary screening of about 100,000 clones yielded twenty positive clones after tertiary screening. Since poly(A)-containing RNAs from ADX and CORT-treated rats were mixed to make the hippocampal cDNA library, both increases and decreases in response to CORT were isolated by this cloning method (TABLE 1). The CORT-responsive clones were subcloned into a plasmid vector for sequencing and further characterization of their glucocorticoid response characteristics. Eleven clones were analyzed by partial or complete DNA sequencing to determine homology with known genes in nucleic acid sequence databases. Sequence analysis indicated that four of these were redundant clones;[25] seven unique sequences and their gene homologies (if found) are listed in TABLE 1.

We also used these clones in hybrid selection/*in vitro* translation experiments to compare them to the previous CORT-responsive mRNAs described by 2-d gel analysis of *in vitro* translation products. By comparing physicochemical characteristics of the polypeptides and comigration on 2-d gels of hybrid-selected and CORT-

TABLE 1. Identification of CORT-Responsive Clones/Translation Products

Clone	Gene[a]	Translation Product Mol. Wt.	pI	Direction of Response to CORT	Intact/ADX Fold-Change
CR1	glutamine synthetase	43 kd[b]	6.7	↑	not done
CR3	glycerol phosphate dehydrogenase	35 kd[b,c]	6.3	↑	2
CR16	unknown	none found		↑	2
CR36	unknown	not done		↑	not done
CR62	unknown	42–46 kd[b]	5.1–5.4	↑	3
	unknown	33 kd[c]	6.5	↑	>10[d]
	unknown	20 kd[c]	6.9	↑	3
CR46	glial fibrillary acidic protein	50 kd[b,c]	5.5	↓	0.4
CR90	transforming growth factor-β1	none found		↓	0.6

[a]DNA sequences were determined using the modified Sanger chain termination method[25] for comparison with known genes in Genbank. Sequence homologies were based on partial sequences except for TGF-β1, which was extensively sequenced,[28] and CR16, which was completely sequenced.[29]

[b]mRNAs were selected from total RNA using immobilized cDNA clones and hybrid selection as described in Nichols *et al.*,[26] translated *in vitro* and translation products were resolved on 2-d gels.[23]

[c]CORT-responsive mRNAs were translated *in vitro* and resolved on 2-d gels.[22]

[d]The 33-kd polypeptide was not detected on 2-d gels of hippocampal mRNA translation products from ADX rats.

responsive *in vitro* translation products, two of four polypeptides were identified as glial fibrillary acidic protein (GFAP)[26] and glycerol phosphate dehydrogenase (GPDH);[27] two others are unidentified (TABLE 1). In addition, a clone for glutamine synthetase (GS) selected a mRNA coding for a 43-kd, pI 6.7 and a clone (CR62) for an unknown gene selected a 42-kd, pI 5.4 polypeptide and 4 additional charge variants, ranging from 43 to 46 kd and pI 5.3 to 5.1 (TABLE 1). Although the clones for transforming growth factor-β1 (TGF-β1) and CR16 were analyzed in similar experiments, no translation products were detected. These results may be due to the low prevalence of TGF-β1 mRNA[28,29] and the very basic pI of the predicted amino acid sequence of CR16 (Masters, Nichols & Finch, in preparation), which could have precluded their detection by the 2-d gel system used for this analysis.[23]

Glucocorticoid Response Characteristics of Hippocampal mRNAs

The clones mRNAs and mRNA translation products were analyzed in a series of experiments to determine the following glucocorticoid response characteristics: an acute time course (2, 8, 32 h after CORT treatment); dose response (0.2, 1.0 and 10.0 mg CORT for 8 h); receptor specificity (RU28362, CORT, aldosterone, and 5α-dihydrotestosterone treatment for 8 h); ADX (3 and 7 days) response compared with a.m. or p.m. intact rats.[23,26,27,29,30] From these data the following profile emerged for most responses: increase or decrease within 2 to 8 h of CORT treatment, maximum response between 1 and 10 mg CORT, GR (type II low affinity glucocorticoid receptor) specificity and opposite response to ADX compared with that to CORT. In addition, all the hippocampal mRNA responses to CORT also occurred in other brain regions, including the cortex, striatum, hypothalamus, and cerebellum. The responses to ADX indicate that these mRNAs are normally under positive or negative regulation by CORT in intact rats (TABLE 1). These results, together with the CORT replacement and receptor specificity data, indicate that the mRNAs are under physiological regulation by glucocorticoids.

A notable exception to this profile is CR16 mRNA which was expressed in select neurons of the forebrain, but not in the cerebellum (Masters, Nichols & Finch, in preparation). CR16 mRNA also increased in a.m. intact rats at the nadir of the circadian variation compared with ADX rats[31] and plateaued between 4 and 80 ng/ml CORT, indicating MR (type I high-affinity mineralocorticoid receptor) response characteristics (Masters, Nichols & Finch, in preparation). Although CR16 mRNA (4374 nt) has been completely sequenced, the identity of this intriguing response is unknown.[32] Functional studies are underway in the laboratory of J. N. Masters at Ohio State University, Columbus, OH.

Glucocorticoid-Regulated mRNAs Exhibit Differential Responses to Stress

Based on the profile of rapid GR-mediated response to high-dose administered CORT, we postulated that the hippocampal mRNAs should also respond to endogenous elevation of CORT in response to stress. Subsequently, we showed that similar increases in two of the mRNA translation products are elicited by the stress of shaking for 2 h in a cage mounted on a metabolic shaker.[30] This stressor maintains glucocorticoid levels in the high physiological range (250–450 ng/ml) for 2 h. GPDH

TABLE 2. Summary of Differential Responses to Shaking Stress

	Fold-Change (Treatment/Control)	
CORT Response	2 h CORT/ADX	2 h Stress/Intact
I. Respond to 2 h CORT and 2 h stress		
GPDH/35 kd	3	3
GS	2	1.5
20 kd	10	3
II. Respond to 2 h CORT but not to 2 h stress.		
TGF-β1	0.6	1
33 kd	27	1
III. Do not respond to 2 h CORT or to 2 h stress.		
CR16	1	1
CR62	1	1
GFAP/50 kd	1	1

and GS mRNAs are also increased in response to shaking stress[27] (TABLE 2, I). Since prior ADX blocked the effects of shaking stress, these mRNA responses to stress are glucocorticoid-dependent. Other CORT-responsive mRNAs did not respond to 2 h stress (TABLE 2, II and III). The lack of response of some of the mRNAs is consistent with their slower response within 8 h, but not within 2 h of CORT treatment (TABLE 2, III). Therefore, these mRNAs may show a delayed response to stress or respond to chronic stress. TGF-β1 mRNA and the mRNA coding for the 33-kd polypeptide responded to 2 h CORT, but not to 2 h stress. The lack of a glucocorticoid-dependent stress effect on preproenkephalin mRNA in rat striatum was attributed to maximum levels of the mRNA in adrenally intact rats.[33] Since TGF-β1 and the 33-kd translation product both responded to CORT treatment of intact rats[23] (unpublished data), the lack of response to stress-induced elevation of glucocorticoids indicates differences between stress-activated and exogenous CORT effects. Collectively, these data suggest that rate- or level-sensitive mechanisms of glucocorticoid action or interactions with a stress-activation pathway determine the differential responses of glucocorticoid-regulated mRNAs to stress.

The hippocampal GPDH mRNA/35-kd response to CORT was of sufficiently large magnitude (10-fold) that gradations of response to different stressors were easily detected by RNA blot hybridization analysis: novelty, 2-fold; restraint and 0.4 M LiCl injection ip, 4-fold; and intermittent tail-shock/restraint, 6-fold.[34] These stressors differ in circulating CORT levels achieved or in neural or peripheral synaptic inputs to the hypothalamus;[12,35,36] therefore, differences between them in the magnitude of GPDH mRNA response could be attributed to level-sensitive mechanisms or specific neural interactions. GPDH mRNA is expressed in oligodendrocytes in both white and gray matter,[27] and the expression and activity of this enzyme are sensitive to neuronal-glial interactions.[37,38] Differential responses to stressors were also shown for a glucocorticoid-regulated 35-kd polypeptide that we subsequently identified by immunoprecipitation and SDS-polyacrylamide gel electrophoresis as GPDH[39] (unpublished data). This reversible enzyme provides substrates for phospholipid synthesis or for energy metabolism. The increase in GPDH in response to various stressors and to nerve injury[40] argues for its role as a stress protein, similar to that of the heat shock-induced family of proteins,[41] with a putative role in cellular protection via changes in oligodendrocyte membrane composition or energy metabolism.

Responses of Glucocorticoid-Regulated mRNAs during Neurodegeneration and Aging

Two glial mRNAs were cloned as decreases in response to CORT: GFAP and TGF-β1 (TABLE 1). GFAP is an intermediate filament protein of astrocytes, and GFAP concentration in the hippocampus and other brain regions is decreased in response to glucocorticoids and increased (astrogliosis) in response to brain injury and during aging.[42–45] TGF-β1 mRNA and peptide increased in response to brain injury in rats.[46] We have shown that TGF-β1 mRNA is expressed in activated microglia in response to both deafferenting and neurotoxic lesions[28] and are presently investigating lesion-induced expression of the peptide. In humans, TGF-β1 and interleukin-1 colocalized in brain disease states,[47] and TGF-β1 was found in amyloid plaques of Alzheimer disease and Down syndrome.[48] However, cloning of TGF-β1 mRNA from the hippocampus was surprising, especially as a decrease in response to CORT, since expression of this cytokine was limited to the choroid plexus and meninges in intact brain.[49] Recently we found that TGF-β1 mRNA is highly expressed in the hippocampal dentate gyrus following ADX. This layer undergoes neurodegeneration in response to ADX in the adult rat; the neurodegeneration is prevented by CORT-replacement.[22] GFAP mRNA is also expressed in a neuroanatomically specific manner in the hippocampus after ADX (FIG. 1). GFAP mRNA is increased 3-fold in the subgranular layer of the dentate gyrus (dg) after 3 d ADX compared with intact rats, and is decreased in response to CORT treatment. The increase in the subgranular layer coincides with the appearance of pyknotic nuclei in the adjacent granule cell layer of the dentate gyrus. In addition to the subgranular layer of the dentate gyrus, the GFAP response to ADX is localized in the molecular layer (ml) and in the hilus (h), where the dendritic fields and axonal projections of the granule cells are found, respectively (FIG. 1). Although other regions of the brain show regulation of GFAP mRNA by glucocorticoids,[26] they do not show this striking regulation of GFAP expression. These data support that the increased expression of GFAP in the hippocampus is a response both to cell death[15,21] and to the reduction of CORT.

We and others have shown that GFAP and TGF-β1 mRNAs increase with brain lesioning in rats,[28,50–53] with advanced age in rats and humans,[54,55] and with neuropathology in humans[54,55] (TABLE 3). Therefore, regulation of mRNA transcription or stability may underlie increases in GFAP protein and TGF-β1 peptide found in these conditions (see above). Increases in GFAP and TGF-β1 mRNAs represent early astrocytic and microglial responses to brain injury and lesioning.[28,51,52,55] Since the direction of the glial responses to aging and/or neurodegeneration is in the opposite direction to that of CORT, we proposed that the inhibitory effects of glucocorticoids were involved in the maintenance of cellular homeostasis. In this context, glucocorticoids mediate a return to basal expression after physiological disturbances, in much the same manner as proposed by Munck *et al.*[1] for the immunosuppressive and antiinflammatory actions of glucocorticoids in keeping natural defense mechanisms from overshooting. If left unchecked, overexpression of these proteins may result in pathophysiology, or contribute to age-related neurodegeneration. Examining brain mRNA responsiveness to CORT during aging in the future will determine whether hormonal regulation is disinhibited, thus contributing

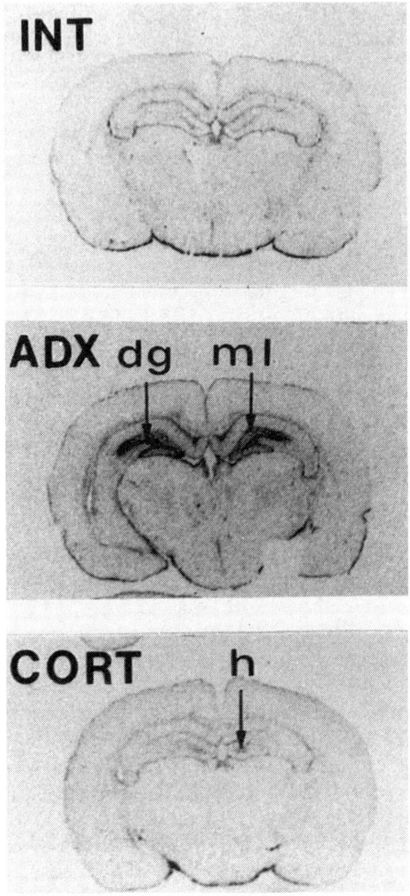

FIGURE 1. Representative autoradiographs of GFAP mRNA *in situ* hybridization in adult rat brain following glucocorticoid manipulation. Male F344 rats were ADX for 3 d (*middle panel*) or ADX and treated with CORT (40 mg/kg/day) for 3 d (*bottom panel*) and compared with p.m. intact (INT, *top panel*). dg = dentate granule cell layer; ml = molecular layer; h = hilus.

TABLE 3. GFAP and TGF-β1 mRNAs Respond Differentially to Glucocorticoids Compared with Neurodegeneration and Aging

	GFAP	TGF-β1
Rat		
CORT	↓	↓
ADX	↑	↑
Lesion	↑	↑
Aging	↑	↑
Human		
Neuropathology	↑	↑
Aging	↑	↑

to the increased gene expression seen in both rodents and humans at advanced ages[54] (TABLE 3).

SUMMARY

We used two methods to examine altered patterns of gene expression in rat hippocampus in response to administered glucocorticoids: analysis of RNA *in vitro* translation products on 2-d gels and cloning of cDNAs from a rat hippocampal library by differential hybridization (± CORT). We determined that two of the CORT-responsive cDNA clones encoded the 35- and 50-kd RNA translation products and identified them as GPDH and GFAP, respectively, by sequence analysis. Cloned mRNAs that increased and decreased in response to CORT were determined to be under positive and negative regulation by glucocorticoids in intact rats.

Despite their similarities in glucocorticoid response characteristics, we found three subsets of hippocampal mRNA responses to CORT and shaking stress which differ in temporal and level-dependent aspects of CORT regulation. In addition, GPDH gene expression represents a glucocorticoid-dependent stress response which is rapidly increased in a dose- and stressor-dependent manner. It is a candidate for a sensitive indicator of stress responsiveness in the brain as a function of neuroendocrine activity.

Mechanisms of adaptation to stress in the brain are likely to involve responses that are both mediated by glucocorticoids and opposed by them.[3] GFAP and TGF-β1 mRNA responses may be examples of the latter, since they are decreased in response to glucocorticoids, are under negative regulation by glucocorticoids in intact rats, and are increased in response to brain injury and disease and during aging. If these astrocytic and microglial responses are involved in cellular defense mechanisms in the brain, then their regulation by glucocorticoids would be important in maintaining and restoring cellular homeostasis in physiological and pathophysiological states. Future studies using these sensitive probes for glucocorticoid-regulated gene expression may identify new mechanisms by which the brain coordinates acute and chronic responses to stress and disease.

ACKNOWLEDGMENTS

The authors thank Jeff Masters, Nick Laping, and Heinz Osterburg for their contributions to these studies.

REFERENCES

1. MUNCK, A., P. M. GUYRE & N. J. HOLBROOK. 1984. Physiological functions of glucocorticoids in stress and their relation to pharmacological actions. Endocr. Rev. **5:** 25–44.
2. JONAT, C., H. J. RAHMSDORF, K. K. PARK, A. C. B. CATO, S. GEBEL, H. PONTA & P. HERRLICH. 1990. Antitumor promotion and antiinflammation: down-modulation of AP-1 (fos/jun) activity by glucocorticoid hormone. Cell **62:** 1189–1204.
3. MCEWEN, B. S. & R. E. BRINTON. 1987. Neuroendocrine aspects of adaptation. Prog. Brain Res. **72:** 11–26.
4. DE KLOET, E. R. 1991. Brain corticosteroid receptor balance and homeostatic control. Front. Neuroendocrinol. **12:** 95–164.
5. SWANSON, L. W., P. E. SAWCHENKO, R. W. LIND & J. H. RHO. 1987. The CRH

motoneuron: differential peptide regulation in neurons with possible synaptic, paracrine, and endocrine outputs. Ann. N.Y. Acad. Sci. **512:** 12–23.
6. Dallman, M. F., S. F. Akana, C. S. Cascio, D. N. Darlington, L. Jacobson & N. Levin. 1987. Regulation of ACTH: variations on a theme of B. Recent Prog. Horm. Res. **43:** 113–173.
7. Dallman, M. F., S. F. Akana, N. Levin, L. Jacobson, C. S. Cascio, D. N. Darlington, S. Suemaru & K. Scribner. 1989. Corticosterone (B) replacement in adrenalectomized rats: insights into the regulation of ACTH secretion. *In* The Control of the Hypothalamic-Pituitary-Adrenocortical Axis. F. C. Rose, Ed. 95–116. International Universities Press. Madison, WI.
8. Besedovsky, H. O., A. del Rey, I. Klusman, H. Furukawa, G. Monge-Arditi & A. Kabiersch. 1991. Cytokines as modulators of the hypothalamic-pituitary-adrenal axis. J. Steroid Biochem. Mol. Biol. **40:** 613–618.
9. Jacobson, L. & R. Sapolsky. 1991. The role of the hippocampus in feedback regulation of the hypothalamic-pituitary-adrenocortical axis. Endocr. Rev. **12:** 118–134.
10. Jones, M. T. & B. Gillham. 1988. Factors involved in the regulation of adrenocorticotrophic/β-lipotrophic hormone. Physiol. Rev. **68:** 744–800.
11. Kovacs, K. J. & E. Mezey. 1987. Dexamethasone inhibits corticotropin-releasing factor gene expression in the rat paraventricular nucleus. Neuroendocrinology **46:** 365–368.
12. Plotsky, P. M., K. V. Thrivikraman & M. J. Meaney. 1993. Central and feedback regulation of hypothalamic corticotropin-releasing factor secretion. Ciba Found. Symp. **172:** 59–75.
13. Ratka, A., W. Sutanto, M. Bloemers & E. R. de Kloet. 1989. On the role of brain mineralocorticoids (type I) and glucocorticoid (type II) receptors in neuroendocrine regulation. Neuroendocrinology **50:** 117–123.
14. McEwen, B. S., R. E. Brinton, H. M. Chao, H. Coirini, M. N. Gannon, E. Gould, J. O'Callaghan, R. L. Spencer, R. R. Sakai & C. S. Woolley. 1990. The hippocampus: a site for modulatory interactions between steroid hormones, neurotransmitters and neuropeptides. *In* Neuroendocrine Perspectives. E. Mueller & R. McCloud, Eds. 93–131. Springer Verlag. New York.
15. Gould, E., H. A. Cameron, D. C. Daniels, C. S. Woolley & B. S. McEwen. 1992. Adrenal hormones suppress cell division in the adult rat dentate gyrus. **12:** 3642–3650.
16. Joels, M. & E. R. de Kloet. 1992. Control of neuronal excitability by corticosteroid hormones. Trends Neurosci. **15:** 25–30.
17. de Kloet, E. R. 1992. Corticosteroids, stress and aging. Ann. N.Y. Acad. Sci. **663:** 357–372.
18. Sapolsky, R. M., L. C. Krey & B. S. McEwen. 1986. The neuroendocrinology of stress and aging: the glucocorticoid cascade hypothesis. Endocr. Rev. **7:** 284–301.
19. Kerr, D. S., L. W. Campbell, M. D. Applegate, A. Brodish & P. W. Landfield. 1991. Chronic stress-induced acceleration of electrophysiologic and morphometric biomarkers of hippocampal aging. J. Neurosci. **11:** 1316–1324.
20. Sloviter, R. S., G. Valiquette, G. M. Abrams, E. C. Ronk, A. L. Sollas, L. A. Paul & S. Neubort. 1989. Selective loss of hippocampal granule cells in the mature rat brain after adrenalectomy. Science **243:** 535–538.
21. Sloviter, R. S., E. Dean & S. Neubort. 1993. Electron microscopic analysis of adrenalectomy-induced hippocampal granule cell degeneration in the rat: apoptosis in the adult central nervous system. J. Comp. Neurol. **330:** 337–351.
22. Gould, E., C. S. Woolley & B. S. McEwen. 1990. Short term glucocorticoid manipulations affect neuronal morphology and survival in the adult rat dentate gyrus. Neuroscience **37:** 367–375.
23. Nichols, N. R., S. P. Lerner, J. N. Masters, P. C. May, S. L. Millar & C. E. Finch. 1988. Rapid corticosterone-induced changes in gene expression in rat hippocampus display type II glucocorticoid receptor specificity. Mol.Endocrinol. **2:** 284–290.
24. Poirier, J. & N. R. Nichols. 1991. Alteration of hippocampal RNA prevalence in response to deafferentation. Methods Neurosci. **7:** 182–199.
25. Nichols, N. R., J. N. Masters & C. E. Finch. 1994. Cloning of steroid-responsive mRNAs by differential hybridization. Methods Neurosci. **22:** 296–313.

26. NICHOLS, N. R., H. H. OSTERBURG, J. N. MASTERS, S. L. MILLAR & C. E. FINCH. 1990. Messenger RNA for glial fibrillary acidic protein is decreased in rat brain following acute and chronic corticosterone treatment. Mol. Brain Res. **7:** 1–7.
27. MASTERS, J. N., C. E. FINCH & N. R. NICHOLS. 1994. Rapid increase in glycerol phosphate dehydrogenase mRNA in adult rat brain: a glucocorticoid-dependent stress response. Neuroendocrinology **60:** 23–35.
28. MORGAN, T. E., N. R. NICHOLS, G. M. PASINETTI & C. E. FINCH. 1993. TGF-β1 mRNA increases in macrophage/microglial cells of the hippocampus in response to deafferentation and kainic acid-induced neurodegeneration. Exp. Neurol. **120:** 291–301.
29. NICHOLS, N. R. & C. E. FINCH. 1991. Transforming growth factor-β1 mRNA in rat brain decreases in response to glucocorticoid treatment. Mol. Cell. Neurosci. **2:** 221–227.
30. NICHOLS, N. R., J. N. MASTERS, P.C. MAY, J. DEVELLIS & C. E. FINCH. 1989. Corticosterone-induced responses in rat brain RNA are also evoked by acute vibratory stress in hippocampus. Neuroendocrinology **49:** 40–46.
31. NICHOLS, N. R., J. N. MASTERS & C. E. FINCH. 1990. Changes in gene expression in hippocampus in response to glucocorticoids and stress. Brain Res. Bull. **24:** 659–662.
32. MASTERS, J. N., W. H. ASHMAN & S. L. COTMAN. 1993. Differential RNA splicing of CR16, a novel gene expressed in hippocampal neurons. Soc. Neurosci. Abst. #461.17, p. 1122.
33. CHAO, H. M. & B. S. MCEWEN. 1990. Glucocorticoid regulation of preproenkephalin messenger ribonucleic acid in the rat striatum. Endocrinology **126:** 3124–3130.
34. NICHOLS, N. R., T. J. SHORS, J. N. MASTERS, S. LEVINE & C. E. FINCH. 1993. Differential responsiveness of glycerol phosphate dehydrogenase (GPDH) mRNA in rat hippocampus to different types of stressors. Abstracts of the 75th Annual Meeting of The Endocrine Society, #666, p. 217.
35. PALKOVITS, M. 1987. Organization of the stress response at the anatomical level. Prog. Brain Res. **72:** 47–55.
36. BEAULIEU, S., T. DI PAOLO, J. COTE & N. BARDEN. 1987. Participation of the central amygdaloid nucleus in the response of adrenocorticotropin secretion to immobilization stress: opposing roles of the noradrenergic and dopaminergic systems. Neuroendocrinology **45:** 37–46.
37. MCGINNIS, J. F. & J. DE VELLIS. 1981. Cell surface modulation of gene expression in brain cells by down regulation of glucocorticoid receptors. Proc. Natl. Acad. Sci. **78:** 1288–1292.
38. FISHER, M. & R. J. MULLEN. 1988. Neuronal influence on glial enzyme expression: evidence from chimeric mouse cerebellum. Neuron **1:** 151–157.
39. SCHLATTER, L. K. & L. A. DOKAS. 1989. Receptor specificity of a glucocorticoid- and stress-induced hippocampal protein. J. Neurosci. **9:** 1134–1140.
40. MEYER, J. S., P. J. LEVEILLE, J. DE VELLIS, J. L. GERLACH & B. S. MCEWEN. 1982. Evidence for glucocorticoid target cells in the rat optic nerve. Hormone binding and glycerolphosphate dehydrogenase induction. J. Neurochem. **39:** 423–434.
41. SCHLESINGER, M. J. 1986. Heat shock proteins: the search for functions. J. Cell. Biol. **103:** 321–325.
42. O'CALLAGHAN, J. P., R. E. BRINTON & B. S. MCEWEN. 1991. Glucocorticoids regulate the synthesis of glial fibrillary acidic protein in intact and adrenalectomized rats but do not affect its expression following brain injury. J. Neurochem. **57:** 860–869.
43. O'CALLAGHAN, J. P. & D. B. MILLER. 1991. The concentration of glial fibrillary acidic protein increases with age in the mouse and rat brain. Neurobiol. Aging **12:** 171–174.
44. HANSEN, L. A., D. M. ARMSTRONG & R. D. TERRY. 1987. An immunohistochemical quantification of fibrous astrocytes in the aging human cerebral cortex. Neurobiol. Aging **8:** 1–6.
45. FREDERICKSON, R. C. 1992. Astroglia in Alzheimer's disease. Neurobiol. Aging **13:** 239–253.
46. LOGAN, A., S. A. FRAUTSCHY, A. GONZALEZ, M. B. SPORN & A. BAIRD. 1992. Enhanced expression of transforming growth factor β1 in the rat brain after a localized cerebral injury. Brain Res. **587:** 216–225.
47. DA CUNHA, A., J. JEFFERSON, R. W. JACKSON & L. VITKOVIC. 1993. Glial cell-specific

mechanisms of TGF-β1 induction by IL-1 in cerebral cortex. J. Neuroimmunol. **42:** 71–86.
48. Van der Wal, E. A., F. Gomez-Pinilla & C. Cotman. 1993. Transforming growth factor-β1 is in plaques in Alzheimer and Down pathologies. NeuroRep. **4:** 69–72.
49. Flanders, K. C., G. Ludecke, S. Engels, D. S. Cissel, A. B. Roberts, P. Kondaiah, R. Lafyatis, M. B. Sporn & K. Unsicker. 1991. Localization and actions of transforming growth factor-βs in the embryonic nervous system. Development **113:** 183–191.
50. Nichols, N. R., N. J. Laping, J. R. Day & C. E. Finch. 1991. Increases in transforming growth factor-beta 1 mRNA in hippocampus during response to entorhinal cortex lesions in intact and adrenalectomized rats. J. Neurosci. Res. **28:** 134–139.
51. Pasinetti, G. M., N. R. Nichols, G. Tocco, T. Morgan, N. Laping & C. E. Finch. 1993. Transforming growth factor β1 and fibronectin mRNA in rat brain: responses to injury and cell-type localization. Neuroscience **54:** 893–907.
52. Laping, N. J., N. R. Nichols, J. R. Day & C. E. Finch. 1991. Corticosterone differentially regulates the bilateral response of astrocyte mRNAs in the hippocampus to entorhinal cortex lesions in male rats. Mol. Brain Res. **10:** 291–297.
53. Laping, N. J., T. E. Morgan, N. R. Nichols, I. Rozovsky, C. Young-Chang, C. Zarow, J. R. Day & C. E. Finch. 1994. Transforming growth factor-β1 induces neuronal and astrocyte genes: tubulin α1, glial fibrillary acidic protein, and clusterin. Neuroscience **58:** 563–572.
54. Nichols, N. R., J. R. Day, N. J. Laping, S. A. Johnson & C. E. Finch. 1993. GFAP mRNA increases with age in rat and human brain regions. Neurobiol. Aging **14:** 421–429.
55. Finch, C. E., N. J. Laping, T. E. Morgan, N. R. Nichols & G. M. Pasinetti. 1993. TGF-β1 is an organizer of responses to neurodegeneration. J. Cell. Biochem. **53:** 314–322.

DISCUSSION OF THE PAPER

Question: In the two conditions, corticol-treated and adrenalectomized animals, have you ever co-stained the hippocampus for TGF-beta or GFAP and some early response genes?

Nichols: Those are planned studies. Both genes have AP1 sites in the promoters and there is some data indicating that AP1s may be involved in the glucocorticoid regulation of GFAP. GFAP and TGF-beta are transcriptionally regulated by glucocorticoids in this system, in the opposite direction. Although for TGF-beta, we are not getting enough of a signal yet to say for sure.

Question: Are you satisfied with the number of genes you have or are you going after others?

Nichols: That is half of them. I have a box, and whenever I get bored I can just go sequence one and start again.

Question: Have you tried other methods like this differential cDNA display?

Nichols: We do not need it. Others in the laboratory are doing that to convince themselves that they have a system that is working. They are using GFAP and TGF-beta as controls and will probably come up with things that we didn't. Nevertheless, I was very pleased that by just using a differential hybridization scheme we were able to pull out something like a cytokine.

B. S. McEwen (*The Rockefeller University*): The increase that you showed after three days of adrenalectomy in GFAP message, specifically in dentate gyrus hilus, is certainly reminiscent of the changes that may be underlying the formation or expression of a radial glia network for the replacement of cells. In a study that we collaborated with Jim O'Callahan on, on measuring not so much message but GFAP proteins, adrenalectomy caused a rather widespread increase in GFAP protein level. But by Northern blot, your data also show a very specific or a larger change in the hippocampus. I was wondering if there were perhaps several relatives of the GFAP gene and you might have just happened to get one of them that may be more specifically expressed in the hippocampus.

NICHOLS: The clone we have is nearly full length, so it might pick up other forms. By an *in situ* analysis I would not be able to tell you that. But we only see one message, 2.9 kB. GFAP will cross-hybridize to vimentin, which is about 2.3 kB under low stringency. But these *in situs* are done with a very high stringency wash and they are RNAse treated. So I do not think you are going to have that kind of a problem. I do intend to look at the vimentin probe just to make sure that I am not picking up vimentin. Radial glia do express GFAP, so I am sure the enhanced expression is in that radial glia layer.

QUESTION: Two hours is already quite quick and I must say I am impressed by what you have seen. But have you tried 30 minutes or something of that sort?

NICHOLS: You see GPDH polypeptide increases in 30 minutes. But you will not see GFAP decrease at that time. You might see TGF-beta respond to glucocorticoids.

A. MUNCK (*Dartmouth Medical School*): I do not know whether it is possible, but have you tried blocking with cycloheximide before you give corticosterone to see if some of the responses are wiped while some of them persist, and which are primary, which are secondary?

NICHOLS: I probably will not do that. I am more interested in looking at aging and at physiological stress circuits. However, one could do that kind of an experiment.

MUNCK: The experiment would be to see if there is some key initiator of a whole cascade of events or whether they are all independent targets of glucocorticoids.

NICHOLS: It is a good idea. That would be a good problem.

I. S. EDELMAN (*Columbia University*): I am interested in the experimental design when you were comparing two hours of stress and two hours of glucocorticoids. At the end of the two-hour exposure period did you immediately sample or did you wait? You sampled immediately. Were the glucocorticoid animals adrenalectomized? The stress animals were not?

NICHOLS: Right. But both GFAP and TGF-beta do increase in response to glucocorticoids in intact rats. I have done those studies.

EDELMAN: I was asking that just to be sure I understood the experimental design. Clearly one of the differences would be the release of catecholamines from the adrenal medulla. There is a wealth of physiological data on catecholamine release in response to stress. It is a very fast acting pathway. But what may be more relevant to your interests is that if you look at the time course of the steroid, with a single bolus injection you will have a brief spike and then an exponential decay. With stress there is going to be a delay in glucocorticoid release, then a build-up to some level, and then when the stress is removed glucocorticoids will decrease. Consequently, the integral average dose is going to be offset in time and if you are sampling rapidly, you could have very similar patterns but the differences might be a consequence of the differences in the turnover time of the messenger pool and the magnitude of the messenger pool before you began. Some of these responses are very sensitive to turnover. So if you did a time course from two hours on, you might see a much closer correspondence between the glucocorticoid response and the stress response.

NICHOLS: I have wanted to do the kinetic experiments, but have been sidetracked by all the different clones.

EDELMAN: The time course is critical if you are going to look at patterns.

NICHOLS: I agree completely.

S. LEVINE (*Stanford University Medical School*): I take issue with the point you make that the stress and the dynamics of the CORT response differ very strikingly depending on which method you are using. If you look at the dynamics of the response, the novelty, or the response to restraint, that response is going down significantly by the end of two hours so that it is barely above base level. The only response in that paradigm that was able to elevate for the full two hours was the intermittent tail shock with the restraint. So that basically if you integrated over time, it is not that you have a chronic level at which point when you sacrifice the animal you have that chronic level persistent. Within the time there is a dynamic where the integration is really different over time. So it is not that the stress and the dynamics of the glucocorticoid response differ very strikingly depending on the procedure that you use.

NICHOLS: I think there are going to be interactions with all different kinds of neurocircuits, including paracrine circuits, which may explain differential responses to glucocorticoids and stress.

S. J. WATSON (*University of Michigan*): This affect of corticosterone on GFAP is quite a paradox. The glucocorticoid hypothesis of brain aging started basically with evidence of a correlation between greater astrocytosis in hippocampus and higher levels of corticosterone and adrenal activity in Fischer rats. So you have elevated corticosterone in these aged animals in many instances, and I do not understand why that did not downregulate the GFAP.

NICHOLS: In aging or degenerative diseases the glucocorticoids are not doing what they are supposed to be doing, so it does not matter if they are high.

QUESTION: What strain did you use?

NICHOLS: These are Fischer 344 males. In my study they did not have elevated corticosterone, but I did not measure free, I only did total.

WATSON: In a lot of studies they tend to be on the high side at least.

NICHOLS: Yes, but I handled them the way a person who does stress experiments would. I did a staggered design, and they were separated and sacrificed very quickly. I did everything I could not to stress any group differentially, because if you do that then you do not see elevations.

WATSON: That is obvious. The point is, just in the spontaneously aging Fischer rat, you have elevated GFAP, at least at the protein level.

NICHOLS: Definitely, and elevated TGF-beta message. It is the opposite of the glucocorticoid response, opposite direction but same magnitude.

WATSON: It is an interesting paradox. Thank you.

Corticosteroid-Induced Proteins in Brain[a]

LINDA A. DOKAS,[b] LAWRENCE K. SCHLATTER,[c]
AND CHRISTINA S. BARR

Departments of Neurology and Biochemistry/Molecular Biology
Medical College of Ohio
Toledo, Ohio 43699-0008

INTRODUCTION

Neuroendocrine factors participate in the regulation of brain development, synaptic plasticity and aging-related degeneration through a number of different mechanisms. Prominent among them in the ability of hormones with nuclear receptors, such as glucocorticoids, to regulate transcription from specific genes. Since the proteins which are coded by these genes mediate the functions of these hormones, their characterization should provide insight into the neuronal and glial mechanisms by which hormonal effects are exerted.

Although glucocorticoid hormones regulate intracellular processes that acutely mediate stress responses, prolongation of such effects becomes maladaptive. In the brain, corticosteroids have an inhibitory influence on development and synaptic plasticity. Treatment of rats with glucocorticoids during brain development reduces brain weight and DNA content, interferes with normal myelination, alters neuronal morphogenesis and results in learning and memory deficits that persist throughout adult life.[1,2] Presumably as a mechanism to protect the developing brain from such deleterious effects, a stress nonresponsive period occurs during the neonatal period.[3] Some growth-related processes, as demonstrated in the rat hippocampus, remain sensitive to high levels of glucocorticoids throughout life. Lesion-induced neuronal sprouting in the dentate gyrus is retarded if rats are concurrently exposed to glucocorticoids, and relatively short exposure to these steroids can alter pyramidal neuron morphology.[4,5] More chronic glucocorticoid exposure, whether as a consequence of old age or of repeated administration, leads to alterations in the rat hippocampus including astroglial proliferation and neuronal degeneration.[6,7]

Steroid-dependent protein expression depends, at the genomic level, upon the nature of the response element with which the hormone receptor binds and its interactions with a number of other transcriptional factors which, in combination, determine the direction in which gene transcription is regulated.[8,9] In the rat hippocampus, corticosteroids have been shown to have both positive and negative effects on protein synthesis.[10,11] The purposes of this review are to describe one example of each type of regulation and to discuss the possible relevance of each to acute and/or chronic effects of glucocorticoids on hippocampal function.

[a]Research in the authors' laboratory was supported by grants from the American Federation for Aging Research, the VanNess-Thompson Foundation, and the National Institutes of Health (NS 17118).

[b]To whom correspondence should be sent at: Department of Neurology, Medical College of Ohio, P.O. Box 10008, Toledo, OH 43699-0008.

[c]Current address: Kemin Industries, Des Moines, IA 50301.

Hippocampal Protein Synthesis in Relation to Short-Term Stress

An experimental system which adapts itself readily to identification of such proteins is analysis of protein synthesis *in vitro* following treatment of rats *in vivo* with steroids or agents that modulate endogenous glucocorticoid levels. In this laboratory, the rat hippocampal slice preparation incubated in the presence of radiolabeled amino acid precursors has been employed for this purpose.[10,12,13] This combined protocol permits a variety of animal treatments, including long-term administration of steroids which would be impossible *in vitro* given the short viability of hippocampal slices. Conversely, incubation of slices with a labeled protein precursor provides an intact system which maintains the normal translational controls on protein synthesis. Measurement of protein labeling in slices can also be combined with other protein-based techniques to more fully define posttranslational factors that may regulate activity of corticosteroid-responsive proteins.

With this approach, the most consistently observed corticosteroid-induced alteration is enhanced synthesis of a hippocampal cytosolic protein with an apparent molecular mass (M_r) of 35,000 (FIG. 1, *left panel*). The increase in synthesis of this protein is directly related to a type II (GR) receptor interaction, since it occurs following injection of corticosterone and the GR receptor agonist RU28362 at relatively low concentrations, but only by pharmacological doses of aldosterone. The response is elicited by either exogenous steroids or by treatment of animals with agents (stressors, ACTH) that raise endogenous plasma levels of corticosterone. The time course of increased synthesis of the 35,000-M_r protein parallels the period of elevated serum corticosterone produced by physiological stressors or by exogenous application. Changes of comparable magnitude in the synthesis of this protein are seen immediately following a single injection of corticosterone or after 14 consecutive days of injection. After 90 consecutive daily steroid injections, enhanced synthesis of the 35,000-M_r protein is still observed although at a diminished level as compared to animals treated with a single injection (FIG. 2), possibly reflecting the downregulation of hippocampal GR receptors which results from chronic steroid administration.[6]

A protein with similar properties was identified as a translation product from mRNA isolated from rats that had been stressed or treated with exogenous corticosterone.[11,14] To establish identity between this protein and the one characterized in hippocampal slices, a direct comparison was made between them (Nichols *et al.*, manuscript in preparation). The protein radiolabeled in hippocampal slices was found to migrate identically on two-dimensional SDS/PAGE gels with the *in vitro* translation product. Because of the close correspondence in steroid sensitivity and time course of response for both the slice protein and the *in vitro* translation product, synthesis of these proteins most likely depends upon a common factor in both experimental systems, that being availability of its messenger mRNA.

The molecular weight (35,000) and isoelectric point (6.3) of the hippocampal cytosolic protein matches that of glycerol 3-phosphate dehydrogenase (GPDH), an oligodendrocyte glucocorticoid-responsive protein.[15] Several experimental measures support co-identity of GPDH and the hippocampal 35,000-M_r protein. Firstly, the radiolabeled hippocampal slice protein co-purifies with GPDH using gel filtration chromatography.[13] Secondly, an antibody to GPDH immunoprecipitates both the hippocampal slice protein and translation product (Nichols *et al.*, manuscript in preparation). Finally, direct measurement of hippocampal mRNA levels for GPDH using RNA blot hybridizations and RNAse protection assays indicate similar changes following stress or steroid administration to those seen in protein labeling studies.[16]

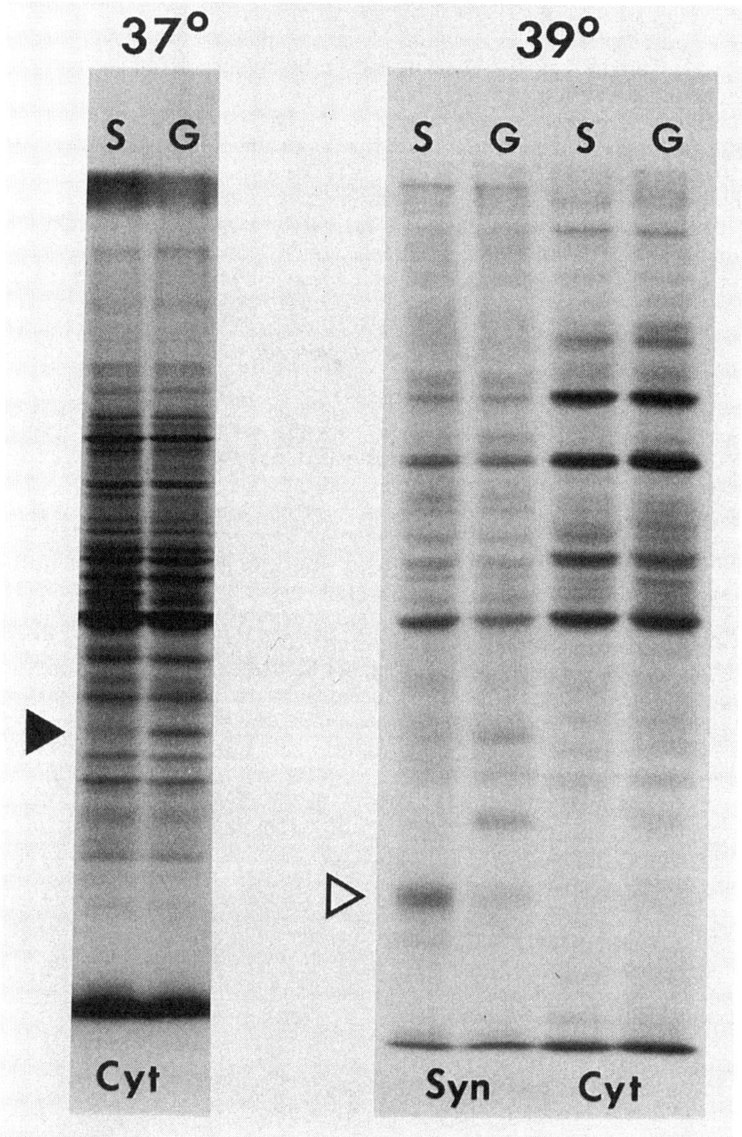

FIGURE 1. Protein synthesis in hippocampal slices following *in vivo* administration of corticosterone. Four hours after injection of rats with 5 mgs of corticosterone in sesame oil (glucocorticoid-injected; G) or sesame oil alone (sham-injected; S), tissue slices were prepared and incubated with [^{35}S]-methionine/cysteine at 37°C or 39°C. Labeling of cytosolic (Cyt) or synaptosomal (Syn) proteins was visualized with fluorography after SDS/polyacrylamide gel electrophoresis. The *open arrowhead* shows the position of a protein with an apparent M_r of 25,000 and the *closed arrowhead* indicates a protein with an apparent M_r of 35,000.

Induction of cytosolic GPDH in the hippocampus in response to stress would appear to be adaptive since it would ensure regeneration of NAD^+ needed for glycolysis and produce glycerol 3-phosphate for both transport into the mitochondria with subsequent production of ATP and also for lipid biosynthesis. However, the relationship between increased synthesis of GPDH in oligodendrocytes and the response of hippocampal neurons during stress remains unclear. Furthermore, it should be noted that the rapid increases in GPDH synthesis which parallel a stress response are not coincident with increases in enzyme activity that can be measured above that of the relatively stable pool of total cytosolic GPDH activity.[13] Thus, stress-induced GPDH may be a posttranslationally modified isozymic form or a subfraction of the total which is selectively localized within cells.

Heat-Shock and Hippocampal Protein Synthesis

The molecular properties of the hippocampal 35,000-M_r protein correspond to those of one of a group of heat-shock proteins that are translationally regulated in thymic lymphocytes.[17] This led us to investigate the possibility that it was expressed as a heat-shock protein in the hippocampus as well. To do so, rats were injected with corticosterone in sesame oil or sesame oil alone and hippocampal slices were prepared in the normal fashion, but incubated at 39°C, instead of the standard 37°C. In the former case, the pattern of radiolabeled proteins changes, becoming more limited, while species with the molecular weights of the major heat-shock proteins (approximately 90,000 and 70,000) become more predominant in the cytosolic fraction (FIG. 1; *right panel*). At the higher slice incubation temperature, the 35,000-M_r protein not only is not prominent, but cannot be detected in the cytosolic fraction from either sham-injected or steroid-treated animals. Instead, a low-M_r protein (25,000) becomes apparent, whose synthesis is decreased in preparations from corticosterone-treated rats (FIG. 1, *right panel*). This protein is not present in the cytosolic fraction, but is seen in the mitochondrial/synaptosomal (P2) fraction. Unlike the 35,000-M_r protein, synthesis of the 25,000-M_r species is affected in hippocampal slices at both 4 and 24 hours after a single injection of corticosterone. Given its molecular weight and the temperature-sensitive nature of its expression,

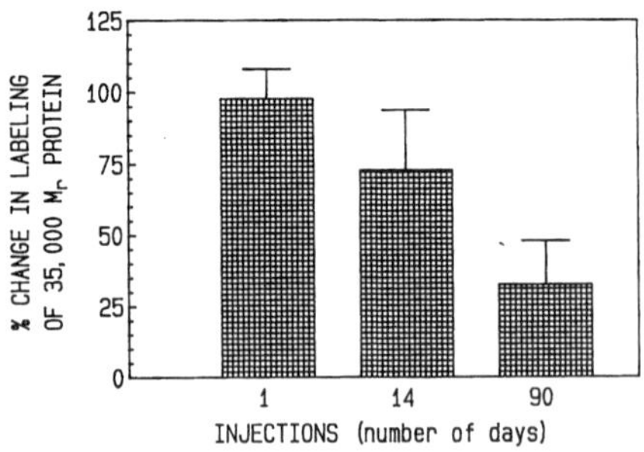

FIGURE 2. Synthesis of a hippocampal protein following cumulative exposure to corticosterone. Rats were injected subcutaneously with corticosterone (5 mgs) in sesame oil or with sesame oil alone. Four hours after a single injection or after the last of consecutive injections of the same type given for 14 or 90 days, hippocampal slices were prepared and incubated with [^{35}S]-methionine/cysteine to label proteins. The cytosolic fraction was prepared, and the synthesis of the 35,000-M_r protein was analyzed with SDS/polyacrylamide gel electrophoresis, fluorography and densitometry. Results are expressed as the mean change in labeling of the protein observed in steroid-treated animals as compared to sham-injected animals (n ≥3).

this may be the small mammalian heat-shock protein. A comparison of the properties of the 35,000 and 25,000 glucocorticoid-sensitive proteins identified in these studies is given in TABLE 1.

CONCLUSIONS

Use of a combined *in vivo/in vitro* approach for the characterization of glucocorticoid-sensitive proteins has indicated two patterns of expression which may be of potential significance for hippocampal function. In the first instance, as exemplified by the 35,000-M_r cytosolic protein, glucocorticoids enhance synthesis in direct relation to the stress state of the animal. Given the close correlation between onset and termination of stress responses and increased synthesis of such proteins, they probably serve a function in the short-term adaptation to stress. Even if the relevance of synthesis of these proteins to a functional response is unclear, as is the case for glycerol 3-phosphate dehydrogenase, they can still serve as valid markers of the hippocampal response to stress. The short-term adaptiveness of this pattern of synthesis implies that the actions of these proteins may become negative only when inappropriately (during critical times of development) or chronically (aging) expressed. Alternatively, long-term endocrine alterations which dampen the responsiveness of the signal for protein expression, such as receptor downregulation, may lead to negative consequences. In this latter model, it is the chronic lack of the protein which becomes maladaptive.

The ability of glucocorticoid hormones to acutely inhibit synthesis of other proteins suggests a different relationship to hippocampal function. By interfering with the synthesis or activation of this class of proteins, these steroids may diminish the ability of neurons to respond appropriately to environmental signals. One corollary of this statement is that this class of glucocorticoid-sensitive proteins is only required at critical levels subsequent to exposure to potentially-injurious stimuli. This model places particular emphasis on the effect of steroids on expression of proteins that maintain neuronal integrity or plasticity, such as heat-shock proteins and those that are growth-associated. Examples of the latter category are the 25,000-M_r protein and the neuronal growth-associated protein GAP-43. An inverse relationship between GAP-43 expression and glucocorticoid levels is indicated by increases of GAP-43 mRNA levels in the rat hippocampus, as measured by *in situ* hybridization[18] or by RNA blot hybridization (Dokas *et al.*, unpublished observations) that follows adrenalectomy.

Although glucocorticoid-inhibited expression of such proteins may have little effect under basal conditions, the lack of their synthesis may place the neuron in a state of endangerment when it encounters stimuli which require elevated synthesis

TABLE 1. Comparison of Hippocampal Corticosteroid-Sensitive Proteins

	Mr	
	35,000	25,000
Change in synthesis	↑	↓
Time course	4 hrs$^+$/24 hrs$^-$	4 hrs$^+$/24 hrs$^+$
Temperature sensitivity	37°C	39°C
Localization	cytosolic	synaptosomal

of the proteins for survival. One specific application of this general statement may be that glucocorticoid-exposed neurons are unable to produce a normal growth response under conditions that require synaptogenesis as an adaptive response. Extended chronically, and particularly in combination with other degenerative conditions, glucocorticoid exposure could exacerbate the degree of neuronal loss. Most studies on glucocorticoid-mediated neuronal degeneration have used morphological or anatomical criteria to measure stages of the process. It is important to identify biochemical markers that could serve as molecular indices of such a response. The assumption that interference by corticosteroids of the synthesis of proteins that mediate plasticity and survival responses underlies loss of neuronal adaptability to injurious stimuli is testable given the methodology that is available for characterization of proteins within this class.

SUMMARY

Cumulative exposure of hippocampal neurons to stress-like levels of corticosterone produces a negative spectrum of cellular alterations from ultrastructural changes to disruption of dendritic morphology and eventual degeneration. An experimental system which adapts itself to characterization of corticosteroid-induced proteins which mediate such effects is the hippocampal slice incubated in the presence of a radiolabeled amino acid following treatment of rats with corticosterone. The most consistently observed response to elevated corticosterone levels produced by exogenous injection is synthesis of a hippocampal cytosolic protein which has characteristics of glycerol phosphate dehydrogenase. Because synthesis of this protein is enhanced with a short latency as serum corticosterone levels are increased and terminated quickly upon re-establishment of basal conditions, it serves as a valid biological marker of the response of the hippocampus to short-term stress. In contrast, alterations in synthesis of other proteins following corticosterone treatment only become apparent under defined conditions or after chronic treatment. For example, steroid-inhibited synthesis of a hippocampal protein with an approximate molecular weight of 25,000 is only observed when slices are incubated at an elevated temperature. Such negative changes may represent loss of adaptive responses that protect the neuron from damage by cellular insults.

ACKNOWLEDGMENTS

Dr. Nancy Nichols is acknowledged for her collaborative efforts and helpful evaluation of this work.

REFERENCES

1. Balazs, R. & M. Cotterrell. 1972. Effect of hormonal state on cell number and functional maturation of the brain. Nature (London) **236:** 348–350.
2. Olton, D. S., C. T. Johnson & E. Howard. 1974. Impairment of conditioned active avoidance in adult rats given corticosterone in infancy. Dev. Psychobiol. **8:** 55–61.
3. Sapolsky, R. M. & M. J. Meaney. 1986. Maturation of the adrenocortical stress response: neuroendocrine control mechanisms and the stress hyporesponsive period. Brain Res. Rev. **11:** 65–76.
4. Scheff, S. W. & S. T. DeKosky. 1983. Steroid suppression of axon sprouting in the

hippocampal dentate gyrus of the adult rat: dose-response relationship. Exp. Neurol. **82:** 183–191.
5. WOOLEY, C. S., E. GOULD & B. S. MCEWEN. 1990. Exposure to excess glucocorticoids alters dendritic morphology of adult hippocampal pyramidal neurons. Brain Res. **531:** 225–231.
6. LANDFIELD, P. W., R. K. BASKIN & T. A. PITLER. 1981. Brain aging correlates: retardation by hormonal-pharmacological treatments. Science **214:** 581–584.
7. SAPOLSKY, R. M., L. KREY & B. S. MCEWEN. 1987. Prolonged glucocorticoid exposure reduces hippocampal number: implications for aging. J. Neurosci. **5:** 1221–1227.
8. DIAMOND, M. I., J. N. MINER, S. K. YOSHINAGA & K. R. YAMAMOTO. 1990. Transcription factor interactions: selectors of positive or negative regulation from a single DNA element. Science **249:** 1266–1272.
9. PEARCE, D. & K. R. YAMAMOTO. 1993. Mineralocorticoid and glucocorticoid receptor activities distinguished by nonreceptor factors at a composite response element. Science **259:** 1161–1165.
10. SCHLATTER, L. K. & L. A. DOKAS. 1987. Glucocorticoid-stimulated [^{35}S]-labeling of a cytosolic protein in the rat hippocampus. Neurosci. Res. Commun. **1:** 71–78.
11. NICHOLS, N. R., S. P. LERNER, J. N. MASTERS, P. C. MAY, S. L. MILLER & C. E. FINCH. 1988. Rapid corticosterone-induced changes in gene expression in rat hippocampus display type II glucocorticoid receptor specificity. Mol. Endocrinol. **2:** 284–290.
12. SCHLATTER, L. K. & L. A. DOKAS. 1989. Receptor specificity of a glucocorticoid- and stress-induced hippocampal protein. J. Neurosci. **9:** 1134–1140.
13. SCHLATTER, L. K., S.-M. TING, L. A. MESERVE & L. A. DOKAS. 1990. Characterization of a glucocorticoid-sensitive hippocampal protein. Brain Res. **522:** 215–223.
14. NICHOLS, N. R., J. N. MASTERS, P. C. MAY, J. DEVELLIS & C. E. FINCH. 1989. Corticosterone-induced responses in rat brain RNA are also evoked in hippocampus by acute vibratory stress. Neuroendocrinology **49:** 40–46.
15. DEVELLIS, J. & D. INGLISH. 1968. Hormonal control of glycerol phosphate dehydrogenase in the rat brain. J. Neurochem. **15:** 1061–1070.
16. NICHOLS, N. R., J. N. MASTERS & C. E. FINCH. 1990. Changes in gene expression in hippocampus in response to glucocorticoids and stress. Brain Res. Bull. **24:** 659–662.
17. COLBERT, R. A. & D. A. YOUNG. 1987. Detection of mRNAs coding for translationally regulated heat-shock proteins in non-heat-shocked thymic lymphocytes. J. Biol. Chem. **262:** 9939–9941.
18. CHAO, H. M., R. L. SPENCER, R. R. SAKAI & B. S. MCEWEN. 1992. The expression of growth-associated protein GAP-43 mRNA in the rat hippocampus in response to adrenalectomy and aging. Mol. Cell. Neurosci. **3:** 529–535.

DISCUSSION OF THE PAPER

QUESTION: Are your GPDH antibodies good enough to do a Western blot to look in the hippocampal slices to see whether or not after 4 hours or 4 days the level of the protein matches what you are looking at in the *in vitro* translation, which is really a synthesis rate measurement?

DOKAS: I would think so. Dr. Nichols has the antibody.

QUESTION: That would be one way to explain the difference between activity and synthesis rate. Are you taking into account that there may be some degradation that may not occur in the *in vitro* experiment? If you could do a Western blot after four hours of corticosterone (CORT) treatment, and in the control and say that the increase you see in synthesis actually will be an increase in protein, it would really tell you a lot.

DOKAS: You could also do quantitative immunoprecipitation in response to various doses.

QUESTION: The key would be to look at total protein as opposed to sythesized protein. I mean, not to do it by labelling, just by accumulated protein.

DOKAS: It is a polyclonal antibody, I do not see why that would not work.

H. AKIL (*University of Michigan*): Did you look at the HSPs themselves in any of your treatments to see if they are at all modulated, whether at 39°C or at 37°C?

DOKAS: We see no changes. I do not know for a fact that it is the small molecular heat-shock protein, but the only one that seems to change consistently in response to steroid treatment is the 25 kilodalton. The 70 and 90 kDa do not change.

AKIL: A major stressor increases a lot of steroids. A lot of things might kick into action which may or may not amount to anything in the end. I would liken this to hearing a hurricane alarm or earthquake alarm depending on where you are, and so you take your goodies and your candles and your stuff and you go hide in your basement or under your door and then nothing happens and you get out, and nothing really comes out of it unless the major disaster actually occurs. Then the alarm becomes relevant. But alarms are not relevant most of the time. The question is with differential strategies, how to tell apart those that are really necessary for more standard stress and coping versus the most extreme cases of stress and coping?

DOKAS: The second type of pattern, where the synthesis of the protein may go on normally but is not critical. It only becomes critical when the neuron needs it in essence. To show functional significance requires long-term studies.

AKIL: The strategy needs an antagonist in it. If you block these things, would anything be different at short-term compared to long-term?

E. R. DE KLOET (*Sylvius Laboratories*): You showed basically two strategies to find glucocorticoid responsive proteins. One is that you give glucocorticoid and look at a set of genes. The other is that you consider it in the context of a stimulus. You and Dr. Nichols see the same proteins or messenger RNA when you give glucocorticoids or when you stress the animal. Intuitively you would not expect that to be the case, because glucocorticoids would interfere or adjust the synthesis of a protein induced by stress. Yet you see the same pattern. Is it because your design is not appropriate or because you are not sensitive enough in the measurements?

DOKAS: It depends on which type of protein—whether it is the one that is stimulated or the one that is inhibited. The other thing is you notice that Dr. Nichols had a grouping of proteins, some of which responded to both, some of which responded to one and not the other. There are variations on that theme in terms of whether stress or CORT injection are synonymous.

NICHOLS: Maybe I am looking at it too simply. In terms of the adaptive and atrophic, or mediated by and opposed by, you are going to see both types of responses. Some things you are going to need. If you need to correlate between anaerobic and aerobic respiration, then turning on GPD is going to work. Or if an oligodendrocyte is changing its membrane phospholipid composition that is protecting the axon during a period of stress, then the stress and the glucocorticoid could be doing the same thing. But if it returns to baseline after everything has happened, so it restores homeostasis, then glucocorticoids are going to oppose the stress response. There are stress pathways that are interacting and that is why I got the different groups. Although if it was a glucocorticoid direct response, you would get the glucocorticoid direct response even in the presence of stress.

DE KLOET: If you challenge a cell with a transmitter it will produce a whole range of responses and you may be able to measure these in terms of messenger RNA synthesis of protein, and then somewhere along the line this is being extinguished or suppressed. In the context of the stimulus you are studying, you may have to see what type of protein is under the control of the corticosteroids. Stress and corticosteroids are not doing the same. I know that. So why do you still think that you would see the same proteins changed by both conditions? Is it not a function of the time course? If you look at five minutes or two minutes you might see the onset of the effect.

DOKAS: That is right. The period of CORT injection certainly differs with regard to GPDH and what happens with stress and what happens with endogenous CORT. But there is actually a similarity with regard to time course of response because as the stress goes up and comes down, the labelling goes up and comes down. With a 5-mg injection it goes up, stays high, and then comes down, and the labelling also is increased over the longer time frame.

S. J. WATSON (*University of Michigan*): You could really break the experiments into three pieces, (1) things that turn on with stress, you take out the adrenals and you do a fast time course, so it is real stress activation. (2) The second class of event might well be an acute set of responses to the first pass of corticosteroids, sort of a termination class event. (3) The third one is the impact chronic exposure. In Dr. Akil's analogy, early into the hurricane, the creature has seen an abundance of corticosteroids as well as stress and now you are looking at those emergency machines turning on in order to compensate.

DOKAS: We do not have the time course data for the 25-kDa protein. It may be an example of a protein that synthesized on demand. Glucocorticoids interfere with that and I suspect that this class of proteins also is one that is affected by glucocorticoids in the more chronic sense. Glucocorticoids are adaptive in the short term and become maladaptive in the long term. These negative effects of this type are going to be a problem when the animals are more chronically exposed.

D. B. DEFRANCO (*University of Pittsburgh*): Getting back to Dr. Akil's question, is there any way to test whether or not there is more oxygen consumption in these slices after adrenalectomy? Does that have any affect on any morphological change? Then you can disrupt oxygen consumption in other ways.

DOKAS: That is part of the reason why we use slices in the first place. That is really getting back to basics. We could get a Warburg apparatus out and measure oxygen consumption.

QUESTION: Yes, that could be relevant to neuronal survival or whatever. *In vivo* that could be what is really happening.

DOKAS: I do suspect that energy metabolism has something to do with it. Certainly with regard to GPH.

Gene-Mediated Control of Hippocampal Neuronal Excitability

M. JOËLS, H. KARST, W. HESEN,
AND W. J. WADMAN
Institute of Neurobiology
University of Amsterdam
Kruislaan 320
1098 SM Amsterdam, The Netherlands

INTRODUCTION

In the late sixties McEwen and co-workers[1] showed that corticosteroid hormones are highly retained in some brain areas, particularly in the hippocampal formation. Since then, several groups have investigated how these hormones can affect, *in vivo*, the generation and propagation of electrical signals in the brain.[2–7] Two factors inherent in the mechanism of action by which steroids act have complicated these investigations.[8] First, steroid actions mediated by intracellular receptors are both slow in onset and also of prolonged duration. Electrophysiological recording of such slow and long-lasting effects is difficult. Second, steroid actions may very well be conditional, in that they will only become apparent when the system is shifted from its 'resting' condition. The latter implies that steroid actions should be studied, preferably, with good voltage control over the cell membrane.

To achieve good voltage control and to study steroid levels at the level of cellular membrane properties, intracellular recording with microelectrodes or whole cell recording with patch electrodes are indicated. However, both techniques can only be reliably performed *in vitro*, particularly when prolonged recordings are involved. Over the past ten years, a number of studies have been initiated which have focussed on steroid-induced changes in electrical properties at the cellular level, in hippocampal slices.[9–13]

These studies have proved, amongst many things, that due to the slow but long-lasting actions of corticosteroids, effects that were initiated *in vivo* (prior to the preparation of the slices) persist *in vitro*. In rats with an intact hypothalamo-pituitary-adrenal axis, circulating levels will range from very low in the morning under rest to very high at the circadian peak and after stress[14] (see paper by Dallman in this volume). Correspondingly, the relative steroid receptor occupation may vary between predominant occupation of the high-affinity mineralocorticoid receptors (MRs) on the one hand, and concurrent activation of MRs and a considerable part of the lower-affinity glucocorticoid receptors (GRs) on the other hand[15] (see paper by de Kloet in this volume). The lack of control over relative MR/GR occupation *in vivo* and the ensuing effects *in vitro* may cause difficulties in interpretation. Therefore, recent electrophysiological studies have employed pharmacological tools to selectively occupy MRs or GRs, in either adrenally intact or adrenalectomized (ADX) animals. As pointed out above, particularly the shift in electrical properties observed with the transition from predominant MR occupation to the situation where both MRs and GRs are activated is physiologically interesting.

Steroid Actions on Electrical Properties of Hippocampal Cells

Potentially, corticosteroid hormones may affect electrical properties of hippocampal cells in many ways, including the following (FIG. 1): 1. They could alter the passive membrane properties (leak current), thereby inducing changes in resting membrane potential and input resistance. This was never observed for hippocampal neurons,[12,13] but does occur in coeliac ganglia cells.[16] 2. The steroids could alter voltage-gated or ion-sensitive currents. 3. They could affect transmitter-induced responses, mediated through ligand-gated ion channels or G protein-coupled receptors. Evidence for the possibilities mentioned under 2 and 3 will be presented below.

Voltage-Gated Currents

Voltage-gated calcium (Ca) currents were found to be very sensitive to changes in corticosteroid levels[17,18] (FIG. 2). Predominant MR occupation results in very small Ca currents in CA1 pyramidal cells.[18] When GRs are occupied in addition to MRs the low- and high-threshold Ca currents are relatively large. Exclusive activation of GRs resulted in large Ca currents only when very high doses of the GR agonist RU 28362 were administered,[17] suggesting that GR-mediated actions are optimally expressed when MRs are also activated. Interestingly, when neither the MRs nor the GRs were occupied (ADX) Ca currents were also found to be large, particularly in the lower voltage range.[18] Voltage characteristics and kinetic properties of the Ca

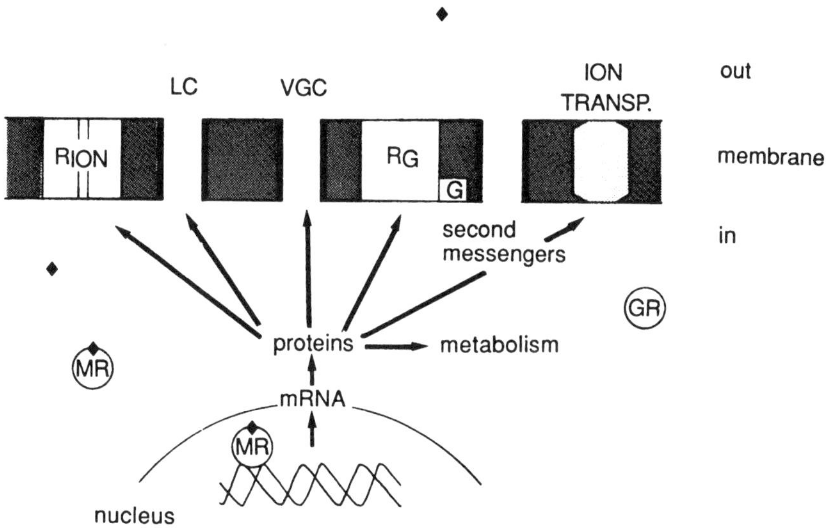

FIGURE 1. Corticosteroid hormones (*diamonds*) pass the cell-membrane and bind to intracellular mineralocorticoid and glucocorticoid receptors (MR and GR respectively). The activated steroid-receptor complex acts as a transcription factor on the genome, resulting in an altered mRNA expression and, thus, protein synthesis. These proteins can affect (amongst other things) 1) characteristics of ion currents through voltage-independent (LC) and voltage-dependent channels (VGC), 2) the responsiveness to transmitters acting via ionotropic (R_{ion}) or G protein-coupled (R_G) receptors or 3) the function of ion transporters.

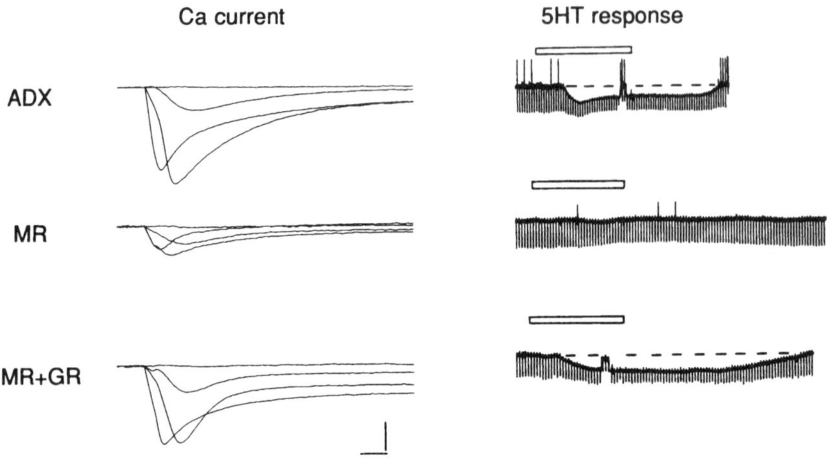

FIGURE 2. In the absence of corticosteroid hormones (ADX) depolarization-induced Ca currents are large (*left*). Depicted are the currents evoked by voltage steps to −100, −60, −30 and 0 mV; currents were recorded with patch electrodes in hippocampal slices from 1 week ADX rats. *In vitro* application of corticosterone in the presence of RU 38486, thus occupying MRs only (MR), results in a considerable reduction of the Ca currents. Additional occupation of GRs (MR + GR) is associated with relatively large Ca currents. Calibration: horizontal = 20 ms; vertical = 0.5 nA. A similar steroid dependency is observed for serotonin (5HT)-induced hyperpolarization of the membrane (*right*). The duration of the 5HT application is depicted by the *open bar*. In this case, recording is performed with microelectrodes in hippocampal slices, under current clamp conditions. Calibration: horizontal = 1 min.; vertical = 13 mV.

currents were not significantly different for the various conditions of steroid receptor occupation.

The conditions during these voltage clamp studies, *e.g.*, the presence of Na and K channel blockers or the application of prolonged depolarizations, are optimal for recording Ca currents in isolation. To what extent steroid-induced modulation of Ca influx will affect neuronal excitability normally is much harder to predict. The increase in low-threshold Ca current observed in the absence of corticosteroids may be particularly important for the Ca influx in dendrites of CA1 pyramidal cells: at least part of the low-threshold current was found to be generated in the distal dendrites.[19] These effects on Ca influx through voltage-gated channels in the dendrites may add to the local Ca influx through NMDA receptors.[20] Corticosteroid-dependent increases of the high-threshold Ca current will mainly affect Ca influx associated with large depolarizations of the membrane. Larger Ca influxes, in turn, are expected to enhance the Ca-dependent inactivation of the current and, more importantly, to activate Ca-dependent K channels. Consequently, the initial enhancement of Ca influx may eventually be reversed. This is corroborated by the observation that occupation of GRs in addition to MRs indeed increases the frequency accommodation and the afterhyperpolarization which are linked to the slow Ca-dependent K conductance.[12,13,21] Also, the duration of the high-threshold Ca spike in CA1 neurons from ADX rats was shortened, suggesting a more pronounced Ca-dependent inactivation of the high-threshold Ca current.[12]

While the effect of increased Ca influx during temporary shifts in corticosteroid receptor occupation may thus be relatively in control, the result of prolonged (24 hrs) exposure to very high corticosteroid levels may be much more dangerous. There are indications that basal Ca levels and depolarization-induced Ca influx are elevated under those conditions,[22,23] to the extent that delayed cell death may occur.[24] Whether these extreme conditions are indicative of the processes that occur during prolonged stress remains to be proved. Studies undertaken until now show that CA1 neurons are relatively resistant to damage induced by prolonged exposure of an animal to high levels of corticosterone[25,26] (see also Ref. 27).

Potassium (K) currents were found to be far less sensitive to steroid modulation.[28] The transient outward current I_A and the delayed rectifier were not appreciably affected by corticosteroid treatment. The inwardly rectifying K current I_Q, however, showed significant differences in amplitude (but not kinetics) with various conditions of MR/GR occupation. Thus, with predominant MR activation the I_Q amplitude is small. Additional GR occupation is associated with a large I_Q amplitude. Intermediate amplitudes were observed for tissue from ADX rats where no steroids were applied. As with the Ca currents, rather small I_Q currents were observed with moderate doses of the GR agonist RU 28362. The functional meaning of the modulation of the I_Q may be limited: the conductance is quite small in the physiological range of membrane potentials (>–90 mV).

Other voltage-gated ion currents, *e.g.*, Na currents, have not been investigated regarding their sensitivity to corticosteroids in the rat. Based on results from current clamp studies, showing that neither the spike threshold nor the shape of the Na spike were significantly altered with various conditions of MR/GR occupation, we assume that corticosteroid modulation of voltage-gated Na currents may be limited. Other steroid hormones such as androgens, though, were found to alter Na current characteristics in fish electric organ considerably.[29]

Transmitter Responses

Pyramidal neurons in the hippocampal CA1 area are recipient of many afferent projections (reviews in Refs. 30,31). The main input flow is carried by excitatory amino acids and mediated by non-NMDA receptors. Local inhibition is mainly supplied by feed-forward and feed-back GABAergic projections. In addition to these amino acids, biogenic amines mediate part of the synaptic input. Afferents containing acetylcholine (ACh), noradrenaline (NA) and serotonin (5HT), and receptors for these compounds have been demonstrated in the CA1 area. Pyramidal cells are also targets for peptidergic projections, *e.g.*, for somatostatin, neuropeptide Y, vasopressin or corticotropin-releasing factor (CRF).

Modulation of transmitter responses may take place at various stages of the signal transduction pathway. The first possibility concerns the modulation of transmitter release. Although some steroid actions on transmitter release in the hippocampus have been described,[32] this subject has received relatively little attention. It should be noted though that steroid actions on the excitability of neurons projecting to the CA1 area will certainly affect neurotransmitter release. Also, the fact that steroids have been found to alter Ca influx in soma opens the possibility that Ca channels in the synaptic region may be similarly affected, potentially leading to altered transmitter release.

A second target for steroid actions involves the neurotransmitter receptor. This may lead to, for instance, changes in the total amount of receptors, in the affinity for the ligand or in desensitization characteristics. Biochemical and molecular biology studies have shown that steroid modulation of receptor characteristics indeed occurs, *e.g.*, for the 5HT1a receptor.[33–36] For G protein-coupled receptors the next step in the cascade from transmitter receptor to eventual conductance change is the activation of intracellular messengers via G proteins. So far, few studies have investigated the possibility that steroid hormones affect these messengers.[37,38] Modulation of second messengers, however, would be a very effective way for steroids to act, since so many transmitter receptors converge on these intracellular compounds. Finally, steroids could interfere with the conductance of ion channels which are linked to transmitter receptors. At the molecular level this has not been investigated yet.

With electrophysiological methods the endpoint of the transmitter-response cascade is recorded. Any steroid-induced change in either the receptor characteristics, the receptor-activated second messenger systems or the ion channel will be reflected in an altered transmitter-evoked potential shift. The combination of electrophysiology with, *e.g.*, molecular or biochemical methods can provide more insight into the precise target for steroid actions.

Several groups have reported that the fast amino acid carried input to CA1 pyramidal neurons is modulated by corticosteroid hormones.[9–11,39–43] In most but not all cases,[39] synaptic responses recorded in tissue from ADX rats were similar to responses in tissue from adrenally intact rats. Application of low doses of corticosteroids, corresponding with a predominant MR occupation, generally maintained or even enhanced synaptic responsiveness.[9,40–42] By contrast, high to very high steroid concentrations resulting in additional GR activation reduced the response to excitatory amino acids[10,11,40–42] and (with very high steroid levels) to inhibitory amino acids.[43] The GR-evoked decrease of excitatory responses depended on the extracellular Ca concentration, in that marked steroid sensitivity was observed in the presence of high [Ca]o.[44]

Not only the amplitude but also the stability of the synaptic potentials depended on steroid receptor occupation. Both in the absence of corticosteroids but also with very high steroid levels the synaptic responses attenuated with repeated stimulation of the afferent fibers.[41,42] Stable responses were observed with predominant MR activation. This phenomenon may contribute to the observation that both in the CA1 area and in the dentate gyrus primed burst potentiation is very effective under conditions of predominant MR activation, while both in the absence of steroids or with very high steroid concentrations primed burst potentiation is poorly expressed.[6,7]

Of the biogenic amines, 5HT responses appear to be very sensitive to steroid treatment. The synthesis of 5HT, the turnover rate and (5HT1a) receptor characteristics are all affected by corticosterone (see elsewhere in this volume). Moreover, the 5HT-induced hyperpolarization of the membrane depends largely on the relative MR/GR occupation (FIG. 2). Thus, predominant MR activation results in a suppression of the 5HT responses.[45] When GRs are occupied in addition to MRs or when both MRs and GRs are unoccupied, 5HT responses are relatively large.[45,46] Responses to 5HT mediated by other receptor subtypes were not affected by steroid administration. It is presently unclear whether or not the steroid-induced altered responsiveness to 5HT is directly linked to the changes in 5HT1a receptor properties.

Steroid modulation of responses to carbachol (a metabolically stable analogue of acetylcholine) followed the same pattern as observed with 5HT.[47] With predominant MR activation responses to carbachol were small, whereas simultaneous activation of MRs and GRs induced large responses (for details see elsewhere in this volume). This was in contrast to the steroid effect on β1-adrenergic receptor-mediated responses.[13] Occupation of GRs by a selective GR-agonist or by the mixed agonist corticosterone suppressed the transmitter responses.

The effect of corticosteroid application on peptidergic responses in the hippocampus has not been investigated yet with electrophysiological methods. Putative interactions with CRF, vasopressin or enkephalin can be expected, based on interactions described in other brain regions. Clearly, by affecting voltage-gated ion conductances (*e.g.*, for Ca) steroids will at least indirectly interfere with hippocampal responses to neuropeptides.

Consequences for Neuronal Excitability

From the above reviewed studies a general picture regarding steroid actions on electrical activity emerges: corticosteroid levels resulting in a predominant activation of MRs are associated with a stable amino acid-mediated transmission; the voltage-gated Ca and K currents are small, as are the responses to most of the biogenic amines. When steroid levels rise, *i.e.*, when GRs become fully occupied in addition to the MRs, amino acid-mediated transmission is reduced, particularly with repetitive stimulation. At the same time voltage-gated Ca influx is enhanced; aminergic responses are relatively large. The shift from predominant MR to concomitant MR/GR occupation (and vice versa) is physiologically relevant, since it probably occurs daily, as a result of circadian variations in circulating steroid levels. Consequently, the intrinsic membrane properties and transmitter responsiveness of CA1 neurons may be expected to alter accordingly. It is important to realize, though, that all studies reviewed above were performed *in vitro*. Extrapolation of these results to functional processes in the whole animal should be done with extreme care. The controlled flow of ions, fuels and oxygen, and the absence of much of the afferent input *in vitro* may enhance or obscure steroid-mediated actions in comparison to the *in vivo* situation.

More extreme variations in steroid levels, which could occur during (semi) pathological conditions, also change electrical signals in the hippocampus. This refers to observations made in the absence of corticosteroids or in the presence of very high amounts of steroids. Interestingly, both in the absence of corticosteroids and with (very) high concentrations the modulation of electrical properties seems to be in the same direction: responses mediated by excitatory and inhibitory amino acid are suppressed, and depolarization of the membrane is associated with a large Ca influx. This implies a U-shaped dose dependency for the steroid-induced changes in electrical signals (FIG. 3). Naturally, the exact shape of the dose-response curve heavily depends on the amount of free corticosterone and on the properties of the MRs and GRs. These factors may be subject to changes during prolonged exposure to very high or to negligible amounts of corticosterone.

The consequences of modulation of a particular membrane characteristic by corticosterone cannot be evaluated regardless of other properties of the neuron. For instance, it was reported that high doses of corticosterone or RU 28362 enhance the amplitude of voltage-gated Ca currents. It is conceivable that excessive depolarization of the membrane, leading to large Ca influxes through voltage-gated and ligand-gated Ca channels, will result in very high local Ca gradients. Such high Ca levels can be responsible for subsequent delayed cell death. However, CA1 neurons were shown to be quite resistant to cell damage caused by prolonged exposure to high levels of corticosteroid hormones, combined with excessive depolarization. Clearly, the properties of, *e.g.*, Ca buffering and Ca extrusion mechanisms, of Ca-dependent K channels and the local balance of excitatory and inhibitory input are equally important in predicting what the consequences of an altered Ca current could be. In addition, morphological features (*e.g.*, cell size, shape of the dendrites) and the

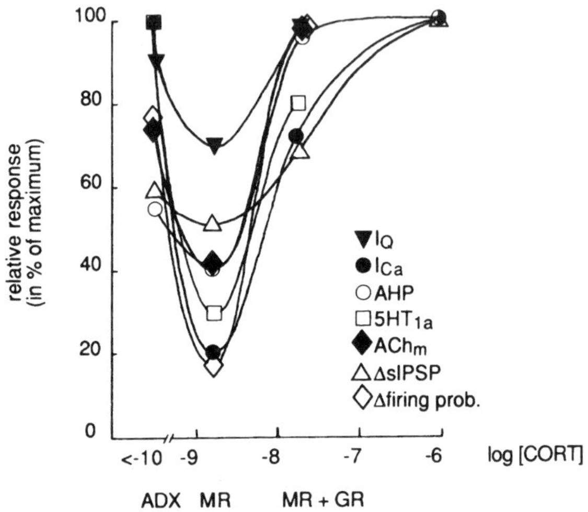

FIGURE 3. Relative ion currents and transmitters responses of CA1 pyramidal neurons as a function of corticosterone concentration or putative MR/GR occupation. The curves show that the inward rectifier I_Q, the voltage-gated Ca current, the afterhyperpolarization (AHP) linked to a Ca-dependent K conductance, the response to 5HT and to carbachol (ACh_m) as well as the stability of synaptically induced IPSPs and firing probability display a U-shaped form: extremely low steroid levels as well as very high levels result in large ion conductances, transmitter responses and unstable synaptic responses, while predominant MR-occupation (occurring with low to moderate levels) induce small ion currents, transmitter responses and stable synaptic transmission. (From Joëls & de Kloet.[50] Reprinted by permission from *Progress in Neurobiology*.)

spatial distribution of ion channels and transmitter receptors will be important determinants.

The mechanism underlying steroid-mediated actions on electrical properties is largely unknown. There are ample indications that in many cases the genome is involved. This is inferred from 1) the considerable time-delay between steroid application and appearance of effects, 2) the effectiveness of synthetic analogues which act specifically on intracellular steroid receptors, and 3) the fact that (when tested) the steroid actions did not develop in the presence of protein synthesis inhibitors.[17,48] However, since steroid actions on gene transcripts and electrical properties have not yet been studied simultaneously in a single cell, direct evidence for a role of the genome has not been provided. Therefore, we cannot exclude the possibility that the actions are partly mediated by membrane receptors for corticosteroids, as they have been identified in amphibian brain tissue.[49]

Which genes form a target for corticosteroid actions in hippocampal cells remains to be evaluated. Analysis of mRNA content has so far mainly provided evidence for effects on transmitter receptors. Clearly, the active membrane properties (*e.g.*, ion channels) and transmitter receptors may be direct targets for gene-mediated steroid actions; however, the actions can also arise secondary to other effects, such as changes in cellular energy metabolism, in the morphology of the cell, or in the activity of intracellular components (*e.g.*, phosphatases or second messengers).

In summary, the presently available data supply evidence that corticosteroids, through their daily fluctuations, affect electrical activity in the brain. The exact nature of the steroid actions will depend on the amount of free steroid, and on the local expression and properties of the two steroid receptor subtypes. The mechanism by which steroids induce their effects is largely unknown, but there are good indications that gene transcription is involved. If so, the corticosteroids provide an essentially new tool to modulate brain function, resulting in slow but persistent changes in signal transduction in the brain.

REFERENCES

1. McEwen, B. S., J. M. Weiss & L. S. Schwartz. 1968. Selective retention of corticosterone by limbic structures in rat brain. Nature **220:** 911–912.
2. Pfaff, D. W., M. T. A. Silva & J. M. Weiss. 1971. Telemetered recording of hormone effects on hippocampal neurons. Science **171:** 394–395.
3. Dafny, N., M. I. Phillips, A. N. Taylor & S. Gilman. 1973. Dose effects of cortisol on single unit activity in hypothalamus reticular formation and hippocampus of freely behaving rats correlated with plasma steroid levels. Brain Res. **59:** 257–272.
4. Ben Barak, Y., M. J. Gutnick & S. Feldman. 1977. Iontophoretically applied corticosteroids do not affect the firing of hippocampal neurons. Neuroendocrinology **23:** 248–256.
5. Saphier, D. & S. Feldman. 1988. Iontophoretic application of glucocorticoids inhibits identified neurones in the rat paraventricular nucleus. Brain Res. **453:** 183–190.
6. Diamond, D. M., M. C. Bennett, M. Fleshner & G. M. Rose. 1992. Inverted-U relationship between the level of peripheral corticosterone and the magnitude of hippocampal primed burst potentiation. Hippocampus **2:** 421–430.
7. Pavlides, C., Y. Watanabe & B. S. McEwen. 1993. Effects of glucocorticoids on hippocampal long-term potentiation. Hippocampus **3:** 183–192.
8. Joëls, M. 1994. Gene-mediated steroid control of neuronal activity. *In* Neurobiology of Steroids. E. R. de Kloet & W. Sutanto, Eds. Methods in Neurosciences. Academic Press.
9. Reiheld, C. T., T. J. Teyler & R. M. Vardaris. 1984. Effects of corticosterone on the electrophysiology of hippocampal CA1 pyramidal cells *in vitro*. Brain Res. Bull. **12:** 349–353.
10. Vidal, C., W. Jordan & W. Zieglgansberger. 1986. Corticosterone reduces the excitability of hippocampal pyramidal cells *in vitro*. Brain Res. **383:** 54–59.
11. Rey, M., E. Carlier & B. Soumireu-Mourat. 1987. Effects of corticosterone on hippocampal slice electrophysiology in normal and adrenalectomized BALB/c mice. Neuroendocrinology **46:** 424–429.
12. Kerr, D. S., L. W. Campbell, S.-Y. Hao & P. W. Landfield. 1989. Corticosteroid modulation of hippocampal potentials: increased effect with aging. Science **245:** 1505–1507.
13. Joëls, M. & E. R. de Kloet. 1989. Effects of glucocorticoids and norepinephrine on the excitability in the hippocampus. Science **245:** 1502–1505.
14. Dallman, M. F., S. F. Akana, C. S. Cascio, D. N. Darlington, L. Jacobson & N. Levin. 1987. Regulation of ACTH secretion: variation on a theme of B. Prog. Horm. Res. **43:** 113–173.
15. Reul, J. M. H. M. & E. R. de Kloet. 1985. Two receptor systems for corticosterone in rat brain: microdissection and differential occupation. Endocrinology **117:** 2505–2512.
16. Hua, S.-Y. & Y.-Z. Chen. 1989. Membrane receptor-mediated electrophysiological effects of glucocorticoid on mammalian neurons. Endocrinology **124:** 687–691.
17. Kerr, D. S., L. W. Campbell, O. Thibault & P. W. Landfield 1992. Hippocampal glucocorticoid receptor activation enhances voltage-dependent Ca conductances: Relevance to brain aging. Proc. Natl. Acad. Sci. USA **89:** 8527–8531.
18. Karst, H., W. J. Wadman & M. Joëls. 1993. Corticosteroid induced effects on voltage

dependent calcium currents in rat hippocampal CA1 neurons. Soc. Neurosci. Abstr. 291.9
19. KARST, H., M. JOËLS & W. J. WADMAN. 1993. Low-threshold calcium current in dendrites of the adult rat hippocampus. Neurosci. Lett. **164:** 154–158.
20. REGEHR, W. G. & D. W. TANK. 1992. Calcium concentration dynamics produced by synaptic activation of CA1 hippocampal pyramidal cells. J. Neurosci. **12:** 4204–4223.
21. JOËLS, M. & E. R. DE KLOET. 1990. Mineralocorticoid receptor-mediated changes in membrane properties of rat CA1 pyramidal neurons *in vitro.* Proc. Natl. Acad. Sci. USA **87:** 4495–4498.
22. ELLIOT, E. M. & R. M. SAPOLSKY. 1992. Corticosterone enhances kainic acid induced calcium elevation in cultured hippocampal neurons. J. Neurochem. **59:** 1033–1040.
23. ELLIOT, E. M. & R. M. SAPOLSKY. 1993. Corticosterone impairs hippocampal neuronal calcium regulation-possible mediating mechanisms. Brain Res. **602:** 84–90.
24. CHOI, D. W. 1990. Cerebral hypoxia: some new approaches and unanswered questions. J. Neurosci. **10:** 2493–2501.
25. SAPOLSKY, R. M. 1985. A mechanism for glucocorticoid toxicity in the hippocampus: increased neuronal vulnerability to metabolic insults. J. Neurosci. **5:** 1228–1232.
26. WOOLLEY, C., E. GOULD & B. S. MCEWEN. 1990. Exposure to excess glucocorticoids alters dendritic morphology of adult hippocampal pyramidal neurons. Brain Res. **531:** 225–231.
27. SAPOLSKY, R. M. 1992. Stress, the Aging Brain and the Mechanism of Neuron Death. MIT Press. Cambridge, MA.
28. KARST, H., W. J. WADMAN & M. JOËLS. 1993. Long-term control by corticosteroids of the inward rectifier in rat CA1 pyramidal neurons, *in vitro.* Brain Res. **612:** 172–179.
29. ZAKON, H. H. 1993. Androgen-dependent changes in the Na^+ current in electric organ. Proc. Symp. Novel Actions of Steroids, November 1993. Washington, DC.
30. LOPES DA SILVA, F. H., M. P. WITTER, P. H. BOEIJINGA & A. H. M. LOHMAN. 1990. Anatomic organization and physiology of the limbic cortex. Physiol. Rev. **70:** 453–511.
31. FROTSCHER, M., P. KUGLER & U. MISGELD, EDS. 1988. Neurotransmission in the Hippocampus. Springer Verlag. Berlin.
32. GILAD, G. M., J. M. HABEY & V. H. GILAD. 1987. Presynaptic effects of glucocorticoids on dopaminergic and cholinergic synaptosomes. Implications for rapid endocrine-neuronal interactions in stress. Life Sci. **40:** 2401–2408.
33. DE KLOET, E. R., H. SYBESMA & J. M. H. M. REUL. 1986. Selective control by corticosterone of serotonin 1 receptor capacity in raphe-hippocampal system. Neuroendocrinology **42:** 513–521.
34. MENDELSOHN, S. D. & B. S. MCEWEN. 1992. Autoradiographic analyses of the effects of adrenalectomy and corticosterone on 5HT1A and 5HT1B receptors in the dorsal hippocampus and cortex of the rat. Neuroendocrinology **55:** 444–451.
35. CHALMERS, D. T., S. P. KWAK, A. MANSOUR, H. AKIL & S. J. WATSON. 1993. Corticosteroids regulate brain 5-HT1A receptor mRNA expression. J. Neurosci. **13:** 914–920.
36. MEIJER, O. C. & E. R. DE KLOET. 1994. Corticosterone suppresses the expression of 5-HT1A receptor mRNA in rat dentate gyrus. Eur. J. Pharmacol. **266:** 255–261.
37. DUMAN, R. S., S. J. STRADA & S. J. ENNA. 1989. Glucocorticoid administration increases receptor-mediated and forskolin-stimulated cyclic AMP accumulation in rat brain cerebral cortical slices. Brain Res. **477:** 166–171.
38. SAITO, N., X. GUITART, M. HAYWARD, J. F. TALLMAN, R. S. DUMAN & E. J. NESTLER. 1989. Corticosterone differentially regulates the expression of Gsa and Gia messenger RNA and protein in rat cerebral cortex. Proc. Natl. Acad. Sci. USA **86:** 3906–3910.
39. DOI, N., S. MIYAHARA & N. HORI. 1991. Glucocorticoids promote localized activity of rat hippocampal CA1 pyramidal neurons in brain slices. Neurosci. Lett. **123:** 99–101.
40. REY, M., E. CARLIER & B. SOUMIREU-MOURAT. 1989. Effects of RU 486 on hippocampal slice electrophysiology in normal and adrenalectomized BALB/c mice. Neuroendocrinology **49:** 120–124.
41. JOËLS, M. & B. FERNHOUT. 1993. Decreased population spike in CA1 hippocampal area

of adrenalectomized rats after repeated synaptic stimulation. J. Neuroendocrinol. **5:** 537–543.
42. Joëls, M. & E. R. de Kloet. 1993. Corticosteroid actions on amino acid-mediated transmission in rat CA1 hippocampal cells. J. Neurosci. **13:** 4082–4090.
43. Zeise, M. L., A. Teschemacher, J. Arriagada & W. Zieglgansberger. 1992. Corticosterone reduces synaptic inhibition in rat hippocampal and neocortical neurons *in vitro*. J. Neuroendocrinology **4:** 107–112.
44. Talmi, M., E. Carlier, M. Rey & B. Soumireu-Mourat. 1992. Modulation of the *in vitro* electrophysiological effect of corticosterone by extracellular calcium in the hippocampus. Neuroendocrinology **55:** 257–263.
45. Joëls, M., W. Hesen & E. R. de Kloet. 1991. Mineralocorticoid hormones suppress serotonin-induced hyperpolarization of rat hippocampal CA1 neurons. J. Neurosi. **11:** 2288–2294.
46. Joëls, M. & E. R. de Kloet. 1992. Coordinative mineralocorticoid and glucocorticoid receptor-mediated control of responses to serotonin in rat hippocampus. Neuroendocrinology **55:** 344–350.
47. Hesen, W. & M. Joëls. 1993. Modulation of carbachol responsiveness in rat CA1 pyramidal neurons by corticosteroid hormones. Brain Res. **627:** 159–167.
48. Karst, H. & M. Joëls. 1991. The induction of corticosteroid actions on membrane properties of hippocampal CA1 neurons requires protein synthesis. Neurosci. Lett. **130:** 27–32.
49. Orchinik, M., T. F. Murray & F. L. Moore. 1991. A corticosteroid receptor in neuronal membranes. Science **252:** 1848–1851.
50. Joëls, M. & E. R. de Kloet. 1994. Mineralocorticoid and glucocorticoid receptors in the brain. Implications for ion permeability and transmitter systems. Prog. Neurobiol. **43:** 1–36.

DISCUSSION OF THE PAPER

P. W. Landfield (*University of Kentucky*): In your intracellular sharp electrode studies you find that the MR tends to increase excitability on most of your measures and the GR tends to reduce it. But with the calcium currents, your patch clamp studies, you did not see the effect that we saw with the GR increasing calcium influx. An increase in calcium influx reduces excitability, since it activates afterhyperpolarization, which reduces the firing of the cell. The whole cell preparation that you are using exchanges the inside of the cell with the pipette solution. You can get some unusual looking currents. Calcium currents are very sensitive. If the receptor-mediated effect requires cytosolic receptors and second messengers, you may not have the substances you need there. I do not think we can explain our differences by a concentration effect, because as you recall in our first study we showed that adrenalectomy reduced the calcium current. There are no hormones there.

Joëls: First of all, concentration is important. When you increase the concentration above 1 micromolar you should be cautious; as we should be cautious with the patch clamp technique. I will be very brief on this technical point because it involves a debate between the electrophysiologists, which may not be too interesting for other people. There is a continuum between sharp electrodes and patch electrodes. In both cases there must be dialysis of an intracellular component. Otherwise you would never be able to record calcium currents in isolation: you put cesium in your pipette and hope that it will get into the cell. People have seen effects when they put EGTA into the sharp electrode. It is not that the one gives dialysis and the other does not. It is a matter of a difference in rate. Now in terms of the intracellular exchange, I am not worried about that. We give the steroid and then let it sit for a couple of hours for the effect to develop and then we record. The final thing is that, of course, we do see clear differences between groups, which gives me a little bit of peace that at least the essential components are still there.

B. S. McEwen (*The Rockefeller University*): Does your progesterone effect to increase calcium currents require estrogen priming? You get it in an ovariectomized animal. If it is blocking MR then it probably would occur whether or not you primed with estrogens unless the synapses that are induced are ones that for some reason have a particular calcium channel.

Joëls: We set out to do the experiment exactly as you described in your papers. Unfortunately, we started thinking only when we got the results.

McEwen: It is a good addition, but it does not dilute the importance of what you have already done for progesterone. There is a potential for synapse formation, and then later it causes the destruction. That destructive phase is blocked by RU486, which implies that it may either be mediated by a glucocorticoid receptor or more likely some kind of a progesterone receptor. We do not know whether the facilitation or potentiation of estrogen induction of synapses is blocked by RU486, and it is conceivable that it might be blocked by an RU318.

Joëls: I totally agree that we should do the experiment with just progesterone.

B. Bohus (*University of California, Irvine*): Could you speculate on the implications of these findings for such fundamental neural processes as learning and memory?

Joëls: I focussed on whether or not steroid treatment could give degeneration. I showed that there are local mechanisms to protect the cell against it, but nevertheless changes in calcium currents occur. We also looked at steroid effects on the response of hippocampal cells to synaptic stimulation. Steroids indeed will affect the efficacy of the local CA1 circuitry. For any function in which this circuitry is involved—*e.g.*, learning, memory, or epileptic processes—this is very important. This agrees with findings that certain behavioral paradigms are sensitive to either MR or GR activation, but the challenge is to bring electrophysiological data together with behavioral.

J. W. Funder (*Baker Institute*): Two things about time. First, how long after adrenalectomy in the ADX animals do effects appear? Second, the intact animals being in a clean cage and being mildly stressed and having circulating levels of B that occupy both receptors. That is carried on for 4 hours and does not wash out. Yet the animals were adrenalectomized for 2 days or 5 days. Are you mixing apples and oranges?

Joëls: Most of the data I showed was in 5 to 7 days ADX animals; when not, I specifically said it developed after usually 3 to 4 days of adrenalectomy, and the change in calcium current is one of those examples. On the one hand, you evoke some effects by taking away the adrenals and waiting for a week. The effects with the OVX rats are in the same category. On the other hand, you seem to be able to restore this within hours. So you can question whether or not this is two sides of the coin or two totally different effects, and I have no ready answer for that.

Question: For example, if you take an intact rat, with that continuing pattern and in your 20 minutes of steroid administration hit it with RU486 and RU318, do you then generate the adrenalectomized pattern?

Joëls: This is something we are working on now. I see as a next step that you treat the animals *in vivo* and look *in vitro* assuming that the effects persist. To investigate this we took the response to serotonin with dependency on steroid levels and this was looked at *in vitro*. What we are doing now is treating animals *in vivo* with steroids and seeing what happens *in vitro*. We know from the *in vitro* experiments with serotonin, when you apply it continuously at very low doses (0.5 nanomolars), it takes indeed two hours for the effect to develop.

L. DOKAS (*Medical College of Ohio*): Was cycloheximide tested in slices from intact animals or ADX animals?

JOËLS: ADX animals.

DOKAS: Was the critical protein synthesis initiated during that 20 minutes or during the time when you switched cages and their receptors were occupied in intact rats?

JOËLS: No, it was done in the ADX rats. What we do is record cells before cycloheximide, then give cycloheximide, and then give a short pulse of CORT; then for the rest of the experiment the cycloheximide is present.

DOKAS: So they really are genomic effects that are happening in that 20-minute period?

JOËLS: I would not say 20 minutes. Some of the effects would take at least 2 hours to develop, too rapid for Dr. McEwen but not for me.

DOKAS: You said you could reestablish things within hours. You can reestablish synthesis of the protein within hours but, if you use GPDH as an example, synthesis really has little significant effect on the pool of enzyme activity. Unless you assume a minor isozyme that is special because it is localized in a particular place within a single cell, *e.g.*, neurons versus glia.

JOËLS: We do not have that kind of interpretational difficulty, because we do see the effects.

DOKAS: The last thing was to try and build a bridge between your talk and my talk. You said something about metabolism and ion channels, and I was wondering if you would just finish your comments on that.

JOËLS: We are looking at very artificial conditions where you control the supply of ions and of glucose and of oxygen. Everything is more ideal than normal. The only time we had a metabolic problem was when we used very high concentrations of glucose (more than twice what the cell really needs) to stimulate slices of adrenalectomized rats. Then over time, the slice has difficulty in giving synaptic responses, full synaptic responses. If you wait then and then start again the cell starts again at a normal level and after a while gets smaller again. This is helped a little bit when you give glucose, more glucose than normal.

General Discussion

D. B. DEFRANCO (*University of Pittsburgh*): Has anyone looked for corticosterone effects on calcium binding proteins in hippocampus?

M. JOËLS (*University of Amsterdam*): After adrenalectomy there was a decrease of calbindin in CA1 pyramidal cells which could be reversed by giving corticosteroids. Also the pumps are very important, the ADPases and calcium ADPases and calcium-sodium exchange.

DEFRANCO: In terms of the long-term effects of hormorne, this would make sense for protecting the cell against increasing calcium influxes.

JOËLS: Yes, it could help the cell to a certain extent, but when there are really devastating situations with high CORT levels for a long period of time and long depolarizations, the combination of those two could give very bad results.

QUESTION: Dr. Dokas, how many proteins are impacted by the presence of glucocorticoid? Is it a tenth of a percent or one percent? Did you have the opportunity to quantitate the shift in signal density on the gels or is it simply presence or absence?

L. DOKAS (*Medical College of Ohio*): No, it is quantitative. All the change in labelling bands that I showed are all based on densimetric readings.

M. C. BOHN (*University of Rochester Medical Center*): We have studied how many proteins are regulated by glucocorticoids in culture astrocytes using 2D gels, and 5 percent of the total cellular proteins and 5 percent of the secreted proteins are significantly changed by long-term glucocorticoid exposure. Only 12 of the messages are changed in the presence of cycloheximide. Some of these codings are the same proteins.

QUESTION: Are you going to promote degradation?

BOHN: No. In the presence of cycloheximide you are only seeing the directly regulated ones, and the others are probably indirectly.

F. A. ANTONI (*University of Edinburgh*): On the calcium binding protein front, calmodulin is rapidly induced in both T cells and AtT20 cells by dexamethasone, and the messenger RNA goes up within 45 minutes. It is the AtT20 cell line that is responsive to dexamethasone and in these cells the intracellular calcium transients are suppressed by pretreatment with dexamethasone.

QUESTION: Have you tried any GR agonist or antagonist? Do you know whether it is specific for the GR?

ANTONI: Sure, it is a type 2 effect. There are only type 2 receptors in AtT20 cells.

RESPONSE: There was a recent report of dexamethasone increasing calcium influx and calcium current in smooth muscle cells, so it may have a lot to do with the type of cell.

QUESTION: Dr. McEwen showed us that aldosterone enhances LTP, whereas we know that LTP is dependent on calcium influx. Of course, the calcium influx may well be through NMDA channels. I would like the electrophysiologist to start a discussion on how the steroid effects on LTP compare with the data on the actions of calcium channels that are represented by Marion Joëls.

M. JOËLS (*University of Amsterdam*): How is the synaptic input affected by steroids? I showed some data done with field potential showing that when you give high corticosterone levels you get a depression of synaptic reponse, but with aldosterone you get either a maintenance or an increase of the synaptic response. The synaptic activation is very important for the development of LTP. It has not only been shown in Dr. McEwen's group, but the first ones who really showed that very nicely were Diamond, Bennett and Rose. The glucocorticoid receptor interacts physically with one calcium binding protein, and cal reticulum will inhibit the effects of glucocorticoid receptor and other steroid receptors. We certainly have preliminary experiments suggesting that both in AtT20 and cardiac cells changes in extract cellular calcium levels affect glucocorticoid response.

QUESTION: Dr. Nemer, either by shift assays or by transfection into the AtT20 cells, is the NGRE equivalently promiscuous to a cannolico response element? If you put progesterone receptor on a glucocorticoid receptor on your gels, would you still get the formation of a dimer? What are the relative binding affinities?

M. NEMER (*Institute of Clinical Research of Montreal*): This is a very interesting question, but we do not have an answer.

Dexamethasone Reverses Adrenalectomy-Induced Neuronal De-differentiation in Midbrain Raphe-Hippocampus Axis[a]

EFRAIN C. AZMITIA AND BAOLANG LIAO

Department of Biology
New York University
1009 Main Building
100 Washington Square East
New York, New York 10003

INTRODUCTION

The first step in the formation of neurons is determination in which ectodermal cells become committed to form neuroblast.[1] During this process a cell loses its ability to develop into other cell types such as muscle or skin. This genetically establishes a cell to become neuronal despite its ability to proliferate. When a neuron differentiates, it undergoes mitosis, migration and maturation. Most of the neuronal phenotypic characteristics are expressed as the neuron loses its ability to proliferate and shows an increase in size and RNA-content in the cell body. During the maturation process a neuron usually shows the following four events: expression of specific neurochemical markers, elongation of the axon, elaboration of the dendritic tree and engagement with other neurons or target cells.

Steroids are strong differentiating agents during the development of the brain. Exposure to sex steroids early in life can produce masculinization or feminization of behavior which is characterized by the unique morphology of neurons in selected brain areas.[2] Glucocorticoids are also differentiating agents on the immature brain. Increased glucocorticoids during early neonatal periods results in decreased proliferation of cells and accelerated maturation.[3,4] Early handling, a mild form of stress, results in long-term changes in hippocampal levels of glucocorticoid receptors and in the rate of hippocampal degeneration seen during aging.[5]

It is generally assumed that once the brain is mature, the differentiating effects of steroids are lost. However, in peripheral tissues, the process of de-differentiation is well known and refers to a cell regaining its previous specific potentiality. The skeletal system is a good example.[6] In this paper we shall present evidence that de-differentiation of neurons occurs in both brainstem and forebrain, and that glucocorticoids are a specific differentiating signal for both serotonergic neurons and hippocampal granule neurons. A model will be presented that brain astroglial cells may participate in the process of both differentiation and de-differentiation. Finally, references will be made as to the generality of this phenomenon to other neural

[a]Support for this work was provided by the National Institute of Aging Program Project AG10108.

systems that appear to share the reversible loss of neuronal phenotypic characteristics. The clinical relevance of this model is also discussed.

The Serotonergic Neurons of the Brainstem

These neurons are among the earliest brain cells to complete their final mitosis. ^{3}H-Thymidine studies show that neurons in the midbrain raphe nuclei undergo their final mitosis between day 12–14 of gestation in the rat.[7] The serotonergic neurons form two groups of cells, a superior group that gives rise to ascending and an inferior group that gives rise to the descending projecting cells of the brainstem.[8,9] The superior group expresses 5-HT biosynthetic enzymes before and during migration from the ventricle while the inferior group does not show their 5-HT phenotype until they have completed their migration. Neurons begin to show both dendritic elaboration and axonal elongation very early and fibers reach all target areas prenatally. The final adult levels of 5-HT and fiber density are reached late in adult life.[10,11] In fact, serotonergic neurons retain the potential for axonal elongation throughout life provided a suitable target is made available.[12] Glucocorticoids exert a positive influence on the differentiation of serotonergic neurons.[13] The neurons in the raphe nuclei have glucocorticoid receptors (GR).[14] Injecting rat pups with ACTH produces an increase in the 5-HT levels and sprouting of 5-HT fibers in both cortex and hippocampus.[15] Conversely, removal of adrenal steroids during the first weeks of life, produces a decrease in the 5-HT innervation of the forebrain as measured by levels of tryptophan hydroxylase.[16]

In the adult brain, glucocorticoids are essential for maintaining the normal levels of 5-HT, 5-HT turnover and tryptophan hydroxylase in the midbrain and in the hippocampus.[13,17–20] These changes initially reflect a change in the phosphorylation (activation) of tryptophan hydroxylase, but with prolonged deprivation the changes are due to amount of protein.[13,21] These changes can be seen either by injections of glucocorticoids or by administering stress-inducing treatments to the animal.[21,22] For example, placing rats in the cold (5°C) for 5 hours increased the activity of tryptophan hydroxylase (FIG. 1). This treatment produced a large increase in tryptophan hydroxylase activity. However, no stress-induced increase in tryptophan hydroxylase was seen if the rats had been adrenalectomized, which deprives the animal of the normal source of corticosterone.

The increase in tryptophan hydroxylase molecules can be demonstrated by administering dexamethasone to adrenalectomized rats.[13] An increase of about 80% in midbrain tryptophan hydroxylase immunoreactivity was found in immunoblots by densitometry measures 24–72 hours after providing dexamethasone (10 μg/ml) to the drinking saline (FIG. 2).

Fluctuation in neurotransmitter enzyme is not a sufficient reason to assume that the differentiated state of a neurons had changed. Therefore, the serotonergic neurons were visualized with a specific antibody against tryptophan hydroxylase in adrenalectomized rats provided saline with or without dexamethasone. The size of the serotonergic neurons was seen to dramatically increase in response to the dexamethasone treatment (FIG. 3). The increase in the size of the cell body (50–80%) was accompanied by an increase in the size of the neuronal processes.

The morphological evidence of increased neuronal size and process thickness coupled with the increased biochemical markers implicate neuronal differentiation. The conclusion of these studies is that serotonergic neurons de-differentiate when corticosterone is removed and can rapidly re-differentiate when the steroid is replaced.

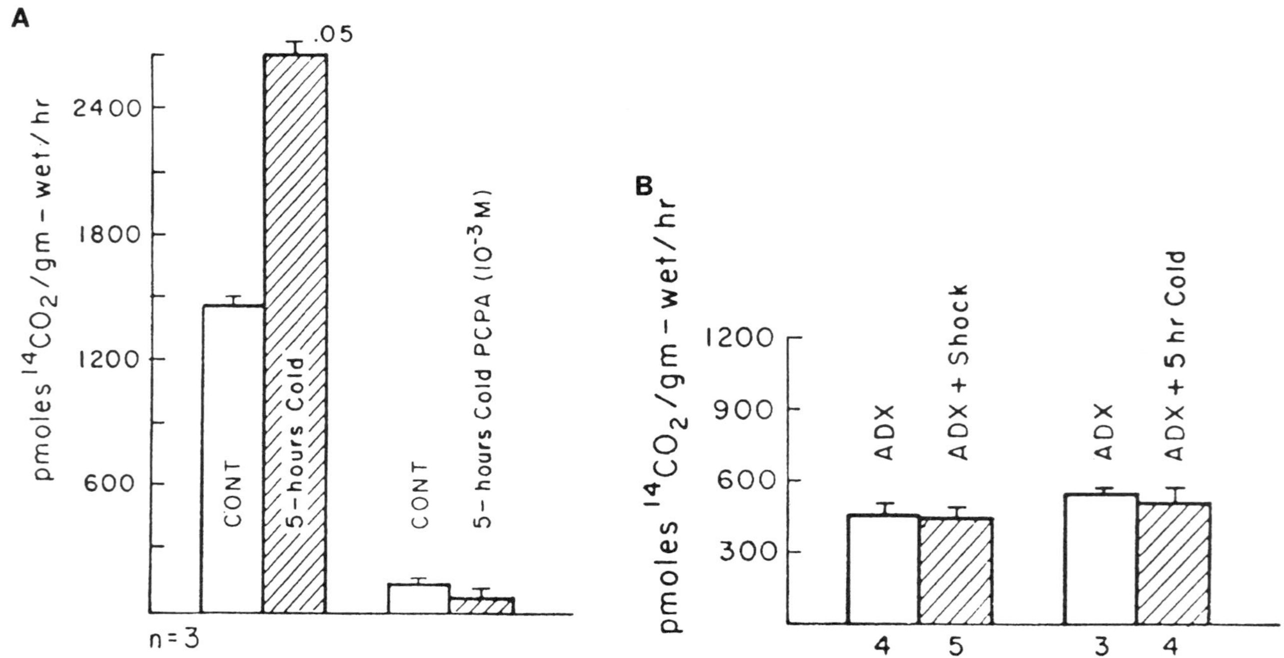

FIGURE 1. The effects of cold stress and PCPA on tryptophan hydroxylase activity in midbrain of normal and adrenalectomized rats. **(A)** Exposure of adult male rats to 5°C for 5 hours in their home cage. The PCPA was added to the reaction mixture. Student *t* test and results presented as mean ± SEM. **(B)** Rats adrenalectomized for 2 weeks exposure to cold stress as described above and to electric foot shock three times a day for 2 days. (From Azmitia.[53])

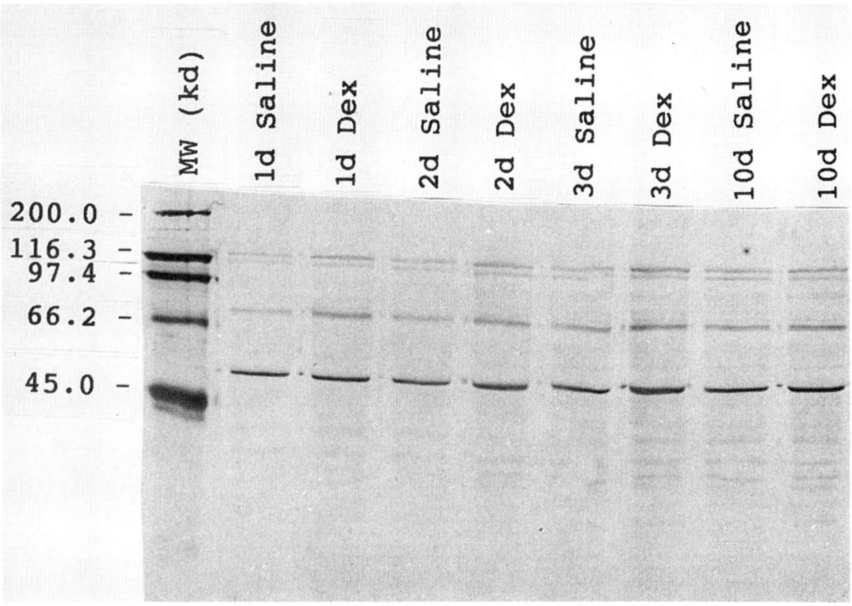

FIGURE 2. Western blots for tryptophan hydroxylase with antibody WH-412 (1/2000) using 10% SDS-PAGE. *Lane 1* is the biotinylated heavy molecular marker. *Lanes 2–9* represent the times of treatment of 0.9% saline or DEX to rats ADX for 5 days. Increase immunoreactivity was seen in the principle band at 49 kDa and the three heavier minor bands after 1 day (lanes 2 and 3). The differences in immunostaining between saline- and DEX-treated brains were maximal at 2–10 days. (From Azmitia *et al.*[13] Reprinted by permission from the *Journal of Neuroscience.*)

The Granule Neurons of the Hippocampal Dentate Gyrus

These neurons are among the last brain cells to complete their final mitosis. ^{3}H-Thymidine studies show that most neurons in the hippocampal dentate gyrus (DG) undergo their final mitosis between day 20 of gestation and postnatal day 10, but occasional neuroblast may continue to proliferate in the adult. Granule neurons may remain alive as small neuroblast for many months in the subgranular zone after they have been determined but before differentiation into granule neurons.[23]

> . . . but the dentate gyrus is less distinct because few of its granule cells have started to differentiate in its ectal arm (the arm facing the cerebral cortex) and few are recognizable as mature granule cells (defined as round-to-oval cells with large, pale nuclei) in the endal arm. A count made in animals ranging in age from newborn to 300 days of age indicated that less than 20% of all the granule cells present at 60–90 days of age could be identified during the first week of life. The scarcity of identifiable granule cells in the infant does not necessarily imply that the cells are absent, it may merely indicate a lack of differentiation of cells already formed (p. 99).

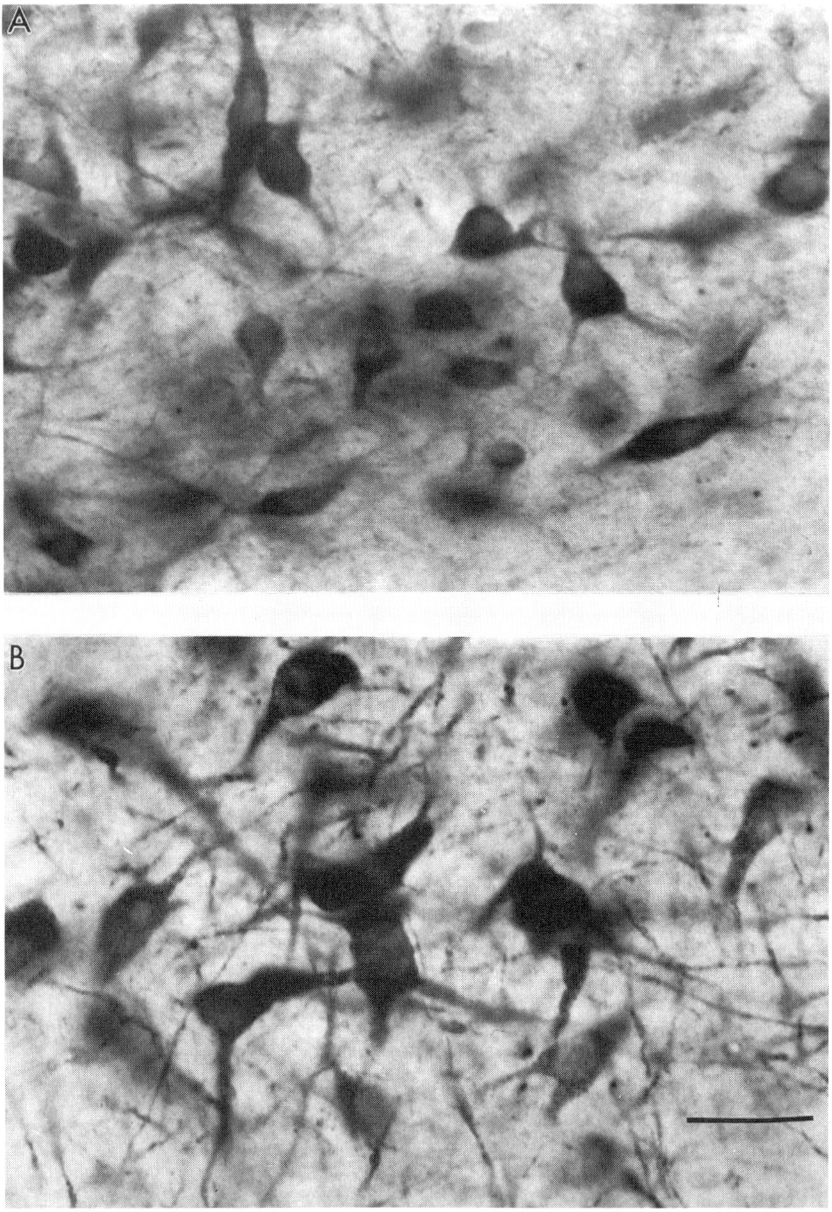

FIGURE 3. Photograph at highest magnification taken at 630 × of WH-66-IR neurons in the lateral wing subdivision of dorsal raphe nucleus (DRN). Individual cellular labelling was easily seen in ADX rats given **(A)** saline alone or **(B)** with DEX. The WH-66-IR neurons in the saline-treated brains were fusiform of oval with few thin processes emanating from the soma. The cell nucleus can be seen to be unlabeled. The neurons of the DEX brain appear more pyramidal in shape and with more pronounced bulbous dendrites extending from the soma. The nucleus was also unlabeled in these neurons. *Scale bar* is 30 microns. (From Azmitia *et al.*[13] Reprinted by permission from the *Journal of Neuroscience.*)

The oldest granule neurons are located on the outer surface of the granule cell layer. The superior blade (ectal arm, closest to CA1) forms before the infragranular blade (endal arm, closest to lateral ventricle). The granule neurons undergo neurogenesis first in the ventricular layer of the lateral ventricle and migrate into the hilar zone of the developing dentate gyrus.[24] This area forms a second and more extensive proliferative zone. The granule neurons use glutamate as a neurotransmitter, express receptors for glucocorticoids (GR), mineralocoids (MR), and 5-HT_{1A} receptors, contain calbindin and send their axons (mossy fibers) exclusively to hippocampal CA3-CA4 pyramidal neurons.[25–28] The granule dendrites in the molecular layer stain with calbindin and receive afferents from the entorhinal cortex, contralateral hippocampus, septal cholinergic neurons, locus coeruleus noradrenergic neurons and the midbrain raphe serotonergic neurons.[29]

Glucocorticoid injections to neonatal rats reduced the proliferation of granule cells in the hippocampal dentate gyrus[3] and increased the expression of GR receptors.[5] The proliferation of astrocytes in the cerebral subependymal zone are decreased by neonatal steroids.[30]

The granule cells in the adult show remarkable plasticity in reorganizing their afferent inputs.[29] After removal of circulating adrenal steroids by adrenalectomy, the granule cells begin to show pyknotic bodies in their cytoplasm within several days.[31,32] After several months, somatodendritic immunolabeling of calbindin and the cell body staining with Nissl are absent.[33] In electron microscopic studies, the granule cells appear to shrink and their cytoplasm becomes empty, although the mitochondria appear normal. Cells with typical granule neuron morphology disappear in the LT-ADX brains and it was concluded that apoptosis killed all these neurons.[34]

In our studies with young female rats (150–180 gm), long-term adrenalectomy produced similar results in DG to those described with certain added features (FIG. 4). As shown in TABLE 1, the granule neurons lose their mRNA for 5-HT_{1A} receptor at the same time they lose Nissl and calbindin staining. In addition, the area of the molecular layer was reduced by nearly 50% in the long-term adrenalectomized rats. These results confirm and extend the studies of Sloviter and establish that granule neurons in the adult dentate gyrus require circulating corticosterone to maintain their adult phenotype (differentiate state).

In the neonatal rat, adrenalectomy accelerates brain growth and increase total DNA content.[35] Since injections of glucocorticoids inhibit neurogenesis in the dentate gyrus, withdrawal of steroids may induce neurogenesis. This hypothesis has been advanced by McEwen (this volume). However, staining of the neurons in the granular layer after long-term adrenalectomy with proliferating cell nuclear antigen (PCNA) antibodies proved negative. In addition, injection of bromo-deoxyuridine (BrdU) into adult adrenalectomized rats and labeling with antibodies against BrdU and calbindin showed no double-labeled neurons. If neurogenesis was accelerated by lack of steroids, we could not detect this process after 2–3 months.

The absence of glucosteroids has a different effect on the astroglial cells. The proliferation of astrocytes was increased after withdrawal of glucocorticoids. Many glial cells in the subgranular layer and in the granular cell layer stained positive for vimentin, a cytoskeletal protein seen predominately in immature glial cells. Glial cells stained positive for PCNA and were labeled after BrdU injections in the DG after long-term adrenalectomy. In addition, the expression of 5-HT_{1A} mRNA in subgranular layer is retained as a narrow band after long-term adrenalectomy (FIG. 4). These 5-HT_{1A} receptors are probably not active since adrenalectomy down-regulates serotonin turnover in the hippocampus.[20]

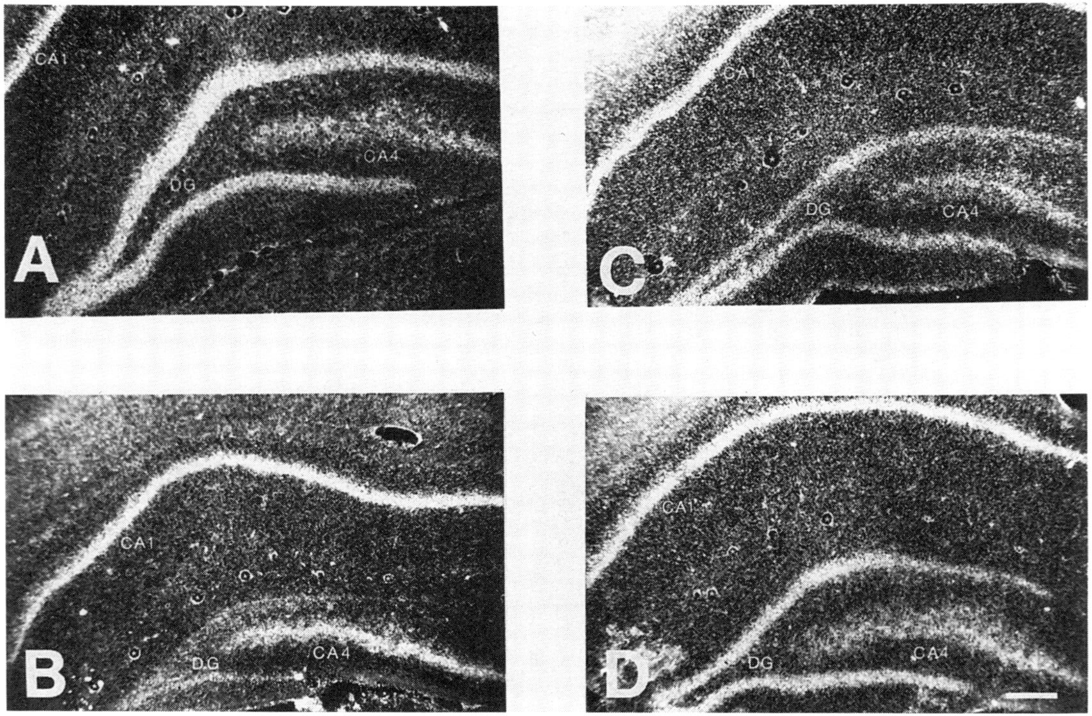

FIGURE 4. Dark-field photomicrographs of radioautography showing 5-HT_{1A} receptor mRNA labelled by ^{35}S-oligoprobes in dorsal hippocampus. (**A**) Sham ADX; (**B**) 2 month ADX; (**C**) 2 month ADX given DEX 24 hr; (**D**) 2 month ADX given DEX 72 hr. DG, dentate gyrus; CA1 and CA4, CA1 and CA4 regions of Ammon's horn. *Bar* = 200 μm. (From Liao *et al.*[25] Reprinted by permission from *Molecular Brain Research*.)

TABLE 1. Change in Dentate Gyrus after Long-Term Adrenalectomy in Adult Female Rats

1. Loss of Nissl staining in the granular cell area
2. Increase of small histone immunostaining cells in granular cell area
3. Loss of mRNA for serotonin 5-HT_{1A} receptor in granular cell area
4. Loss of calbindin immunostaining in the cell bodies of granule neurons
5. Loss of calbindin immunostaining in the dendrites of granule neurons (retraction of dendrites)
6. Decrease in area of the molecular layer
7. Increase of vimentin immunostaining glial cells in the subgranular zone

The loss of neuronal staining in the granular layer may represent death or a de-differentiation of the neurons (FIG. 5). One way to distinguish between these two possibilities is to determine if steroid replacement in the LT-ADX can quickly restore the mature granule neurons. If the neurons are dead, dexamethasone administration should have no effect on DG morphology. However, if the neurons are de-differentiated, then dexamethasone administration should restore normal morphology of the DG. This latter effect would be consistent with the effects of steroids on midbrain serotonergic neurons after ADX. Dexamethasone (10 μg/ml) was placed in the drinking saline and the LT-ADX rats were perfused after 24 or 72 hr. A dramatic increase in Nissl and 5-HT_{1A} mRNA labeling occurred after 24 hr and reached nearly normal levels by 72 hr.[25] The area of the DG molecular layer showed a progressive return to normal. In addition, calbindin-positive cell bodies and dendrites were again seen in the DG. Staining with PCNA and BrdU after BrdU injections gave no indications that massive neurogenesis had been induced by the steroid. The rapid return of these differentiated granule neurons supports the idea that glucocorticoids may be necessary to maintain the differentiated state of granule neurons.

Neuronal-Glial Hypothesis: Serotonin and S-100

The mechanism of this steroid-induced differentiation remains unclear. However, it would not be unreasonable to suspect that glial cells and serotonin play a role. The vimentin-IR radial glial cells seen in the LT-ADX were "matured" by the short-term treatment with dexamethasone. This is reminiscent of what occurs during development where steroids are known to suppress glial proliferation and induce their differentiation. In our studies, the dexamethasone treatment to LT-ADX rats produced a pronounced increase in branching and somal size of the vimentin-IR radial glial cells, and in certain studies actually produced a marked decrease in the amount of vimentin immunoreactivity. Vimentin is replaced in mature glial cells by a glial acidic fibrillary protein (GFAP).[36] All the evidence therefore suggests that dexamethasone treatment after LT-ADX performed the same functions in adults as seen during development.

5-HT_{1A} receptors are located on radial glial cells and astrocytes during gestation.[37] Stimulation of cultured astrocytes with a 5-HT_{1A} receptor agonist leads to a release of S-100β.[38] This soluble glial protein is a known neurotrophic factor which promotes maturation of serotonergic and cortical neurons.[39] In developmental stud-

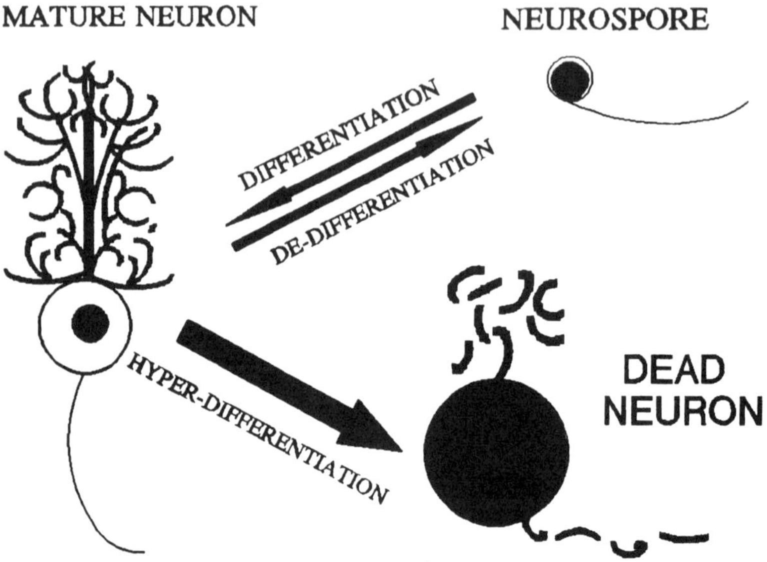

FIGURE 5. Schematic model of the plasticity of neurons in the adult brain. The model assumes differentiation and de-differentiation are reversible processes governed by epigenetic factors such as neurotransmitters (*e.g.*, serotonin, NMDA), neurotrophic factors (*e.g.*, S100, NGF) and steroids (*e.g.*, glucocorticoids, estrogen). In the presence of an appropriate differentiating factor, the neuron assumes a mature phenotype. In the absence of these same differentiating factors, the mature neuron can de-differentiate into a stable neurospore. It is assumed that the axons of these neurospores retract from the synapse but are maintained in the vicinity with the support of oligodendroglial cells. When the differentiating agent is administered, the neurospore can quickly assume the phenotype of a mature neuron and its axon re-engage with its target cells. However, if the mature neuron is continuously exposed to the actions of differentiating agents it will undergo hyperdifferentiation, which leads irreversibly to neuronal death. The neurodegenerative cascade would be consistent with the prolonged actions of glucocorticoids, serotonin and NMDA drugs.

ies, when 5-HT is reduced in the forebrain, cortical neurons and glial cells are arrested in an immature state.[40–42] The glial cells fail to migrate properly into the cortex, the neurons show less branching and synaptic density and cortical thickness are reduced. The glial migration and cortical thickness, parameters of an immature brain, are rapidly reversed by the injection of a 5-HT_{1A} agonist.[40] The hypothesis has been advanced that 5-HT interaction with the 5-HT_{1A} receptor on astrocytes promotes the maturation of brain.[43]

In adults, 5-HT levels maintain the levels of S-100β in the brain. Mature glial cells express 5-HT_{1A} receptor.[44] Removal of adult 5-HT fibers from the hippocampus by either 5,7-dihydroxyltryptamine (a 5-HT fiber toxin) or para-chlorophenylalanine (an inhibitor of tryptophan hydroxylase) reduces S-100β levels and immunoreactivity.[45,46] Interesting, para-chlorophenylalanine (PCPA, a serotonin synthesis inhibitor) treatment to adult rats for 1 week results in a loss of synaptic density in cortex between 30–50%.[42] We have found that a similar treatment in adult rats results in the loss of cortical and hippocampal immunolabeling with antibodies against synaptophysin and MAP-2.[47] Thus, 5-HT may be essential for maintaining the adult differentiated states of neurons and glial cells in the cortex and hippocampus.

Clinical Implication of De-differentiated Neurons

In our studies, the withdrawal of circulating glucocorticoids in the adult rat produces a loss of adult neuronal-phenotypic staining in both the midbrain raphe nuclei and in the dentate gyrus. In the DG, loss of steroids was associated with an increase in immature (vimentin-positive) glial cells. Similar conclusions were drawn from studies in which the trophic factors were withdrawn from adult cholinergic neurons. In the septal-hippocampal system, cutting the fornix-fimbria results in a loss of choline acetyltransferase and nerve growth factor receptor immunocytochemistry and Nissl staining in the septum.[48] Finally, in an adult Addison's patient who was not treated with cortisol, the DG showed numerous necrotic neurons with karyorrhexis throughout the granular layer with no evidence of gliosis.[49] We propose that these "lost" neurons are not dead, but de-differentiated with loss of their adult phenotype.

In our animal studies, providing dexamethasone after 2–3 months of ADX led to a rapid return of the normal mature neuronal phenotype.[50] When NGF is infused into the ventricles of rats up to 95 days after the transection of the fornix-fimbria, the mature cholinergic phenotype quickly emerges within 3–7 days.[48] Finally, in adult Addison's patients of several years' duration, even if the patient had several episodes of Addisonian crisis, providing cortisone treatment resulted in a hippocampus showing no nerve cell loss.[49]

How long can de-differentiated neurons survive? Hibernating animals show neuronal de-differentiation for several months with a rapid return to normal phenotype within hours after awakening.[51] The changes in mature neuron morphology of granule neurons without steroids[50] or cholinergic neurons without NGF[48] can be reversed after 3 months within days. Addison's patients of several years' duration who have experienced multiple crises do have normal hippocampus at autopsy if given cortisone.[49] During development, granule cells can remain undifferentiated for up to 3 months.[23] It may be proposed that a neuron that has de-differentiated can remain in suspended animation as a "neurospore" for the life of the animal. In this state, the dendrites and cell body shrink but the axon appears to be maintained. Such a situation would have major implications in the study of aging and neurodegenerative diseases where the etiology is due to loss of a differentiating agent, such as hormone, trophic factor or neurotransmitters (transtrophins).

In conclusion, 5-HT and glucocorticoids may represent either two parallel and redundant differentiating processes or may function by the same process. Studies to date have not allowed a choice between these two alternatives. However, it is interesting to note that the loss of synaptophysin and MAP-2 by PCPA injections in the adult hippocampus is reversed by treatment with dexamethasone.[52]

SUMMARY

Differentiation leads to specific morphological and biochemical characteristics. We examined whether epigenetic factors (*e.g.*, glucocorticoids) are required to maintain neuronal differentiation in the adult brain. In the midbrain, adrenalectomy (ADX) (1–2 wk) reduced the size of tryptophan hydroxylase (WH)-immunoreactive (IR) neurons. ADX rats exposed to short-term (24–72-h) dexamethasone (ST-DEX) in the drinking saline (10 mg/l) showed an increase in WH protein, somal area and dendritic size of WH-IR neurons.

In the hippocampus, ADX for 2–3 mo (long-term; LT) reduced Nissl staining, calbindin (CBD)-IR and 5-HT_{1A} receptor mRNA in the granular cell layer, and the size of the molecular layer and its CBD-IR dendrites. Small vimentin (Vim)-IR glial cells emerged in the granular layer. ST-DEX after LT-ADX rapidly induced a recovery of 5-HT_{1A} mRNA, Nissl labeling and CBD-IR in the granule cell layer. In the molecular layer, there was an increase in the area and in the number of CBD-IR dendrites. Furthermore, the Vim-IR glial cells were enlarged in size and branching.

The rate of cell proliferation was studied in these animals. Immunostaining with antibodies against proliferating cell nuclear antigen (PCNA) and use of bromouridine argue against enhanced neurogenesis after ST-DEX in LT-ADX.

We propose that glucocorticoids induce and maintain differentiation of serotonergic and CBD-IR neurons in the midbrain-hippocampal axis. A neuronotrophic role for the glial 5-HT_{1A} receptor is suggested.

ACKNOWLEDGMENTS

The authors wish to thank Xiao Ping Hou, Yun-Shuo Chen and Barbara Miesak for their technical assistance in various aspects of this work.

REFERENCES

1. Jacobson, M. 1991. Developmental Neurobiology. Third edit. Plenum Press. New York and London.
2. Beach, F. A. 1975. Hormonal modification of sexually dimorphic behavior. Psychoneuroendocrinology **1:** 3–23.
3. Bohn, M. C. 1980. Granule cell genesis in the hippocampus of rats treated neonatally with hydrocortisone. Neuroscience **5:** 2003–2012.
4. Bohn, M. C. 1984. Glucocorticoid induced teratologies of the nervous system. *In* Neurobehavioral Teratology. J. Yanai, Ed. 365–387. Elsevier. Amsterdam.
5. Meaney, M., D. Aitken, S. Bhatnagar, C. Van Berkel & R. Sapolsky. 1988. Post-natal handling attenuates neuroendocrine, anatomical and cognitive impairments related to the aged hippocampus. Science **238:** 766–768.
6. Reddi, A. H. 1982. Local and systemic mechanisms regulating bone formation and remodeling: an overview. *In* Current Advances in Skeletogenesis: Development, Biomineralization, Mediators, and Metabolic Bone Disease. M. Silbermann & H. C. Slavkin, Eds. 77–86. Experta Medica. Amsterdam.
7. Lauder, J. M. & F. E. Bloom. 1974. Ontogeny of monoamine neurons in the locus coeruleus, raphe nuclei and substantia nigra of the rat. J. Comp. Neurol. **155:** 469–482.
8. Lidov, H. G. W. & M. E. Molliver. 1982. Immunohistochemical study of the development of serotonergic neurons in the rat CNS. Brain Res. Bull. **9:** 559–604.
9. Wallace, J. A. & J. M. Lauder. 1983. Development of the serotonergic system in the rat embryo: an immunocytochemical study. Brain Res. Bull. **10:** 459–479.
10. Lidov, H. G. W. & M. E. Molliver. 1982. An immunohistochemical study of serotonin neuron development in the rat: ascending pathways and terminal fields. Brain Res. Bull. **8:** 389–430.
11. Azmitia, E. C., M. J. Brennan & D. Quartermain. 1983. Adult development of the hippocampal-serotonin system of C57BL/6N mice; analysis of high-affinity uptake of ^{3}H-5HT. Intl. J. Neurochem. **5:** 39–44.
12. Azmitia, E. C., M. J. Perlow, M. J. Brennan & J. M. Lauder. 1981. Fetal raphe and hippocampal transplants into adult and aged C57BL/6N mice: a preliminary immunocytochemical study. Brain Res. Bull. **7:** 703–710.

13. AZMITIA, E. C., B. LIAO & Y.-S. CHEN. 1993. Increase of tryptophan hydroxylase enzyme protein by dexamethasone in adrenalectomized rat midbrain. J. Neurosci. **13:** 5041–5055.
14. HÄRFSTRAND, A., K. FUXE, A. CINTRA, L. F. AGNATI, I. ZINI, A.-C. WIKSTROM, S. OKRET, Z.-Y. YU, M. GOLDSTEIN, H. STEINBUSCH, A. VERHOFSTAD & J.-A. GUSTAFSSON. 1986. Glucocorticoid receptor immunoreactivity in monoaminergic neurons of rat brain. Proc. Natl. Acad. Sci. USA **83:** 9779–9783.
15. ALVES, S. E. 1994. Neonatal ACTH administration alters the hypothalamic-pituitary-gonadal (HPG) and adrenal (HPA) aves in female rats: evidence from monoamine innervation reproductive parameters and the stress response. Ph.D. dissertation. New York University.
16. SZE, P. Y. 1980. Glucocorticoids as a regulatory factor for brain tryptophan hydroxylase during development. Dev. Neurosci. **3:** 217–223.
17. AZMITIA, E. C. & B. S. MCEWEN. 1969. Corticosterone regulation of tryptophan hydroxylase in midbrain of the rat. Science **166:** 1274–1276.
18. AZMITIA, E. C., S. ALGERI & E. COSTA. 1970. *In vitro* conversion of ^{3}H-L-tryptophan into ^{3}H-serotonin in brain areas of adrenalectomized rats. Science **169:** 201–203.
19. BOADLE-BIBER, M. C., K. C. CORLEY, L. GRAVES, T. PHAN & J. ROSECRANS. 1989. Increase in the activity of tryptophan hydroxylase from cortex and midbrain of male Fischer 344 rats in response to acute or repeated sound stress. Brain Res. **482:** 306–316.
20. DE KLOET, E. R., G. L. KOVACS, G. SZABO, G. TELEGDY, B. BOHUS & D. H. G. VERSTEEG. 1982. Decreased serotonin turnover in the dorsal hippocampus of rat brain shortly after adrenalectomy: selective normalization after corticosterone substitution. Brain Res. **239:** 659–663.
21. SINGH, V. B., K. C. CORLEY, H. PHAN & M. C. BOADLE-BIBER. 1990. Increases in the activity of tryptophan hydroxylase from rat cortex and midbrain in response to acute or repeated sound stress are blocked by adrenalectomy and restored by dexamethasone treatment. Brain Res. **516:** 66–76.
22. AZMITIA, E. C. & B. S. MCEWEN. 1974. Adrenalcortical influence on rat brain tryptophan hydroxylase activity. Brain Res. **78:** 291–302.
23. ALTMAN, J. & S. A. BAYER. 1975. Postnatal development of the hippocampal dentate gyrus under normal and experimental conditions. *In* The Hippocampus. R. L. Isaacson & K. H. Pribram, Eds. Vol. 1: 95–122. Plenum Press. New York.
24. ALTMAN, J. & S. A. BAYER. 1990. Migration and distribution of two populations of hippocampal granule cells precursors during the perinatal and postnatal periods. J. Comp. Neurol. **301:** 365–381.
25. LIAO, B., B. MIESKA & E. C. AZMITIA. 1993. Loss of 5-HT_{1A} receptor mRNA in the dentate gyrus of long-term adrenalectomized rats and rapid reversal by dexamethasone. Mol. Brain Res. In press.
26. ARONSSON, M., K. FUXE, Y. DONG, L. F. AGNATI, S. OKRET & J. A. GUSTAFSSON. 1988. Localization of glucocorticoid receptor mRNA in the male rat brain by *in situ* hybridization. Proc. Natl. Acad. Sci. USA **85:** 9331–9335.
27. ISAACSON, R. L. & K. H. PRIBRAM. 1975. The Hippocampus. Vol. 2. Neurophysiology and Behavior. Plenum Press. New York & London.
28. CHAN-PALAY, V. & C. KOHLER. 1989. The Hippocampus: New Vistas. Alan R. Liss, Inc. New York.
29. LYNCH, G., G. ROSE, C. GALL & C. W. COTMAN. 1973. The response of the dentate gyrus to partial deafferentation. *In* The Golgi Centennial Symposium. R. Santini, Ed. Raven Press. New York.
30. BOHN, M. C. 1979. Effects of hydrocortisone on neurogenesis in the neonatal rat brain: a morphological and autoradiographic study. Ph.D. dissertation. University of Connecticut at Storrs.
31. WOOLLEY, C. S., E. GOULD, R. R. SAKAI, R. L. SPENCER & B. S. MCEWEN. 1991. Effects of aldosterone or RU28362 treatment on adrenalectomy-induced cell death in the dentate gyrus of the adult rat. Brain Res. **554:** 312–315.
32. GOULD, H., C. S. WOOLLEY & B. S. MCEWEN. 1990. Short-term glucocorticoid man-

ipulations affect neuronal morphology and survival in the adult dentate gyrus. Neuroscience **37:** 367–375.
33. Sloviter, R. S., G. Valiquette, G. M. Abrams, E. C. Ronk, A. L. Sollas, L. A. Paul & S. Neubort. 1989. Selective loss of hippocampal granule cells in the mature rat brain after adrenalectomy. Science **243:** 535–538.
34. Sloviter, R. S., E. Dean & S. Neubort. 1993. Electron microscopic analysis of adrenalectomy-induced hippocampal granule cell degeneration in the rat: apoptosis in adult central nervous system. J. Comp. Neurol. **330:** 337–351.
35. Meyer, J. S. 1985. Biochemical effects of the corticosteroids on neural tissues. Physiol. Rev. **65:** 946–1020.
36. Dahl, D., D. C. Rueger, A. Bignami, K. Weber & M. Osborn. 1981. Vimentin, the 57000 molecular weight protein of fibroblast filaments, is the major cytoskeletal component in immature glia. Eur. J. Cell Biol. **24:** 191–196.
37. Hellendall, R. P., U. Schambra, J. Liu, G. R. Bresse, D. E. Millhorn & J. Lauder. 1992. Detection of serotonin receptor transcripts in the developing nervous system. J. Chem. Neuroant. **5:** 2799–310.
38. Whitaker-Azmitia, P. M., R. Murphy & E. C. Azmitia. 1990. Stimulation of astroglial 5-HT_{1A} receptors releases the serotonergic growth factor, protein S-100, and alters astroglial morphology. Brain Res. **528:** 155–158.
39. Azmitia, E. C., W. S. T. Griffin, D. R. Marshak, L. J. Van Eldik & P. M. Whitaker-Azmitia. 1992. S-100β and serotonin: a possible astrocytic-neuronal link to neuropathology of Alzheimer's disease. Prog. Brain Res. **94:** 459–473.
40. Akbari, H. M., P. M. Whitaker-Azmitia & E. C. Azmitia. 1994. *In utero* exposure to cocaine results in decreased availability of the tropic factor S-100β: a possible mechanism for cocaine-induced microcephaly. Neurosci. Lett. **170:** 141–144.
41. Daugherty, J. A. & J. H. Haring. 1989. Effects of neonatal serotonin depletion upon development of rat somatosensory cortex. Soc. Neurosci. Abstr. **15:** 1050.
42. Okado, N., L. Cheng, S.Tanatsugu & K. Hamaguchi. 1993. Synaptic loss following removal of serotonergic fibers in newly hatched and adult chickens. J. Neurobiol. **24:** 687–691.
43. Azmitia, E. C. & P. M. Whitaker-Azmitia. 1991. Awaking the sleeping giant: anatomy and plasticity of the brain serotonergic system. J. Clin. Psychiatry **52:** 4–16.
44. Whitaker-Azmitia, P. M., C. Clarke & E. C. Azmitia. 1993. Localization of 5-HT_{1A} receptors to astroglial cells in adult rats: implication for neuronal-glial interactions and psychoactive drug mechanism of action. Synapse **14:** 201–205.
45. Azmitia, E. C., D. R. Marshak & P. M. Whitaker-Azmitia. 1990. Functional interactions between glial S-100β and CNS 5-HT neurons. J. Cell Biochem. **8:** 14F.
46. Haring, J. H., A. Hagan, J. Olson & B. Rodgers. 1993. Hippocampal serotonin levels influence the expression of S-100β detected by immunocytochemistry. Brain Res. **631:** 119–123.
47. Whitaker-Azmitia, P. M., A. Borella, N. Raio, N. Okado & E. C. Azmitia. 1994. Serotonin as a synapse maintenance factor in the adult hippocampus: loss of synaptic markers after PCPA treatment. Soc. Neurosci. Abstr. **20:** 200.
48. Hagg, T., B. Fass-Holmes, H. L. Vahlsing, M. Manthrope, J. M. Conner & S. Varon. 1989. Nerve growth factor (NGF) reverses axotomy-induced decreases in choline acetyltransferase, NGF receptor and size of medial septum cholinergic neurons. Brain Res. **505:** 29–38.
49. Maehlen, J. & A. Torvik. 1990. Necrosis of granule cells of hippocampus in adrenalcortical failure. Acta Neuropathol. **80:** 85–87.
50. Liao, B. & E. C. Azmitia. 1994. Differential effects of corticosteroids on neuronal and glial cells in the dentate gyrus of adult rat hippocampus. Soc. Neurosci. Abstr. **20:** 600.
51. Popov, V. I., L. S. Bocharova & A. G. Bragen. 1992. Repeated changes of dendritic morphology in the hippocampus of ground squirrels in the course of hibernation. Neuroscience **48:** 45–51.
52. Azmitia, E. C., J. Huang & P. M. Whitaker-Azmitia. Steroid interactions with serotonin as a synaptic maintenance factor in adult hippocampus: reversal of synaptic loss by dexamethasone. Soc. Neurosci. Abstr. **20:** 200.

53. AZMITIA, E. C. 1973. Rapid alterations in rat tryptophan hydoxylase activity: effects of adrenal steroids, drugs and stressors. Ph.D. dissertation. The Rockefeller University. New York.

DISCUSSION OF THE PAPER

QUESTION: It did not look like you saw any changes in the pyramidal neurons after ADX and yet those have high steroid receptors.

AZMITIA: We did not really see much of a change in the CA neurons. In certain animals with calbindin staining, it looked like the calbindin staining was reduced, but by and large the CA neurons were spared the effects of long-term adrenalectomy.

QUESTION: Could you speculate, if the loss of cells is a direct effect of glucocorticoids, why granule neurons would be more susceptible?

AZMITIA: If neuronal plasticity is being regulated by astrocytes, there is an increased number of astrocytes in the granular cell layer compared to the CA regions. This means that maybe these granular neurons are much more dependent on products coming from granular glial cells than CA neurons might be. So they may in fact have different ways of maintaining their differentiated states.

QUESTION: It is really an amazing effect, and it is unlikely that you would have restoration of that number of lost cells in 24 or 72 hours after long-term adrenalectomy. I am not yet convinced by the studies of bromo-deoxyuridine. I wonder why you chose embryonic day 21 rather than postnatal periods for labeling the granule neurons?

AZMITIA: We tried postnatal day 4, and the number of labelled cells that we could detect in the adult granular layer was less than when we were able to label at E21. Whether that is because we were injecting the mother in one case and the pups in the other case, I am not sure.

QUESTION: That is surprising, because the peak of neurogenesis is postnatal of the granular cells.

AZMITIA: It could be how long the cells have been exposed to the bromodeoxyuridine rather then the actual rate of neurogenesis.

QUESTION: The phenomena you described in terms of cell shrinkage and a gain in cell size is really not restricted to the brain. In fact people studying direct effects of steroids a number of years ago in fibroblast and our own work in HeLa cells have shown that glucocorticoids in particular really have profound effects on the sizes of cells and HeLa cells can almost double in volume in response to glucocorticoids. So this occurs in your physiological system. This may be a more general and widespread type of steroid effect that perhaps others should look at.

AZMITIA: I agree.

P. W. LANDFIELD (*University of Kentucky*): It is a very nice series of studies and probably very important and may help to resolve a number of ongoing issues that have been brought up with regard to these kinds of studies. It really emphasizes the point you made that just because you do not see it, does not mean it is not there. I would like to make an advertisement for somebody doing a semithin section or a low-power light electronmicroscopic analysis of this phenomenon. It is important to use multiple methods so that you get some sort of triangulation and you do not miss something. Many phenomena, such as aging, as well as a number of other things, can alter stainability of neurons.

AZMITIA: The classic example that I know of was the studies that Rusty Gauge did where he cut the fornix-fimbria and showed that the cholinergic neurons in the septum degenerated. He could prevent that degeneration by giving NGF. So his idea was that NGF maintained the cells alive. Sylvio Varon also cut the fornix-fimbria and let the cells "die," waited two weeks, then gave NGF, and the cholinergic cells cam back again. So this is not just a situation restricted to granule cells and it may in fact be a much more widespread phenomenon.

QUESTION: Since the neurons appear to be coming back, what do you think about their own suvival factors or differentiation factors? I am thinking of the neurotrophins NT3 and BDNF. Have you looked at their expression or their regulation?

AZMITIA: We have looked at the trophic factor S100, and S100 seems to go down following adrenalectomy. S100 is located in the glia cells.

QUESTION: Right. If the neurons are coming back, if they are there, and they come back, maybe not through a cell division process.

AZMITIA: These cells show no evidence of being new neurons. We have tried several methods for detecting neurogenesis.

QUESTION: Right, survival factors are re-regulated when you put the steroids back.

AZMITIA: Yes, a lot of people when they study survival factors will use specific stains to in fact determine whether a substance is a survival factor. In many cases it is not that the neuron has died but that the neuron has stopped making the protein which allows you to identify it. NGF has as one of its main affects, to increase the cholinergic cells markers, not to keep those cells alive.

B. S. MCEWEN (*The Rockefeller University*): One technical question about the ADX animals that you studied. Did you follow their salt intake, say, in a two-bottle preference test to find out if the animals you were studying at the end of two months showed an elevated salt preference as they would if there was still an inadequate amount of accessory adrenal tissue regrowth?

AZMITIA: No, we did not. But that is a good idea. It might be a good idea if we all do that in the future to keep track of it.

Glucocorticoids and Neurotrophin Gene Regulation in the Nervous System[a]

DAN LINDHOLM, MAIJA CASTRÉN, BASTIAN HENGERER,[b] AXEL LEINGÄRTNER, EERO CASTRÉN, AND HANS THOENEN

Department of Neurochemistry
Max Planck Institute for Psychiatry
Am Klopferspitz 18 A
D-82152, Martinsried, Germany
and
[b]Ciba-Geigy AG
CH-40002 Basel, Switzerland

INTRODUCTION

Glucocorticoid hormones play an important role in mediating the stress response in brain, but the hormones also influence survival of specific neuronal populations.[1,2] Adrenal steroids exert their gene effects by interacting with specific hormone receptors belonging to the superfamily of steroid receptors.[3] In the brain, two types of glucocorticoid receptors, type I or mineralocorticoid receptor (MR) and type II or classical glucocorticoid receptor (GR), are present with a partly overlapping distribution.[4–6] Corticosterone, which is the physiological glucocorticoid in rats, interacts with both receptors in brain cells albeit with different affinities.[7] Interestingly, the GRs are widely distributed in rat brain,[4] while the MR, which binds mineralocorticoids and natural glucocorticoids with high affinity, is highly expressed particularly in the hippocampus.[5,6,8] Hippocampus is also the site of synthesis of many of the neurotrophic factors which have important functions in promoting survival and differentiation of certain populations of neurons.[9] Thus, nerve growth factor (NGF), brain-derived neurotrophic factor (BDNF), and neurotrophin-3 (NT-3) are all expressed in the rat hippocampus although in different relative proportions.[10,11] It seems that at least in the intact rat brain the expression of the neurotrophins is confined largely to neurons. In addition, the levels of neuronal NGF and BDNF mRNAs are increased by enhanced neuronal activity following glutamate[12] and muscarinic receptor stimulation.[13]

Besides neurotrophins, hippocampal neurons also express the TrkB and TrkC tyrosine kinase receptors, which mediate the effects of BDNF and NT-3, respectively (for a review see Ref. 14). Recently it was shown that BDNF and NT-3 are able to elevate intracellular calcium in hippocampal neurons.[15] This finding supports the view that neurotrophins might have autocrine or paracrine actions within this brain region. On the other hand, adrenal steroids have been shown to influence survival of hippocampal neurons in addition to having effects on neuronal excitability.[1,16] In order to study whether the actions of glucocorticoids involve alterations in neurotrophic gene expression, we have analyzed the effects of the hormones on the levels of mRNA

[a]MC is supported by a grant from the Finnish Academy of Science.

encoding different neurotrophins. In the present paper we would like to discuss the effects of glucocorticoid hormones on neurotrophin gene expression in brain and periphery in more detail and present some evidence indicating that some of the hormonal effects on neurotrophin expression levels involve transcriptional events.

Effects of Glucocorticoid Hormones on NGF Expression in the Peripheral Nerve

Transection of a peripheral nerve such as the sciatic nerve leads to upregulation of NGF synthesis in non-neuronal cells surrounding the axons.[17] This was shown to be mainly due to the invasion of the lesioned nerve by activated macrophages,[17] and interleukin-1 (IL-1) was identified as one of the major factors involved.[18] However, it also became clear that the lesion- and IL-1-mediated increase in NGF production could be inhibited by the synthetic glucocorticoid, dexamethasone.[19,20] In addition, by using sciatic fibroblasts in culture it was shown that dexamethasone rapidly decreases the levels of NGF mRNA in these cells.[19] The downregulating effect of dexamethasone was also observed in transfection assay employing the mouse NGF promoter[21] linked to a choline acetyl transferase (CAT) reporter gene.[19] This data suggested that dexamethasone decreases NGF levels in the sciatic fibroblast via a direct effect on the NGF gene transcription. Likewise studies from other groups have shown that dexamethasone as well as aldosterone are able to downregulate NGF mRNA and protein levels in the fibroblasts cell line, L-929.[22,23] However, the physiological significance of the strong downregulating effect of glucocorticoid hormones on NGF expression in non-neuronal cells is not fully understood. It is also not clear whether the decreased levels of NGF in the lesioned peripheral nerve affect subsequent regeneration. Preliminary data on the transection of the sciatic nerve in dexamethasone-treated rats indicated to impairment of outgrowth of the regenerating nerve fibers (Bresjanac, Sketelj, and Lindholm, unpublished) which could be due to the presence of other growth factors. Recently, it was shown that BDNF is also upregulated following this nerve lesion paradigm.[24] However, it is not known whether BDNF levels are influences by glucocorticoids in the peripheral nervous system.

Effects of Glucocorticoids on Neurotrophins in Brain

In contrast to the periphery, administration of glucocorticoids was found to increase NGF mRNA levels in the rat brain.[25,26] As shown in FIGURE 1, there was a rapid increase in hippocampal NGF mRNA in adult mice after injection of corticosterone. Following the maximal rise in NGF mRNA at 3-h, the mRNA levels in hippocampus declined subsequently to below those of controls (FIG. 1). In cortex, NGF mRNA increased more slowly before returning to control levels within 12 h. A similar pattern of change in NGF mRNA levels in the hippocampus and cortex has also been found following administration of dexamethasone to adult rats.[25] In these studies the decrease below control levels in NGF mRNA by dexamethasone lasted more than 48 h suggesting a long-term inhibition of NGF protein levels by an excess of glucocorticoids. The reason for the biphasic response in brain NGF mRNA to glucocorticoid treatments is not clear, but it might be due to a differential effect of the hormones on different brain cells, *i.e.*, neuron and glia, or alternatively it might reflect changes in neuronal excitability with time. These studies clearly show that NGF is a target for glucocorticoid action in brain. However, the issue on the relationship between circulating levels of glucocorticoid hormones and neurotro-

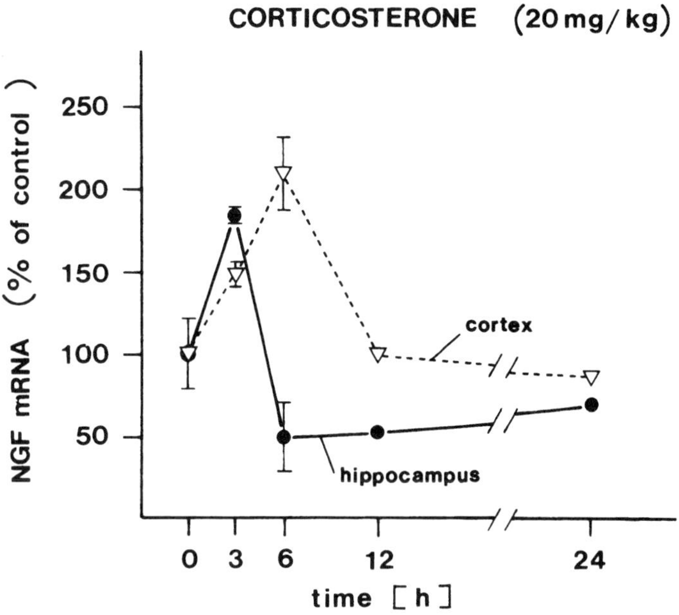

FIGURE 1. Effect of corticosterone on NGF mRNA levels in hippocampus and cortex of adult rats.

phins is even more complex, since adrenalectomy decreases the levels of NGF in the rat hippocampus. It was originally shown by Aloe[27] that adrenalectomy of young rats within 12 days reduced the levels of NGF protein in the hippocampus. Adrenalectomy also decreased NGF mRNA in the hippocampus and cortex of adrenalectomized rats were found to be reduced.[25] In the control rats corticosteroids had no short-term effects on brain BDNF-mRNA levels but the levels significantly decreased at later time-points. In keeping with these observations, BDNF mRNA was not changed in cultured hippocampal neurons after 4 h of dexamethasone treatments.[26] Previous studies also suggest that NT-3 mRNA in hippocampus is decreased by adrenalectomy, whereas dexamethasone transiently elevated this mRNA in control rats. Again, as for NGF, the response of NT-3 was biphasic and there was a pronounced inhibition of NT-3 mRNA by dexamethasone after 24 h.[25]

The results of previous studies have shown that levels of NGF and BDNF expression are regulated by various neurotransmitters in an activity-dependent manner (for a review see Lindholm *et al.*[28]). An interesting issue to consider was therefore whether there is an interaction between the glucocorticoid and the transmitter system in controlling neurotrophin production in brain. Indeed, adrenalectomy was found to attenuate the kainic acid-mediated increase in NGF and BDNF mRNA in the hippocampus.[29] In addition, an additive[26] or even synergistic effect[29] of dexamethasone and kainic acid in elevating NGF mRNA in the hippocampus has been reported. On the other hand, there seemed to be no short-term potentiation by dexamethasone of the hippocampal BDNF mRNA levels induced by kainic acid. However, dexamethasone strongly decreased the response of BDNF to kainic acid administered 24 h later.[29] Taken together, these studies show that there is a functional interaction between the action of adrenal steroids and the activation of glutamate receptors in regulation of neurotrophin levels in the rat brain.

Cell-Specific Effects of Glucocorticoids on NGF Expression

As discussed above glucocorticoid hormones inhibited NGF expression in primary fibroblasts and the L-929 cell line. However, corticosterone or dexamethasone *in vivo* elevated NGF mRNA in the rat hippocampus and cortex. To study whether there is a cell-specific response of brain cells to glucocorticoids, neuronal and glia cell cultures were established. Since NGF is a low-abundant mRNA in neurons a quantitative polymerase chain reaction (PCR) was employed for the analysis of neurons. Addition of dexamethasone to the cultures elevated NGF mRNA in the hippocampal neurons in a concentration-dependent manner.[26] In addition, mineralocorticoid aldosterone increased neuronal NGF mRNA levels suggesting that MR might also be involved in NGF regulation in the hippocampal neurons. In contrast, dexamethasone downregulated both the basal and growth factor-stimulated expression of NGF in cultured astrocytes.[26] *In vivo* NGF is expressed mainly by neurons but cultured astrocytes also produce considerable amounts of NGF.[30] An interesting possibility emerging from these studies is that the low levels of NGF in astrocytes *in vivo* might be partly controlled by the normal circulating levels of glucocorticoids.

The present results also suggest that there might exist cell-specific transcriptional factors differentially affecting NGF expression in neurons and non-neuronal cells. In order to study whether the glucocorticoid-mediated effects on NGF mRNA are reflected by changes in transcription, we analyzed the effect of adrenal steroids on the activity of the mouse NGF promoter[21] in different neuronal cell lines. An immortalized hippocampal cell line, HN10e[31] and the mouse N2A neuroblastoma cell line were transiently transfected with various NGF promoter constructs in front of the luciferase reporter gene. As shown in FIGURE 2, all the hormones tested, corticosterone, aldosterone and dexamethasone, enhanced NGF promoter activity in the hippocampal cell line. Unexpectedly, the same hormones reduced the NGF promoter activity in transfected N2A cells (FIG. 2). The reason for the differential response of the NGF promoter in these two neuronal cell lines towards glucocorticoids is at the moment unknown. Interestingly, the HN10e cell line has been shown to recapitulate many of the properties of the hippocampal cells from which they have been derived,[31] and thus they might represent a more adequate model system to study neuronal transcription.

The stimulatory effect of dexamethasone on the NGF promoter was observed using different deletion constructs of the NGF promoter in transient transfection assays (FIG. 3). This data suggest that the positive enhancement of dexamethasone on the NGF promoter observed here is probably not mediated by a single 5′-element in the NGF promoter. Interestingly, there are sequences resembling conserved glucocorticoid receptor binding half-sites in the upstream region of the NGF promoter which may contribute to the gene activity. Previous studies in the 3T3 fibroblast cell line indicated the existence of a repressive element for dexamethasone action within a 162 base pair region of the NGF promoter (nucleotides −275 to −437, see Lindholm *et al.*[19]). Deleting this region from the NGF promoter abolished the inhibition of NGF gene activity observed in the N2A cell line (FIG. 3). However, this deletion did not potentiate the dexamethasone-induced transcriptional stimulation of NGF in the HN10e cells (FIG. 3). Glucocorticoid receptor has been shown to interact with c-fos and in the collagenase gene, the active glucocorticoid receptor was found to inhibit the AP-1 activity and repress transcription. An active AP-1 site for the transcription

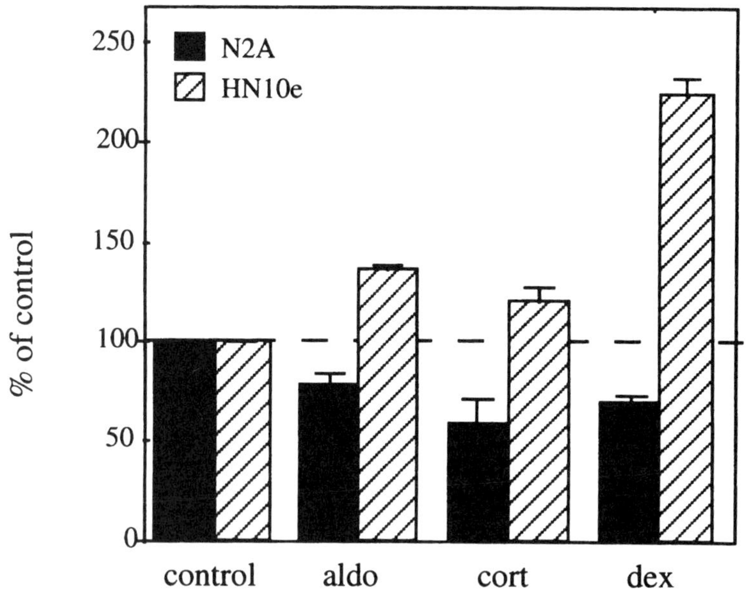

FIGURE 2. Effects of aldosterone (aldo), corticosterone (cort) and dexamethasone (dex) on NGF promoter activity in N2A and HN10e cells. The cells were transiently transfected with pNP-LUC(-1900) and hGR/hMR plasmids by calcium phosphate method. Hormones (1×10^{-8} M) were added 16 h after transfection and the cells were harvested for luciferase assay after additional incubation for 24 h. Relative luciferase activity is determined for each experiment by dividing the luciferase activity by β-galactosidase activity for the same extract.

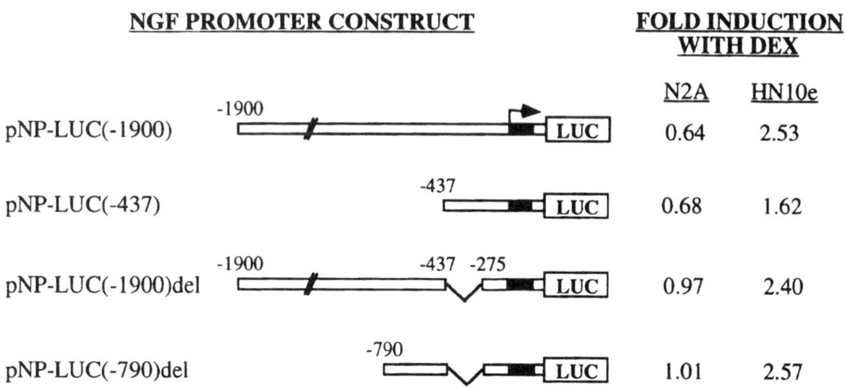

FIGURE 3. Influence of deletions in the NGF promoter on effects of dexamethasone on the promoter activity. Neuro 2A and HN10e cells were transfected with NGF promoter constructs containing the luciferase reporter gene. After the transfection cells were incubated for 24 h in the absence or presence of dexamethasone (1×10^{-8} M).

factors c-fos and c-jun dimer, exists in the first intron of the NGF gene.[21] Thus, the downregulation of the NGF gene expression might involve an interaction between glucocorticoid receptors and AP-1 complex. However, the mechanisms by which glucocorticoids in some cell types stimulate while in other cells repress NGF gene activity are presently unknown.

Functional Significance of the Neurotrophin Regulation by Adrenal Steroids

FIGURE 4 depicts the effects of different levels of corticosteroid on neurotrophin expression and proposes a hypothesis about the relationship between corticosteroid levels, neurotrophins and neuronal survival in the hippocampus. Adrenalectomy has been shown to preferentially reduce the number of dentate granule neurons.[32] On the other hand, adrenalectomy has been found to decrease the basal levels of BDNF and NGF mRNA as well as to attenuate the activity-dependent increase in these mRNAs in the dentate gyrus.[29] Likewise, the levels of NT-3 mRNA were found to be decreased in the dentate gyrus and CA2 region in adrenalectomized animals.[33] Because hippocampal neurons have been shown to possess functional trkB and trkC receptors and are therefore responsive to BDNF and NT-3, it is possible that the decreased levels of BDNF and NT-3 after adrenalectomy might render the granule neurons more vulnerable to other insults by providing them with a diminished neurotrophic support.

Excessive or prolonged exposure to glucocorticoids adversely affects neurons mainly in the CA3 region of the hippocampus (for a review see McEwen and Gould[2]). As discussed above, extended exposure to dexamethasone reduced BDNF mRNA in the hippocampus. This, together with the known action of glucocorticoids in increasing neuronal excitability[1,16] might lead to vulnerability of some neurons towards different insults in the presence of high levels of these hormones. BDNF has recently been shown to protect hippocampal neurons against ischemic cell death *in vivo*,[34] suggesting that endogenously produced BDNF also has a survival effect.

Although the hypothesis presented in FIGURE 4 emphasizes the role of neurotrophins in the cell death produced by reduced or excess levels of corticosteroid, it is probable that this phenomenon is more complex and involves other factors as well. To better understand the interaction of neurotrophins and glucocorticoid hormones in the physiology of hippocampal neurons more data are needed, *e.g.*, by studying mice lacking endogenous glucocorticoid receptors or analyzing animals having altered levels of neurotrophic factors.

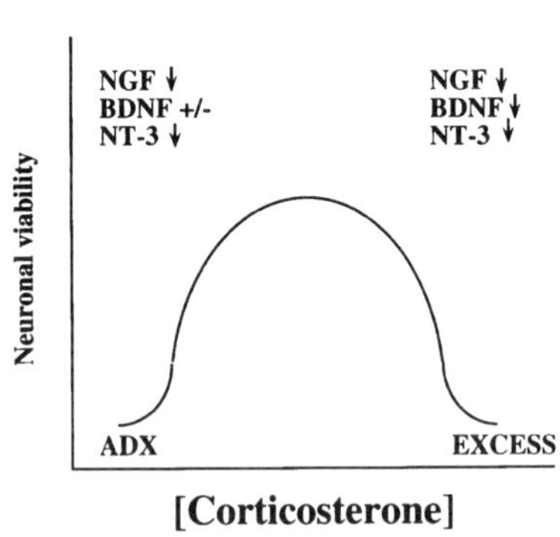

FIGURE 4. Hypothetical model of the relationship between corticosteroid levels, neurotrophin expression and neuronal viability in hippocampus. For references, see text.

ACKNOWLEDGMENT

We thank Dr. Bruce Wainer for the kind gift of the hippocampal cell line.

REFERENCES

1. DE KLOET, E. R. 1991. Brain corticosteroid balance and homeostatic control. Front. Neuroendocrinol. **12:** 95–164.
2. MCEWEN, B. S. & E. GOULD. 1990. Adrenal steroid influences on the survival of hippocampal neurons. Biochem. Pharm. **40:** 2393–2402.
3. TRUSS, M. & M. BEATO. 1993. Steroid hormone receptors: interaction with deoxyribonucleic acid and transcription factors. Endocrinol. Rev. **14:** 459–479.
4. ARONSSON, M., K. FUXE & Y. DONG. 1988. Localization of glucocorticoid receptor mRNA in the male rat brain by in situ hybridization. Proc. Natl. Acad. Sci. USA **85:** 9331–9335.
5. ARRIZA, J. L., R. B. SIMERLY, L. W. SWANSON & R. M. EVANS. 1988. The neuronal mineralocorticoid receptor as a mediator of glucocorticoid response. Neuron **1:** 887–900.
6. VAN EEKELEN, J. A. M., W. JIANG, E. R. DE KLOET & M. C. BOHN. 1988. Distribution of the mineralocorticoid and glucocorticoid receptor mRNAs in the rat hippocampus. J. Neurosci. Res. **21:** 88–94.
7. REUL, J. M. H. & E. R. DE KLOET. 1985. Two receptor systems for corticosterone in rat brain: microdistribution and differential occupation. Endocrinology **117:** 2505–2511.
8. CASTRÉN, M. & K. DAMM. 1993. A functional promoter directing expression of a novel type of rat mineralocorticoid receptor mRNA in brain. J. Neuroendocrinol. **5:** 461–466.
9. THOENEN, H. 1991. The changing scene of neurotrophic factors. Trends Neurosci. **14:** 165–170.
10. ERNFORS, P., C. WETMORE, L. OLSON & L. PERSSON. 1990. Identification of cells in rat brain and peripheral tissues expressing mRNA for members of the nerve growth factor family. Neuron **5:** 511–526.
11. PHILLIPS, H. S., J. M. HAINS, G. R. LARAMEE, A. ROSENTHAL & J. W. WINSLOW. 1990. Widespread expression of BDNF but not NT3 by target areas of basal forebrain cholinergic neurons. Science **250:** 290–294.
12. ZAFRA, F., B. HENGERER, J. LEIBROCK, H. THOENEN & D. LINDHOLM. 1990. Activity-dependent regulation of BDNF and NGF mRNAs in the rat hippocampus is mediated by non-NMDA glutamate receptors. EMBO J. **9:** 3545–3550.
13. BERZAGHI, M. P., J. D. COOPER, E. CASTRÉN, F. ZAFRA, M. V. SOFRONIEW, H. THOENEN & D. LINDHOLM. 1993. Cholinergic regulation of brain-derived neurotrophic factor (BDNF) and nerve growth factor (NGF) but not neurotrophin-3 (NT-3) mRNA levels in the developing rat hippocampus. J. Neurosci. **13:** 3818–3826.
14. CHAO, M. V. 1992. Neurotrophin receptors: a window into neuronal differentiation. Neuron **9:** 583–593.
15. BERNINGER, B., D. E. GARCÍA, N. INAGAKI, C. HAHNEL & D. LINDHOLM. 1993. BDNF and NT-3 induce intracellular Ca^{+} elevation in hippocampal neurones. NeuroReport **4:** 1303–1306.
16. JOELS, M. & E. R. DE KLOET. 1992. Control of neuronal excitability by corticosteroid hormones. Trends Neurosci. **15:** 25–30.
17. HEUMANN, R., D. LINDHOLM, C. BANDTLOW, M. MEYER, M. J. RADEKE, T. P. MISKO, E. SHOOTER & H. THOENEN. 1987. Differential regulation of mRNA encoding nerve growth factor and its receptor in rat sciatic nerve during development, growth and regeneration: role of macrophages. Proc. Natl. Acad. Sci. USA **84:** 8735–8739.
18. LINDHOLM, D., R. HEUMANN, M. MEYER & H. THOENEN. 1987. Interleukin-1 regulates synthesis of nerve growth factor in non-neuronal cells of rat sciatic nerve. Nature **330:** 658–659.
19. LINDHOLM, D., B. HENGERER, R. HEUMANN, P. CARROLL & H. THOENEN. 1990. Glu-

cocorticoid hormones negatively regulate nerve growth factor expression *in vivo* and in cultured rat fibroblasts. Eur. J. Neurosci. **2:** 795–801.
20. LINDHOLM, D., R. HEUMANN, B. HENGERER & H. THOENEN. 1988. Interleukin-1 increases stability and transcription of mRNA encoding nerve growth factor in cultured rat fibroblasts. J. Biol. Chem. **263:** 16348–16351.
21. HENGERER, B., D. LINDHOLM, R. HEUMANN, U. RUTHER, E. F. WAGNER & H. THOENEN. 1990. Lesion-induced increase in nerve growth factor mRNA is mediated by c-fos. Proc. Natl. Acad. Sci. USA **87:** 3899–3903.
22. SIMINOSKI, K., R. A. MURPHY, P. RENNERT & G. HEINRICH. 1987. Cortisone, testosterone, and aldosterone reduce levels of nerve growth factor messenger ribonucleic acid in L-929 fibroblasts. Endocrinology **121:** 1432–1437.
23. WION, D., R. HOULGATTE & P. BRACHET. 1986. Dexamethasone rapidly reduces the expression of the β-NGF gene in mouse L-929 cells. Exp. Cell Res. **162:** 562–565.
24. MEYER, M., I. MATSUOKA, C. WETMORE, L. OLSON & H. THOENEN. 1992. Enhanced synthesis of brain-derived neurotrophic factor the lesioned peripheral nerve: different mechanisms are responsible for the regulation of BDNF and NGF mRNA. J. Cell Biol. **119:** 45–54.
25. BARBANY, G. & H. PERSSON. 1992. Regulation of neurotrophin mRNA expression in the rat brain by glucocorticoids. Eur. J. Neurosci. **4:** 396–407.
26. LINDHOLM, D., E. CASTRÉN, B. HENGERER, B. ZAFRA, B. BERNINGER & H. THOENEN. 1992. Differential regulation of nerve growth factor (NGF) synthesis in neurons and astrocytes by glucocorticoid hormones. Eur. J. Neurosci. **4:** 404–410.
27. ALOE, L. 1989. Adrenalectomy decreases nerve growth factor in young adult rat hippocampus. Proc. Natl. Acad. Sci. USA **86:** 5636–5640.
28. LINDHOLM, D., E. CASTRÉN, M. DA PENHA BERZAGHI & H. THOENEN. 1993. Effects of transmitters and hormones on neuronal production of neurotrophins. Sem. Neurosci. **5:** 279–283.
29. BARBANY, G. & H. PERSSON. 1993. Adrenalectomy attenuates kainic acid-elicited increases of messenger RNAs for neurotrophins and their receptors in the rat brain. Neuroscience **54:** 909–922.
30. SPRANGER, M., D. LINDHOLM, C. BANDTLOW, R. HEUMANN, H. GNAHN, M. NÄHER-NOÉ & H. THOENEN. 1990. Regulation of nerve growth factor (NGF) synthesis in rat central nervous system: comparison between the effects of interleukin-1 and various growth factors in astrocyte culture and *in vivo.* Eur. J. Neurosci. **2:** 69–76.
31. LEE, H. J., D. N. HAMMOND, T. H. LARGE, J. D. ROBACK, A. J. SIM, D. A. BROWN, U. H. OTTEN & B. H. WAINER. 1990. Neuronal properties and trophic activities of immortalized hippocampal cells from embryonic and young adult mice. J. Neurosci. **16:** 1779–1787.
32. SLOVITER, R. S., G. VAIGUETTE, G. M. ABRAMS, E. C. RONK, A. I. SOLLAS, L. A. PAUL & S. L. NEUBORT. 1989. Selective loss of hippocampal granule cells in the mature rat brain after adrenalectomy. Science **243:** 535–538.
33. LAUTERBORN, J. C. & C. M. GALL. 1992. Adrenalectomy disrupts aspects of neurotrophin expression in hippocampus. Abstr. Soc. Neurosci. 22nd Annual Meeting, Anaheim, p. 775.
34. BECK, T., D. LINDHOLM, E. CASTRÉN & A. WREE. 1994. Brain-derived neurotrophic factor protects against ischemic cell damage in rat hippocampus. J. Cereb. Blood Flow Metab. In press.

DISCUSSION OF THE PAPER

E. C. AZMITIA (*New York University*): How long after removal of the target area can the neurons remain alive and then be brought back with the NGF in your experiments?

LINDHOLM: I am talking about years. We have been following this out to 500 days. This is published in the December 1993 volume of *Neuroscience*. We had 34 pictures; they cut it down to 17.

QUESTION: You mentioned all the tracks. What about the low-affinity NGF receptor? It plays a role in apoptosis, I think.

LINDHOLM: Yes, there is a paper in *Nature* about this, but as far as we know we do not see an expression of the P75, the low-affinity NGF receptor, as we did in hippocampal neurons. There is some misconception about this.

QUESTION: With regard to the ischemic damage in CA1 where you can rescue the cells with BDNF, which types of cells did you look at?

LINDHOLM: We did an overall stain, but it is the principal neurons—they do die. But you are right that some of the large interneuron cells in the oriens layer are alive.

QUESTION: Again, do you think these glucocorticoid effects are direct effects on the promoters or are they indirectly mediated through glutamate or GABA changes which would then change the growth factors?

LINDHOLM: Difficult question. For NGF at least, this receptor expression in the hippocampus cell lines, it is clearly a direct effect. But on the BDNF, we do not know, and we just recently finished the cloning of the promoter. There are only three promoters. These promoters are driven by steroids, but when we do a CAT assay for the dexamethasone, we do not see anything.

QUESTION: Do you need very high levels of corticosterone or glucocorticoids to decrease BDNF?

LINDHOLM: We see a small decrease, 10 or 15 percent with 5 mg/kg in a rat.

Steroid Regulation of Cytokines

Relevance for TH1-to-TH2 Shift?[a]

FRANCESCO CHIAPPELLI,[b] ERCOLANO MANFRINI, CLAUDIO FRANCESCHI, ANDREA COSSARIZZA, AND KEITH L. BLACK

[b]*Laboratory of Human Oral and Molecular Immunology*
Diagnostic Sciences, School of Dentistry
Department of Neurobiology, School of Medicine
Brain & Dental Research Institutes
University of California, Los Angeles, California 90024-1668
and
Laboratory of Human Immunology and Psychoneuroimmunology
West Los Angeles VAMC

Steroid Regulation of Cytokine Production: TH1 and TH2 Patterns of Response

The ability of T cells to engage in the appropriate immune response depends upon their competence to be adequately activated following exposure to a pathogen. T cell activation is monitored as transmembrane and transcytosolic signaling events, overall cellular metabolic activity, production of cytokines and of their receptors, and clonal expansion and proliferation. The activation of T cells is modulated by a variety of endocrine products, including steroids. The immunomodulatory effect of steroids is mediated largely by their effects upon the expression of certain cytokines. Depending upon the antigenic and endocrine environment during immune activation, either of two polarized end-stages of T cell maturation can result, which are distinguished by the pattern of cytokines they produce. This dichotomy contributes to regulate and determine the type of cellular immune process in several given clinical conditions. When the T helper cell type-1 (TH1) pattern predominates, cytokine responses promote the generation and activation of natural killer and lymphokine activated killer cells and of cytotoxic T lymphocytes for the elimination of virally-infected and tumor cells; when the TH2 pattern predominates, then cellular immune events converge to favor the generation, maturation and regulation of B cells for the production of immunoglobulins. TH1 responses favor the elimination of antigens from intracellular origin, and TH2 responses favor the elimination of extracellular antigens in a manner reminiscent of murine T helper type-1 and type-2 clones.[1–25]

Antigenic stimulation of recent thymic emigrate naive T cells leads to the generation of primary effector cells. These cycling cells have a rather large cytoplasm and produce a range of cytokines that includes interleukin-1 (IL-1), gamma-interferon (γ-IFN), IL-3, IL-4, IL-5 and IL-10, the THO pattern of cytokine response. Further activation of THO cells in the presence of γ-IFN and absence of IL-4 leads to the generation of CD4+ cells that secrete large amounts of IL-2 and γ-INF, but little IL-3, IL-4 or IL-5 upon further stimulation. This is the TH1 pattern of response. By contrast, THO cells activated in the presence of IL-4 and no γ-IFN become

[a]We thank NIH (AI07126; DA07683) and the UCLA Program in Psychoneuroimmunology for support.

capable of a TH2 pattern of cytokine response, characterized by the predominant secretion of IL-3, IL-4, IL-5 and newly identified IL-13 (FIG. 1). IL-12, an accessory cytokine produced by activated antigen-presenting cells, plays an important regulatory role in the generation of TH1 and TH2 patterns of response and favors the TH1 pathway. Metabolic inhibitors, designed to blunt certain events during lymphocyte activation (*e.g.*, pentoxifylline, which blocks phosphodiesterase) permit the study of these maturational events because they selectively blunt the generation of one or the other pattern of cytokine response.[15,18–27]

The patterns of T cell cytokine production produced in response to an antigenic challenge are modulated by the regulatory influences on gene expression exerted by steroid hormones. Glucocorticoids (GC) are potent anti-inflammatory agents that suppress IL-2 and γ-IFN production, but lead to enhanced production of IL-4.[5,10,14,15] These affects are mediated predominantly, albeit not exclusively[28,29] by the GC receptor.[11,30,31]

The GC receptor characteristically consists of a hetero-tetramer: one receptor polypeptide (MW ≈ 100 kDa), two heat shock proteins (MW ≈ 90 kDa), and one protein of the immunophilin family (MW ≈ 59 kDa). The human GC receptor peptide contains 777 amino acids: the NH_2-end, "modulation domain," is a stretch of 450 amino acids, which interacts with transcription factors, perhaps the RNA polymerase itself; the polypeptide's middle portion consists of approximately 70 amino acids and represents the DNA binding domain of the molecule, it has strong homology with the corresponding region of other members of the steroid receptor family, including the typical zinc finger arrangement (8 cysteine residues involved in capturing 2 zinc ions in a tetrahedral folding that permit effective interaction with GC responsive

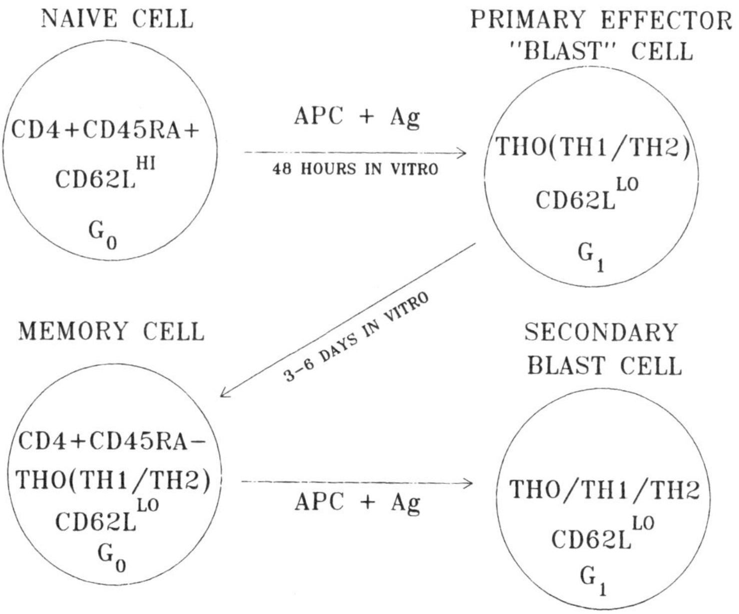

FIGURE 1. Diagrammatic representation of the generation of TH1 and TH2 patterns of response to indicate potential immunodulatory sites of steroids.

elements of DNA); the molecule's COOH-end is the hormone-binding domain that consists of a stretch of about 250 amino acids characterized by specific folds that permit to bind the steroid with high specificity. One-point mutations in that region (*e.g.*, ASP_{641}→VAL) considerably decrease steroid binding affinity.[30–33] We and others have characterized the number and binding affinity of the GC receptor in established human T and B cell lines, as well as in peripheral blood mononuclear cells. Human macrophages are richer in GC receptors (≈7000 sites/cell) compared to T or non-T lymphocyte populations (≈3000 sites/cell). The number of GC receptor increases 2–6-fold upon stimulation of human T cells with mitogens or in a mixed leukocyte reaction, and this reflects the surge in receptor content during S and G_2 compared to G_1.[11,34–37] We have confirmed the reported heterogeneity of T cell subpopulations for GC receptor content:[38,39] human acute lymphoblastic T cell leukemia cells, Molt, have an average of 31000 sites/cell, cutaneous human T cell lymphoma cells, Hut-78, have an average of 22000 sites/cell, but immunoglobulin-secreting human lymphoblast cells, IM-9, have only about 2000 sites/cell; in other experiments, we found that PMBC from normal adult subjects can vary from as low as 4000 sites/cell to as high as 22000 sites/cell. Treatment of these immune cells with 10^{-9} M dexamethasone (DEX) for 16 h leads to a reduction in the cytosolic form of the receptor (GC-CR) by 37.5% ± 2.0. The characterization of other steroid receptors, including that for the weak androgen dehydroepiandrosterone (DHEA) in human immune cells[40] is now ongoing: with the aid of our newly characterized protocol for the purification of subpopulations of CD4+ T cells,[41,42] we now are poised to study the biochemistry of their GC and DHEA receptors.

In contrast to GC, DHEA favors IL-2 production over IL-4,[5–8,13–15] counters GC-induced thymic involution, and antagonizes GC-mediated suppression of lymphocyte proliferation *in vitro.* T cells obtained from healthy adults, pretreated with 10^{-9} M DHEA and stimulated with mitogens, generally generate more IL-2 and are more cytotoxic than untreated cells. These outcomes, which appear to be DHEA-receptor mediated, appear to be specific to the CD4+ subpopulation, and to be dependent, at the molecular level, upon the enhancement of IL-2 mRNA transcription.[7,13] This is of profound potential implications for clinical conditions that require immunotherapy protocols, as noted below. Our data show that, at the cellular and the molecular immune levels, DHEA normalizes the depression in activation response of human T cells following treatment with DEX, alcohol or cocaine.[42]

In summary, immune competence depends largely upon the appropriate generation of TH1 or TH2 patterns of cytokine response for the effective surveillance of antigens of intrinsic or extrinsic origin. Expectedly, the pathogen plays a determinant role in directing the generation of TH1 and TH2 patterns of response. Unexpectedly perhaps if one fails to consider neuroendocrine-immune modulatory phenomena, the endocrine environment also plays a significant role in this context. Steroids, particularly GC and DHEA (and, although not reviewed here, dihydrotestosterone) and steroid receptor-mediated modulation of cytokine production are crucial for the regulation of T cell maturation into specific subpopulations capable of these specialized patterns of cytokine production. The relevance of these observations for immune competence in certain clinical situations that involve distinct aspects of brain pathology is outlined below.

Clinical Implications

Severe chronic illness, surgery and other exacting modes of clinical intervention are powerful stressors to the patient. A significant rise in cortisol production, and a parallel decrease in circulating DHEA and DHEA-S can ensue. This reflects a shift in the metabolism of adrenocortical pregnenolone toward the GC pathway (17α-OH-pregnenolone→17α-OH-progesterone→11-deoxycortisol→cortisol). Concurrently, the efficiency of the androgen pathway (17α-OH-pregnenolone→DHEA → DHEA-S) decreases (FIG. 2). This metabolic shift may benefit the patient to the extent that increased levels of GC could maximize the chances of survival, while decreased levels of DHEA may be less critical to survival.[43–47]

A rise in plasma cortisol and a drop in plasma DHEA levels occur in aging, and to a greater extent in patients with Alzheimer's disease.[48] Pharmacological administration of DHEA (30–90 mg/day PO) normalizes plasma DHEA levels in these patients. DHEA levels are generally inversely correlated ($r = 0.78$, $p < 0.05$) with mid-afternoon cortisol,[48] perhaps because DHEA both depresses ACTH production by the anterior pituitary corticotrophs and blunts GC metabolic enzymes.[49]

Alzheimer's disease is a cerebral degenerative disease, which afflicts primarily, albeit not exclusively, aged individuals. Its hallmark is a progressive deterioration of neurocognitive performance, which includes neuropathologic (*e.g.*, cortical atrophy, neuronal loss, neurofibrillary tangles, senile plaques, vascular deposits of β-amyloid peptide complexes) and psychopathologic outcomes (*e.g.*, severe loss of memory, depression, personality changes). The observations that amyloid complexes are proteinaceous deposits composed of several protein moieties that include certain complement components and immunoglobulins suggest that peripheral and/or central immune events contribute to trigger cerebral acute phase responses that lead to increased amyloidogenesis and neuronal toxicity. This, taken together with the observation that T cells from patients with Alzheimer's disease express high levels of the membrane activation marker CD25, suggests that Alzheimer's disease may be a form of autoimmune disease.[50–53] Associated psycho- and physiopathologies commonly occur in these patients, which include frequent bouts against bacterial infections.

Cellular immune competence decreases with age.[10,54,55] This may be in part consequential to the drop in DHEA levels.[56–59] DHEA administration to aged mice normalizes T cell responses and IL-2 production to levels attained by young adult mice, and reverses the phenotypic alterations associated with cellular immunose-

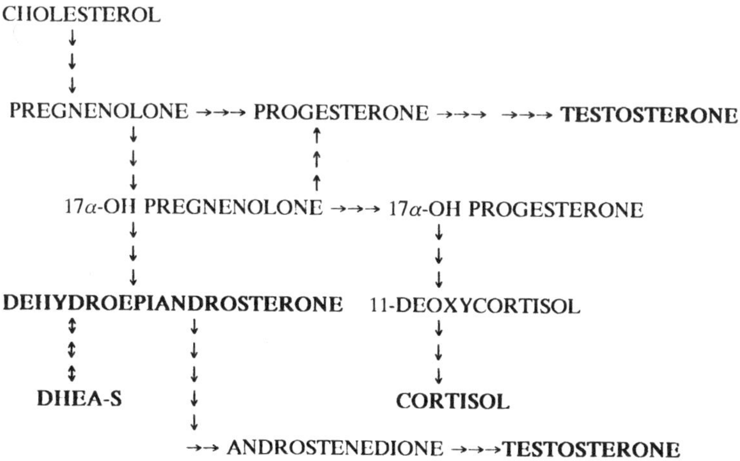

FIGURE 2. Principal pathways of adrenal steroid metabolism.

nescence.[5,15,60] We have confirmed these findings in human subjects: in collaboration with B. Miller, M.D., (Harbor-UCLA Medical Center), we have administered DHEA to a group of healthy aged subjects, and examined phenotypic characteristics of circulating T cells. We note a drop in naive T cell subpopulations (*i.e.*, CD4+CD45RA+, CD8+CD45RA+) by about 50%, an equivalent rise in the memory T cell subpopulations, no difference in total T cells, and slight, albeit not significant drop in CD4/CD8 ratio. These data suggest that DHEA administration favors a shift from naive to memory T cells. Whether or not these effects reflect a preferential rise in cells capable of TH1 or TH2 patterns of response, whether or not similar outcomes will occur in patients with Alzheimer's disease, and whether *in vitro* models can be effectively designed to elucidate if the basic mechanisms for these outcomes involve steroid receptor-mediated events is presently under study.

Altered regulation of the expression of certain cytokines occurs in normal aging. The production of the pleiotropic cytokine, IL-6, is elevated.[8,61,62] Peripheral blood mononuclear cells from healthy elderly subjects produce significantly more IL-1β, IL-6 and TNF-α, but not of other cytokines (*e.g.*, γ-IFN), following stimulation with phytohemagglutinin and the phorbol ester, phorbol-methyl acetate. These functional changes, which are independent from previously reported phenotypic alterations in normal aging,[10,55,63] may be indicative of subtle changes in TH1 and TH2 balance. It is possible, even probable that these outcomes reflect circulating steroid- and steroid receptor-mediated neuroendocrine-immunomodulation.

Patients with Alzheimer's disease may be at greater immunological risk than healthy elderly because the drop in T cell competence observed in healthy aging is exacerbated in these patients.[64–67] These deficiencies in T cell immunity in Alzheimer's disease may reflect a more significant shift in the balance between the TH1 and TH2 patterns of cytokine response to antigenic challenges, which may in turn be associated with a more severe steroid deregulation. These patients have much decreased peripheral GC sensitivity compared to healthy elderly control subjects: a large proportion of these patients maintain elevated levels of plasma cortisol following DEX administration.[58,59,68] The rate of DEX nonsuppression is comparable to that of patients with major depression.[68] Rise in plasma cortisol levels actually predicts progression of Alzheimer's disease severity[70] Contemporaneously with the rise in plasma cortisol and the onset of GC resistance, plasma levels of DHEA and of DHEA-S decrease in normal aging and to a greater extent in patients with Alzheimer's disease.[48,71,72] The DHEA-S/cortisol ratio drops in healthy elderly and to a significantly greater extent in Alzheimer's patients compared to age-matched control subjects.[72] These lines of evidence converge to suggest that such systemic manifestations of Alzheimer's disease as altered steroid plasma levels (*i.e.*, cortisol, DHEA, DHEA-S) contribute to the observed decrease in T cell competence in these patients, and that DHEA-based clinical protocols could benefit them.

Similarly, the prognosis of patients with intracranial tumors (*e.g.*, astrocytoma multiforme, glioblastoma) remains poor notwithstanding concerted efforts by neurosurgeons, clinicians and basic scientists.[73,74] Despite surgical resection of the malignancy, radiation, immuno- or chemotherapy, or combinations of these modes of intervention, the median survival curves fall below 10% within 24 months of the original diagnosis. The 9th International Conference on Brain Tumor Research and Therapy has emphasized the need for interactive basic and clinical research to uncover the fundamental biologic and molecular processes that underlie the aggressive nature of these cancers, and to develop effective treatment modalities for successfully preserving the patients' physical and mental health.

Edema and inflammation of the brain parenchyma are significant complications of intracranial tumors, which clinicians attempt to control with the administration of pharmacological doses of DEX and other GC. These interventions, while beneficial in terms of reduction of peritumoral edema and of cerebral blood volume, have a wide spectrum of effects within and outside of the central nervous system. These complex and multifaceted outcomes could contribute to the disappointing observation that high doses of GC, for instance, methylprednisone, as an adjuvant to surgery and radiation therapy often fail to enhance their survival.[75]

Pharmacological doses of GC alter several fundamental physiological processes.[76] Plasma levels of DEX drop below 5 μM within 24 h of intravenous administration of 1.5–2.5 mg:[77] We have observed that 178 ng/dl ± 7% (3.56 μM) DEX remains in plasma 10 h and 35 ng/dl ± 11% (0.7 μM) 18 h following intravenous administration of 1 mg DEX in normal subjects.[78,79] Persistent suppression of the pituitary-adrenocortical axis, significant drop in GC-CR number in peripheral blood lymphocytes, and significant reduction in the number and proliferative response of circulating helper T cell subpopulations are noted at 10 h.[78–80] The impaired functional response of T cells obtained from patients with brain tumors treated with massive doses of GC may reflect in part a GC-induced and GC-CR-mediated drop in the proportion of responding peripheral blood T cells, particularly in terms of their ability to produce IL-2.[75] Neuroendocrine-immune modes of interventions that involve DHEA administration and/or *in vitro* treatment of lymphocytes with DHEA may prove beneficial for these patients and should be tested.

Experimental treatment with the biologic response modifier ImuVert, derived from ribosomes and membrane vesicles isolated from the bacterium *Serratia marcescens,* prolongs survival to 78 weeks.[73] ImuVert may act by stimulating γ-IFN production, thus enhancing natural killer cytotoxic activity. By the same token, sustained ImuVert administration (*e.g.*, once weekly in 5-week cycles with a 2-week rest period for a period of 12 months[73]), probably favors the generation of a TH1 pattern of response, which should favor immune-mediated tumor surveillance. Thus, it is possible, even probable that ImuVert therapy protocols supplemented with neuroendocrine-immune therapy protocols that involve DHEA, or related structural analogs endowed with established cancer chemopreventive properties,[81] could significantly benefit these patients. That steroid therapy may prove advantageous for brain tumor patients is supported by the recent report that the secosteroid $1,25(OH)_2D_3$, the major metabolite of vitamin D, is powerfully cytotoxic to cultured human gliomas.[82]

Centrally, intracranial tumors are associated with substantial immune activation.[83] Several immune events associated with the surveillance of tumors, such as the cytokine cascade, are impeded by elevated circulating GC: the production of several cytokines, such as IL-1 and tumor necrosis factor, by brain and peripheral macrophages as well as by astrocytes is blunted by GC. Production of the neutrophil chemotactic factor IL-8 by tumor cells and by cells of the surrounding reactive tissue is also blunted.[84–86] By contrast, the production of members of the transforming growth factor family (*e.g.*, TGF-β2), which appear to play a central role in orchestrating a variety of neurodegenerative processes including those seen in Alzheimer's disease patients,[87] and which are among the principal immunosuppressive factors produced by glioma cells, is enhanced by GC.[9] TGFs blunt DHEA production.[88]

Malignant cell populations isolated from the brain of patients exhibit specificity in terms of the predominant type of steroid receptor they express. Meningiomas, for example, appear to have hormonal dependence properties similar to other solid tumors (*e.g.*, breast cancer); they express elevated levels of progesterone, androgen and GC receptors.[89] This may explain in part the role of GC receptors in the proliferative stimulation of certain human gliomas by DEX *in vitro.*[77] The presence of steroid receptors also plays an important role in the biologic behavior of these tumors since it appears to correlate with the production of arachidonic acid metabolites by these cells.[89] Arachidonate cicosanoid metabolites (prostaglandins and leukotrienes) participate in blunting IL-2-dependent immune proliferation *in vitro.*[90] Animal models of intracranial tumors have proved useful to show that experimental infusion of leukotrienes, powerful metabolites of arachidonic acids through the 5-lipoxygenase pathway, significantly increase the permeability of the blood-brain barrier. Leukotriene LTE_4 is most potent and increases capillary permeability by an average 355%. Taken together, these data suggest that vascular permeability within and adjacent to intracranial tumors is modulated by arachidonic metabolism, particularly leukotrienes produced endogenously, that localized concentrations of GC may modulate the production of these metabolites, and that these conditions could play an important role in brain tumor-associated edema.[91,92] Thus, GC therapy of brain tumor patients, while beneficial for the control of inflammation, may alter arachidonate metabolism, in some tumors at least. Certain of these metabolites may in turn increase permeability of the blood-brain barrier, and contribute to aggravate peritumoral edema and to exacerbate the state of immune suppression that afflicts these patients.

In summary, a complex picture emerges that suggests that GC have both beneficial (*i.e.*, decreased inflammation and edema) and detrimental outcomes (*e.g.*, potential decrease in leukocyte influx, altered arachidonic acid metabolism) for patients with intracranial tumors. While the importance of GC in the palliative treatment of brain tumor patients remains, for now at least, unquestionable, emerging evidence points to the strong potential of DHEA in this clinical context.

CONCLUSION

Based upon the observations outlined above, the question now arises as to whether or not DHEA and DHEA-S may come to play a new key role in improving the immune competence of a variety of immunocompromised subjects, including alcoholics, addicts to cocaine and other drugs of abuse, patients with acute or chronic forms of oral pathology (*e.g.*, oral lichen planus) and, as outlined above, patients with Alzheimer's disease or with intracranial tumors. Promising new research should now focus to confirm that these androgens act to enhance immune surveillance mechanisms,[81] that their administration is safe and has minimal side effects,[47,81,92] and that their mode of action involves largely steroid receptor-mediated modulation of cytokine expression.

ACKNOWLEDGMENTS

We thank the many collaborators who have contributed to these studies, and who are acknowledged in the papers cited.

REFERENCES

1. Jefferies, W. McK. 1990. Cortisol and immunity. Med. Hyp. **34:** 198–208.
2. Compton, M. M., L. A. M. Caron & J. A. Cidlowski. 1987. Glucocorticoid action on the immune system. J. Steroid Biochem. **27:** 201–208.

3. GUYRE, P. M., M. T. GIRARD, P. M. MORGANELLI *et al.* 1988. Glucocorticoid effect on the production and actions of immune cytokines. J. Steroid Biochem. **30:** 89–93.
4. CUPPS, T. R. & A.S. FAUCI. 1982. Corticosteroid-mediated immunoregulation in man. Immunol. Rev. **65:** 133–155.
5. DAYNES, R. A., D. J. DUDLEY & B. A. ARANEO. 1990. Regulation of murine lymphokine production *in vivo.* II. Dehydroepiandrosterone is a natural enhancer of interleukin-2 synthesis by helper T cells. Eur. J. Immunol. **20:** 793–802.
6. DAYNES, R. A., A. W. MEIKLE & B. A. ARANEO. 1991. Locally active steroid hormones may facilitate compartmentalization of immunity by regulating the types of lymphokines produced by helper T cells. Res. Immunol. **142:** 40–45.
7. SUZUKI, T., N. SUZUKI, R. A. DAYNES *et al.* 1991. Dehydroepiandrosterone enhances IL-2 production and cytotoxic effector function of human T cells. Clin. Immunol. Immunopathol. **61:** 202–211.
8. DAYNES, R. A., B. A. ARANEO, W. B. ERSHLER *et al.* 1993. Altered regulation of IL-6 production with normal aging: possible linkage to the age-associated decline in dehydroepiandrosterone and its sulfate derivative. J. Immunol. **150:** 5219–5230.
9. AYANLAR BATUMAN, O., A. P. FERRERO, A. DIAZ *et al.* 1991. Regulation of transforming growth factor-β1 gene expression by glucocorticoids in normal human T lymphocytes. J. Clin. Invest. **88:** 1574–1580.
10. CHIAPPELLI, F., C. FRANCESCHI, E. OTTAVIANI *et al.* 1994. Neuroendocrine modulation of the immune system. *In* Human Physiology: from Cellular Mechanisms to Integration. R. Gregor, H.-P. Koepchen, W. Mommaerts & U. Winhorst, Eds. Section L, Chapter 86. Springer. New York. In press.
11. SMITH, K. A., G. R. CRABTREE, S. J. KENNEDY *et al.* 1977. Glucocorticoid receptors and glucocorticoid sensitivity of mitogen stimulated and unstimulated human lymphocytes. Nature **267:** 523–526.
12. BLAUER, K. L., M. POTH, W. M. ROGERS *et al.* 1991. Dehydroepiandrosterone antagonizes the suppressive effects of dexamethasone on lymphocyte proliferation. Endocrinology **129:** 3174–3179.
13. DAYNES, R. A. & B. A. ARANEO. 1989. Contrasting effects of glucocorticoids on the capacity of T cells to produce the growth factors interleukin-2 and interleukin-4. Eur. J. Immunol. **19:** 2319–2325.
14. STAM, W. B., A. J. M. VAN OOSTERHOUT & F. P. NIJKAMP. 1993. Pharmacologic modulation of TH1- and TH2-associated lymphokine production. Life Sci. **53:** 1921–1934.
15. ROOK, G. A., P. ONYEBUJOH & J. L. STANFORD. 1993. TH1/TH2 switching and loss of CD4+ T cells in chronic infections: an immunoendocrinological hypothesis not exclusive to HIV. Immunol. Today **14:** 568–569.
16. CHIAPPELLI, F. 1991. Neuroimmunological perspectives on lymphocyte homing and adhesion. Int. J. Immunopathol. Pharmacol. **4:** 115–121.
17. MOSMANN, T. R. & R. L. COFFMAN. 1987. Two types of mouse helper T cell clone. Immunol. Today **8:** 223–227.
18. MOSMANN, T. R., H. CHEZWINSKI, M. W. BOND *et al.* 1986. Two types of muzine helper T-cell clone. Definition according to profiles of lymphokine activities and secreted proteins. J. Immunol. **136:** 2348–2351.
19. TRINCHIERI, G. 1993. Interleukin-12 and its role in the generation of TH1 cells. Immunol. Today **14:** 335–3358.
20. VAN DER POUW-KRAAN, T., R. DE JONG & L. AARDEN. 1993. Development of human TH1 and TH2 cytokine responses: the cytokine production profile of T cells is dictated by the primary *in vitro* stimulus. Eur. J. Immunol. **23:** 1–5.
21. ROMAGNANI, S. 1994. Human TH1 and TH2 subsets. Eur. Cytokine Netw. **5:** 7–12.
22. ROTT, O., E. CASH & B. FLEISCHER. 1993. Phosphodiesterase inhibitor pentoxifylline, a selective suppressor of T helper type-1- and not type-2-associated lymphokine production, prevents induction of experimental autoimmune encephalomyelitis in Lewis rats. Eur. J. Immunol. **23:** 1745–1751.
23. DEL PRETE, G., A. TIRI, E. MAGGI *et al.* 1989. Defective *in vitro* production of gamma-interferon and tumor necrosis factor-α by circulating T cells from patients with the hyperimmunoglobulin E syndrome. J. Clin. Invest. **84:** 1830–1835.

24. Del Prete, G., M. de Carli, C. Mastromauro *et al.* 1991. Purified protein derivative of mycobacterium tuberculosis and excretory-secretory antigens of *Toxocara canis* expand *in vitro* human T cells with stable and opposite (type 1 T helper or type 2 T helper) profile of cytokine production. J. Clin. Invest. **88:** 346–350.
25. Del Prete, G., E. Maggi, P. Parronchi *et al.* 1988. IL-4 is an essential factor for the IgE synthesis induced *in vitro* by human T cell clones and their supernatants. J. Immunol. **140:** 4193–4198.
26. Bradley, L. M., M. Croft & S. L. Swain. 1993. T cell memory: new perspectives. Immunol. Today **14:** 197–199.
27. Zurawski, G. & J. E. de Vries. 1994. Interleukin-13, an interleukin-4-like cytokine that acts on monocytes and B cells, but not on T cells. Immunol. Today **15:** 19–26.
28. Paliogianni, F., S. S. Ahuja, J. P. Balow *et al.* 1993. Novel mechanism for inhibition of human T cells by glucocorticoids. J. Immunol. **151:** 4081–4089.
29. Chiappelli, F., C. Franceschi, E. Ottaviani *et al.* 1993. Phylogeny of the neuroendocrineimmune system: fish and shellfish as a model system for social interaction stress research in humans. Annu. Rev. Fish. Dis. **3:** 327–346.
30. Distelhorst, C. W. 1993. Steroid hormone receptors. J. Lab. Clin. Med. **122:** 241–244.
31. Litwach, G. 1989. The glucocorticoid receptor at the protein level. Cancer Res. **48:** 2636s–2640s.
32. Ku Tai, P.-K., M. W. Albers, H. Chang *et al.* 1992. Association of a 59 kDa immunophilin with the glucocorticoid receptor complex. Science **256:** 1315–1318.
33. Gehring, U. 1993. The structure of glucocorticoid receptors. J. Steroid Biochem. Mol. Biol. **45:** 1–3.
34. Lippman, M. & R. Barr. 1977. Glucocorticoid receptors in purified subpopulations of human peripheral blood lymphocytes. J. Immunol. **18:** 1977–1981.
35. Werb, Z., R. Foley & A. Munck. 1978. Interaction of glucocorticoids with macrophages: identification of glucocorticoid receptors in monocytes and macrophages. J. Exp. Med. **22:** 1684–1694.
36. Cidlowski, J. A. & N. B. Cidlowski. 1982. Glucocorticoid receptors and the cell cycle: evidence that the accumulation of glucocorticoid receptors during the S phase of the cell cycle is dependent on ribonucleic acid and protein synthesis. Endocrinology **118:** 1653–1662.
37. Martins, V. R., M. I. Birolli, J. S. Duarte *et al.* 1987. Glucocorticoid receptors in subpopulations of human lymphocytes defined by monoclonal antibodies. Cell. Immunol. **105:** 443–446.
38. Distelhorst, C. W. & B. M. Benutto. 1981. Glucocorticoid receptor content of T lymphocytes: evidence for heterogeneity. J. Immunol. **126:** 1630–1634.
39. Homo-Delarche, F. & D. Duval. 1986. Glucocorticoid receptors in lymphoid tissue. *In* Hormones and Immunity. I. Berczi & K. Kovacs, Eds. 1–19. MTP Press Limited. New York.
40. Meikle, A. W., R. W. Dorchuck, B. A. Araneo *et al.* 1992. The presence of a dehydroepiandrosterone-specific receptor binding complex in murine T cells. J. Steroid Mol. Biol. **42:** 293–304.
41. Chiappelli, F., C. Garcia, E. Manfrini *et al.* 1993. Regulation of migration of human naive helper T lymphocytes by glucocorticoids: implications for cocaine and alcohol abusers. Proc. Int. Cong. Lymphol. M. H. Witte & C. L. Witte, Eds. 206–211.
42. Chiappelli, F., P. Frost & T. Wylie. 1993. Alcohol and alcohol/cocaine mixtures blunt the activation of human CD4+ T cells. Alcologia **5:** 221–229.
43. Parker, L. N., E. R. Levin & E. T. Lifrak. 1985. Evidence for adrenocortical adaptation to several illness. J. Clin. Endocrinol. Metab. **60:** 947–952.
44. Ozasa, H., M. Kita, T. Inoue *et al.* 1990. Plasma dehydroepiandrosterone-to-cortisol ratios as an indicator of stress in gynecologic patients. Gynecol. Oncol. **37:** 178–182.
45. Welle, S., R. Jozefowicz & M. Statt. 1990. Failure of dehydroepiandrosterone to influence energy and protein metabolism in humans. J. Clin. Endocrinol. Metab. **71:** 1259–1264.
46. Mulder, J. W., P. H. J. Frissen, P. Krijnen *et al.* 1992. Dehydroepiandrosterone as

predictor for progression to AIDS in asymptomatic human immunodeficiency virus-infected men. J. Infect. Dis. **165:** 413–418.
47. Jacobson, M. A., R. E. Fusaro, M. Galmarini *et al.* 1991. Decreased serum dehydroepiandrosterone is associated with an increased progression of human immunodeficiency virus infection in men with CD4 cell count of 200–499. J. Infect. Dis. **164:** 864–868.
48. Wolkowitz, O. M., V. I. Reus, F. Manfredi *et al.* 1992. Dr. Wolkowitz and associates reply to Antiglucocorticoid effects of DHEA-S in Alzheimer's disease. Am. J. Psychiatry **149:** 1126.
49. Parker, L. N. 1989. Adrenal Androgens in Clinical Medicine. Academic Press. San Diego.
50. Fundenberg, H. H., H. D. Whitten, P. Arnand *et al.* 1984. Is Alzheimer's disease an immunological disorder? Observations and speculations. Clin. Immunol. Immunpathol. **32:** 127–131.
51. Ikeda, T., K. Yamamoto, K. Takahashi *et al.* 1991. Interleukin-2 receptor in peripheral blood lymphocytes of Alzheimer's disease patients. Acta Psychiatr. Scand. **84:** 262–265.
52. Bauer, J., S. Strauss, B. Volk *et al.* 1991. IL-6-mediated events in Alzheimer's disease pathology. Immunol. Today **12:** 422.
53. Vandenabeele, P. & W. Fiers. 1991. Is amyloidogenesis during Alzheimer's disease due to an IL-1-/IL-6-mediated "acute phase response" in the brain? Immunol. Today **12:** 217–219.
54. Chopra, R. 1990. Mechanisms of impaired T cell function in the elderly. Rev. Biol. Res. Aging **4:** 83–104.
55. Cossarizza, A., C. Ortolani, R. Paganelli *et al.* 1992. Age related imbalance of virgin (CD45RA+) and memory (CD45RO+) cells between CD4 and CD8 T lymphocytes in humans: a study from newborns to centenarians. J. Immunol. Res. **3:** 118–126.
56. Del Ponte, A., M. G. Di Monte, D. Graziani *et al.* 1990. Changes in plasma DHEA-S circadian rhythm in elderly men. Prog. Clin. Biol. Res. **341A:** 791–796.
57. Armanini, D., M. Scali, G. Vittadello *et al.* 1993. Corticosteroid receptors and aging. J. Steroid Biochem. Mol. Biol. **45:** 191–194.
58. Martignoni, E., F. Petraglia, A. Costa *et al.* 1990. Dementia of the Alzheimer type and hypothalamus-pituitary-adrenocortical axis: changes in cerebrospinal fluid corticotropin releasing factor and plasma cortisol levels. Acta Neurol. Scand. **81:** 452–456.
59. Swerdloff, R. S. & C. Wang. 1993. Androgen deficiency and aging in men. West J. Med. **159:** 579–585.
60. Araneo, B. A., M. L. Woods & R. A. Daynes. 1993. Reversal of the immunosenescent phenotype by dehydroepiandrosterone: hormone treatment provides an adjuvant effect on the immunization of aged mice with recombinant hepatitis B surface antigen. J. Infect. Dis. **167:** 830–840.
61. Ershler, W. B., W. H. Sun, N. Binkley *et al.* 1993. Interleukin-6 and aging: blood levels and mononuclear cell production with advancing age and *in vitro* production is modifiable by dietary restriction. Lymphokine Cytokine Res. **12:** 225–230.
62. Wei, J., H. Xu, J. L. Davies *et al.* 1992. Increase of plasma IL-6 concentration with age in healthy subjects. Life Sci. **51:** 1953–1956.
63. Fagiolo, U., A. Cossarizza, E. Scala *et al.* 1993. Increased cytokine production in mononuclear cells of healthy elderly people. Eur. J. Immunol. **23:** 2375–2378.
64. Leonardi, A., L. Arata, G. Bino *et al.* 1988. Functional study of T lymphocyte responsiveness in patients with dementia of the Alzheimer type. J. Neuroimmunol. **22:** 19–22.
65. MacDonald, L. M., A. H. Goldstone, J. E. Morris *et al.* 1982. Immunologic parameters in the aged and in Alzheimer's disease. Clin. Exp. Immunol. **49:** 123–128.
66. Skias, D., M. Bania, A. T. Reder *et al.* 1985. Senile dementia of Alzheimer's type (SDAT): reduced T8+ cell mediated suppressor activity. Neurology **35:** 1635–1638.
67. Wayne, S. J., L. R. Rhyne, P. J. Garry *et al.* 1990. Cell-mediated immunity as a predictor of morbidity and mortality in subjects over 60. J. Gerontol. **45:** M45–M48.

68. MOLCHAN, S. E., J. L. HILL, A. M. MELLOW *et al.* 1990. The dexamethasone suppression test in Alzheimer's disease and major depression: relationship to dementia severity, depression and CSF monoamines. Int. Psychogeriat. **2:** 99–122.
69. LINDER, J., P. NOLGRAD, B. NASMAN *et al.* 1993. Decreased peripheral glucocorticoid sensitivity in Alzheimer's disease. Gerontology **39:** 200–206.
70. WEINER, M. F., S. VOBACH, D. SVETLIK *et al.* 1993. Cortisol secretion and Alzheimer's disease progression: a preliminary report. Biol. Psychiatry **34:** 158–161.
71. LEBLHUBER, F., E. WINDHAGER, C. NEUBAUER *et al.* 1992. Antiglucocorticoid effects of DHEA-S in Alzheimer's disease. Am. J. Psychiatry **149:**8 1125–1126.
72. SUNDERLAND, T., C. R. MERRILL, M. G. HARRINGTON *et al.* 1989. Reduced plasma dehydroepiandrosterone concentrations in Alzheimer's disease. Lancet **2:** 570.
73. BLACK, K. L., J. D. CIACCI, M. AMMIRATI *et al.* 1993. Clinical trial of *Serratia marcescens* extract and radiation therapy in patients with malignant astrocytoma. J. Clin. Oncol. **11:** 1746–1750.
74. ROSZMAN, T. L., L. ELLIOTT & W. BROOKS. 1991. Modulation of T cell function by gliomas. Immunol. Today **12:** 370–374.
75. MCVICAR, D. W., D. F. DAVIS & R. E. MERCHANT. 1992. *In vitro* analysis of the proliferative potential of T cells from patients with brain tumor: glioma-associated immunosuppression unrelated to intrinsic cellular effects. J. Neurosurg. **76:** 251–260.
76. MUNCK, A., D. B. MENDEL, L. I. SMITH *et al.* 1990. Glucocorticoid receptors and actions. Am. Rev. Respir. Dis. **141:** S2–S10.
77. LANGEVELD, C. H., M. P. VAN WAAS, J. C. STOOF *et al.* 1992. Implication of glucocorticoid receptors in the stimulation of human glioma cell proliferation by dexamethasone. J. Neurosci. Res. **31:** 524–531.
78. CHIAPPELLI, F., H. GWIRTSMAN, M. LOWY *et al.* 1991. Pituitary-adrenal-immune system in normal subjects and in patients with anorexia nervosa: the number of circulating helper T lymphocytes (CD4) expressing the homing receptor Leu8 is regulated in part by pituitary-adrenal products. Psychoneuroendocrinology **16:** 423–432.
79. CHIAPPELLI, F., H. GWIRTSMAN, G. GORMLEY *et al.* 1992. Effects of dexamethasone on selective lymphocyte subpopulations in hypercortisolemic patients with anorexia nervosa and with bulimia nervosa: preliminary report. Int. J. Eating Dis. **12:** 37–47.
80. CHIAPPELLI, F. & S. TRIGNANI. 1993. Neuroendocrine-immune interactions: implications for clinical research. *In* Primary and Secondary Eating Disorders: a Psychoneuro-Endocrine and Metabolic Approach. E. Ferrari, F. Brambilla & S. B. Solerte, Eds. 185–198.
81. SCHWARTZ, A. G., J. M. WHITCOMB, J. W. NYCE *et al.* 1988. Dehydroepiandrosterone and structural analogs: a new class of cancer chemopreventive agents. Adv. Cancer Res. **51:** 391–424.
82. NAVEILHAN, P., F. BERGER, K. HADDAD *et al.* 1994. Induction of glioma cell death by 1,25$(OH)_2$ vitamin D_3: toward an endocrine therapy of brain tumor? J. Neurosci. Res. **37:** 271–277.
83. BLACK, K. L., K. CHEN, D. P. BECKER *et al.* 1992. Inflammatory leukocytes associates with increased immunosuppression by glioblastoma. J. Neurosurg. **77:** 120–126.
84. NITTA, T., M. ALLEGRETTA, K. OKUMARA *et al.* 1992. Neoplastic and reactive human astrocytes express interleukin-8 gene. Neurosurg. Rev. **15:** 203–207.
85. TADA, M., K. SUZUKI, Y. YAMAKAWA *et al.* 1993. Human glioblastoma cells produce 77 amino acids interleukin-8 (IL-8_{77}). J. Neuro-Oncol. **16:** 25–34.
86. GAUTHIER, T., M.-F. HAMOU, L. MONOD *et al.* 1993. Expression and release of interleukin-1 by human glioblastoma *in vitro* and *in vivo.* Acta Neurochir. **121:** 199–205.
87. FINCH, C. E., N. J. LAPING, T. E. MORGAN *et al.* 1993. TGF-β1 is an organizer of responses to neurodegeneration. J. Cell. Biochem. **53:** 314–322.
88. SORDOILLET, C., M. A. CHAUVIN, J. C. HENDRICK *et al.* 1992. Sites of interaction between epidermal growth factor and transforming growth factor-β1 in the control of steroidogenesis in cultured porcine Leydig cells. Endocrinology **130:** 1352–1358.
89. BUTTI, G., P. GAETANI, C. CHIARABRANDO *et al.* 1991. A study on the biological behavior of human tumors. Part II. Steroid receptors and arachidonic acid metabolism. J. Neuro-Oncol. **10:** 241–246.

90. Borofsky, M. A., R. B. Zurier, H. Rosenbaum *et al.* 1992. Effects of polyunsaturated fatty acids on interleukin-2-dependent T cell growth. Immunol. Res. **11:** 154–164.
91. Baba, T., C.-C. Chio & K. L. Black. 1992. The effect of 5-lipoxygenase on blood-brain barrier permeability in experimental brain tumor. J. Neurosurg. **77:** 403–406.
92. Chio, C.-C., T. Baba & K. L. Black. 1992. Selective blood-tumor barrier disruption by leukotrienes. J. Neurosurg. **77:** 407–410.

DISCUSSION OF THE PAPER

Question: To get the DHEA effects that you are talking about, what kind of concentrations of hormone do you have to use?

Chiappelli: We used between 1 and 5 milligrams weekly.

Question: What kind of blood levels does that lead to?

Chiappelli: We normalized the levels in normal elderly to the blood levels found in normal healthy young adults.

Question: Does DHEA have any direct effects on lymphocytes in culture?

Chiappelli: Yes and no. The literature will say no. When you start to be very specific in terms of the immune outcomes, such as specific cytokines, then you begin to find some effects. In terms of the general response to a mitogen PHA, you do not see effects.

Question: In the 10-minute effect that seemed to be nonreceptor mediated, what concentration of dexamethasone is involved?

Chiappelli: 10^{-9} M dexamethasone for 10 minutes will significantly blunt phosphoinositol phosphate metabolism. We have published those data.

Question: You showed dexamethasone killing of CD4 cells. Is there any difference between the naive cells and memory cells in that respect?

Chiappelli: Yes. The dexamethasone effect is seen on the naive cells, not on the memory cells.

Question: Glucocorticoids kill the naive cell and simultaneously inhibit their growth; do they also inhibit the growth of the memory cells?

Chiappelli: This is a difficult question. Both naive cells and memory cells are resting cells and therefore will not grow on their own unless you stimulate them. The experiments of testing the role of glucocorticoids in the generation of memory cells has not been done. We are now trying to define naive cells based on the expression of CD45RA and we are in the process of testing whether glucocorticoids in fact alter or blunt the maturation of these cells. In terms of the killing, yes, the glucocorticoids do kill the naive cells at high concentrations. At 10^{-9} M there is an effect, but it is not a dramatic one, only a 15% decrease. It does occur and it is there. But there are enough cells in the population to continue growing when you stimulate them.

Question: I had a question about the cytokines. You had a scheme which I assume is taken from somebody else.

Chiappelli: Yes, from G. Trinchieri (Ref. 19).

Question: Glucocorticoids inhibit IL-2 and not IL-4. Have you yourself measured some of those cytokines?

Chiappelli: Yes.

Question: But with dexamethasone you have not looked, for instance, to see whether it does not inhibit the production of IL-4?

Chiappelli: No. We are looking at the expression of IL-2 and IL-4.

Downregulation of the Glucocorticoid Receptor

A Mechanism for Physiological Adaptation to Hormones[a]

JEFFREY C. WEBSTER[b,d] AND JOHN A. CIDLOWSKI[c]

[b]*Department of Physiology*
and
[c]*The Lineberger Comprehensive Cancer Center*
University of North Carolina at Chapel Hill
Chapel Hill, North Carolina 27599-7545.

The glucocorticoid receptor (GR) is a member of a large family of ligand-inducible transcription factors including all of the steroid hormone receptors, as well as the thyroid hormone and retinoic acid receptors. Current dogma asserts that a steroid molecule binds to its receptor with high affinity and as a result, a protein DNA interaction ensues, causing an alteration in the transcription in those genes that contain the appropriate hormone response elements.[1] Recent work suggests that the steroid receptor, once bound to its cognate DNA response element, interacts directly or indirectly with components of the basal transcription complex via protein-protein interactions.[2] These studies demonstrate the importance of the steroid hormone receptor in hormone-induced transcription. The pivotal role that the receptor plays in hormone action is further underscored by physiological studies that reveal a direct correlation between receptor number and cellular responsiveness to hormones. Such a relationship has been demonstrated in a number of cell lines and intact animals including humans.[3–5] For these reasons, the mechanisms governing the tissue specific regulation of the steroid receptors themselves are critically important for our understanding of steroid hormone action.

Our laboratory, Cidlowski and Cidlowski,[6] and Svec and Rudis,[7] first demonstrated that the glucocorticoid receptor could be homologously downregulated by prolonged exposure to its own ligand. This process was shown to be readily reversible, following short-term glucocorticoid treatment. In addition to GR, homologous downregulation has been documented for other receptors, including the many other members of the steroid receptor super-family (for a comprehensive review see Ref. 8). Such homologous downregulation, which is also associated with peptide hormone receptors and growth factor receptors, probably reflects a mechanism to attenuate hormone responsiveness during excess hormone production or therapeutic administration. This acclimatization by cells and tissues to a new environment (a stress stimulus in the case of corticosteroids) can be potentiated by this downregulation mechanism. TABLE 1 displays several modes whereby a cell's glucocorticoid receptor number can be regulated. These potential regulation sites for the glucocorticoid receptor include an increase in receptor degradation, a decrease in receptor synthesis, and receptor modification to an inactive form. For receptor degradation, the receptor

[a]Supported in part by National Institutes of Health Grants DK 32460 and Fellowship 1F32DK08905 to JCW.

[d]Corresponding author: 460 Medical Science Research Building, Department of Physiology, CB#7545, University of North Carolina at Chapel Hill, Chapel Hill, NC 27599-7545.

TABLE 1. Potential Mechanisms Involved in Glucocorticoid Receptor Downregulation

1. Increased Receptor Degradation
   A. General protein degradation
   B. Ubiquitin-dependent degradation
2. Decreased receptor synthesis
   A. Altered mRNA stability
   B. Reduced transcription initiation
   C. Transcription attenuation
3. Receptor inactivation/modification
   A. Non-steroid-binding forms
   B. Non-DNA-binding receptors

may be specifically targeted for increased turnover. A decrease in receptor synthesis can take place if there is an alteration of hGR mRNA stability, a decrease in transcription initiation from the hGR promoter, or an attenuation of transcription of the hGR gene. Receptor modification could include forms of the receptor which no longer bound steroid or failed to bind to the DNA response element.

One well documented mechanism whereby glucocorticoids modulate GR levels in cells is at the point of GR gene expression. In the past several years different laboratories, including our own, have demonstrated that glucocorticoids markedly decrease GR mRNA levels by 50–80% in a variety of cell types and tissues. Transcriptional analyses have further shown that a decrease in transcription rate accounts for a majority of the reduction in steady state levels of GR mRNA. However, the precise mechanism responsible for this transcriptional effect remain elusive. Interestingly, GR/DNA binding experiments performed in our laboratory utilizing Southwestern blotting and sucrose density gradient analysis revealed that the GR had a high affinity toward its intact cDNA even under stringent conditions. Computer analysis (unpublished observation, D. B. Tully) of the human GR cDNA disclosed that the cDNA contains a number of GRE-like elements although no perfect GRE is present. These observations taken together strongly suggest that the GR may be playing a direct role in glucocorticoid-mediated downregulation.

To test if these sequences were of functional significance in down regulation, the human glucocorticoid receptor (hGR) cDNA (whose expression is controlled by an RSV promoter which is not responsive to glucocorticoids) was transiently transfected into COS I cells or stably transfected into CHO cells to produce a functional receptor. Surprisingly, it was found that the GR cDNA contained sufficient information to achieve downregulation at both the mRNA and protein levels in both transient and stably transfected cell lines. This effect was found to be independent of the expression vector used (*i.e.*, independent of promoter context). The kinetics of the mRNA downregulation were rapid, dose dependent for dexamethasone, and restricted only to those ligands which naturally bind GR. Furthermore, studies using transcriptional and translational inhibitors indicate that the glucocorticoid receptor is solely responsible for hormone-mediated downregulation via interactions with the transcription initiation complex on the GR gene.

An observation that presents an interesting insight as to the molecular mechanisms involved in downregulation is that the glucocorticoid antagonist RU486 can also cause GR downregulation with the same dose response as dexamethasone. Since RU486 is a complete antagonist at low concentrations, this finding suggests that previously proposed models of transcriptional initiation by steroid receptors may not apply to GR downregulation. Together, the above information shows that the process of homologous downregulation of the GR gene is mediated by the hormone-bound glucocorticoid receptor complex but perhaps not in a classical inductive manner.

The question arises as to why the glucocorticoid receptor has intragenic elements to autoregulate its own expression? FIGURE 1 shows a potential model of how the glucocorticoid receptor can act as an attenuator of a stress response. This model proposes that via the intrageneic elements, glucocorticoid receptor concentration can be coordinately downregulated regardless of the various combinatorial associations of transcription factors (tissue specific) which may be present on the 5′ flanking region of the GR gene in different tissues. In FIGURE 1, the set of transcription factors binding to the 5′ flanking region of hGR from a liver cell could be quite different from those binding the 5′ flanking region of hGR in a hippocampal-neuron. Furthermore, the ability of the intragenic elements to interact with the basal transcription complex seems to be a dominant event. This mechanism potentially allows for a coordinate effect in different tissues where their respective promoters may vary with regard to various transcriptional factors. On a broader level, these results suggest a common mechanism whereby steroid-resistant states may simply be an acclimatization to a new environmental state. By utilizing the glucocorticoid receptor in a rate-limiting manner, cells and tissues have a shared avenue available in which to carry out responses to various environmental stimuli.

An interesting question that arises is what happens to the glucocorticoid receptor following chronic exposure to glucocorticoids. Glucocorticoid receptor downregulation in HeLa cells (which contain endogenous GR) treated up to 26 days in hormone-containing media, was freely reversible after removal of ligand. This result suggests that the remaining receptors are in a quiescent state but are capable of initiating a reversal in receptor phenotype. However, our laboratory[9] recently showed that after chronic administration (2 years) of dexamethasone, HeLa cells contained no measurable amounts of glucocorticoid receptor protein or mRNA. Even upon a return to control media (no hormone present) for up to six weeks these cells remained unable to express detectable levels of glucocorticoid receptor. Southern blot analysis of those cells chronically treated with dexamethasone revealed that the loss of receptor mRNA and protein was not due to gene loss or rearrangement. However, the loss seems to be irreversible, generating a cell population which is truly incompetent in their ability to recognize a glucocorticoid signal. Interestingly, when an expression vector for the hGR was cotransfected with a reporter plasmid containing a glucocorticoid response element (GRE) into the chronically treated cells, there was a glucocorticoid-inducible response of the reporter plasmid when compared to control (no transfected hGR). This would argue that the defect is at the receptor level only and not with any of the ancillary factors of the transcriptional complex which are necessary for glucocorticoid regulation of gene expression.

One possible model to explain this phenomenon is that initially, the receptor instigates hormone-mediated downregulation of expression of the GR gene. However, after long-term exposure to hormone, the factors responsible for constitutive expression of the glucocorticoid receptor gene become disassociated from the transcriptional initiation complex resulting in an alteration of chromatin structure which no longer allows transcription initiation to take place, even when the ligand is removed. Alternatively or concurrently, the glucocorticoid receptor gene may become methylated, which could render the gene in an "inactive" state. Once in the methylated state the receptor would not be constitutively expressed and therefore would not be responsive to future hormone stimulation. However, an introduced

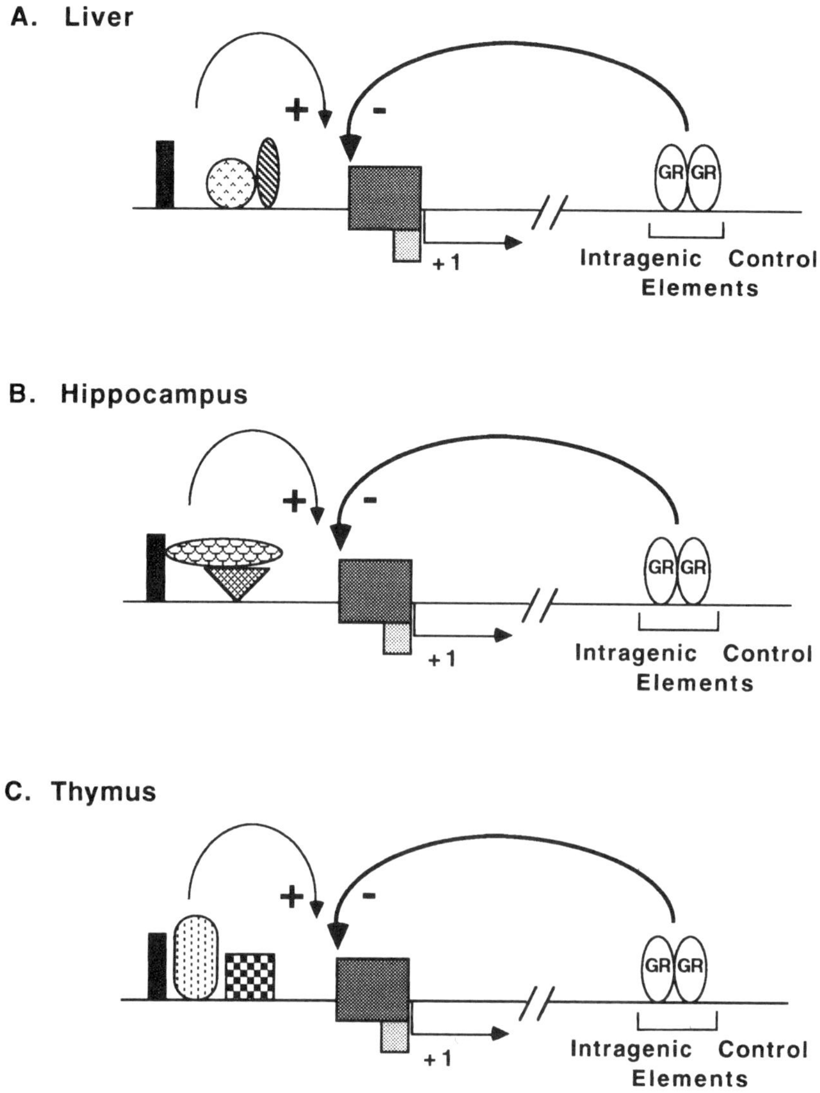

FIGURE 1. Coordinate control of GR gene expression in different tissues.

copy of the glucocorticoid gene (via transient transfection) would produce functional receptors which would respond with a normal physiological modality. These ideas are currently being evaluated by our laboratory.

The long-term hormone-treated HeLa cells may represent a bonifide physiological system to study the effects of chronic exposure of hormone in cases such as stress-related disorders (chronic depression) or long-term therapeutic regimes such as treatment of systemic lupus erythematosus (SLE). This example presents a viable model to study how genes may be switched from a transcriptionally active to a transcriptionally inactive state. For these reasons further investigations on the mechanisms by which long-term irreversible downregulation of glucocorticoid receptors occurs will be carried out.

In summary, the ability of cells and tissues to respond to a stress stimulus is of critical importance to the physiological maintenance of all higher organisms. Therefore, the importance of steroid receptors and how they carry out their physiological processes must not be underestimated. The ubiquitous nature of hormone-mediated downregulation suggests that this phenomenon represents a common negative feedback mechanism enabling a cell to attenuate the continued signal evoked by chronic exposure to ligand. Our proposed models show this process of hormone-mediated downregulation to be a complex event with marked differences from transcriptional activation or even other forms of transcriptional repression. For these reasons, the mechanisms of hormone-mediated downregulation will be the subject of intense study, and the clinical benefits derived from our understanding of this process should be significant.

ACKNOWLEDGMENTS

The authors would like to thank Dr. Douglas B. Tully for his critical reading of the manuscript and for helpful suggestions.

REFERENCES

1. YAMAMOTO, K. R. 1985. Steroid receptor regulated transcription of specific genes and gene networks. Annu. Rev. Genet. **19:** 209–252.
2. BEATO, M. 1989. Gene regulation by steroid hormones. Cell **56:** 335–344.
3. BELLINGHAM, D. L., M. SAR & J. A. CIDLOWSKI. 1992. Ligand-dependent downregulation of stably transfected human glucocorticoid receptor. Mol. Endocrinol. **6:** 2090–2102.
4. BOURGEOIS, S. & R. F. NEWBY. 1979. Correlation between glucocorticoid receptor and cytolytic response of murine lymphoid cell lines. Cancer Res. **39:** 4749–4751.
5. VANDERBILT, J. N., R. MIESFIELD, B. A. MALER & K. R. YAMAMOTO. 1987. Intracellular receptor concentration limits glucocorticoid-dependent enhancer activity. Mol. Endocrinol. **1:** 68–74.
6. CIDLOWSKI, J. A. & N. B. CIDLOWSKI. 1981. Regulation of glucocorticoid receptors by glucocorticoids in cultures HeLa S3 cells. Endocrinology **109:** 1975–1882.
7. SVEC, F. & M. RUDIS. 1981. Glucocorticoids regulate the glucocorticoid receptor in the AtT-20 cell. J. Biol. Chem. **256:** 5984–5987.
8. OAKLEY, R. H. & J. A. CIDLOWSKI. 1993. Homologous downregulation of the glucocorticoid receptor: the molecular machinery. Crit. Rev. Eukaryotic Gene Exp. **3**(2): 63–88.
9. SILVA, C. M., F. POWELL-OLIVER, C. M. JEWELL, M. SAR, V. E. ALLGOOD & J. A. CIDLOWSKI. 1994. Regulation of the human glucocorticoid receptor by long-term and chronic treatment with glucocorticoid. Steroids **59:** 436–442.

General Discussion

QUESTION: Dr. Bob Watson and I are preparing a manuscript looking at calcium binding protein in the hippocampus of fetal sheep. In the last 20 days of gestation fetal cortisol goes from a few nanograms per mL and peaks at term, it could be 90 to 100 nanograms. This seemed a natural system where cortisol goes from very low to very high, and we thought to look in hippocampus to see what calcium binding protein is doing. When we stained for it, we saw no calcium binding protein in Ammon's horn and in dentate gyrus. We did see it in the subiculum, but we did not see any change between the animals that have very low cortisol and those at term. We thought it would be a good way for the hippocampus to protect itself against insults at parturition like ischemia or hypoxemia. But we in fact did not see changes in calbindin that accompany cortisol changes.

E. C. AZMITIA (*New York University*): Does anyone have any thoughts on how a neuron can remain quiescent for 500 days?

ANSWER: This potential model from someone who studies apoptosis for most of our life could really be a very exciting one, and I would encourage any of you who are interested in apoptosis to follow this up. I suspect both Drs. Azmitia and Sloviter are right. These cells shrink, and shrinkage is one of the earliest steps in the process of apoptosis. But we know from studies in our own laboratories with lymphocytes that the shrinkage step in apoptosis is reversible. So you can shrink a cell which is the earliest phase of an apoptotic process at least in lymphocytes, and you can rescue that cell. It is only once the cell has gone farther along during the apoptotic process and fragmented its DNA that you can no longer rescue that cell and recoup it. You have a fantastic model in this dentate gyrus to make major contributions to the study of programmed cell death, as I would call it, in the immune system.

QUESTION: Dr. Sloviter made the point that some cells died within hours and others were alive for five months and continued to die. Is there a mechanistic explanation for that? Is that the answer to the discrepancy between the two views, that shrinkage may be happening on the way to death? That could still be a point where it could be rescued by glucocorticoids.

ANSWER: In all other examples of apoptosis that I know, except for this model system that we are working with now in our laboratory, when the cell shrinks it commits to apoptosis and continues to go. I am excited about this neural model, because perhaps you can shrink a population of cells and hold them in that state, and we might be able to learn more about the signal transductions in the apoptotic process. Whereas the half time of apoptosis in most cells is 4 hours. So once a cell shrinks, in four hours it is usually gone and engulfed by a neighboring cell. That is an average from two hours in the lymphocyte to six hours in a hepatocyte. So here you may have something unique and that could be very powerful in terms of understanding long-term plasticity.

QUESTION: Dr. Sloviter's data tell you that there is a point that the cell needs glucocorticoids, and if it does not have them then it commits. That long period is what I am asking about.

ANSWER: In the thymus where we studied the process, apoptosis is a very asynchronic process. Unless you have synchronized cells you cannot get them to die in a uniform manner. It all depends on their state of glucose that they have in the cell. It depends on how happy they are, and even in highly synchronized populations of cells you cannot get them to die all at once. At least that has been our experience in lymphocytes.

QUESTION: You said you rescued the cells. Presumably there are some specific manipulations you are doing. How actually are you inducing apoptosis? For instance, can you do it with steroids?

ANSWER: It is being done with ions. We can ionically shrink cells and then rescue them. We cannot do that with steroids, because it takes a long time to get the steroid receptor out of the nucleus.

QUESTION: Dr. Sloviter made the statement that he could actually protect his cells with 20 nanograms per mL of corticosteroids in the drinking water, which otherwise would have catastrophically died following adrenalectomy. That is lower then you can measure, because it is not a great assay. But when you think of that being largely bound to transcortin, which it will be, have you repeated it by infusion, or are we looking at the corticosteroid acting on the nasal organ or in the stomach, especially if you put that amount of corticosterone throughout a 150, 200 gram rat? You can do it with the same concentration of aldosterone in the drinking water. People have been talking about low doses of corticosterone as a milligram. Dr. McEwen is right. A microgram an hour will give an animal hypertension. That is not going to drain into the circulation. But this is 20 nanograms per mL, which means on a thirsty day the rat will drink 1 microgram per day. It is going to be 96 percent bound to CBG. ALDO is less than 50 percent bound to plasma protein. If you take cortisol, you get a systemic effect, whereas putting ALDO in hourly will not do anything. I am just concerned about that dose. A dose of 10 μgms is 500 times more than Dr. Sloviter is maintaining the hippocampus with.

R. S. SLOVITER (*Helen Hayes Hospital*): We used to give that concentration every week just to wean them off. In the last week they got 5 nanograms per mL in the drinking water and that protected it. It is amazing.

QUESTION: Then I think we are really looking at it being awfully concentrated and doing something in the mouth. If you have to put that amount past the liver in 150 mLs of extracellular fluid it should not reach the brain.

SLOVITER: Lipid soluble drugs will certainly be absorbed in nitroglycerin from the mucosa of the mouth very well. So I can easily see that that is the route of distribution absorption. That is no problem.

QUESTION: But not just distribution absorption, it has to produce a reaction. That is why I asked if you put it in, in a pump to get it into the body. Can you do it with the same dose? I think that is a live question.

SLOVITER: I have not done it; I do no know.

QUESTION: Have you looked at nuclear extractability of the DEX receptor complex that went into the nucleus versus RU486 receptor complex? Are these extractable with high salt?

ANSWER: We are just now beginning that.

QUESTION: So it is conceivable that the RU486 complex is in the nuclear plasma but not actually bound in a way comparable to the DEX receptor complex?

ANSWER: That is conceivable. We have done some confocal microscopy with RU486 versus DEX, and we really do not see any difference doing immunohistochemistry. But we have no real information about binding. What I can tell you is that *in vitro* with gel shift assay, with sucrose density gradient shifts, with virtually anything, we cannot distinguish the affinity of the RU486 complex for naked DNA from that of dexamethasone at all.

QUESTION: What about HSP90 association with the receptor? RU486 at least in some systems does not cause dissociation of HSP90, whereas dexamethasone does.

ANSWER: There is no direct evidence that HSP90 binds to GR in an intact living cell. We have looked at activation on sucrose gradients, and we cannot reproduce the findings that HSP90 does not dissociate as well, at least in HeLa cells.

Pharmacology of Neurosteroid Biosynthesis

Role of the Mitochondrial DBI Receptor (MDR) Complex

E. COSTA,[a] D. L. CHENEY,[a] D. R. GRAYSON,
A. KORNEYEV, P. LONGONE,[a] L. PANI, E. ROMEO,
E. ZIVKOVICH, AND A. GUIDOTTI[a]

Fidia-Georgetown Institute for the Neurosciences
Georgetown University Medical School
3900 Reservoir Road, NW
Washington, DC 20007

DBI: Discovery and Biological Activities

DBI is the acronym for diazepam binding inhibitor and refers to a 9-kd endogenous polypeptide first purified from rat brain by monitoring this peptide's ability to displace [^{3}H]-diazepam specifically bound to high-affinity recognition sites located in crude synaptic membrane preparations.[1] DBI is endowed with multiple biological actions: it modulates γ-aminobutyric acid (GABA)-mediated transmission, mitochondrial steroidogenesis, insulin secretion, conflict behavior and Acyl CoA binding and intracellular transport.[1] This peptide is present in brain (mostly in glial cells), in endocrine steroidogenic tissues and in several other organs (liver, spleen, kidney etc.).[1] Because of the method used in monitoring DBI purification, it was natural after its purification to homogeneity that the first DBI biological activity investigated was related to its ability to modulate GABA action on $GABA_A$ receptors.[2] These experiments revealed that DBI is a negative allosteric modulator of Cl^- channel GABA-gating in a manner that is inhibited by flumazenil,[2] the pure antagonist of positive and negative allosteric modulators of GABA action, which bind to a specific recognition site located in the ectodomain of the α subunits of $GABA_A$ receptors.[3]

The amino terminus of DBI purified from rat brain was found to be blocked. In order to obtain an amino acid sequence, DBI cyanogen bromide fragments were prepared, purified and sequenced.[4] The DBI_{33-50} sequence was used to prepare the probes to clone human[5] and rat[6] DBI cDNAs. The putative DBI amino acid sequences deduced from these two cDNAs were found to differ in length. The human cDNA encoded 104 amino acids whereas the rat cDNA encoded only 86 amino acids. Thus, human DBI appears to be a 5′ slightly elongated variant of rat DBI but the 86 amino acid sequence shared by the two peptides has an amino acid homology of about 75%. DBI was isolated, purified and directly sequenced also from porcine intestine.[7] This DBI was found to have 88% amino acid homology with murine DBI. Independently, DBI was also sequenced from bovine brain[8] and found to have 92% amino acid homology with human brain DBI.[5] Shoyab *et al.*[8] in their report refer to bovine DBI as a benzodiazepinoid (endozepine). Similar terminology des (Gly-Ile) endozepine was adopted later by Besman *et al.*[9] In 1989 they identified a protein that

[a]Present address: Center for Neuropharmacology, N. S. Kline Institute for Psychiatric Research, Orangeburg, NY 10962.

is essentially identical to DBI except that it lacks only the last two amino acid residues which were probably lost during its purification. They reported that this protein functions as an endogenous effector of the corticotropin-mediated increase of adrenal cortex steroidogenesis. Based on the similar and equally potent biological activity of DBI and des (Gly-Ile) endozepine, the authors[9] concluded that very likely the endogenous effector of the corticotropin action on adrenal steroidogenesis is DBI. In addition, a recent report by Goujrad *et al.*[10] shows that the steroidogenesis stimulated by a trophic hormone for MA-10 Leydig tumor cells is decreased following treatment with a phosphorothioate oligodeoxy nucleotide antisense to DBI cDNA. The antisense concentration that depletes the DBI content of MA-10 Leydig tumoral cells also potently reduces the increase of progesterone biosynthesis caused by saturating amounts of human chorionic gonadotropin. In addition to mediating steroidogenesis elicited by specific pituitary hormones in endocrine tissues, DBI and DBI_{17-50}, but not DBI_{33-50}, in nM concentrations, stimulate the conversion of cholesterol into pregnenolone by mitochondria prepared from glial cells.[11]

DBI is expressed in pancreatic islets. When it is added to isolated islets or perfused pancreas it inhibits the insulin secretion stimulated by high concentrations of glucose.[12] In the intestinal tract DBI is present in D cells and colocalizes with somatostatin.[13] Very likely these cells are identical to the glucagon-producing cells (A cells) which are located at the periphery of the islets and represent about 10% of the endocrine cells of an islet.[14] Another putative biological role of DBI is to bind and transport acyl CoA.[15]

Before considering the role of DBI in brain steroidogenesis directly, we should mention that the term endozepine should not be used to refer to DBI, because it is now reserved for endogenous ligands of benzodiazepine recognition sites which do not include peptidic bonds[16] in their chemical structure.

Functional Studies of DBI Genes

Chromosomal localization of DBI genes was studied using *in situ* hybridization in human cells[17] and Southern blot analysis of rat[6] and human[5] genomic DNA. This work has shown that DBI is encoded by a multigene family. The sequence of the rat DBI gene and that of four processed pseudogenes have been reported.[18] The lack of TATA and CCAAT boxes and other considerations have suggested that the DBI gene might be a housekeeping gene.[18] However, a differential cell type specific expression[19] and the presence of two different DBI forms in human cells[5] might favor a tissue-restricted regulation of DBI transcription. During the investigation of the regulation of DBI gene expression, three RNA polymerase initiation sites were detected by RNase protection and primer extension assays. Transient expression experiments[20] using DNA fragments of different lengths located upstream of the 5′-untranslated region of the DBI gene in glioma cell lines showed that basal promoter activity requires 146 bp of the proximal DBI sequence, whereas full activation is achieved with 423 bp of the 5′-untranslated region. DNase I protection experiments with liver nuclear proteins demonstrates three protected regions at nucleotides −387 to −333, −295 to −271 and −176 to −139 relative to the ATG initiation codon; in other tissues the pattern of protection is different. The region −176 to −139 was found to bind several general transcription factors and cell type-restricted nuclear proteins. Computer analysis has detected consensus recognition sequences for GR(−417 and −389), SP_1 (−340), AP_2 (−321) and AP_1 (−153). Because previous studies[20] had shown maximal promoter activity with a construct from −423 bp to −84 bp, Exo III deletions were generated, fused to a CAT reporter

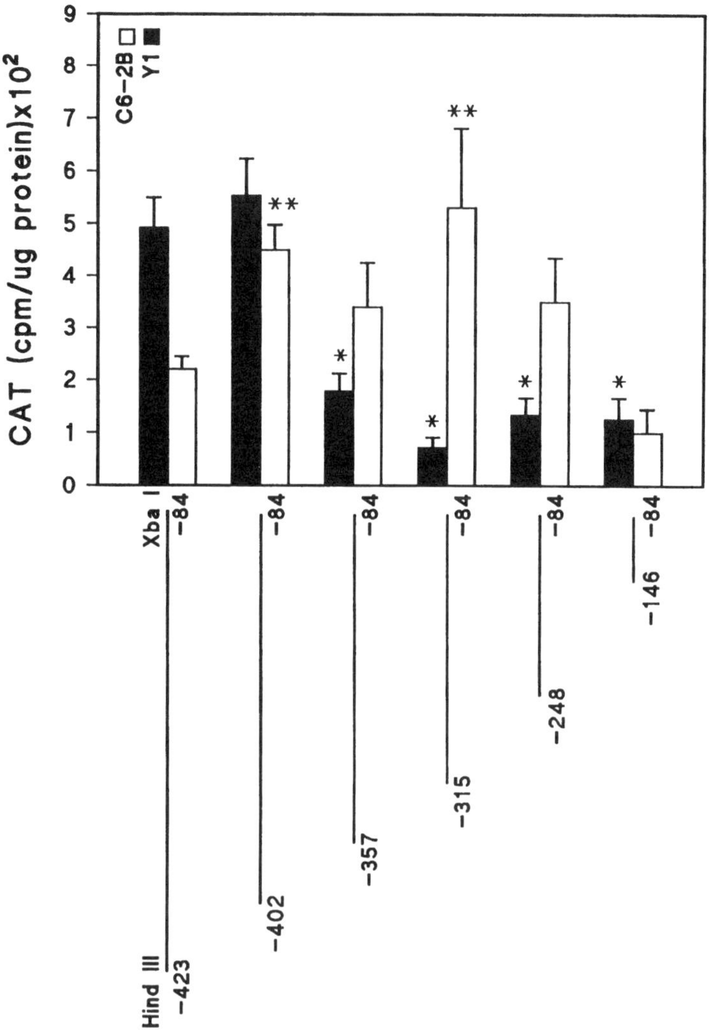

FIGURE 1. Functional analysis of the rat DBI promoter by fine point deletion mapping. Exo III deletions of the DBI promoter between positions −423 bp to −84 were generated and fused to a CAT reporter gene construct and transiently transfected into C6-2B glioma cells and Y-1 adrenal cells. DNA transfections in Y-1 cells were performed using the calcium-phosphate method[65] while the C6-2B transfections were performed using the lipofection method.[66] Cells were harvested 48 hours (Y-1) or 72 hours (C6-2B) after transfection and CAT activity was analyzed in cell extracts (prepared by three freeze-thaw cycles) following the Promega CAT Enzyme Assay System. The butylated products were measured by scintillation spectrometry. Results were corrected for the amount of protein and for the activity of β-galactosidase. Data are shown as the mean ± SEM of four transfections in triplicate for Y-1 and two transfections in triplicate for C6-2B * = Significantly different from transcription with −423 to −84 and −402 to −84. (p <0.05 with Duncan Multiple Range Test). ** = Significantly different from transcription with −146 to −84.

gene and transfected into C6-2B glioma cells and Y-1 cells in parallel. As shown in FIGURE 1, in Y1 cells a 50% loss in transcriptional activity (as a percentage of CAT conversion) occurs with a deletion between −402 and −357 without any recovery in activity with shorter constructs. In contrast, in C6-2B cells, maximal transcription begins after a deletion between −423 and −402, and, in fact, as shown in FIGURE 1, the transcription activity is significantly different in constructs excluding the −423 to −402 region fragments. This maximal transcription decays after a deletion of the region between −248 and −146. These data suggest the possible regulatory involvement in promoter activation of one putative GRE (located at −389) for Y-1, but other regulatory element transcripts are equally important for transcription in C6-2B cells. The SP_1 (−340) and multiple AP_2 (−321 to −295) sites appear to be involved in the DBI promoter regulation in C6-2B, but not in Y-1, cells. Thus, the regulation of the DBI gene is cell dependent on, and appears to change with, the different signalling of steroidogenesis regulation. In Y-1 cells where this regulation is mediated by pituitary corticotropic hormones, a GRE consensus (−389), but not SP_1 or AP_2, appears to regulate the DBI promoter. In contrast, in C6-2B cells, where the pituitary hormones are not operative in transcriptional regulation (probably in glial cells multiple extracellular signals can regulate steroidogenesis), GRE, SP_1 and AP_2 consensi binding activate DBI expression. Of course, these conclusions are mentioned here as working hypotheses that require verification with gel shift and footprint experiments that are now underway in our laboratory.

Mitochondrial DBI Receptors: a Subclass of "Peripheral" Receptors for Benzodiazepines

The discovery that anxiolytic benzodiazepines bind with high affinity to crude synaptic membrane preparations[21,22] raises the possibility that this high-affinity binding site is located on the $GABA_A$ receptors through which benzodiazepine might exert their anxiolytic and anticonvulsant actions.[23,24] However, it is also possible that the action of anxiolytic benzodiazepines on $GABA_A$ receptors is indirect and the high-affinity binding site of these drugs is located in the receptors for another as yet unknown transmitter. Successive work has clarified that the benzodiazepine recognition site is located in the α subunit of $GABA_A$ receptors.[25] This finding, on the one hand, stopped the search for the possible new transmitter system where benzodiazepines might be acting and, on the other hand, focused the work on the search for putative endogenous ligands of the allosteric modulatory center located in the α subunits of $GABA_A$ receptors. In fact, this site is endowed with high-affinity recognition sites for positive (anxiolytic) and negative (anxiogenic) modulators of the Cl^- channel gated by GABA. In the search for a clarification it became evident that the benzodiazepine binding sites located in the crude synaptic membranes prepared from brain homogenates are not structurally uniform; they include a "peripheral type" of benzodiazepine recognition site which, incidentally, is also present in brain[26] and very likely is abundantly expressed in glial cells. In addition, these "peripheral" benzodiazepine recognition sites may not be structurally homogeneous. In fact, subcellular fractionation has demonstrated the existence of a mitochondrial site, a nuclear site and a cell membrane site, but the site located in the mitochondrial outer membrane appears to be quantitatively predominant.[11] Since DBI has been proposed by several laboratories[11] to be a putative endogenous ligand for mitochondrial recognition sites, we have followed the traditional nomenclature procedure that names a receptor after its putative ligand, and have suggested that this site be named mitochondrial DBI receptor (MDR). The advantage of such nomenclature is to clarify that this recognition site is not only a drug binding site but it is also a putative effector site for the pituitary regulation of adrenal steroidogenesis.[10] Moreover, the term we suggest does not rule out that various drugs can also activate mitochondrial steroidogenesis by binding to MDR. Thus far, one of the drugs with the highest affinity for MDR is alpidem, an anxiolytic imidazopyridine which has a pM affinity for MDR.[11] However, this drug is not very selective because it also has a sub μM affinity (one to two orders of magnitude smaller than that for MDR) for the allosteric modulatory center of $GABA_A$ receptors were anxiolytic benzodiazepines bind. Since MDR represents the most abundant and widespread recognition site for alpidem, this site is persistently saturated following dose regimens of alpidem that saturate the alpidem recognition site located in $GABA_A$ receptors. The dual site mediation of alpidem action can be easily detected *in vivo* with appropriate pharmacological experiments. Evidence for such a dual mechanism stems from the observation that alpidem modifies rat behavior in the elevated plus maze, *i.e.*, it increases the time the rats spend in the open arm maze. The data of TABLE 1 show that this anxiolytic response to alpidem can be inhibited by both flumazenil and PK 11195 (an isoquinoline carboxamide acting as a weak partial agonist of MDR that *in vivo* functions as an antagonist by virtue of its sluggish intrinsic activity and its high affinity for MDR[1]). In contrast, the anxiolytic action of a chemically related imidazopyridine derivative, zolpidem, which has a low affinity for MDR and is virtually devoid of steroidogenic action, modifies the rat plus maze behavior similarly to alpidem, but this action exclusively derives from a direct allosteric modulatory activity on $GABA_A$ receptor. This is demonstrated by the antagonistic action of flumazenil (TABLE 1) against zolpidem and by the lack of antagonism of PK 11195. In this respect the action of zolpidem compares to that of clonazepam (a benzodiazepine devoid of affinity for MDR). The action of the neurosteroid tetrahydrodeoxycorticosterone (THDOC) differs from that of the other drugs reported in TABLE 1 because

TABLE 1. Influence of PK 11195 or Flumazenil Pretreatment on the Percentage Change of Time Spent in the Elevated Plus Maze Open Arm by Rats Receiving Zolpidem (Z), Alpidem (A), THDOC (T), or Clonazepam (C)[a]

PK 11195 (μmole/kg, i.v.)	Flumazenil (μmole/kg, i.v.)	Anxiolytic Drug (μmole/kg, i.v.)	Time in Open Arm with vs. without Antagonist Pretreatment (Percentages)	Time in Open Arm without Antagonist Pretreatment (Sec)
4.2	—	(A) 1.23	40*	17 ± 4.2
4.2	—	(C) 0.003	92	19 ± 3.2
4.2	—	(Z) 0.20	100	14 ± 3.5
4.2	—	(T) 2.76	98	13 ± 2.8
—	6.5	(A) 1.23	48**	17 ± 4.2
—	6.5	(C) 0.003	18**	19 ± 3.2
—	6.5	(Z) 0.20	34**	14 ± 3.5
—	6.5	(T) 2.76	100	13 ± 2.8

[a]PK 11195 or flumazenil was injected i.v. 5 min before the respective anxiolytic drug which was injected i.v. 10 min before a 5-min elevated plus maze trial. Statistical analysis: Duncan Multiple Range Test. (Compiled from data published by Auta *et al.*[54])

*p <0.05 when compared to PK 11195 pretreatment ± (C) or (Z) or (T).

**p <0.05 when compared to flumazenil + (T).

it is resistant to both flumazenil and PK 11195 inhibition. It is important to keep in mind that in a tumoral cell line derived from adrenal cortex (Y-1) pM concentrations of alpidem stimulate mitochondrial conversion of cholesterol into pregnenolone[11,27] (the parent steroid for every steroid hormone synthesized in mammalian tissues).

MDR is a heteroligomeric protein very likely formed by several integral membrane proteins spanning the mitochondrial ectomembranes;[11] probably one of them is a mitochondrial voltage-dependent anionic channel (VDAC). It is presumed that MDR activation by appropriate endogenous ligands (DBI?) increases the cholesterol influx from cytosolic storage sites in lipid vacuoles to the inner mitochondrial membrane where the cytochrome P_{450scc} contributes to the conversion of cholesterol into pregnenolone. Radioligand binding studies revealed that MDRs are not present in every mitochondria even though they are present in almost all tissues[11] including liver where they might be involved in biosynthesis regulation of biliary acids. Perhaps in different tissues the structure of MDR complex may differ. This diversity might be designed to accommodate its putative ability to participate in a variety of different metabolic patterns of cholesterol utilization and even the mitochondrial utilization of other metabolic substances. However, the functional connection of mitochondrial steroidogenesis with MDR is indirectly supported by the demonstration that a selective decrease of adrenal cortex MDR content occurs following hypophysectomy.[28]

Structure of a MDR Protein That Recognizes Endogenous DBI and Other Synthetic MDR Ligands

In 1983, Le Fur and collaborators[29,30] developed a series of isoquinoline carboxamides with high affinity for MDR. This series includes PK11195, the above-mentioned weak partial agonist endowed with strong antagonistic activity against the stimulation of mitochondrial steroidogenesis elicited by various chemical classes of MDR ligands.[1] A pair of quinoline propanamide enantiomers synthesized by Le Fur *et al.*[29] demonstrated a stereospecific binding for MDR. In addition, the synthesis of a nitrophenyl derivative of PK11195 (PK14105)[31] allowed for the development of a photoaffinity probe for the purification[32,33] of the MDR structural protein on which is located the high affinity site that binds various steroidogenic drugs including isoquinolin-carboxamides, imidazopyridines, and 2-aryl-3-indolacetamides. The purification of this protein from adrenal glands allowed the molecular cloning and expression of a cDNA encoding for this protein which presumably is included in the supramolecular structure of MDR.[34] The primary sequence deduced from this cDNA predicted a 169 amino acid chain. These data were confirmed by successive studies conducted in several animal species which indicate a 79% amino acid homology with the 18-kd MDR rat protein.[27] The gene encoding for this protein appears to be located in the 13,3 region of the long arm of human chromosome 22.[35] This protein is hydrophobic and includes five segments that appear to be putative transmembrane domains presumably spanning the mitochondrial membrane.[34] The expression of the rat cDNA encoding for this 18-kd protein in an embrionic kidney cell line (HEK 293) increased the density of the MDR binding domain of these cells for isoquinoline carboxamides, imidazopyridines and 4′ Cl diazepam, that is, for MDR-specific ligands.[34] Recently, the cDNA for this 18-kd protein present in the human MDR complex was expressed in yeast devoid of constitutional 18-kd protein; these studies convincingly confirmed that this protein contains binding sites for a specific benzodiazepine ligand which binds to MDR but not to $GABA_A$ receptors.[36]

McEnery *et al.*[37] presented evidence consistent with the view that in the MDR complex the 18-kd protein discussed above is associated with two proteins of 30 and 32 kd, which provisionally were identified with the mitochondrial adenine nucleotide carrier and the mitochondrial voltage-dependent anion channel (VDAC), respectively. This channel is presumably formed by the interaction of a single β sheet with the mitochondrial external lipid bilayer, has a complex voltage-dependence characteristic, is located in the contact points between inner and outer mitochondrial membranes and forms a large diameter pore which has a large conductance.[37] The adenine nucleotide carrier is an inner membrane protein which, in the solubilized MDR complex, was identified by immunoblot analysis. McEnery *et al.*[37] proposed the existence of a three-protein complex formed by the 18-kd protein, VDAC and the adenine nucleotide carrier which, probably, are functionally interrelated. Currently, Dr. Krueger's work in our Institute envisages that in adrenal cortex a 10-kd and 80-kd protein also participate with the 18-kd protein to form a complex of possible functional significance in the regulation of cholesterol mitochondrial influx. Since these prtoeins have not been sequenced it is fair to say that though we do not know for sure the molecular aspects regulating cholesterol mitochondrial influx, it is likely that DBI, by binding to the 18-kd mitochondrial protein, is operative in the regulation of adrenal steroidogenesis. It is important to recall that MDR and DBI are also present in liver, where, perhaps in a molecular arrangement different from that of endocrine tissue and glia cells, they might function also in the regulation of the hepatic mitochondrial transformation of cholesterol into biliary acids. With regard to the role of DBI in the regulation of mitochondrial steroid biosynthesis via MDR, several lines of independent investigation[38–43] suggest that this peptide and/or its processing product DBI_{17-50} play a role in glial cells and in endocrine steroidogenesis. The stimulation of adrenal (cortical cells) steroidogenesis by MDR ligands fails to occur *in vivo* in intact animals where steroidogenesis is mainly regulated by pituitary hormones that use DBI as an effector;[9] whereas in hypophysectomized rats,[44] MDR ligands promote adrenal steroidogenesis *in vivo*. In contrast, in the glia cells the steroidogenesis appears to be independent from a control by pituitary hormones[45] and can be increased by MDR ligands also in gonadectomized and adrenalectomized rats with intact pituitary glands.[45]

2-Aryl-3-Indolacetamide Derivatives: a New Class of MDR Synthetic Ligands

The 2-aryl-3-indolacetamide derivatives (FGIN-1) potently and selectively bind to glial MDR and increase mitochondrial steroidogenesis.[46] From the data[46] summarized in TABLE 2 the high affinity binding of the novel 2-aryl-3-indolacetamides to mitochondrial preparations from olfactory bulb or adrenal gland depends on the following characteristics of their chemical structure: (1) dialkylation of the amide nitrogen; (2) length of the amide alkyl chain (n-hexyl is optimal); and (3) halogenation of aryl groups appended to the indole nucleus. The specificity of the 2-aryl-3-indolacetamide derivatives listed in TABLE 2 was documented by showing that these compounds fail to bind in concentrations of 10^{-5} M to $GABA_A$, $GABA_B$, glycine, glutamate (NMDA, kainic acid and AMPA), dopamine, serotonin, opiate, cholecystokinin, β-adrenergic, cannabinoid or sigma receptors.[46] FGIN-1-27 enters the brain and, presumably by binding to brain MDR, increases glial cell steroidogenesis as it does *in vitro* (TABLE 3). In TABLE 3 the steroidogenic potency and efficacy *in vitro* of one of the most potent derivatives (FGIN-1-27) are compared to a very weak derivative (FGIN-1-7) and three other derivatives: a N-dipentyl (FGIN-1-41) and two compounds halogenated in the Z and X positions (FGIN-1-45; FGIN-1-13) (see TABLE 2). These data allow us to make two major points: (1) the n-hexyl

TABLE 2. Displacement of [^{3}H] 4′ Chlordiazepam by Various 2-Aryl-3-Indolacetamide Derivatives

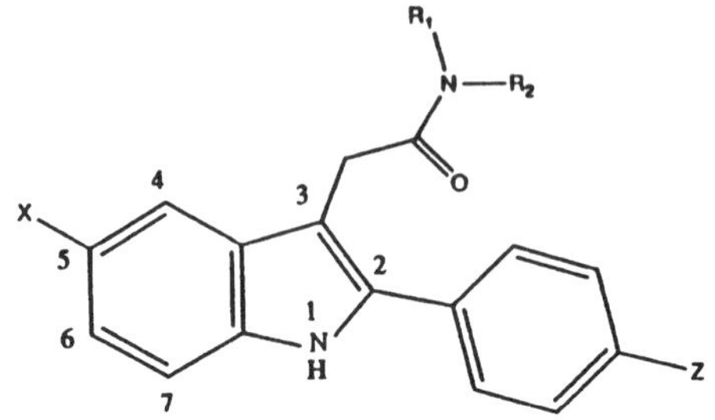

FGIN-1-	R[1]	R[2]	Z	X	K_i (nM)[a]
7	Methyl	Methyl	H	H	170 ± 16
38	Ethyl	Ethyl	H	H	160 ± 17
2	Propyl	Propyl	H	H	100 ± 10
51	Hexyl	Hexyl	H	H	9.7 ± 0.6
52	Octyl	Octyl	H	H	>1000
5	Propyl	Propyl	Cl	H	66 ± 6.6
20	Propyl	Propyl	F	H	100 ± 17
40	Butyl	Butyl	F	H	45 ± 2.1
41	Pentyl	Pentyl	F	H	15 ± 1.3
27	Hexyl	Hexyl	F	H	4.4 ± 0.15
46	Hexyl	Hexyl	H	Cl	4.1 ± 0.10
43	Hexyl	Hexyl	Cl	Cl	3.6 ± 0.080
45	Hexyl	Hexyl	F	Cl	3.9 ± 0.090
6	Propyl	Propyl	Cl	Cl	3.9 ± 0.080

[a]The concentrations of (^{3}H) 4′ Cl diazepam to obtain 50% saturation of MDR binding sites in mitochondria prepared from C6-2B glioma cells was 2.5 nM. Values are means ± SE of at least 3 experiments.

TABLE3. Stimulation of Pregnenolone Formation in C6-2B Glioma cell Mitochondria

FGIN-1-"X" X =	Maximal Stimulation % of Controls	EC_{50}[a] nM
7	130 ± 19	>100
27	240 ± 25	3.3 ± 0.30
41	180 ± 20	10 ± 1.2
45	220 ± 20	3.0 ± 0.28
13	180 ± 10	100 ± 12

[a]Drug concentration which increases the rate of pregnenolone formation over the value of vehicle which was 5.4 ± 0.4 (ng/mg protein/15 min). (For details see Ref. 46.)

amidation in the N residue maximizes the steroidogenic action of the FGIN series; and (2) halogenation in X or Z position contributes to maintain the high efficacy of these compounds. Like the neurosteroid THDOC, FGIN-1-27 also delays the onset of isoniazid-induced convulsions, but fails to change the bicuculline-induced convulsions.[46] However, unlike THDOC the antagonism of FGIN-1-27 against isoniazid convulsions is blocked by PK 11195.[46] In the elevated plus maze test, FGIN-1-27 inhibits neophobia in a manner that is antagonized by PK 11195, but not by flumazenil.[46] Because FGIN-1-27 binds to MDR and it does not bind to $GABA_A$ receptors,[46] it is likely that this drug "*in vivo*" may positively modulate $GABA_A$ receptors indirectly via the stimulation of brain pregnenolone biosynthesis and metabolism.[11]

Neurosteroids Modulate GABA Action at $GABA_A$ Receptors

The modulation of GABA activity by THDOC, allopregnanolone (3αOHDHP) and pregnenolone sulfate (PS) has been studied in native $GABA_A$ receptors of rat cortical and cerebellar neurons in primary culture and in structurally different recombinant $GABA_A$ receptors expressed in the HEK 293 cell line.[47–49]

In cortical neurons 3αOHDHP and THDOC positively modulate GABA-elicited Cl^- currents while 10 μM PS negatively modulates the GABA-gated Cl^- current (50% decrease). However, 10 nM PS positively modulates this GABA current (40% increase). This positive modulation by neurosteroids is equally potent in various types of recombinant $GABA_A$ receptors, except for receptors containing $\alpha_6\beta_1\gamma_2$ subunits which are less sensitive to the positive allosteric modulation by neurosteroids. In contrast, in receptors which include the γ_1 subunit, the efficacy of 3αOHDHP is doubled[48] (THDOC was not tested). The negative modulation by PS is similarly potent in recombinant $GABA_A$ receptors including various molecular forms of α or γ subunits. Thus, the receptor requirements for inhibition differ from those for positive allosteric modulation.[47,48] An activation of Cl^- current in absence of GABA by various concentrations of 3αOHDHP is observed in native and recombinant receptors but the current intensity on the various molecular forms of $GABA_A$ receptor tested is always smaller than that elicited by equimolar concentrations of GABA (10 μM).[48]

Brain Steroidogenesis Activation in Vivo by 2-Aryl-3-Indolacetamide Derivatives

In order to postulate that FGIN-1-27 is changing behavior via an activation of brain steroidogenesis we tested whether the treatment with FGIN-1-27 could increase the concentration of brain pregnenolone in animals with reduced steroidogenesis in peripheral endocrine tissues.[45]

Male rats were adrenalectomized and castrated (A-C) in order to exclude the possibility that the increase in brain pregnenolone content following the activation of the conversion of cholesterol into pregnenolone by the injection of FGIN-1-27 was due to a blood-borne contribution related to an increase in steroid synthesis and secretion by these endocrine tissues. Moreover, these A-C rats were treated with trilostane to minimize the further metabolism of pregnenolone to progesterone.

The data reported in TABLE 4 show that FGIN-1-27 in a dose-related manner increases the forebrain content of pregnenolone. It should be noted that in A-C rats

TABLE 4. Dose-Dependence of FGIN-1-27-Induced Increase of Pregnenolone Forebrain Content in A-C Rats[a]

FGIN-1-27[b] μmol/kg p.o.	Forebrain Pregnenolone Content ng/g ± SE
Vehicle	9 ± 1.5
200	7 ± 2.0
400	19 ± 4.0*
800	25 ± 6.5*

[a]The orchyectomy and adrenalectomy were performed two to three days before the FGIN-1-27 injection. (For additional details see Ref. 45.) $p < 0.05$ when drug-treated rats were compared to vehicle-treated rats. (Data from Korneyev *et al.*[45])

[b]Rats were given FGIN-1-27 1 hr after 600 μmol/kg i.p. trilostane. Rats were killed 1 hr after FGIN-1-27 or vehicle.

*Duncan multiple analysis (n = 5).

the blood pregnenolone content fails to increase as a result of the combined treatment with FGIN-1-27 and trilostane when compared to A-C rats receiving trilostane alone.[45] Moreover, it was observed that the increase in brain pregnenolone content elicited by 800 μmol/kg, p.o., of FGIN-1-27 lasts longer than 2 hrs.[45]

The term neurosteroid was introduced by Baulieu and Robel[50] to indicate that certain steroids possess an action on behavior which is prompt and independent from an activation of gene expression. Although the pharmacological significance of the behavioral actions of neurosteroids is well established,[51] the physiological significance of their actions requires further clarification.

As a preliminary step to elucidate the physiological role of neurosteroids we studied whether the behavioral effects of FGIN-1-27 could be abolished by a specific MDR antagonist in doses which could also reduce the stimulation of the cholesterol conversion into pregnenolone elicited by FGIN-1-27. Unfortunately, such a demonstration is not yet unquestionable. The data of TABLE 5 show that pretreatment with PK 11195 can abolish the increase of forebrain pregnenolone content elicited by a large dose of FGIN-1-27 (800 μmol/kg, p.o).[52] For technical reasons a similar oral treatment with FGIN-1-27 could not be used to evaluate the anticonflict activity of FGIN-1-27 using the Vogel test.[53] Thus, in the behavioral study we had to inject

TABLE 5. Effect of FGIN-1-27 and PK 11195 on the Pregnenolone Forebrain Content of A-C Rats

	Pregnenolone (ng/g)	
Treatment	Vehicle	PK 11195*
Vehicle	6.0 ± 1.2	6.5 ± 2.5
FGIN-1-27 (800 μmol/kg p.o.)*	14 ± 4.0**	6.5 ± 2.0[+]

*FGIN-1-27 was administered 1 hr before killing, while PK 11195 (100 μmol/kg, i.p.) was given 10 min before FGIN-1-27. Each value is the mean ± SEM of six rats. (For additional details see Ref. 52.)

**$p < 0.05$ vs. vehicle (Student t test).

+ $p < 0.05$ vs. FGIN-1-27 (Student t test).

TABLE 6. The Anticonflict Activity of FGIN-1-27 Is Blocked by PK 11195 in the Vogel Conflict Test

FGIN-1-27[a] μmol/kg, i.v.	PK 11195[a] μmol/kg, i.v.	Licking Periods in 3 Min (Mean ± SE)[b]
Vehicle	vehicle	9 ± 1.2
2.9	vehicle	13 ± 2.4
5.7	vehicle	16 ± 1.7**
11.7	vehicle	17 ± 3.2**
17.2	vehicle	22 ± 4.1**
17.2	4.2	10 ± 1.8***
17.2	2.1	10 ± 1.8***

[a]PK 11195 was injected 5 min before FGIN-1-27, which was injected 10 min before test.
[b]Each licking period value is the averge of 10 rats. (For additional details see Ref. 54.)
**p <0.05 compared to rats receiving only vehicle.
***p <0.05 compared to rats receiving 17.2 μmol/kg of FGIN-1-27 only.

intravenously smaller doses of FGIN-1-27 and PK 11195 in order to limit the interference by possible side effects associated with intravenous administration of large doses of these drugs.[54] Hence, the experiments on the inhibition by PK 11195 of FGIN-1-27 induced brain steroidogenesis (TABLE 5) cannot be directly compared with the behavioral experiments (TABLE 6), nevertheless these data suggest a similar action of PK 11195 on the two parameters.

TABLE 6 reports a dose-dependent anticonflict action of FGIN-1-27, expressed as a blockade of the behavioral inhibition induced by punishment using the Vogel paradigm, according to a procedure described by Corda *et al.*[53] The MDR antagonist PK 11195 inhibits the anticonflict action of FGIN-1-27 but fails to change the anticonflict action elicited by anxiolytic benzodiazepines or THDOC.[54] The anticonflict action of both THDOC and FGIN-1-27 is resistant to flumazenil inhibition.[52,54] These experiments reveal that there are two mechanisms whereby GABA mediates anticonflict actions, one epitomized by benzodiazepine, imidazopyridines and β-carboline derivatives which is inhibited by flumazenil, another directly or indirectly mediated by neurosteroids which is resistant to flumazenil. THDOC, which acts with a direct mechanism, is resistant to flumazenil and PK 11195 inhibition; FGIN-1-27 which acts with an indirect mechanism, is resistant to flumazenil but not to PK 11195.[52,54] Similar to results obtained in the elevated plus maze experiments reported in TABLE 1, results obtained in the Vogel test show that the action of low doses of alpidem is selectively inhibited by PK 11195 but not by flumazenil. However, flumazenil will antagonize the anticonflict action of higher doses of alpidem.[54]

It must be added that FGIN-1-27 also causes an anticonflict action in A-C rats, indicating that its action is likely to depend on the drug stimulation of brain steroidogenesis via high-affinity binding to brain MDR receptors. In order to elucidate the putative physiological role of neurosteroids it is important to understand why PK 11195, which inhibits the behavioral[54] steroidogenic action of MDR ligands,[52] fails *per se* to modify the behavior of rats in conflict situations, or to decrease brain neurosteroid content (Cheney, D. L. unpublished). In normal rats peripherally produced pregnenolone is taken up by the brain, where it can be converted into 3αOHDHP and THDOC. Since PK 11195 fails to change adrenal production of steroids in normal animals, it is possible that when PK 11195 is given to a rat, the brain steroidogenesis related to brain uptake of blood-borne pregnenolone proceeds unabated but this rate of metabolism obscures the inhibition of the brain mitochondrial steroidogenesis elicited by PK 11195. Clearly, the mechanism that we propose to explain the lack of behavioral and metabolic actions of PK 11195 in normal rats requires further experimental documentation. Perhaps, experiments with hypophysectomized rats performing on the elevated plus maze test after receiving PK 11195 may help to verify the validity of our inferences.

Neurosteroids Can Antagonize Experimentally Induced Cognitive Deficits

The systemic administration of appropriate doses of pregnenolone increases the brain content of several neurosteroids (Cheney, D. L. unpublished) and the administration of pregnenolone and pregnenolone sulfate into brain ventricles enhances memory and learning in male mice.[55] Though the molecular mechanisms of this hormonal action is not understood, an action of steroids on $GABA_A$,[56–58] NMDA[59] or glycine receptor[57] modulation has been documented. Specifically, as mentioned earlier, the potency (nM) of 3αOHDHP and THDOC in modulating $GABA_A$ receptors[47–49] has attracted special attention. Because of this background we have investigated whether MDR ligands that enhance brain pregnenolone content[45,52] reverse or attenuate the cognition deficit elicited by dizocilpine[60] and scopolamine,[61] antagonists of NMDA and muscarinic receptors, respectively.

Using a two-compartment box with one side illuminated and the other side dark and including an electric shock device located in the floor of the dark compartment as the avoidance stimulus, it was found that oral pretreatment with FGIN-1-27 antagonizes the dizocilpine disruption of passive avoidance with an EC_{50} of 57 μmol/kg. The peak effect is reached with 114 μmol/kg.[62] This dose of FGIN-1-27 is active when given 15 min after dizocilpine.[62] Moreover, the MDR antagonist PK 11195 (28 μmol/kg, i.p.) fails to modify the retention for passive avoidance response either when given alone (data not shown) or when given before dizocilpine. A pretreatment with this MDR antagonist reverses the FGIN-1-27 inhibition of the dizocilpine induced retention deficit. In considering these results in the light of a putative physiological role of neurosteroids, the lack of effect of PK 11195 should not be surprising since the electric shock elicits a strong response by the pituitary-adrenal axis with a marked increase of brain uptake of blood-borne pregnenolone.

The interaction between dizocilpine and FGIN-1-27 was tested in radial and water maze paradigms to study this drug interaction on spatial and working memory in rats. In both tests FGIN-1-27 fails to modify performance but reverses the performance impairment elicited by dizocilpine.[62]

As shown in TABLE 7 FGIN-1-27 antagonizes the dizocilpine disruption of the passive avoidance retention in sham-operated (SO) and A-C rats, suggesting that very likely a stimulation of brain steroidogenesis by FGIN-1-27 might be operative. This protection by FGIN-1-27 is inhibited by SKF 105111, a 5α-reductase inhibitor, in both A-C and SO rats supporting the concept that the formation of 3αOHDHP via 5α-reductase is required to achieve an antagonism of the dizocilpine disruption of the passive avoidance retention by FGIN-1-27. This action is not due to a primary action of SKF 105111 *per se* because this drug neither blocks the dizocilpine effects nor antagonizes the inhibition of the dizocilpine disruption of the passive avoidance retention elicited by THDOC. That the similar antagonism of the dizocilpine disruption of passive avoidance by pregnenolone and FGIN-1-27 is mediated by a

TABLE 7. FGIN-1-27, Pregnenolone and THDOC Abolish Dizocilpine Disruption of Passive Avoidance Retention. A 5α-Reductase Inhibitor (SKF 105111) Selectively Curtails the Action of FGIN-1-27 or Pregnenolone but Not That of THDOC[a]

Steroid or FGIN-1-27 μmol/kg	Dizocilpine 0.3 μmol/kg, i.p.	Retention Trial (Sec to Enter Dark Chamber)			
		Vehicle		SFK 105111**	
		SO	A-C	SO	A-C
Vehicle	—	140 ± 6.1	150 ± 7.3	140 ± 9.8	150 ± 8.3
Vehicle	Yes	21 ± 3.2	18 ± 2.1	28 ± 5.0	20 ± 2.1
FGIN-1-27 (114 p.o.)	Yes	120 ± 9.5	140 ± 8.3	23 ± 6.2*	16 ± 3.1*
Pregnenolone sulfate (48 i.p.)	Yes	125 ± 15	—	35 ± 6.0	—
THDOC (15 i.v.)	Yes	140 ± 13	—	125 ± 15	—

[a]SO = sham operated. A-C = adrenalectomized-castrated. SKF 105111 (11.7 μmol/kg, i.v.) 5 min after FGIN-1-27 or pregnenolone or THDOC. (40 min before trial/the acquisition). Dizocilpine was given 15 min later and rats were exposed to the acquisition trial 30′ after dizocilpine. The retention trial was recorded 24 h after acquisition trial. (For additional details see Ref. 62.)

*$p < 0.01$ when SKF 105111 group is compared with respective control. Anova with Dunnett test.

metabolite formed by 5α-reductase can be inferred from the data of TABLE 7 which shows a strong inhibition of the pharmacological action of pregnenolone and FGIN-1-27 by the 5α-reductase inhibitor. Before investigating more closely the chemical nature of the metabolic process that is operative we had to carry on experiments to investigate whether FGIN-1-27, pregnenolone and its neurosteroid metabolites change the binding of dizocilpine.[62] These experiments were carried out by maximizing the [^{3}H]-dizocilpine binding capacity of crude synaptic membranes with the addition of 10 μM glutamate in presence of low Mg^{2+} and 10 μM glycine. In such conditions, because of the opening of NMDA operated cationic channels by glutamate and glycine, the [^{3}H[-dizocilpine binding was increased by 30%; however, neither the membranes' maximal binding capacity nor their basal binding capacity for [^{3}H]-dizocilpine were changed by FGIN-1-27 up to 10^{-4} M. Furthermore, the addition of pregnenolone, pregnenolone sulfate, THDOC or 3αOHDHP up to 10^{-4} M also fails to change the basal or the glutamate-stimulated [^{3}H]-dizocilpine binding.

We then investigated[62] whether the injection of pregnenolone sulfate (48 μmol/kg i.p.) increases the pregnenolone, progesterone and allopregnanolone brain content. We found that the brain content of pregnenolone sulfate, pregnenolone, progesterone and allopregnanolone increased by about 10-, 9-, 2- and 7-fold, respectively, after the injection of pregnenolone sulfate. These findings corroborate the view that the action of pregnenolone on the cognitive tests mentioned above which is antagonized by the 5α-reductase inhibitor is mediated by 3αOHDHP formation from blood-borne pregnenolone. Indirectly, these findings suggest that FGIN-1-27 not only activates the conversion of cholesterol into pregnenolone in discrete brain areas important for the regulation of cognition but by a mechanism still poorly understood also promotes the activation of neurosteroid biosynthesis in these areas which, in turn, very likely mediates the antagonism of the cognitive impairment elicited by dizocilpine.

SPECULATIVE CONCLUSION

The experiments summarized above fail to provide unequivocal evidence that brain steroidogenesis plays a physiological role in normal brain function. Nevertheless, they contributed information with regard to the important role played by neurosteroids synthesized by MDR pharmacological stimulation in mediating the behavioral modification of cognition elicited by dizocilpine. Keeping in mind the desire to clarify whether neurosteroids play a physiological role in brain function, it is important to reflect on the finding that either the injection of pregnenolone, a 3αOHDHP precursor that is usually present in blood, or the increase of brain pregnenolone biosynthesis via brain MDR receptor stimulation in A-C rats not only can elicit a similar antagonism of the disruption of passive avoidance retention elicited by dizocilpine, but also very likely the action of these two different treatments is mediated by the same putative effector: 3αOHDHP. Thus, blood-borne pregnenolone and endogenously synthesized pregnenolone, are not metabolized similarly in brain but can also, apparently, reach functionally equivalent neuronal targets. Several lines of independent investigation suggest that these two different stimuli act on endogenous metabolic mechanisms that physiologically may activate the positive modulation of GABAergic synaptic activity via 3αOHDHP. If the response to MDR receptor stimulation can reach the same neuronal targets in brain as an increase of blood-borne pregnenolone, taken up by the brain, it appears that we might have two different metabolic mechanisms to subserve the same synaptic physiological modulation. Should this be the case, we have to consider that a brain uptake mechanism of a blood borne precursor of neurosteroids and a localized increase in the neurosteroid biosynthesis stimulated by specific MDR agonists may represent two alternative physiological mechanisms for the physiological modulation of GABAergic synaptic strength. One might also consider that these two different regulatory signaling systems can operate in different physiological situations: one is part of a more complex and generalized response mediated by the hypothalamic hypophyseal system and the other can be triggered, perhaps even physiologically, by local feedback mechanisms regulated by a signaling system that, via receptors very likely located on glial cell membranes, uses cyclic AMP as the second messenger. In fact, it is known that the addition of cAMP or forskolin to brain minces can increase pregnenolone formation within minutes.[63]

It is important to keep in mind that, while the injection of high doses of progesterone can cause a 3αOHDHP-dependent deep sedation, a similar response was never obtained with the maximal dose of FGIN-1-27 given to a rat (800 μmol/kg, p.o.). Perhaps the rate of 3αOHDHP biosynthesis that can be increased via MDR stimulation operates physiologically in specifically located functional circuits, while the increase of the 3αOHDHP steady state that can be elicited by systemic administration of progesterone involves almost every brain area where the appropriate enzymes for the 3αOHDHP biosynthesis from progesterone are located. Since these enzymes are quite abundant in every brain area, perhaps the resultant deep sedation and sleep depends on the simultaneous facilitation of GABAergic transmission in large neuronal populations.

In light of these considerations the physiological surge of the blood levels of progesterone and its rapid drop in the late luteal period that occur during the menstrual cycle may also be followed by changes in brain neurosteroid content. Progesterone has been frequently considered to have a significance in causing premenstrual dysphorias in a large population of women. Though it is not certain whether a temporal correlation exists between the decrease in progesterone blood

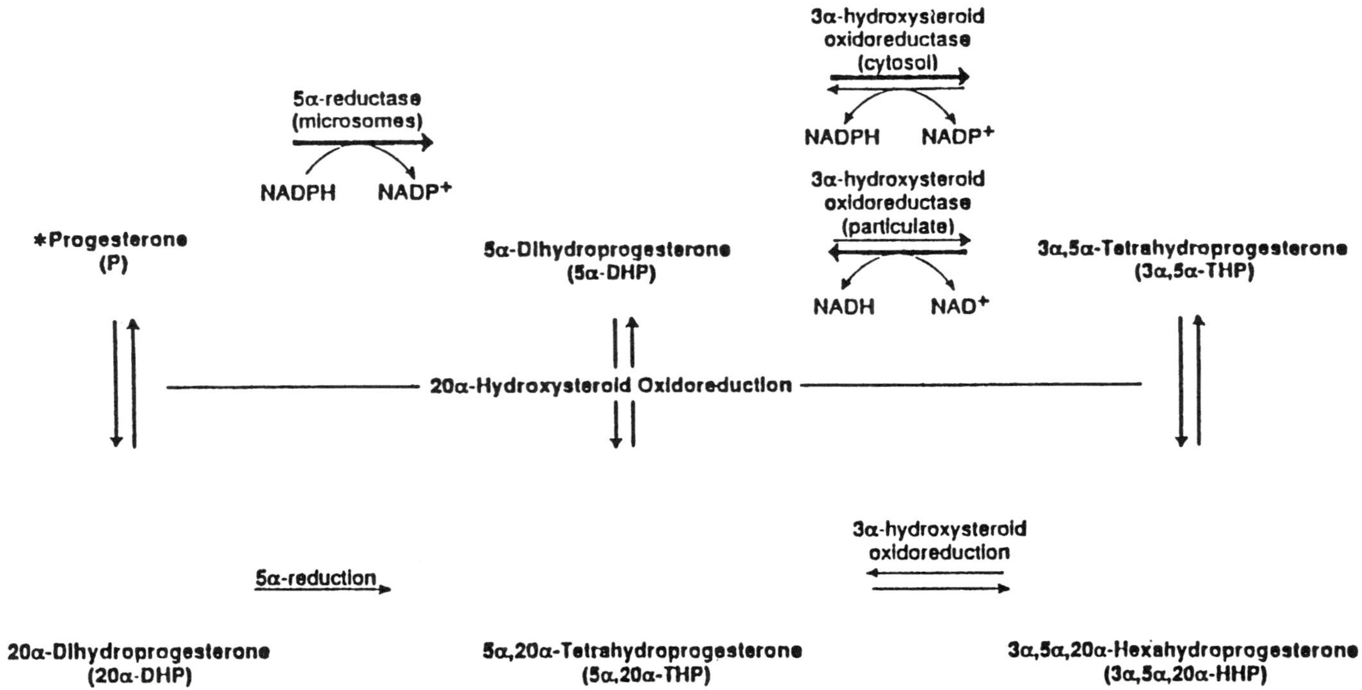

FIGURE 2. Schematic representation of progesterone procession in CNS. (Adapted from Karavolas and Hodges.[64])

levels and the severity of dysphorias in the late luteal phase of the menstrual cycle, it is possible that low levels of progesterone may somehow limit the endogenous production of 3αOHDHP. Alternatively during the late luteal phase of the menstrual cycle and an altered genetic substratum may shift neurosteroid biosynthesis from synthesis of steroids that positively modulate $GABA_A$ receptors to synthesis of steroids that negatively modulate $GABA_A$ receptors creating an imbalance in glutamatergic-GABAergic tone. This may nurture the pathological shift between the allosteric modulation of inhibitory and excitatory transmitter-regulated ion channels that appear operative in progesterone surge-related dysphorias.

In fact, as shown in FIGURE 2, the 5α-reductase that converts progesterone into 5α-dihydroprogesterone could be, in selected brain circuits, a rate-limiting event in the brain synthesis of 3αOHDHP. The metabolism of an excess of progesterone reaching the brain during several hours could shift toward a secondary pathway that, hypothetically, may result in an accumulation of a putative negative modulator of GABAergic or glutamatergic ion channels. Since we know that negative allosteric modulation of either of these two ligand-gated channels by β carbolines or phencyclidine can cause severe dysphorias and psychotomimetic symptoms, this hypothetical possibility finds some support in the experiments reported in TABLE 7. In fact, these data show that an inhibitor of 5α-reductase reduces the antagonism of both pregnenolone and FGIN-1-27 against the dizocilpine disruption of passive avoidance retention. Moreover, an inhibition of progesterone metabolism to 3αOHDHP via the 3α-hydroxysteroid reductase, could trigger an alternative progesterone metabolism leading to the possible accumulation of metabolites (FIG. 2) that change the balance between GABAergic and glutamatergic tone.

Before closing we would like to add a word of caution in future studies of the pharmacology of MDR agonists. If MDR were to play a generalized role in the liver metabolism of cholesterol via biliary acid formation, then one should study the affinity of MDR ligands for the liver and brain MDR and investigate their efficacy in promoting the enhancement of pregnenolone or biliary acid production. Presumably, a selectivity could be found even though both functions appear to have DBI as an effector. As mentioned earlier different DBI processing products may be operative in steroidogenesis and in the regulation of biliary acid synthesis. Very likely the differences in chemical structure of the DBI processing products which act as effectors is different in a variety of MDR recognition sites. This diversity could be used to provide a specific activation of brain steroidogenesis versus liver production of biliary acids.

Recently alpidem was withdrawn from the market because it caused a number of liver function problems. Is this a sign that the hypothesis formulated above with regard to possibly liver toxicity for MDR ligands is operative?

REFERENCES

1. COSTA, E. & A. GUIDOTTI. 1991. Diazepam binding inhibitor (DBI): a peptide with multiple biological actions. Life Sci. **49:** 325–344.
2. BORMANN, J., P. FERRERO, A. GUIDOTTI & E. COSTA. 1985. Neuropeptide modulation of GABA receptor Cl^- channels. Regul. Pept. (Suppl.) **4:** 33–45.
3. COSTA, E., G. PUIA, P. GIUSTI, J. AUTA, I. DUCIC, S. VICINI & A. GUIDOTTI. 1994. Mechanistic and pharmacological implications in the partial allosteric modulation of $GABA_A$ receptors by imidazenil. *In* The Challenge of Neuropharmacology: a Tribute to the Memory of Willy Haefely 19 March 1930 to 19 April 1993. H. Möhler & M. Da Prada, Eds. 45–53. Editiones Roche. Basel, Switzerland.
4. FERRERO, P., M. R. SANTI, B. CONTI-TRONCONI, E. COSTA & A. GUIDOTTI. 1986. Study

of an octadecaneuropeptide derived from diazepam binding inhibitor (DBI): biological activity and presence in brain. Proc. Natl. Acad. Sci. USA **83:** 827–831.

5. Gray, P. W., D. Gleister, P. H. Seeburg, A. Guidotti & E. Costa. 1986. Cloning and expression of a cDNA for human diazepam binding inhibitor, a natural ligand of the allosteric regulatory site of the γ aminobutyric acid type A receptor. Proc. Natl. Acad. Sci. USA **83:** 7547–7551.
6. Mocchetti, I., R. Einstein & J. Brosius. 1986. Putative diazepam binding inhibitor peptide cDNA clones from rat. Proc. Natl. Acad. Sci. USA **83:** 7221–7225.
7. Chen, Z., B. Agerberth, K. Gell, M. Andersson, V. Mutt, C.-G. Östenson, S. Efendic, J. Barros-Söderling, B. Persson & H. Jörnvall. 1988. Isolation and characterization of porcine diazepam-binding inhibitor, a polypeptide not only of cerebral occurrence but also common in intestinal tissues and with effects on regulation of insulin release. Eur. J. Biochem. **174:** 239–245.
8. Shoyab, M., L. E. Gentry, L. H. Marquardt & G. J. Todaro. 1986. Isolation and characterization of a putative endogenous benzodiazepinoid (endozepine) from bovine and human brain. J. Biol. Chem. **261:** 11968–11973.
9. Besman, M. J., K. Yanagibashi, T. D. Lee, M. Kawamara, P. F. Hall & J. E. Shively. 1989. Identification of des (Gly-Ile) endozepine as an effector of corticotropine dependent adrenal steroidogenesis: stimulation of cholesterol delivery is mediated by the peripheral benzodiazepine receptor. Proc. Natl. Acad. Sci. USA **86:** 4897–4901.
10. Boujrad, N., J. R. Hudson & V. Papadopoulos. 1993. Inhibition of hormone stimulated steroidogenesis in cultured Leydig tumor cells by cholesterol-linked phosphorothioate oligodeoxynucleotide antisense to diazepam binding inhibitor. Proc. Natl. Acad. Sci. USA **90:** 5728–5731.
11. Papadopoulos, V. 1993. Peripheral type benzodiazepine/diazepam binding inhibitor receptors: biological role in steroidogenic cell functions. Endocr. Rev. **14:** 227–240.
12. Östenson, C. G., B. Ahren, S. Karlsson, E. Sandberg & S. Efendic. 1990. Effects of porcine diazepam binding inhibitor on insulin and glucagone secretion *in vitro* from the rat endocrine pancreas. Regul. Pept. **29:** 143–151.
13. Höög, A., E. Sandberg & C. G. Östenson. 1989. Porcine diazepam binding inhibitor (p DBI) occurs in the D-cells of porcine and human gastrointestinal tract and pancreas, and modulates pancreatic hormone secretion. Pathol. Res. Pract. **185:** 73A.
14. Johansson, D., M. Hilliges, C. G. Östenson, E. Sandberg, S. Efendic & V. Mutt. 1991. Immunohistochemical localization of porcine diazepam binding inhibitor (DBI) to rat endocrine pancreas. Cell Tissue Res. **263:** 395–398.
15. Knudsen, J. 1991. Aryl-Co A-binding and transport, an alternative function for diazepam binding inhibitor (DBI) which is identical with acyl Co A binding protein. Neuropharmacology **30:** 1405–1410.
16. Rothstein, J. & A. Guidotti. 1993. Endozepines: nonbenzodiazepine endogenous allosteric modulators of $GABA_A$ receptors. *In* Natural Occurring Benzodiazepines: Structure, Distribution and Function. I. Izquierdo & J. Medina, Eds. 115–130. Ellis Horwood Ltd. Chichester, UK.
17. De Bernardi, M. A., R. R. Growe, I. Mocchetti, T. B. Shows, R. L. Eddy & E. Costa. 1988. Chromosomal localization of the human diazepam binding inhibitor gene. Proc. Natl. Acad. Sci. USA **85:** 6561–6565.
18. Mandrup, S., R. Hummel, S. Ravn, G. Jensen, P. H. Andreasen, N. Gregersen, J. Knudsen & K. Kristiansen. 1992. Acyl-CoA-binding protein/diazepam binding inhibitor gene and pseudogenes. A typical housekeeping gene family. J. Mol. Biol. **228:** 1011–1022.
19. Alho, H., R. Fremeau, H. Tiedge, J. Wilcox, P. Bovolin, J. Brosius, J. Roberts & E. Costa. 1988. Diazepam binding inhibitor gene expression: location in brain and peripheral tissues of rat. Proc. Natl. Acad. Sci. USA **85:** 7018–7022.
20. Kolmer, M., H. Alho, E. Costa & L. Pani. 1993. Cloning and tissue-specific functional characterization of the promoter of the rat diazepam binding inhibitor, a peptide with multiple biological actions. Proc. Natl. Acad. Sci. USA **90:** 8439–8443.
21. Möhler, H. & T. Okada. 1977. Demonstration of benzodiazepine receptors in the central nervous system. Science **198:** 848–851.

22. SQUIRES, R. F. & C. BRAESTRUP. 1977. Benzodiazepine receptors in rat brain. Nature **266:** 732–734.
23. COSTA, E., A. GUIDOTTI & C. C. MAO. 1975. Evidence for GABA involvement in the action of benzodiazepines: studies on rat cerebellum. *In* Advances in Biochemical Psychopharmacology. E. Costa and P. Greengard, Eds. Vol. 14: 113–130. Raven Press. New York.
24. HAEFELY, W., A. KULCSAR, H. MÖHLER, L. PIERI, P. POLC & R. SCHAFFNER. 1975. Possible involvement of GABA in the central actions of benzodiazepines. *In* Advances in Biochemical Psychopharmacology. E. Costa and P. Greengard, Eds. Vol. 14: 131–152. Raven Press. New York.
25. OLSEN, R. W. & A. J. TOBIN. 1990. Molecular biology of $GABA_A$ receptors. FASEB J. **4:** 1469–1480.
26. ANHOLT, R. R. H., P. L. PEDERSEN, E. P. DE SOUZA & S. H. SNYDER. 1986. The peripheral-type benzodiazepine receptor on outer mitochondrial membranes. J. Biol. Chem. **261:** 576–583.
27. KRUEGER, K. E. & V. PAPADOPOULOS. 1993. Cellular role and pharmacological implications of mitochondrial P-sites recognizing diazepam binding inhibitor (DBI) as a putative endogenous ligand. *In* Imidazophyridines in Anxiety Disorders: a Novel Experimental Approach. G. Bartholini, M. Gerreau, P. L. Morselli & B. Zivkovic, Eds. 39–48. Raven Press. New York.
28. ANHOLT, R. R. H., E. B. DE SOUZA, M. J. KUHAR & S. H. SNYDER. 1985. Depletion of peripheral-type benzodiazepine receptors after hypophysectomy. Eur. J. Pharmacol. **110:** 41–46.
29. LE FUR, G., M. L. PERRIER, N. VAUCHER, F. IMBAULT, A. FLAMIER, A. UZAN, C. RENAULT, M. C. DUBROEUCQ & C. GUEREMY. 1983. Peripheral benzodiazepine binding sites: effects of PK 11195, a-(2-chlorophenyl)-*N*-(1-methylpropyl)-3-isoquinolinecarboxamide. I. *In Vitro* Studies. Life Sci. **32:** 1839–1847.
30. DUBROEUCQ, M. C., J. BENAVIDES, A. DOBLE, F. GUILLOUX, D. ALLAM, N. VAUCHER, P. BERTRAND, C. GUEREMY, C. RENOULT, A. UZAN & G. LE FUR. 1986. Stereoselective inhibition of the binding of (^{3}H) PK 11195 to peripheral-type benzodiazepine binding sites by a quinoline propanamide derivative. Eur. J. Pharmacol. **128:** 269–272.
31. DOBLE, A., O. FERRIS, M. C. BURGEVIN, J. MENAGER, A. UZAN, M. C. DUBROEUCQ, C. RENAULT, C. GUEREMY & G. LE FUR. 1987. Photoaffinity labelling of peripheral-type benzodiazepine binding sites. Mol. Pharmacol. **31:** 42–49.
32. ANTKIEWICZ-MICHALUK, L., A. GUIDOTTI & K. E. KRUEGER. 1988. Molecular characterization and mitochondrial density of a recognition site for peripheral-type benzodiazepine ligands. Mol. Pharmacol. **34:** 272–278.
33. ANTKIEWICZ-MICHALUK, L., A. G. MUKHIN, A. GUIDOTTI & K. E. KRUEGER. 1988. Purification and characterization of a protein associated with peripheral-type benzodiazepine binding sites. J. Biol. Chem. **263:** 17317–17321.
34. SPRENGEL, R., P. WERNER, P. H. SEEBURG, A.G. MUKHIN, M. R. SANTI, D. R. GRAYSON, A. GUIDOTTI & K. E. KRUEGER. 1989. Molecular cloning and expression of cDNA encoding a peripheral-type benzodiazepine receptor. J. Biol. Chem. **264:** 14082–14087.
35. RIOND, J., M. G. MATTEI, M. KAGHAD, X. DUMONT, J. C. GUILLEMOT, G. LE FUR, D. CAPUT & P. FERRARA. 1991. Molecular cloning and chromosomal localization of a human peripheral-type benzodiazepine receptor. Eur. J. Biochem. **195:** 305–311.
36. RIOND, J., P. LEPLATOIS, P. LAURENT, G. LE FUR, D. CAPUT, G. LOISON & P. FERRARA. 1991. Expression and pharmacological characterization of the human peripheral type benzodiazepine receptor in yeast. Eur. J. Pharmacol. **208:** 307–312.
37. MCENERY, M. W., A. M. SNOWMAN, R.R. TRIFILETTI & S. H. SNYDER. 1992. Isolation of the mitochondrial benzodiazepine receptor: association with the voltage dependent anionic channel and the adenine nucleotide carrier. Proc. Natl. Acad. Sci. USA **89:** 3170–3174.
38. PAPADOPOULOS, V., P. GUARNERI, B. PAN, K E. KRUEGER & E. COSTA. 1991. Steroid synthesis by glial cells. *In* Neurosteroid and Brain Function. E. Costa & S. M. Paul, Eds. Vol. 8: 165–170. Thieme Medical Publishers. New York.

39. Papadopoulos, V., A. G. Mukhin, E.Costa & K. E. Krueger. 1990. The peripheral type benzodiazepine receptor is functionally linked to Leydig cell steroidogenesis. J. Biol. Chem. **265:** 3772–3779.
40. Papadopoulos, V., P. Guarneri, K. E. Krueger, A. Guidotti & E. Costa. 1992. Pregnenolone biosynthesis in C6-2B glioma cell mitochondria: regulation by a diazepam binding inhibitor mitochondrial receptor. Proc. Natl. Acad. Sci. USA **89:** 5118–5122.
41. Krueger, K. E. & V. Papadopoulos. 1990. Peripheral type benzodiazepine receptors mediate translocation of cholesterol from outer to inner mitochondrial membranes in adrenocortical cells. J. Biol. Chem. **265:** 15015–15022.
42. Papadopoulos, V., A. Berkovich, K. E. Krueger, E. Costa & A. Guidotti. 1991. Diazepam binding inhibitor (DBI) and its processing products stimulate mitochondrial steroid biosynthesis via an interaction with mitochondrial benzodiazepine receptors. Endocrinology **129:** 1481–1488.
43. Guarneri, P., V. Papadopoulos & E. Costa. 1992. Regulation of pregnenolone synthesis in C6 glioma cells by 4′ chlorodiazepam. Proc. Natl. Acad. Sci. USA **89:** 5113–5117.
44. Cavallaro, S., A. Korneyev, A. Guidotti & E. Costa. 1992. Diazepam binding inhibitor (DBI)-processing products, acting at the mitochondrial DBI receptor, mediate adrenocorticotropic hormone-induced steroidogenesis in rat adrenal gland. Proc. Natl. Acad. Sci. USA **89:** 10598–10602.
45. Korneyev, A., B. S. Pan, A. Polo, E. Romeo, A. Guidotti & E. Costa. 1993. Stimulation of brain pregnenolone synthesis by mitochondrial DBI receptor ligands *in vivo.* J. Neurochem. **61:** 1515–1524.
46. Romeo, E., J. Auta, A. P. Kozikowski, D. Ma, V. Papadopoulos, G. Puia, E. Costa & A. Guidotti. 1992. 2-Aryl-3-indoleacetamides (FGIN-1): a new class of potent and specific ligands for the mitochondrial DBI receptor (MDR). J. Pharmacol. Exp. Ther. **262:** 971–978.
47. Puia, G., M. R. Santi, S. Vicini, D. B. Pritchett, R. H. Purdy, S. M. Paul, P. H. Seeburg & E. Costa. 1990. Neurosteroids act on recombinant human $GABA_A$ receptors. Neuron **4:** 759–765.
48. Puia, G., I. Ducic, S. Vicini & E. Costa. 1993. Does neurosteroid modulatory efficacy depend on $GABA_A$ receptor subunit composition? Recept. Channels **1:** 135–142.
49. Majewska, M. D., N. L. Harrison, R. D. Schwartz, J. L. Barker & S. M. Paul. 1986. Steroid hormone metabolites are barbiturate-like modulators of the GABA receptor. Science **232:** 1004–1007.
50. Baulieu, E. E. & P. Robel. 1990. Neurosteroids: a new brain function. J. Steroid Biochem. Mol. Biol. **37:** 397–403.
51. Paul, S. M. & R. H. Purdy. 1992. Neuroactive steroids. FASEB J. **6:** 2311–2322.
52. Romeo, E., S. Cavallaro, A. Korneyev, A. P. Kozikowski, D. Ma, A. Polo, E. Costa & A. Guidotti. 1993. Stimulation of brain steroidogenesis by 2-aryl-indole-3-acetamide derivatives acting at the mitochondrial diazepam binding inhibitor receptor complex. J. Pharmacol. Exp. Ther. **267:** 462–471.
53. Corda, M. G., W. D. Blaker, W. B. Mendelson, A. Buidotti & E. Costa. 1983. Betacarbolines enhance shock induced suppression of drinking in rats. Proc. Natl. Acad. Sci. USA **80:** 2070–2076.
54. Auta, J., E. Romeo, A. Kozikowski, D. Ma, E. Costa & A. Guidotti. 1993. Participation of mitochondrial diazepam binding inhibitor receptors in the anticonflict, antineophobic and anticonvulsant action of 2-aryl-3-indolacetamide and imidazophyridine derivatives. J. Pharmacol. Exp. Ther. **265:** 649–656.
55. Flood, J. F., J. E. Morley & E. Roberts. 1992. Memory enhancer effects in male mice of pregnenolone and steroids metabolically derived from it. Proc. Natl. Acad. Sci. USA **89:** 1567–1571.
56. Majewska, M. D., J. M. Mienville & S. Vicini. 1988. Neurosteroid pregnenolone sulfate antagonizes electrophysiological responses to GABA neurons. Neurosci. Lett. 279–284.
57. Mienville, J. M. & S. Vicini. 1989. Pregnenolone sulfate antagonizes $GABA_A$ receptor-

mediated currents via a reduction of channel opening frequency. Brain Res. **489:** 190–196.

58. Wu, F. S., T. T. Gibbs & D. H. Farb. 1991. Inverse modulation of γ aminobutyric acid and glycine-induced currents by progesterone. Mol. Pharmacol. **37:** 397–602.
59. Wu, F. S., T. T. Gibbs & D. H. Farb. 1990. Pregnenolone sulfate: a positive allosteric modulator of the N-methyl-D-aspartate receptor. Mol. Pharmacol. **40:** 333–336.
60. Collingridge, G. L. & R. A. J. Lester. 1989. Excitatory amino acid receptors in the vertebrate central nervous system. Pharmacol. Rev. **40:** 167–174.
61. Alpern, H. P. & R. A. Marriott. 1973. Short term memory: facilitation and disruption with cholinergic agents. Physiol. Behav. **11:** 571–575.
62. Romeo, E., D. L. Cheney, I. Zovkovic, E. Costa & A. Guidotti. 1994. Mitochondrial DBI receptor complex (MDRC) agonists antagonize dizocilpine amnesia: putative role for allopregnanolone. J. Pharmacol. Exp. Ther. **270:** 89–96.
63. Barbaccia, M. L., G. Roscetti, M. Trabucchi, C. Ambrosio & M. Massotti. 1992. Cyclic AMP dependent increase of steroidogenesis in brain cortical minces. Eur. J. Pharmacol. **219:** 485–486.
64. Karavolas, H. J. & D. R. Hodges. 1991. Metabolism of progesterone and related steroids by neural and neuroendocrine structures. *In* Neurosteroids and Brain Function. E. Costa & M. M. Paul, Eds. 135–145. Thieme. New York.
65. Chen, C. & H. Okayana. 1987. High efficiency transformation of mammalian cells by plasmid DNA. Mol. Cell. Biol. **7:** 2745–2752.
66. Felgner, P. L., T. R. Gadek, M. Holm, R. Roman, H. W. Chan, M. Wenz, J. P. Northrop, G. M. Ringold & M. Danielson. 1987. A highly efficient lipid mediated DNA-transelection procedure. Proc. Natl. Acad. Sci. USA **87:** 7413–7417.

In Vitro Studies of Glucocorticoid Effects on Neurons and Astrocytes[a]

M. C. BOHN,[b] M. K. O'BANION, D. A. YOUNG,[c]
R. GIULIANO, S. HUSSAIN, D. O. DEAN,[d]
AND L. A. CUNNINGHAM[e]

Department of Neurobiology and Anatomy
and
[c]Department of Medicine
University of Rochester School of Medicine and Dentistry
Rochester, New York 14642

INTRODUCTION

Adrenal glucocorticoid hormones exert remarkably diverse effects on development, function, aging and regeneration in the nervous system. For example, these steroids have been implicated in such developmental processes as the determination of cell fate, neurogenesis, gliogenesis, the ontogeny of neuronal and glial phenotypic characters, myelination and development of behavior.[1–10] In mature animals, glucocorticoids are known to play a critical role in modulating the stress response; they are also critical for some more specific processes that include regeneration in the CNS following lesioning, expression of neurotrophic factors, neuronal cell death in the hippocampus and behaviors involved in learning, memory and adaptation.[11–18]

Glucocorticoids influence many of these processes by acting through two intracellular receptors that are known to function as transcriptional regulators, the glucocorticoid receptor (GR) and the mineralocorticoid receptor (MR). The expression of GR occurs in most neurons and glia throughout the brain, whereas high levels of MR expression are confined mainly to limbic and olfactory brain regions.[19–26] A particularly interesting aspect of GR and MR expression in the brain is the high level of co-expression of these receptors exhibited by hippocampal pyramidal and granule neurons.[19,23–25] As shown below, GR and MR are also co-expressed in astrocytes although at much lower levels than in neurons. The co-expression is interesting because both receptors appear to use the same ligand in brain and both are capable of activating the same glucocorticoid response element in a reporter construct.[27] It is known that some neurons express high levels of both receptors while other neurons express high levels of one receptor and low levels of the other receptor. However, the functional significance of this differential expression is not known. In addition, although the genes for both receptors are expressed in the fetal hippocampus, the precise molecular actions of glucocorticoids during brain development have not been defined.

[a]Supported by National Institutes of Health Grant NS-20832.

[b]Corresponding author: M. C. Bohn, Dept. of Neurobiology and Anatomy, Box 603, 601 Elmwood Avenue, Rochester, NY 14642.

[d]Present address: Dept. of Neuroscience and Cell Biology, Robert Wood Johnson Medical School, Piscataway, NJ 08854.

[e]Present address: Dept. of Pharmacology, University of New Mexico Medical School, Albuquerque, NM 87131.

The goal of the studies described here was to determine whether cultures of neurons or astrocytes might be used to simplify dissecting the molecular actions of glucocorticoids in the brain. To approach this goal, it was first necessary to determine whether GR and MR are expressed in neurons and astrocytes grown in defined conditions and, if so, whether glucocorticoids exert cell-specific effects on gene expression *in vitro*. Reviewed here are the results of our studies of GR and MR expression in cultured neurons and astrocytes and the identification of glucocorticoid-responsive genes in astrocytes.

Expression of GR and MR in Cultured Hippocampal Neurons

To study the expression of GR and MR in limited amounts of tissue or in cultured cells, a sensitive quantitative RNase protection assay was developed to measure the levels of GR mRNA and MR mRNA in the same sample, as well as the level of the constitutively expressed cyclophilin mRNA for the purpose of normalizing RNA levels among samples, quantifying absolute mRNA levels and comparing between experiments (FIG. 1).[28] This assay was first applied to developing hippocampus *in vivo* to study possible sex differences in expression of GR or MR mRNA. As shown in FIGURE 2, the levels of MR mRNA are 2–3 times higher than those of GR mRNA throughout development and in the adult.[28] With the exception of GR mRNA levels in the adult male rat hippocampus which are somewhat lower than those in the female, no significant sex differences were observed in levels of either receptor mRNA throughout ontogeny. These observations of message abundance are inter-

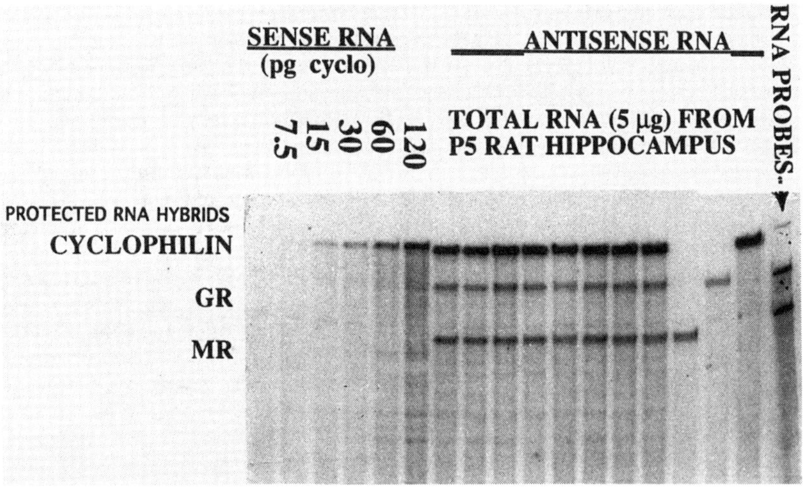

FIGURE 1. A representative RNase protection assay gel using 5 µg of hippocampal RNA from P5 rat. *Left five lanes* show standards of cyclophilin sense cRNA hybridized with antisense cRNA without tissue RNA. *Far right lane* shows undigested antisense cRNA probes without tissue RNA. *Middle lanes show* protected bands for all three probes added together to tissue RNA as well as added separately. (From Bohn *et al.*[28] Reprinted by permission from *Developmental Brain Research.*)

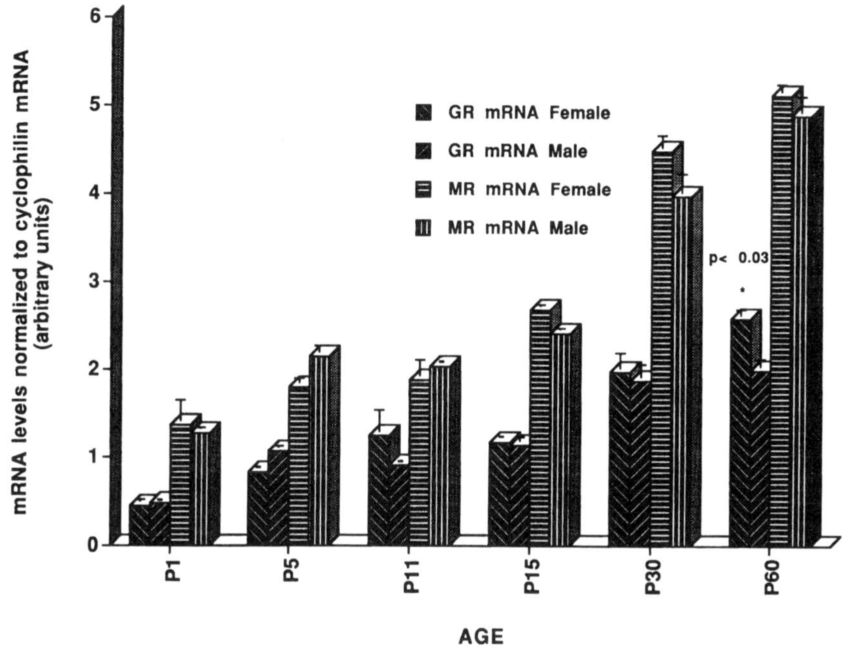

FIGURE 2. Postnatal ontogeny of GR mRNA and MR mRNA in hippocampus of male and female rat as determined by RNase protection assay. Data are corrected for amount of RNA/lane by normalizing to levels of cyclophilin mRNA and are shown as mean ± SEM. $p \leq 0.03$. (From Bohn *et al.*[28] Reprinted by permission from *Developmental Brain Research.*)

esting when considered in conjunction with data from binding studies suggesting that there are more GR than MR binding sites available in the hippocampus at all postnatal ages[29–32] and reports of sex differences in levels of receptor binding.[33] Thus, the levels of GRs and MRs available for binding glucocorticoids are not strictly paralleled by the levels of their respective mRNAs during ontogeny. Processes controlling either translation or posttranslational modification, *e.g.*, phosphorylation, may be crucial to receptor regulation as suggested previously.[34] On the other hand, occupation of receptors by nonglucocorticoid steroids, such as progesterone, in the adrenalectomized rat could lead to an underestimation of receptor protein in binding assays.

We also studied expression of GR and MR in cultures of dissociated embryonic day-E18 rat hippocampus established according to previously described methods.[35] After 7 days *in vitro,* the majority of neurons are immunoreactive for both GR and MR using the Bugr 2 and MinRec 4 antisera, respectively (FIG. 3). Preliminary studies using confocal microscopy suggest that, as in neurons *in vivo,*[22,36,37] GR is mainly cytoplasmic in the absence of ligand and nuclear in the presence of ligand.[38] In contrast, MR appears to be distributed in cytoplasm of the soma and neurites, as well as in the nucleus, with no translocation of MR evident in the presence or absence of ligand.[38] It is not clear whether this represents a real difference in subcellular

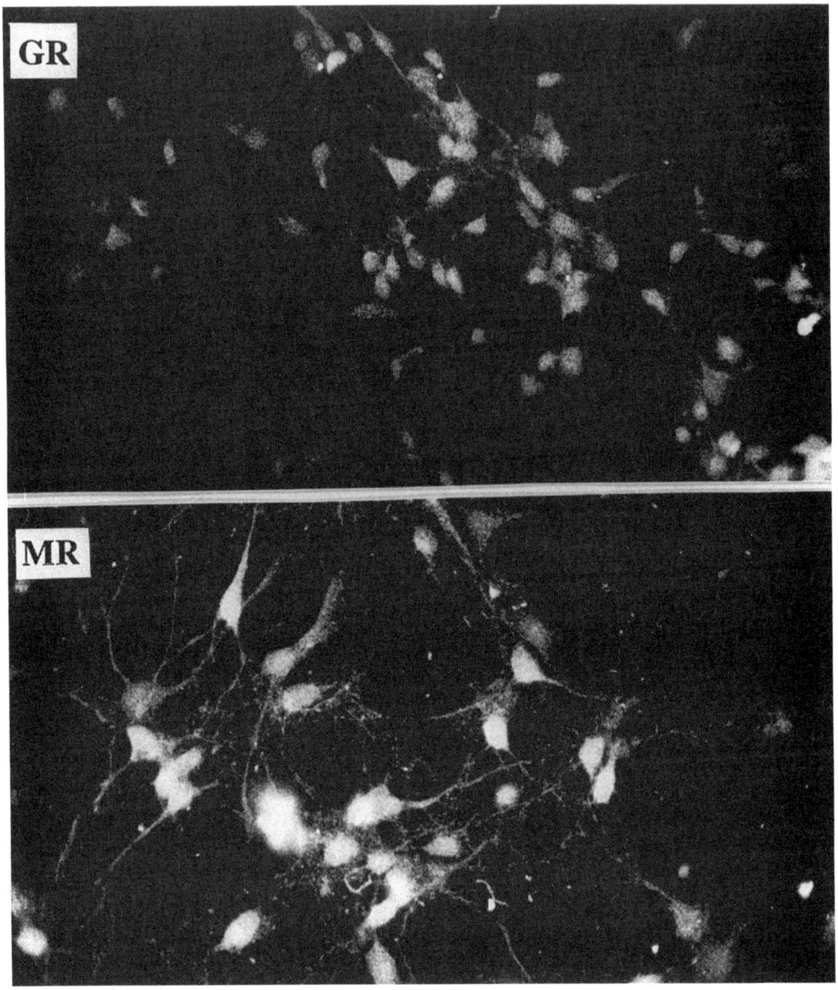

FIGURE 3. Immunofluorescence for GR and MR in embryonic day 18 hippocampal neurons grown in serum-free, steroid-free defined medium for 7 days and treated with 10^{-6} M corticosterone for 48 hours prior to fixation.

behavior of GR and MR in neurons, or whether other factors play a role in the subcellular localization of these receptors in embryonic neurons grown in culture.

In any case, RNase protection assays show that the levels of GR and MR mRNA expression in hippocampal neurons cultured for 7 days are approximately the same as those in the hippocampus *in vivo* on postnatal day 5 (data not shown) and that the ratio of MR mRNA/GR mRNA is also similar to that *in vivo* (FIG. 4). Growth of the neurons in low levels of serum or in medium conditioned by astrocytes significantly enhances this ratio with both increases being due to increased MR mRNA expres-

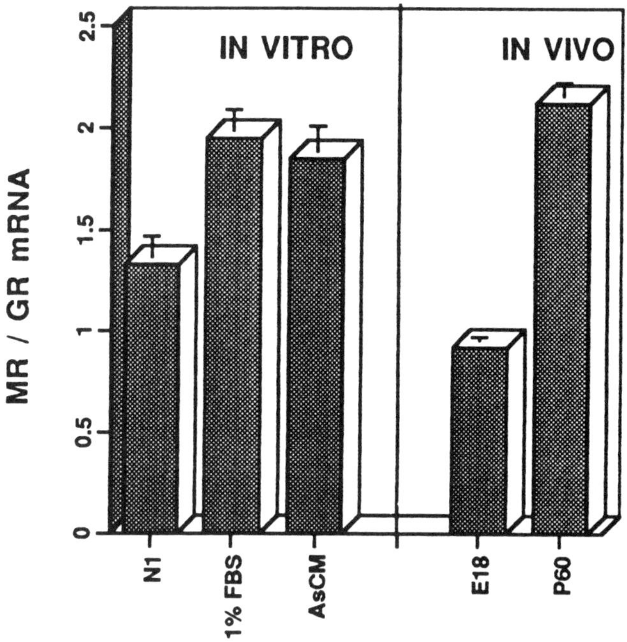

FIGURE 4. Results of RNase protection assay comparing *in vitro* to *in vivo* expression of hippocampal GR and MR. *Left panel*: ratio of MR mRNA/GR mRNA in hippocampal neurons cultured for 7 days in serum-free, steroid-free defined medium (N1) compared to growth in N1 plus 1% fetal bovine serum (FBS) or N1 pre-conditioned by astrocytes (AsCM) for 24 hours. *Right panel*: ratio of MR mRNA/GR mRNA levels in the embryonic day 18 (E18) and postnatal day 60 (P60) hippocampus. Data are expressed as mean ± SEM.

sion, and no significant change in GR mRNA expression (data not shown). In addition, the expression of both receptors was augmented by the addition of micromolar concentrations of corticosterone during the last 48 hours of culture (FIG. 5).

The positive regulation of receptor expression by ligand in the cultured embryonic neurons is in distinct contrast to the negative downregulation of receptor expression by ligand that occurs in most cells.[39] As shown in FIGURE 6, the levels of GR and MR mRNA are increased in the hippocampus after 3 days of adrenalectomy and are decreased by corticosterone replacement. Studies of other investigators using *in situ* hybridization and radioligand binding have also reported negative regulation by glucocorticoids of receptor expression in adult rat hippocampus,[23,29,40,41] while no increase in GR or MR mRNA after adrenalectomy has also been reported.[31] On the other hand, the levels of GR mRNA in cultured cells from cerebral cortex and hypothalamus are reported to be regulated in a complex, biphasic manner.[42]

There are several possible explanations for the apparent paradoxes raised by these diverse results. Corticosterone might be required for full expression of receptor genes during early ontogeny prior to the onset of a negative regulatory mechanism. A study of Meaney and co-workers is interesting in this respect. They reported that the negative glucocorticoid regulation of its own receptor does not occur until after

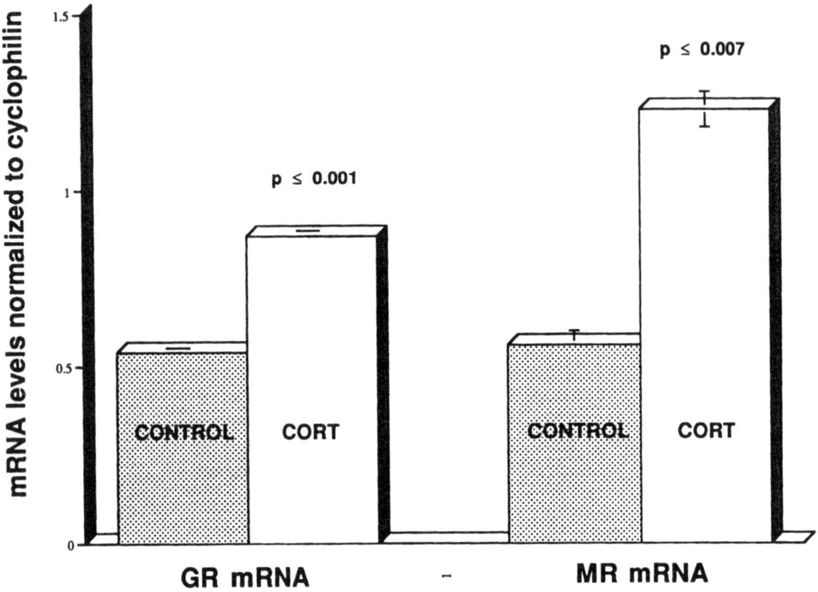

FIGURE 5. Results of RNase protection assay showing effects of corticosterone 10^{-6} M added for the last 2 days of culture on GR and MR mRNA levels in hippocampal neurons grown for 7 days in serum-free, steroid-free defined medium. Data are expressed as mean ± SEM.

puberty, a phenomenon that would permit the ontogenic increase of receptors to occur in parallel with increasing glucocorticoid levels.[43] It is also possible that corticosterone affects receptor expression by acting on different aspects of cell differentiation *in vitro* than *in vivo* or that corticosterone has different effects in different neuronal cell types. A further consideration is the possibility that assays of the effects of glucocorticoids on receptor expression *in vivo* represent the summation of direct effects and of possible indirect effects of steroids occurring in other neurons that are synaptically integrated with the hippocampus. In any case, it is unlikely that this paradox can be explained by the inclusion of glial receptor expression in the *in vivo* measurements since, as discussed below, the levels of receptor expression in astrocytes are at least two orders of magnitude lower than those in neurons.

Expression of GR and MR in Astrocytes

In early studies, we reported that every type of glial cell studied in culture stained for GR-immunoreactivity (IR), including type 1 and type 2 astrocytes from cerebral cortex and cerebellum, oligodendrocytes, Schwannoma and microglia.[26] To determine whether astrocytes also co-express MR, we studied astrocytes cultured from hippocampus by immunocytochemistry and from cerebral cortex by RNase protection assay. As shown in FIGURE 7, hippocampal astrocytes contain both GR-IR and

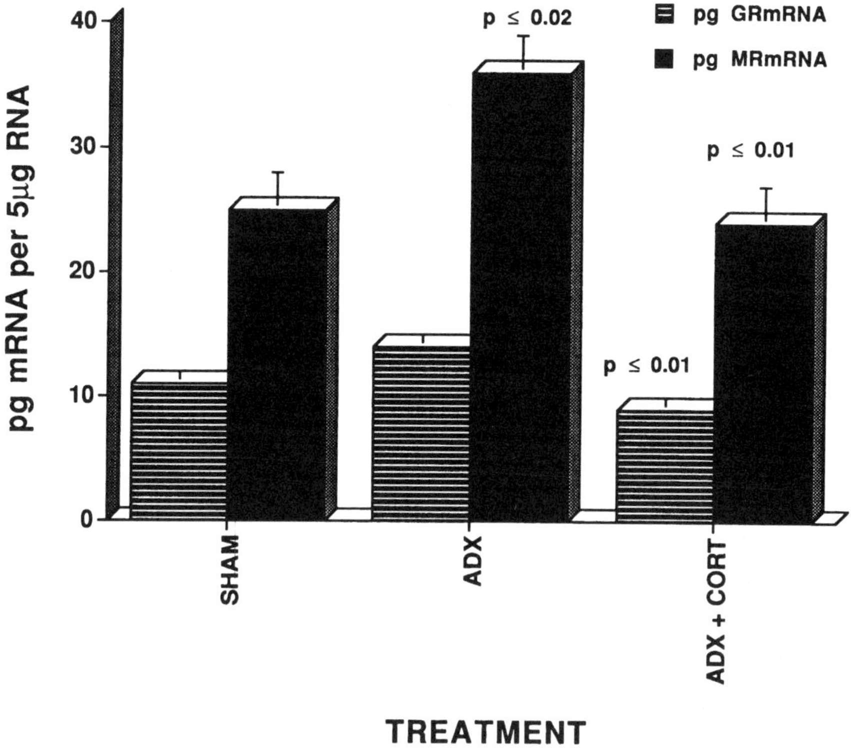

FIGURE 6. RNase protection assay showing effects of 3 days of adrenalectomy (ADX) and corticosterone replacement on GR and MR mRNA levels in adult rat hippocampus. Corticosterone replacement was administered by placing solidified pellets of Silastic adhesive containing 100 mg corticosterone subcutaneously in the cervical region at the time of surgery. The ADX group received pellets of silastic alone, whereas the sham group received no pellets. Data are expressed as mean ± SEM.

MR-IR which is predominantly nuclear in cells pretreated with corticosterone for 30 minutes prior to fixation. However, the intensity of staining in astrocytes is much less than that observed in neurons, and MR-IR is generally less than GR-IR. As reported previously, astrocyte nuclei are devoid of GR-IR in cells grown in the absence of glucocorticoid.[26] In contrast, no evidence of a ligand-dependent translocation of MR in cultured astrocytes has been observed either with the light or confocal microscope.[38,44]

The low levels of GR and MR mRNA expression in astrocytes were confirmed by RNase protection assay. In contrast to neuronal cultures in which both GR and MR mRNAs are easily detected using 5 μg of total RNA, it was necessary to use 1 μg of poly A RNA to detect GR and MR mRNAs in astrocytes. Consequently, expression of these receptors in astrocytes is estimated to be about two orders of magnitude lower than in hippocampal neurons. Moreover, the regulation of GR and MR expression in cultured astrocytes is further downregulated by treatment with corticosterone (FIG. 8).

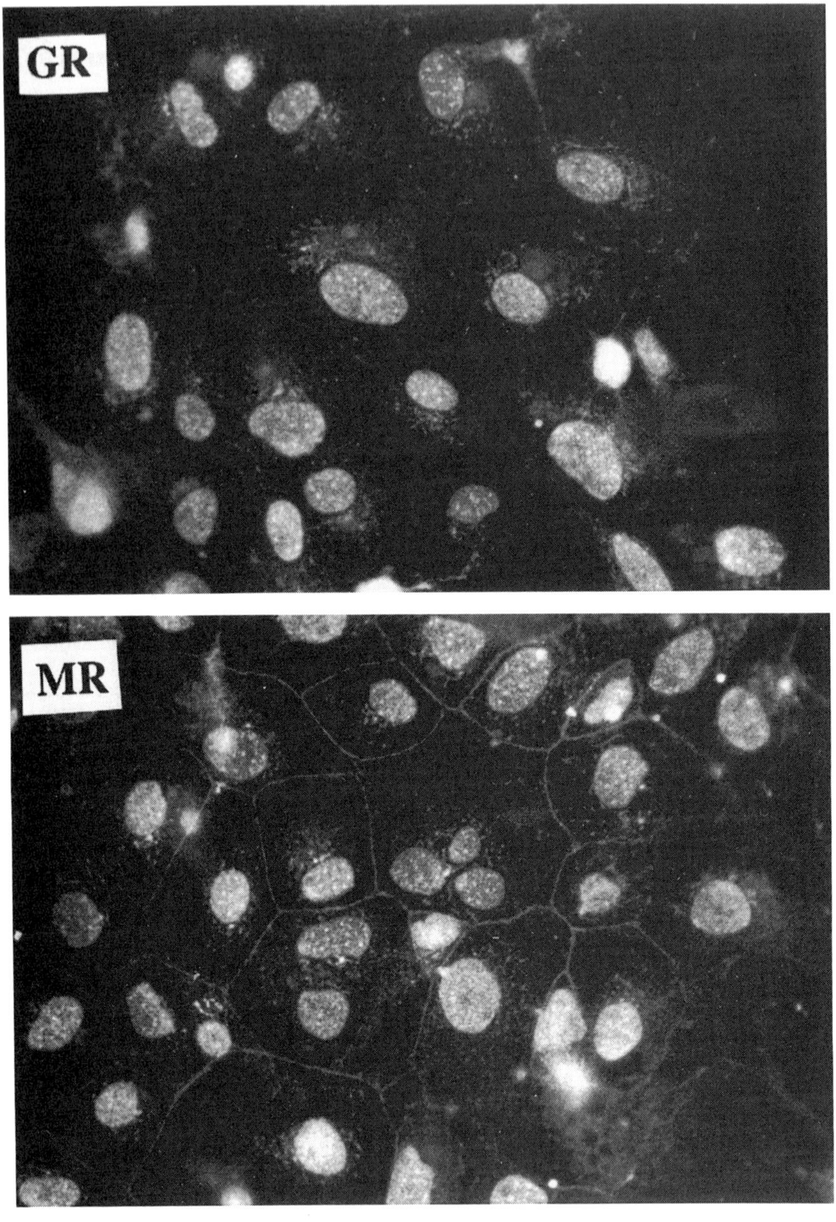

FIGURE 7. Immunofluorescence for GR and MR in astrocytes in embryonic day 18 hippocampal cultures grown in serum-free, steroid-free defined medium for 7 days and treated with 10^{-6} M corticosterone for 48 hours prior to fixation.

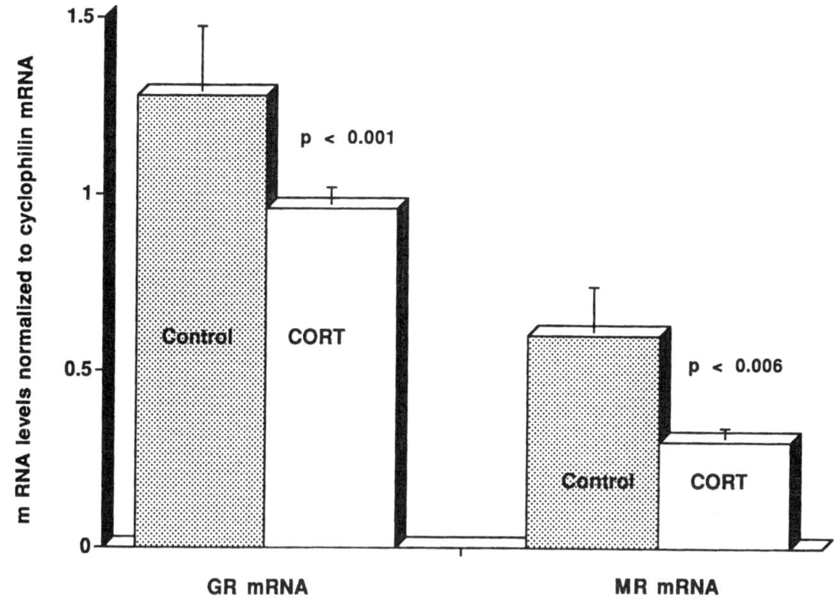

FIGURE 8. RNase protection assay showing effects of corticosterone 10^{-6} M added for 48 hours on GR and MR mRNA levels in astrocytes purified from postnatal day 1 rat cerebral cortex. ± SEM.

Effects of Glucocorticoids on Gene Expression in Astrocytes

To begin to identify glucocorticoid target genes in specific cell types in the nervous system, we studied the effects of treating cultures of pure type 1 astrocytes with corticosterone using ultra-high resolution giant two-dimensional (2D) gel electrophoresis.[45] Since we are primarily interested in those proteins that are regulated by glucocorticoids at the level of mRNA, the effects of glucocorticoids on astrocytes were studied by treating cultures with corticosterone for 3 or 24 hours, isolating poly A^+ RNA, translating this *in vitro* using rabbit reticulocyte lysate in the presence of ^{35}S-methionine, and separating the *in vitro* translation products by 2D gel electrophoresis.[46] The resulting patterns were analyzed using laser densitometric scanning and the PDQuest™ software package. Cycloheximide was added to some cultures prior to corticosterone treatment to discern those mRNAs that respond to glucocorticoid treatment in the absence of protein synthesis. The analysis of approximately 1500 *in vitro* translation products ranging in isoelectric point (pI) from 7.5–4.0 and in apparent molecular weight from 100 kDa to 14 kDa revealed 11 glucocorticoid inductions and 1 repression that occurred in repeated experiments (Fig. 9).[47] Most of these changes were observed after 3 hours of corticosterone treatment, although the changes were of greater magnitude after 24 hours of treatment. All of the changes appeared to be primary effects at the level of transcription, since pretreatment of cultures with cycloheximide, a treatment which inhibits protein synthesis by >97%,[47] did not prevent these changes in mRNAs and, in most cases, resulted in an even

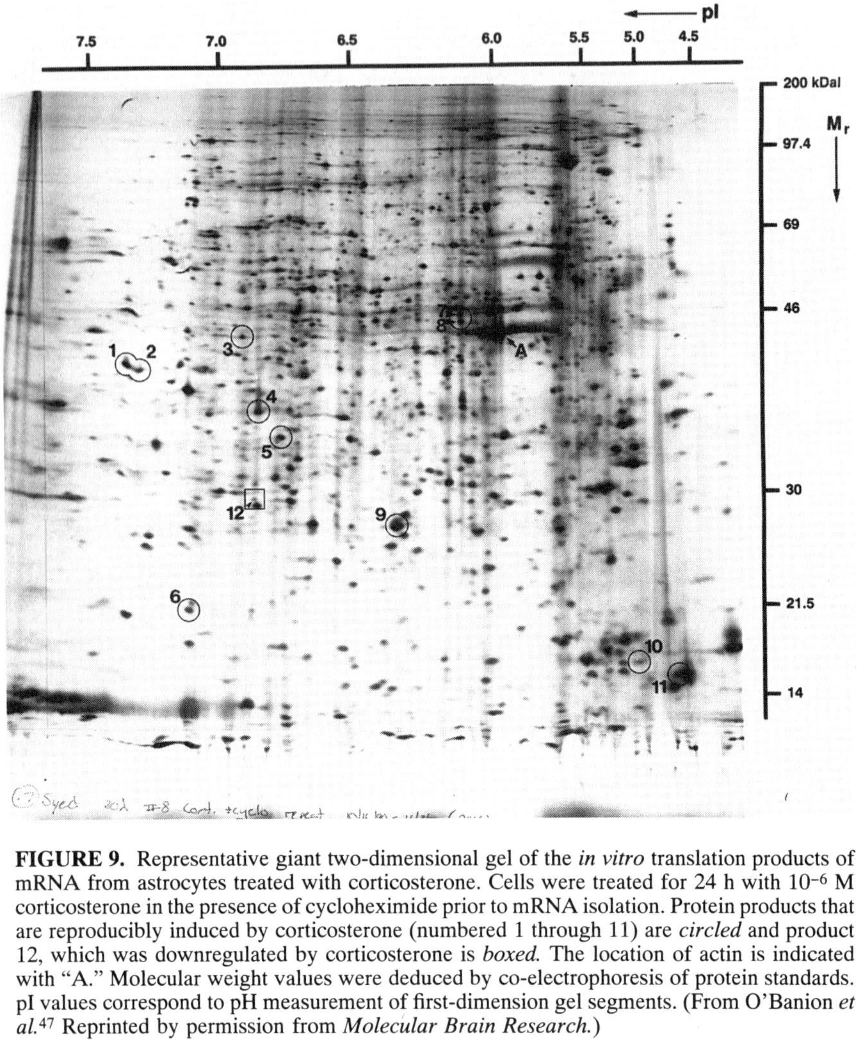

FIGURE 9. Representative giant two-dimensional gel of the *in vitro* translation products of mRNA from astrocytes treated with corticosterone. Cells were treated for 24 h with 10^{-6} M corticosterone in the presence of cycloheximide prior to mRNA isolation. Protein products that are reproducibly induced by corticosterone (numbered 1 through 11) are *circled* and product 12, which was downregulated by corticosterone is *boxed.* The location of actin is indicated with "A." Molecular weight values were deduced by co-electrophoresis of protein standards. pI values correspond to pH measurement of first-dimension gel segments. (From O'Banion *et al.*[47] Reprinted by permission from *Molecular Brain Research.*)

greater magnitude of change (TABLE 1). However, it is not possible to rule out an effect of corticosterone on mRNA stability. A comparison between gels of *in vitro* translation products and metabolically labeled cellular proteins showed that most, but not all, of the *in vitro* translation products have cognate proteins that also show similar changes in response to glucocorticoid treatment (FIG. 10). However, no cognate protein was identified for the glucocorticoid-repressed *in vitro* translation product (spot 12) on gels of whole cell proteins, suggesting that this may represent a secreted protein. In this regard, glucocorticoids markedly alter the pattern of proteins secreted by glial cells.[44,48,49] Immunoprecipitation identified spot 3 as gluta-

TABLE 1. Induction of *in Vitro* Translation Products by 24-h Corticosterone Treatment[a]

Spot	MW	pI	Control[b]	+ Cort[b]	24-h Induction[c]	24 h + Cx[d]
1	42	7.4	1698 ± 348	2752 ± 293	1.62	3.43
2	42	7.35	589 ± 12	1449 ± 240	2.46*	5.86
3	44	6.9	1487 ± 264	1735 ± 230	1.17	1.97
4	37.5	6.8	605 ± 143	1717 ± 38	2.83*	2.3
5	35	6.75	380 ± 150	1763 ± 209	4.64*	8.11
6	22	7.1	296 ± 139	476 ± 113	1.61	9.67
7	47.5	6.15	578 ± 68	945 ± 213	1.64	2.88
8	46	6.15	104 ± 24	365 ± 21	3.51*	10.6
9	27.5	6.3	1198 ± 246	2575 ± 325	2.15*	3.62
10	17.5	5	145 ± 31	262 ± 33	1.8	1.35
11	16.5	4.7	—	689 ± 252	>65.0	>100.0
12	29	6.8	492 ± 220	206 ± 114	−2.66	−6.4

[a]From O'Banion *et al.*[47] Reprinted by permission from *Molecular Brain Research.*
[b]Arbitrary optical density units, mean of three separate samples (n = 3) from one experiment ± SEM.
[c]Induction ratio defined as (value in the presence of hormone/control value) or the negative reciprocal of this formula in the case of repressions.
[d]Induction ratios were calculated from the means of two determinations per condition (cycloheximide (cx) alone and cycloheximide + corticosterone) within one experiment.
*p <0.05 by independent Student t test.

mine synthase,[47] a previously identified glucocorticoid-regulated protein.[50,51] Spot 5 is tentatively identified as glycerol phosphate dehydrogenase (N. R. Nichols, personal communication). Spot 11 was identified as glucocortin, a protein whose mRNA has been shown to be rapidly induced *de novo* in a variety of cell types[52] and which is not present on 2D gels run on astrocyte proteins unless cells have been treated with corticosterone.[47] The other spots remain to be identified. Our data also suggest that all of the changes observed in this experiment were mediated by the GR, and not the MR, since they occurred only after cells were treated with 1 or 10 nM dexamethasone and not 100 nM aldosterone.[47]

Unresolved Issues

Even though GR and MR are apparently activated by the same ligand in the same cell and can drive the same promoter element,[24] a number of studies suggest that selective activation of GR or MR differentially affects animal behavior and neuronal electrophysiological properties in slice cultures of hippocampus.[17,53–55] The molecular mechanisms underlying these differential effects need to be defined. Since both GR and MR are expressed *in vitro* at levels representative of those *in vivo,* neuronal cultures of dissociated embryonic hippocampus, which can be easily grown under defined, steroid free conditions,[56] should facilitate elucidation of the precise role that these two receptors play in gene regulation in the hippocampus.

Although our studies have found a number of glucocorticoid-regulated proteins on 2D gels, the identity of some of these is unknown. In addition, glucocorticoid-

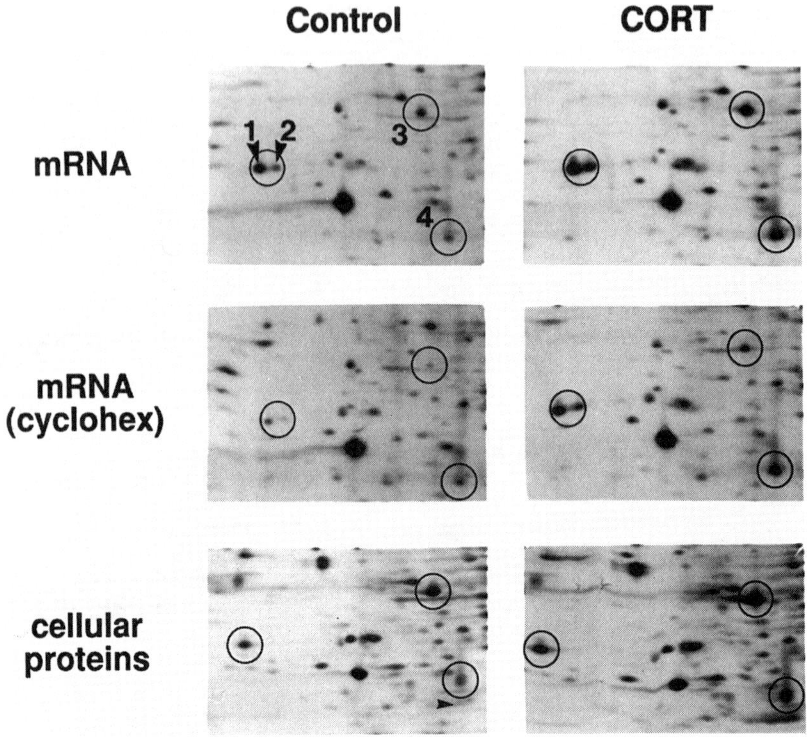

FIGURE 10. Corticosterone-responsive gene products are induced in the presence of cycloheximide and in metabolic labeling experiments. Astrocytes were treated with 1 μM corticosterone for 24 h (*in vitro* translations) or for 6 h with 0.1 μM dexamethasone (metabolic labeling). Portions of autoradiographs from gels of *in vitro* translation products (mRNA; mRNA (cyclohex)) or from metabolically labeled cells (cellular proteins) are shown. Protein product 2 was not observed in metabolically labeled cells. The mobility of product 4 increases in metabolically labeled cells; its position in *in vitro* translations is indicated with an *arrowhead* in the control metabolically labeled panel. (From O'Banion *et al.*[47] Reprinted by permission from *Molecular Brain Research.*)

induced alterations in proteins secreted by astrocytes may represent proteins that have functional significance during development or regeneration. Thus, the sequencing of the protein products of the glucocorticoid-regulated mRNAs in astrocytes may lead to the identification of proteins that play a pivotal role in glial-neuronal interactions. A similar strategy may lead to the identification of previously unrecognized mRNAs that are regulated by glucocorticoids in hippocampal neurons.

Molecular studies in nonneuronal cells suggest that glucocorticoid activation of GR and MR can be synergistic or antagonistic depending on the levels of other transcriptional factors, such as c-*fos* and c-*jun* acting at the AP1 element.[57–59] This observation suggests that the co-expression of GR and MR at conspicuously high levels in hippocampal neurons may be relevant to modulating the activity of promoter elements other than, or in addition to, the glucocorticoid response elements.

Consequently, the precise action of glucocorticoids on hippocampal neurons may depend on the state of the neuron and the integrative effects of GRs and MRs with other transcription factors whose levels are changes in response to extracellular signals other than steroids, such as neurotransmitters or neurotrophic factors. If such interesting molecular mechanisms underlie glucocorticoid action in the hippocampus, these could be demonstrated more easily in neurons grown *in vitro* than *in vivo.*

SUMMARY

Studies using immunocytochemistry and RNase protection assay demonstrate that glucocorticoid and mineralocorticoid receptors (GR, MR) and their corresponding mRNAs are co-expressed in hippocampal neurons cultured in serum-free, defined medium and at lower levels in cultured astrocytes. Addition of serum or medium conditioned by astrocytes increases the levels of MR mRNA, but has little effect on the levels of GR mRNA. Cellular levels of both GR mRNA and MR mRNA are upregulated by growth of embryonic hippocampal neurons in corticosterone. This is in distinct contrast to regulation of receptor expression *in vivo* where mRNAs for these receptors are downregulated in the rat hippocampus by corticosterone treatment of the adult adrenalectomized rat. However, in cultured astrocytes, GR and MR mRNAs are also downregulated by corticosterone. To begin to define the role of glucocorticoids in gene expression in astrocytes, we have used giant two-dimensional (2D) gel electrophoresis to separate astrocyte cellular proteins and translation products synthesized *in vitro* from astrocyte poly A^+ RNA. Analysis of approximately 1,500 *in vitro* translation products by giant 2D gel electrophoresis reveals 11 protein inductions and 1 repression that occur at the level of mRNA in the absence of protein synthesis following treatment of astrocytes with corticosterone. Interestingly, these changes appear to be mediated by GR, but not by MR. The *in vitro* studies described here are relevant to identifying the role of GR and MR in gene expression in specific cell types in the hippocampus.

ACKNOWLEDGMENTS

The authors thank Dr. R. Harrison and Dr. Z. Krozowski for their generous gifts of the Bugr and MinRec 4 antisera, respectively.

REFERENCES

1. ANDERSON, D. J. 1993. Cell fate determination in the peripheral nervous system: the sympathoadrenal progenitor. J. Neurobiol. **24:** 185–198.
2. BOHN, M. C. 1984. Glucocorticoid induced teratologies of the nervous system. *In* Neurobehavioral Teratology. J. Yanai, Ed. 365–387. Elsevier Biomedical Press. Amsterdam.
3. BOHN, M. C. 1993. Glucocorticoids: roles in development of the autonomic nervous system. *In* Autonomic-Endocrine Interactions. *In* The Autonomic Nervous System Series. G. Burnstock, Ed. Harwood Acad. Publ. Chur, Switzerland. In press.
4. DOUPE, A. J. & P. H. PATTERSON. 1982. Glucocorticoids and the developing nervous system. *In* Adrenal Actions on Brain. D. Ganta & D. Pfaff, Eds. 23–43. Springer-Verlag. Berlin.
5. MCEWEN, B. S., E. R. DE KLOET & W. ROSTENE. 1986. Adrenal steroid receptors and action in the nervous system. Physiol. Rev. **66:** 1121–1188.

6. MEYER, J. S. 1985. Biochemical effects of corticosteroids on neural tissue. Physiol. Rev. **65:** 946–1020.
7. VICEDOMINI, J. P., A. J. NONNEMAN, S. T. DE KOSKY & S. W. SCHEFF. 1986. Perinatal glucocorticoids disrupt learning: a sexually dimorphic response. Physiol. Behav. **36:** 145–149.
8. GOULD, E., C. S. WOOLLEY, H. A. CAMERON, D. C. DANIELS & B. S. MCEWEN. 1991. Adrenal steroids regulate postnatal development of the rat dentate gyrus. I. Effects of glucocorticoids on cell death. J. Comp. Neurol. **313:** 479–485.
9. GOULD, E., C. S. WOOLLEY, H. A. CAMERON, D. C. DANIELS & B. S. MCEWEN. 1991. Adrenal steroids regulate postnatal development of the rat dentate gyrus. II. Effects of glucocorticoids and mineralocorticoids on cell birth. J. Comp. Neurol. **313:** 486–493.
10. MEANEY, M. J. 1994. This volume.
11. SCHEFF, S. W. & S. T. DE KOSKY. 1989. Glucocorticoid suppression of lesion-induced synaptogenesis: effect of temporal manipulation of steroid treatment. Exp. Neurol. **105:** 260–264.
12. SCHEFF, S. W., J. K. MORSE & S. T. DE KOSKY. 1988. Hydrocortisone differentially alters lesion-induced axon sprouting in male and female rats. Exp. Neurol. **100:** 237–241.
13. SLOVITER, R. S., G. VALIQUETTE, G. M. ABRAMS, E. C. RONK, A. I. SOLLAS, L. A. PAUL & S. L. NEUBORT. 1989. Selective loss of hippocampal granule cells in the mature rat brain after adrenalectomy. Science **243:** 535–538.
14. SAPOLSKY, R. M. 1990. Glucocorticoids, hippocampal damage and the glutamatergic synapse. Prog. Brain Res. **86:** 13–23.
15. DE KLOET, E. R. 1991. Brain corticosteroid receptor balance and homeostatic control. Front. Neuroendocrinol. **12:** 95–164.
16. SAPOLSKY, R. M. 1994. This volume.
17. DE KLOET, E. R. 1994. This volume.
18. AZMITIA, E. 1994. This volume.
19. REUL, J. M. H. M. & E. R. DE KLOET. 1985. Two receptor systems for corticosterone in rat brain: microdistribution and differential occupation. Endocrinology **117:** 2505–2512.
20. DE KLOET, E. R., A. RATKA, J. M. H. M. REUL, W. SUTANTO & J. A. M. VANEEKELEN. 1987. Corticosteroid receptor types in brain: regulation and putative function. Ann. N.Y. Acad. Sci. **521:** 351–361.
21. FUNDER, J. W. 1993. Mineralocorticoids, glucocorticoids, receptors and response elements. Science **259:** 1132–1133.
22. FUXE, K., A. C. WILKSTRØM, S. OKRET, L. F. AGNATI, A. HÅRFSTRAND, Z. Y. YU, L. GRANHOLM, M. ZOLI, W. VALE & J. Å. GUSTAFSSON. 1985. Mapping of glucocorticoid receptor immunoreactive neurons in the rat tel- and diencephalon using a monoclonal antibody against rat liver glucocorticoid receptors. Endocrinology **117:** 1803–1812.
23. HERMAN, J. P., P. D. PATEL, H. AKIL & S. J. WATSON. 1989. Localization and regulation of glucocorticoid and mineralocorticoid receptor messenger RNAs in the hippocampal formation of the rat. Mol.Endocrinol. **3:** 1886–1894.
24. ARRIZA, J. L., L. W. SWANSON & R. M. EVANS. 1988. Neuronal mineralocorticoid as a mediator of glucocorticoid response. Neuron **1:** 887–900.
25. VAN EEKELEN, J. A. M., W. JIANG, E. R. DE KLOET & M. C. BOHN. 1988. Distribution of the mineralocorticoid and the glucocorticoid receptor mRNAs in the rat hippocampus. J. Neurosci. Res. **21:** 88–94.
26. VIELKIND, U., A. WALANCEWICZ, J. LEVINE & M. C. BOHN. 1990. Type II glucocorticoid receptors are expressed in oligodendrocytes and astrocytes. J. Neurosci. Res. **27:** 360–373.
27. ARRIZA, J. L., G. WEINBERGER, T. M. CERELLI, B. L. GLASER, B. L. HENDELIN, D. E. HOUSEMAN & R. M. EVANS. 1987. Cloning of human mineralocorticoid receptor complementary DNA: structural and functional kinship with the glucocorticoid receptor. Science **237:** 268–275.
28. BOHN, M. C., D. DEAN, S. HUSSAIN & R. GIULIANO. 1994. Development of mRNAs for glucocorticoid (GR) and mineralocorticoid receptors (MR) in rat hippocampus. Dev. Brain Res. **77:** 157–162.

29. REUL, J. M. H. M., F. R. VAN DEN BOSCH & E. R. DE KLOET. 1987. Differential responses of type 1 and type 2 corticosteroid receptors to changes in plasma steroid level and circadian rhythmicity. Neuroendocrinology **45:** 407–412.
30. ROSENFELD, P., W. SUTANTO, S. LEVINE & E. R. DE KLOET. 1989. Ontogeny of mineralocorticoid (type 1) receptors in brain and pituitary: an *in vivo* autoradiographical study. Dev. Brain Res. **52:** 57–62.
31. CHAO, H. M., P. H. CHOO & B. S. MCEWEN. 1989. Glucocorticoid and mineralocorticoid receptor mRNA expression in rat brain. Neuroendocrinology **50:** 365–371.
32. SPENCER, R. L., A. H. MILLER, M. STEIN & B. S. MCEWEN. 1991. Corticosterone regulation of type I and type II adrenal steroid receptors in brain, pituitary and immune tissue. Brain Res. **549:** 236–246.
33. TURNER, B. B. & D. A. WEAVER. 1985.Sexual dimorphism of glucocorticoid binding in rat brain. Brain Res. **343:** 16–23.
34. DONG, Y., L. POELLINGER, J.-Å. GUSTAFSSON & S. OKRET. 1988. Regulation of glucocorticoid receptor expression: evidence for transcriptional and posttranslational mechanisms. Endocrinology **129:** 2166–2174.
35. BANKER, G. A. & W. M. COWAN. 1977. Rat hippocampal neurons in dispersed cell culture. Brain Res. **126:** 397–425.
36. LIPOSITS, Z., R. M. UHT, R. W. HARRISON, F. P. GIBBS, W. K. PAULL & M. C. BOHN. 1987. Ultrastructural co-localization of glucocorticoid receptor and corticotropin-releasing factor in hypothalamic neurons. Histochemistry **87:** 407–412.
37. UHT, R. M., J. F. MCKELVY, R. W. HARRISON & M. C. BOHN. 1988. Demonstration of glucocorticoid receptor-like immunoreactivity in glucocorticoid-sensitive vasopressin and corticotrophin-releasing factor neurons in the hypothalamic paraventricular nucleus. J. Neurosci. Res. **19:** 405–411.
38. BARKER, D. L., R. GIULIANO, Z. KROZOWSKI & M. C. BOHN. 1992. Confocal microscopy of glucocorticoid and mineralocorticoid receptors (GR, MR) in cultured hippocampal neurons and astrocytes. Soc. Neurosci. Abst. **18:** 478.
39. ROSEWICZ, S., A. R. MCDONALD, B. A. MADDUX, I. D. GOLDFINE, R. L. MIESFELD & D. C. LOGSDON. 1988. Mechanism of glucocorticoid receptor down-regulation by glucocorticoids. J. Biol. Chem. **263:** 2581–2584.
40. MCEWEN, B. S., G. WALLACH & C. MAGNUS. 1974. Corticosterone binding to hippocampus: immediate and delayed influences of the absence of adrenal secretion. Brain Res. **70:** 321–334.
41. PEIFFER, A., B. LAPOINTE & N. BARDEN. 1991. Hormonal regulation of type II glucocorticoid receptor messenger ribonucleic acid in rat brain. Endocrinology **129:** 2166–2174.
42. PEPIN, M.-C., S. BEAULIEU & N. BARDEN. 1990. Differential regulation by dexamethasone of glucocorticoid receptor messenger RNA concentrations in neuronal cultures derived from fetal rat hypothalamus and cerebral cortex. Cell. Mol. Neurobiol. **10:** 227–235.
43. MEANEY, M. J., R. M. SAPOLSKY & B. S. MCEWEN. 1985. The development of the glucocorticoid receptor system in the rat limbic brain. II. An autoradiographic study. Dev. Brain Res. **18:** 165–168.
44. Unpublished data.
45. YOUNG, D. A., B. P. VORIS, E. V. MAYTIN & R. A. COLBERT. 1983. High-resolution two-dimensional gel electrophoresis of proteins. Methods Enzymol. **91:** 190–214.
46. COLBERT, R. A. & D. A. YOUNG. 1986. Electrophoretic separation of *in vitro* translation products on giant two-dimensional gels allows detailed analysis of cellular mRNAs. J. Biol. Chem. **261:** 14733–14739.
47. O'BANION, M. K., D. M. YOUNG & M. C. BOHN. 1994. Corticosterone responsive mRNAs in primary rat astrocytes. Mol. Brain Res. **22:** 57–68.
48. ARENANDER, A. T. & J. DE VELLIS. 1981. Glial-released proteins. II. Two-dimensional electrophoretic identification of proteins regulated by hydrocortisone. Brain Res. **224:** 105–116.
49. ARENANDER, A. T. & J. DE VELLIS. 1980. Glial-released proteins in clonal cultures and their modulation by hydrocortisone. Brain Res. **200:** 401–419.
50. VARDIMON, L., L. L. FOX, L. DEGENSTEIN & A. A. MOSCONA. 1988. Cell contacts are

required for induction by cortisol of glutamine synthetase gene transcription in the retina. Proc. Natl. Acad. Sci. USA **85:** 5981–5985.

51. Kumar, S., E. Holmes, S. Scully, B. W. Birren, R. H. Wilson & J. de Vellis. 1986. The hormonal regulation of gene expression of glialmarkers: glutamine synthetase and glycerol phosphate dehydrogenase in primary cultures of rat brain and in C6 cell line. J. Neurosci. Res. **16:** 251–264.
52. Colbert, R. A. & D. A. Young. 1986. Glucocorticoid-mediated induction of glucocortin: a rapid primary response common to major target tissues. Proc. Natl. Acad. Sci. USA **83:** 72–76.
53. Joëls, M. 1994. This volume.
54. Joëls, M. & E. R. de Kloet. 1989. Effects of glucocorticoids and norepinephrine on the excitability in the hippocampus. Science **245:** 1503–1505.
55. Joëls, M. & E. R. de Kloet. 1990. Mineralocorticoid receptor-mediated changes in membrane properties of rat CA1 pyramidal neurons *in vitro.* Proc. Natl. Acad. Sci. USA **87:** 4495–4498.
56. Bohn, M. C. 1994. *In vitro* approaches to studying glucocorticoid effects on gene expression in neurons and glia. *In* Neurobiology of Steroids. E. R. de Kloet & W. Sutano, Eds. *In* Methods in Neurosciences. P. M. Conn, Ed. Academic Press, Inc. Orlando, FL. In press.
57. Pearce, D. & K. R. Yamamoto. 1993. Mineralocorticoid and glucocorticoid receptor activities distinguished by nonreceptor factors at a composite response element. Science **259:** 1161–1165.
58. Yang-Yen, H., J. Chambard, Y. Sun, T. J. Schmidt, J. Drouin & M. Karin. 1990. Transcriptional interference between c-jun and the glucocorticoid receptor: mutual inhibition of DNA binding due to direct protein-protein interaction. Cell **62:** 1205–1215.
59. Diamond, M. I., J. N. Miner, S. K. Yoshinaga & K. R. Yamamoto. 1990. Transcriptional factor interaction: selectors of positive or negative regulation from a single DNA element. Science **249:** 1266–1272.

DISCUSSION OF THE PAPER

Question: Do you have any idea about the labelling that is occurring in these long processes of neurites? Is it dendritic or axonal?

Bohn: We have not done double labelling with a dendritic or axonal marker and receptors.

Question: Obviously, if there is going to be some kind of signal, some kind of function for those receptors, it probably is not going to involve translocation back to the nucleus.

Bohn: Actually I did not mention the EM work that we did on the GR localization. After adrenalectomy, in hippocampal neurons and paraventricular nuclear neurons, you see labelling way out in the dendrites and dendritic spines for GR. Then after replacement of corticosterone you drive the labelling right back in. Therefore, the receptors in the processes do seem capable of translocation, at least for GR. If I had to guess what those processes are in culture I would guess dendrites, because we have done a lot of MAP 2 staining and most of the processes are dendrites. It is hard to see the axons, but in principle we could do that. However, we have never seen receptor labelling in axons at the EM level.

Question: Also, you mentioned that a number of proteins are being secreted; considering that astrocytes contain so many neurotropic factors, a number of which have been characterized, have you identified any of those proteins that are secreted?

BOHN: We are trying to. Most of the neurotropic factors are very basic and we probably would not see them on these gels. They are also mostly very low molecular weight and near the limits of our gels.

QUESTION: S100 is acidic.

BOHN: I am not sure exactly where that runs. We should probably look for S100.

E. R. DE KLOET (*Sylvius Laboratories*): I noticed in the confocal laser microscopy images that the MR activity was not homogeneously distributed in the nucleus. Is there some comarker you have applied to see whether the heterogenous distribution has some significance related to transcription or to the matrix?

BOHN: We have never done anything like that. Are you speaking of GR? Little hot spots of receptor staining are very evident for GR in the nucleus. We have not seen this for MR where the most intense staining is perinuclear.

A. MUNCK (*Dartmouth Medical School*): Can you do the dual staining with the anti-GR and anti-MR to visualize a single cell with the two different antibodies?

BOHN: We did quite a bit of double staining, but I did not trust the results. There is enough bleed-through in the fluorescence that I was not convinced that we were seeing specifically only one receptor and not the other in specific subcellular locations.

MUNCK: It would be nice to see the same cell with both receptors behaving differently when they are being treated with hormone.

BOHN: We did try to do this, but to be truthful, gave up on it for technical reasons.

B. S. MCEWEN (*The Rockefeller University*): Along the same line, even though the MR staining does not seem to concentrate in the nucleus with hormone, if you gave radioactive aldosterone would you get a translocation of at least some of the ligand?

BOHN: We have treated these cultures with all different steroids at different doses and have never seen an indication of translocation. We have never done this with labeled steroid though, which would be interesting. Note, however, there is already substantial nuclear staining for MR even in the absence of ligand.

Environmental Regulation of the Development of Glucocorticoid Receptor Systems in the Rat Forebrain

The Role of Serotonin

M. J. MEANEY,[a] J. DIORIO, D. FRANCIS, S. LaROCQUE, D. O'DONNELL, J. W. SMYTHE, S. SHARMA, AND B. TANNENBAUM

Developmental Neuroendocrinology Laboratory
Douglas Hospital Research Centre
Departments of Psychiatry, and Neurology and Neurosurgery
McGill University
Montreal, Canada H4H 1R3

Several years ago Levine, Denenberg, Ader, and others[1–5] described the effects of postnatal "handling" on the development of behavioral and endocrine responses to stress. As adults, handled rats exhibited attenuated fearfulness in novel environments and a reduced adrenal glucocorticoid response to a variety of stressors. These findings demonstrated that the development of rudimentary, adaptive responses to stress could be modified by environmental events. In addition the handling paradigm provided a marvelous opportunity to examine how subtle variations in the early environment alter the development of specific neurochemical systems, leading to stable individual differences in biological responses to stimuli that threaten homeostasis.

The effects of handling and other environmental manipulations, such as maternal separation or early immune challenge, on the development of HPA responses to stress have been reviewed recently in detail.[6–8] In this paper the focus is on how these early environmental events alter the differentiation of the forebrain neurons which regulate HPA function, and thus we will present only a brief summary of the neuroendocrine differences between adult animals that were handled vs nonhandled early in life. The important point to be gleaned here is that the handling effect on pituitary-adrenal responses to stress is mediated, in part at least, by changes in hippocampal glucocorticoid receptor levels.

In these studies rat pups are handled from the day following birth until weaning. The animals are then housed normally in groups and tested as fully mature adults. With the exception of some subtle differences (see Ref. 7) the handling effect on HPA function is the same for both males and females. Briefly, postnatal handling reduces adult HPA responses to stress and this is apparent in plasma levels of either ACTH or corticosterone.[9–12] Interestingly, there is no difference (as best as we can detect) between handled (H) and nonhandled (NH) animals in basal levels of ACTH or corticosterone.[10,11]

At the level of the paraventricular n. of the hypothalamus, H animals how lower levels of the nRNAs for corticotropin releasing-hormone (CRH) and arginine vaso-

[a]Correspondence to: Michael J. Meaney, Developmental Neuroendocrinology Lab., Douglas Hospital Research Centre, 6875 Boul. LaSalle, Montreal, Quebec H4H 1R3, Canada.

pressin as well as decreased levels of these peptides,[12–14] both of which serve to stimulate the release of ACTH. In addition, CRH release during stress is decreased in the H animal.

The inhibition of HPA activity is achieved through a negative-feedback process,whereby the elevated GC levels act on various neural targets and the pituitary to decrease the synthesis of hypothalamic CRN and AVP, eventually resulting in decreased pituitary ACTH secretion.[15–20] H animals show increased negative-feedback sensitivity to elevated GC levels (either endogenous or synthetic GCs, see Ref. 10). These differences, in turn, appear to be mediated by differences in GC receptor density in the hippocampus and frontal cortex, critical neural sites for inhibition of HPA activity.[21–22] Thus, H animals show increased GC receptor binding sites in the hippocampus and frontal cortex.[9–11,23–26] This effect is tissue specific: GC receptor populations in the pituitary, hypothalamus, septum, and amygdala are unaffected, as are hippocampal mineralocorticoid receptor levels. Handling also increases GC receptor nRNA levels in the hippocampus.[27] When the difference in hippocampal GC receptors between H and NH animals is experimentally eliminated, the difference in stress-induced GC secretion is no longer apparent,[10] arguing for the importance of the receptors in these regions in determining differences in pituitary-adrenal responses to stress.

Interestingly, the differences between H and NH animals in the synthesis of hypothalamic CRH and AVP, as well as those in ACTH release during stress, occur in response to a basal glucocorticoid signal.[12] Since H and NH animals do not differ in basal corticosterone levels, this underscores the importance of the differences in glucocorticoid receptor binding capacity for the handling-induced increase in negative feedback sensitivity (see FIG. 1 for a summary).

Mechanism of Action of Handling on Glucocorticoid Receptor Development

The handling studies provided a marvelous paradigm with which to examine how variations in the early environment alter the development of specific neurochemical systems, leading to stable individual differences in biological responses. The effect of handling on the hypothalamic-pituitary-adrenal (HPA) stress response shares two important features with other known early environmental effects on CNS development. First, there is a 'critical period' during which specific brain regions which regulate HPA function are sensitive to the effects of handling. Second, the effects of handling persist throughout the life of the animal, despite the fact that the handling manipulation is limited to the first weeks of life. We are now examining the neurochemical pathways which mediate the effects of handling and feel that these studies address certain fundamental questions in developmental neurobiology: How do environmental stimuli influence certain aspects of neural differentiation (in this case, determining the sensitivity of certain neurons to a class of hormones) and, in doing so, alter the development of a fundamental CNS response to stress? Why is the occurrence of such long-term effects limited to specific periods during early development (what cellular events underlie the onset and the termination of such critical periods)? The differences in hippocampal glucocorticoid receptor levels appear to be a critical feature for the effect of handling on HPA responses to stress. Thus, we are examining the way in which the stimuli associated with the handling procedure actually serve to increase hippocampal glucocorticoid receptor gene expression.

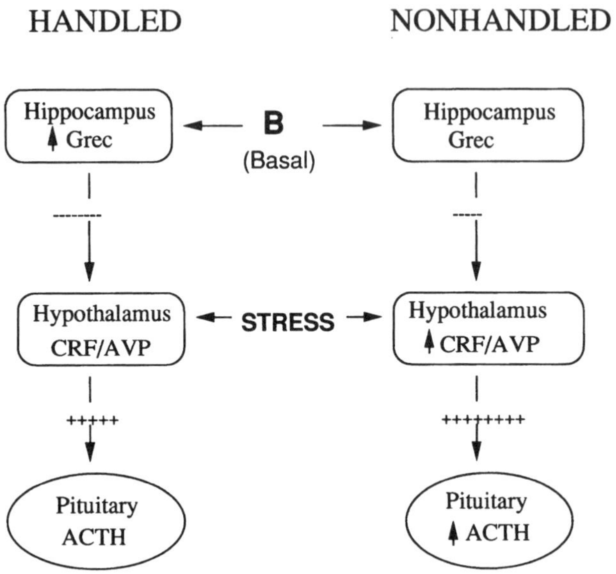

FIGURE 1. A summary of our current understanding of the mechanisms underlying the differences in HPA responses to acute stress in handled and nonhandled animals. The increased glucocorticoid receptor levels confer greater sensitivity to corticosterone (B) in hippocampal tissue from handled animals. Therefore, under even under basal corticosterone levels (which do not differ in young, adult handled and nonhandled rats) there is a greater tonic inhibitory signal (–) on hypothalamic corticotropin-releasing factor (CRF) and vasopressin (AVP) synthesis in handled rats. In response to the neural signals associated with stress and impinging upon the hypothalamus, there is a greater release of CRF/AVP into portal circulation (+) in the nonhandled rats, resulting in greater ACTH release.

Temporal Features of the Handling Effect

Initially animals were handled for the first 21 days of life. However, Levine[28] found that handling during the first week of life was as effective as handling during the entire first three weeks of life in reducing HPA responses to stress. Similarly, handling between Days 1 and 7 of life was as effective in increasing hippocampal glucocorticoid receptor density as handling over the entire first 3 weeks.[24] Handling over the second week of life (*i.e.,* between Days 8 and 14) was somewhat less effective, whereas animals handled between Days 15 and 21 did not differ from NH animals in glucocorticoid receptor binding. Thus, in terms of both HPA activity and glucocorticoid receptor binding, the sensitivity of the system to environmental regulation wanes progressively over the first three weeks of life. Moreover, in comparison to same-aged NH animals, H animals exhibited significantly increased hippocampal glucocorticoid receptor density as early as Day 7 of life and that magnitude of the effect did not increase thereafter.[24] Thus, glucocorticoid receptor binding capacity appears to be especially sensitive to environmental regulation during the first week of life.

Note, however, that this does not preclude the possibility of other periods of environmental regulation. Recently Abdul *et al.*[29] reported that environmental enrichment during the postnatal period results in increased glucocorticoid receptor mRNA expression. Ader[1] had previously reported that handling during the postweaning period also reduced the magnitude of HPA responses to stress. There may well exist a period during the postweaning period when HPA development is, once more, sensitive to environmental events. If this is indeed true, it will be interesting to compare the biological bases for the handling effects during these two periods.

The temporal pattern reported in our studies and those of Levine corresponds to the normal developmental changes in glucocorticoid receptor density occurring over the early postnatal life (see Ref. 30 for a review). Glucocorticoid receptor density is low on postnatal Day 3 (~30% of adult values) and increases steadily towards adult values which are achieved by about the third week of life. A comparable pattern occurs for glucocorticoid receptor mRNA levels (Refs. 31,32 and see M. Bohn, this volume). It is during this period of ontogenic variation that environmental events can influence the development of the receptor population. In contrast to the glucocorticoid receptor, mineralcorticoid receptor density does not vary significantly with age: hippocampal mineralocorticoid receptor density in early postnatal life is largely indistinguishable from that of adult rats (Refs. 23, 33 and see Ref. 30; although the pattern for mRNA may be different depending upon the isoform, see H. Akil and M. Bohn, this volume) and, as mentioned above, handling has no effect on hippocampal mineralocorticoid receptor binding.[10,23] Thus, it is tempting to consider the relationship between the developmental pattern in hippocampal glucocorticoid receptor density and both 1) the sensitivity of this receptor system to environmental stimuli, and 2) the timing of the critical period for these stimuli on glucocorticoid receptor development. However, the developmental pattern for glucocorticoid receptor binding in regions not affected by handling, such as the hypothalamus, amygdala, and septum[30,34] is identical to that of the hippocampus (Refs. cited above) and the frontal cortex.[35] Thus, it is unlikely that the handling effect on glucocorticoid receptor density in the hippocampus and the frontal cortex can be explained simply by the status of the glucocorticoid receptor system during the first weeks of life.

The Role of Thyroid Hormones

Handling during the first week of life activates the hypothalamic-pituitary-thyroid axis leading to increased levels of circulating thyroxine (T_4) and increased intracellular levels of the biologically more potent T_4 metabolite, triiodothyronine (T_3). Handling results in a modest, but reliable (~30%) increase in plasma thyroid hormones.[36] Thus, we[26] examined whether the effects of handling might be mediated by increased exposure to these hormones. Neonatal rat pups were administered T_4, T_3, or the vehicle over the first week of life. Both T_4 and T_3 treatment resulted in significantly increased glucocorticoid receptor binding capacity in the hippocampus in animals examined as adults. Like the handling manipulation, neither T_4 nor T_3 treatment affected hypothalamic or pituitary glucocorticoid receptor density. Moreover, administration of the thyroid hormone synthesis inhibitor, propylthiouracil (PTU), to H pups for the first 2 weeks of life completely blocked the effects of handling on hippocampal glucocorticoid receptor binding capacity. These data are consistent with the idea that the thyroid hormones might mediate, in part at least, the effects of neonatal handling on the development of the hippocampal glucocorticoid receptor system.

Systemic injections of neonatal rat pups represent a rather crude manipulation, particularly procedures involving thyroid hormones. While these data might implicate the thyroid hormones, there is no indication that the hippocampus is the actually critical site of action for the thyroid hormone effect. In order to examine whether thyroid hormones might act directly on hippocampal cells we turned to an *in vitro* system, using primary cultures of dissociated hippocampal cells.[37] The hippocampal cells are taken from embryonic rat pups (E20) and beginning on the fifth day after plating the cultures were exposed to 0, 1, 10, or 100 nM T_3. These cells express both mineralocorticoid and glucocorticoid receptor binding (Ref. 38; FIG. 2). Indeed, both receptors as well as their mRNAs (on Northern blots) can be detected using material from a 60-mm dish. The cells were cultured in 10% fetal calf serum stripped of thyroid hormones. The results of several experiments have failed to detect any effect of thyroid hormones on glucocorticoid receptor density in cultured hippocampal cells.[36] These *in vitro* data suggest that 1) the effects of the thyroid hormones on the glucocorticoid receptor binding occurs at some site distal to the hippocampal cells or 2) thyroid hormones interact at the level of the hippocampus with some other hormonal signal that is obligatory for the expression of the thyroid hormone effect.

The Role of Serotonin (5-HT)

Thyroid hormones have pervasive effects throughout the developing CNS and one such effect involves the regulation of central serotonergic neurons (*e.g., Ref.* 39). Thyroid hormones increase 5-HT turnover in the hippocampus of the neonatal rat.[40] Handling also increases hippocampal 5-HT turnover,[40,41] and thus both manipulations increase serotonergic stimulation of hippocampal neurons. There is also direct evidence for an effect of 5-HT on glucocorticoid receptor density in the neonatal rat. 5,7-Dihydroxytyrptamine (5,7-DHT) lesions of the raphe 5-HT neurons dramatically reduce the ascending serotonergic input into the hippocampus. Rat pups administered 5,7DHT on Day 2 of life showed reduced hippocampal glucocorticoid receptor density as adults.[40] Similar lesions of ascending catecholaminergic systems, using 6-hydroxydopamine, had no effect on the development of glucocorticoid receptor density in the hippocampus (Meaney, unpublished), although these lesions do appear to alter hippocampal GR binding in adult animals.[42]

The next question is whether 5-HT might be mediating the effects of handling at the level of the hippocampal cells. Indeed, 5-HT had a profound effect on glucocor-

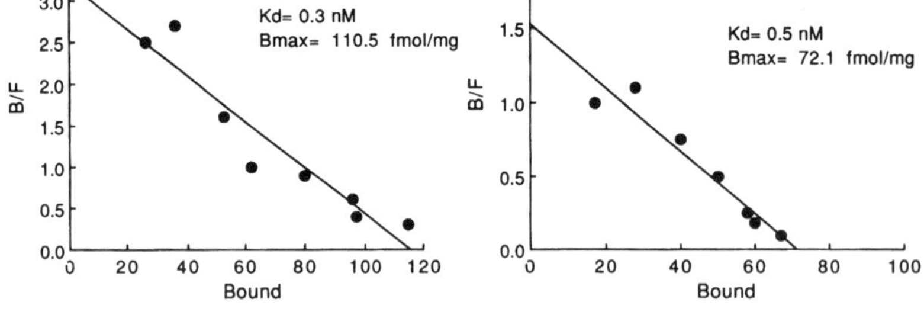

FIGURE 2. Representative Scatchard plots of glucocorticoid ([^{3}H]dexamethasone ± cold RU28362) and mineralocorticoid ([^{3}H]corticosterone + RU28362 ± cold aldosterone) receptor binding in primary hippocampal cell cultures.

ticoid receptor density in cultured hippocampal cells.[38,43] In hippocampal cells cultured in the presence of increasing concentrations of 5-HT there was as much as a twofold increase in glucocorticoid receptor binding (FIG. 3). The effect of 5-HT had a profound effect on glucocorticoid receptor density in cultured hippocampal cells.[38,43] In hippocampal cells cultured in the presence of increasing concentrations of 5-HT there was as much as a twofold increase in glucocorticoid receptor binding (FIG. 3). The effect of 5-HT was dose-related, with an EC_{50} of 4–5 nM and the maximal effect was achieved at 10-nM concentrations. Interestingly, a minimum of 4 days of treatment were required for the maximal effect of 5-HT to occur. Shorter periods of exposure were ineffective, suggesting that the effect of 5-HT may involve the increased synthesis of receptors. In support of this idea, we (O'Donnell and Meaney, unpublished) found that the effect of 5-HT on glucocorticoid receptor density in cultured hippocampal cells was blocked by either actinomycin D or cyclohexamide and was paralleled by an increase in glucocorticoid receptor mRNA levels.

The effect of 5-HT on glucocorticoid receptor binding appears to occur uniquely in the neuronal cell population. We found that there was no effect of 5-HT on glucocorticoid receptor binding in hippocampal glial-enriched cell cultures. This finding is not surprising, since our initial studies were performed with cultures comprised largely (~75–85%) of neuron-like cells;[40] in these studies we used neuron-specific enolase as neuronal marker and GFAP as a glial marker. Moreover, the composition of the cultures was unaffected by 5-HT treatment. We also examined the potential involvement of the glial cells by using a conditioned-medium experiment in which glial-enriched cultures were treated for 5 days with 5-HT and the medium was then used to feed the neuronal cultures. This procedure had no effect on glucocorticoid receptor density, suggesting that the effect was not due to a 5-HT-induced glial secretory product.

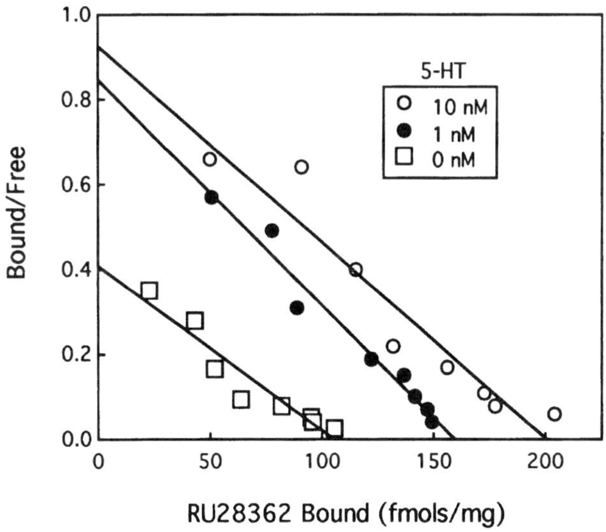

FIGURE 3. Representative Scatchard plots of glucocorticoid receptor binding (see FIG. 2) in primary hippocampal cell cultures treated for 4 days with various concentrations of 5-HT. (Adapted from Mitchell *et al.*[38])

The effects of 10 nM 5-HT on glucocorticoid receptor density in cultured hippocampal cells was completely blocked by the 5-HT_2 receptor antagonists, ketanserin and mianserin.[38,43] Moreover, the 5-HT_2 agonists 1-(2,5-dimethoxy-4-iodophenyl)-2-amino-propane (DOI), m-trifluromethyl-henylpiperazine (TFMPP), and quipazine were also effective in increasing glucocorticoid receptor binding in hippocampal culture, although not as effective as 5-HT. Selective agonists or antagonists of the 5-HT_{1A} or 5-HT_3 receptors had no effect on glucocorticoid receptor binding. Using [^{125}I]7-amino-8-iodo-ketanserin as radioligand, we confirmed the presence of high-affinity 5-HT_2 binding sites in our cultured hippocampal cells (FIG. 4). Teitler *et al.*[44,45] have provided evidence that the ketanserin-labeled 5-HT_2 site may exist in two states: an agonist state (5-HT_{2H}) with a high, nanomolar affinity for 5-HT and an antagonist state (5-HT_{2L}) with a low, micromolar affinity for 5-HT. In both states the receptor shows a high affinity for antagonists, such as ketanserin. The 5-HT_{2H} site binds with high affinity to DOI, TFMPP, and quipazine, and Teitler *et al.*[44] reported a K_d of ~5 nM for 5-HT. This K_d closely approximates the EC_{50} (4–5 nM) for the effect of 5-HT on glucocorticoid receptors in cultured hippocampal cells. Taken together, these findings suggest that this effect of 5-HT appears to be mediated via high-affinity, 5-HT_2-like receptor.

At this point in time it is difficult to pharmacologically discriminate between the 5-HT_2 and 5-HT_{1C} receptors. To the best of our knowledge there is no compound which can, with any certainty, distinguish 5-HT_2 and 5-HT_{1C} receptor sites, and the mRNAs encoding both receptors are expressed in the hippocampus (see Ref. 46). Both receptors are reported to mediate an increase in intracellular calcium and to activate inositol phosphate second messenger systems.[47-49] For the present, we assume that the effects described here could occur at either site and for this reason we refer to a 5-$HT_{2/1C}$ receptor effect.

We have begun to examine the nature of the secondary messenger systems involved in this serotonergic effect on glucocorticoid receptor binding. Mitchell *et al.*[43] found that low nanomolar concentrations of 5-HT (EC_{50} = 7.2 nM) produce a fourfold increase in cAMP levels in cultured hippocampal cells, with no effect on cGMP levels. This increase in cAMP is blocked by ketanserin and at least partially mimicked by quipazine, TFMPP and DOI. Indeed, there is a rather good correlation between the effects of these 5-HT receptor agonists on cAMP and glucocorticoid receptor levels (FIG. 5). Moreover, treatment with the stable cAMP analog 8-bromo-cAMP or with 10 μM forskolin produces a significant increase in glucocorticoid

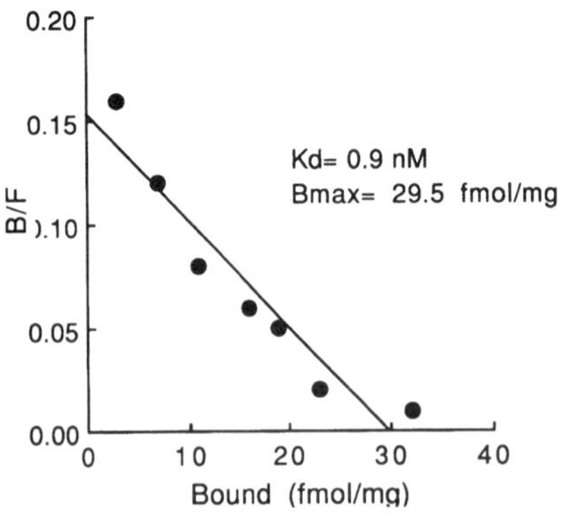

FIGURE 4. Scatchard plot of [I^{125}]AMIK (ketaserin) binding in primary hippocampal cell cultures.

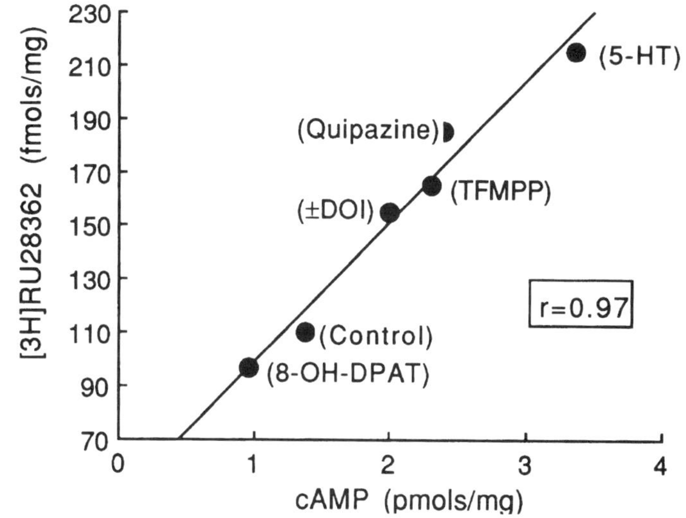

FIGURE 5. Correlation between the potency of various 5-HT receptor agonists in stimulating cAMP formation and effects on glucocorticoid receptor binding. The effect of 5-HT on cAMP is apparent after one hour incubation and maximal by 24 hours. The effect then remains quite stable for at least 5 days. The samples used here for cAMP determination and glucocorticoid receptor binding were obtained 4 days following treatment.

receptor density in cultured hippocampal neurons. The effect of 8-bromo cAMP is concentration-related, and the maximal effect of 8-bromo-cAMP (~181%) is comparable to that for 5-HT (~195%). Interestingly, as with 5-HT, the effects of 8-bromo cAMP on glucocorticoid receptor density were not apparent until at least 4 days of treatment.

Taken together, these findings suggest that changes in cAMP concentrations may mediate the effects of 5-HT on glucocorticoid receptor synthesis in hippocampal cells. We[43] also found that the cyclic nucleotide-dependent protein kinase inhibitor, H8, completely blocked the effects of 10 nM 5-HT on glucocorticoid receptor binding in hippocampal cell cultures. In contrast, the protein kinase C inhibitor, H7, had no such effect. These data suggest that activation of protein kinase A might, at some point, be involved in the serotonergic regulation of hippocampal glucocorticoid receptor development.

We are working with intact cells over long incubation periods and there is ample possibility for an interaction between second messenger systems. This issue also arises here, since the 5-$HT_{2/1C}$ receptor has been linked not to cyclic nucleotide, but inositol phosphate second messenger systems. *In vivo,* 5-HT increases both diacylglycerol (DAG) levels and inositol phosphate metabolism (notably IP_1) in hippocampal membranes from neonates.[50] However, in contrast with what is often seen with other compounds, such as glutamate or carbachol, stimulation of IP metabolism by 5-HT is muted in animals during the first week of life. Interestingly, the effect of 5-HT on IP metabolism is somewhat greater in the handled animals on Day 7 of life. Thus, repeated handling of pups increases the amount of stimulation with 5-HT. We also have data suggesting that treatment with a phorbol ester also increases glucocorticoid receptor binding in the hippocampal cell cultures (O'Donnell & Meaney, unpublished). However, the effect here is apparently smaller than that for either 5-HT or 8-bromo-cAMP (about 50 vs 100%), and the toxic effects of phorbol esters on cultured neurons are difficult to account for in these studies. These findings suggest that 5-HT effects on glucocorticoid receptor density might, at some level, involve the mobilization of intracellular Ca^{++} stores. Interestingly, Ca-calmodulin activation has been found to increase intracellular cAMP levels.

In Vivo *5-HT Effects on Glucocorticoid Receptor Expression*

The dorsal hippocampus of the neonatal rat shows readily detectable $5\text{-HT}_{2/1C}$ receptor sites (FIG. 6). Using [^{3}H]ketanserin as radioligand, we found $5\text{-HT}_{2/1C}$ binding evident throughout the hippocampus. However, handling does not alter hippocampal $5\text{-HT}_{2/1C}$ binding (Smythe *et al.,* in press). Indeed, in adult H animals $5\text{-HT}_{2/1C}$ binding in the hippocampus and frontal cortex is decreased in comparison to NH rats. Nevertheless, the sensitivity of the hippocampus and frontal cortex to neonatal handling may well be associated with the comparatively high levels of 5-HT_2 receptors in these regions which are apparent as early as Day 1 of life.

Recent studies (Smythe *et al.,* in press) of 5-HT turnover have provided some insight into why the hippocampus and frontal cortex are selectively affected by handling. In rat pups handled for the first seven days of life, and sacrificed immediately following handling on Day 7, 5-HT turnover was significantly increased in the hippocampus and frontal cortex, but not in the hypothalamus or amygdala (regions where handling has no effect on glucocorticoid receptor density). Interestingly, the increase in 5-HT turnover was greater following 7 days of handling than after only one handling episode, suggesting that some sensitization might be occurring along the ascending 5-HT pathways.

These data suggest that handling selectively activates certain ascending 5-HT pathways and that it is this feature of the handling effect, together with the existence of a developing glucocorticoid receptor system, that underlies the sensitivity of this receptor system in specific brain regions to regulation by environmental events during the first week of life. Whether or not handling alters different ascending 5-HT neuronal populations is just now being studied.

The *in vivo* data suggest that 5-HT acts at the $5\text{-HT}_{2/1C}$ receptor to alter hippocampal glucocorticoid receptor density. We found that effects of postnatal handling on glucocorticoid receptor binding are blocked by concurrent administration of the 5-HT_2 receptor antagonist, ketanserin.[40] Moreover, ketanserin treatment also blocks the effects of T3 on hippocampal glucocorticoid receptor density.[36] This finding supports the idea that thyroid hormones mediate the handling effect by serving to increase 5-HT activity.

There are two very intriguing features of this effect that bear directly on the question of the hippocampal cell cultures as a model for neural differentiation. First, the effects of 5-HT on glucocorticoid receptor binding in hippocampal cell cultures are restricted to the first 3 weeks in culture. Thus, cultures treated with 10 nM 5-HT for 7 days at any time during the first 3 weeks in culture show a significant increase in glucocorticoid receptor density; however, the effect is lost after this point. Cultures treated with 10 nM 5-HT for 7 days during the 3rd to the 4th week following plating show no increase in glucocorticoid receptor binding. Second, and perhaps most exciting, the increase in glucocorticoid receptor binding capacity following exposure to 10 nM 5-HT persists as long as 40 days following 5-HT removal from the

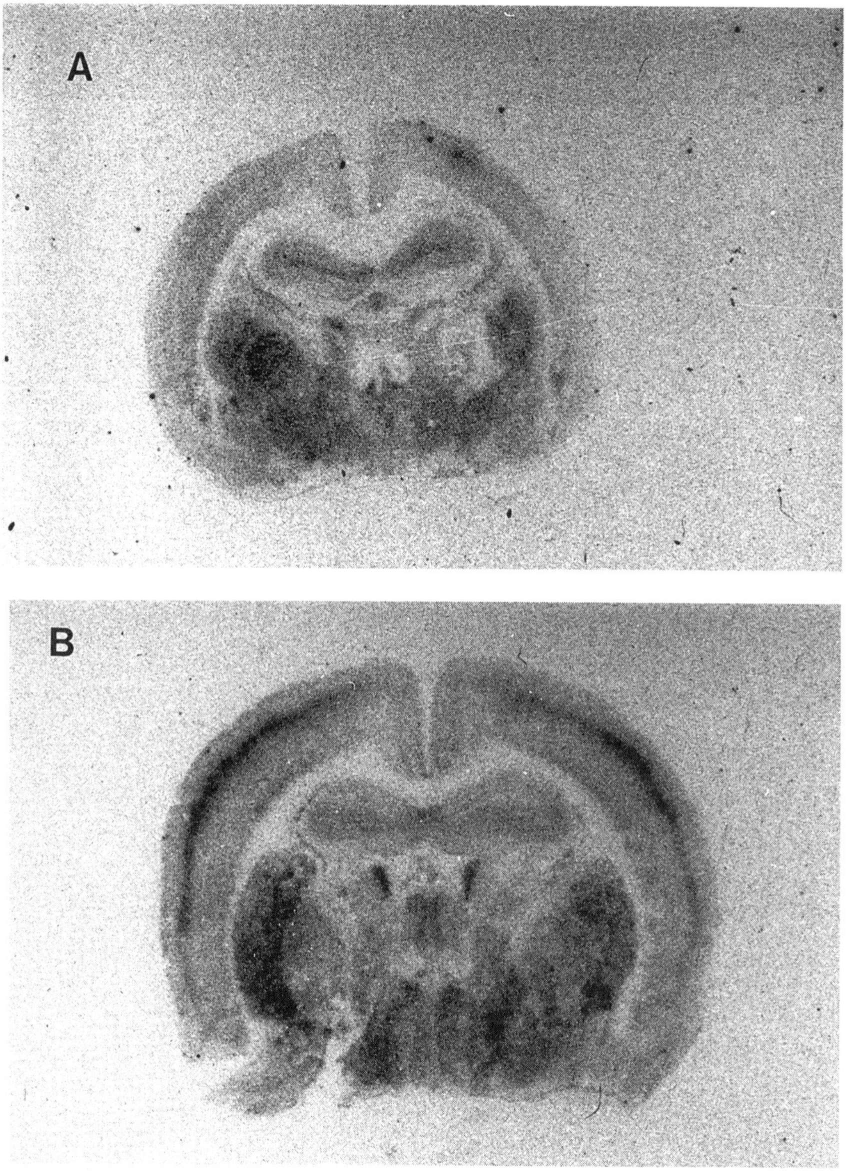

FIGURE 6. Autoradiograms of [^{3}H]ketaserin binding in coronal sections of **(A)** Day 3 and **(B)** Day 7 rat brain. Note the prominent labelling in the hippocampus at both ages (Larocque and Meaney, unpublished).

medium.[51] Thus, the effect of 5-HT on glucocorticoid receptor density observed in hippocampal culture cells mimics the long-term effects of neonatal handling.

Working Model

A model of the way in which neonatal handling might increase glucocorticoid receptor density in the hippocampus and frontal cortex is provided in FIGURE 7. Levine[52] has mobilized a convincing argument for the role of mother-pup interactions in mediating the handling effect on HPA development. Handling, although a brief interlude in the routine of mother-pup interactions, provides for an increase in sensory stimulation which, in turn, elicits a set of physiological responses in the pup. These responses and their relevance for the handling effect remain as a black box at the present. In addition, the mother is also subject to the manipulation. We have found that handling the mother alone (leaving the pups undisturbed) has no effect on hippocampal glucocorticoid receptor density (Aitken and Meaney, unpublished). However, handling the pups does alter the behavior of the mother towards the offspring, and the effect persists well beyond the immediate period following the return of the mother and pups to the nest. H pups show increased ultrasonic vocalizations,[53] and such calls elicit maternal care (*e.g.*, retrieval). Immediately following the return of the pups, mothers of H litters show increased care of their offspring. Over the long term, handling also increases maternal licking[54] and alters nest-building.[55] Several aspects of the mother-pup interactions are of known importance in neural development, especially maternal licking.[56] We (Woodside, Jans and Meaney, unpublished) have examined the effect of handling on the amount of time the mother spends with her offspring. Typically, a nesting bout begins when the mother approaches the litter, gathers the pups under her, suckles, and terminates when the

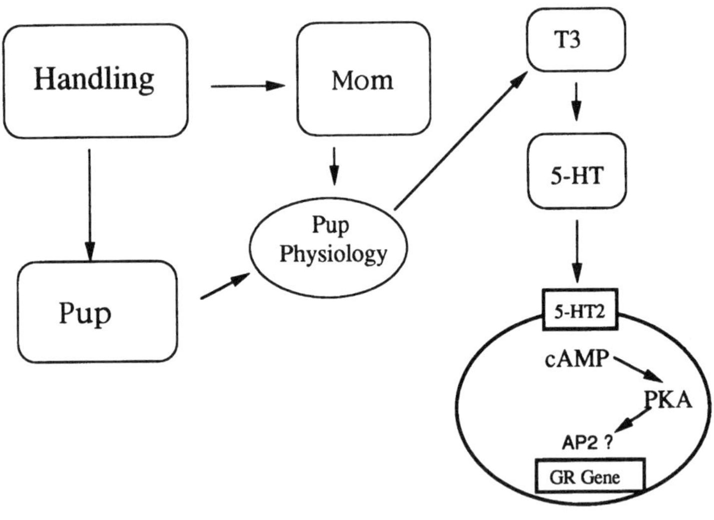

FIGURE 7. A speculative, working model describing the cascade of events involved in the handling effect on hippocampal glucocorticoid receptor expression. See text for a description.

mother licks the pups and leaves the nest (the termination of a nest bout is associated with a buildup in the mother's body temperature). Overall, mothers of H pups spend the same amount of time with their litters as mothers of NH pups. However, mothers of H litters have shorter but more frequent nest bouts and spend significantly more time manipulating their pups, especially licking of pups (Ref. 55; Meaney, unpublished) than mothers of NH pups. This latter effect is likely due to the fact that mothers of H litters are getting on and off their pups more frequently, and licking is associated with the termination of each bout. In addition, handling of pups also elicits endocrine responses from the mother that do not occur as a function of simply being separated from her pups.[57] The question that arises is whether some aspect of this altered pattern of maternal behavior/physiology may serve, in part at least, as the critical stimulus for the effects of handling on the development of the HPA axis and glucocorticoid receptor systems (see Ref. 58).

The handling procedure is associated with increased plasma thyroid hormone levels. Whether this is due to alterations in mother-pup interactions or to the actual manipulation of the pups is not clear. Thyroid hormones serve to increase 5-HT turnover in the hippocampus (see above). Both *in vitro* and *in vivo* experiments suggest that the effects of 5-HT on glucocorticoid receptor binding is mediated through a 5-$HT_{2/1C}$ receptor. Importantly, the effects of T3 on hippocampal glucocorticoid receptor binding are blocked by concurrent treatment with ketanserin.[36] This finding strengthens the proposed link between thyroid hormones and 5-HT in the regulation of glucocorticoid receptor density in the hippocampus.

The post-receptor events involved in the effect of 5-HT appear to involve increases in cAMP levels and the activation of cycle nucleotide-dependent protein kinases. Whether this cascade is initiated or facilitated by changes in inositol phosphate or diacylglycerol metabolism is not clear; however, there are numerous examples in the literature of such 'cross-talk' between second messenger systems (*e.g.*, Ref. 59). The regulation of gene transcription by cAMP has been sown to be mediated by at least two transcription factors: the cyclic nucleotide response element binding protein (CREB) and activator protein-2 (AP2).[60–62] Both CREB and AP2 are induced by both cAMP and forskolin. The promoter region of the human glucocorticoid receptor gene has been at least partially sequenced and contains numerous binding sites for the AP2 transcription factor. We have recently found that handling increases AP2 mRNA expression throughout the dorsal hippocampus (Diorio, Seckl, and Meaney, unpublished). The effect occurs within 30 to 60 minutes and persists for at least 2 hours following handling. Therefore, at this time it is interesting to speculate that the cascade of intracellular events may involve AP2 induction.

Then we arrive at the most interesting feature of these effects: the finding that the handling effect, as well as the *in vitro* effect of 5-HT, persists well beyond the period of the treatment. There are at least two possible explanations for this finding. First, the increase in 5-HT turnover associated with the handling procedure might be accompanied by an increase in 5-HT innervation of the hippocampus which persists throughout the life of the animal. The increased 5-HT innervation could then serve to maintain the handling effect. Seckl *et al.*[64] have shown that 5,7-DHT lesions in the hippocampus decrease glucocorticoid receptor mRNA level. However, such lesions also decrease mineralocorticoid receptor mRNA levels, which are unaffected by handling. This possibility seems unlikely for more direct reasons. Handling does not seem to permanently alter 5-HT innervation into the dorsal hippocampus. Thus, HPLC analysis suggests comparable levels of hippocampal 5-HT in adult H and NH rats (Smythe *et al.*, in press), while immunohistochemistry with 5-HT antibodies also indicates comparable levels of 5-HT innervation (Desjardins and Meaney, unpublished). Alternatively, handling could result in a 'structural' change in the glucocorticoid receptor gene (and/or the promoter region) which sustains a difference in basal transcription rates throughout the life of the animal. There is considerable research to be done on this topic, including a verification of differences in transcription rates. However, this remains an interesting possibility, and one that might explain the persistence of the handling effect, an effect which may represent a rather intriguing example of receptor imprinting occurring in response to an environmental signal.

REFERENCES

1. Ader, R. & L. J. Grota. 1969. Physiol. Behav. **4:** 303-305.
2. Hess, J. L., V. H. Denenberg, M. X. Zarrow & W. D. Pfeifer. 1969. Physiol. Behav. **4:** 109-112.
3. Levine, S. 1957. Science **126:** 405-406.
4. Levine, S. 1962. Science **135:** 795-796.
5. Levine, S., G. C. Haltmeyer, G. G. Karas & V. H. Denenberg. 1967. Physiol. Behav. **2:** 55-63.
6. Meaney, M. J., S. Bhatnagar, C. M. McCormick, N. Shanks, J. W. Smythe & V. Viau. 1993. Ann. N. Y. Acad. Sci. **697:** 70-86.
7. Meaney, M. J., S. Bhatnagar, J. Diorio, S. Larocque, D. Francis, D. O'Donnell, N. Shanks, S. Sharma, J. W. Smythe & V. Viau. 1993. Cell. Mol. Neurobiol. **13:** 321-347.
8. Meaney, M. J., B. Tannenbaum, D. Francis, S. Bhatnagar, N. Shanks, V. Viau & D. O'Donnell. Semin. Neurosci. In press.
9. Meaney, M. J., D. H. Aitken, S. Bhatnagar, C. Van Berkel & R. M. Sapolsky. 1988. Science **238:** 766-768.
10. Meaney, M. J., D. H. Aitken, S. Sharma, V. Viau & A. Sarrieau. 1989. Neuroendocrinology **50:** 597-604.
11. Meaney, M. J., D. H. Aitken, S. Sharma & V. Viau. 1992. Neuroendocrinology **55:** 205-213.
12. Viau, V., S. Sharma, P. M. Plotsky & M. J. Meaney. 1993. J. Neurosci. **13:** 1097-1105.
13. Plotsky, P. M. & M. J. Meaney. 1993. Mol. Brain Res. **18:** 195-200.
14. Zarrow, M. X., P. S. Campbel & V. H. Denenberg. 1972. Proc. Soc. Exp. Biol. Med. **356:** 141-143.
15. Dallman, M. F., S. Akana, C. S. Cascio, D. N. Darlington, L. Jacobson & N. Levin. 1987. Recent Prog. Horm. Res. **43:** 113-173.
16 Jones, M. T., B. Gillham, B. D. Greenstein, U. Beckford & M. C. Homes. 1982. *In* Current Topics in Neuroendocrinology. D. Ganten & D. Pfaff, Eds. Vol. 2:45-68. Springer. New York.
17. Keller-Wood, M. & M. F. Dallman. 1984. Endocr. Rev. **5:** 1-24.
18. Plotsky, P. M. & W. W. Vale. 1984. Endocrinology **114:** 164-169.
19. Plotsky, P. M., S. Otto & R. M. Sapolsky. 1986. Endocrinology **119:** 1126-1130.
20. Sawchenko, P. E. 1987. J. Neurosci. **7:** 1093-1106.
21. Jacobson, L. & R. M. Sapolsky. 1991. Endocr. Rev. **12:** 118-134.
22. Diorio, D., V. Viau & M. J. Meaney. 1993. J. Neurosci. **13:** 3839-3847.
23. Sarrieau, A., S. Sharma & M. J. Meaney. 1988. Dev. Brain Res. **43:** 158-162.
24. Meaney, M. J. & D. H. Aitken. 1985. Dev. Brain Res. **22:** 301-304.
25. Meaney, M. J., D. H. Aitken, S. R. Bodnoff, L. J. Iny, J. E. Tatarewicz & R. M. Sapolsky. 1985. Behav. Neurosci. **99:** 760-765.
26. Meaney, M. J., D. H. Aitken & R. M. Sapolsky. 1987. Neuroendocrinology **45:** 278-283.
27. Larocque, S., D. O'Donnell, C. Gianoulakis, J. R. Seckl & M. J. Meaney. 1992. Soc. Neurosci. Abst. **18:** 479.
28. Levine, S. & G. W. Lewis. 1959. Science **129:** 42-43.

29. Mohammed, A. H., B. G. Henriksson, S. Soderstrom, T. Ebendal, T. Olsson & J. R. Seckl. 1993. Behav. Brain Res. **57:** 183-191.
30. Meaney, M. J., D. O'Donnell, V. Viau, S. Bhatnagar, A. Sarrieau, J. W. Smythe, N. Shanks & C.-D. Walker. *In* Receptors and the Developing Nervous System. P. McLaughlin & I. Zagon, Eds., 163-202. Chapman and Hall. London.
31. Van Eekelen, J. . M., M. C. Bohn & E. R. De Kloet. 1992. Dev. Brain Res.
32. Larocque, S., D. O'Donnell, J. R. Seckl & M. J. Meaney. 1993. Soc. Neurosci. Abst. **19:** 881.
33. Rosenfeld, P., W. Sutanto, S. Levine & E. R. de Kloet. 1988. Dev. Brain Res. **42:** 113-118.
34. Meaney, M. J., R. M. Sapolsky & B. S. McEwen. 1985. Dev. Brain Res. **18:** 159-164.
35. Meaney, M. M. & D. H. Aitken. 1985. Brain Res. **328:** 176-180.
36. Sharma S. & M. J. Meaney. Submitted.
37. Banker, G. A. & W. M. Cowan. 1977. Brain Res. **126:** 397-425.
38. Mitchell, J. B., W. Rowe, P. Boksa & M. J. Meaney. 1990. J. Neurosci. **10:** 1745-1752.
39. Savard, P., Y. Merand, T. Di Paolo & A. Dupont. 1984. Brain Res. **292:** 99-108.
40. Mitchell, J. B., W. Rowe, P. Boksa & M. J. Meaney. 1990. J. Neurosci. **10:** 1745-1752.
41. Smythe, J. W., W. Rowe & M. J. Meaney. Dev. Brain Res. In press.
42. Maccari, S., P. V. Piazza, H. Simon, L. Angelucci & M. Le Moal. 1992. Soc. Neurosci. Abst. **18:** 478.
43. Mitchell, J. B., K. Betito, W. Rowe, P. Boksa & M. J. Meaney. 1992. Neuroscience **48:** 631-639.
44. Teitler, M., R. A. Lyon, K. H. Davis & R. A. Glennon. 1987. Biochem. Pharmacol. **36:** 3265-3271.
45. Leonhardt, S., E. Gorospe, B. J. Hoffman & M Teitler. 1992. Mol. Pharmacol. **42:** 328-335.
46. Peroutka, S. 1993. J. Neurochem. **60:** 408-416.
47. Prichett, D. B., A. W. J. Bach, M. Wozny, O. Taleb, R. DalToso, J. C. Shih & P. H. Seeburg. 1988. EMBO J. **7:** 4135-4140.
48. Julius, D., A. B. MacDermott, R. Axel & T. Jessel. 1988. Science **241:** 558-564.
49. Apud, J. A., D. R. Grayson, E. DeErausquin & E. Costa. 1992. Neuropharmacology **31:** 1-8.
50. Parent, A. R., S. Sharma, R. Quirion & M. J. Meaney. 1993. Soc. Neurosci. Abst. **19:** 307.
51. Meaney, M. J., J. B. Mitchell, K. Betito, D. O'Donnell, W. Rowe & J. Diorio. Submitted.
52. Levine, S. 1975. *In* Society, Stress and Disease. (L. Levi, Ed.) 43-50. Oxford University Press. London.
53. Bell, R. W., W. Nitschke, T. H. Gorry & T. A. Zachman. 1971. Dev. Psychobiol. **4:** 181-191.
54. Lee, M. H. S. & D. I. Williams. 1974. Anim. Behav. **22:** 679-681.
55. Lee, M. H. S. & D. I. Williams. 1975. Dev. Psychobiol. **8:** 91-95.
56. Kuhn, C. M., J. Pauk & S. M. Schanberg. 1990. Dev. Psychobiol. **23:** 395-410.
57. Smotherman, W. P., S. G. Wiener, S. P. Mendoza & S. Levine. 1977. Dev. Psychobiol. **10:** 113-122.
58. Hennessy, M. B., J. Vogt & S. Levine. 1982. Physiol. Psychol. **10:** 153-157.
59. Yoshimasa, T., D. R. Sibley, M. Bouvier, R. J. Lefkowitz & M. G. Caron. 1987. Nature **d327:** 67-70.
60. Habener, J. 1990. Mol. Endocrinol. **4:** 1087-1094.
61. Imagawa, M., R. Chiu & M. Karin. 1987. Cell **51:** 251-260.
62. Montminy, M. R., G. A. Gonzalez & K. K. Yamamoto. 1990. Trends Neurosci. 184-188.
63. Roesler, W. J., G. R. Vandenbark & R. W. Hanson. 1988. J. Biol. Chem. **263:** 9063-9066.
64. Seckl, J. R., K. L. Dickson & G. Fink. 1990. J. Neuroencrinol. **2:** 911-916.

DISCUSSION OF THE PAPER

QUESTION: Is lactation a necessary part of this mother-pup interaction? Has somebody shown that?

MEANEY: I really would like to have Dr. Levine comment on some of the work that he had done with Michael Hennessey. One of the things that has been shown is that if you handle the pups, you see changes in the mother-pup interaction. We have seen (in collaboration with Barbara Woodsite, who is more the behavioral expert in these areas) that you change the pattern of maternal care of the offspring. The mother normally gets on her pups, feeds, licks the pups, stimulates anogenital activity with licking and then moves off and probably dissipates body heat. This has a rather ordered sequence so that it occurs during 20 to 25 minutes. Handled mothers are on the litters for a shorter period of time, off them for a shorter period of time and then back on again. So the frequency of the bouts is increased. Intriguingly, this would and does produce greater amounts of tactile stimulation and licking of the pups. Cindy Kuhn and Shonberg at Duke have shown that maternal licking is something that is absolutely critical in regulating neuroendocrine activity, immediately and over the course of development. So that is one aspect of it. But there is better evidence that indeed looking at the mother and maternal behavior can be the critical feature here. Stephanie McCowia has done studies cross-fostering the pups, which produces comparable changes in mother-pup interactions and can produce a neuroendocrine sequela that is similar to the handling experiment.

QUESTION: Is there any reciprocal effect, or is there an effect on the 5HT receptors themselves?

MEANEY: No. Jim Smite said you might be altering the 5HT input into the hippocampus and subsequently, it could just be a recurrence of increased serotinergic activity. We worked on that particular hypothesis for some time. Jim did 5HT2/1C receptor binding in the hippocampus and frontal cortex. In fact, in the frontal cortex there is a decrease that is a consequence of handling, while in the hippocampus there is no change. We also looked at the levels of 5HT using HPLC in the hippocampus and frontal cortex and found no permanent changes. We did immunocytohistochemistry in the hippocampus and frontal cortex and found no differences. So it does not appear as though you are enhancing the serotinergic signal over the course of a lifetime.

QUESTION: What is TSH and T_4 and hormone binding doing in the animals that are handled? Is this a peripheral effect, the increase in T_3, or is it from the central TRH?

MEANEY: I have no idea. In fact all we have done, I am afraid, is the quick and dirty radioimmunoassay. We do not know exactly what the pituitary-adrenal output is. Cindy Kuhn has been measuring TSH. She finds slight increases in TSH but not quite as impressive as we would like to see. So it could very well have something to do with the metabolism of T_4 and T_3.

The Ontogeny of the Hypothalamic-Pituitary-Adrenal Axis

The Influence of Maternal Factors[a]

SEYMOUR LEVINE

Department of Psychiatry and Behavioral Science
Stanford University Medical School
Stanford, California 94305

In adult organisms, glucocorticoids (GCs) serve a wide variety of regulatory and permissive functions.[1] These functions are aimed basically at controlling the organism's response to stress and regulating certain circadian activities. Most of the effects of GCs, in the adult, are readily reversible. However, during development, GC administration has been demonstrated to have permanent effects on growth and differentiation in a number of systems, including the CNS. In rats, high doses of GCs administered neonatally cause decreased mitosis, myelination, and altered neuromorphogenesis.[2] Animals treated neonatally with GCs show reduced DNA content and brain size, as well as impaired neuroendocrine function as adults[3] as well as behavior.[4] Conversely, the absence of GCs following adrenalectomy also affects CNS development.[5] Thus, relatively constant low levels of GCs are indispensable for normal maturation. It appears that during development a delicate and critical balance of GCs is essential.

The dynamics of the HPA system show a characteristic developmental pattern. During the final days of gestation and immediately after birth, basal levels of CORT in the rat are declining but are still high.[6] However, during the first two postnatal days, CORT concentrations decrease dramatically and remain at low levels until approximately day 14. In addition, concentrations of pituitary ACTH and hypothalamic CRF are also low during the first two postnatal weeks.[7] The system is further characterized in the neonatal period by the fact that stimuli which normally elicit CORT elevations in adults are unable to do so in the infant. This neonatal period which covers, in the rat, the period from postnatal day 4 to 14, has been referred to as the "stress hyporesponsive period" (SHRP). During the SHRP, there is also a diminution of adrenal sensitivity, evidenced by the fact that the infant fails to show a significant elevation of CORT following administration of high doses of exogenous ACTH.[8] The HPA system does not develop uniformly; different components of the system have different ontogenetic patterns. Thus, although the infant rat is capable of responding to stressful stimuli by day 15, circadian rhythmicity and feedback regulation are not yet fully developed until much later in development.[9–11]

One proposed mechanism for the suppression of the HPA system during this period of development is that most stressors are unable to evoke a sufficiently large CRF signal to overcome an "enhanced CORT feedback signal" at the brain and pituitary levels.[12,13] There are several experiments which question the validity of this hypothesis.[14] The overall picture that emerges, however, is that there are multiple differences in the neonatal neuroendocrine system that ensure that plasma CORT

[a]This research was supported by Grants HD-02881 from the National Institute of Child Health and Human Development and MH-45006 from the National Institute of Mental Health (NIMH), and by U.S. Public Health Service Research Scientist Award MH-19936 from NIMH to the author.

levels in the developing infant are maintained at relatively low and unperturbable levels. In view of the permanent deleterious effects which have been reported to follow perinatal administration of high levels of GCs, this suppression of the HPA system appears highly adaptive.

There is a growing body of evidence which indicates that physiological processes in the developing infant are regulated by the mother in subtle ways.[15,16] Studies, mostly on developing rodents, have demonstrated that the mother regulates physiological responses in the infant as diverse as heart rate, sleep/wake cycles, and growth hormone production. Maternal regulation is made manifest by separating mother and infant. Physiological changes in the infant, which develop slowly over a relatively protracted period of separation, and which can be tightly linked to specific features of mother-infant interaction, are thought to reflect the absence of maternal regulation. For example when two-week-old rat pups are separated from their mothers for 24 hr there is a 40% decrease in cardiac rate. This decrease results specifically from the absence of mother's milk, as opposed to lack of maternal contact or other aspects of maternal care. Thus, when milk is infused into the pup directly, in the absence of the mother, normal cardiac rate can be maintained or restored.

Another physiological response of the infant rat pup to separation from the mother is a decline in growth hormone secretion and in levels of a tissue enzyme regulating growth and differentiation: ornithine decarboxylase (ODC). This decline can be traced precisely to the lack of vigorous tactile stimulation from the mother, as opposed to deprivation of milk, passive maternal contact, or other stimuli emanating from the mother. Vigorous stroking of a separated rat pup by an experimenter with a moistened artist brush can normalize growth hormone and ODC activity.[17,18]

Recent work has demonstrated that there appears to be a profound influence of maternal behavior on the maintenance of the SHRP. We have found (see Ref. 14 for review) that following 24 hr of maternal deprivation the neonatal rat responds with significant increases in ACTH and CORT when exposed to novelty, injection of isotonic or hypertonic saline, and ether vapors. Thus, the HPA system appears to be yet another physiological system which is subject to maternal regulation.[19]

The initial study of maternal regulation of the HPA axis was designed to investigate whether the infant rat is capable of using consummatory behavior to suppress HPA activity as was demonstrated in the adult. Thus, rats at post-natal day 12, 16, and 20 were maternally deprived for 24 hr and then exposed to a novel environment.[20] The two variables that were manipulated were the presence or absence of the mother and the presence or absence of food. The animals were fed through a previously implanted tongue cannula. The results indicated that a brief feeding bout was not capable of suppressing the CORT response following exposure to novelty. However, the presence of an anesthetized mother completely inhibited the CORT response. There was one surprising finding in this study, namely 12 day-old pups also showed significant elevation of both basal and stress-induced CORT secretion. This was unexpected in view of all the data which indicated that at this age the rat pup is in the midst of the (SHRP).[13] Although the purpose of the deprivation procedure was to increase the level of food-motivated behavior, so that the infants would consume the infused milk, the results indicated that maternal factors might be critical in the regulation of the neonatal HPA axis. Subsequent experiments revealed that maternal deprivation seemed to produce a stress-responsive HPA system during that period in development when nondeprived infants are nonresponsive.[21] These data suggested that the HPA axis was yet another physiological system that was influenced by maternal behavior.

Hofer[15] has designated the processes by which maternal factors regulate physiological development as "hidden regulators." One of the criteria for a hidden regulator is that it requires a long period of maternal deprivation to become evident, in contrast to the immediate responses to separation. An examination of the time course of HPA activity following deprivation reveals that the maternal influences on the developmental patterns of this system do indeed meet this criterion established for hidden regulators. Evidence for this conclusion comes from a study in which pups were maternally deprived for varying lengths of time (*i.e.*, 0, 2, 4, 8, and 24 hr). At the end of each deprivation period CORT secretion in response to stress (novelty or novelty plus a saline injection and ACTH) was measured. Basal levels of CORT increased progressively over time in day 7 and 11 (but not day 3) pups. CORT release in response to stress followed a similar pattern. In contrast, ACTH injection resulted in marked increases in CORT secretion regardless of the length of deprivation in 3-day-old pups. At older ages, 24 hr of deprivation led to even larger increases. Thus, for animals that are in the midst of the (SHRP) prolonged maternal deprivation appears to result in the disinhibition of the HPA axis.[22]

In view of the evidence which convincingly demonstrated that maternal factors are important in the regulation of the neonatal HPA axis, we began to focus on the issue of what specific aspects of the mother-infant interaction were responsible for the nonresponsiveness characteristic of this system. In our early studies we found that contact with a lactating dam suppressed the CORT response in maternally deprived pups who were either exposed to novelty or injected with saline. A lactating dam was the most effective stimulus for producing this suppression. The presence of a nonlactating female, a littermate, or an adult male were not capable of suppressing the CORT response.[21] Thus, maternal contact has been shown to regulate at least some aspect of the pup's HPA responsiveness during the SHRP. Kuhn *et al.*[23] have shown that the increase in CORT levels seen following two hr of maternal separation is partially reversed when the pup is allowed to contact a nipple-cauterized dam. Thus, we designed an experiment[24] to investigate whether prolonged (24-hr) nonnutritional contact with a maternally behaving dam would also inhibit the stress response observed in maternally deprived pups. Seven-, 11- and 15-day-old pups were left with a dam that exhibited maternal care but was unable to lactate (*i.e.*, thelectomized, nipples removed) for 24 hr, or were maintained with their lactating dams. At the end of this period, basal and stress (novelty) levels of CORT were compared between different treatments. The results provided little evidence that maternal contact in the absence of suckling and/or feeding was capable of downregulating the infants' HPA system. Pups that remained in contact with thelectomized dams showed CORT levels equivalent to deprived infants (with the exception of a small but significant modulation of the CORT response to novelty at day 12): in all cases however, these levels were higher than those of the nondeprived pups. It should be noted that thelectomized dams have been described not to exhibit an active nursing posture.[25] Thus, pups in contact with these dams did not experience this specific aspect of normal maternal behavior. To ensure that food-deprived pups received normal maternal behavior we conducted an experiment[26] in which normal lactating dams were in contact with pups who were unable to feed. Feeding was prevented by implanting a polyethylene cannula through the tongue. Although this cannula did not appear to prevent suckling, all pups lost significant weight over the 24-hr period indicating either the absence of feeding, or at least a marked reduction in milk intake. Observation of maternal behavior did not reveal any major influences on maternal behavior as a consequence of the pups being cannulated and the pups were observed to be attached to the nipple while the dam was in its nursing postures.

The CORT results showed once again only a small reduction in CORT levels compared to maternally deprived pups in response to novelty. However, food-restricted pups still showed a significant elevation in basal levels and a response to novelty which does not occur in nondeprived pups. The findings obtained in the above-mentioned pups that were permitted to remain in contact with their mothers but not able to ingest milk differed only slightly from totally deprived pups.

These findings suggest that the processes responsible for maintaining the SHRP differ from those that modulate the stress response in an infant that has been rendered responsive by 24 hr of maternal deprivation. Thus, we have discriminated two seemingly independent events: 1) an inhibitory effect that maintains hyporesponsiveness, and 2) a suppressing effect that prevents the CORT response from being activated once the pup is capable of responding to mild stressors.

It is frequently assumed that elevations in CORT are under the direct regulation of ACTH. However, there is evidence both in adults[27] and in infants[28] that adrenal sensitivity to ACTH can vary. For example, in adult animals adrenal sensitivity varies as a function of circadian rhythms, and in the infant there is evident that one of the indicators of the SHRP is a reduction in adrenal sensitivity to ACTH. Several studies conducted in this laboratory have shown that following maternal deprivation CORT levels in the adrenal and in plasma are significantly elevated in response to exogenous ACTH, in contrast to nondeprived pups, which show little or no CORT response when exposed to high levels of ACTH at all ages tested (days 4, 8, 12, and 16). At day 4 it was demonstrated not only that maternally deprived pups elicit a greater CORT response to ACTH but also that extremely low levels of ACTH (in the order of magnitude of 0.001 IU of ACTH) are capable of inducing a CORT response in these deprived pups.

To determine which component of the HPA axis is affected by feeding, it was essential to demonstrate that the ACTH response to either the deprivation procedure and/or following exposure to novelty and/or a saline injection was also under maternal regulation. Thus, the following experiments (FIG. 1) examined the ACTH response to a saline injection and novelty in both maternally deprived and non-deprived pups at days 6, 9 and 12. Further, it has been reported that neonatal rat pups do not have the capacity to efficiently thermoregulate. Therefore, pups were tested either at room temperature (23°C) or at nest temperature (33°C). Body temperature was also measured in deprived and nondeprived infants. An initial blood sample was obtained after 24 hr deprivation and at 10, 20, and 30 min following the saline injection. The data[19] demonstrate that the ACTH response to the mild stress used in these experiments was significantly higher in deprived pups. However, the magnitude of the response was dependent upon both ambient temperature and age. At day 6, only those deprived pups tested at nest temperature showed elevated ACTH levels. At days 9 and 12, all deprived pups showed elevated circulating ACTH, but once again the animals tested at nest temperature were further elevated. Nondeprived pups did not elicit an ACTH response at any of the ages tested. The magnitude of the ACTH response was greater in the older deprived pups (day 12). The nondeprived pups were capable of maintaining their body temperature through the 30 min of exposure to room temperature, whereas deprived pups showed a significant drop in body temperature. This was most likely due to the loss of body weight and probably a reduction in brown adipose tissue, an important thermoregulatory factor in the rat.

In addition to demonstrating that maternally deprived pups showed an elevation of ACTH following exposure to mild stress the data also indicated that once the system is activated there appears to be a deficiency in the negative feedback mechanism. In adult animals the ACTH response to a control injection returns to basal

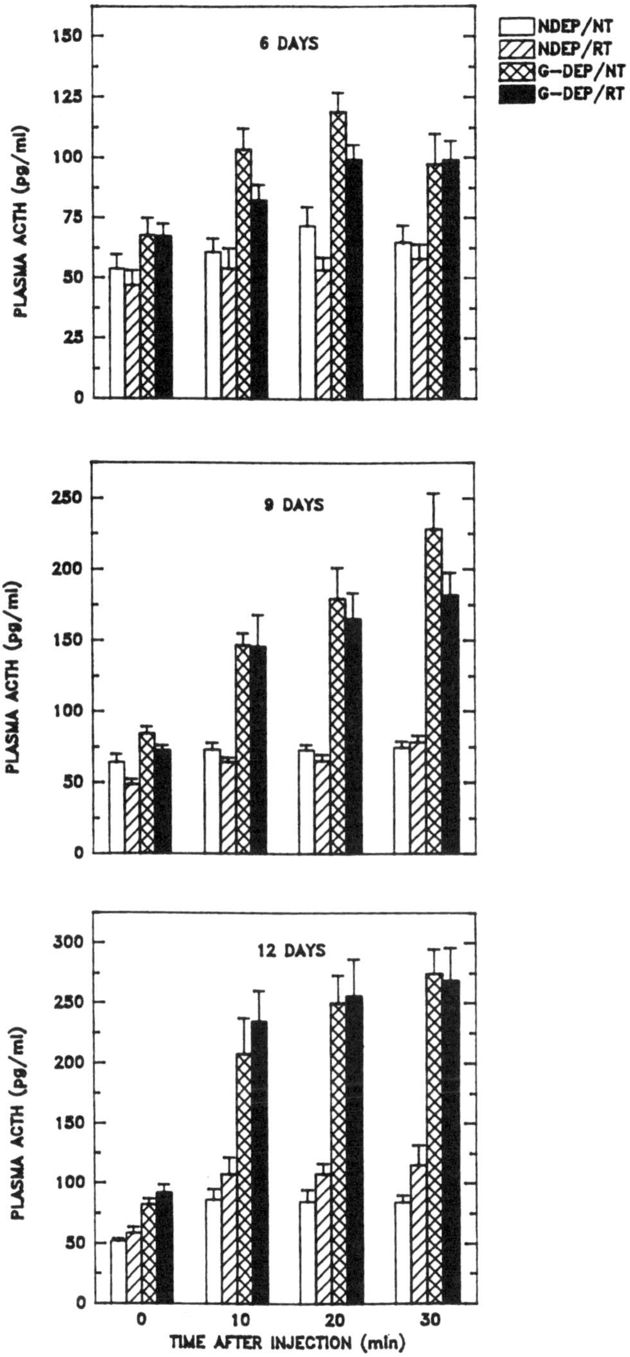

FIGURE 1. Time course and effects of nest (NT) versus room temperature (RT) on the ACTH response (pg/ml) to a saline injection in 6-, 9-, and 12-day-old nondeprived (NDEP) and maternally deprived in a social group (G-DEP) neonates. Values represent the mean ± SEM of 6–10 pups/group.

values by 30 min. At days 9 and 12, we have demonstrated that ACTH remains elevated for at least two hr following an acute injection with saline. At day 15 ACTH is still elevated in maternally deprived pups, but appears to return slowly to basal levels.

Given this body of data we took up again the question of the possible inhibitory role of feeding in the maintenance of the SHRP. In order to accomplish this experiment (FIG. 2) the use of a technique for feeding the pup in the absence of the dam was required. Such a technique was developed initially by Hall and Rosenblatt[29] and involved implanting a cannula through which milk could be delivered at regular intervals throughout the 24 hr of maternal deprivation. For this study we used a modification of the Hall and Rosenblatt procedure which was reported by Spear *et al.*[30] The cannula was placed in the cheek of the pup. The placement of the cannula did not affect swallowing and the pups were observed to swallow and ingest the milk. At one-hr intervals 0.24–0.45 ml of milk were delivered by an infusion pump attached to the cannula. The amount of milk delivered was calibrated to maintain the normal weight gain that was seen in the nondeprived pups. One of the more pervasive maternal behaviors is anogenital licking. The primary purpose of this behavior is to stimulate the eliminative processes in the pup. In order to prevent accumulation of waste products in the infant the anogenital area of the pups was stimulated manually by the experimenter on three different occasions separated by 7–8 hr. This stimulation was also performed on nonfed deprived pups. At the end of 24 hr, blood samples were obtained from fed and nonfed pups at days 9, 12, and 15. For half the animals, blood was taken immediately following the 24 hr period of maternal separation (nontreated, NT). The remaining infants were injected with saline, and blood was taken at 30 min post-injection. The results[19] showed that nonfed NT deprived pups had elevated CORT levels when compared to the fed NT deprived animals. Thus feeding apparently reduced the deprivation-induced elevations in basal levels. This effect was present at days 9 and 12. Although the saline injection produced elevations in CORT above those observed in NT pups, feeding significantly reduced the CORT response at all ages tested. However, the ACTH response did not appear to be affected by feeding. At all ages the ACTH levels were not significantly different between the fed and nonfed deprived infants. ACTH levels were elevated following saline injection, but not differentially as a consequence of feeding.

These data suggest that there are at least two different components of maternal behavior that regulate distinct elements of the HPA system. Based on the results of the experiment described above it would appear that feeding acts to reduce the sensitivity of the adrenal to ACTH. Thus, even under conditions which elicit a significant ACTH response *i.e.*, IL-1,[31] the adrenal is still incapable of releasing large amounts of CORT. In contrast to the adult, where feeding appears to regulate CRF and ACTH secretion,[32] in the infant ACTH secretion does not appear to be influenced by feeding. Several mechanisms have been explored to account for the decreased adrenal sensitivity during the SHRP. 1) A limited store of cholesterol in the adrenal gland of the neonate. This may be due to a decrease response to the action of peptides other than ACTH which may play a role in the activation of the biosynthetic pathway of CORT. One candidate is Lys-y-3-MSH, which is another product of the POMC molecule cleavage. Receptors for this peptide have been identified in the rat adrenal and this peptide has been shown to be active in the adrenal of hypophysectomized rats.[33] 2) A decrease in steroidogenic activity of the adrenal, possibly due to a reduction in ACTH secretion during this period in development. Doering and Clayton[34] have demonstrated that cholesterol side-chain cleavage is high on day 1, but then is markedly reduced on days 6 and 12, increasing again by the third week of

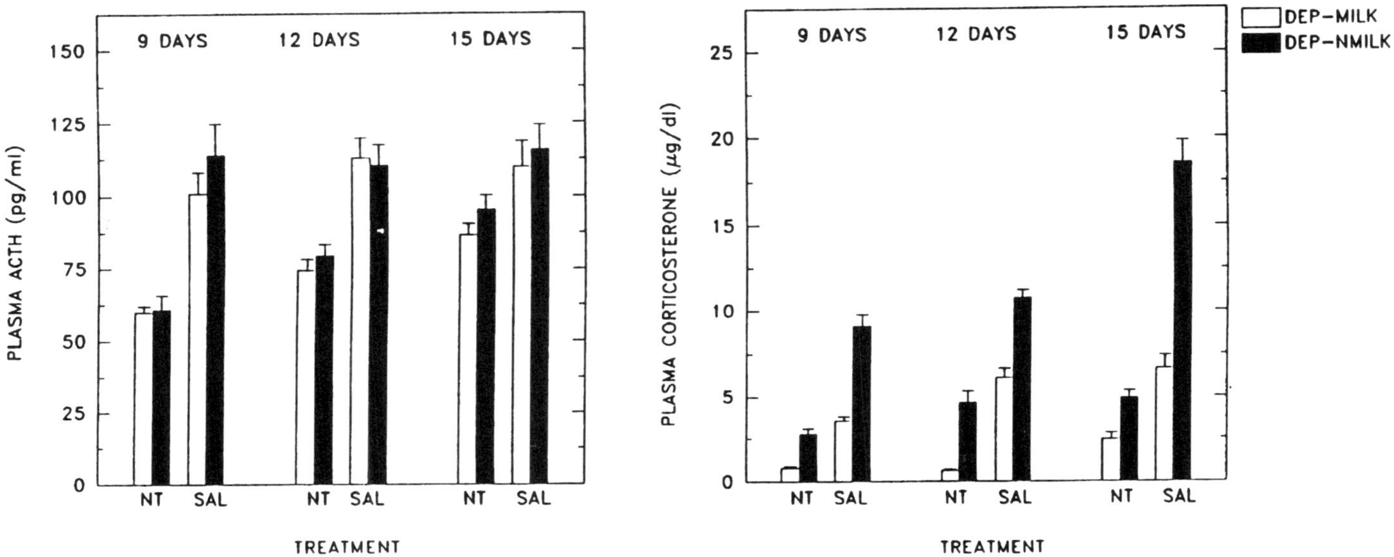

FIGURE 2. Effects of feeding on ACTH (pg/ml) and CORT (µg/dl) plasma levels in 9-, 12-, and 15-day-old pups. Pups were either provided (DEP-MILK) or not (DEP-NMILK) with a milk solution delivered episodically during the deprivation period. Values represent the mean ± SEM of 7–13 pups/group.

life. One function of this enzyme is to catalyze the conversion of cholesterol to pregnenolone and is thought to be the rate-limiting step in steroidogenesis.[35] It is not possible at this time to specify how feeding interacts with any of these proposed mechanisms. We can conclude however, that the absence of feeding does render the neonatal adrenal more sensitive to ACTH. We can further conclude that some other aspect of maternal behavior is involved in the regulation of ACTH secretion, and presumably CRF and possibly AVP.

When we examined the results obtained in the experiment described above with those presented in the initial study which investigated the ACTH response in maternally deprived pups, it was evident that the magnitude of the ACTH response was significantly lower than what we had obtained in the previous experiment. The only change in the procedure between the two experiments was that in the feeding experiment all animals had been stroked three times during the 24-hr deprivation period. Therefore, we conducted an experiment (FIG. 3) to specifically investigate the role of stroking in the regulation of the neonatal HPA axis. CORT and ACTH secretion was measured in four groups of pups at two ages (days 9 and 12). Three groups of pups were maternally deprived. One group received no treatment during the deprivation period (DEP), the second group was stroked in the anogenital region using a fine brush (DEP+STROKED), the third group was picked up at the same times as the stroked group but received no stroking (DEP+HANDLED). These animals were compared to a nondeprived group (NDEP). Basal and stress levels of CORT and ACTH were measured. None of the treatments elevated basal levels of ACTH. The results[19] showed that stroking reduced the ACTH response to stress to the levels of the nondeprived pups. Handling did not prevent the elevation of ACTH levels induced by deprivation at day 9. At day 12, this manipulation lead to a further increase of the peptide secretion. The CORT results, once again, reflected the absence of feeding. Thus, although there were no differences between the nondeprived and stroked infants in ACTH levels, both basal and stress levels of CORT were significantly elevated in the stroked pups indicating that adrenal sensitivity to the same endogenous levels of ACTH was increased. These findings help resolve the paradox created by earlier findings which seemed to suggest that contact with a nonlactating dam was not effective in preventing activation of the HPA system in nonfed pups. Those findings were based solely on the CORT response. In view of the fact that different properties of maternal behavior appear to differentially regulate specific components of the developing HPA axis it is likely that if ACTH was measured during 24 hr of contact with the nonlactating dam we might have observed a suppression of the ACTH response, although the adrenal remains responsive due to an increase in sensitivity.

As stated earlier in this paper during development, glucocorticoids have been shown to produce permanent effects on growth and differentiation of a number of systems, including the CNS. It thus appears that intrinsic factors such as low levels of CRF and AVP, deficient POMC processing in the pituitary, and alterations in steroid biosynthesis and metabolism prevent the infant rat from releasing high circulating levels of CORT. In addition it also appears that there are extrinsic factors, specifically certain facets of maternal behavior which also act to maintain the developing HPA axis in an unresponsive state. It would appear as though the down-regulation of the HPA system during this period of development is so critical that there are numerous fail-safe mechanisms that ensure that it will either not respond or respond minimally. Even stimuli that are capable of eliciting a vigorous ACTH response in nondeprived pups, such as immune signals[31] and excitatory amino acids,[36] fail to stimulate marked increases in CORT secretion. All the evidence leads to the conclusion that the major effect of feeding is to reduce the sensitivity of the adrenal, preventing the adrenal from secreting high levels of endogenous CORT, whereas stroking appears to actively inhibit the neural systems which regulate ACTH.

Although the SHRP appears to terminate around day 14 postnatally the emergence of GC negative feedback occurs much later in development.[10,11] It was observed that once the ACTH and CORT response was elicited in deprived pups (ages 9, 12, and 15) the levels of ACTH and CORT remained elevated for at least two hr in contrast to adult rats, which return to basal levels within 30 min. We initiated a series of studies (FIG. 4) designed to investigate the ontogeny of negative feedback using adrenalectomized (ADX) pups. It has been reported[37] that ADX results in an acute elevation of ACTH in 5-day-old pups (three hr following ADX). We have replicated these findings and extended them to include older aged pups. At all ages during the SHRP, ACTH was markedly elevated three hr after ADX. There were no differences in ACTH levels between deprived and nondeprived pups. Replacing CORT, by injection, immediately following ADX did not result in a suppression in the ADX induced ACTH elevations (unpublished results). However, implants of a CORT pellet immediately following ADX[37] maintain the normally low levels of ACTH. It appears that chronic exposure to CORT exerts a negative feedback action on basal ACTH secretion. Once ACTH is elevated a bolus injection of CORT is no longer capable of suppressing ACTH. Thus during development low levels of circulating CORT appear capable of suppressing basal levels of ACTH but are unable to prevent stress-induced ACTH once it is activated.

It has been postulated that the different GC receptors in the brain (type 1, MR and type 2, GR) act to regulate different functions of the HPA system.[38] According to de Kloet, limbic system type 1 receptors mediate the control of basal activity of the HPA axis, whereas the negative feedback action of CORT on stress-induced HPA activity is mediated primarily by type 2 receptors in the PVN and the pituitary corticotropes. Studies on the ontogeny of these receptors[39] indicate that during the SHRP the number of available type 2 receptors is less than 50% of those available in the adult. In contrast binding studies indicate that type 1 receptors achieve adult levels by postnatal day 8. The results of the experiments investigating CORT negative feedback in the ADX neonate would support the view that the type 1 receptors mediate the basal activity of the HPA system and the presence of low levels of CORT in the neonate are required to maintain basal levels of ACTH. However, once ACTH is activated either by stress in the maternally deprived pup, or by ADX, negative feedback that is under the control of the type 2 receptor is deficient by virtue of the fact that the available type 2 receptors are reduced during development.

The existence of the SHRP has been challenged by several investigators. The fact that the neonatal HPA system is different in many ways from that of the adult has been well documented. Whether it is stress hyporesponsive is dependent on which component of the HPA is examined and the specific stimuli used to activate this system. The concept of the SHRP is clearly applicable to neonatal CORT secretion. Although feeding appears to inhibit CORT synthesis and/or secretion there are additional as yet undetermined rate-limiting factors which result in reduced CORT levels during the SHRP. Thus nonfed pups at days 4 and 8 show a lower CORT response to exogenous ACTH than 12-day-old maternally deprived pups. The reasons for the suppression of the CORT response early in ontogeny are unknown. However, the capacity of the neonate to release ACTH and presumably CRF is not absolutely impaired, but does exhibit much greater specificity than observed in adults and is age dependent. Thus immune signals, LPS,[40] IL-1,[31,41] and histamine;[42] ether

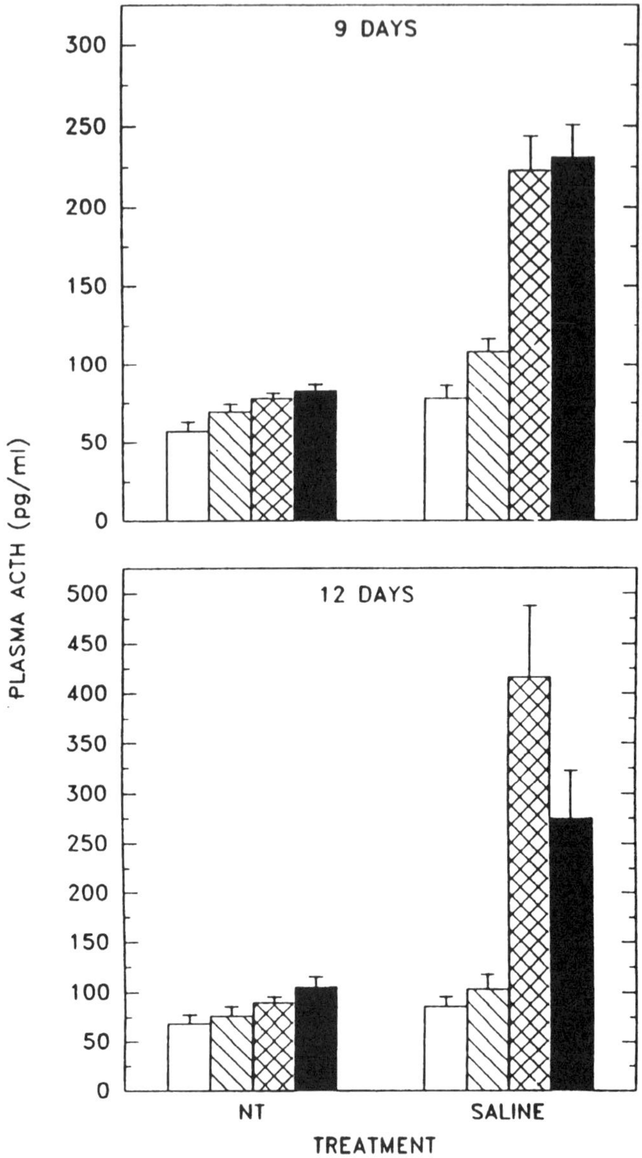

FIGURE 3. Effects of stroking on ACTH (pg/ml) and CORT (μg/dl) plasma levels in 9- and 12-day-old pups. Rat pups were either left with their dams until testing (NDEP), or were maternally-deprived and either stroked (DEP+STROKED), Handled (DEP+HANDLED) or

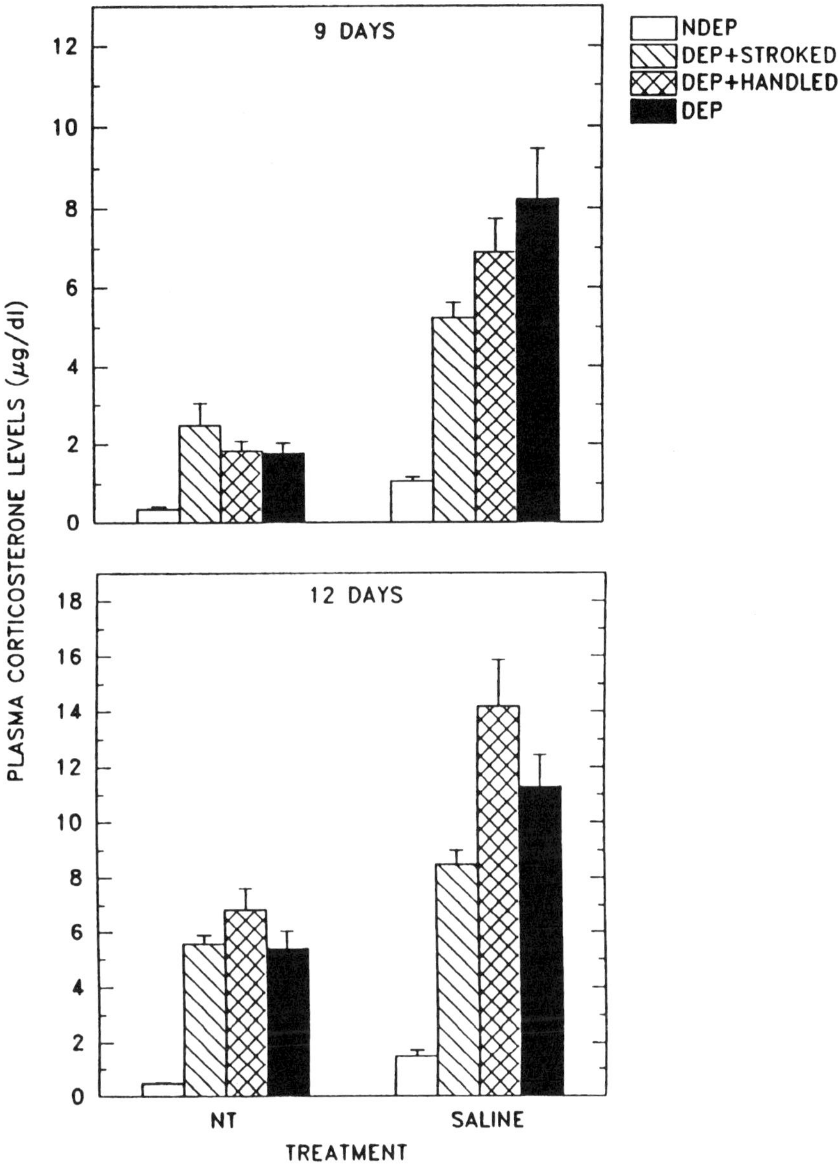

left undisturbed during the deprivation period (DEP). Values represent the mean ± SEM of 9-12 pups/group.

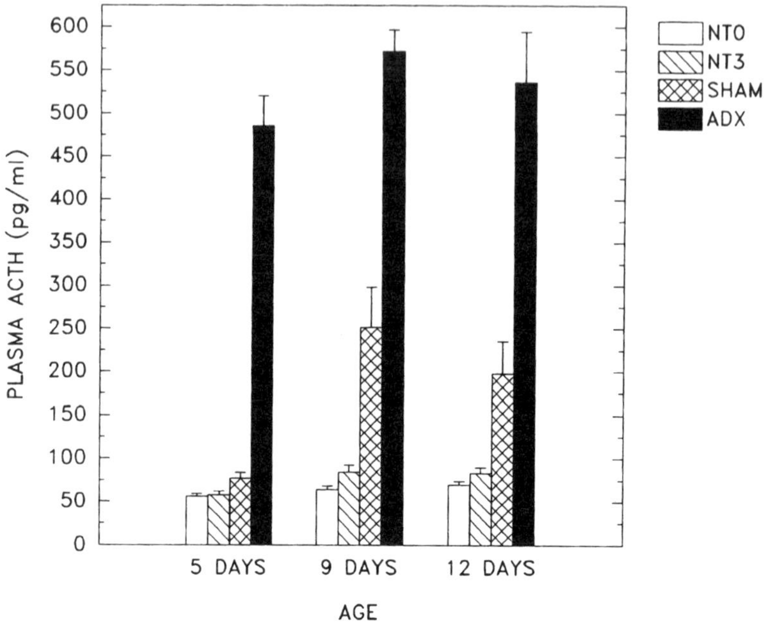

FIGURE 4. Plasma ACTH concentrations (pg/ml) in response to 3 hr of maternal separation (NT3), sham operation (SHAM), or adrenalectomy (ADX) in pups 5, 9, and 12 days old. Values represent the mean ± SEM of 6–10 pups/group.

vapors,[42] EEA;[36] and ADX[31,37] elicit a significant increase in ACTH. The stimulating effects of EEA are blocked after passive immunization with an anti-CRF antiserum,[36] indicating that CRF release is involved in the ACTH-stimulating effect of EEA agonists. However, the magnitude of the ACTH response is age dependent. Thus the response to IL-1[31] and ether vapors[42] is significantly lower early in development. This reduction in ACTH secretion does not appear to be a consequence of a reduced capacity of the pituitary to secrete ACTH, insofar as the ACTH response to ADX is equivalent throughout ontogeny. However, the normally reared pup does not exhibit an ACTH response to stimuli such as novelty, saline or vehicle injections, and acute maternal deprivation, which are observed only in pups following maternal-deprivation. The ontogenetic changes in the stimulus-specificity of the stress response may reflect concurrent maturation in the specific neural pathways that mediate the effects of each particular stimulus. It could also be argued that maternal factors are capable of inhibiting only those specific neural pathways which involve the processing of psychogenic stimuli and most likely involve limbic system mediation of CRF.[43]

REFERENCES

1. Munck, A., P. M. Guyre & N. J. Holbrook. 1984. Physiological functions of glucocorticoids in stress and their relation to pharmacological actions. Endocr. Rev. **5**(1): 25–44.
2. Bohn, M. C. 1980. Granule cell genesis in the hippocampus of rats treated neonatally with hydrocortisone. Neuroscience **5:** 2003–2012.

3. Erskine, M. S., E. Geller & A. Yuwiler. 1979. Effects of neonatal hydrocortisone treatment on pituitary and adrenocortical response to stress in young rats. Neuroendocrinology **29:** 191–199.
4. Bohn, M. C. 1984. Glucocorticoid induced teratologies of the nervous system. *In* Neurobehavioral Teratology. J. Yanai, Ed. 365–387. Elsevier Science Publishers. New York.
5. Meyer, J. S. 1983. Early adrenalectomy stimulates subsequent growth and development of the rat brain. Exp. Neurol. **82:** 432–446.
6. Martin, C. E., M. H. Cake, P. E. Hartmann & I. F. Cook. 1977. Relationship between foetal corticosteroids, maternal progesterone and parturition in the rat. Acta Endocrinol. **84:** 167–176.
7. Walker, C. D., M. Perrin, W. Vale & C. Rivier. 1986. Ontogeny of the stress response in the rat: role of the pituitary and the hypothalamus. Endocrinology **118:** 1445–1451.
8. Levine, S., D. Glick & P. K. Nakane. 1967. Adrenal and plasma corticosterone and vitamin A in rat adrenal glands during postnatal development. Endocrinology **80:** 1177–1179.
9. Levin, R. & S. Levine. 1975. Development of circadian periodicity in base and stress levels of corticosterone. Am. J. Physiol. **229:** 1397–1399.
10. Goldman, L., C. Winget, G. W. Hollingshead & S. Levine. 1973. Postweaning development of negative feedback in the pituitary-adrenal system of the rat. Neuroendocrinology **12:** 199–211.
11. Vazquez, D. M. & H. Akhil. 1993. Pituitary-adrenal response to ether vapor in the weanling animal: characterization of the inhibitory effect on glucocorticoids on adrenocorticoid secretion. Pediatr. Res. **34:** 646–653.
12. Sakly, M. & B. Koch. 1983. Ontogenetical variations of transcortin modulate glucocorticoid receptor function and corticotropic activity in the pituitary gland. Horm. Metab. Res. **15:** 92–96.
13. Sapolsky, R. M. & M. J. Meaney. 1986. Maturation of the adrenocortical stress response: neuroendocrine control mechanisms and the stress hyporesponsive period. Brain Res. Rev. **11:** 65–76.
14. Rosenfeld, P., D. Suchecki & S. Levine. 1992. Multifactorial regulation of the hypothalamic-pituitary-adrenal axis during development. Neurosci. Biobehav. Rev. **16:** 553–568.
15. Hofer, M. A. 1978. Hidden regulatory processes in early social relationships. *In* Perspectives in Ethology: Social Behavior. P. P. G. Bateson & P. H. Klopfer, Eds. 135–166. Plenum Press. New York.
16. Hofer, M. A. 1984. Early stages in the organization of cardiovascular control. Proc. Soc. Exp. Biol. Med. **175:** 147–157.
17. Kuhn, C. M., S. R. Butler & S. M. Schanberg. 1978. Selective depression of serum growth hormone during maternal deprivation in rat pups. Science **201:** 1034–1036.
18. Schanberg, S. M., G. Evoniuk & C. M. Kuhn. 1984. Tactile and nutritional aspects of maternal care: specific regulators of neuroendocrine function and cellular development. Proc. Soc. Exp. Biol. Med. **175:** 135–146.
19. Suchecki, D., P. Rosenfeld & S. Levine. 1993. Maternal regulation of the hypothalamic-pituitary-adrenal axis in the infant rat: the role of feeding and stroking. Dev. Brain Res. **75:** 185–192.
20. Stanton, M. E., J. Wallstrom & S. Levine. 1987. Maternal contact inhibits pituitary-adrenal stress responses in preweanling rats. Dev. Psychobiol. **20:** 131–145.
21. Stanton, M. E. & S. Levine. 1988. Maternal modulation of infant glucocorticoid stress responses: role of age and maternal deprivation. Psychobiology **16**(3): 223–228.
22. Levine, S., D. M. Huchton, S. G. Wiener & P. Rosenfeld. 1991. Time course of the effect of maternal deprivation on the hypothalamic-pituitary-adrenal axis in the infant rat. Dev. Psychobiol. **24**(i): 547–558.
23. Kuhn, C. M., J. Pauk & S. M. Schanberg. 1990. Endocrine responses to mother-infant separation in developing rats. Dev. Psychobiol. **23:** 395–410.
24. Cirulli, F., S. L. Gottlieb, P. Rosenfeld & S. Levine. 1992. Maternal factors regulate stress responsiveness in the neonatal rat. Psychobiology **20**(2): 143–152.

25. STERN, J. 1990. Sensory and dopaminergic regulation of maternal behavior in Norway rats: critical role of trigemunal and veneral somatosensory inputs (abstract). Paper presented at the Annual Meeting of the International Society for Developmental Psychobiology. Cambridge, UK.
26. ROSENFELD, P., J. EKSTRAND, E. OLSON, D. SUCHECKI & S. LEVINE. 1993. Maternal regulation of adrenocortical activity in the infant rat: effects of feeding. Dev. Psychobiol. **26**(5): 261–277.
27. DALLMAN, M. F., S. F. AKANA, C. S. CASCIO, D. N. DARLINGTON, L. JACOBSON & N. LEVIN. 1987. Regulation of ACTH secretion: variations on a theme of B. Recent Prog. Horm. Res. **43:** 113–173.
28. GRINO, M., J. M. BURGUNDER, R. L. ESKAY & L. E. EIDEN. 1989. Onset of glucocorticoid responsiveness of anterior pituitary corticotrophs during development is scheduled by corticotropin-releasing factor. Endocrinology **124:** 2686–2692.
29. HALL, W. G. & J. S.ROSENBLATT. 1977. Suckling behavior and intake control in the developing rat pup. J. Comp. Physiol. Psychol. **91:** 1232–1247.
30. SPEAR, L. P., S. M. SPECHT, C. L. KIRSTEIN & C. M. KUHN. 1989. Anterior and posterior, but not cheek, intraoral cannulation procedures elevate serum corticosterone levels in neonatal rat pups. Dev. Psychobiol. **22:** 401–411.
31. LEVINE, S., F. BERKENBOSCH, D. SUCHECKI & F. J. G. TILDERS. 1994. Pituitary-adrenal and interleukin-6 responses to recombinant interleukin-1 in neonatal rats. Psychoneuroendocrinology **19:** 143–153.
32. DALLMAN, M. F. 1984. Viewing the ventromedial hypothalamus from the adrenal gland. Am. J. Physiol. **15:** R1–R12.
33. PEDERSEN, R. C. & A. C. BROWNIE. 1987. Gamma$_3$-melanotropin promotes mitochondrial cholesterol accumulation in the rat adrenal cortex. Mol. Cell. Endocrinol. **50:** 149–156.
34. DOERING, C. H. & R. B. CLAYTON. 1969. Cholesterol side-chain cleavage activity in the adrenal gland of the young rat: development and responsiveness to adrenocorticotropic hormone. Endocrinology **85:** 500–511.
35. FEVOLD, H. R. 1983. Regulation of the adrenal and gonadal microsomal mixed function oxygenase of steroid hormone biosynthesis. Annu. Rev. Physiol. **45:** 19–36.
36. CHAUTARD, T., F. BOUDOURESQUE, V. GUILLAUME & C. OLIVER. 1993. Effect of excitatory amino acid on the hypothalamo-pituitary-adrenal axis in the rat during the stress-hyporesponsive period. Neuroendocrinology **57:** 70–78.
37. WALKER, C.-D., S. F. AKANA, C. S. CASCIO & M. DALLMAN. 1990. Adrenalectomy in the neonate: adult-like adrenocortical system responses to both removal and replacement of corticosterone. Endocrinology **127:** 832–842.
38. DE KLOET, E. R. 1991. Brain corticosteroid receptor balance and homeostatic control. Front. Neuroendocrinol. **13**(2): 95–164.
39. ROSENFELD, P., W. SUTANTO, S. LEVINE & E. R. DE KLOET. 1988. Ontogeny of type I and type II corticosteroid receptors in the rat hippocampus. Dev. Brain Res. **42:** 113–118.
40. WITEK-JANUSEK, L. 1988. Pituitary-adrenal response to bacterial endotoxin in developing rats. Am. J. Physiol. **255:** E525–E530.
41. O'GRADY, M. P., N. R. S. HALL & R. A. MENZIES. 1993. Interleukin-1β stimulates adrenocorticotropin and corticosterone release in 10-day-old rat pups. Psychoneuroendocrinology **18:** 241–247.
42. WALKER, C.-D., K. SCRIBNER, C. S. CASCIO & M. F. DALLMAN. 1991. The pituitary-adrenocortical system of neonatal rats is responsive to stress throughout development in a time-dependent and stressor-specific fashion. Endocrinology **128:** 1385–1395.
43. SAWCHENKO, P. E. The final common path: issues concerning the organization of central mechanisms controlling corticotropin secretion. *In* Stress: Neurobiology and Neuroendocrinology. M. R. Brown, G. F. Koob & C. Rivier, Eds. 55–71. Marcel Dekker, Inc. New York.

General Discussion

S. LEVINE (*Stanford University Medical School*): At some point, I would like to discuss the amount of licking contact, which is really a phenomenal regulatory aspect. It not only affects the growth hormone but is suppresses ACTH secretion so that it effects the whole HPA axis during the so-called SHRP. The licking is very selective to CRF and ACTH.

QUESTION: Have you looked at the peripheral vasopressin responses to some of your stressors, and if the ACTH and corticosteroid responses are suppressed in these animals, is there any sort of overall implication for the animals' survival of stress in the adult period?

ANSWER: No, we have not looked at vasopressin mechanisms much beyond what I showed you. In terms of implications, the one I would like to underscore, as Martin Dayley wrote in a review several years ago is that these animals are different. One is not more adaptive or better fit than the other, but they are quite different. Ideally you can think of a context in which one would be at an advantage with respect to a particular pathology. We did do some studies in collaboration with Robert Sapolsky a number of years ago in which we looked at glucocorticoid sensitive neuropathology in the hippocampus in later life. Our argument here was simple: that if negative feedback is enhanced in the handled animals then perhaps they would be able to ward off and regulate more tightly and efficiently pituitary-adrenal activity in the later phase of life and thus not see levels that affect hippocampal morphology. This is in fact exactly what we saw. The frequency of what we would now call "neurologically impaired aging" in the nonhandled animals is about 85 percent, whereas it was about 10 percent in the handled rats. The hippocampal morphology lined up rather nicely with that. So that was one scenario where you saw an advantage for the handled animals. Fair and Fantonie and Stanford Lightman and a number of people in discussions have suggested another scenario which might be quite different, related to, as Dr. Munck would suggest, the immunoregulatory activities of glucocorticoids. If you look at the fact that the glucocorticoid response is absolutely necessary for the survival of a particular immune challenge, then it may well be that the nonhandled animals would be better at surviving say an EAA challenge than the handled animals. I hope we will get enough animals to study that. Thus, you could probably think of circumstances under which both are advantageous. Dr. Dallman's group has found that, at least in an acute challenge, you do get greater suppression of plaque-forming cells and antibody responses in the nonhandled animals. So there are a number of aspects of physiology which are quite different in the animals. Whether the differences are related to pituitary-adrenal output is not entirely clear at this point.

LEVINE: There really is no evidence, at least from Dr. Meaney's work or any other work, that the handling effect is occurring through glucocorticoids. Also quite remarkable is that if you look at the pattern of development, at least in the rodent, there is a period in which the adrenal essentially becomes unresponsive. You can get it to respond, but in the normal state with its mother when the pup is adequately fed, you can give it gallons of ACTH and barely get it to budge. You can get an ACTH response with interleukin during that period, but no CORT response. The CORT response simply is absent. There are a lot of things that are downregulating, and feeding is clearly one, because if you remove food you can then get a CORT response. So the biosynthetic mechanisms are there, but there is something apparently adaptive or fundamental about maintaining low, stable CORT levels. They are stable, and there is no rhythmicity. They do not budge, for at least a period of about 10 days. If you look at Dr. Bohn's data on some of the detrimental effects of high glucocorticoids during this critical period of development, it makes sense that there are a lot of protective mechanisms.

QUESTION: A lot of biological changes that you have observed are very similar to what we have seen in posttraumatic stress disorder. Is early handling a stressful thing that has happened to these animals? In earlier studies it used to be called the stressor, and now people are backing away from that.

M. J. MEANEY (*McGill University*): I back away from the semantic argument that emerges, but let me draw one distinction. We have used at least two other manipulations and I am sure there are others which can be distinguished from handling in their consequences. We have used early endotoxin exposure so as to activate the immune system, with low doses, and a longer period of maternal separation of about 180 to 360 minutes. Both of those manipulations produce effects which are diametrically opposite to handling. They produce animals which even by comparison to nonhandled controls show exaggerated expression of CRH, exaggerated ACTH or increased ACTH during glucocorticoid responses. So at least functionally you can distinguish between what we are doing with the handling manipulation and intuitively what we may think of as more extreme forms of stress. Is it a stressor? I will let Dr. Levine handle that one.

LEVINE: I am older and wiser than you and I ducked it a long time ago. It is very hard to say, depending on which component of the system you look at. If you give a long period of maternal deprivation you will see a rise in glucocorticoids but you will not see a change in ACTH. Probably the rise in glucocorticoids is a function of a more sensitive adrenal. It becomes very difficult to deal with whether these are stress or not stress issues, partly because you have an animal that is unresponsive to begin with. It is hard to know what is changing acutely in these animals.

QUESTION: Is there any other manipulation that you know of that can be employed later in life in an adult animal that would cause the same kind of profile of an upregulation of glucocorticoid receptors?

ANSWER: Abdul Mohammad, Jonathan Seckl in Edinburgh, and Robert have been collaborating looking at the effects of enriched environments in the postweaning phase and have found increases at least in glucocorticoid receptor messenger RNA throughout the hippocampus. You see them at least in CA1 and you see them in other areas as well. That is one other later mainipulation that at least at the level of GR expression has comparable effects.

QUESTION: I was fascinated by the 5HT experiment in tissue culture. Can that be reproduced by cyclic AMP or do you know if the 5HT exposure for a couple of days in tissue culture leads to a permanent reset in the cyclic AMP level in that cell?

ANSWER: No, we do not know about that. We have not come back and looked at cyclic AMP mechanisms following 5HT exposure.

QUESTION: Have you been able to discriminate between turning on GR gene expression in a cell versus selecting for cells that simply have a normally higher rate of GR gene expression in these experiments? I do not know if there is any cell birth or cell death at this time. You probably do.

ANSWER: Yes, there is, as Dr. McEwen described, particularly within the granular cell layer. You see the effect throughout the pyramidal cell layer as well as the granular cell layer and those cells show very different patterns of neurogenesis, one being primarily prenatal, the pyramidal cells, and the other being somewhat more postnatal, the granular cells. So despite having very different neurogenic profiles, the effect seems to occur in both of them. The only other relevant thing I can tell you is that we know we are not changing the overall density of hippocampal neurons in any field with the handling manipulation, based on very gross neurodensity estimates. As to whether we may be selecting for the survival of a particular subpopulation, that is really intriguing, but I can tell you no more than that.

QUESTION: Given what Dr. Levine said, the other thing that is important to consider is the tissue-specific regulation of GR, as Dr. Akil showed, throughout the course of development. The one thing we need to keep in mind is that the adrenal is being actively inhibited, so presumably other tissues in the body are expressing very high levels of GR, including the pancreas, the pituitary and the lung. So these are highly sensitive tissues, whereas the hippocampus is being undersensitive. As your earlier work showed the activation of GR during this period would probably be somewhat detrimental, do you think that hippocampal GR is being actively inhibited during this period of time?

ANSWER: You mean, not the expression, but the function of GR.

QUESTION: No, I mean the expression. There is a tendency somehow to look at the type of developmental profile you showed and say that is an autogenetic pattern, that it starts off low and it gets high as if there was something programmed about cells to do that.

ANSWER: If it follows the *in vitro* data, you know you need glucocorticoids to increase expression of GR and MR in the hippocampal neurons. That would make sense because your glucocorticoid levels are low during that period.

QUESTION: But then, of course, GR expression in other tissues at the same time is pretty high.

ANSWER: There were some data a long time ago from Jean Lauder suggesting that serotonin plays a role in regulating the timing of neurogenesis. So if you have only very subtle changes in the timing of genesis of particular granular cells, that may change the synaptogenesis also.

QUESTION: Would you say that activating 5HT systems might have something to do with the differentiation of the hippocampal neurons as well? Could you be prematurely maturing these neurons?

M. C. BOHN (*University of Rochester Medical Center*): We are referring to the handling model. I cannot see where a change in that short a period of time is really going to have a long-lasting effect on synaptic plasticity. I think the plasticity is such a robust phenomenon that you may dampen it down for a while, but it is going to bounce right back.

RESPONSE: That may be indeed what is occurring here.

BOHN: In most developmental studies, you can see changes up to day 15, but by day 30 the system is back to normal. In Jean Lauder's report where she gave the PCPA, she retarded development, but I think by day 30 everything was back up to normal.

RESPONSE: The last cell division can be changed.

BOHN: Yes, but I think eventually everything gets put in the right place.

MEANEY: It is hard to know how that would change subtle aspects of differentiation. I was going to comment on your question of the possibly functional relevance of this low trough in glucocorticoid levels. The cell cycle is very sensitive to glucocorticoid levels and there is a lot of neurogenesis going on in that period, both in the cerebellum and the hippocampus. In our early studies when we raised glucocorticoid levels there were two things that happened: one, you can turn neurogenesis off early in some cells and then when you remove the steroids, you find that in some cells you have turned it off permanently and early, while others come back and rebound and are generated late. So that you change the temporal pattern over which certain neurons are generated. That may play a role. You may have to have very controlled levels of glucocorticoids as well as other hormones.

E. R. DE KLOET (*Sylvius Laboratories*): I would like to discuss the exclusiveness of the serotonin and hippocampal–frontal cortex-GR connection. I am doing this because in our studies where we use genetically selected lines for apomorphine responsiveness, we obviously have rat lines with different dopamine receptors. The surprising thing was that if you measure the receptors which you like so much, the glucocorticoid receptors, then the frontal cortex also showed an increase in glucocorticoid receptors in rats sensitive for dopamine. Also there were changes in the hippocampus and there was a change in pituitary-adrenal responsiveness throughout life which is remarkably similar to the one you were showing. I can imagine that if you had looked for dopamine effects, since there is a lot of data on dopamine regulation of glucorticoid receptors in neuroendocrine tissues like the pituitary, you may have gotten a story which looks remarkably similar, except that serotonin is exchanged for dopamine. I am saying this because there is some clinical literature on neurophychiatric disorder which very much focusses on the role of dopamine in later life and which refers this to early life events. So I am bringing this up as a comment, and would like to know your ideas on the exclusiveness of serotonin. I understand that you do not have data on this; otherwise you would have obviously showed us some slides.

ANSWER: Yes, I did leave some slides at home, you are absolutely right. Stephanie Maccari and others have data showing 6-hydroxydopamine lesions can regulate glucocorticoid receptor expression. There are at least two possibilities. First, it is clear that in the course of the handling manipulation, which is a fairly substantial thing to perform with young pups, you are altering more than pituitary, thyroid and serotinergic systems. You are probably activating other systems in addition to that. We have focussed on serotonin because it provides a way of looking at the more molecular mechanisms, but at no time do we delude ourselves into thinking it is the only neurotransmitter or neuromodulator that is relevant here. Second, there are probably many different ways of arriving at a similar neuroendocrine profile. I certainly saw the fascinating poster that your lab presented with the apomorphine-susceptible rats and immediately thought I have to go home and see how these handled and nonhandled animals respond to apomorphine. It is entirely feasible that there are many different environmental inputs which can act to produce a similar profile and by different neuromodulatory neurotransmitter systems. However, there is no doubt that other systems are involved in the handling as well.

QUESTION: Certainly in the hippocampus dopamine is not a major player. Maybe in frontal cortex but not in hippocampus.

ANSWER: But it would be a good candidate for the frontal cortex.

QUESTION: We need to remember that the rat may have a different physiology than many species, including the human being. We heard about the stress nonresponsive period and we have seen some very striking data; however, primates and human beings do not have a stress nonresponsive period.

ANSWER: No, I disagree with you. There is more and more evidence that primates and humans may indeed have a stress nonresponsive period.

RESPONSE: Right, but the postnatal first few days in the rat may be much more comparable to the last two or three weeks prenatally in the human being and some large mammals. We also need to remember that primates certainly are exposed prenatally to very high levels of endogenous corticosteroids due to the normal development of the hypothalamus-pituitary-adrenal axis; not only high levels of DHAS but also high levels of cortisol. Also we must not forget that clinically women are commonly treated with very, very high doses of glucocorticoids and an antenatal treatment for pulmonary development in babies that are destined to be born early. Therefore, though this is extremely important data, we need a word of caution that not everything in the physiology of the rat may be applicable to the human being.

QUESTION: You might have pointed out that you do have stress-specific changes after handling. Can you comment on the possible regional differences in the activation profiles of those various stressors that you find behave differently in this way?

ANSWER: Yes. The phrase we used to use was that handling alter HPA responses to a wide range of stressors. We now know thanks to Semous's work in our group that not all stressors, such as endotoxin exposure and cold exposure, elicit exactly the same HPA response from both groups of animals. Presumably this has something to do with reflecting the roles of the hippocampus and the frontal cortex under some conditions and perhaps different roles or perhaps not participating in regulating pituitary adrenal responses under other conditions.

The Physiological Relevance of Glucocorticoid Endangerment of the Hippocampus[a]

ROBERT M. SAPOLSKY

Department of Biological Sciences
Stanford University
Stanford, California 94305-5020

INTRODUCTION

A theme that permeates the papers in this volume is the deleterious nature of glucocorticoid (GC) excess. Sometimes the adverse effects of the steroid are implied indirectly. For example, when considering the extraordinary complexity of the regulation of GC secretion (see the paper by Dallman, for example), one is reminded of the need for such tight control—*i.e.,* the pathologic consequences when such regulation fails. Or one must consider the catabolic effects of GCs when asking what the ultimate purpose is of GC secretion in response to stress (as in the work of Munck).

Other papers consider the pathogenic consequences of GCs more directly. One of the most striking examples of such pathogenicity is the ability of GCs and of stress to damage the hippocampus, a structure in the limbic system of the brain that is critical to learning and memory. Moreover, GCs are able to endanger the hippocampus, impairing its capacity to survive a variety of coincident neurological insults. Such GC neurotoxicity and neuroendangerment have become relevant to understanding aspects of neural development (see the paper by Meaney), brain aging (see Landfield), a number of neuropsychiatric and endocrine disorders (see Gottfries), as well as some of the most common neurological diseases that damage the hippocampus.

In this paper, I review the current knowledge concerning the cellular and molecular mechanisms by which GCs endanger the hippocampus. Most of these findings emerged only just recently, and they make it possible to begin asking some rather subtle questions about the phenomenon. I shall emphasize two such questions: a) Do GC endanger the hippocampus under physiological, as well as pharmacological circumstances? b) Are the mechanisms by which GCs act over the course of hours to days to endanger the hippocampus during a neurological insult related to the mechanisms by which GCs act over the course of weeks to months to directly damage the aging hippocampus?

Glucocorticoid Neurotoxicity and Neuroendangerment

The first experimental demonstration of GC neurotoxicity occurred nearly thirty years ago, with the report that pharmacological concentrations of the steroid would damage the central nervous system, with preferential injury to the hippocampus.[1] It

[a]These studies were made possible by an RCDA Grant and an RO1 Grant from the National Institute on Aging and by grants from the Adler Foundation and the Alzheimer's Association.

was approximately a decade later that the first hints emerged that this phenomenon was of physiological relevance. Circulating GC concentrations tend to rise with age in the rat (reviewed in Ref. 2), and Landfield and colleagues[3] showed that the magnitude of GC hypersecretion in old age predicted the extent of hippocampal degeneration. Following that, a series of studies more directly implicated GC exposure as a pacemaker of hippocampal aging.

First, it was shown that prolonged diminution of GC exposure decelerated hippocampal aging. In one variant of this, Landfield *et al.* adrenalectomized mid-aged (12-month) rats and maintained them on minimal GC replacement thereafter. They observed that such rats were spared the neuron loss and compensatory glial reactivity and infiltration typical of aging in the hippocampus, as well as some of the hippocampal-dependent cognitive deficits.[4] In work inspired by these observations, Meaney and colleagues showed that behavioral manipulations which lowered cumulative GC exposure were neuroprotective as well. Specifically, they took advantage of the well-established phenomenon in developmental psychobiology, "neonatal handling."[5,6] A prior literature had shown that a particular regime of daily handling of neonatal rats caused permanent reductions in GC secretion in the adult (cf Ref. 7). It was shown that a consequence of this imprinting was decelerated hippocampal aging. Neonatally handled rats, when aged, were spared the neuron loss, cognitive deficits and neuroendocrine abnormalities typical of hippocampal aging.

During that same period, a body of studies showed the converse of this protection, namely, that elevated exposure to GCs accelerated many of the indices of hippocampal aging. First it was shown that elevation of GC concentrations into the upper physiological range (*i.e.*, levels typical of major stressors) for three months caused selective loss of pyramidal neurons in the Ammon's horn, accompanied by glial hypertrophy,[8] patterns identical to that seen in normal aging of the rat hippocampus (reviewed in Ref. 9). If stress levels of exogenous GCs were damaging, stress should be as well, and in an important follow-up, it was shown that prolonged stressors caused similar degeneration as well, in this case including accelerated electrophysiological indices of aging.[10] Importantly, the older the rat, the more vulnerable it was to such stress-induced damage. A series of more recent studies have replicated the general features of stress and/or GCs damaging the rodent hippocampus,[11,17] including the enhanced vulnerability to the phenomenon with age. Notably, they have shown that as little as three weeks of elevated GC exposure will cause what might arguably be the first steps in the process, namely, atrophy of dendritic processes in pyramidal neurons.[14,16]

Following these early studies, it was hypothesized that under circumstances where exposure to GCs was not yet sufficient to be directly toxic, there would instead be "endangerment" of the hippocampus, making its neurons less likely to survive coincident challenges. This idea generated the prediction that neurologic insults to the hippocampus should become more damaging when occurring in the presence of increased GC concentrations.[18] This has now been shown to be the case for a variety of insults by close to a dozen independent groups. One striking example is hypoxia-ischemia. Transient cessation of blood flow to the brain (following cardiac arrest) causes selective damage to the CA1 region of the hippocampus, and it was shown that the magnitude of such damage in the adult was increased by GC exposure around the time of the ischemia and decreased by adrenalectomy;[19-23] in contrast, GCs have fairly opposite effects with neonatal hypoxia-ischemia, which differs from the adult model in a number of ways.[24,25] As another example, status epilepticus seizures preferentially damage the CA3 region of the hippocampus and again, toxicity is modulated by GCs.[17,18,26,27] Hypoglycemia and or administration of an antimetabolite neurotoxin both damage the dentate gyrus region, with damage being sensitive to GC presence.[18,28] Finally, GCs worsen the neuropathology induced by sodium azide,[29] oxygen radical generators such as paraquat[30] or adriamycin (unpublished), glial-specific toxins,[31] NMDA,[32] cholinergic[33] or serotomergic toxins.[34] In some of these models, the extent of damage could be modulated an order of magnitude by GC manipulation, attesting to the potency of this effect.

The Endocrine and Metabolic Features of Glucocorticoid Endangerment

Thus, GCs can exacerbate the toxicity of a variety of hippocampal insults, suggesting that the steroid-induced endangerment is a broad one. Subsequent work turned to undercovering the underlying mechanisms. In an important advance, it was shown that GCs could endanger hippocampal neurons *in vitro* in primary neuronal tissue culture. Specifically, the *in vitro* toxicity of a number of excitotoxins, of aglycemia and anoxia were shown to be exacerbated by exposure to high nM to uM GC concentrations for a few days bracketing the *in vitro* insult.[30,35,36] These findings were important for two reasons. First, by establishing the endangerment *in vitro,* it made possible some of the more reductionist studies needed to understand the mechanisms of the endangerment. Second, and of great heuristic value, the occurrence of such endangerment in the absence of the rest of the brain, the cerebrovascular circulation, peripheral organs, and so on, demonstrated that the endangerment was not secondary to one of the vast number of peripheral GC actions. Instead, the endangerment appeared to be direct.

Subsequent *in vitro* work defined some of the parameters of the endangerment. GCs did not increase the toxicity of insults to cerebellar, cortical or hypothalamic cultures,[30] supporting the *in vivo* observations that GC endangerment was either exclusive to or most pronounced in the hippocampus.[18,19] Furthermore, non-GC steroids such as estrogens, progestins or androgens did not exacerbate damage to hippocampal cultures.[37] The tissue- and steroid-specificity suggested a classical, receptor-mediated effect; in support of this, the GC endangerment could be blocked by antagonists to, and mimicked by agonists to the type II GC receptor.[15,20,23,32,37]

Both *in vivo* and *in vitro* studies have indicated that the GC endangerment is strongly related to disruptive effects of GCs upon hippocampal energetics. The first evidence for this was indirect.[38] A few paragraphs above, it was emphasized how varied the different neurological insults are whose hippocampal toxicities are sensitive to GCs. Yet, they all share a common theme, namely, of constituting energy crises for brain tissue. Such insults either involve disruption of energy production (*e.g.,* the cessation of delivery of oxygen and/or glucose to neurons) or pathologic elevation of energy utilization (*e.g.,* the demands of repeated seizure activity, the costs of repair of oxidative damage). All involve profound suppression of levels of high-energy substrates in brain tissue (reviewed in Refs. 39–41). It was hypothesized that GCs were able to potentiate the damage of all of these varied insults by inducing some manner of broad energetic vulnerability in the hippocampus.[38] This generated a simple prediction: if the GC endangerment was energetic in nature, supplementing neurons with excess energy should "substract" out the endangerment. It has now been shown both in vivo and in vitro that glucose, mannose or ketones can protect hippocampal neurons from GC endangerment.[30,35,38]

Subsequent work has shown some of the underpinnings of the energetic disruption. Glucocorticoids suppress glycogen levels in the hippocampus.[35] While glycogen storage in the brain is overwhelmingly glial, such a depletion is of considerable significance for neurons, given that glycogenolysis makes possible the fueling of neuronal metabolism by glial transfer of lactate.[42] Furthermore, GCs make hippocampal ATP concentrations and cytochrome oxidase activity more vulnerable to insults.[29,43,44] This last finding is subtle and informative. GC administration does not suppress basal ATP concentrations in cultured hippocampal neurons or in glia, nor does it alter cytochrome oxidase activity. Instead, the decline in the levels of ATP and in the activity of the enzyme that follows various insults is accelerated or magnified by GCs. This seems the metabolic equivalent of GCs endangering, rather than damaging outright.

Recent work has identified a mechanism by which GCs might cause this energetic vulnerability. GCs have long been recognized for their capacity to inhibit glucose transport in various peripheral tissues. This is accomplished by GC-induced translocation of glucose transporter molecules from the cell surface[45] and, over a longer time span, decreases in levels of mRNA for the transporter.[46] These actions are one way by which GCs halt superfluous energy storage in peripheral tissues during a physical stressor (along with mobilizing energy already stored from such sites).[47] Reminiscent of those peripheral actions, GCs inhibit local cerebral glucose utilization in the hippocampus,[48,49] and inhibit glucose uptake in cultured neurons and glia from the hippocampus.[31,50] Such inhibition shows the same receptor and steroid specificity as does the endangerment.[50] Furthermore, reiterating the theme of GCs inducing a vulnerability, rather than damaging outright, the inhibition of transport is relatively mild (on the order of 25–30%, in contrast to the 75% inhibition that occurs in some peripheral tissues).

Glucocorticoids and Excitotoxic Cascades in the Hippocampus

Insofar as GCs induce an energetic vulnerability in the hippocampus, they could potentially worsen any of the vast number of perturbations caused by insults such as hypoxia-ischemia, seizure or hypoglycemia. Recent work suggests that the GC endangerment centers on the effects of the steroids on the trafficking of glutamate and calcium in the hippocampus following an insult.

Excitatory amino acids (EAAs) such as glutamate and aspartate are the predominate excitatory neurotransmitters in the brain. EAAs are particularly important in the hippocampus, where the potent and nonlinear features of their excitation are critical to long-term potentiation. All of the major necrotic insults to the hippocampus cause loss of control over the trafficking of EAAs, producing elevated concentrations of them in the synapse. EAAs, via interaction with both NMDA and non-NMDA ionotropic receptors as well as with metabotropic receptors, mobilize free cytosolic calcium in the post-synaptic neuron. Such mobilization is normally tightly-regulated and critical to EAA-induced plasticity; during a necrotic insult, the pathologic accumulation of synaptic EAAs leads to a pathologic overabundance of free cytosolic calcium. This, in turn, causes an array of calcium-dependent events throughout the neuron, including activation of proteases, lipases and nucleases; the most damaging consequence of this probably is the subsequent generation of oxygen radicals (reviewed in Refs. 39–41, 51). The elucidation of this pathway is, in my opinion, the major achievement in neuropathology in the last decade.

Critically, the potency of this pathway is highly sensitive to energy (reviewed in Refs. 39–41). With energy depletion, neurons will depolarize, enhancing EAA release. Furthermore, removal of EAAs from the synapse depends on costly high-affinity uptake pumps, and these fail or even reverse during an energy crisis. Thus, EAA accumulation in the synpase is greater and more persistent. This produces a greater mobilization of post-synaptic calcium, worsened by depolarization-induced opening of voltage-gated calcium channels and release of NMDA-gated channels from voltage-dependent magnesium blockades. Furthermore, all of the mechanisms for calcium sequestering or extrusion are disrupted during energy deprivation, producing waves of calcium that are thus larger and more persistent. As the most striking manifestation of the energy-dependency of the EAA/calcium cascade of damage, energy availability determines whether EAAs act as neurotransmitters or neurotoxins.[52–56]

These observations generate an obvious prediction. To the extent that GCs exacerbate the toxicity of insults that activate this cascade, that the potency of this cascade is energy-dependent, and that the GC exacerbation revolves around producing an energetic vulnerability in the hippocampus, GCs should worsen some of the steps in this cascade. This has now been shown to be the case. As indirect evidence, GC endangerment is blocked with NMDA receptor antagonists.[57] Furthermore, GCs contribute to stress- and seizure-induced activation of glutamatergic events in the hippocampus.[58] A body of recent studies show this GC interaction with the EAA/calcium cascade even more explicitly:

First, GCs increase the extracellular levels of EAAs in the hippocampus after a seizure,[59,60] and enhance cyanide-induced extracellular accumulation of EAAs in hippocampal cultures.[61] This GC effect seems to be due to impaired removal of EAAs from the synapse, rather than enhanced release.[62] Furthermore, stress causes EAA accumulation in the hippocampus.[62] This effect is prevented by adrenalectomy[63] and reinstated with GCs.[64]

Second, as would thus be expected, GCs increase the magnitude and duration of calcium mobilized in hippocampal neurons in response to EAAs.[65,66] This has been shown with calcium-sensitive dyes *in vitro.* Moreover, GCs enhance calcium-dependent currents in hippocampal neurons.[67] Studies of the kinetics of the calcium mobilization suggest indirectly that the GC effect is more pronounced in disrupting calcium removal, rather than enhancing calcium influx.[66]

Third, not surprisingly, GCs also worsen calcium-dependent degenerative events in the hippocampus following EAA exposure. The two endpoints examined have been proteolysis of the cytoskeletal protein spectrin, and accumulation of the abnormally phosphorylated tau microtubule protein.[17,68]

As evidence of the energetic nature of these GC actions, the hormone's effects on EAA accumulation, on calcium mobilization, and on calcium-dependent degenerative events can be prevented by energy supplementation.[59,65,68]

Thus, by inducing a mild energy vulnerability in neurons at the time of a major and costly necrotic crisis, GCs impair the capacity of the hippocampus to contain potentially damaging EAAs and calcium. It is worth noting that this model does not require that inhibition of glucose uptake be the sole route by which GCs induce an energetic vulnerability, nor that the vulnerability induced by them be solely energetic in nature.[41] At present, however, this appears to be the most plausible sequence of events underlying the GC endangerment. This model also predicts that GCs should also impair the capacity of the hippocampus to survive insults which, while not involving the EAA/calcium cascade, are nonetheless energetic crises. This has been shown.[33]

Do Physiologic Concentrations of Glucocorticoids Endanger the Hippocampus?

Should the findings regarding GC endangerment in the rodent hippocampus apply to the human, they suggest some immediate implications concerning the pharmacologic administration of GCs. Exogenous GCs are often used to control post-stroke edema. This was prompted by the great efficacy with which GCs inhibit tumor edema in the brain. In contrast, thirty years of research has shown that GCs are generally less effective at lessening post-stroke edema than are nonsteroidal alternatives (reviewed in Ref. 41, chapter 11). The present findings caution further against the use of GCs at such times and may explain the observation that GCs often worsen post-stroke sequelae when given to control edema. Furthermore, these findings suggest caution in the use of GCs for any acute neurological insult involving the hippocampus. These data also suggest that hippocampal damage may ensue when endogenous secretion of GCs rises to the pharmacological range (*i.e.,* during Cushing's syndrome). In support of this, Cushingoid patients have selective hippocampal atrophy, as assessed by MRI, and there is a positive correlation between circulating cortisol concentrations and the extent of atrophy.[70]

But are these findings concerning GC endangerment of physiological relevance? Humans and experimental animals who have suffered a necrotic neurological crisis typically secrete massive amounts of GCs.[71–74] Separate from the issue of the exogenous administration of GCs, is such endogenous, physiological secretion sufficient to increase hippocampal damage? In effect, is what we have come to view as "typical" hippocampal damage following a neurologic crisis damage due in part to the adrenocortical stress-response triggered at such times?

Apparently, this is the case. As the most direct evidence for this, adrenalectomy immediately following a necrotic insult reduces the subsequent hippocampal damage (cf Ref. 19). Moreover, similar protection can be shown with chemical, rather than surgical adrenalectomy, using the adrenocortical steroidogenesis inhibitor metyrapone.[22,71]

A number of additional studies suggest that the GC endangerment is a physiological as well as pharmacological phenomenon. Many of the demonstrations that GCs exacerbate the toxicity of hippocampal insults utilized GC concentrations within the physiological range (cf Ref. 18), employed relatively brief periods of GC exposure,[28] or showed stress itself to be endangering.[17] Moreover, many of the studies showing that GCs worsen steps in the EAA/calcium have utilized physiological GC concentrations.[59,68]

How little of an elevation of GC exposure is required to exact a neuropathologic price? Surprisingly little, apparently. In a pair of recent studies, rats were adrenalectomized and replaced with corticosterone pellets with a variety of secretion rates. This generated rats with stable circulating GC concentrations over the entire physiologic range. In one study, rats were subjected to kainic-acid induced status epilepticus seizures while extracellular EAA concentrations were measured by microdialysis.[60] In the second study, the extent of tau immunoreactivity and spectrin proteolysis were quantified following kainic acid seizures.[17] In both cases, there was a dose-response relationship between increments in GC exposure and worsening of the neuropathological endpoint. To give some sense of this relationship, raising of GC concentrations from the low basal to high basal range caused an approximate doubling of glutamate accumulation, while elevation of GC concentrations into the stress range caused an 4-fold increase in EAA accumulation.[60]

Are the Mechanisms That Underlie Glucocorticoid Endangerment Related to Glucocorticoid Toxicity?

The recent literature concerning GC effects on hippocampal energetics and on trafficking of EAAs and calcium might well explain how GCs, over the course of hours to days, augment the toxicity of massive and acute insults to the hippocampus.

But do these GC actions help explain how the steroids can directly damage the hippocampus over the course of weeks to months?

This is an extremely difficult question, but an essential one to explore if we are to understand GC actions in the brain in the context of neurogerontology, rather than neurology. A recent experiment suggests that the progressive and direct ability of GCs to damage the hippocampus involves this same EAA/calcium cascade. As discussed previously, three weeks of stress or GC exposure will cause atrophy of dendritic processes in the pyramidal neurons.[14,16] In an important follow-up, Watanabe *et al.*[69] have shown that this damage can be prevented by the antiseizure drug phenytoin, which prevents EAA release.

How might direct GC toxicity over time involve the EAA/calcium cascade? One possibility is that there are microinsults occurring with some regularity, and with their mild toxicities being augmented by GCs. An additional or alternative possibility is that GCs can activate the EAA/calcium cascade in the absence of a coincident neurological insult, mild or otherwise.

The first possibility, of GCs exacerbating the toxicities of microinsults would, in effect, be the story of GC endangerment writ small but repeatedly. This goes to the heart of many facets of gerontology research, in which investigators seek to understand how much a senescent impairment is due to true intrinsic decline versus the cumulative effects of repeated exogenous insults over the lifetime. Current knowledge about the neurology of aging suggests that repeated microinsults do occur over time. For example, in cases of cerebrovascular disease, small infarcts occur frequently and are often impossible to detect at the time, appearing afterward as neuropsychological impairments or foci of necrosis in MRIs. Furthermore, as scarification accumulates around necrotic lesions, small foci of transient epileptiform activity often emerge. In theory then, should repeated microinsults be occurring over time and, in each case, incurring a small degree of neuron loss, GCs might well be exacerbating this process. This is obviously a difficult arena in which to carry out experiments. One must either be able to detect these mild neurologic insults, or impose them repeatedly, all the while studying the effects of GCs upon the subsequent slow emergence of damage; such experiments are in progress.

As the second possibility, GCs might activate the EAA/calcium cascade in the absence of a coincident insult. In the demonstrations that GCs enhance EAA accumulation in the rat hippocampus following kainic acid seizures, it was also shown that glutamate concentrations were already elevated by GCs *prior* to the kainic acid infusion.[59] Moreover, the studies showing that GCs increase and prolong the wave of free cytosolic calcium provoked by kainic acid also showed GC-induced increases in basal, pre-kainic acid calcium concentrations.[65] Moreover, the studies showing that GCs enhance calcium currents in the hippocampus were carried out in the absence of a coincident necrotic insult.[67]

Thus, GCs activate the EAA/calcium cascade, even without an accompanying insult. However, technical problems limit this interpretation. First, one must voice the obligatory but very valid caveat anytime *in vitro* studies are carried out. The studies of calcium currents[67] were carried out with acutely prepared hippocampal slices which, by definition, have recently undergone the neuropathologic insult of having been generated from a living animal. Moreover, the studies concerning calcium concentrations[65] were carried out with long-term primary hippocampal cultures, once again far from a physiological preparation. The studies showing that GCs augment extracellular EAA concentrations even prior to the insult of kainic acid exposure[59] were, nonetheless, carried out shortly after the insult of inserting a massive microdialysis probe into the hippocampus. As an alternative strategy, microdialysis probes can be inserted permanently and animals allowed to recover from the acute trauma of the surgery. In one such study, stress levels of GCs no longer elevated "pre-insult" EAA concentrations.[64] In contrast, in the other relevant study, stress alone elevated EAA concentrations.[58] Thus, because of the invasive nature of most techniques needed to study the EAA/calcium cascade, it has been difficult to determine whether GCs or stress activate this cascade in the absence of a coincident insult, with the minimal amount of data currently available being inconclusive.

CONCLUSIONS

Synthetic GCs are prescribed in vast quantities, among other reasons, for the treatment of asthma or arthritis, the suppression of autoimmunity, or to prevent rejection of transplanted tissue. Their beneficial effects in those cases are considerable but, unfortunately, their deleterious side effects can also often be numerous. Those side effects have long been recognized to include altering of cognition and affect, and work in the last decade suggests that sustained exposure to exogenous GCs might also cause or worsen degeneration in the hippocampus. The recent work reviewed here suggests that some of the mechanisms thought to underlie these damaging GC actions might be activated under physiological circumstances, even by only moderate elevations in GC concentrations.

REFERENCES

1. Aus der Muhlen, K. & H. Ockenfels. 1979. Morphologische Veranderung en im Diencephalon und Telencephalon nach Storngen des regelkreises Adenohypophyse-Nebennierenrinde III. Ergebnisse beim Meerschweinchen nach Verabreichung von Cortison und hydrocortison. Z. Zellforsch. **93:** 126–133.
2. Sapolsky, R. 1991. Do glucocorticoid concentrations rise with age in the rat? Neurobiol. Aging **13:** 171–175.
3. Landfield, P. J. Waymire & G. Lynch. 1978. Hippocampal aging and adrenocorticoids: a quantitative correlation. Science **202:** 1098–1101.
4. Landfield, P., R. Baskin & T. Pitler. 1981. Brain-aging correlates; retardation by hormonal-pharmacological treatments. Science **214:** 581–585.
5. Meaney, M., D. Aitken, S. Bhatnager, C. van Berkel & R. Sapolsky. 1988. Effects of neonatal handling on age-related impairments associated with the hippocampus. Science **239:** 766–769.
6. Meaney, M. D. Aitken & R. Sapolsky. 1991. Postnatal handling attenuates neuroendocrine, anatomical and cognitive dysfunctions associated with aging in female rats. Neurobiol. Aging 12, 31–40.
7. Levine, S. 1962. Plasma-free corticosteroid response to electric shock in rats stimulated in infancy. Science **135:** 795–799.
8. Sapolsky, R., L. Krey & B. McEwen. 1985. Prolonged glucocorticoid exposure reduces hippocampal neuron number; implications for aging. J. Neurosci. **5:** 1221–1227.
9. Coleman, P. & D. Flood. 1987. Neuron numbers and dendritic extent in normal aging and Alzheimer's disease. Neurobiol. Aging **8:** 521–543.
10. Kerr, D., L. Campbell, M. Applegate, A. Brodish & P. Landfield. 1991. Chronic stress-induced acceleration of electrophysiologic and morphometric biomarkers of hippocampal aging. J. Neurosci. **11:** 1316–1321.
11. Levy, A., S. Dachir, I. Arbel & T. Kadar. 1994. Aging, stress and cognitive function. Ann. N. Y. Acad. Sci. In press.
12. Levy, A., S. Dachir, T. Kadar, B. Shukitt-Hale, M. Stillman & H. Lieberman. 1993. Nimodipine, stress, and central cholinergic function. Drugs Dev. **2:** 223–235.

13. Talmi, M., E. Carlier & B. Soumireu-Mourat. 1993. Similar effects of aging and corticosterone treatment on mouse hippocampal function. Neurobiol. Aging **14:** 239–245.
14. Woolley, C., E. Gould & B. McEwen. 1990. Exposure to excess glucocorticoids alters dendritic morphology of adult hippocampal pyramidal neurons. Brain Res. **531:** 225.
15. Mizoguchi, K., T. Kunishita, D. Chui & T. Tabira. 1992. Stress induces neuronal death in the hippocampus of castrated rats. Neurosci. Lett. **138:** 157–163.
16. Watanabe, Y., E. Gould & B. McEwen. 1992. Stress induces atrophy of apical dendrites of hippocampal CA3 neurons. Hippocampus **2:** 431–439.
17. Stein-Behrens, B., M. Mattson, I. Chang, M. Yeh & R. Sapolsky. 1994. Stress exacerbates neuron loss and cytoskeletal pathology in the hippocampus. J. Neurosci. In press.
18. Sapolsky, R. 1985. A mechanism for glucorticoid toxicity in the hippocampus; increased neuronal vulnerability to metabolic insults. J. Neurosci. **5:** 1228–1234.
19. Sapolsky, R. & W. Pulsinelli 1985. Glucocorticoids potentiate ischemic injury to neurons: therapeutic implications. Science **229:** 1397–1401.
20. Koide, T., T. Wieloch & B. Siesjo. 1986. Chronic dexmethasone pretreatment aggravates ischemic neuronal necrosis. J. Cereb. Blood Flow Metab. **6:** 395–408.
21. Miller, G. & J. Davis. 1991. Post-ischemic surge in corticosteroids aggravates ischemic damage to gerbil CA1 pyramidal cells. Soc. Neurosci. Abstr. **17:** 302.4.
22. Morse, J. & J. Davis. 1990. Regulation of ischemic hippocampal damage in the gerbil: adrenalectomy alters the rate of CA1 cell disappearance. Exp. Neurol. **110:** 86–94.
23. Hall, E. 1990. Steroids and neuronal destruction or stabilization. *In* Steroids and Neuronal Activity. Ciba Foundation Symposium 153. 206–215. Wiley. Chichester.
24. Barks, J., M. Post & U. Tuor. 1991. Dexamethasone prevents hypoxic-ischemic brain damage in the neonatal rat. Pediatr. Res. **29:** 558–563.
25. Tuor, U., C. Simone, R. Arellano, K. Tanswell & M. Post. 1993. Glucocorticoid prevention of neonatal hypoxic-ischemic damage; role of hyperglycemia and antioxidant enzymes. Brain Res. **604:** 65–72.
26. Sapolsky, R. 1985. Glucocorticoid toxicity in the hippocampus: temporal aspects of neuronal vulnerability. Brain Res. **339:** 300–307.
27. Theoret, Y., J. Caldwell-Kenkel & M. Krigman. 1985. The role of neuronal metabolic insult in organometal neurotoxicity. Toxicologist **6:** Abstract 491.
28. Sapolsky, R. 1986. Glucocorticoid toxicity in the hippocampus: synergy with an excitotoxin. Neuroendocrinology **43:** 440.
29. Bennett, M., J. Lehman, G. Mlady & G. Rose. 1993. Synergy between corticosterone and sodium azide in producing a place learning deficit in the Morris water maze. Soc. Neurosci. Abstr. **19:** 78.12.
30. Sapolsky, R., D. Packan & W. Vale. 1988. Glucocorticoid toxicity in the hippocampus: *in vitro* demonstration. Brain Res. **453:** 367–373.
31. Virgin, C., T. Ha, D. Packan, G. Tombaugh, S. Yang & R. Sapolsky. 1991. Glucocorticoids inhibit glucose transport and glutamate uptake in hippocampal astrocytes: implications for glucorticoid neurotoxicity. J. Neurochem. **57:** 1422–1428.
32. Supko, D. & M. Johnston. 1994. Dexamethasone enhances NMDA receptor-mediated injury in the postnatal rat. Eur. J. Pharmacol. **270:** 105–113.
33. Hortnagl, H. M. Berger, L. Havelec & O. Hornykiewicz. 1993. Role of glucocorticoids in the cholinergic degeneration in rat hippocampus induced by ethylcholine aziridinum (AF64A). J. Neurosci. **13:** 2939–2945.
34. Johnson, M., D. Stone, L. Bush, G. Hanson & J. Gibb. 1989. Glucocorticoids and 3,4-methylenedioxymethamphetamine (MDMA)-induced neurotoxicity. Eur. J. Pharmacol. **161:** 181–187.
35. Tombaugh, G., S. Yang, R. Swanson & R. Sapolsky. 1992. Glucocorticoids exacerbate hypoxic and hypoglycemic hippocampal injury *in vitro;* biochemical correlates and a role for astrocytes. J. Neurochem. **59:** 137–143.
36. Tombaugh, G. & R. Sapolsky. 1993. Endocrine features of glucocorticoid endangerment in hippocampal astrocytes. Neuroendocrinology **57:** 7–11.

37. PACKAN, D. & R. SAPOLSKY. 1990. Glucocorticoid endangerment of the hippocampus: tissue, steroid and receptor specificity. Neuroendocrinology **51:** 613–620.
38. SAPOLSKY, R. 1986. Glucocorticoid toxicity in the hippocampus: reversal by supplementation with brain fuels. J. Neurosci. **6:** 2240–2248.
39. AUER, R., & B. SIESJO. 1988. Biological differences between ischemia, hypoglycemia, and epilepsy. Ann. Neurol. **24:** 699–715.
40. BEAL, M. 1992. Does impairment of energy metabolism result in excitotoxic neuronal death in neurodegenerative illnesses? Ann. Neurol. **31:** 119–130.
41. SAPOLSKY, R. 1992. Stress, the Aging Brain, and the Mechanisms of Neuron Death. MIT Press. Cambridge, MA.
42. DRINGEN, R., R. GEBHARDT & B. HAMPRECHT. 1993. Glycogen in astrocytes: possible function as lactate supply for neighboring cells. Brain Res. **623:** 208–214.
43. TOMBAUGH, G. & R. SAPOLSKY. 1992. Corticosterone accelerates hypoxia-induced ATP loss in cultured hippocampal astrocytes. Brain Res. **588:** 154–159.
44. LAWRENCE, M. & R. SAPOLSKY. 1994. Glucocorticoids accelerate ATP loss cultured hippocampal neurons. Brain Res. **646:** 303–308.
45. HORNER, H., MUNCK A. & G. LIENHARD. 1987. Dexamethasone causes translocation of glucose transporters from the polasma membrane of an intracellular site in human fibroblasts. J. Biol. Chem. **262:** 17696–17703.
46. GARVEY, W., T. HUECKSTEADT, F. LIMA & M. BIRNBAUM. 1989. Expression of a glucose transporter gene cloned from brain in cellular models of insulin resistance: dexamethasone decreases transporter mRNA in primary cultured adipocytes. Mol. Endocrinol. **3:** 1132–1138.
47. MUNCK, A. 1971. Glucocorticoid inhibition of glucose uptake by peripheral tissues. Old and new evidence, molecular mechanisms and physiological significance. Perspect. Biol. Med. **14:** 265–283.
48. KADEKARO, M., I. MASANORI & P. GROSS. 1988. Local cerebral glucose utilization is increased in acutely adrenalectomized rats. Neuroendocrinology **47:** 329–335.
49. FREO, U., H. HOLLOWAY, K. KALOGERAS, S. RAPOPORT & T. SONCRANT. 1992. Adrenalectomy or metyrapone-pretreatment abolishes cerebral metabolic responses to the serotonin agonist 1-(2,5-dimethoxy-4-iodophenyl)-2-aminopropane (DOI) in the hippocampus. Brain Res. **586:** 256–264.
50. HORNER, H., D. PACKAN & R. SAPOLSKY. 1990. Glucocorticoids inhibit glucose transport in cultured hippocampal neurons and glia. Neuroendocrinology **52:** 57–63.
51. CHOI, D. 1994. Excitotoxic neuron death. Semin. Neurosci. In press.
52. LYSKO, P., J. COX, M. ALESSANDRA & R. HENNEBERRY. 1989. Excitatory amino acid neurotoxicity at the NMDA receptor in cultured neurons: pharmacological characterization. Brain Res. **499:** 258.
53. NOVELLI, A. J. REILLY, P. LYSKO & R. HENNEBERRY. 1988. Glutamate becomes neurotoxic via the NMDA receptor when intracellular energy levels are reduced. Brain Res. **451:** 205.
54. COX, J., P. LYSKO & R. HENNEBERRY. 1989. Excitatory amino acid neurotoxicity at the NMDA receptor in cultured neurons: role of the voltage-dependent magnesium block. Brain Res. **499:** 267.
55. ZEEVALK, G., & W. NICKLAS. 1991. Mechanisms underlying initiation of excitotoxicity associated with metabolic inhibition. J. Pharmacol. Exp. Ther. **257:** 870.
56. SIMPSON, J. & O. ISACSON. 1993. Mitochondrial impairment reduces the threshold for *in vivo* NMDA-mediated neuronal death in the striatum. Exp. Neurol. **121:** 57.
57. ARMANINI, M., C. HUTCHINS, B. STEIN & R. SAPOLSKY. 1990. Glucocorticoid endangerment of hippocampal neurons is NMDA-receptor mediated. Brain Res. **532:** 7–15.
58. KRUGERS, H. D. JAARSMA & J. KORF. 1992. Rat hippocampal lactate efflux during electroconvulsive shock or stress is differentially dependent on entorhinal cortex and adrenal integrity. J. Neurochem. **58:** 826.
59. STEIN-BEHRENS, B., E. ELLIOTT, C. MILLER, J. SCHILLING, R. NEWCOMBE & R. SAPOLSKY. 1992. Glucocorticoids exacerbate kainic acid-induced extracellular accumulation of excitatory amino acids in the rat hippocampus. J. Neurochem. **58:** 1730–1735.
60. STEIN-BEHRENS, B., W. LIN & R. SAPOLSKY. 1994. Physiological elevations of glucocor-

ticoids potentiate glutamate accumulation in the hippocampus. J. Neurochem. **63:** 596–603.
61. Chou, Y., W. Lin & R. Sapolsky. 1994. Glucocorticoids exacerbate cyanide-induce asparate accumulation in hippocampal cultures. Brain Res. In press.
62. Moghaddam, B. 1993. Stress preferentially increases extraneuronal levels of excitatory amino acids in the prefrontal cortex: comparison to hippocampus and basal ganglia. J. Neurochem. **60:** 1650–1657.
63. Lowy, M., L. Gault & B. Yamamoto. 1993. Adrenalectomy attenuates stress-induced elevations in extracellular glutamate concentrations in the hippocampus. J. Neurochem. **61:** 1957.
64. Moghaddam, B., M. Boliano, B. Stein-Behrens & R. Sapolsky. 1994. Glucocorticoids mediate the stress-induced accumulation of extracellular glutamate. Brain Res. In press.
65. Elliott, E. & R. Sapolsky. 1993. Corticosterone impairs hippocampal neuronal calcium regulation: possible mediating mechanisms. Brain Res. **602:** 84–90.
66. Elliott, E. & R. Sapolsky. 1992. Corticosteroid enhances kainic acid-induced calcium mobiliztion in cultured hippocampal neurons. J. Neurochem. **59:** 1033–1038.
67. Kerr, D., L. Campbell, O. Thibault & P. Landfield. 1992. Hippocampal glucocorticoid receptor activation enhances voltage-dependent calcium conductances: relevance to brain aging. Proc. Natl., Acad. Sci. USA **89:** 8527–8531.
68. Elliott, E., M. Mattson, P. Vanderklish, G. Lynch, I. Chang & R. Sapolsky. 1993. Corticosterone exacerbates kainate-induced alterations in hippocampal tau immunoreactivity and spectrin proteolysis *in vivo.* J. Neurochem. 61: 57–65.
69. Watanbe, Y., E. Gould, H. Cameron, D. Daniels, & B. McEwen. 1992. Phenytoin prevents stress- and corticosterone-induced atrophy of CA3 pyramidal neurons. Hippocampus **2:** 431–436.
70. Starkman, N., S. Gebarski, S. Berent & D. Schteingart. 1992. Hippocampal formation volume, memory dysfunction, and cortisol levels in patients with Cushing's syndrome. Biol. Psychiatry **32:** 756–764.
71. Stein, B. & R. Sapolsky. 1988. Chemical adrenalectomy reduces hippocampal damage induced by kainic acid. Brain Res. **473:** 175–181.
72. Feibel, J., P. Hardi, M. Campbell, N. Goldstein & R. Joynt. 1977. Prognostic value of the stress response following stroke. J. Am. Med. Assoc. **238:** 1374–1380.
73. Oka, M. 1956. Effect of cerebral vascular accident on the levels of 17-hydroxycorticoids in plasma. Acta Med. Scand. **156:** 221–226.
74. Olsson, T. 1990. Urinary free cortisol excretion shortly after ischemic stroke. J. Intern. Med. **20:** 1685–1690.

DISCUSSION OF THE PAPER

Question: Your model, you argued, hinges on the excitatory amino acids. You get an increase in the release of excitatory amino acids and a larger effect on the amino acid receptors and in the calcium influx. Now, if you look at high CORT levels, both *in vivo* and *in vitro*, it has been shown that the response to excitatory amino acids is actually depressed by high CORT. This is not only through other receptors, but probably through NMDA receptors as well, because things like LTP are also depressed. There are a couple of defense mechanisms induced by high CORT levels, like this depression of excitatory amino acids and the activation of calcium-dependent potassium conductance, which help the cell maintain control under normal conditions. This CORT effect also influences other things like the increase in calcium influx and the depression of inhibitory current. Those mechanisms are perhaps even more important and only come into play when the cell is, for instance, depolarized for a prolonged period of time or when there is repetitive firing. Then, the depression of the IPSP becomes very important. So I would like to know what you think about the idea that perhaps not only excitatory amino acids but particularly things like the depression of inhibitory amino acids may be something that the whole devastating effect hinges on.

SAPOLSKY: Two answers to that. I agree completely. It does not hinge at all on elevated levels of glutamate in the synapse. What the general model predicts is whatever the insult is to the hippocampus; if it is energetically costly, the hippocampus is going to be in trouble under the circumstances. The nicest demonstration of that was a paper in the *Journal of Neuroscience* a few months ago from Hortnagl's group in Austria showing an insult to the hippocampus with a cholinergic toxin which does not activate this cascade in the slightest but nevertheless is an energetic insult that is also made worse by glucocorticoids. That is the common theme. The effects could very readily sidestep this particular glutamate cascade, which happens to be the hot cascade when you come to ischemia, seizure, and so on; but other insults that are completely separate from the cascade are also made worse. In the broader sense of your question, I totally agree as well. The field is at the point where we are pretty comfortable about what the things are that insult the hippocampus after a neurological disaster: the roles of glutamate, oxygen radicals, and so forth. What people are just getting to think about in a much more organized way is what the neuronal defenses are: the heat shock proteins, the GRPs, the means of changing receptor sensitivity, and the neurotransmitters like adenosine or GABA that can be released, which is an incredibly fertile area to look at. One could just as readily guess, to the extent that glucocorticoids make the insulting components more insulting, that glucocorticoids might very well be impairing some of the defenses.

QUESTION: Your work also brings up the possibility that the *in vivo* activity of glucocorticoids to antagonize insulin action and thereby raise serum glucose may actually be a compensatory protective effect. The insults that you talk about that occur through raising glucocorticoids and not feeding the animal might be offset by the glucocorticoid effect on raising serum glucose. It also explains an old clinical study in which elderly patients who had serum glucoses above 200 mg percent had fewer short-term memory deficits than elderly patients with normal serum glucose.

SAPOLSKY: That certainly fits into work from Paul Gold's group, University of Virginia, who for years have been showing that glucose availability into the bloodstream has dramatic affects on cognition, and an entire literature showing, for example, the amount of glucose in the bloodstream will modulate the amount of seizure-induced damage to the hippocampus. There is a general theme—excitotoxicity is a very expensive thing for a neuron, and there are lots of ways to get into more trouble that way.

QUESTION: On the timing of high glucocorticoid levels in relation to the insult, does it make a big difference if they are before, during, or after?

SAPOLSKY: We have tried to define the minimal time window of endangerment. In the first studies, we exposed the hippocampus to glucocorticoids 7 days before and after the insult, and then cut it down to as little as 24 hours bracketing the insult. Both approaches worsened damage. Another way of asking that is by restricting the glucocorticoid exposure only to the time before and doing it this way, 7 days before, 6 days before and so on, or doing it in the either direction. We find that most of what goes on is in the aftermath. That certainly fits with the picture in neuropathology these days. What goes on here in response to glutamate calcium, etc. is a fairly slowly emerging process and one the the clichés in the field is delayed neuron death. The remarkable thing is if you look at neurons at the end of the 10-minute ischemic crisis, there is lactic acid all over the place, highly acidic, no ATP, no activity, all sorts of problems. A few hours later the neurons are relatively normal again. It is only over the next 72 hours that neurons die. For most of what can be done pharmacologically at present—blocking calcium channels, blocking NMDA receptors—there is a time window to intervene afterward, which seems to be where most of the arena is. It is a slowly emerging process. That is when vulnerability occurs.

QUESTION: In relation to your measurement of the energy state of the cell, you mentioned ATP. Don Young many years ago when we were working with rat thymus cells as targets for glucocorticoids looked at energy charge which is some relationship of ATP to ADP. Have you looked at that, and is it a more sensitive measure?

SAPOLSKY: We are collaborating with David McDougal and Oliver Lowry at Washington University who pioneered most of those techniques, doing micropunches on different hippocampal cell fields after insults with and without glucocorticoids, basically trying to see how far down the glycolytic chain you see this echo of glucocorticoid effects.

QUESTION: In correlating the proteolysis spectrum or the breakdown products, these things also correlate with just basic plasticity, synaptic reorganization, memory, etc. To correlate those with cell death may be stretching the point. In your studies in tissue culture, you are using levels of corticosterone and of glucose which exceed that necessary for saturating either the receptor in the one case or the transport site in the other. We find in our studies and culture that if we get a specific effect of corticosterone or dexamethasone at a physiological concentration and then start pushing it to the micromolar concentration range, very often we will get the exact opposite effect. Too much of a compound can in fact inhibit the response that you get when you use that compound at the concentration that is close to its K_d for its receptor. With glucose, usually the glucose concentration in tissue culture media is close to its transport K_d, which means you get maximal uptake efficiency. We find that if we increase glucose we tend to slow down the growth of the cultures. Do you see that at all or do you, as you described, see a strictly linear relationship between corticosterone responses at 10^{-9} to 10^{-5}? Do you see increased cell growth as you increase glucose or do you see these bell-shaped curves that we observed in our laboratory?

SAPOLSKY: Regarding the spectrin proteolysis, for one thing all of the studies are *in vivo* with spectrin. I totally agree that one of the puzzling things in the field is that if you get a bunch of people together who do hippocampl plasticity, PTP, and the like, then in a sense glutamate is your friend and calcium is good news and spectrin proteolysis is part of what Lynch has speculated involves necessary morphological changes for plasticity to occur. In contrast, in terms of neuropathology, glutamate, calcium excess, cytoskeletal breakdown are central to neuron death. Essentially, the problem is one of scale, and that speaks to the whole issue of why the hippocampus is so constantly involved in neuropathology. It lives in the fast lane. In terms of its percentage synapses it is more glutamatergic, it is more excitatory than any other part of the brain and it is a part of the brain that, so to speak, lives fast and dies young. You cannot have a nonlinear excitation that is essential to LTP and not be very vulnerable to the energetic costs of losing control over glutamate and calcium. It is mostly a matter of scale. As for the effects of high glucose, we see that when you are getting into the 20-mM glucose range that is bad news for the cells. We do see a bell-shaped curve and it points to all the reasons why, as soon as you are working *in vitro*, you have endless reasons to wonder about the physiological relevance. You notice half of what we do is *in vitro*, but half *in vivo*, and that is exactly why at the end we feel like we have to do the physiological studies, not zero verus high, but over the entire physiological range. That is exactly what makes us nervous.

QUESTION: What is so special about the CA3 cells that make them so vulnerable? Is it the great amount of glutamate or the glucocorticoids doing something different there to glucose transport than anywhere else?

SAPOLSKY: From the neurological perspective, there is nothing that fancy in CA3. Essentially, whatever cell field in the hippocampus is most vulnerable to a particular neurological insult is most vulnerable to the synergy between glucocorticoids and that insult. Hypoxia-ischemia is CA1, excitotoxicity is CA3; for energetic insults, hypoglycemia and suchlike, it is dentate gyrus. In general, there is enough glucocorticoid sensitivity in almost any part of the hippocampus that when coupled with one of these other disasters you get in trouble. The metaphor we have used throughout is that of a Venn diagram: you need both the sensitivity of the glucocorticoids and whatever the insult is to have the synergy occur.

The Glucocorticoid Hypothesis of Age-Related Hippocampal Neurodegeneration: Role of Dysregulated Intraneuronal Calcium

PHILIP W. LANDFIELD[a] AND J. CHARLES ELDRIDGE[b]

[a]Department of Pharmacology
University of Kentucky College of Medicine
Chandler Medical Center
MS-305
Lexington, Kentucky 40536-0084
and
[b]Department of Physiology and Pharmacology
Bowman Gray School of Medicine of Wake Forest University
Winston-Salem, North Carolina 27157-1083

INTRODUCTION

The hippocampal formation is among the brain structures most vulnerable to normal brain aging in animals[1-3] and humans.[3,4] The hippocampus also demonstrates severe damage in Alzheimer's disease (AD) and related disorders,[5,6] and damage to this region is believed to contribute substantially to the cognitive/mnemonic manifestations of that disease. A number of hypotheses for mechanisms underlying the age- and/or AD-dependent pathologic degeneration of the hippocampus have been presented (cf. Refs. 3,7,8 for discussion), but one continuing hypothesis suggests a central role for glucocorticoid (GC) steroids of the adrenal gland.[9] Although originally proposed during the late 1970s,[10,11] the GC hypothesis has been subjected to extensive testing and still appears to be one of the few currently viable explanations of mammalian brain decline with aging. However, the hypothesis has not remained unmodified since it was initially proposed. Ongoing experimentation on the role of corticosteroids in brain aging has led to a progression of evolving concepts regarding specific mechanisms underlying interactions of corticosteroids and the brain aging process. The present paper discusses briefly some of the developmental history of the GC hypothesis and summarizes recently obtained evidence that appears to warrant further modifications of the hypothesis. The new evidence suggests a greater emphasis on (a) a need to consider an apparently important age-related susceptibility to GC effects on the brain, and (b) the role of altered intracellular calcium homeostasis as a possible key mechanism mediating GC-related neurotoxicity and brain aging.

Evolving Concepts of Glucocorticoid Interactions with Brain Aging

The concept that disturbances in the adrenal-pituitary axis might be a factor in aging and age-related disease was formulated many decades ago, based on several lines of evidence, including clinical observations,[12,13] the classic studies of Selye on the "stress" response (reviewed in Ref. 14) and studies of GC-mediated accelerated aging in Pacific salmon and in mammals (reviewed in Ref. 15). However, those observations did not address aging of the brain. Subsequently, it was suggested, based on studies by Finch and associates,[16] Meites, Simpkins and colleagues,[17,18] Everitt[19] and others, that the initial sites of the observed endocrine disturbances with aging were in the hypothalamus and related brain systems that regulate peripheral endocrine activity. This "neuroendocrine cascade" hypothesis was notable for its emphasis upon alterations in the homeostatic balance among extrinsic circulating factors, rather than upon intrinsic cellular changes, as a fundamental factor in aging.[16] Of course, cellular-level changes in at least some of the neurons in the brain regulatory systems must first precede systemic alterations and, except for the suggestion of programmed neural changes,[16] the altered neuroendocrine concept did not address specific causes of aging in brain neurons.

The proposal of the hypothesis of glucocorticoid-mediated brain aging[9–11] attempted to address the initial causes of changes in brain cells (which could lead in turn to neuroendocrine regulatory alterations as well as cognitive decline). The hippocampal formation was proposed as a key target site for GCs, because it had recently been discovered that this brain region is particularly rich in GC receptors (reviewed in Refs. 20,21) and because, as noted above, it is particularly vulnerable to age-related decline. Among the key predictions of this view were that blood glucocorticoid concentrations and other measures of chronic adrenocortical activity should correlate with indices of brain age, and that long-term interventions to alter GC action would similarly alter biomarkers of hippocampal aging.[10,11] In addition, since the hippocampus had been reported to participate in negative feedback regulation of the adrenal-pituitary axis (reviewed in Ref. 22), it was suggested that gradual deterioration of the hippocampus with aging might lead to rising titers of corticosteroids, thereby resulting in a positive feedback-mediated acceleration of hippocampal aging.[9–11] An initial test of these predictions found that plasma corticosterone and adrenal weights of healthy Fischer-344 rats increased with aging, and that these measures correlated positively in individual animals with quantitative biomarkers of hippocampal aging.[23] Further, when mid-aged rats were adrenalectomized (ADX) and allowed to age an additional 9 months, these same markers of hippocampal aging (pyramidal neuron loss, astrocyte reactivity) were markedly reduced, compared to age-matched, intact control animals[9,24] (FIG. 1A). The latter finding was particularly important to the hypothesis, because it demonstrated that normal levels of endogenous GCs appeared sufficient to drive the hippocampal aging process. The study also showed for the first time that it was possible to delay or reduce at least some biomarkers of brain aging through hormonal manipulation. Similar results have since been obtained by early handling of animals, a procedure that reduces corticosterone titers later in life.[25]

The converse study, namely, to administer high doses of GCs, has proved much more difficult to carry out due to toxic effects of GC treatment in rats (cf. discussion in Ref. 9). However, it has been possible to treat animals for shorter periods of time with high doses of corticosterone[26] and with a synthetic GC agonist,[27] and also to apply a mild anxiety-sustaining stress for 6 mo.[28] Each study found that signs of brain aging were enhanced by elevated GC exposure (*e.g.,* FIG. 1B). Additional studies have shown that GC administration reduced axon sprouting in lesioned rat hippocampus,[29] that plasma levels of corticosterone correlated with learning and memory impairment of aging animals,[30] and that direct infusion of GCs into monkey hippocampus produced early signs of neurotoxicity.[31] Several recent studies have also found that chronic GC administration impairs learning/memory performance in rodents (*e.g.,* Ref. 32). Thus, these findings, along with others, form a relatively

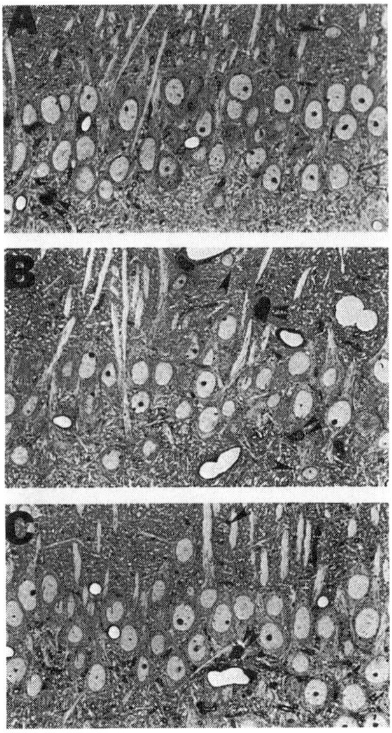

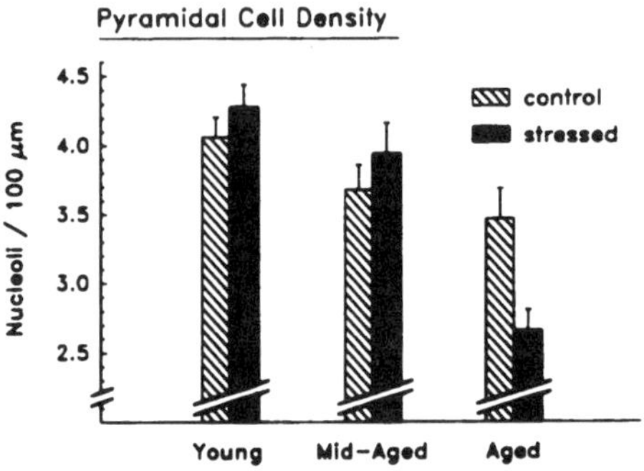

FIGURE 1. Effect of aging, adrenalectomy and chronic stress on CA1 pyramidal cell density in rat hippocampus. *Left panel.* Examples of semithin sections from **(A)** young intact rats, **(B)** aged intact rats, and **(C)** rats of the same age as (B) but adrenalectomized 9 mo earlier. Neuronal nuclei and major glia are easily recognizable: astrocytes with lucent cytoplasm *(arrowheads)* and darker microglia and oligodendrocytes, with chromatin clumps in the nucleus *(double arrowheads).* Long-term adrenalectomy protected against decreased neuronal density and increased astrocyte reactivity. (From Landfield *et al.*[24] Reprinted by permission from *Science.*) *Right panel.* Pyramidal cell density of young, mid-aged and aged rats (non-stressed vs stressed). Values are expressed as no. of nucleoli (mean ± SEM) per 100 μm length of stratum pyramidale. Main effects of age were observed and chronic stress resulted in an increased cell loss in the aged group. (From Kerr *et al.*[28] Reprinted by permission from the *Journal of Neuroscience.*)

strong line of evidence consistent with the concept that the brain, particularly the hippocampus, is cumulatively sensitive to toxic effects of GCs, even at normal levels, and that conditions which enhance GC secretion or activity over prolonged periods can further exacerbate the concomitants of brain aging. (A separate line of studies, beyond the scope of this paper, has provided evidence of analogous effects of estrogens on the hypothalamus,[33] although other recent evidence suggests that estrogens may have beneficial actions on brain function in aging.[34])

An important development in the GC hypothesis arose from studies by Sapolsky and McEwen and colleagues (reviewed in Ref. 35), emphasizing the role of the brain's own GC receptor population. These investigators[36] and several other groups, including ourselves,[37–39] reported that GC receptor populations are lower in hippocampus of aged rats. This observed decline of receptors led Sapolsky to propose that, since hippocampal GC receptors presumably modulate negative feedback control by influencing systems that regulate CRF and ACTH secretion, a reduced receptor population would impede effective control and permit elevated and/or prolonged GC secretion following stressful events. This enhanced GC secretion should prove increasingly toxic, reduce hippocampal receptor levels still further (by downregulation), and produce a positive feedback of enhanced decline and age-dependent neurodegeneration[35] similar to that described above.

These concepts provided an important impetus in that they directly integrated receptor studies into ongoing investigations of the hypothesis. However, the evidence of declining GC receptor content with age also presented, at least on the surface, something of a paradox for the GC-mediated brain aging hypothesis: if rising GC levels induced progressive declines in the GC receptor population, this diminishing population would be expected to be less sensitive to the toxic actions of GCs on the hippocampus (*i.e.,* downregulation should serve as a protective mechanism). In contrast, however, evidence has begun to accumulate that GC toxicity instead is exacerbated with aging. For example, in the 6-mo chronic stress study (FIG. 1B), stress increased hippocampal neuronal loss in comparison to age-matched controls only in the oldest of the 3 age groups.[28] This apparent paradox has stimulated a number of additional studies, by ourselves and others, and the results now suggest that the role of GC receptors in brain aging is more complex than was realized and appears to warrant reconsideration.

Recent Findings and Proposed Adjustments to the Glucocorticoid Hypothesis

Changes in Receptor Plasticity with Aging May Increase the Impact of Glucocorticoid Neurotoxicity

The hippocampus is now well recognized to contain two distinct, but closely related, corticosteroid receptors. Type I is similar to the renal mineralocorticoid receptor (MR) and type II is similar to the peripheral glucocorticoid receptor (GR) (reviewed in Ref. 40). Both hippocampal receptors can bind endogenous corticosterone (CORT). Type I has the higher affinity and is probably saturated at even low diurnal CORT levels, while type II, with a lower affinity for CORT (but a higher affinity for synthetic GCs such as dexamethasone), becomes substantially occupied only at high blood CORT levels (*e.g.,* during stress episodes). The different degrees of occupation have led to proposals that the two-receptor systems make separate use of the single hormone (cf. Ref. 40 and reviews in this volume). If this is the case, it seems more reasonable that the type II receptor would mediate the neurotoxic effects of stress and elevated GC secretion, because the type I receptor is almost fully and continually occupied at all CORT levels.

New insights that may contribute to resolution of the apparent paradox described above came out of studies designed to examine regulation of the type II GC receptor in hippocampus. Although it had been reported earlier that binding capacity for the type II receptor declined in aged rat hippocampus,[38,39] as did overall GC binding,[36] a subsequent examination of hippocampal type II binding for radiolabeled dexamethasone (DEX) found that receptor capacity did not decline, and was possibly increased, with aging in both stressed and unstressed rats.[41] Moreover, the type II receptors were found to be more resistant to stress-induced downregulation in the aged animals (FIG. 2A). It was notable, however, that this study was conducted in intact rats,[41] whereas most earlier studies had examined animals that were recently adrenalectomized (ADX) to clear endogenous steroid from the receptor population. One issue that arises when using intact rats is whether binding data have been confounded by the presence of endogenous ligand. Although this did not appear to be the case in this study,[41] we have also studied receptor content using an anti-GR antibody. This confirmed the high density of GR in the CA1 region (cf. Fuxe, this volume), and also suggested a higher density of immunoreactive GR in the hippocampus of aged rats (unpublished data).

The finding of reduced hippocampal type II binding in aged ADX but not intact animals might be explained by an observation we had made indicating that the process of upregulation occurring after adrenalectomy also appeared to be impaired in aged rats (FIG. 2B).[37] Thus, in studies of GC receptor binding conducted after ADX of aged rats, the observations of lower binding in hippocampus may not have reflected a decline of receptor population from normal levels but rather a decreased ability to upregulate binding in response to ADX.

The observations of lesser up- and downregulation in aged animals (FIG. 2) may have identified a fundamental and perhaps critical change with aging, namely, an overall diminished capacity for receptor plasticity. A reduction of plasticity could exacerbate GC-mediated neurotoxicity in that, if downregulation of GC binding normally follows excessive or chronic exposure to CORT and acts as a protective measure, the inability to downregulate with aging could render the tissue more susceptible to GC-mediated toxicity. Therefore, this loss of plasticity could explain why the impact of GC exposure apparently is increased with aging (FIG. 1B), and why neurodegenerative effects appear to accelerate with advancing age. Interestingly, analogous aging effects may occur for other steroid receptors in that impairment of relatively rapid up- and downregulating changes in brain estrogen receptors has been found to accompany reproductive senescence in rodents.[42] Reduced GC receptor plasticity might also account in part for the age-related impairment of GC-mediated negative feedback control of CRH/ACTH secretion: perhaps the dynamic feedback process involves a flexible, shifting series of events that becomes less flexible, and therefore less effective, in aged animals.

Thus, at least for the hippocampus, it may be concluded that the impact of GCs may increase rather than decrease with aging, as suggested by diminished dynamic change or plasticity of GC receptor regulation (FIG. 2), and by the greater susceptibility of hippocampus of aged animals to potentially toxic actions of GCs (FIG. 1B). It should be emphasized, however, that the enhanced susceptibility to neurodegeneration with aging is probably not limited to the toxic effects of GCs, and it seems likely that multiple adaptive mechanisms, including those involved in other toxic pathways, may be impaired by aging. It is also highly conceivable that some final common pathway of toxic influences (*e.g.*, intraneuronal Ca^{2+} regulation, cf. below) is altered by aging.

Alterations of Intraneuronal Calcium Homeostasis

In recent years, a number of studies have tested the possibility that aging is associated with changes in voltage-activated calcium (Ca^{2+}) currents in brain neurons. One electrophysiological test of Ca^{2+} influx into hippocampal neurons is to measure a well characterized Ca^{2+}-dependent, K^+-mediated afterhyperpolarization (AHP) that follows depolarizing action potentials in hippocampal neurons.[43] In studies that tested the AHP quantitatively in neurons from hippocampal slices of adult and aged rats, the AHP was found to be larger and prolonged with aging[44,45]

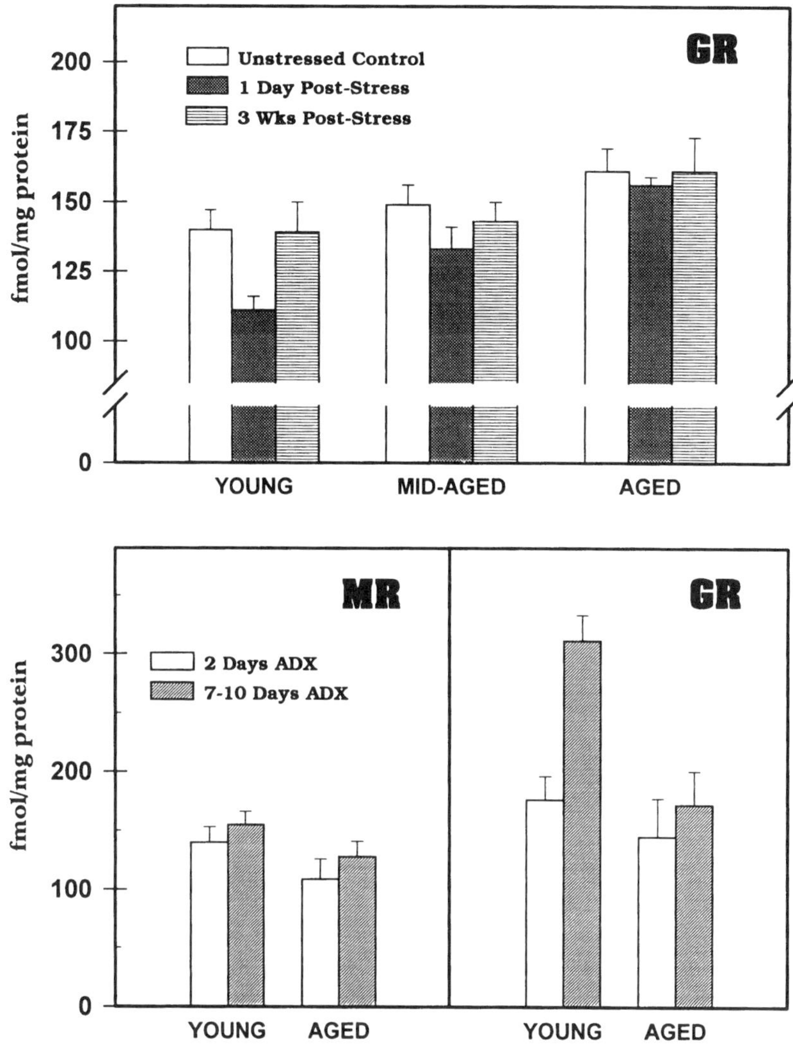

FIGURE 2. Effects of aging, adrenalectomy and chronic stress exposure on corticosteroid receptor levels in hippocampus of *intact* rats. *Upper panel.* Glucocorticoid receptor content following 6 mo mild daily escape stress. Ages at beginning of training: young = 6 mo, mid-aged = 12 mo, aged = 18 mo. Each *bar* represents mean ± SEM of 8 animals sacrificed at 1 day or 3 wks after the final training session. Note absence of downregulation at 1 day in the older groups. Significant overall effects of age and stress were observed. (Redrawn from Eldridge *et al.*[41]) *Lower panel.* Measures of mineralocorticoid (MR or type I) and glucocorticoid (GR or type II) steroid receptors in rat hippocampus following adrenalectomy. Receptor content was measured in young animals (3–4 mo of age) at 2 days ADX (n = 10) and 7–10 days ADX (n = 13); and in aged animals (24–26 mo of age) at 2 days ADX (n = 9) and 7–10 days ADX (n = 5). For MR binding, no significant effects of age or ADX were observed; for GR, significant overall effects of both age (lower in the aged) and ADX (upregulation in young) were observed. (Redrawn from Eldridge *et al.*[37])

(FIG. 3A). Elevating Ca^{2+} in the extracellular media prolonged the AHP, whereas elevating Mg^{2+} reduced the AHP.[44] Therefore, the size of the AHP was directly dependent on the influx of Ca^{2+} following depolarization, and, in turn, the observed age-related increases in the AHP strongly suggested an elevation of intracellular Ca^{2+} following depolarization. However, because the AHP is more a measure of final Ca^{2+} concentration than a direct index of influx, it was not clear from the AHP studies whether influx or buffering, among other possibilities, was primarily affected by aging.

Subsequent studies, using slice neurons treated with tetrodotoxin and tetraethylammonium, plus a cesium-loaded recording pipette to block Na^+ and K^+ currents, have clearly demonstrated aging-related increases in Ca^{2+} action potentials and voltage-sensitive Ca^{2+} currents in hippocampal neurons.[46–48] These are more direct measures of influx, and although the findings do not preclude changes in buffering or extrusion mechanisms (which in fact apparently occur[49,50]), they indicate that voltage-gated Ca^{2+} influx is increased in hippocampal pyramidal cells of aging rats. Importantly, a similar pattern of results has been observed in a separate mammalian species (rabbits).[51]

Linkage of GCs and Ca^{2+} Homeostasis in Aging

To test whether glucocorticoids might interact with Ca^{2+} influx, recordings were made from neurons in slices from hippocampi of ADX and intact rats. Direct application of CORT to the slice medium induced a larger and prolonged AHP, beginning 15–30 min postapplication[45] (FIG. 3B). Adrenalectomy was found to diminish the AHP and Ca^{2+} action potential duration in hippocampal neurons, particularly those from aged rats (FIG. 3C), whereas injection of CORT to ADX rats several hours prior to recording increased the recorded AHP, even in young rats[45] (FIG. 3C). Further studies have shown that the GC effect on Ca^{2+} influx is mediated by the type II receptor; the effect can be blocked by cycloheximide.[52] Very similar observations on type II receptor mediation of the AHP have been obtained independently by another group.[53]

A particularly interesting observation was that the GC effect on the AHP appeared to be greater in aged neurons since, after ADX, the young-aged difference in the AHP disappeared (FIG. 3C). Taken together, these observations indicate that CORT is able to increase Ca^{2+} influx into neurons during or after depolarization (*i.e.*, by voltage-activated Ca^{2+} currents) and moreover, that aging enhances this linkage. This disruption of Ca^{2+} homeostasis with aging may have considerable implications for neuronal function and survival, as it is now well recognized that Ca^{2+} is critical to a wide range of cellular functions and that elevation of intracellular Ca^{2+} can be toxic to excitable cells. Thus, a number of lines of evidence have come together to suggest that gradual alterations in Ca^{2+} homeostasis may play a key role in brain aging and Alzheimer's disease (cf. reviews and discussions in Refs. 47,49–51,54–56).

Of some relevance to the mechanisms underlying the GC-Ca^{2+} homeostasis linkage may be that a novel mode of activity of L-type Ca^{2+} channels, which could underlie generation of the AHP, was recently found.[57] Although several types of Ca^{2+} channels have been recognized,[58,59] the high voltage-activated, L-type channel appears to be associated with the AHP.[51,60] In our recent studies, single Ca^{2+}-channel activity was examined in cell-attached patches of cultured hippocampal neurons. When near-physiological levels of divalent cations were used as the charge carrier, the L-type channels were found to continue to open extensively after the completed

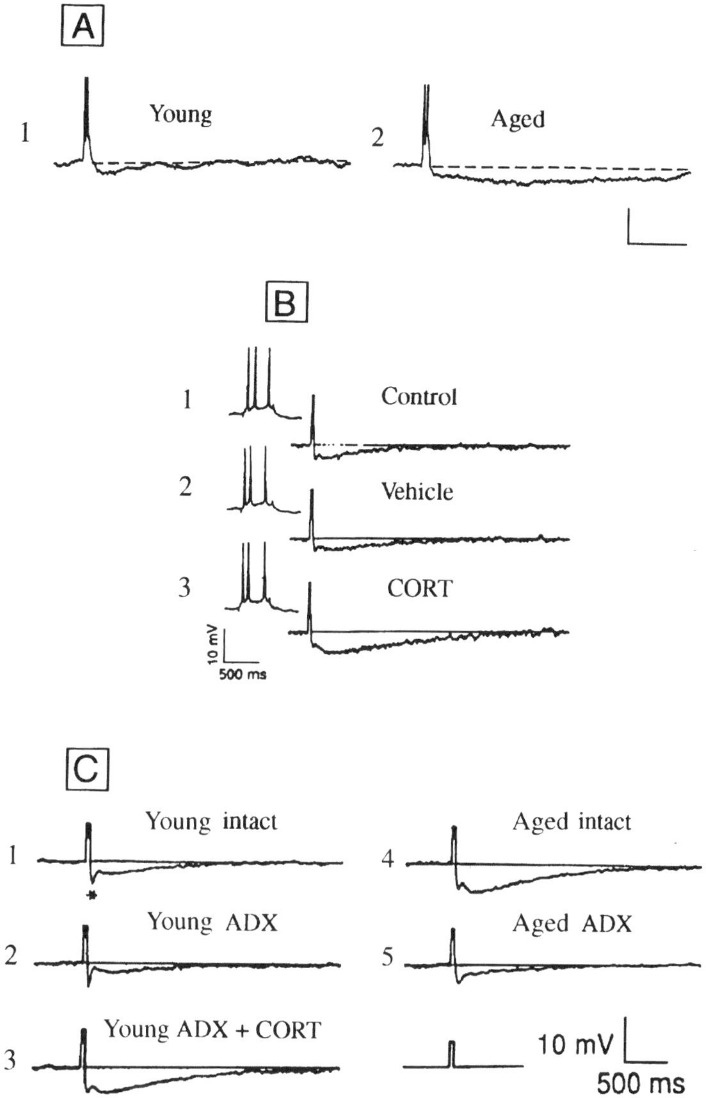

FIGURE 3. Intracellular current bursts of action potentials and subsequent afterhyperpolarizations (AHPs) in CA1 neurons of hippocampal slices. AHPs and concomitant conductance increases were recorded following a 0.4 nA current-induced burst of 3 spikes. **(A)** AHPs recorded from (1) young and (2) aged rats. AHPs were consistently prolonged with aging. (From Landfield & Pitler.[44] Reprinted by permission from *Science*.) **(B)** Intracellular AHP recordings from the same CA1 neuron (slice from young rat). (1) Baseline period, (2) after application of CORT vehicle to slice medium, and (3) after application of CORT to slice medium. Note that the AHP was increased when steroid was added. (From Kerr *et al.*[45] Reprinted by permission from *Science*.) **(C)** Representative examples of AHPs recorded from CA1 neurons of (1) young intact rat, (2) young ADX rat, (3) young ADX rat injected with CORT several hr prior to sacrifice, (4) aged intact rat, and (5) aged ADX rat. AHPs were diminished by ADX, and the normally prolonged AHP of the aged rat was reduced to the size seen in young rats following ADX. The AHP response was also extended in young ADX rats injected with CORT. (From Kerr *et al.*[45] Reprinted by permission from *Science*.) Thus, AHPs were prolonged in aged intact rats or in neurons of young ADX rats exposed to CORT.

depolarization-repolarization sequence.[57] These openings, termed "repolarization openings" (FIG. 4) could last for hundreds of milliseconds, or the approximate duration of the AHP. Previously, it had been thought that Ca^{2+} channels deactivate rapidly following depolarization, but this conclusion had been based primarily on studies using very high concentrations of divalent cations. Further, it may be that brain (hippocampal) neurons are more likely to exhibit this prolonged repolarization opening mode. The demonstration that Ca^{2+} channels can open extensively post-repolarization raises the interesting possibility that this novel activity pattern may mediate the AHP and, moreover, may be a key target for both aging changes and GC actions. Studies to test those possibilities are currently underway.

The evidence outlined above traces a series of observations that combine to suggest that altered function of the aging hippocampal neuron may result from interaction of GCs with systems controlling Ca^{2+} homeostasis. That both steroids and Ca^{2+} homeostasis were important factors in brain aging had been indicated in separate lines of study, but the new evidence suggests that the two processes interact and may in fact be two aspects of the same brain aging neurodegenerative process.

CONCLUSIONS

Susceptibility Factors and Enhanced Linkage of GCs with Ca^{2+} Homeostasis

Results above suggest that GCs have the ability to increase Ca^{2+} influx and AHPs in hippocampus, and that this linkage is enhanced with aging. Consequently, long-term exposure to hormone activation could drive Ca^{2+}-associated deterioration of neuronal structure and function. Aging is the major risk factor for many neurodegenerative diseases and neuropathologic responses to GCs appear to be greater in aged animals (FIG. 1B). Therefore, one critical age-related susceptibility factor may be an enhanced sensitivity of intraneuronal calcium homeostasis to glucocorticoid action. It has been found that hippocampus of aged animals exhibits less plasticity of GC receptor modulation (up- and downregulation) (FIG. 2). If the Ca^{2+} homeostatic system is GC modulated in hippocampal neurons, and loses dynamic responsiveness with advancing age, the result could be a cell that permits excessive Ca^{2+} influx during or after depolarization (FIG. 5). Of course, changes in other factors that regulate Ca^{2+} channels also might occur, resulting in a hypersensitivity of Ca^{2+} channels to a variety of influences.

Thus, further modification of the glucocorticoid hypothesis of neuronal aging appears to be necessary, to include the following additions and revisions: (1) GC-mediated neurotoxic effects likely are mediated primarily via the type II intracellular receptor (GR); (2) the impact of GCs on hippocampal neurons may actually be increased, rather than decreased, with aging; (3) this increased impact may result from a loss of plasticity in control of receptor function, or from hypersensitivity of Ca^{2+} channels, or both; (4) one adversely-affected GC-mediated function may be homeostatic control of intraneuronal Ca^{2+} homeostasis, which is clearly altered with aging, and which can be neurotoxic. Conceivably, overactivation of this system by stress exposure, or with cumulative activation from the passage of time, gradually impairs important aspects of plasticity, which then leads to enhanced susceptibility to neuropathology from subsequent GC-mediated actions.

It should be noted that another recent review also emphasized the need to reexamine the GC hypothesis of brain aging, in light of recent findings.[61] In addition

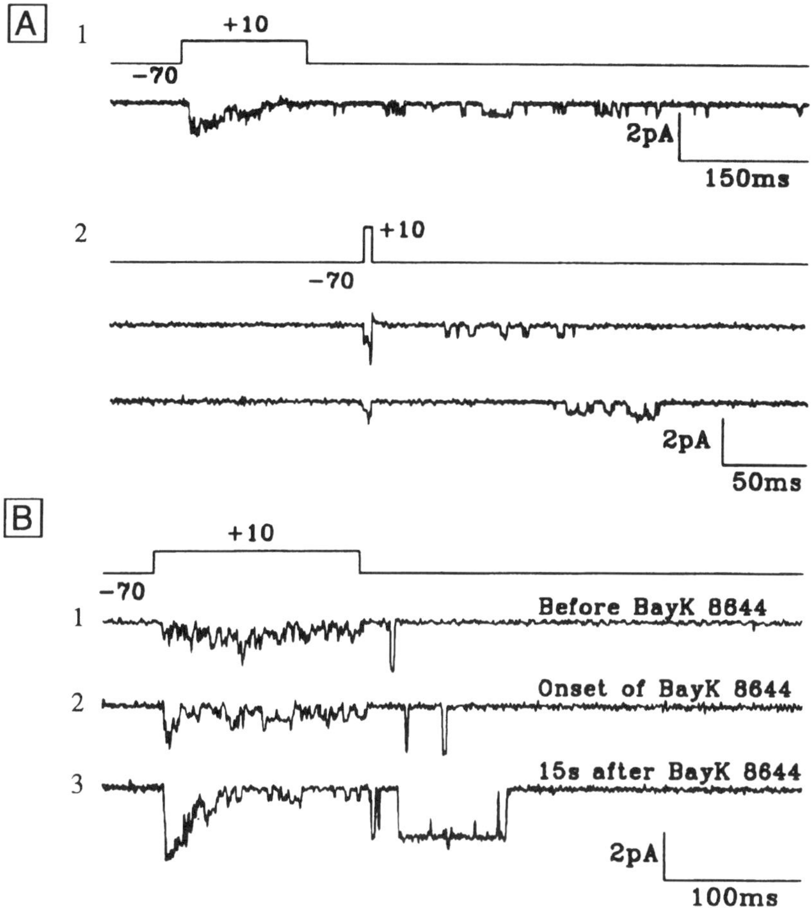

FIGURE 4. Single Ca^{2+} channel recordings from multichannel patches in primary cultures of fetal rat hippocampus, using on-cell patch-clamp methodology. **(A)** Repolarization openings (ROs) under relatively physiological conditions. (1) Channel activity recorded with a pipette containing 5 mM Ca^{2+} as the charge carrier; openings during the step are poorly resolved because of the low conductance at this concentration, but ROs are clearly seen extending for substantial periods after repolarization. (2) Examples of ROs recorded, with 5 mM Ca^{2+} in the pipette, in response to brief (5-ms) depolarizing steps to +10 mV that stimulate an action potential. ROs can be seen following many of the brief depolarizations. **(B)** L-type dihydropyridine sensitivity of ROs recorded with 20 mM Ba^{2+} as the charge carrier, without Bay K 8644 in the recording pipette. *Top trace.* A single RO is seen following the depolarizing step. *Middle trace.* The onset of application of Bay K 8644 (100 nM) onto the cell with a puffer pipette. ROs do not yet exhibit dihydropyridine effects, although longer openings during the step are apparent. *Bottom trace.* 15 sec after the application of Bay K 8644, clear increases in the open time of the amplitude-identified RO can be seen, confirming that ROs are generated by L-type Ca^{2+} channels. Increased simultaneous channel activity also is seen during the step. (From Thibault *et al.*[52] Reprinted by permission from the National Academy of Sciences.)

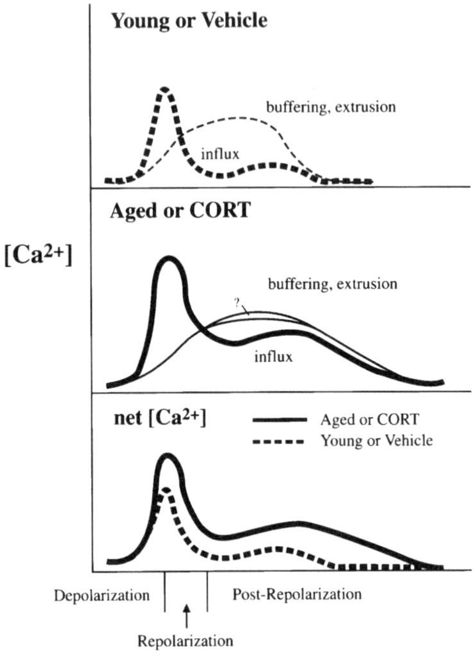

FIGURE 5. Schematic illustrations of neuropathology mediated by glucocorticoid dysregulation of intraneuronal Ca^{2+} homeostasis. In a typical neuron, depolarization is accompanied by a rapid rise of free intracellular Ca^{2+}, caused largely by opening of voltage-sensitive membrane Ca^{2+} channels. Following repolarization, intracellular levels of Ca^{2+} fall as the ion is buffered and extruded. Glucocorticoids appear to enhance the Ca^{2+} elevation that occurs during and after depolarization. Enhancement could occur by increasing the number of voltage-gated channels which open during depolarization, by increasing channel open probability, by increasing repolarization openings of channels, or by diminishing buffering and outward transport of Ca^{2+}. With aging, there appears to be increased efficacy of GCs in elevating intraneuronal Ca^{2+}.

to some of the points raised here, that review also focuses on some different lines of evidence, including short-term effects of hormones on brain structure, which are beyond the scope of the present paper.

REFERENCES

1. LANDFIELD, P. W. 1982. Measurement of brain age: conceptual issues and neurobiological indices. *In* Endocrine and Neuroendocrine Mechanisms of Aging. R. Adelman & G. Roth, Eds. 183–207. CRC Press. Boca Raton, FL.
2. COLEMAN, P. D. & D. G. FLOOD. 1987. Neuron numbers and dendritic extent in normal aging and Alzheimer's disease. Neurobiol. Aging **8:** 521–545.
3. WISNIEWSKI, H. M. & R. D. TERRY. 1973. Morphology of the aging brain, human and animal. *In* Progress in Brain Research. Vol. 240. D. M. Ford, Ed. 167–186. Elsevier Science Publishers. Amsterdam.
4. TOMLINSON, B. L. & G. HENDERSON. 1976. Some quantitative cerebral findings in normal

and demented old people. *In* Neurobiology of Aging. R. D. Terry & S. Gershon, Eds. 183–204. Raven Press. New York.
5. MANN, D. M. A. 1985. The neuropathology of Alzheimer's disease: a review with pathogenetic, etiological and therapeutic considerations. Mech. Ageing Dev. **31:** 213–255.
6. WISNIEWSKI, H. M. & G. S. MERZ. 1985. Neuropathology of the aging brain and dementia of the Alzheimer type. *In* Our Health Care Destiny. Vol. 1. Biomedical Issues. C. M. Gaitz & T. Samorajski, Eds. 231–243. Springer-Verlag. New York.
7. KHACHATURIAN, Z. S., T. S. RADENBAUGH & A. A. MONJAN. 1990. Alzheimer's disease: causes, controls and cures. *In* Alzheimer's Disease: Treatment and Long-Term Management (Neurological Diseases and Therapy, No. 4). J. Cummings & J. Miller, Eds. 275–285. Marcel Dekker. New York.
8. KATZMAN, R. & T. SAITOH. 1991. Advances in Alzheimer's disease. FASEB J. **5:** 278–286.
9. LANDFIELD, P. W. 1987. Modulation of brain-aging correlates by long-term alterations of adrenal steroids and neurally-active peptides. *In* Progress in Brain Research. Vol. 72. E. R. de Kloet, V. M. Wiegant & D. de Wied, Eds. 279–300. Elsevier Science Publishers. Amsterdam.
10. LANDFIELD, P. W. 1978. An endocrine hypothesis of brain aging and studies on brain-endocrine correlations and monosynaptic neurophysiology during aging. *In* Parkinson's Disease. Vol. 2. Aging and Neuroendocrine Relationships. C. E. Finch, Ed. 179–199. Plenum Press. New York.
11. LANDFIELD, P. W., D. K. SUNDBERG, M. S. SMITH, J. C. ELDRIDGE & M. MORRIS. 1980. Mammalian aging: theoretical implications of changes in brain and endocrine systems during mid- and late-life. Peptides **1** (Suppl. 1): 185–196.
12. FINDLAY, T. 1949. Role of the neurohypophysis in the pathogenesis of hypertension and some allied disorders associated with aging. Am. J. Med. **7:** 70–84.
13. SOLEZ, C. 1952. Aging and adrenal cortical hormones. Geriatrics **7:** 241–245.
14. SELYE, H. & B. TUCHWEBER. 1976. Stress in relation to aging and disease. *In* Hypothalamus, Pituitary and Aging. A. F. Everitt & J. A. Burgess, Eds. 553–569. Charles C. Thomas. Springfield, IL.
15. WEXLER, B. C. 1976. Comparative aspects of hyperadrenocorticism and aging. Ibid. 333–361.
16. FINCH, C. E. 1976. The regulation of physiological changes during mammalian aging. Q. Rev. Biol. **51:** 49–83.
17. MEITES, J., H. H. HUANG & J. W. SIMPKINS. 1978. Recent studies on the neuroendocrine control of reproductive senescence in rats. *In* Aging. Vol. 4. The Aging Reproductive System. E. L. Schneider, Ed. 213–235. Raven Press. New York.
18. SIMPKINS, J. W., G. P. MUELLER, H. H. HUANG & J. MEITES. 1977. Evidence for depressed catecholamine and enhanced serotonin metabolism in aging male rats: possible relation to gonadotropin secretion. Endocrinology **100:** 1672–1678.
19. EVERITT, A. V. 1976. Hypophysectomy and aging in the rat. *In* Hypothalamus, Pituitary and Aging. A. V. Everitt & J. A. Burgess, Eds. 68–85. Chas. C. Thomas. Springfield, IL.
20. MCEWEN, B. S., E. R. DE KLOET & W. ROSTENE. 1986. Adrenal steroid receptors and actions in the nervous system. Physiol. Rev. **66:** 1121–1188.
21. STUMPF, W. E., C. HEISS, M. SAR, G. E. DUNCAN & C. CRAVER. 1989. Dexamethasone and corticosterone receptor sites: differential topographic distribution in rat hippocampus revealed by high resolution autoradiography. Histochemistry **92:** 201–210.
22. BOHUS, B. 1975. The hippocampus and the pituitary-adrenal system hormones. *In* The Hippocampus. R. L. Isaacson & K. H. Pribram, Eds. 323–354. Plenum Press. New York.
23. LANDFIELD, P. W., J. L. WAYMIRE & G. S. LYNCH. 1978. Hippocampal aging and adrenocorticoids: quantitative correlations. Science **202:** 1098–1102.
24. LANDFIELD, P. W., R. K. BASKIN & T. A. PITLER. 1981. Brain aging correlates: retardation by hormonal-pharmacological treatments. Science **214:** 581–584.
25. MEANEY, M. J., D. H. AITKEN, S. BHATNAGAR, C. H. VAN BERKEL & R. M. SAPOLSKY.

1988. Postnatal handling attenuates neuroendocrine, anatomical and cognitive impairments related to the aged hippocampus. Science **238:** 766–768.
26. Sapolsky, R. M., L. C. Krey & B. S. McEwen. 1985. Prolonged glucocorticoid exposure reduces hippocampal neuron number: implications for aging. J. Neurosci. **5:** 1222–1227.
27. Landfield, P. W., J. D. Lindsey & G. Lynch. 1978. Apparent acceleration of brain aging pathology by prolonged administration of glucocorticoids. Soc. Neurosci. Abstr. **4:** 325.
28. Kerr, D. S., L. W. Campbell, M. D. Applegate, A. Brodish & P. W. Landfield. 1991. Chronic stress-induced acceleration of electrophysiologic and morphometric biomarkers of hippocampal aging. J. Neurosci **11:** 1316–1324.
29. Scheff, S. W., L. S. Bernardo & C. W. Cotman. 1980. Hydrocortisone administration retards axon sprouting in the rat dentate gyrus. Exp. Neurol. **68:** 195–201.
30. Issa, A. M., W. Rowe, S. Gauthier & M. J. Meaney. 1990. Hypothalamic-pituitary-adrenal activity in aged, cognitively impaired and cognitively unimpaired rats. J. Neurosci. **10:** 3247–3254.
31. Sapolsky, R. M., H. Uno, C. S. Rebert & C. E. Finch. 1990. Hippocampal damage associated with prolonged glucocorticoid exposure in primates. J. Neurosci. **10:** 2897–2902.
32. Dachir, S., T. Kadar, B. Robinzon & A. Levy. 1993. Cognitive defects induced in young rats by long-term corticosterone administration. Behav. Neural Biol. **60:** 103–109.
33. Finch, C. E., L. S. Felicio, C. V. Mobbs & J. F. Nelson. 1984. Ovarian and steroidal influences on neuroendocrine aging processes in female rodents. Endocr. Rev. **5:** 467–497.
34. Sherwin, B. B. 1988. Estrogen and/or androgen replacement therapy and cognitive functioning in surgically menopausal women. Psychoneuroendocrinology **13:** 345–357.
35. Sapolsky, R. M., L. C. Krey & B. S. McEwen. 1986. The neuroendocrinology of stress and aging: the glucocorticoid cascade hypothesis. Endocr. Rev. **7:** 284–301.
36. Sapolsky, R. M., L. C. Krey & B. S. McEwen. 1983. Corticosterone receptors decline in a site-specific manner in the aged brain. Brain Res.2 **289:** 235–240.
37. Eldridge, J. C., D. G. Fleenor, D. S. Kerr & P. W. Landfield. 1989. Impaired upregulation of type II corticosteroid receptors in hippocampus of aged rats. Brain Res. **478:** 248–256.
38. Landfield, P. W. & J. C. Eldridge. 1989. Increased affinity of type II corticosteroid binding in aged rat hippocampus. Exp. Neurol. **106:** 110–113.
39. Reul, J. M. H. M., H. A. D. M. Tonnaer & E. R. de Kloet. 1988. Neurotrophic ACTH analogue promotes plasticity of type I corticosteroid receptor in brain of senescent rats. Neurobiol. Aging **8:** 253–261.
40. de Kloet, E. R. & J. M. H. M. Reul. 1987. Feedback action and tonic influence of corticosteroids on brain function: a concept arising from the heterogeneity of brain receptor systems. Psychoneuroendocrinology **12:** 83–105.
41. Eldridge, J. C., T. E. Kute, A. Brodish & P. W. Landfield. 1989. Apparent age-related resistance of hippocampal type II corticoid receptors to down-regulation by chronic escape training. J. Neurosci. **9:** 3237–3242.
42. Wise, P. M. & B. Parsons. 1984. Nuclear estradiol and cytosol progestin receptor concentrations in the brain and pituitary gland and sexual behavior in ovariectomized estrogen-treated middle-aged rats. Endocrinology **115:** 810–816.
43. Alger, B. E. & R. A. Nicoll. 1980. Epileptiform burst afterhyperpolarization: calcium-dependent potassium potential in hippocampal CA1 pyramidal cells. Science **210:** 1122–1124.
44. Landfield, P. W. & T. A. Pitler. 1984. Prolonged Ca^{2+}-dependent afterhyperpolarization in hippocampal neurons of aged rats. Science **226:** 1089–1092.
45. Kerr, D. S., L. W. Campbell, S-Y. Hao & P. W. Landfield. 1989. Corticosteroid modulation of hippocampal potentials: increased effect with aging. Science **245:** 1505–1509.

46. Campbell, L. W., S-Y. Hao & P. W. Landfield. 1989. Aging-related increases in L-like calcium currents in rat hippocampal slices. Soc. Neurosci. Abstr. **15:** 234.
47. Landfield, P. W., L. W. Campbell, S-Y. Hao & D. S. Kerr. 1989. Aging-related increases in voltage-sensitive, inactivating calcium currents in rat hippocampus: implications for mechanisms of brain aging and Alzheimer's disease. Ann. N.Y. Acad. Sci. **568:** 95–105.
48. Pitler, T. A. & P. W. Landfield. 1990. Aging-related prolongation of calcium spikes in rat hippocampal slice neurons. Brain Res. **508:** 1–6.
49. Michaelis, M. L., K. Johe & T. E. Kitos. 1984. Age-dependent alterations in synaptic membrane systems for Ca^{2+} regulation. Mech. Ageing Dev. **25:** 215–225.
50. Gibson, G. E. & C. Peterson. 1987. Calcium and the aging nervous system. Neurobiol. Ageing **8:** 329–343.
51. Disterhoft, J. F., J. R. Moyer, Jr., L. T. Thompson & M. Kowalska. 1993. Functional aspects of calcium-channel modulation. Clin. Neuropharmacol. **16** (Suppl. 1): S12–S24.
52. Kerr, D. S., L. W. Campbell, O. Thibault & P. W. Landfield 1992. Hippocampal receptor activation enhances voltage-dependent Ca^{2+} conductances: relevance to brain aging. Proc. Natl. Acad. Sci. USA **89:** 8527–8531.
53. Joëls, M. & E. R. de Kloet. 1989. Effects of glucocorticoids and norepinephrine in the hippocampus. Science **245:** 1502–1505.
54. Khachaturian, Z. S. 1984. Towards theories of brain aging. *In* Handbook of Studies on Psychiatry and Old Age. D. S. Kay & G. W. Burrows, Eds. 7–30. Elsevier Science Publishers. Amsterdam.
55. Khachaturian, Z. S. 1989. The role of calcium regulation in brain aging: reexamination of a hypothesis. Aging **1:** 17–34.
56. Scriabine, A., T. Schuurman & J. Traber. 1989. Pharmacological basis for the use of nimodipine in central nervous system disorders. FASEB J. **3:** 1799–1806.
57. Thibault, O., N. M. Porter & P. W. Landfield. 1993. Low Ba^{2+} and Ca^{2+} induce a sustained high profitability of repolarization openings of L-type Ca^{2+} channels in hippocampal neurons: physiological implications. Proc. Natl. Acad. Sci. USA **90:** 11792–11796.
58. Tsien, R. W., D. Lipscombe, D. V. Madison, K. R. Bley & A. P. Fox. 1988. Multiple sites of neuronal calcium channels and their selective modulation. Trends Neurosci. **11:** 431–437.
59. Regan, L. J., D. Y. W. Sah & B. P. Bean. 1991. Ca^{2+} channels in rat central and peripheral neurons: high-threshold current resistant to the dihydropyridine blockers and w-conotoxin. Neuron **6:** 269–280.
60. Mazzanti, M. L., O. Thibault & P. W. Landfield. 1991. Dihydropyridine modulation of normal hippocampal physiology in young and aged rats. Neurosci. Res. Commun. **9:** 117–126.
61. McEwen, B. S. 1992. Re-examination of the glucocorticoid hypothesis of stress and aging. *In* Progress in Brain Research. Vol. 93. D. F. Swaab, M. A. Hofman, M. Mirmiran & F. W. van Leeuwen, Eds. 365–383. Elsevier Science Publishers. Amsterdam.

DISCUSSION OF THE PAPER

Question: Do the calcium channels have the characteristics of a known calcium channel type?

Landfield: Yes, They are L-type channels; they are blocked by nimodipine completely.

Question: Has this prolonged activation been seen in any other tissue?

Landfield: This phenomenon may reflect a brain neuron-specific state of the calcium channel, because there have been some previous studies showing low frequencies of these openings by investigators using high barium in the pipettes. They noted these repolarization openings as infrequent phenomena, but only in hippocampal preparations and only in cerebellum cells. This activity has never been seen in peripheral neurons or the cardiac tissue that so many laboratories use.

QUESTION: Do you see any calcium-dependent inactivation?

LANDFIELD: We see it all the time. One of our major thrusts of study is how these channels are regulated. They are strongly regulated, in part, I think, by calcium-dependent inactivation.

QUESTION: I was struck by the lack of downregulation in the older animals in terms of glucocorticoid receptors. Do you have any information on the mechanism of this? Is there a difference in the size of the messenger RNA in the older animals?

LANDFIELD: We are just now looking at the mechanisms of this process and have not yet looked at that level.

QUESTION: Am I right in assuming that under the conditions you record the calcium currents, when you block potassium channels and sodium channels, that the gradient from the outside to the inside of the cell would be an important determinant of the size of the rate of the calcium influx?

LANDFIELD: Absolutely.

QUESTION: So, if you are using a calcium buffer would you expect this, since in effect it reduces intracellular calcium and you would then end up with an increased calcium influx?

LANDFIELD: The effects of buffers like BAPTA and EGTA may be minimal in this regard, because the internal concentration of calcium is so low (it is submicromolar to begin with) that the driving force is very large in any case. However, we do not use a buffer in the sharp electrode intracellular studies of the AHP and the calcium action potential. In whole cell, patch clamp studies, which I did not show here, we use buffers, but the main effect of those buffers is to reduce calcium-dependent inactivation and rundown rather than to change the driving force.

QUESTION: The critical experiment is where you measure voltage in the presence of cesium and TTX.

LANDFIELD: Cesium blocks potassium channels and TTX affects sodium, so we are not affecting the calcium channels.

QUESTION: Under those conditions have you tried to buffer intracellular calcium?

LANDFIELD: We have done some preliminary comparison studies. We did not see much effect on the buffer. We have not done this systematically in that particular preparation.

QUESTION: With regard to GR regulation in the aged rat and the plasticity and the relative failure of the aged rat to move the receptor levels up or down in response to changes in CORT, the data that you looked at was two days and eight to ten days following ADX. Though the plasticity argument is really useful, it would not resolve the discrepancies that exist in the literature on GR binding following ADX, because they are done generally within about 16 to 24 hours.

LANDFIELD: There was a tendency for there to be an age difference even at the two-day point. I do not think it was quite significant, but there was a strong tendency for that effect to be there. One could imagine that these changes begin almost immediately upon removal of the hormones and that it just takes a longer time for our inefficient or insensitive types of measurement to detect a significant effect.

QUESTION: You end up with the old notion of unmasking receptors versus actually producing new ones. Was the antibody that was used generated against the activated form of the receptor?

LANDFIELD: We got it from John Cidlowski. It was made against the activated form, but I believe it measures both the occupied and the unoccupied forms of the receptor.

QUESTION: Is there any evidence that naturally occurring variations amongst aged animals in serum calcium might be predictive of neurological damage?

LANDFIELD: From what we have seen so far in our electrophysiology I doubt that the kinds of serum changes that might be expected under physiological conditions would modulate the calcium currents very much, although we do not know for sure.

QUESTION: Working in Huda Akil's laboratory, we are studying 150 Fischer rats, and we find that intact animals and ADX animals when they are in the 27-month-old range have a decrease in the levels of GR, IM-GR, and MR in the hippocampus. However, after two days of adrenalectomy you can increase the levels of RNA, of the GR and MR, to exactly the same levels that the young ones have. Yet the binding capacity of GR and MR is not upregulated to the levels that you have in the young ones. There are no age differences in the size of the RNA levels. So I do not know exactly what is happening here, because we have many differences in working across different laboratories.

LANDFIELD: The problem is that your sample size of 150 animals was not large enough.

QUESTION: Also we studied every one of these animals for corticosterone levels and ACTH levels and we saw completely superimposable curves and circadian rhythms in young and aged rats. I never saw a difference in the circulated levels of corticosterone. Even in the face of all the types of compounds and receptor changes, we saw absolutely normal levels of circulating corticosterone. It is true when we stressed the aged animals, their recovery to base level after stress was delayed. But we need to use a 60-minute restraint stress to get that difference. If we do it for a shorter time, there is no effect.

LANDFIELD: Are all your mRNA studies done in adrenalectomized animals?

RESPONSE: In both cases, in intact animals and in ADX animals.

LANDFIELD: You do find an age-related decrease in the GR?

RESPONSE: We found a decrease in intact animals, but after two days of adrenalectomy the non-ADX animals and all ADX animals have exactly the same levels of RNA.

LANDFIELD: That is the opposite of what we find at the receptor binding level. As far as the circulating levels of corticosterone go, lately I have heard from several groups that they are not finding the elevated levels in aged Fischer rats. When we first looked at this, it was very easy to find an age effect, especially under stress conditions. Possibly the data we showed in our 1978 paper did reflect some small amount of stress, and the animals who were most responsive to stress had the highest levels of brain astrocytosis. There is considerable genetic drift in the Fischer strain. Maybe some changes occur in the adrenal system, because it used to be easy to see changes in corticosterone with aging in the Fischer rat strain, and in other rat strains for that matter. We have not looked at this for a while. Whether there is some change occurring or not, I do not know.

QUESTION: This particular phenomenon appears in some places, it does not appear in other places. Within a laboratory there seems to be a lot of individual variability. The fact of the matter is that there are populations of rats that do not show this phenomenon, others that do, and some that show it in various degrees of gradation. No one addresses that particular issue except perhaps Drs. Meaney and Sapolsky where they did handling and showed individual differences. But the aging issue, even the cell death issue, seems to be very, very idiosyncratic and has a lot of variability. I do not see people in general addressing that issue of the variability.

LANDFIELD: Variability is a major problem, I agree. But I want to emphasize that the glucocorticoid hypothesis of brain aging, if we can call it that, has never really relied on there being an elevation. Long-term adrenalectomy studies show that you can remove endogenous levels and obtain a cumulative protective effect on brain aging. Also, in some human studies of normal aging there are elevations in cortisol. But there again, such studies have been highly controversial and variable. This is a sensitive system that is extremely difficult to measure reliably and to obtain results that do not vary from one group to another. You have to study the adrenal system under a variety of conditions and consider the overall weight of the evidence.

RESPONSE: To respond to your point, Dr. Meaney has graciously consented to allow me to speak about his work. The best evidence that people are dealing with the variability of the aging research these days is the whole field of successful aging. Though there is tremendous heterogeneity, there is not all bad news, because there are substantial subsets of aged organisms that do well in aging. Probably the nicest example of that was the study that Dr. Meaney did a couple of years ago in the *Journal of Neuroscience* showing there is a substantial group, about a third, of the old rats, who as defined in the study, showed no cognitive impairments at all. He termed these unimpaired aged rats who also did not have the neuroendocrine problems or extensive neuron death. This suggests that what occurs in your first week of life, may be a major factor in whether or not you have successful aging. Hopefully, some things that happen after the first week are also important.

QUESTION: The early handling animal has higher levels of GR in the hippocampus. You are predicting that high levels of GR during aging may in fact enhance the aging process. If you could somehow activate these receptors by giving exogenous steroids or exposing them to stress, would you predict then that animals that had been handled earlier on would in fact show a greater susceptibility to aging?

LANDFIELD: The handling condition may reflect a specific instance or unusual circumstance, but I would say that a very simple prediction of our view is that, if you enhance glucocorticoid receptor activation in a neuron, you should have a greater neurotoxic impact. Now, maybe these upregulated receptors are less efficacious in inducing some kinds of cellular actions which could influence the effects of activation.

RESPONSE: I would agree completely. You end up with this paradox where you have high GRs. Is this protective or is it actually running a risk? I think that depends on the level of the ligand that you would see. If there is high GR, then this certainly would render the hippocampal neuron vulnerable either to calcium trafficking disorders or worse, or to the metabolic vulnerability that Dr. Sapolsky talked about. There is a study that Jonathan Seckl's group has done recently that is rather interesting in that sense. What they did was to look at the spatial memory ability of a set of aged rats and then they correlated that with plasma corticosterone levels and also with hippocampal GR and MR messenger RNA levels. What they found was that amongst the aged rats, the higher the hippocampal GR level, the lower the plasma CORT levels. Under that particular set of conditions then, high levels of GR predicted good spatial memory performance. But on the other hand if you combine the high levels of the ligand with the high levels of the receptors, then you end up with a possible toxicity and a different result.

LANDFIELD: That is one of the unresolved questions in our studies, in that we find a bigger impact of CORT with age, in terms of chronic stress and of glucocorticoids on the AHP. But, of course, the aged animals generally have higher circulating ligand. So that is one possible confounding factor in evaluating receptor sensitivity. However, when we get to studies of hormone action at the single channel level, we can test definitively whether or not the hormones have a greater effect specifically on the calcium channels.

RESPONSE: I would just like to add the possibility that the way both of you could have your cake and eat it too is for the handled rats to have higher GR as they age but to show a continuation of the downregulation in response to a chronic stress which would, according to your idea, protect them. So you could have a tighter regulation and continued plasticity of the system. That is something, if you have enough money and enough animals, that could be tested.

LANDFIELD: Yes, an interesting idea. I should also give Dr. McEwen credit for having suggested that the downregulation should be protective. We discussed that point years ago.

QUESTION: I, like you, worry about the drift that we see in the changes of HPA activity in the aged rats. I know that your data, and the work of Steve Scheff and Steve DeKosky, and the studies of Dr. Sapolsky in the Fischer rats had consistently seen these changes in corticosterone levels. Now, one of the questions that arises is we often house our animals under very different conditions. There is variability in that. But you also wonder about different background levels of immunological activity that might have been occurring in aged rats, and whether or not the differences in the way in which we house rats over the course of aging may be contributing to the kind of drift that we are talking about here.

LANDFIELD: That is possible, but also in many of these studies people are not measuring adrenal weight as we did. Consistently, adrenal weight used to be elevated in the aged rats. If it is, it is hard to imagine how you would not get a greater hormone output, unless there is substantial deterioration of the glands. Actually, Wexler did show that at the final stage of life the adrenal gland fails, so it might be that in some cases there has been more of an output, and then output is on the way down when it is measured.

E. R. DE KLOET *(Sylvius Laboratories):* In old animals with elevated corticosterone levels one actually may even assume that for the larger part of the day the mineralocorticoid receptor and the glucocorticoid receptor are occupied. Now, the Surgeon General says, "The glucocorticoid receptor is dangerous for your health." However, I am glad to have the glucocorticoid receptor and to have adrenals. By the way we have shown that the mineralocorticoid receptor is an extremely important receptor for maintaining the stability of the hippocampal neurons and for determining sensitivity to excitatory stimuli. If you talk about glucocorticoid receptor activation, there happens to be the second receptor that also is simultaneously activated. In other words, the mineralocorticoid receptors are not only activated at low concentrations but also at high concentrations of corticosteroids. Would it not be justified to have a role for this very important mineralocorticoid receptor in the glucocorticoid toxicity hypothesis?

LANDFIELD: Your point is well taken. I am looking at it from a simplified point of view. We use the specific GR agonist and we find our effects, and we are looking at the effects of elevated concentrations that occur in stress. But, you and Dr. Dallman have been two key players in pointing out that the GR is the receptor that would be affected most or changed most by elevated levels, because the MR system is already occupied at the lower level. So you have hurt your own argument now by showing that earlier. But your point is well taken, and what you and Dr. Dallman are studying and emphasizing is that the joint occupation of the MR and the GR may be a very different business from occupying only the MR.

M. F. DALLMAN *(University of California, San Francisco):* I was going to back up Dr. Meaney and maybe suggest it is not genetic drift, but instead what has happened in the last 15 years is that there has been a huge thrust for barrier bred rats and for rats that are not exposed to any pathogens. SPF rats. This I am convinced makes a very different animal with a whole set of properties that scare me, because you want the variability of an animal that has been exposed to the slings and arrows of usual life. I am worried about it.

LANDFIELD: It is true that a lot of the early aging studies used animals that were not barrier reared and many of them were retired breeder animals. On the other hand, from day 1 in my laboratory we have used specific pathogen-free animals. We have used the Fischers from the National Institute on Aging colony all along. A lot of the other studies showing elevated corticosterone with aging have also been done in those animals.

DALLMAN: That is correct. But one thing that has changed from your initial studies to subsequent ones is that the animals are now studied in the facility. At least for the chronic stress study they were studied in the facility rather than being moved. No, they were studied in the laboratory but were covered on their way from the facility to the laboratory.

LANDFIELD: We have taken precautions in that regard, but you are right: maybe people are being much more compulsive about being sure that they have low nonstress levels. Perhaps the age difference is not apparent unless there is a mild stress.

Hypothalamic-Pituitary-Adrenal Axis Regulation and Human Aging[a]

MURRAY A. RASKIND, ELAINE R. PESKIND,
AND CHARLES W. WILKINSON

Department of Psychiatry and Behavioral Sciences
University of Washington School of Medicine
Seattle, Washington 98195
and
Seattle/American Lake Veterans Affairs Medical Centers

INTRODUCTION

Studies in the rat suggest that hypothalamic-pituitary-adrenal (HPA) axis responsivity to stress increases with aging[1,2] and that resultant increased circulating glucocorticoids lower the threshold for the hippocampal pyramidal cell damage and loss which accompanies aging.[3,4] If these phenomena also occur in human aging, they could contribute to the pathophysiologic processes underlying the modest but potentially clinically important loss of hippocampal pyramidal neurons in normal human aging[5] and the far more extensive loss of hippocampal pyramidal neurons which is a central neuropathologic feature of Alzheimer's disease.[6] The hypotheses that enhanced HPA axis responsivity to stress occurs with human aging and that resultant episodic increases of circulating cortisol contribute to neuronal degeneration in normal aging and Alzheimer's disease have potentially important therapeutic implications. If correct, they suggest that reduction of HPA axis responsivity might form part of a treatment strategy to delay the onset or slow the progression of aging associated memory impairment and/or Alzheimer's disease. The rationale for such a treatment strategy depends, in part, on demonstrating that human aging is associated with enhanced HPA axis responsivity.

Earlier studies suggested that HPA axis regulation does not change with human aging (see Ref. 2 for review). However, if one focuses on recent human studies which address aging effects on HPA axis responsivity to stimuli acting at or above the level of the pituitary, or on HPA axis responsivity to glucocorticoid feedback inhibition, a case for enhanced HPA axis responsivity with human aging in some circumstances begins to emerge. Here we shall critically review these more recent studies and shall present data from our laboratory addressing aging effects on HPA axis responses to suprapituitary cholinergic stimulation and to manipulation of glucocorticoid feedback inhibition.

Unstimulated Plasma Cortisol and ACTH Concentrations in Human Aging

Resting concentrations of cortisol generally have not been reported to differ between old and young normal humans.[7–9] However, one recent study reported

[a]This investigation was supported by the Department of Veterans Affairs and NIH Grants AG-8419, AG-5136, and NS-20311.

increased resting morning plasma cortisol concentrations in old humans.[10] The diurnal rhythm of plasma cortisol is retained with aging,[11] although a phase advance of cortisol peak, nadir and acrophase in old men has been described.[12] Resting plasma ACTH concentrations appear unaffected by aging.[13] Resting concentrations of beta endorphin (βE), a proopiomelanocortin product released with ACTH from the anterior pituitary corticotroph,[14] also are unaffected by aging.[15]

Effects of Aging on Adrenocortical Responses to ACTH

There appears to be no effect of aging on cortisol responses to a supraphysiologic dose of ACTH.[16] More recent studies[17,18] have extended these findings by demonstrating no consistent aging effects on cortisol responses to graded physiologic doses of ACTH. One of these studies[17] must be interpreted cautiously because of small subject number (5 young and 6 old subjects), the relatively young age of the "old" subjects (mean age 65 years) and the fact that the majority of subjects were either being treated for thyroid disease or had chronic medical disorders. These shortcomings were addressed by a subsequent study[18] which included larger numbers of healthy subjects, an older group of "old" subjects (range 67 to 83 years) and a protocol which used successive injections of ACTH in a manner which controlled for possible differences in cortisol clearance among groups. These investigators reported no consistent effects of either age or sex on cortisol responses to ACTH, but did find a steeper dose-response curve in older women than in the other subject groups.

Effects of Aging on Anterior Pituitary Responses to Direct Stimulation

Recent studies of aging effects on anterior pituitary responsivity to ACTH secretagogues are less easily interpreted than the above described studies of aging effects on adrenocortical responsivity. Although several investigators have failed to demonstrate marked effects of aging on the ACTH response to ovine corticotropin releasing hormone (oCRH), a close reading of these studies suggests that aging may subtly enhance corticotrope responsivity. For example, in one study plasma ACTH responses were measured over 120 minutes following 100 μg oCRF in six young and six old men.[19] Although the mean ACTH area under the curve in old subjects was approximately twice that of the young subjects, analysis of variance failed to demonstrate that the responses significantly differed between groups. Other investigators reported similar results in a study which examined the ACTH response to 1 μg/kg oCRH in three groups of healthy men (mean ages ± SEM; 35 ± 2 years, 61 ± 1 years, 77 ± 1 years).[10] Although ACTH responses were significantly higher at 30 minutes in the oldest group, there was only a trend toward higher overall ACTH responses with aging. Because basal cortisol concentrations were significantly elevated in the oldest group (who also tended to have higher ACTH responses to oCRH), these findings were interpreted as compatible with diminished sensitivity of ACTH secretion to glucocorticoid feedback inhibition in old men.

Vasopressin is an important ACTH secretagogue, which modestly stimulates ACTH release when administered alone and markedly enhances the ACTH response to CRH when both secretagogues are administered together.[20] A recent study demonstrated substantial aging effects on the ACTH response to combined administration of vasopressin and CRH.[21] In this study a bolus injection containing both 50 μg human CRH (hCRH) and 0.5 IU lysine vasopressin was administered intravenously to healthy young men (age range 21 to 25 years) and healthy old men and women (age range 66 to 90 years). ACTH responses were substantially and significantly higher in the old subjects and return to baseline was delayed in the old subjects. A potential problem with this study, however, was the health status of the old subjects. Although all old subjects were described as free of any symptoms and showing normal physical activity, they all were awaiting discharge from an inpatient medical service having convalesced from a diagnostic procedure. The possibility that illness or the stress of inpatient hospitalization explained the enhanced ACTH responses in the old subjects has to be considered particularly in light of a recent report that the presence of a stable chronic illness (diabetes mellitus or hypertension) enhanced HPA axis responsivity to CRH.[22]

Effects of Aging on HPA Axis Responsivity to Suprapituitary Stimulation

Studies examining aging effects on HPA axis response to suprapituitary stimuli are particularly relevant to the present discussion. This is so because the original rat studies demonstrating enhanced glucocorticoid responses with aging utilized a restraint stressor presumably acting at a suprapituitary level.[1] Unfortunately, the number of studies addressing aging effects on HPA axis responsivity with quantifiable stimuli acting at suprapituitary sites are few. In an earlier study,[23] the cortisol responses to insulin-induced hypoglycemia were compared among three groups of subjects with mean ages 27 years, 41 years, and 68 years, respectively. Each group contained approximately equal numbers of men and women. Overall cortisol responses to intravenous insulin 0.10U/kg did not differ. However, the achievement of hypoglycemia was slower, the glucose nadir was less marked, and the cortisol increase tended to be more rapid in the old group. Thus, the old subjects appeared to have a more rapid cortisol response despite a weaker stimulus than the young subjects. Inclusion of an even older group of "old" subjects and measurement of plasma ACTH would have been useful to evaluate further this suggestion of enhanced HPA axis responsivity to hypoglycemia. A recent study[24] reported an even clearer aging effect on the cortisol response to hypoglycemia in young (20-42 years) and old (66-72 years) normal volunteers. Despite a trend toward a less marked glucose nadir in the old subjects, they achieved significantly higher peak cortisol responses.

We recently examined the effects of aging on the cortisol and βE responses to the central nervous system (CNS) active cholinesterase inhibitor physostigmine in 9 normal young (25 ± 1 years) men and 12 normal old (68 ± 2 years) men.[25] Physostigmine easily crosses the blood brain barrier and increases the intrasynaptic availability of acetylcholine in the CNS.[26] Physostigmine thus enhances cholinergic stimulation of the HPA axis at CNS sites. Subjects were given a 10-minute i.v. infusion of physostigmine at 0.0125 mg/kg and blood samples were obtained over a 90-minute period for measurements of cortisol and βE. For technical reasons at the time the study was performed, βE rather than ACTH was used to assess the responsivity of the anterior pituitary corticotrope. Old subjects had substantially greater plasma cortisol and βE responses than did young subjects (FIG. 1). Because physostigmine concentrations were not measured in this study, the possibility remained that aging-associated pharmacokinetic factors could have led to higher plasma and brain physostigmine concentrations. If this were the case, enhanced cortisol and βE

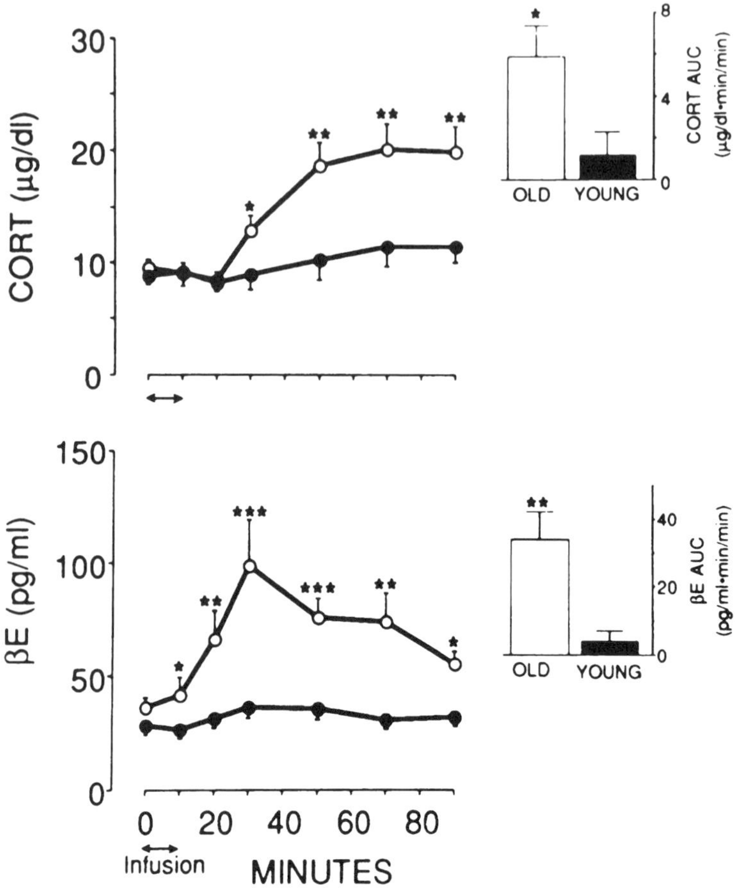

FIGURE 1. Plasma cortisol (CORT) and beta endorphin (βE) responses to 10-minute physostigmine infusion (0.0125 mg/kg) in 12 old (68 ± 2 yrs) and 9 young (25 ± 1 yrs) normal volunteers. Responses greater in old than young: $*p < 0.05$, $**p < 0.01$, $***p < 0.001$. (Modified from Raskind *et al.*[25])

responses could be the result of a more intense stimulus rather than enhanced HPA axis responsivity. To address this issue and to replicate the initial findings, the physostigmine protocol was repeated in new groups of young (10 men, age 27 ± 2 years) and old (5 men, 5 women, age 71 ± 2 years) normal healthy subjects.[27] In this second study, plasma physostigmine concentrations were determined and plasma ACTH was directly measured. Again, HPA axis responsivity to physostigmine as indicated by plasma cortisol, βE and ACTH concentrations was markedly and significantly greater in the old subjects than in the young subjects. In contrast, plasma physostigmine concentrations did not differ between groups. Therefore, aging effects on physostigmine pharmacokinetics did not explain the twice observed enhanced HPA axis responsivity to physostigmine in old subjects.

Effects of Aging on HPA Axis Sensitivity to Glucocorticoid Feedback Inhibition

In an early study,[11] our laboratory was unable to demonstrate an aging effect of 1 mg dexamethasone administered at 2300 hours on plasma cortisol concentrations sampled at 0800, 1600 and 2300 hours the following day. However, the oldest subject (age 82 years) also was the only young or old subject with a post-dexamethasone cortisol at 1600 hours higher than 5 μg/dl. In contrast, in a recent study comparing normal old persons (mean age 69 years), and normal young persons (mean age 34 years) afternoon plasma cortisols following 1.5 mg oral dexamethasone at 2300 the previous evening were significantly higher in the old subjects.[28] Higher cortisol concentrations were particularly prominent among older women. Afternoon plasma ACTH concentrations tended to be higher in old subjects but differences between groups were not statistically significant. In a second phase of this study, 100 μg hCRH was administered to these dexamethasone-pretreated subjects. Cortisol responses to the CRH challenge were significantly higher and ACTH responses tended to be higher in the old subjects. Again, old female subjects demonstrated particularly robust cortisol responses to hCRH. These results support enhanced HPA axis activity with human aging, at least in the afternoon phase of the diurnal cycle. They also suggest that female gender may add to these aging effects. Although these investigators interpreted the combined dexamethasone/CRH results as reflecting decreased sensitivity to glucocorticoid feedback inhibition, the data also are compatible with increased intrinsic anterior pituitary responsivity to CRH with aging.

In another carefully designed study,[13] morning plasma ACTH and cortisol concentrations in old men (66–78 years) and young men (21–38 years) were measured frequently between 0400 and 0800 hours following dexamethasone at doses of 0.15 mg, 0.3 mg, or 0.6 mg administered at 2400 hours the night before. These early AM cortisol and ACTH concentrations following dexamethasone failed to differ significantly between old and young men. Subjects then were administered a bolus injection of 1 μg/kg oCRH and blood was collected every 30 minutes for the subsequent two hours for ACTH and cortisol measurements. Although inspection of the data suggested less inhibition by dexamethasone pretreatment of the ACTH response to oCRH in old subjects, differences between age groups as analyzed failed to reach statistical significance. The increase in cortisol following oCRH in the 0.3 mg dexamethasone pretreatment condition was significantly greater in the old subjects. Differences between results in this study and that of Heuser *et al.*[28] may be partially attributable to the inclusion of women in the latter study, and the very different times of the day during which the two studies were performed. It is possible that feedback inhibition of the HPA axis induced by dexamethasone is more prominently affected by aging in the late afternoon and evening than in the early morning.

We recently took another approach to assessing the effects of aging on the sensitivity of the HPA axis to glucocorticoid feedback inhibition. In this experiment, endogenous cortisol feedback inhibition was substantially reduced by administering metyrapone, an 11-β hydroxylase inhibitor which greatly reduces the adrenocortical production of cortisol and corticosterone. Tonic HPA axis inhibition by endogenous cortisol was estimated by measuring the rise of ACTH following blockade of cortisol production by metyrapone. Glucocorticoid feedback inhibition was subsequently reinstituted by an exogenous cortisol infusion. The sensitivity of the axis to cortisol feedback inhibition was estimated by measuring the rate of ACTH decline toward baseline levels during the exogenous cortisol infusion.

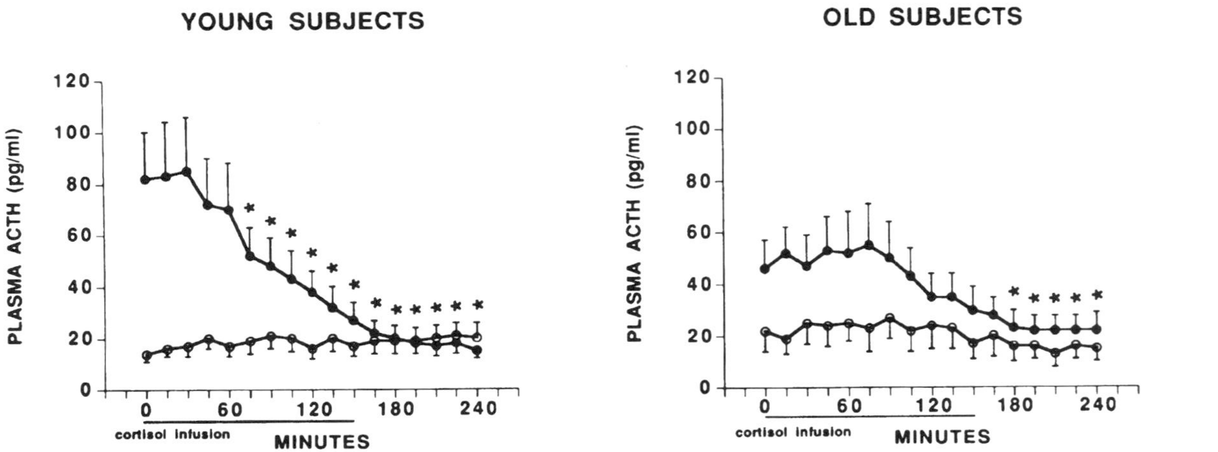

FIGURE 2. Plasma ACTH responses to exogenous cortisol infusion following metyrapone pretreatment in 6 young (23 ± 4 yrs) and 6 old (74 ± 3 yrs) normal volunteers. •-•, subjects given metyrapone 750 mg at minus 180 minutes and again at 0 minutes, then given 150-minute cortisol infusion (0.06 mg/kg/hr). O-O, subjects given placebo tablets at minus 180 minutes and again at 0 minutes, then given 150-minute normal saline infusion. *p <0.05 compared to time 0.

This metyrapone/cortisol infusion protocol was administered to six old subjects (72 ± 3 years; 3 men, 3 women) and six young subjects (25 ± 1: 3 men, 3 women). All young women were in the follicular phase of the menstrual cycle. All subjects were in excellent health and free of drugs known to affect the HPA axis. Subjects underwent two study conditions in random order. In the "active drug" condition, metyrapone 750 mg p.o. was administered at 0600 hrs and again at 0900 to maintain 11-β hydroxylase inhibition during the exogenous cortisol infusion. After baseline blood samples for determination of plasma ACTH were withdrawn from an indwelling intravenous catheter, an infusion of cortisol, 0.6 mg/kg/hr was begun through a second intravenous catheter and continued for 150 minutes during the cortisol infusion. The "placebo" study was identical to the active drug study except that placebo tablets were substituted for metyrapone and a normal saline infusion was substituted for the cortisol infusion.

The plasma ACTH concentrations during both the active drug and placebo conditions in the old and young subject groups are presented in FIGURE 2. Analysis of variance with repeated measures revealed an overall group by time by treatment interaction (p <0.001). The fall of ACTH during exogenous cortisol infusion was significantly blunted in the old subjects compared to the young subjects (group by time interaction, p = 0.02). Plasma cortisol concentrations achieved during the cortisol infusion did not differ between old and young subjects (data not shown). These findings support decreased HPA axis sensitivity to glucocorticoid feedback inhibition with aging. Old subjects also demonstrated a trend toward a blunted increase in ACTH following metyrapone, suggesting an aging effect on tonic glucocorticoid feedback inhibition as well. Specifically the difference in ACTH at baseline following metyrapone compared to baseline following placebo tended to be smaller in old than in young subjects (21 ± 7 pg/ml vs 62 ± 19 pg/ml; p = 0.07).

CONCLUSION

Although human aging does not appear substantially to alter resting cortisol and ACTH concentrations, a growing body of evidence suggests enhancement with aging of HPA axis responsivity to stimulation at or above the level of the anterior pituitary. The increased prevalence of chronic medical illness in later life may further enhance HPA axis responsivity in old persons.

The pathophysiological significance of enhanced HPA axis responsivity with human aging remains unclear. The hippocampal neurotoxicity of glucocorticoids reported in animal studies,[2–4] the impairment of cognitive function in hypercortisolemic humans afflicted with Cushings' syndrome or in normal volunteers administered exogenous glucocorticoids (reviewed in Ref. 29), and a recent report of an association between high HPA axis activity and subsequent cognitive decline in old normal humans[30] all support the possibility that enhanced HPA axis activity contributes to CNS events underlying cognitive deterioration in late life. Careful prospective studies are necessary to clarify the potential role of HPA axis activity in the development of aging-associated memory impairment and Alzheimer's disease in late life.

REFERENCES

1. SAPOLSKY, R., L. KREY & B. MCEWEN. 1986. The adrenocortical axis in the aged rat: impaired sensitivity to both fast and delayed feedback. Neurobiol. Aging **7:** 331–335.

2. Sapolsky, R., M. Armanini, D. Packan & G. Tombaugh. 1987. Stress and glucocorticoids in aging. Endocrinol. Metab. Clin. **16:** 965–980.
3. Landfield, P. W., J. L. Waymire & G. S. Lynch. 1978. Hippocampal aging and adrenocorticoids: quantitative correlations. Science **202:** 1098–1102.
4. Kerr, S. D., L. W. Campbell, M. D. Applegate, A. Brodish & P. W. Landfield. 1991. Chronic stress-induced acceleration of electrophysiologic and morphometric biomarkers of hippocampal aging. J. Neurosci. **11:** 1316–1324.
5. Coleman, P. & D. Flood. 1987. Neuron numbers and dendritic extent in normal aging and Alzheimer's disease. Neurobiol. Aging. **8:** 521–546.
6. Ball, M., M. Fishman, V. Hachinski, W. Blume, A. Fox, V. A. Kral, A. J. Kirshen & H. Merskey. 1985. A new definition of Alzheimer's disease: a hippocampal dementia. Lancet **1:** 14–16.
7. Jensen, J. K. & M. Blichert-Toft. 1971. Serum corticotrophin, plasma, cortisol, and urinary excretion of 17-ketogenic steroids in the elderly (age group: 66–94 years). Acta Endocrinol. (Copenhagen) **66:** 25–34.
8. Friedman, M., M. F. Green & D. E. Sharland. 1969. Assessment of hypothalamic-pituitary-adrenal function in the geriatric age group. J. Gerontol. **24:** 292–297.
9. Blichert-Toft, M., B. Blichert-Toft & H. Kaalund-Jensen. 1970. Pituitary-adrenocortical stimulation in the aged as reflected in levels of plasma cortisol and compound S. Acta Chir. Scand. **136:** 665–670.
10. Pavlov, E. P., S. M. Harman, G. P. Chrousos, D. L. Loriaux & M. R. Blackman. 1986. Responses of plasma adrenocorticotropin, cortisol, and dehydroepiandrosterone to ovine corticotropin-releasing hormone in healthy aging men. J. Clin. Endocrinol. Metab. **62:** 767–772.
11. Rivard, M. A., M. A. Raskind & D. Rivard. 1981. The dexamethasone suppression test in an elderly population. Biol. Psychiatry **16:** 1177–1184.
12. Sherman, B., C. Whysham & B. Pfohl. 1985. Age-related changes in the circadian rhythm of plasma cortisol in man. J. Clin. Endocrinol. Metab. **61:** 439–443.
13. Waltman, C., M. R. Blackman, G. P. Chrousos, C. Riemann & S. M. Harman. 1991. Spontaneous and glucocorticoid-inhibited adrenocorticotropic hormone and cortisol secretion are similar in healthy young and old men. J. Clin. Endocrinol. Metab. **73:** 495–502.
14. Risch, S., N. Kalin, D. Janowsky, R. Cohen, D. Pickar & D. Murphy. 1983. Corelease of ACTH and beta-endorphin immunoreactivity in human subjects in response to central cholinergic stimulation. Science **222:** 77.
15. Forman, L., T. Cavalieri, J. Fitzharris, A. Chopra, D. Marquis & S. Estlow. 1987. Plasma levels of beta-endorphin in young and aged human males during the morning hours. Horm. Metab. Res. **19:** 38–39.
16. Vermeulen, A., J. P. Deslypere, W. Schelfhout, L. Verdonck & R. Rubens. 1982. Adrenocortical function in old age: response to acute adrenocorticotropin stimulation. J. Clin. Endocrinol. Metab. **54:** 187–191.
17. Ohashi, M., K. Kato, H. Nawata & H. Ibayashi. 1986. Adrenocortical responsiveness to graded ACTH infusions in normal young and elderly human subjects. Gerontology **32::** 43–51.
18. Roberts, N. A., R. N. Barton & M. A. Horan. 1990. Ageing and the sensitivity of the adrenal gland to physiological doses of ACTH in man. J. Endocrinol. **126:** 507–513.
19. Ohashi, M. N. Fujio, K. Kato, H. Nawata & H. Ibayashi. 1986. Aging is without effect on the pituitary-adrenal axis in men. Gerontology **32:** 335–339.
20. Negro-Villar, A., C. Johnston, E. Spinedi, M. Valencia & F. Lopez. 1987. Physiological role of peptides and amines on the regulation of ACTH secretion. Ann. N.Y. Acad. Sci. **512:** 218–236.
21. Dodt, C., J. Duttmann & J. Hruby. 1991. Different regulation of adverse corticotropin and cortisol secretion in young, mentally healthy elderly, and patients with senile dementia of Alzheimer's type. J. Clin. Endocrinol. Metab. **72:** 272–276.
22. Greenspan, S. L., J. W. Rowe, L. A. Maitland, M. McAloon-Dyke & D. Elahi. 1993. The pituitary-adrenal glucocorticoid response is altered by gender and disease. J. Gerontol. Med. Sci. **48:** M72–M77.

23. Muggeo, M., D. Fedele, A. Tiengo, M. Molinari & G. Crepaldi. 1986. Human growth hormone and cortisol responses to insulin stimulation in aging. J. Gerontol. **30:** 546–551.
24. Meneilly, G. S., K. L. Minaker, J. B. Young, L. Landsberg & J. W. Rowe. 1985. Counterregulatory responses to insulin-induced glucose reduction in the elderly. J. Clin. Endocrinol. Metab. **61:** 178–182.
25. Raskind, M. A., E. R. Peskind, R. C. Veith, C. W. Wilkinson, D. Federighi & D. M. Dorsa. 1990. Differential effects of aging on neuroendocrine responses to physostigmine in normal men. J. Clin. Endocrinol. Metab. **70:** 1420–1425.
26. Taylor, P. 1990. Anticholinesterase agents. *In* Goodman and Gilman's The Pharmacological Basis of Therapeutics, 8th edit. A. G. Gilman, T. W. Rall, A. S. Nies & P. Taylor, Eds. 131–140. Pergamon Press. New York.
27. Peskind, E. R., M. A. Raskind, D. Wingerson, M. Pascualy, L. J. Thal, D. J. Dobie, R. C. Veith, D. M. Dorsa, S. Murray, C. Sikkema, S. A. Galt & C. W. Wilkinson. Enhanced hypothalamic pituitary adrenocortical axis responses to physostigmine in normal aging. J. Gerontol. In press.
28. Heuser, J., U. Gotthardt, U. Schweiger, J. Schmider, C. H. Lammers, M. Dettling & F. Holsboer. 1994. Age-associated changes of pituitary-adrenocortical hormone regulation in humans: importance of gender. Neurobiol. Aging **15:** 227–231.
29. Wolkowitz, O. M. 1994. Prospective controlled studies of the behavioral and biological effects of exogenous corticosteroids. Psychoneuroendocrinology **19:** 233–255.
30. Lupien, S., A. R. Lecours, I. Lussier, G. Schwartz, N. P. V. Nair & M. J. Meaney. Basal cortisol levels and cognitive deficits in human aging. J. Neurosci. In press.

DISCUSSION OF THE PAPER

Question: Your physostigmine data would suggest that there is an increased responsivity to stress, but your metyrapone data suggest that there is a decreased responsivity to corticosteroids. Do you see this as a paradox, and if so, how do you resolve it?

Raskind: I do not see a paradox in our separate observations of increased HPA axis responsiveness with aging in the physostigmine stimulation protocol and decreased sensitivity with aging to glucocorticoid feedback inhibition of the axis in the metyrapone/cortisol infusion protocol. In fact, decreased sensitivity to cortisol feedback inhibition is a possible mechanism contributing to the observed enhanced ACTH and cortisol responses to physostigmine with aging. Although the contribution of decreased glucocorticoid feedback inhibition in older humans to their greater HPA axis response to physostigmine cannot be determined from our data, decreased sensitivity to glucocorticoid feedback inhibition appears to contribute substantially to increased HPA axis responsiveness to stress in the old rat.

Question: I should like to commend you for claiming that 70 year olds also show what is being considered an aging response. When we call people aged, we have to be rather careful about how we define it.

Raskind: Physiologic aging is a complex construct which varies greatly among individuals and is only roughly approximated by chronologic age. Of particular importance to human aging studies is the high likelihood that "older" old individuals in their ninth and tenth decades differ from "younger" individuals in their seventh and eighth decades. These "older" old individuals often are excluded from clinical studies because of their high frequency of coexistent medical illnesses and prescribed medications.

Regulation of the Hypothalamic-Pituitary-Adrenal Axis in Dementia Disorders

C. G. GOTTFRIES, JAN BALLDIN, KAJ BLENNOW,
GÖREL BRÅNE, INGVAR KARLSSON, BJÖRN REGLAND,
AND ANDERS WALLIN

Department of Clinical Neuroscience
Section of Psychiatry and Neurochemistry
University of Göteborg
Mölndal Hospital
S-431 80 Mölndal, Sweden

INTRODUCTION

Alzheimer-type dementia is often classified into two subgroups, one with early onset, Alzheimer's disease (AD), and one with late onset, senile dementia of the Alzheimer type (SDAT). Vascular dementia (VAD) is looked upon as more or less the consequence of thromboembolism with multiple cerebral infarcts. Our findings suggest that subcortical white matter changes are another important VAD substrate. When infarcts are present, they may be the endpoint manifestations of the vascular pathology rather than the cause of the disease. Based on these findings, we consider the white matter substrate, which possibly causes a state of subcortical-cortical disconnection, more important than the mechanical destruction of the brain parenchyma by infarction (for review see Ref. 1). Results of epidemiological studies indicate that at least 5% of people above the age of 65 are assumed to suffer from dementia disorders, 10% if mild dementia is included. Around 60% of all dementias are of the Alzheimer type, while around 20% are VAD.

Multiple neurochemical changes have been observed in grey matter from patients with AD/SDAT. A cholinergic dysfunction is well established, but monoamine metabolism is also disturbed, especially that of serotonin (5-HT) and noradrenaline (NA). Neurotransmitter dysfunction occurs in several areas, among them the hippocampus (for review see Refs. 2, 3). The hippocampal neurons are glucocorticoid-sensitive and involved in the regulation of the hypothalamic-pituitary-adrenal (HPA) axis.[4] It is of interest that, in AD/SDAT, reduced concentrations of vasopressin are reported from the cortex, hippocampus and basal ganglia and from CSF investigations.[5–8] However, in studies by Wallin *et al.*[9] increased concentrations in the hypothalamus were reported in patients with AD/SDAT. Corticotropin-releasing factor has also been investigated in postmortem human brain material, and decreased concentrations are reported.[10,11] Reduced concentrations of corticotropin-releasing factor have also been reported in CSF from AD/SDAT patients.

Several studies have also found a disturbed neurotransmitter function in VAD.[12–14] It is of interest that Alzheimer patients with late onset, the SDAT group, exhibited white matter degeneration on CT scans and blood-brain barrier disturbances.[15] A significantly larger number of vascular risk factors was also noted in SDAT patients than in AD patients, suggesting that vascular disease may be part of the etiopathological factors for the dementia. Not only brain imaging but also postmortem studies have shown subcortical white matter degeneration in VAD patients.[16] Thus, subcortical white matter lesions may be an important pathogenetic factor in a subset of VAD patients and to some extent also in SDAT patients.

The dexamethasone suppression test (DST) has been shown to be pathological in different types of dementia disorders (for review see O'Brien[17]). Hypercorticolism revealed by the DST is not only seen in AD/SDAT but also in VAD and in patients with acute ischemic stroke.[18] It is of interest that in the investigation of DST in demented patients no correlation is usually seen between the degree of depressed mood and the frequency of nonsuppressors.

We earlier proposed the hypothesis that an abnormal DST may reflect disturbances of the central neurotransmission in the CNS[19] and that the treatment with specific 5-HT reuptake blockers in demented patients may reduce the overactivity in the HPA axis. The aim of this study was to investigate more closely the relationship between mental impairment in dementia disorders and the frequency of pathological DST and post-DST cortisol levels.

MATERIAL AND METHODS

As shown in TABLE 1, 163 patients with dementia disorders according to DSM-III-R formed the basis of study. The patients were admitted to the psychogeriatric diagnostic unit for diagnosis and evaluation of the severity and profile of their dementias. All the patients underwent psychiatric and physical examinations, including a comprehensive laboratory investigation, ECG, EEG and CT scans of the brain to exclude other causes of the development of dementia symptoms. The estimates of the duration of the disease were based on interviews with patients, key persons and relatives. AD (n = 40) was diagnosed in patients with onset of disease before 65 years of age, while SDAT (m = 56) was diagnosed in patients with onset of disease at the age of 65 or after. The diagnoses were made according to the criteria outlined in DSM-III-R and the NINCDS-ADRDA diagnostic manual.[20] The VAD diagnosis (n = 45) was made in patients with TIA/stroke episodes at the time of dementia onset and/or early in the course of the disease. This diagnosis was also given to patients with pronounced vascular disease that could be assumed to have contributed substantially to the manifestation of dementia. Thus, the VAD group

TABLE 1. Characterization of Patients in Four Diagnostic Groups: Alzheimer's Disease (AD), Senile Dementia of the Alzheimer Type (SDAT), Vascular Dementia (VAD), and Unspecified Dementia (NUD)

					Degree of Dementia		
	n	Age	Male/Female	Duration	Mild (%)	Moderate (%)	Severe (%)
AD	40	64 ± 5 (54–74)	22/18	7 ± 3 (2–16)	8 (20)	22 (55)	10 (25)
SDAT	56	77 ± 7 (68–87)	41/15	4 ± 2 (1–11)	10 (18)	31 (55)	15 (27)
VAD	45	74 ± 6 (62–86)	21/24	4 ± 4 (1–20)	8 (18)	21 (47)	16 (35)
NUD	22	74 ± 5 (62–82)	10/12	5 ± 3 (1–10)	0 (0)	19 (86)	3 (14)

included not only multiinfarct dementia but also dementia in which small vessel disease or disorders of the systemic circulation were assumed to be of pathogenetic importance.

The diagnosis NUD (n = 22) was given to patients who could not be assigned to the AD, SDAT or VAD groups, *i.e.,* cases with mixed dementia, or in whom confounding factors, such as alcohol abuse, exposure to organic solvents, were present.

The degree of dementia was evaluated as mild, moderate or severe according to DSM-III-R by the physician in charge. The patients were assessed by ratings on the GBS scale.[21] This scale gives a quantitative measure and a profile of dementia from different aspects. Three subscales measure impairment of motor performance (ADL) and intellectual and emotional impairment. The scale also includes six symptoms that are common in dementia: confusion, irritably, anxiety, fear-panic, depressed mood and restlessness. Each item is given a score between 0 and 6; 0 represents absence of impairment or symptomatology and 6 represents total impairment or severe symptoms.

The patients received 1 mg dexamethasone (Decadron®) at 10 p.m. The tablets were administered with particular attention paid to the compliance of the patients. Venous blood samples for determination of cortisol concentrations were drawn at 8.00 a.m. on the day of dexamethasone administration (pre-DST cortisol concentration) and at 8.00 a.m. and 3.00 p.m. on the following day. Because of the strong relationship between the post-DST cortisol levels taken at 8.00 a.m. and at 3.00 p.m. (r = 0.75, n = 155, p <0,0001), only the values at 3.00 p.m. are presented. Serum cortisol concentrations were measured using a radioimmunoassay, as described in detail by Balldin *et al.*[22]

Forty-two of the patients received small amounts of neuroleptics (daily dose 2 mg or less of Haldol or equivalent doses of other neuroleptics) and/or antidepressants (daily dose 70 mg or less of Lofepramin or equivalent doses of other antidepressants) before admission to the unit. These treatment regimens remained unchanged over a three-week period before admission and during the study.

STATISTICS

Kruskal-Wallis test was used to identify differences between groups. Correlations between serum cortisol concentrations and scores on the GBS scale were calculated using the Spearman rank correlation test. Differences in GBS scores between patients with pathological and nonpathological DST were evaluated for each diagnostic group using the Kruskal-Wallis chi-square approximation.

RESULTS

TABLE 1 shows number of cases, age, sex, duration of disorder and degree of dementia according to DSM-III-R.

Basal cortisol levels correlated significantly with post-DST cortisol levels (r = 0.28; p <0.0005; n = 147). There were no significant differences in basal cortisol levels between the diagnostic groups. The percentage of pathological DST (escapers) was lowest in the AD group (40%), somewhat higher in the VAD group (49%), still higher in the SDAT group (54%) and highest in the NUD group (59%), but the differences between the groups were not significant. Basal cortisol levels did not vary significantly with the degree of dementia, as assessed according to DSM-III-R (mild, moderate and severe), but post-DST cortisol levels varied from 100 ± 106 (mean ± SD) in the mildly demented through 155 ± 130 in the moderately demented to 195 ± 141 in the severely demented patients (p <0.01) (FIG. 1). Increases in post-DST cortisol levels with the degree of dementia were also noted in all diagnostic groups, but reached a significant level only in the VAD group (FIG. 1). In the total group (n = 163), 7/26 (27%) with mild, 47/93 (51%) with moderate and 27/144 (61%) with severe dementia had pathological DST, which is a significant increase (p <0.05). In the diagnostic groups the increase with the degree of dementia was significant (p <0.05) only in the VAD group in which 1/8 (12%) with mild, 10/21 (48%) with moderate and 11/16 (69%) with severe dementia had pathological DST.

In TABLE 2 GBS scores are related to post-DST cortisol levels. As shown in the table, there were no significant correlations between GBS scores and post-DST cortisol levels in the AD and NUD groups. In the SDAT group impairment of motor performance, intellectual impairment, confusion, depressed mood and restlessness correlated with post-DST cortisol levels, if a probability level of 0.05 is accepted as significant. The strongest correlations were found in the VAD group in which all GBS variables but depressed mood correlated significantly with post-DST cortisol levels. With significance at a probability level of 0.01, intellectual impairment, fear-panic and restlessness were found to correlate significantly with post-DST cortisol levels in the VAD group. In the total group intellectual impairment, confusion, anxiety, fear-panic, depressed mood and restlessness all correlated significantly with the post-DST levels.

The differences in GBS scores between subgroups with pathological and non-pathological DST were significant in the VAD group for intellectual impairment, anxiety, fear-panic and restlessness. In the SDAT group the subgroup with pathological DST had significantly more depressed mood than the subgroup with non-pathological DST.

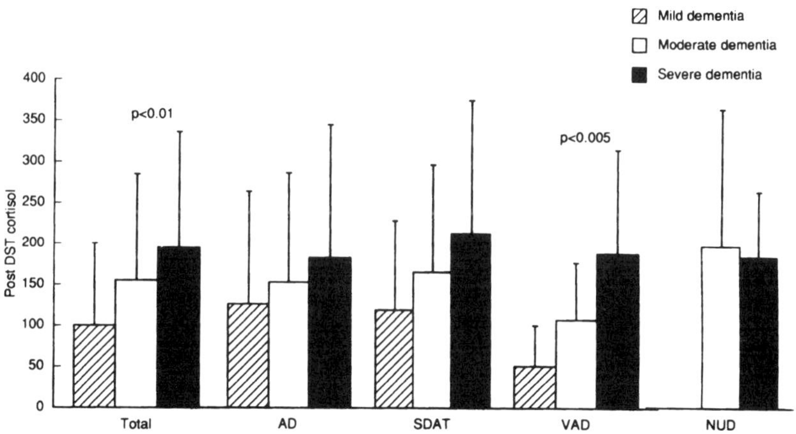

FIGURE 1. Post-DST cortisol levels in subgroups of mild, moderate and severe dementia in the diagnostic groups Alzheimer's disease (AD), senile dementia of the Alzheimer type (SDAT), vascular dementia (VAD) and dementia of unspecified type (NUD).

TABLE 2. Correlations between Post-DST Cortisol Levels and GBS Scores in Diagnostic Subgroups of Demented Patients and in the Total Group According to the Spearman Rank Correlation Test.

	AD (n = 34)	SDAT (n = 49)	VAD (n = 40)	NUD (n = 20)	Total (n = 143)
Motor impairment	0.00	0.35*	0.31*	0.02	0.20
Intellectual impairment	0.08	0.29*	0.43***	0.10	0.29***
Emotional impairment	−0.04	0.04	0.37*	0.06	0.14
Confusion	0.07	0.35*	0.38*	0.26	0.30***
Irritability	−0.12	0.24	0.36*	−0.19	0.12
Anxiety	0.25	0.25	0.42**	−0.17	0.24***
Fear-panic	0.25	0.15	0.49***	−0.18	0.23***
Depressed mood	0.16	0.34*	0.18	0.19	0.24***
Restlessness	0.15	0.28	0.49***	0.09	0.30***

*$p < 0.05$.
**$p < 0.01$.
***$p < 0.005$.

DISCUSSION

As reported by several investigators and also confirmed in this study, 40–60% of demented patients give insufficient response to DST, indicating either that their feedback systems are out of order or that their HPA axis activity is no longer influenced by the feedback signals.

Based on the finding of relatively high basal cortisol levels (505 ± 143) in this study of demented patients and the high frequency of pathological DST, together with the clinical picture of irritability, anxiety, fear-panic and restlessness, we conclude that there is continuous overactivity in the hypothalamus in a substantial proportion of demented patients, a notion that is supported by results of postmortem human brain studies.[23] To some extent in agreement with Sapolsky and McEwen[24] we hypothesize that the high HPA activity in demented patients is attributable to a reduced central inhibitory tone over the hypothalamus that is secondary to a neurotransmitter deficit. We believe that, in demented patients, pathological DST is not associated (at least not exclusively) with the presence of depression, but with the severity of dementia (the degree of disturbed central neurotransmitter function). To test our hypothesis we studied correlations between degree of dementia and symptomatology in demented patients and their pathological DST and post-DST cortisol levels. Depression was assessed by ratings on one item of the GBS scale with seven defined scale steps (0–6). Degree of dementia was assessed according to DSM-III-R as mild, moderate or severe and also by ratings on GBS subscales. In the total group there were significant positive correlations between scores on the GBS intellectual subscale or degree of dementia according to the DSM-III-R and the number of pathological DST or post-DST cortisol levels. These correlations were especially strong in the VAD group; they bordered on significance in the SDAT group, but were not significant in the AD and NUD groups.

VAD is a dementia of the subcortical type, with moderate to severe white matter lesions, with or without lacunas. White matter lesions—mild to moderate—are also present in SDAT, as are various vascular risk factors.[25,26] In VAD and SDAT the long axons between the cortex and the subcortical centers may be deranged as a consequence of the white matter lesions. The disconnection between the cortex (including hippocampus) and the subcortical centers, such as the hypothalamus, may explain the uncontrolled activity in the HPA axis.

The correlations between scores on the item of depression and post-DST cortisol levels were significant in the SDAT group but not in the VAD group. Thus, at least in the VAD group, the presence of depression can hardly explain the high frequency of pathological DST. It is probable that the connection between the hippocampus and hypothalamus is disturbed in subcortical dementias. In postmortem investigations significant correlations were found between disturbed 5-HT activity in the hippocampus and neuropeptide activity in the hypothalamus.[27] It is possible that the feedback system that is tested by the administration of dexamethasone is partly mediated by 5-HT axons. The pharmacological strategy to treat some behavioral disturbances with selective serotonin reuptake blockers is based on this hypothesis which has been supported by results of clinical trials.[12,28,29]

When we tried to characterize VAD patients with pathological DST by analyzing GBS scores, we found that these patients were significantly more intellectually impaired and showed a higher degree of anxiety, restlessness and fear-panic than VAD patients with normal DST.

Overactivity in the HPA axis is attributable to stress experienced by the demented patients. An insufficient feedback system that cannot switch off the increased activity leads to chronic stress adaptation failure. To restore the feedback system therapeutic interventions with psychological, as well as pharmacological measures are called for.

SUMMARY

In 163 patients with dementia disorders, subdivided into Alzheimer's disease with early onset (AD; n = 40), senile dementia of the Alzheimer type (SDAT; n = 56), vascular dementia (VAD; n = 45) and dementia of unspecified type (NUD; n = 22) the dexamethasone suppression test (DST) was performed. The patients were rated according to the DSM-III-R criteria as having mild, moderate or severe dementia and were also assessed using the GBS scale which gives a profile of the dementia syndrome. In the total group of dementia there were significant correlations between severity of dementia and post-DST levels. The frequency of pathological DST also correlated significantly with the severity of dementia. In the subgroups of dementia a strong correlation between severity of dementia and high post-DST cortisol levels was found only in the VAD group. Between the subgroups of dementia disorders there were no significant differences in basal cortisol levels. The percentage of pathological DST was lowest in the AD group (40%). It was somewhat higher in the VAD group (49%), still higher in the SDAT group (54%) and highest in the NUD group (59%). When the relationship between post-DST cortisol levels and GBS scores was analyzed, significant correlations were found mainly in the VAD group. There intellectual impairment, anxiety, fear-panic and restlessness correlated significantly with post-DST cortisol levels. The results indicate hypothalamic overactivity in a substantial number of demented patients. In VAD and to a certain extent also in SDAT a disconnection between cortical areas, including the hippocampus, and the hypothalamus is assumed. Overactivity in the hypothalamic-pituitary-adrenal (HPA) axis is due to stress, and an insufficient feedback system leads to chronic stress adaptation failure.

REFERENCES

1. WALLIN, A. & K. BLENNOW. 1991. Pathogenetic basis of vascular dementia. Alz. Dis. Assoc. Dis. **5:** 91–102.
2. COWBURN, R. F., J. A. HARDY & P. J. ROBERTS. 1989. Neurotransmitter deficits in Alzheimer's disease. *In* Alzheimer's Disease: Towards an Understanding of the Etiology and Pathogenesis. D. C. Davies, Ed. 9–32. John Libbey. London and Paris.
3. GOTTFRIES, C. G. 1992. Clinical and neurochemical aspects on diseases with cognitive impairment. Rev. Neurosci. **3:** 191–206.
4. SAPOLSKY, R. M., L. C. KREY & B. S. MCEWEN. 1984. Glucocorticoid-sensitive hippocampal neurons are involved in terminating the adrenocortical stress response. Proc. Natl. Acad. Sci. USA **81:** 6174–6177.
5. ROSSOR, M. N., L. L. IVERSEN, C. Q. MOUNTJOY, M. ROTH, M., J. HAWTHORN, W. Y. ANG & J. S. JENKENS. 1980. Arginine vasopressin and cholineacetyl transferase in brains of patients with Alzheimer-type senile dementia. Lancet **ii:** 1367–1368.
6. SUNDQUIST, J., M. L. FORSLING, J. E. OLSSON & M. ÅKERLUND. 1983. Cerebrospinal fluid arginine vasopressin in degenerated disorders and other neurological diseases. J. Neurol. Neurosurg. Psychiatry **46:** 14–17.
7. MAZUREK, M. F., J. H. GROWDON, M. F. BEAL & J. B. MARTIN. 1986. CSF vasopressin concentration is reduced in Alzheimer's disease. Neurology **36:** 1133–1137.
8. FUJIYOSHI, K., H. SUGA, K. OKAMOTO, S. NAKAMURA & M. KAMEYAMA. 1987. Reduction of arginine-vasopressin in the cerebral cortex in Alzheimer-type senile dementia. J. Neurol. Neurosurg. Psychiatry **50:** 929–932.
9. WALLIN, A., A. CARLSSON, R. EKMAN, C. G. GOTTFRIES, I. KARLSSON, L. SVENNERHOLM & E. WIDERLÖV. 1991. Hypothalmic monoamines and neuropeptides in dementia. Eur. J. Neuropsychopharmacol. **1:** 165–168.
10. BISSETTE, G., G. P. REYNOLDS, C. D. KITT, E. WIDERLÖV & C. B. NEMEROFF. 1985. Corticotropin-releasing factor-like immunoreactivity in senile dementia of the Alzheimer type. J. Am. Med. Assoc. **254:** 3067–3069.
11. NEMEROFF, C. B., S. J. KIZER, G. P. REYNOLDS & G. BISSETTE. 1989. Neuropeptides in Alzheimer's disease: a postmortem study. Regul. Pept. **25:** 123–130.
12. CARLSSON, A. & C. G. GOTTFRIES. 1986. Neurotransmitter abnormalties in old age dementias. *In* Proceedings of the Vth South-East European Neuropsychiatric Conference, Graz 1983. H. Leichner & A. Paraschos, Eds. 634–645. University Studio Press. Salonica, Greece.
13. WALLIN, A., I. ALAFUZOFF, A. CARLSSON, S. Å. ECKERNÄS, C. G. GOTTFRIES, I. KARLSSON, L. SVENNERHOLM & B. WINBLAD. 1989. Neurotransmitter deficits in a non-multi-infarct category of vascular dementia. Acta Neurol. Scand. **79:** 397–406.
14. GOTTFRIES, C. G., I. ALAFUZOFF, A. CARLSSON, S.-Å. ECKERNÄS, I. KARLSSON, L. ORELAND, L. SVENNERHOLM & A. WALLIN. 1991. Neurochemical changes in brains from patients with vascular dementia. *In* Cerebral Ischemia and Dementia. A. Hartman, W. Kuchinsky & S. Hoyer, Eds. 94–102. Springer-Verlag. Berlin.
15. BLENNOW, K., A. WALLIN, P. FREDMAN, C. G. GOTTFRIES, I. KARLSSON & L. SVENNERHOLM. 1990. Blood-brain barrier disturbance in patients with Alzheimer's disease is related to vascular factors. Acta. Neurol. Scand. **81:** 323–326.
16. ENGLUND, E. 1988. A white matter disease in dementia. A study with special reference to Alzheimer's disease. Academic dissertation. Lund U. Lund, Sweden.
17. O'BRIEN, J., D. AMES & I. SCHWEITZER. 1993. HPA axis function in depression and dementia: a review. Int. J. Geriatr. Psychiatry **8:** 887–898.
18. OLSSON, T., M. ÅSTRÖM, S. ERIKSSON & Å. FORSELL. 1989. Hypercortisolism revealed by the dexamethasone suppression test with acute ischemic stroke. Stroke **20**(12): 1685–1690.
19. BALLDIN, J., C. G. GOTTFRIES, I. KARLSSON, G. LINDSTEDT, G. LÅNGSTRÖM & L. SVENNERHOLM. 1988. Relationship between DST and the serotonergic system. Results from treatments with two 5-HT reuptake blockers in dementia disorders. Int. J. Geriatr. Psychiatry **3:** 17–26.
20. MCKHANN, G., D. DRACHMAN, M. FOLSTEIN, R. KATZMAN, D. PRICE & E. M. STAD-

LAN. 1984. Clinical diagnosis of Alzheimer's disease: report of the NINCDS-ADRDA Work Group under the auspices of Department of Health and Human Services Task Force on Alzheimer's Disease. Neurology **34:** 939–944.
21. GOTTFRIES, C. G., G. BRÅNE, B. GULLBERG & G. STEEN. 1982. A new rating scale for dementia syndromes. Arch. Gerontol. Geriatr. **1:** 311–330.
22. BALLDIN, J., C. G. GOTTFRIES, I. KARLSSON, G. LINDSTEDT, G. LÅNGSTRÖM & J. WÅLINDER. 1983. Dexamethasone suppression test and serum prolactin in dementia disorders. Br. J. Psychiatry **143:** 277–281.
23. WALLIN, A., A. CARLSSON, R. EKMAN, C. G. GOTTFRIES, I. KARLSSON, L. SVENNERHOLM & E. WIDERLÖV. 1991. Hypothalamic monoamines and neuropeptides in dementia. Eur. J. Neuropsychopharmacol. **1:** 165–168.
24. SAPOLSKY, R. M. & B. S. MCEWEN. 1986. Stress, glucocorticoids, and their role in degenerative changes in the aging hippocampus. *In* Treatment Development Strategies for Alzheimer's Disease. T. Crook, R. Bartus, S. Ferris & S. Gershon, Eds. 151–171. Mark Powley. Canaan, CT.
25. BLENNOW, K., A. WALLIN, C. UHLEMANN & C. G. GOTTFRIES. 1991. White matter lesions in Alzheimer's disease—relation to clinical symptomatology and vascular factors. Acta Neurol. Scand. **83:** 187–193.
26. SCHELTENS, P. H., F. BARKHOF, J. VALK, P. R. ALGRA, G. VAN DER HOOP, J. NAUTA & C. H. WOLTERS. 1992. White matter lesions on magnetic resonance imaging in clinically diagnosed Alzheimer's disease. Brain **115:** 735–748.
27. GOTTFRIES, C. G. 1991. Disturbance of the 5-hydroxytryptamine metabolism in ageing and in Alzheimer's and vascular dementias. *In* Psychiatry, a Spectrum of Ideas. M. Sandler, A. Coppen & S. Harnett, Eds. 309–323. Oxford University Press. Oxford, New York and Tokyo.
28. NYTH, A. L. & C. G. GOTTFRIES. 1990. The clinical efficacy of citalopram in treatment of emotional disturbances in dementia disorders. A Nordic multicentre study. Br. J. Psychiatry **157:** 894–901.
29. NYTH, A. L., C. G. GOTTFRIES, K. LYBY, L. SMEDEGAARD-ANDERSEN, J. GYLDING-SABROE, M. KRISTENSEN, H. E. REFSUM, E. ÖFSTY, S. ERIKSSON & S. SYVERSEN. 1992. A controlled mulicenter clinical study of citalopram and placebo in elderly depressed patients with and without concomitant dementia. Acta Psychiatr. Scand. **86:** 138–145.

DISCUSSION OF THE PAPER

QUESTION: In your citalopram study, did you measure the DST in the placebo group also before and after?

GOTTFRIES: Yes. We made a study where we measured one control group and one placebo group. The control group consisted of elderly people on a ward that was very calm, that is, the environment did not disturb or stress the patients at all. There only 30% of the demented had a pathological DST. The placebo group consisted of patients who were admitted to the hospital for 3 days. This means that they probably felt stressed. We tested them before and after placebo treatment and obtained approximately the same number of pathological tests both times, *i.e.,* 50% had pathological DST. There was a variance in that some of those who had a pathological test on the first occasion had a normal test on the second. Our conclusion is that stress is an important factor in pathological HPA activity.

QUESTION: Was there a statistical difference between the pre- and post-dexamethasone in the citalopram group vis-à-vis the placebo group?

GOTTFRIES: We did not measure that; we measured only before and after treatment. My guess would be yes.

QUESTION: Is there a relationship between body weight or food intake and the dexamethasone suppressibility in the demented group?

GOTTFRIES: I am sorry I cannot answer that question. We have the necessary data, but we have performed no such analysis. Why do you ask?

QUESTION: People with poor food intake are inclined to be resistant to dexamethasone suppression.

GOTTFRIES: This may be relevant only for a subgroup of Alzheimer-demented people, those who lose weight.

QUESTION: Regarding DST and nonsuppression, did you find increased basal cortisol levels in your demented populations?

GOTTFRIES: Yes, we did. They were around 550, while they mostly were at least 100 nmol less in the control group. Thus, the levels were increased, but there were no significant differences between Alzheimer-demented subgroups (AD, SDAT) or the subgroup with vascular dementia.

QUESTION: The problem is with the interpretation of this suppression, that is, the failure to suppress beyond a particular point. If these people start off somewhat elevated, if you given them dexamethasone they may not suppress underneath. Five micrograms would be the magical number, but they still may be suppressing to the same extent. So if you just did a preimposed treatment, do they suppress to the same degree or are there big differences in the percentage of baseline?

GOTTFRIES: I cannot answer that question.

QUESTION: In our 1982 study there were no differences in basal cortisol concentrations. I think the Mt. Sinai group in New York has demonstrated modest elevations in cortisol over a 24-hour period in Alzheimer's patients compared to controls. However, the percent depression one would be extrapolating from those data would be less than the Alzheimer's subjects with severe disease in a normal elderly person.

GOTTFRIES: The longer demented patients stay on a ward with a peaceful atmosphere, the more closely their DST will come to normal values. After acute admittance to hospital, the cortisol levels can be assumed to be high and it takes a long time for the demented patients to normalize their HPA activity. If you have a short observation period, you will not record these changes.

QUESTION: In the citalopram study do the effects outlast the treatment? If you test the patients at eight weeks or sixteen weeks after cessation of the drug are they better, worse, the same?

GOTTFRIES: In our program an investigation was included in which DST was made a couple of days after cessation of the pharmacological treatment. The result of the DST was the same if the patients were on the drug or if the drug had been withdrawn a couple of days before DST. We have made no DST in patients in whom citalopram treatment has been withdrawn for 8 or 16 weeks. Clinical experience indicates that there is a relapse of emotional disturbances after cessation of citalopram treatment, but so far we have no scientific data to show this.

QUESTION: Dexamethasone has known actions at the anter pituitary level and in various other conditions where you get pathological suppression by dexamethasone. It has been attempted to determine whether the defect is at the pituitary level or above. Has that been done for demented patients?

GOTTFRIES: I do not know any such studies. They would be interesting.

COMMENT: I know of one paper with CRH infusion and measuring the ACTH response in Alzheimer's patients. As in people with a primary depression who did not have a dementia, those in this study had a blunted ACTH response to CRH. This was interpreted as downregulation of CRH receptors on the corticotrope. But there were other possible explanations.

Steroid Effects on Central Neurons and Implications for Psychiatric and Neurological Disorders

FLORIAN HOLSBOER,[a] ANNETTE GRASSER,
ELISABETH FRIESS, AND KLAUS WIEDEMANN

Max Planck Institute of Psychiatry
Clinical Institute
Kraepelinstrasse 2-10
D-80804 Munich, Germany

INTRODUCTION

Stressful stimuli produce a variety of physiological changes, including activation of the pituitary-adrenocortical system, in order to maintain homeostasis. In normal controls this neuroendocrine activation is counteracted by glucocorticoid receptor-mediated negative feedback, which promptly curtails heightened plasma corticosteroid levels. The mechanisms involved in this negative feedback regulation include corticosteroid receptor action not only at the anterior pituitary but also at the hypothalamus, the hippocampus and possibly also other sites inside and outside the limbic system. The dynamic changes in corticosteroid receptor capacity and function are of considerable interest for studies on the causes and treatment of psychiatric and neurological disorders. There are a number of reasons for this: (1) In patients with psychiatric disorders, particularly those with depression, altered function of the hypothalamic-pituitary-adrenocortical (HPA) system is the leading neuroendocrine sign; (2) antidepressants appear to act directly on various regulatory elements of the HPA system; (3) neurological diseases are frequently accompanied by changes in the HPA system; (4) some of these changes have serious implications for treatment with corticosteroids; and (5) naturally occurring or synthetic steroids act on brain cells in specific ways, resulting in behavioral changes.

This review summarizes some of the clinical findings related to disturbed corticosteroid receptor function among patients with affective disorder, multiple sclerosis and HIV infection and demonstrates how steroids derived from the adrenals or brain cells and synthetic steroids can be studied at the systemic level by sleep electroencephalogram (EEG) analysis.

The HPA System in Affective Disorder

Patients with depression exhibit increased HPA activity, as demonstrated by an increased ACTH pulse frequency and an increased magnitude of cortisol bursts (for review see Ref. 1). This HPA overactivity is further reflected in elevated urinary "free" cortisol (UFC) levels, which appear to be about twice as high in depressives as in normal controls, but which are still lower than in patients with Cushing's syndrome. Furthermore, salivary and cerebrospinal fluid (CSF) concentrations of

[a]Corresponding author.

cortisol, which represent the "free" (*i.e.,* unbound) fraction of cortisol, are increased in depression. Finally, Nemeroff *et al.*[2] found that the concentration of corticotropin-releasing hormone (CRH) in the CSF is elevated in depressives.

One method for assessing the activity of the HPA system is intravenous administration of CRH and measurement of the ensuing ACTH and corticosteroid response. In drug-free depressed patients ACTH responses to CRH are blunted, which is probably due to chronic hypersecretion of CRH from the nerve terminals in the median eminence, resulting in a reduced number of anterior pituitary CRH receptors. This reduced CRH receptor capacity together with the elevated levels of plasma cortisol in these patients results in a decreased responsiveness to CRH. The desensitization of CRH receptors in the pituitary in response to excessive CRH secretion has been documented both in laboratory animals[3] and in humans (for review see Refs. 1, 4). In a study of postmortem tissue from suicide victims Nemeroff and co-workers found the density of CRH receptors to be decreased in the frontal cortex, supporting the notion that excessive CRH secretion downregulates CRH receptors among depressed or chronically stressed humans.[4]

That circulating corticosteroids suppress ACTH in response to CRH is indicated by the finding that metyrapone-pretreated depressed patients do not have a blunted ACTH response.[5] Metyrapone inhibits the biosynthesis of cortisol by blocking C_{11}-steroid hydroxylase, thus depriving the system of its major feedback signal. In light of the reciprocal interaction between baseline cortisol and the amount of ACTH released after CRH stimulation and the enhanced ACTH response to CRH among metyrapone-treated subjects, the following finding is surprising. When patients with depression are given a dexamethasone suppression test (DST) and are challenged with CRH during the afternoon of the next day, one would expect that the combination of inadequately suppressed cortisol levels and dexamethasone would be additive in blunting ACTH release. Normal volunteers show an ACTH and cortisol response to CRH that gradually decreases with increasing dosages of dexamethasone after CRH[6] (FIG. 1).

In contrast, depressives show a paradoxical pattern insofar as dexamethasone pretreatment results in increased ACTH and cortisol responses to CRH despite combined endogenous (elevated cortisol) and exogenous (dexamethasone) glucocorticoid levels[7] (FIG. 2). This abnormality disappears after successful antidepressant treatment.[8] The mechanism underlying this phenomenon is not yet fully elucidated, but vasopressin appears to play a role as a possible synergizer of the CRH effects on corticotrophic cells. In dexamethasone-pretreated normal controls even high dosages of CRH do not result in plasma ACTH and cortisol levels like those seen among depressed DST nonsuppressors who were not CRH-stimulated. This suggests that CRH alone may not account for DST nonsuppression in depressives. When dexamethasone-pretreated normal controls received vasopressin and CRH in combination, ACTH and cortisol secretory patterns emerged that strongly resemble those seen in depressives.[9]

From these data and a vast number of preclinical investigations, suggesting a profound synergy between CRH and vasopressin at the pituitary level,[10] the following hypothesis was derived:[1] In depression as in chronic stress there is a shift to a gradually intensifying vasopressinergic regulation of the pituitary adrenocortical system. Dexamethasone, which does not bind to corticosteroid-binding globulins, thus causes the HPA suppression primarily at the level of the pituitary corticotrophs and does not suppress hypothalamic CRH and vasopressin as effectively as endogenous corticosteroids. A transient corticosteroid receptor desensitization gradually develops and vasopressin expression, which reacts more sensitively to changes in

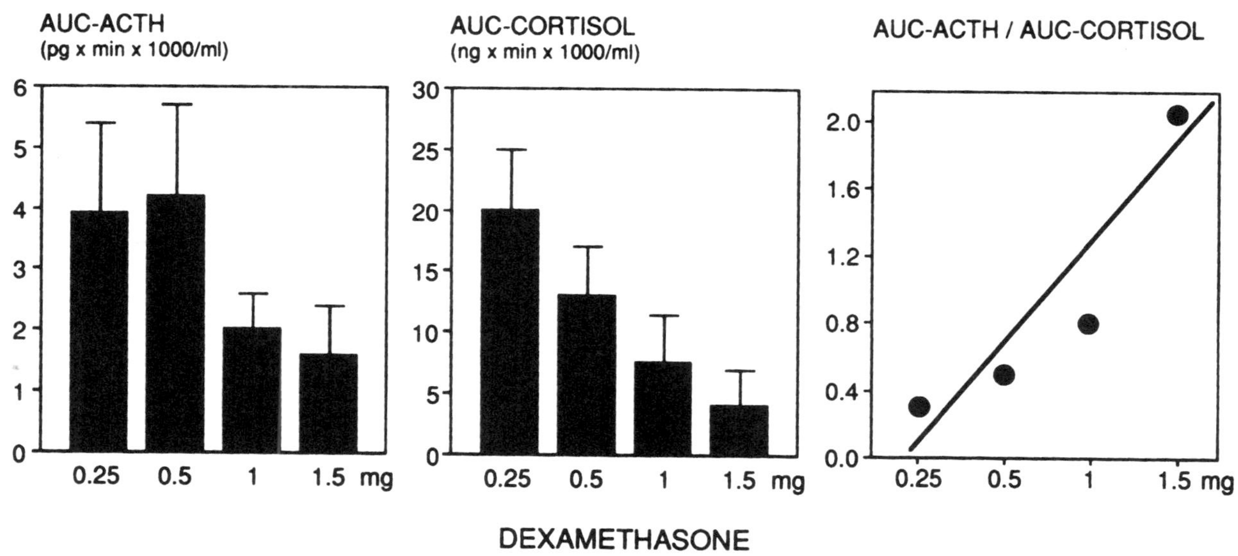

FIGURE 1. ACTH and cortisol response to CRH in controls pretreated with increasing doses of dexamethasone. With increasing doses of oral dexamethasone (23.00 h) plasma ACTH and cortisol response to 100 μg CRH (15.00 h) become gradually reduced while the ACTH/cortisol ratios increase. (From Heuser *et al.*[6] Reprinted by permission from the *Journal of Psychiatric Research.*)

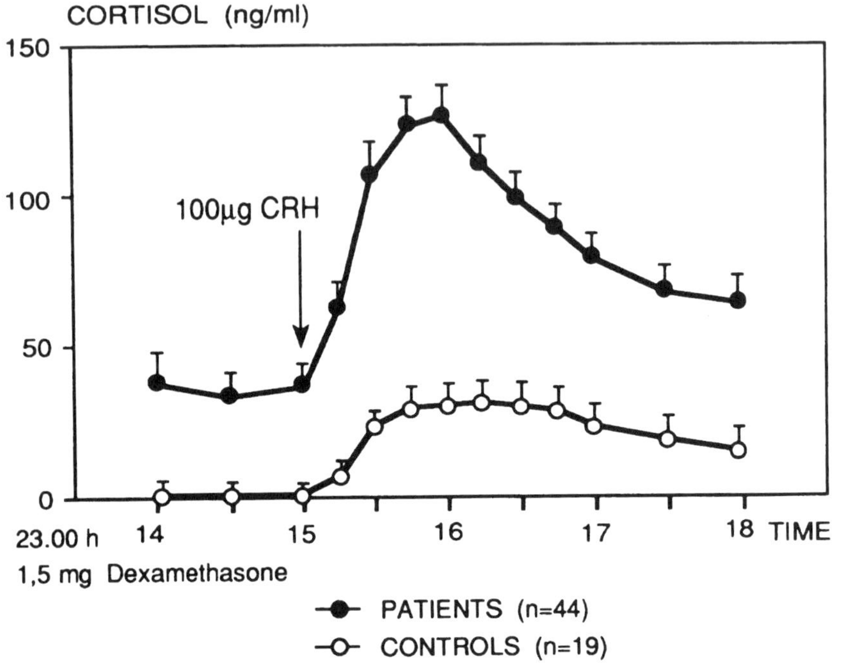

FIGURE 2. DEX/CRH test in depression. In depressives plasma cortisol secretion is enhanced after CRH injection (100 µg; 15.00 h) despite elevated plasma cortisol concentrations and pretreatment with 1.5 mg dexamethasone the night before. (Data from Von Bardeleben & Holsboer[7] and Heuser *et al.*[12])

glucocorticoid concentration than does CRH expression, becomes less effectively suppressed than CRH by circulating corticosteroids. As a result, in a depressed dexamethasone-pretreated patient secretion of ACTH in response to exogenous CRH would be greater than in a control because of the synergistic interaction of the administered CRH bolus with the larger amount of vasopressin present at corticotrophs due to insufficient suppression by dexamethasone.

Recently the above hypothesis received substantial support from preclinical studies, which showed that CRH neurons that project to the median eminence are multimessenger neurons with the capacity to produce arginine vasopressin (AVP) in neurons that under normal conditions belong to the non-AVP-producing CRH phenotype.[10] This remarkable plasticity allows an increase in the AVP/CRH ratio of the signal produced and may explain why dexamethasone-pretreated depressed patients show excessive ACTH and cortisol release after CRH stimulation. We believe that enhanced AVP release synergizes with exogenous CRH to override dexamethasone suppression at the pituitary in these patients. The mechanisms mediating this switch to increased AVP/CRH ratios imply a severe disturbance of negative feedback in these patients. In a series of studies Young *et al.*[11] showed a decrease in β-endorphin and β-lipotropin release concomitantly with rapidly increasing plasma cortisol levels among normal controls, but not among depressed patients. From these findings they concluded that the fast feedback effects of cortisol upon β-endorphin and β-lipotropin are disturbed at the level of the corticosteroid receptor-mediated inhibition of synthesis and release of ACTH secretagogues.

HPA Changes and Therapeutic Response

Whatever the exact mechanism underlying the enhanced cortisol release in the DEX-CRH test may be, the test would appear to be a very useful tool for characterizing patients with depression in order to monitor their response to therapeutic intervention. Heuser *et al.*[12] compared the effects of a six-week regimen of amitriptyline upon HPA regulation in elderly depressed patients and matched healthy controls. DEX-CRH tests were administered before treatment initiation and three times during treatment. The exaggerated plasma cortisol response among depressed patients began to disappear after one week of treatment, thus preceding antidepressant effects upon psychopathology. This effect was more pronounced among responders than among nonresponders, supporting the view that antidepressants act through reinstatement of corticosteroid receptor function.

These findings prompted us to study the effects of amitriptyline and moclobemide (a reversible inhibitor of MAO-A) in rats similarly to what we did in depressed patients. Both amitriptyline and moclobemide transiently elevated hippocampal mineralocorticoid receptor (MR) by 40–70% between 2 and 5 weeks after the start of treatment.[13] At 7 weeks the increments in MR had largely disappeared. Hypothalamic glucocorticoid receptor (GR) was increased by 20–25% at 5 weeks of treatment, suggesting that antidepressant-induced changes in brain corticosteroid capacity may underlie the observed decrease in circulating ACTH and corticosterone levels and the decreased adrenal size. Furthermore, when challenged by a stressor, antidepressant-treated rats showed a decreased ACTH and corticosterone response, which is understandable in terms of an attenuated locus coeruleus-CRH reagibility, possibly through enhanced effectiveness of negative corticosteroid feedback due to reestablished GR and MR capacity. Moreover, CRH binding and POMCmRNA were reduced among moclobemide-treated animals, which also indicates that antidepressant action involves decreased activation of the HPA system at various levels (Reul, unpublished observation).

Effects of Trauma on HPA Activity

The long-term effects of physiological and psychological trauma on neuroendocrine function have been the subject of recent research efforts because patients who fulfill the diagnostic criteria of post-traumatic stress disorder (PTSD) have highly variable cognitive symptoms (intrusive recollections, nightmares, pronounced startle reactions, etc.), suggesting an autonomic nervous system disturbance and altered stress hormone regulation. Studies exploring different cortisol measures at baseline did not provide a clear indication of heightened HPA activity.[14,15] In studies of patients with combat PTSD some authors, but not all, found decreased urinary-free cortisol concentrations. In addition, DST and plasma cortisol concentrations in these patients were decreased even if PTSD was accompanied by depression.

In the CRH test PTSD patients showed a blunted ACTH response but a normal cortisol response in the presence of only slightly elevated plasma cortisol concentrations at baseline.[16] This pattern is similar to that of patients with panic disorder and suggests that acute episodes of PTSD such as flashbacks and nightmares may have desensitizing effects on cortiocotrophic CRH receptors similar to those in episodic

HPA excess, which through trophic effects of ACTH renders the adrenal gland hypersensitive in both disorders.

One possible explanation for the low cortisol secretion at baseline was provided by Yehuda *et al.*,[17] who observed increased lymphocyte glucocorticoid receptor density in patients with PTSD and suggested that if such a receptor increase were also present in the hippocampus, the hypothalamus and the pituitary, then an increased negative feedback sensitivity would develop. Although glucocorticoid receptor changes are not necessarily uniform in all tissues, this finding points to a rather specific HPA alteration in these patients.

Other types of extreme stressors or trauma have effects that are clearly different from those of combat-related PTSD. Excessive physical stress such as that experienced by elderly marathon runners produces neuroendocrine signs of decreased brain corticosteriod receptor capacity and thus weakened negative feedback.[18] In a small group of women who had experienced a violent sexual assault and fulfilled the criteria for PTSD, no signs of neuroendocrine abnormality were detected. Apparently, gender, the type of trauma and possibly also consumption of psychoactive substances (which is common among combat veterans) may be critical determinants of HPA function in PTSD.

HPA Function in Multiple Sclerosis

Multiple sclerosis (MS) is characterized by focal perivascular inflammatory cell infiltrates and demyelination in the central nervous system. In the beginning the disease is usually episodic, but every recurrence is accompanied by new patches of demyelination. The cause of MS is unknown and there is no effective treatment available as yet. If the MS is active, treatment with synthetic corticosteroids is often initiated, and some patients respond favorably. The variance in response to corticosteroid treatment raises the question of differences in corticosteriod receptor capacity. This led us to conduct a combined DEX-CRH test in 19 patients with MS.[19] We found that as a group these patients had increased cortisol responses similar to those of patients with depression (FIG. 3). When we excluded those patients whose cortisol secretion was in the normal range, we were left with a group of excessive responders (hyperresponders) and hyporesponders (FIG. 4). The hyperresponse to DEX-CRH tests was not associated with gender, age, degree of affective symptoms or course of illness (relapsing-remitting or chronic progressive).

In summary, a hyperresponse of ACTH and cortisol after DEX-CRH testing is believed to be indicative of decreased neuronal corticosteriod receptor capacity. If such a mechanism is also involved in hyperresponsive patients with MS this would have therapeutic implications. The corticosteroid treatment of MS is targeted at the active immune response and involves cytokines, whose production is partly controlled by corticosteriod receptors. Future studies will delineate whether the outcome of the DEX-CRH test is a good predictor of the response to corticosteroids. Furthermore, the cognitive impairment frequently observed in patients with MS may be related to changes in corticosteroid receptor function.

HPA Function in Patients with HIV Infection

HIV infection can affect virtually any endocrine gland, but the adrenal glands have received the most attention because adrenal insufficiency can be deleterious when complications occur. Indeed, decreased secretory capacity may develop in the

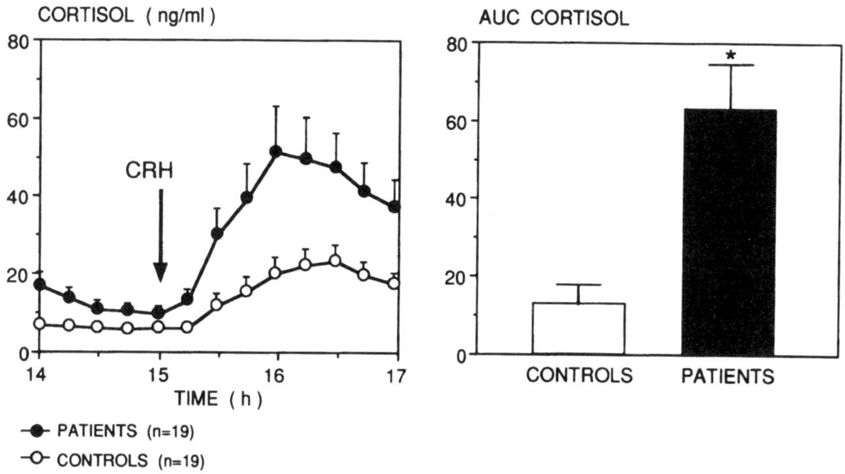

FIGURE 3. Combined DEX/CRH test in patients with multiple sclerosis. Patients with acute exacerbation of multiple sclerosis have an enhanced maximal plasma cortisol response to a combined DEX/CRH-test. (Data from Grasser *et al.*[19])

adrenal gland secondary to certain infections (*e.g.*, cytomegalovirus), and Membreno *et al.*[20] found that 15% of patients with HIV exhibit a decreased cortisol response to synthetic ACTH stimulation. This decreased adrenocortical reserve is not clinically relevant, but it may portend the future development of frank adrenal insufficiency.

In contrast, subclinical hypercortisolism among men with HIV infection has also been reported and has received particular attention because of the effects of corticosteriods upon immune function.[21,22] Specifically, the enhancement of HIV expression by cortisol has led to consideration of using therapeutic strategies that involve reducing stress or depression-induced HPA activation with psychiatric treatments. Other possible explanations for mild hyperactivity of the HPA system among HIV-infected patients include immunocyte activation and cortisol resistance. Cytokines are known to stimulate the HPA system at all levels. The lack of systematically elevated cytokines in patients with HIV infection does not exclude such a possibility because activation of the HPA system by cytokines probably involves a paracrine mechanism. Norbiato *et al.*[23] recently suggested that hypercortisolism in HIV-infected patients who display Addisonian-like clinical features may result from an acquired form of glucocorticoid resistance. These authors based their conclusion upon measurement of [^{3}H]-dexamethasone binding and the effect of dexamethasone on [^{3}H]-thymidine incorporation, which served as an estimate of glucocorticoid receptor activation. Among HIV-infected patients with Addisonian symptoms glucocorticoid receptor density was increased and the affinity was decreased. As a result of this receptor impairment [^{3}H]-thymidine incorporation was decreased. Such changes were absent in HIV-infected patients without Addisonian changes. Furthermore, Merenich *et al.*[24] reported that as a group HIV-infected asymptomatic subjects have lower plasma cortisol concentrations at baseline and following ACTH stimulation than symptomatic subjects. We measured plasma ACTH and cortisol levels serially between 1400 h and 1900 h in HIV-infected men at various stages of the

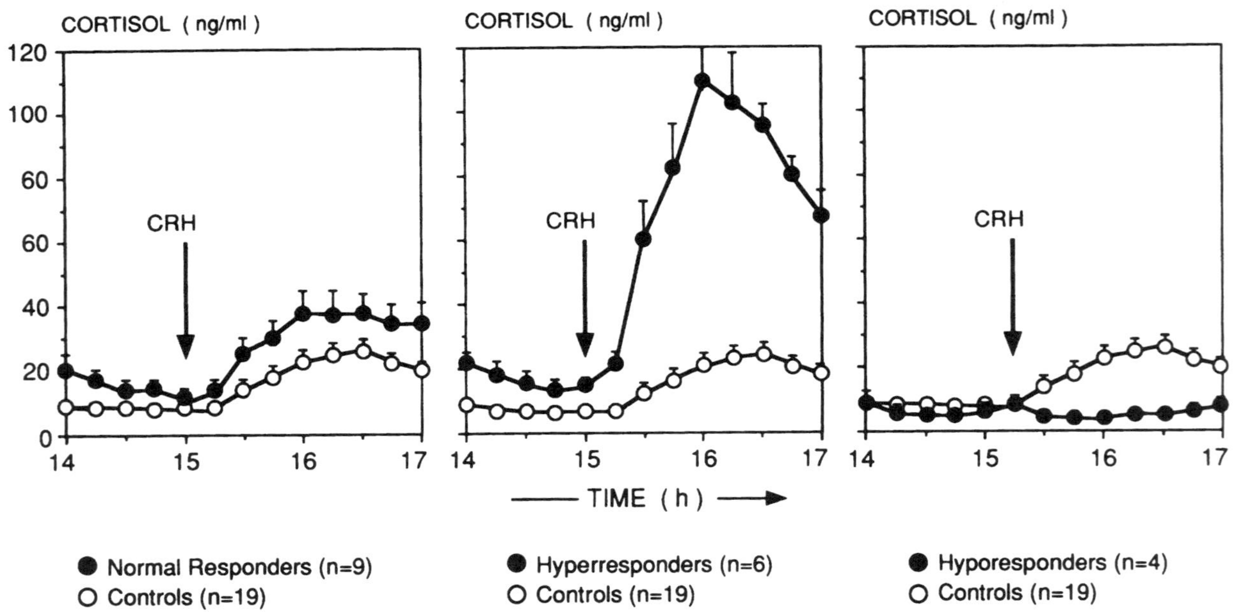

FIGURE 4. Combined DEX/CRH test in patients with multiple sclerosis. Subdivision of the patient sample with acute multiple sclerosis reveals three discernible patterns of response following CRH (100 μg; 15.00 h) which are unrelated to the dexamethasone (1.5 mg; 23.00 h) suppressed plasma cortisol levels prior to CRH injection. (Data from Grasser *et al.*[19])

disease and in healthy controls and failed to find any differences between the two groups (FIG. 5).[25] Further, following a challenge with CRH, the plasma ACTH and cortisol responses of the two groups were indistinguishable. These findings are inconsistent with enhanced HPA activity or decreased adrenocortical responsivity as a common endocrine manifestation of HIV infection.

Effects of Neuroactive Steroids on EEG Sleep in Man

Corticosteroids, their biosynthetic precursors and their naturally occurring metabolites can exert their neuroactivity through genomic (cytosolic GR and MR) and/or nongenomic (membrane sites) mechanisms. Both mechanisms can be studied at the systemic level by analyzing steroid-induced effects on the sleep EEG. More than 20 years ago, Gillin *et al.*[26] showed an effect of glucocorticoid receptor activation on the sleep EEG by demonstrating that prednisone increases the amount of slow wave sleep (SWS) in a dose-dependent fashion in man. Moreover, other glucocorticoids proved to enhance SWS in conjunction with a decreased amount of rapid eye movement (REM) sleep.[27] The interaction among hormones of the HPA system has been extensively studied and an attempt has been made to relate the secretion of adrenocortical hormones with certain sleep stages. In aggregate, these studies suggest that sleep inhibits activation of ACTH and cortisol secretion, but the effect is limited to the early part of sleep, when SWS prevails. Follenius[28] recently rejected the possibility that SWS might have an inhibitory action on cortisol secretion and suggested instead that enhanced adrenocortical activity prevents the transition from shallow sleep to deep sleep. This would support a sleep-regulatory role of CRH, the main central neuropeptide activating ACTH and thus secretion from the adrenal gland. We infused CRH in a repetitive mode prior to and following sleep onset and documented a suppression of SWS, which was more pronounced in the second half

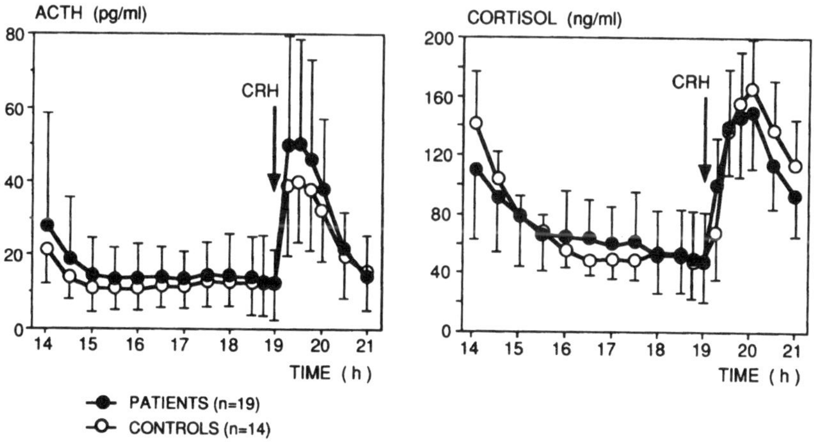

FIGURE 5. Plasma ACTH and cortisol secretion at baseline and after CRH in patients with HIV infection. Baseline and CRH (100 μg; 19.00 h) stimulated plasma ACTH and cortisol concentrations were indistinguishable among 19 males in different stages of HIV infection from that in 19 age-matched controls. (Data from Möller *et al.*[25])

of the sleep period.[29] Furthermore, the nocturnal GH surge was suppressed by CRH, which is interesting because growth hormone-releasing hormone (GHRH) is a stimulator not only of GH but, when injected during sleep, of SWS, too. This effect seems to depend on the repetitive mode of administration because continuous hormone administration produces less pronounced effects.[30]

Because activated GR and MR in the brain are believed to mediate different behavioral and endocrine effects, it is of interest that Born *et al.*[31] were able to show that canrenoate, an MR antagonist, suppressed SWS, whereas REM sleep remained unaffected. In contrast, cortisol increased SWS but suppressed REM sleep, and dexamethasone reduced both SWS and REM sleep. From these data it was concluded that the SWS-stimulatory effect of cortisol involves effects via MR, whereas changes in REM sleep are mediated through GR. This interesting hypothesis cannot be generalized as yet because Steiger *et al.*[32] found that neither the MR agonist deoxycorticosterone nor the MR antagonist spironolactone induced any measurable effect on the sleep EEG or concurrent endocrine activity. A recent study by Frieß *et al.*[33] showed that repetitive administration of cortisol throughout the sleep period enhances SWS in conjunction with activation of GH release, supporting the notion that a corticosteriod receptor-mediated effect on the somatotrophic system is involved in glucocorticoid-induced sleep enhancement.

The probable involvement of glucocorticoid receptors in human sleep was recently underscored by Wiedemann *et al.*,[34] who showed that mifepristone (RU 486) completely abolishes the physiological sleep EEG structure (FIG. 6). In a further study, Wiedemann *et al.*[35] pretreated healthy controls with a low dose of dexameth-

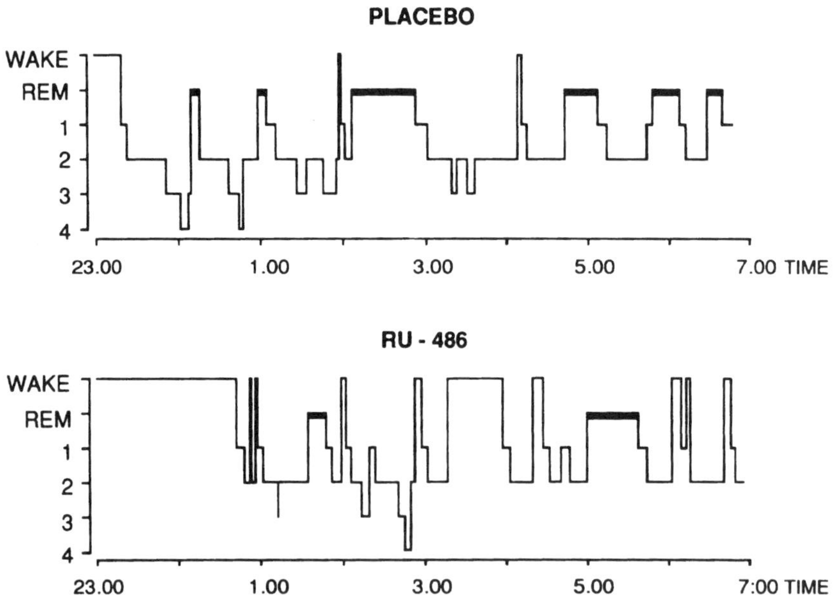

FIGURE 6. GR- and PR-antagonist RU-486 deteriorates sleep in men. An oral dose (22.00 h) of 200 mg RU-486 induces profound alterations of sleep-EEG pattern in healthy men. (Data from Wiedemann *et al.*[34])

asone and then administered spironolactone or mifepristone 24 h later. In these subjects, where the secretion of endogenous corticosteroids had been suppressed through the corticotrophic action of dexamethasone, spironolactone reduced REM sleep, whereas mifepristone reduced both REM sleep and SWS. Thus, effects of cytosolic steroid receptors upon sleep regulation can indeed be differentiated. Considering the highly complex nature of the process by which MR and GR alone and together mediate cellular and systemic effects, it would be surprising if activation of either one were causally related to a crude neurophysiological measure such as sleep EEG stages and more refined spectral analysis is needed for clarification.

Baulieu *et al.* (see review see Ref. 36) suggested the presence of steroids in the brain, independent of peripheral steroidogenic glands. The first such steroids discovered were pregnenolone and dehydroepiandrosterone (DHEA), which were found in various species at concentrations higher than those in plasma, suggesting a functional role in the central nervous system. The exact route by which brain-derived steroids are formed is not yet clear, but it appears that glia cells are the primary site of biosynthesis. These cells can be regarded as paracrine glands, which produce certain steroids that act at neurons to elaborate a variety of effects. Electrophysiological studies suggest that sulfated conjugates of pregnenolone and DHEA act as noncompetitive antagonists at the $GABA_A$ receptor, thus inhibiting GABA-induced chloride currents. At the systemic level, Flood *et al.*[37,38] showed that both of these steroids have memory-enhancing effects, and they suggested that these substances may have a potential for reducing neuronal degeneration. Steiger *et al.*[39] administered pregnenolone orally and showed that this steroid increases the amount of SWS without affecting peripheral secretion of cortisol and growth hormone. Spectral analysis of the sleep EEG revealed decreased power between 8.5 and 19.5 Hz, with the most pronounced effects occurring between 11 and 14 Hz (approximately the sigma band) (FIG. 7). The latter frequency range represents a common "spectral benzodiazepine signature" because benzodiazepines specifically increase spectral power in this frequency range. In addition to their specific effects on spectral power, benzodiazepines also decrease SWS. The effects observed under pregnenolone are compatible with an inverse agonistic effect at the $GABA_A$ benzodiazepine receptor complex. This view has recently been amplified by Lancel *et al.*[40] who administered pregnenolone and midazolam, a benzodiazepine with hypnotic action, to rats and compared the sleep EEGs. The major finding of this study was an increase in sigma-activity within nonREM sleep, which is in accordance with the findings in humans. Pregnenolone is the major precursor of many endogenous steroids including progesterone, acting through cytosolic receptors. In addition, progesterone can be metabolized into 3α-ring A-reduced allopregnanolone (tetrahydroprogesterone), which can act as an allosteric agonist at $GABA_A$ receptors.

Moreover, 3α-hydroxy-ring A-reduced pregnanes such as allopregnanolone can be oxidized intracellularly into 5α-pregnan steroids, which act at progesterone receptors.[41] With the involvement of this hitherto unknown receptor action, it remains open to what degree the progesterone receptor is involved in the observed effects. Whatever the exact mechanism is by which pregnenolone acts on sleep and other behaviors, its supposed inverse agnostic effect on the $GABA_A$ benzodiazepine receptor complex is compatible with the reported effect as a memory enhancer. Moreover, drugs under study for the treatment of dementia, such as β-carbolines, are also inverse agonists at the GABA benzodiazepine receptor complex.

The effects of DHEA upon the sleep EEG also form a basis for understanding its suggested therapeutic use. After oral administration of DHEA EEG power is increased in the higher frequency range (α and β band between 12 and 18 Hz) during

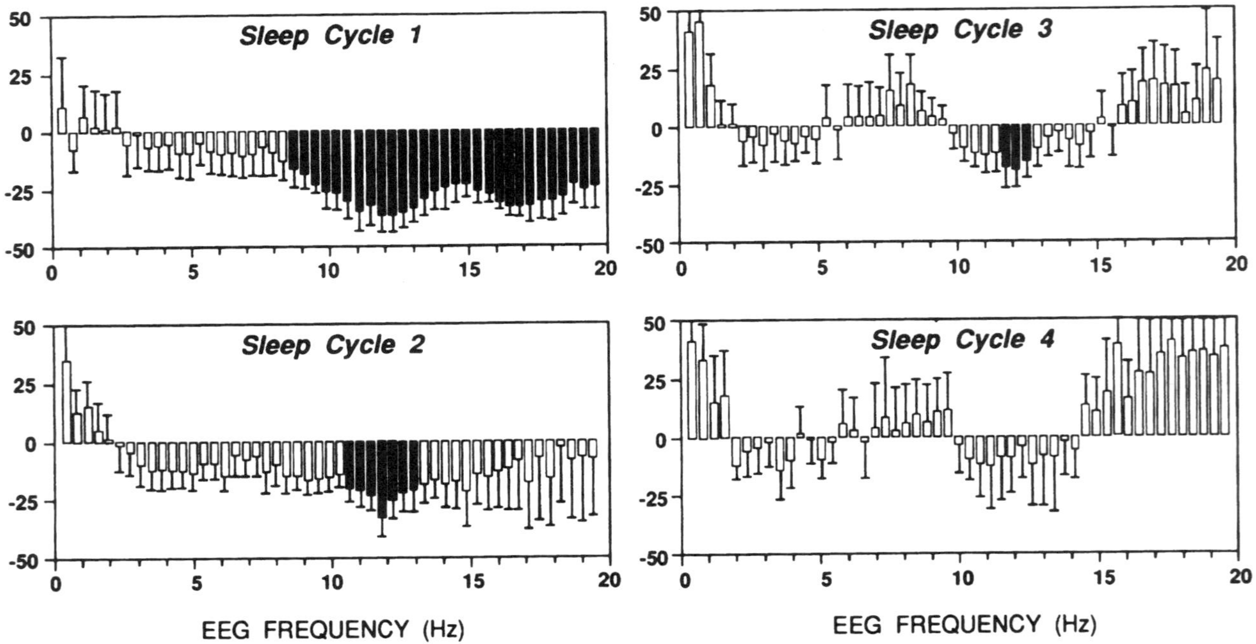

FIGURE 7. Pregnenolone EEG-power spectra (0–19.3 Hz) for sleep cycle 1–4 in men. EEG power specture (0–19.3 Hz) for sleep stages 1–4 in healthy men reveal that 1 mg pregnenolone orally administered at 22.00 h produces strongest decrease of EEG power at 11–13 Hz (approx. −40% from control levels). *Solid bars* denote significant differences between placebo and pregnenolone. The pregnenolone-induced effect becomes less pronounced at later sleep cycles. (Data from Steiger *et al.*[39])

REM sleep and the duration of REM sleep is also increased, whereas non-REM sleep EEG activity is not affected.[42] Given the well established association between the amount of REM sleep and consolidation of memory, the reported memory-enhancing effect of DHEA[38] may be related to its effect on REM sleep. The immediate onset of the DHEA action on the sleep EEG and the absence of peripheral endocrine effects together with the reported electrophysiological findings indicate that this steroid acts in a nongenomic fashion at membranous sites.

SUMMARY AND CONCLUSION

Acute and chronic stress as well as a number of psychiatric and neurological disorders are accompanied by profound disturbances of the HPA system. These neuroendocrine alterations act back on the central nervous tissue mainly via corticosteroids—affecting glucocorticoid and mineralocorticoid receptors. The major conclusions drawn from studies probing these receptors in clinical investigations are: (1) In many such conditions central corticosteroid receptors are weakened in their capacity to curtail spontaneous and stress-elevated corticosteroid levels; (2) the combined DEX-CRH test is the best neuroendocrine tool currently available for identifying HPA abnormalities in psychiatric patients; (3) in depression the decreased corticosteroid receptor capacity is transient, and antidepressants act through reinstatement of GR and MR function probably resulting in reduced hypothalamic CRH and AVP production; (4) several neurological disorders such as MS and HIV infection are often accompanied by altered HPA function, which has therapeutic implications; and (5) various corticosteriods, their biosynthetic precursors and their metabolites have differentiable effects on the sleep EEG, which can be attributed to their mode of action; specifically, steroids such as pregnenolone and DHEA most likely are produced in glia cells and act in a paracrine fashion at neurons, thus modifying the sleep EEG in humans in a manner that suggests their potential as memory enhancers.

REFERENCES

1. HOLSBOER, F., D. SPENGLER & I. HEUSER. 1992. The role of corticotropin-releasing hormone in the pathogenesis of Cushing's disease, anorexia nervosa, alcoholism, affective disorders and dementia. Prog. Brain Res. **93:** 385–417.
2. NEMEROFF, C. B., E. WIDERLÖV & G. BISSETTE. 1984. Elevated concentrations of CSF corticotropin-releasing factor-like immunoreactivity in depressed patients. Science **226:** 1342–1344.
3. HOLMES, M. C., K. J. CATT & G. AGUILERA. 1987. Involvement of vasopressin in the down-regulation of pituitary corticotropin-releasing factor receptors after adrenalectomy. Endocrinology **121:** 2093–2098.
4. OWENS, M. J. & C. B. NEMEROFF. 1991. Physiology and pharmacology of corticotropin-releasing factor. Pharmacol. Rev. **43:** 425–473.
5. VON BARDELEBEN, U., G. K. STALLA, O. A. MÜLLER & F. HOLSBOER. 1988. Blunting of ACTH response to human CRH in depressed patients is avoided by metyrapone pre-treatment. Biol. Psychiatry **24:** 782–786.
6. HEUSER, I. J. E., A. YASSOURIDIS & F. HOLSBOER. 1994. The combined dexamethasone CRH-test: a refined laboratory test for psychiatric disorders. J. Psychiatr. Res. In press.
7. VON BARDELEBEN, U. & F. HOLSBOER. 1989. Cortisol response to a combined dexamethasone-human corticotropin-releasing hormone challenge in patients with depression. J. Neuroendocrinol. **1:** 485–488.
8. HOLSBOER-TRACHSLER, E., R. STOHLER & M. HATZINER. 1991. Repeated administration

of the combined dexamethasone/hCRH stimulation test during treatment of depression. Psychiatry Res. **38:** 163–171.

9. Von Bardeleben, U., F. Holsboer, G. K. Stalla & O. A. Müller. 1985. Combined administration of human corticotropin-releasing factor and lysine vasopressin induces cortisol escape from dexamethasone suppression in healthy subjects. Life Sci. **37:** 1613–1618.
10. Tilders, F. J. H., E. D. Schmidt & D. C. E. De Goeij. 1993. Phenotypic plasticity of CRF neurons during stress. Ann. N. Y. Acad. Sci. **697:** 39–52.
11. Young, E. A., R. F. Haskett, V. Murphy-Weinberg, S. J. Watson & H. Akil. 1991. Loss of glucocorticoid fast feedback in depression. Arch. Gen. Psychiatry **48:** 693–699.
12. Heuser, I. J. E., U. Gotthardt, U. Schweiger, J. Schmider, C. H. Lammers, M. Dettling, A. Grasser & F. Holsboer. 1994. Effects of amitriptyline on pituitary-adrenal-axis regulation in elderly depressed patients with age-matched normal controls. Submitted.
13. Reul, J. M. H. M., I. Stec, M. Söder & F. Holsboer. 1993. Chronic treatment of rats with the antidepressant amitriptyline attenuates the activity of the hypothalamic-pituitary-adrenocortical system. Endocrinology **133:** 312–320.
14. Pitman, R. K. & S. P. Orr. 1990. Twenty-four hour urinary cortisol and catecholamine excretion in combat-related posttraumatic stress disorder. Biol. Psychiatry **27:** 245–247.
15. Yehuda, R., S. M. Southwick, G. Nussbaum *et al.* 1990. Low urinary cortisol excretion in patients with posttraumatic stress disorder. J. Nerv. Ment. Dis. **178:** 366–369.
16. Smith, M. A., J. Davidson, J. C. Ritchie, H. Kudler, S. Lipper, P. Chappell & C. B. Nemeroff. 1989. The corticotropin-releasing hormone test in patients with posttraumatic stress disorder. Biol. Psychiatry **26:** 349–355.
17. Yehuda, R. M. T. Lowy, S. M. Southwick, D. Shaffer & E. L. Giller, Jr. 1991. Lymphotcyte glucocorticoid receptor number in posttraumatic stress disorder. Am. J. Psychiatry **148:** 499–504.
18. Heuser, I. J. E., H. J. Wark, J. Keul & F. Holsboer. 1991. Hypothalamic-pituitary-adrenal-axis function in elderly endurance athletes. J. Clin. Endocrinol. Metab. **73:** 485–488.
19. Grasser, A., A. Möller, H. Backmund, A. Yassouridis & F. Holsboer. 1994. Defective hypothalamic-pituitary-adrenal system responsiveness to a combined dexamethasone-corticotropin-releasing hormone challenge in multiple sclerosis. Submitted.
20. Membreno, L., I. Irony, W. Dere, R. Klein, E. G. Biglieri & R. Cobb. 1987. Adrenocortical function in acquired immunodeficiency syndrome. J. Clin. Endocrinol. Metab. **65:** 482–487.
21. Gorman, J. M., R. Kertzner, T. Cooper, R. R. Goetz, I. Lagomasino, H. Novacenko, J. B. W. Williams, Y. Stern, R. Mayeux & A. A Ehrhardt. 1991. Glucocorticoid level and neuropsychiatric symptoms in homosexual men with HIV infection. Am. J. Psychiatry **148:** 41–45.
22. Markham, P. D., S. Z. Salahuddin, K. Veren, S. Orndorff & R. C. Gallo. 1986. Hydrocortisone and some other hormones enhance the expression of HTLV-III. Int. J. Cancer **37:** 67–72.
23. Norbiato, G., M. Bevilacqua, T. Vago, G. Baldi, E. Chebat, P. Bertora, M. Moroni, M. Galli & N. Oldenburg. 1992. Cortisol resistance in acquired immunodeficiency syndrome. J. Clin. Endocrinol. Metab. **74:** 608–612.
24. Merenich, J. A., M. T. McDermott, A. A. Asp, S. M. Harrison & G. S. Kidd. 1990. Evidence of endocrine involvement early in the course of human immunodeficiency virus infection. J. Clin. Endocrinol. Metab. **70:** 566–571.
25. Möller, A., A., Grasser, A. Fuchs, K. Wiedemann & F. Holsboer. 1994. Neuroendocrine function in HIV infection. Submitted.
26. Gillin, J. C., L. S. Jacobs, D. H. Fram & F. Snyder. 1972. Acute effect of a glucocorticoid on normal human sleep. Nature **237:** 398–399.
27. Fehm, H. L., R. Benkowitsch, W. Kern, G. Fehm-Wolsdorf, P. Pauschinger & J. Born. 1986. Influences of corticosteroids, dexamethasone and hydrocortisone on sleep in humans. Neuropsychobiology **16:** 198–204.
28. Follenius, M., G. Brandenburger, J. J. Bandesapt, J. P. Libert & J. Ehrhart.

1992. Nocturnal cortisol release in relation to sleep structure. Sleep **15:** 21–27.
29. Holsboer, F., U. von Bardeleben & A. Steiger. 1988. Effects of intravenous corticotroph-releasing hormone upon sleep-related growth hormone surge and sleep EEG in man. Neuroendocrinology **48:** 32–38.
30. Born, J., Späth-Schwalbe, H. Schwakenhofer, W. Kern & H. L. Fehm. 1989. Influences of corticotropin-releasing hormone, adrenocorticotropin, and cortisol on sleep in normal man. J. Clin. Endocrinol. Metab. **68:** 904–910.
31. Born, J. E. R. de Kloet, H. Wenz, W. Kern & H. L. Fehm. 1991. Gluco- and antimineralocorticoid effects on human sleep: a role of central corticosteroid receptors. Am. J. Physiol. **260:** E183–E188.
32. Steiger, A., R. Rupprecht, D. Spengler, J. Guldner, U. Hemmeter, B. Rothe, K. Damm & F. Holsboer. 1993. Functional properties of deoxycorticosterone and spironolactone: molecular characterization and effects on sleep-endocrine activity. J. Psychiatr. Res. **27:** 275–284.
33. Friess, E., U. von Bardeleben, K. Wiedemann, C. J. Lauer & F. Holsboer. 1994. Effects of pulsatile cortisol infusion on sleep-EEG and spontaneous release of growth hormone in healthy men. J. Sleep Res. **3:** 73–79.
34. Wiedemann, K., C. Lauer, A. Loycke, Ph. Durst, J. P. Machér & F. Holsboer. 1992. Antiglucocorticoid treatment disrupts endocrine cycle and nocturnal sleep pattern. Eur. Arch. Psychiatry Clin. Neurosci. **241** 372–375.
35. Wiedemann, K., C. J. Lauer, T. Pollmächer & F. Holsboer. 1994. Sleep endocrine effects of antigluco- and antimineralocorticoids in healthy males. Am. J. Physiol. **267:** E109–E114.
36. Baulieu, F. E., P. Robel, O. Vatier, A. Haug, C. Le Goascogne & E. Bourreau. 1987. Neurosteroids: pregnenolone and dehydroepiandrosterone in the rat brain. *In* Receptor-Receptor Interaction. A New Intramembrane Integrative Mechanism. K. Fuxe & L. F. Agnati, Eds. 89–104. Plenum Press. New York, London.
37. Flood, J. F., J. E. Morley & E. Roberts. 1992. Memory-enhancing effects in male mice of pregnenolone and steroids metabolically derived from it. Proc. Natl. Acad. Sci. **89:** 1567–1571.
38. Flood, J. F., G. E. Smith & E. Roberts. 1988. Dehydroepiandrosterone and its sulfate enhance memory retention in mice. Brain Res. **447:** 269–278.
39. Steiger, A., L. Trachsel, J. Guldner, U. Hemmeter, B. Rothe, R. Rupprecht, H. Vedder & F. Holsboer. 1993. Neurosteroid pregnenolone induces sleep-EEG changes in man compatible with inverse agonistic $GABA_A$-receptor modulation. Brain Res. **615:** 267–274.
40. Lancel, M., T. A. M. Crönlein, P. Müller-Preuss & F. Holsboer. 1994. Pregnenolone enhances EEG delta activity during non-rapid eye movement sleep in the rat, in contrast to midazolam. Brain Res. **646:** 85–94.
41. Rupprecht, R., J. M. H. M. Reul, T. Trapp, B. van Steensel, C. Wetzel, K. Damm, W. Zieglgänsberger & F. Holsboer. 1993. Progesterone receptor-mediated effects of neuroactive steroids. Neuron **11:** 1–20.
42. Friess, E., L. Trachsel, J. Guldner, T. Schier, A. Steiger & F. Holsboer. 1994. Dehydroepiandrosterone administration increases rapid eye movement sleep and EEG power in the spindle frequency range. Am. J. Physiol. In press.

DISCUSSION OF THE PAPER

Question: When you do the combined dexamethasone CRH test, you give 1.5 mg of DEX and then you get a very different response between the normals and depressed patients. Then you give them CRH. Is the purpose of giving the dexamethasone before the CRH to shut off negative feedback?

Holsboer: The original finding with the CRH test by Phil Gold and us was that while the ACTH response to CRH is blunted, cortisol levels are elevated at baseline and are indistinguishable in response to CRH. So the simplistic interpretation Phil and I embarked on was that the negative feedback, the integrity of pituitary cortical function, is proved by this test and the locus of pathology is somewhere above. This is okay, but things are more difficult than that. Then there is the other interpretation that the CRH receptors are reduced in their function. This is an overly simplistic interpretation. With the dexamethasone I did not want to find this, I can tell you. I was very surprised when we had that. I thought I could titrate down the peak, and that did not work. What I learned from that is that dexamethasone is not just another cortisol. It is a completely different compound which produces the negative feedback and asymmetry which is direct to the pituitary. A very simplistic interpretation would be that while acting with this very low dose of dexamethasone we mainly act at the pituitary depriving the brain of its major negative feedback signals. The brain does not see any cortisol anymore and it is not substituted by dexamethasone. Therefore, the brain striving to get more cortisol pushes on everything it has, including vasopressin. Vasopressin then comes via the portal blood to the pituitary. Then you give your CRH shot, which synergizes and does not see the dexamethasone anymore. That is my interpretation. In studies with normal controls first, we gave CRH and wondered why we never achieved the ACTH and cortisol levels which we see under other kinds of stimulation like anxiolytic support. So we tried to suppress with the dexamethasone at the normal controls, and then we infused CRH and vasopressin and got in the normal controls treated with the same dosage of dexamethasone exactly the same cortisol and ACTH curves we saw in the antidepressants without vasopressin. That is why we reasoned that vasopressin is involved.

QUESTION: Why not use a bigger dose of dexamethasone so that the normals and depressed patients would be comparable to start with? Then you would not have an issue of negative feedback at all. Alternatively, why would doing this be different than using metyrapone? Could you explain the difference in approach?

HOLSBOER: Metyrapone is a rather tricky compound in this respect. It is not just a compound that absorbs cortisol, it blocks an enzyme that hydroxylates at C11. So it has a lot of substance and you see a lot of oxycorticosterone, which by itself has a certain amount of physiology. For example, the oxycorticosterone is a ligand at the MR. This does not impede the conclusions drawn by Dr. Raskind, though it is something which has to be taken into consideration. You may think that you just filter out the negative feedback signal, but a lot of other physiology and pharmacology comes up. It is a kind of complementary information you get with metyrapone and I wish we could do a metyrapone and DEX CRH test within the same patient and the same condition. But this is hard to do, because you have to wait for a while, and the patient might not be the same, aside from the ethical considerations.

QUESTION: Could you tell us what the sleep pattern is in depressed patients? Have you tried treating them with some of these steroids?

HOLSBOER: The major finding among depressants regarding sleep is that they have a decreased slow wave sleep pressure, if you will, in the first half of the night, which reduces slow wave sleep in the first cycles and pushes the REM sleep closer to sleep onset, which is secondary to this reduced slow wave sleep. They have a reduced latency between sleep onset and rapid eye movements. Secondly, they have an increased rapid eye movement density. The driving force for that is the hyperexpression of CRH, because when CRH is administered to a rat, it has a decreased slow wave sleep. If you do the same thing in the pulsar pill mode to normal controls you have two things: a decrease of slow wave sleep and a decrease of growth hormone in the first portion of sleep. Reduced growth hormone in the first portion of sleep is also a regular concomitant of the sleep of depressants. Thus, if you consider sleep as one correlate of hormonal dysfunction and depression, which is not the whole story, then CRH hyperdrive reduces slow wave sleep, which then reduces the REM latency, and cortisol opposes this effect. So the CRH effect is not mediated by a peripheral induction of cortisol, because CRH reduces slow wave sleep while cortisol increases it.

QUESTION: When you are pulsing with CRH, these are people, and you are not giving it IVT. Or are these rats, and you are giving it IVT?

HOLSBOER: We give it to normal controls intravenously.

QUESTION: Are you assuming the CRH that you are giving is entering brain and acting to some extent in brain?

HOLSBOER: I do not know how that happens. I grew up with a neurology textbook which taught me that the blood brain barrier is something made of steel, 6 cm thick. This is not true. There are leakages. I do not want to overuse the ventricular organs. They are only one fifth of the whole blood brain barrier. But still it does not need much peptide in comparison to the ordinary transmitters. They have a completely different pharmacology regarding receptor binding, so I trust that what we have is a real finding. We have replicated that so often, as have others, and we have done that with the evoked potentials in the waking state and so forth. So we are quite confident that there is a mechanism, whether directly or indirectly I do not know. But there is a direct effect.

Glucocorticosteroid Resistance in Humans

Elucidation of the Molecular Mechanisms and Implications for Pathophysiology

CONSTANTINE A. STRATAKIS,[a] MICHAEL KARL,[a]
HEINRICH M. SCHULTE,[b] AND GEORGE P. CHROUSOS[a]

[a]Developmental Endocrinology Branch (DEB)
National Institute of Child Health & Human Development (NICHD)
National Institutes of Health (NIH)
Building 10, Room 10 N 262
9000 Rockville Pike
Bethesda, Maryland 20892
and
[b]Institute for Hormone and Fertility Research (IHF)
University of Hamburg
22529 Hamburg, Germany

INTRODUCTION

Glucocorticosteroids play major roles in human psychology, affecting almost every tissue in the human body.[1] Glucocorticosteroids are crucial regulators of affect and behavior, metabolism, cardiovascular function, inflammation and immunity.[2,3] Complete inability of glucocorticosteroids to exert their effects on target tissues would be incompatible with life in primates. Indeed, only syndromes of partial or incomplete glucocorticosteroid resistance exist. They are caused by defects of the ubiquitous, classic glucocorticosteroid receptor (GR) (type II), which mediates most of glucocorticoid actions and is the prototypic member of the family of ligand-activated transcription factors homologous to the v-erb A oncogene.[4] Familial glucocorticoid resistance (FGR) or primary cortisol resistance was the last nuclear hormone resistance syndrome to be described, as late as 1976,[5] and to date has been reported in eight unrelated kindreds[5–12] and more recently, in five individual subjects.[13] The clinical phenotype of patients with cortisol resistance is extremely variable and is dependent not only on the described as the GR genotypes but also many factors that participate in the pathophysiology of FGR.

Glucocorticoids regulate development and homeostasis in complex eukaryotes by modulating a wide range of physiological actions. Their day-to-day production, therefore, is tightly controlled by an elaborate feedback system, in which glucocorticoids exert negative feedback on hypothalamic secretion of corticotropin-releasing hormone and arginine vasopressin and on pituitary secretion of corticotropin (ACTH). In states of resistance, this complex system is insensitive to concentrations of cortisol considered normal for the general population,[14] and the hypothalamic-pituitary-adrenal (HPA) axis is reset to a higher level because of compensatory increases of ACTH and cortisol secretion. Since the defect of the receptor is partial, adequate compensation for the end-organ insensitivity appears to be achieved by the elevated circulating cortisol; however, the excess ACTH secretion results in increased production of adrenal steroids with salt-retaining or androgenic activity (FIG. 1).[14] Because in FGR the peripheral tissues are presumably normally sensitive to mineralcorticoids and androgens, the clinical characteristics reflect the increased production of these hormones. Corticosterone, deoxycorticosterone and even cortisol, through their interaction with the mineralocorticoid receptor, cause symptoms and signs of mineralocorticoid excess, such as hypertension and/or hypokalemic alkalosis.[6–8,10,13] The excess androgens, such as Δ_4-androstenedione, dehydroepiandrosterone, and dehydroepiandrosterone sulfate, on the other hand, lead to signs and

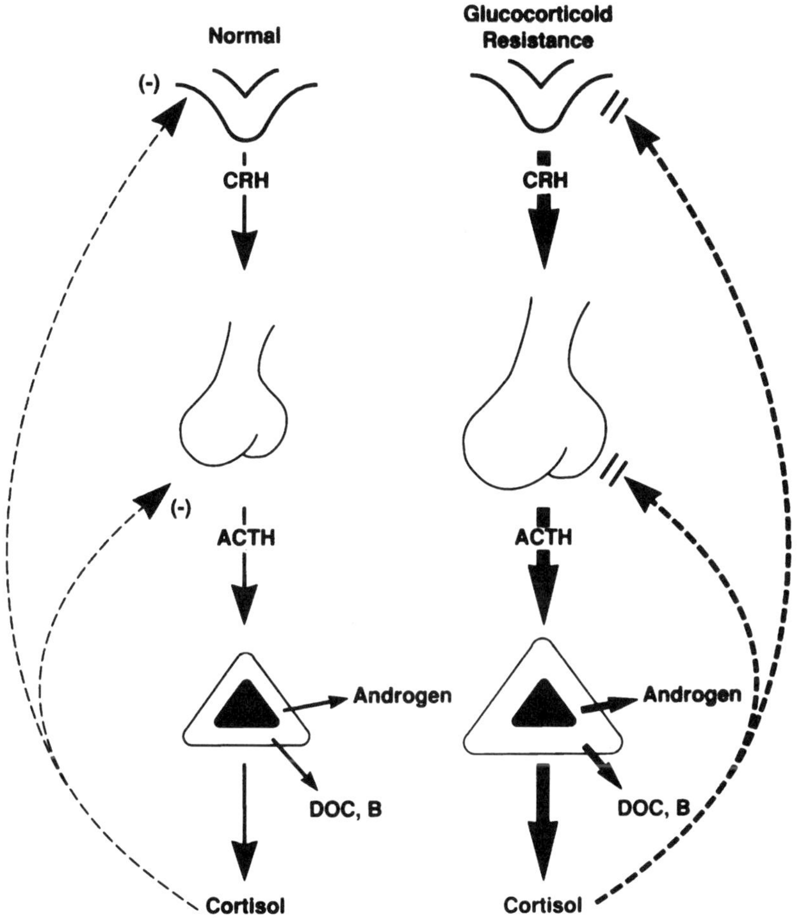

FIGURE 1. Pathophysiology of the clinical manifestations of glucocorticoid resistance in humans. The hypothalamic-pituitary-adrenal axis is reset at higher levels of corticotropin (ACTH) and cortisol. The latter compensate for the insensitivity of tissues to glucocorticoids; however, the increased ACTH levels result in increased secretion of glucocorticoid precursors with mineralocorticoid activity (DOC, deoxycorticosterone; B, corticosterone) and adrenal androgens. (From Chrousos *et al.*[14] Reprinted by permission from the *Annals of Internal Medicine.*)

symptoms of hyperandrogenism. The latter, in women with FGR presents with acne, hirsutism, male pattern baldness, menstrual irregularities, oligo-anovulation, amenorrhea and infertility.[7,11,13] In children, the excessive and early prepubertal adrenal androgen secretion can cause sexual precocity.[12] Finally, the interference of adrenal androgens with the regulation of pituitary secretion of follicle stimulating hormone (FSH) and/or ACTH-induced intratesticular growth of adrenal rests, appear to be responsible for abnormal spermatogenesis and infertility in men with FGR.[13] These clinical manifestations were not reported in all patients with symptomatic glucocorticoid resistance and presentation varied even within families. A large number of subjects had biochemical but not clinically overt resistance and were detected in the course of family studies.[7,16]

Inherited or spontaneous mutations of the GR gene causing quantitative and/or qualitative alterations in the function of the receptor protein had been suspected in FGR, since receptor studies in cells from affected individuals and their relatives had shown decreases in the receptor concentration and ligand binding affinity, thermolability, and/or abnormal interactions with DNA.[6,7,11,13,15] The cloning of the complementary DNA (cDNA) of the human GR in 1985[17] allowed the deduction of its primary structure, definition of its functional domains,[18] and cloning of its gene[19] and permitted further studies on the molecular mechanisms of FGR and other states of glucocorticoid resistance. The gene is located on chromosome 5[20] and is quite complex, consisting of 10 different exons.[14,19] The structure of its 3 kilobases (kb)-long regulatory region is similar to that of other housekeeping genes of the human genome[14] (FIG. 2A). The 777 amino acid GR has three main functional domains: the ligand-binding domain (which lies in the carboxy terminal portion of the protein), the DNA-binding domain (which is in the center), and the aminoterminal domain, which was originally characterized by its property of providing antigenic epitopes for the generation of antireceptor antibodies and, hence, was called the "immunogenic domain." Additional work with construction of chimeric GRs and *in vitro* expression studies showed a number of domains representing other functions involved in the cascade of events leading from ligand binding to regulation of gene transcription in target cells[21–24] (FIG. 2B). The inactivated ligand-free glucocorticoid receptor is cytoplasmic and associated with other proteins, such as heat-shock proteins (hsp) –90[24] and –70[25] and immunophillin of the FK506/rapamycin binding type,[26] in the form of a hetero-oligomer. Binding with the hsp-90 is secured by sites in the steroid-binding domain and this special interaction appears to facilitate the GR's activation by the ligand and its effectiveness in eliciting a glucocorticoid response.[27] When the GR is bound to a ligand, the hsp-90 is released and the domains responsible for nuclear localization, DNA binding and transactivation are unmasked.[24] The activated GR molecules form homodimers as a result of the interaction of dimerization sequences present in the ligand- and DNA-binding domains and with the help of hsp-70, intracellular fibrils and possibly other proteins, cross into the nucleus through recognition of its nuclear localization domains by proteins of a nuclear pore.[28] Inside the nucleus, the activated GR-dimers bind to the DNA using its "zinc finger" motif and selectively interact with the glucocorticoid-responsive elements (GREs)[29] altering the transcription rates of the target genes. Through different or even the same GREs, or by interacting directly with other transcription factors, such as c-jun and c-fos, the ligand-activated GR can repress other genes.[30] The receptor, subsequently, is recycled into the cytoplasm by an as yet undefined mechanism, which may involve phosphorylation and dephosphorylation steps.[31]

We have investigated the molecular basis of FGR in a large kindred (kindred-A), in which the syndrome was described by Vingerhoeds *et al.*[5] and another one reported later by Lamberts *et al.* (kindred-B).[7] The propositus in kindred-A had no features of glucocorticoid excess despite a sevenfold elevation of free cortisol in serum and a 30–40-fold increase in urinary-free cortisol excretion. The dose of dexamethasone needed to suppress serum cortisol levels by 50% in this patient was seven times greater than that required in normal subjects. Less severe but unequivocal cortisol excess and dexamethasone resistance were found in one son and one nephew.[16] Borderline cortisol excess was detected in six of 27 additional relatives from three generations consistent with an autosomal co-dominant mode of inheritance, in which heterozygoes were mildly or biochemically-only affected. Studies of the GR in intact mononuclear leukocytes and cultured skin fibroblasts from the patients showed normal concentrations of the GR. The hGR of Epstein-Barr virus (EBV)-transformed lymphoblast lines from the propositus and his mildly affected relatives, on the other hand, showed GR mRNA of the expected size but decreased ligand binding affinity, increased thermolability and reduced binding to DNA-cellulose after thermal activation.[32] It was concluded from these studies that the most likely molecular defect was a mutation of the GR gene affecting the ability of the receptor to bind efficiently with the ligand.[15,32] In kindred-B, the propositus was an adult woman with symptoms and signs of hyperandrogenism and elevated plasma cortisol concentrations. She, her father and two brothers were biochemically affected to the same degree, with cortisol production indices within the Cushing's syndrome range; the male members, however, were completely asymptomatic.[7] Glucocorticoid binding analysis in circulating mononuclear leukocytes from affected members demonstrated a 50% decrease of receptor concentrations and normal binding affinity for dexamethasone.[7,13] We used the polymerase chain reaction (PCR) to amplify and sequence the GR cDNA and the nine exons of the GR gene in several members of families with FGR[33,34] and glucocorticoid-resistant ACTH-secreting pituitary tumors (unpublished information).

METHODS AND RESULTS

The 2,3-kb coding region of the GR cDNA, obtained by specifically primed reverse transcription (RT) of RNA from the EBV-transformed lymphoblast lines established from the three patients of kindred-A, was amplified in two overlapping segments by PCR. Both amplified products were of the predicted size excluding major deletions or extended rearrangements in the GR cDNA of the patients. Subsequently, overlapping segments of double-stranded cDNA were amplified by cDNA and then reamplified in asymmetric PCR to yield single-stranded cDNA for sequencing. There was complete concordance with the published normal GR cDNA nucleotide sequence for all three patients, except at position 2054, where a thymidine (T) residue was present in place of the normal adenosine (A), resulting in a nonconservative amino acid substitution at position 6451 of the GR protein (FIG. 2) and changing the aspartic acid codon (GAC) to the valine codon (GTC). The propositus was homozygous for the Val-641 mutation, whereas the amplified GR cDNAs from the mildly affected son and nephew contained both A and T residues at position 2054, suggesting heterozygosity. Allele-specific hybridization with amplified genomic DNA from the patients confirmed the sequencing findings, and the oligonucleotide-probe bearing the mutation did not hybridize with PCR-amplified genomic DNAs from 24 normal subjects, suggesting that this base change is not a common polymorphism.

The functional significance of the identified mutation was tested by a series of

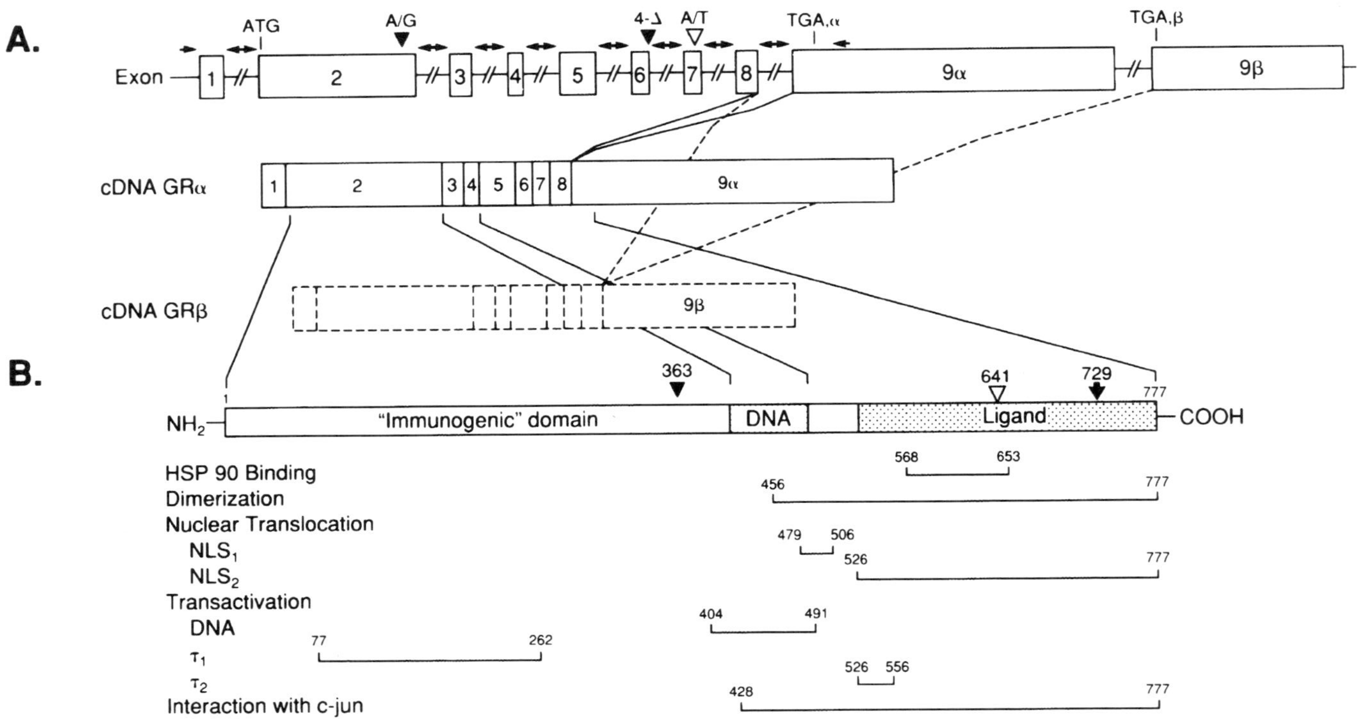
A.
B.
Exon
ATG
A/G
4-∆
A/T
TGA,α
TGA,β
1
2
3
4
5
6
7
8
9α
9β
cDNA GRα
cDNA GRβ
NH2
"Immunogenic" domain
DNA
Ligand
COOH
363
641
729
777
HSP 90 Binding
Dimerization
Nuclear Translocation
NLS1
NLS2
Transactivation
DNA
τ1
τ2
Interaction with c-jun
568
653
456
479
506
526
404
491
77
262
556
428

C.

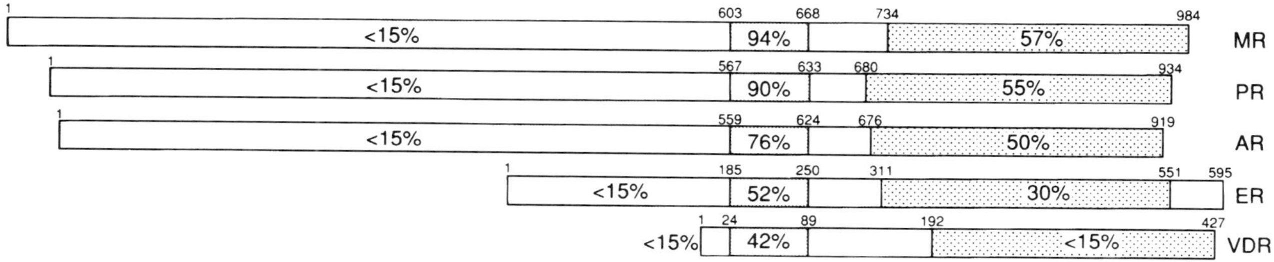

FIGURE 2. Genomic, cDNA and protein structure of the human glucocorticoid receptor (hGRα). **(A)** The gene consists of 10 exons of variable length. The hGRβ is the product of alternative splicing and does not bind glucocorticoids. **(B)** The main domains of the hGR are represented in a linear model and the three identified mutations in kindreds-A and -B (see text) and the one reported by Malchoff *et al.*[41] are shown at positions 641, 363 and 729 of the hGR protein, respectively. (HSP = heat shock protein; NLS = nuclear localization sequences). **(C)** Homologies of the five other classes of steroid and sterol receptors to the hGR expressed as percent identity in primary sequence (MR, PR, AR, ER, and VDR = mineralocorticoid, progesterone, androgen, estrogen and 1,25$(OH)_2$-vitamin D_3 receptor, respectively). (From Chrousos *et al.*[14] Reprinted by permission from the *Annals of Internal Medicine.*)

experiments starting by its introduction by PCR-site directed mutagenesis into the wild-type GR expression vector pRShGRa and, thus, constructing the recombinant plasmid pRShGR-Val 641. The wild-type or the mutant plasmid were separately cotransfected into COS-7 cells with the reporter plasmid pMTVCAT. Dexamethasone-induced CAT gene expression was then assessed by measuring the acetylation of chloramphenicol to acetyl-chloramphenicol. There was no stimulation in cells which had been transfected with the pRShGR-Val 641 even though western blots using monoclonal anti-GR antibody demonstrated similar levels of expression of both wild-type and Val-641-mutant GRs in the transfected cells. Specific high affinity binding [^{3}H]-dexamethasone was demonstrated in cytosol prepared from COS-7 cells transfected with pRShGRa and pRShGR-Val 641 and the equilibrium dissociation constant (Kd) of the Val-641 hGR was 3-fold higher in the latter than that of the wild-type receptor, giving similar to the *in vivo* data of the patients.[33,35]

In kindred-B, the GR gene of the proband, her two affected brothers, and two unaffected siblings was studied also in EBV-transformed lymphoblast lines. The PCR-amplified genomic DNA of the patient and her affected brothers was first analyzed by Southern blotting using GR cDNA probes. There were no major deletions, insertions, or other gross rearrangements in the GR gene. The nine exons of the GR gene were then amplified by PCR, using primers located in the 5′- and 3′-flanking intron regions and directly sequenced. A 4-basepair deletion (Δ_4) was identified at the 3′-boundary of exon and intron 6 (FIG. 2). The PCR amplified exon 6 and the exon 6-intron 6 boundary in the two unaffected siblings of the proband were normal. Therefore, the newly found deletion segregated with the presence of glucocorticoid resistance, and all three affected members of the kindred were heterozygous for it. Interestingly, a single nucleotide substitution that was, in addition, identified at the second position of codon 363 in exon 2 (G_{1220}) of the proband and one of her affected brothers, was also present in an unaffected sister, whereas it was absent in the second of the affected brothers. Thus, this mutation which is causing an amino acid substitution from Asn to Ser at position 363 of the GR protein (GR-Ser 363) did not segregate with the biochemical manifestations of the disorder: the proband and one of her affected brothers, as well as one of her unaffected sisters were heterozygous for the GR sequence bearing the mutation, while one of the affected brothers was homozygous for the wild-type sequence in exon 2.

To confirm that the G_{1220} substitution is a genetic polymorphism of no functional significance, a pRShGR-Ser 363 plasmid was constructed and used in a co-transfection assay as described above for the Val-641 hGR. No difference in the ability of the wild-type and the Ser-363 mutant receptor to activate CAT gene expression was observed. The frequency of this polymorphism was studied by screening PCR-amplified genomic DNA from 23 unrelated caucasian subjects using allele-specific slot-blot hybridization and was found to be lower than 5%.[34] We then employed the Ser-363 mutation as an allelic marker, in order to detect and quantitate transcripts derived from either allele. GR cDNA was derived from the patients' mRNA and amplified using RT-PCR and when the entire coding cDNA sequence was determined, only guanine could be detected at the second position of codon 363 in the proband and in one of the affected brothers demonstrating the presence of transcripts from the G_{1220} allele and lack of expression of the Δ_4 allele for which this patient is heterozygous. The conclusions from the above studies in kindred-B, were that the G_{1220} and Δ_4 mutations of the genomic DNA were on different alleles, the former having no effect on the function of the GR and the latter being responsible for the phenomenon of glucocorticoid resistance in the three affected individuals by either premature termination of the transcription of the mutant gene or nuclease degradation of its aberrant mRNA-product[34] (FIG. 1). The lack of transcription of the Δ_4 allele in the affected individuals is compatible with the reduction of the GR number by 50% in the patients' cells that was demonstrated before.[7,13]

We have studied two additional families with FGR (unpublished information). In one of them a mutation was identified in the ligand-binding domain of the GR. In the other no mutation was found within the structure of the GR gene.

Recently, we also used PCR to investigate the GR gene in the pituitary tumors of four patients with Cushing's syndrome that had undergone surgical and/or medical treatment for their disease and subsequently developed Nelson's syndrome (unpublished information). DNA was extracted from the tumors obtained by transsphenoidal surgery. Amplification of each of eight exons (2–9) of the GR yielded fragments of expected size excluding major rearrangements or deletions within the gene. Direct sequencing of the entire coding region revealed the insertion of an additional nucleotide in the sequence of exon-2 of the GR gene in the pituitary tumor of one of the four patients studied, causing a frameshift and apparently premature termination of one of the GR alleles. It is noteworthy, that this patient had an extremely resistant to treatment form of Cushing's disease, requiring adrenalectomy after two transsphenoidal surgeries, cranial irradiation and a craniotomy for a large suprasellar tumor. The same mutation was not detected in the sequence of the GR gene of DNA obtained from the patient's peripheral blood leukocytes. The other tumors studied displayed the wild-type GR sequence. These results constitute the first evidence that somatic mutations of the GR can account for glucocorticoid resistance at the tissue level and may be involved in the pathogenesis of at least some ACTH-producing tumors.

DISCUSSION

Familial Glucocorticoid Resistance (FGR), also known as primary cortisol resistance, can be caused by mutations of the human GR gene. In the two kindreds studied at the National Institutes of Health (NIH), we identified two alterations in the GR gene that are responsible for the syndrome. The identified Val-461 mutation in the GR of the first reported kindred with FGR lies within the cDNA region that encodes the hormone binding domain of the receptor and is absolutely conserved in the homologous positions of all species studied to date.[4] The allele bearing the base substitution segregated with the clinical phenotype of the patients and the reduced affinity to glucocorticoids of their receptor, and there was no other change in the sequence of the coding region of the GR cDNA of all three patients. The identified mutation was not present in genomic DNAs of normal subjects, suggesting that this base substitution is not a common polymorphism and the *in vitro* expression studies provided definitive confirmation that the Val-641 receptor has reduced ligand-binding affinity and impaired ability to activate gene expression.[33,35] In the second kindred investigated at the NIH, the receptor studies had shown normal binding affinity for glucocorticoids but a 50% decrease of GR concentrations in the patients' leukocytes.[7] Analysis of the GR gene revealed a four-base pair deletion at the exon-intron boundary of exon 6, removing a donor splice site in one allele of the GR gene in all three affected siblings studied. This genetic alteration segregated with the clinical phenotype of FGR in kindred-B and was not present in the unaffected siblings. The Δ_4 deletion presumably caused transcription of unstable precursor-mRNA in all three affected patients, resulting eventually in lowering the first two nucleotides of an intron have been shown to disrupt normal splicing.[36] Aberrant mRNA variants generated in such cases are thought to be more susceptible to nuclease degradation, precluding production of mature, translatable mRNA.[37] Lamberts *et al.*[13] had demonstrated that mitogen-stimulated lymphocytes from affected members of kindred-B had decreased responsiveness to dexamethasone-induced suppression of [^{3}H]thymidine incorporation. In addition, Vanderbilt *et al.*[38] have shown that the magnitude of several transcriptional response elicited by the GR are roughly proportional to the number of receptor molecules per cell. These studies indicate that the splice site deletion identified was responsible for the syndrome of FGR in this family.

Elucidation of the molecular basis of various genetic diseases in recent years has revealed great variability of the identified genotypes with little or no correlation with the known clinical phenotypes. In cystic fibrosis no predictions can be made about the occurrence of common complications or the severity or course of pulmonary disease based on the genotype[39] and in osteogenesis imperfecta more than 70 mutations of the two structural genes coding for the proα-1 and -2 chains result in a variety of clinical phenotypes with no apparent correlation with the identified genetic defects.[40] The syndrome of FGR is no exception to the rule: in the two families studied at the NIH two completely different genetic alterations seem to account for the same disease. A third family studied elsewhere[41] had another defect of the GR gene: a guanine-to-adenine point mutation in cDNA position 2317, leading to an amino acid substitution at position 729 of the GR protein within the ligand-binding domain. The mutation also led to decreased affinity and apparently compromised the functional ability of the GR. The propositus in this family is the only patient with FGR reported to present with isosexual precocious puberty, adding not only genetic but also clinical variability to the syndrome.[12]

The variety of phenotypes and abnormal receptor studies described in FGR cases as well as the complexity of the GR structure and its multitude of interactions within the cell, can with certainty predict that many more genetic defects will be identified as causes of glucocorticoid resistance in humans. The latter can include not only alterations of the GR gene structure and function, but also of other DNA sequences that are important for mediating glucocorticoid responsiveness, such as the GREs and those that code for the c-fos and c-jun transcription factors, the hsp-90 and -70 proteins and the immunophillins. It is noteworthy that in our expression studies of the Val-641 mutant GR, there was complete lack of CAT induction by dexamethasone in cells transfected with the mutant plasmid, even in the presence of high concentrations of the ligand.[33] That was unexpected in light of adequate expression of the defective receptor in the *in vitro* system and the *in vivo* situation of the studied patients where high doses of exogenous glucocorticoids resulted in clinical resolution of the FGR symptomatology. Nevertheless, our finding was not the first showing that *in vitro* studies of the GR differ from *in vivo* situations[35] and underlines the importance of the cellular environment for the function of the GR. It is possible that mutations of the proteins that constitute this environment lead to milder forms of glucocorticoid resistance than those caused by alterations of the GR gene. Alternatively, identical GR-genotypes can result in two distinct receptor phenotypes, as it has been described in the dexamethasone-resistant clones originating from the human leukemic cell line CCRF-CEM-C7.[42]

The above are suggestive of an actually higher prevalence of the phenomenon of generalized glucocorticoid resistance in humans. The latter is at present most likely underestimated and believed to be rare because of the limited awareness of the syndrome by clinicians and its variable clinical presentation. However, in a recent study, 7 of 420 consecutive patients presenting to an adrenal disorders clinic had *bonafide* generalized glucocorticoid resistance with functional abnormalities of their GRs in cultured skin fibroblasts.[43] Their symptoms varied from chronic fatigue and obesity to hirsutism and/or amenorrhea and the abnormalities of the receptor included decreased receptor concentration and/or affinity, thermolability, and defective downregulation of GR mRNA by glucocorticoids. This study suggests that screening women with hyperandrogenism by obtaining measurements of urinary free cortisol excretion per 24 hours may be cost effective, since it reported an incidence of glucocorticoid resistance of 14% among the hirsute women that attended this clinic, a number which is higher than the reported 5 to 7% of the nonclassic forms of congenital adrenal hyperplasia.[14]

In addition to the generalized, congenital syndrome of glucocorticoid resistance, acquired and/or localized forms of deficient glucocorticoid responsiveness may also exist. It is noteworthy, that among the first studies of glucocorticoid resistance were those of human leukemia and mouse lymphoma cell lines that developed unresponsiveness to glucocorticoids cultured under selective conditions.[44] These cells were found to have either decreased specific affinity binding to the ligand or defective interaction with the DNA, whereas their progeny responded normally to glucocorticoids, indicating that the defect of the GR was acquired and potentiated through the selection mechanisms imposed by the culture conditions. The mechanisms of alteration of the GR function in these cell lines varied and included abnormal splicing of the GR mRNA, "activation labile" GRs and even a proposed downregulation of receptor levels.[42,44] More recently, patients with the acquired immunodeficiency syndrome (AIDS) were described as having symptoms similar to patients with adrenal insufficiency (weakness, weight loss, hypotension) and concurrently increased plasma ACTH and cortisol levels, as well as peripheral mononuclear leukocyte GRs with markedly reduced affinity for dexamethasone.[45]

In the present study, we report our preliminary results of the investigation of the GR gene in four patients with Nelson's syndrome. The latter, is a clinical condition caused by the appearance and/or the progression of ACTH-secreting pituitary adenomas, in patients who previously had bilateral medical or surgical adrenalectomy. The increased and often invasive tumor growth is presumably caused by the removal of the restraint of hypercortisolism, which is also the cause of the increased ACTH levels. Glucocorticoids normally repress ACTH production by direct binding of the GR to negative GREs in the proximal region of the proopiomelanocortin (POMC) gene promoter. Since the latter has been found to be intact in ACTH-producing pituitary adenomas,[46] we hypothesized that a defective GR might explain the lack of appropriate negative feedback in these tumors. In one out of the four pituitary tumors investigated, a heterozygous mutation was disclosed. A nucleotide insertion was altering the open reading frame introducing premature termination codons probably excluding the expression of a functioning receptor protein from the mutated allele. The identified genetic alteration of the GR was most likely a somatic mutation, since it was not present in the patient's lymphocytes. The tissue with the mutant GR was part of a large and unusually aggressive suprasellar tumor of a patient with a long and complicated history of Nelson's syndrome. By perhaps lowering the receptor number in the tumor cells, the defect identified might have caused local resistance to the negative glucocorticoid feedback and, thus, promoted tumor growth (unpublished data).

Alterations of the GR function causing local resistance may not be a phenomenon related to tumorigenesis only. It is possible that local glucocorticoid resistance exists in noncancerous tissue cells participating in the pathogenesis or the clinical manifestations of other disease states. Altered GR sensitivity at the various tissue levels may explain the observed variability of the severity of the clinical manifestations of Cushing's syndrome. A deficient GR at the dopaminergic mesocortical or mesolimbic system (responsible for pleasure and reward phenomena) or the norepinephrine-locus ceruleus system (responsible for proper arousal) may result in emotional disorders characterized, respectively, by stimulus or drug-seeking behaviors and chronic fatigue despite normal cortisol production.[14] Decreased GR concentration in circulating peripheral leukocytes to about 50% of control levels has been shown in patients with rheumatoid arthritis, who have normal or even decreased cortisol levels, indicating that decreases of GR function in immunity-mediating cells may be involved in excessive immune responses, and, thus, the pathogenesis of autoimmune diseases.[47] Finally, in glucocorticoid-resistant asthma there is abnormal suppression of several cellular functions by glucocorticoids both *in vitro* and *in vivo*[48] and T-cell glucocorticoid resistance mediated in part by the cytokines interleukin (IL)-2 and -4, which cause a reversible decrease of GR affinity *in vitro*.[49]

We conclude that glucocorticoid resistance may be more prevalent than previously believed and may have important implications for human pathophysiology. Primary cortisol resistance is caused by defects of the GR gene with significant molecular and clinical heterogeneity. It appears to be relatively frequent among patients with hyperandrogenism and should be ruled out in these cases. States of resistance to glucocorticoid action include not only the generalized congenital form observed in FGR but also local tissue unresponsiveness to glucocorticoids observed in a wide array of disease states such as lymphoid and pituitary cell neoplasia, autoimmune diseases and probably psychiatric disorders.

SUMMARY

Familial glucocorticoid resistance (FGR) is a rare hereditary disorder characterized by hypercortisolism and the absence of stigmata of Cushing's syndrome. The inability of glucocorticoids to exert their effects on target tissues is compensated for by increases in circulating corticotropin (ACTH) and cortisol, the former causing excess secretion of both adrenal androgens and adrenal steroid-biosynthesis intermediates with salt-retaining activity. There is considerable variability in the clinical presentations of FGR ranging from asymptomatic, to isolated chronic fatigue and to hypertension with or without hypokalemic alkalosis or to hyperandrogenism, or both. In women, hyperandrogenism can result in acne, hirsutism, menstrual irregularities, oligoanovulation, and infertility; in men it may lead to infertility and in children to precocious puberty. The reported molecular defects in FGR, such as point mutations and a microdeletion of the glucocorticoid receptor (GR) gene, cause partial resistance by, respectively, compromising the function of the GR or decreasing its intracellular concentration in glucocorticoid target tissues. Complete glucocorticoid resistance is believed to be incompatible with life in humans. Hence, the glucocorticoid resistance cases reported have been partial and of variable degree. The extreme variability in the clinical manifestations of the disorder can, additionally, be explained by differing sensitivity of target tissues to mineralocorticoids or androgens or both, and perhaps by different biochemical defects of the glucocorticoid receptor, causing selective resistance of certain glucocorticoid responses in specific tissues. Isolated tissue-resistance from a somatic mutation of the GR in a corticotropinoma from a patient with Nelson's syndrome was also found, suggesting that this may be a mechanism of tumorigenesis. There is additional evidence that defects of GR function can appear surreptitiously in a variety of clinical conditions, suggesting that glucocorticoid resistance in humans may be involved in the pathogenesis and/or clinical picture of a plethora of disease states, of which FGR is the archetype.

REFERENCES

1. Baxter, J. D. & G. G. Rousseau, Eds. 1979. Glucocorticoid hormone action. Springer Verlag. New York, NY.
2. Orth, D. N., W. J. Kovacs, & C. R. DeBold. 1992. The adrenal cortex. *In* Williams Textbook of Endocrinology. 8th edit. J. D. Wilson & D. W. Foster, Eds. 489–620. W.B. Saunders Co. Philadelphia, PA.
3. Munck, A. & P. M. Guyre. 1986. Glucocorticoid physiology, pharmacology and stress. *In* Steroid Hormone Resistance: Mechanisms and Clinical Aspects. Advances in Experimental Medicine and Biology. G. P. Chrousos, D. L. Loriaux & M. B. Lipsett, Eds. Vol. 196: 81–96. Plenum Press. New York, NY.
4. Evans, R. M. 1988. Science **240:** 889 895.
5. Vingerhoeds, A. C. M., J. H. H. Thijssen & F. Schwarz. 1976. J. Clin. Endocrinol. Metab. **43:** 1128–1133.
6. Iida, S., M. Gomi, K. Moriyaki *et al.* 1985. J. Clin. Endocrinol. Metab. **60:** 967–971.
7. Lamberts, S. W. J., D. Poldermans, M. Zweens & F. H. J. DeJong. 1986. J. Clin. Endocrinol. Metab. **63:** 1328–1333.
8. Nawata, H., K. Sediya, K. Higuchi *et al.* 1987. J. Clin. Endocrinol. Metab. **65:** 219–226.
9. Bronnegard, M., S. Werner & J. A. Gustafson. 1989. J. Clin. Invest. **78:** 1270–1278.
10. Vescei, P., K. Frank, D. Haack *et al.* 1989. Cancer Res. (Suppl) **49:** 2220s–2221s.
11. Jaursch-Hancke, C., B. Allolio, M. Reinicke & W. Winkelmann. 1989. Acta Endocrinol. (Copenhagen) (Suppl) **120:** 113.
12. Malchoff, C. D., E. C. Javier, D.M. Malchoff *et al.* 1990. J. Clin. Endocrinol. Metab. **70:** 503–507.
13. Lamberts, S. W. J., J. W. Koper, P. Biemond *et al.* 1992. J. Clin. Endocrinol. Metab. **74:** 313–321.
14. Chrousos, G. P., S. Detera-Wadleigh & M. Karl. 1993. Ann. Intern. Med. **119:** 1113–1124.
15. Chrousos, G. P., A. Vinerhoeds, D. Brandon *et al.* 1982. J. Clin. Invest. **69:** 1261–1269.
16. Chrousos G. P., A. C. M. Vingerhoeds, D. L. Loriaux & M. B. Lipsett. 1983. J. Clin. Endocrinol. Metab. **56:** 1243–1245.
17. Hollenberg, S. M., C. Weinberger, E. S. Ong *et al.* 1985. Nature **318:** 635–641.
18. Giguere, V., S. M. Hollenberg, M. G. Rosenfeld & R. M. Evans. 1986. Cell **46:** 645–652.
19. Encio, I. J. & S. D. Detera-Wadleigh. 1991. J. Biol. Chem. **266:** 7182–7188.
20. Francke, U. & B. E. Foelmer. 1989. Genomics **4:** 610–612.
21. Picard, D. & K. R. Yamamoto. 1987. EMBO J. **6:** 3333–3340.
22. Hollenberg, S. M. & R. M. Evans. 1988. Cell **55:** 899–906.
23. Pratt, W. B. 1990. Prog. Clin. Biol. Res. **322:** 119–132.
24. Dalman, F. C., L. C. Scherrer, L. P. Taylor, H. Akil & W. B. Pratt. 1991. J. Biol. Chem. **266:** 3482–3490.
25. Sanchez, E. R., L. E. Faber, W. J. Henzel & W. B. Pratt. 1990. Biochemistry **29:** 5145–5152.
26. Kuk Tai, P. K., M. W. Albers, H. Chang, L. E. Faber & S. L. Schreiber. 1992. Science **256:** 1315–1318.
27. Picard, D., B. Khursheed, M. J. Garabedian *et al.* 1990. Nature **348:** 166–168.
28. Nigg, E. A., P. A. Bauerle & R. Luhrmann. 1991. Cell **66:** 15–22.
29. Drouin, J., Y. L. Sun, S. Tremblay *et al.* 1992. Mol. Endocrinol. **6:** 1299–1309.
30. Schule, R., P. Rangarajan, S. Kliewer *et al.* 1990. Cell **62:** 1217–1226.

31. Orti, E., D. B. Mendel, L. I. Smith, J. E. Bodwell & A. Munck.1989. J. Steroid Biochem. Mol. Biol. **34:** 85–96.
32. Brandon, D. D., A. J. Marwick, G. P. Chrousos & D. L. Loriaux. 1989. Cancer Res. (Suppl) **49:** 2203s–2213s.
33. Hurley, D. M., D. Accili, C. A. Stratakis *et al.* 1991. J. Clin. Invest. **87:** 680–686.
34. Karl, M., S. W. Lamberts, S. Detera-Wadleigh *et al.* 1993. J. Clin. Endocrinol. Metab. **76:** 683–689.
35. Stratakis, C. A., M. Karl, D. M. Hurley *et al.* 1991. *In* Horizons in Endocrinology. M. Maggi & V. Geenen, Eds. Serono Symposia. Vol. 76: 03—08. Raven Press. New York, NY.
36. Krainer, A. R. & T. Maniatis. 1988. *In* Frontiers in Molecular Biology: Transcription and Splicing. IRL Press. Oxford, Washington, DC.
37. Rees, D. J. G., C. R. Rizza & G. G. Brownlee. 1985. Nature **316:** 799–803.
38. Vanderbilt, J. N., R. Miesfield, B. A. Maler & K. R. Yamamoto. 1987. Mol. Endocrinol. **1:** 68–74.
39. The Cystic Fibrosis Genotype-Phenotype Consortium. 1993. N. Engl. J. Med. **329:** 1308–1313.
40. Prockop, D. J. 1992. N. Engl. J. Med. **326:** 540–545.
41. Malchoff, D. M., A. Brufsky, G. Reardon. *et al.* 1993. J. Clin. Invest. **91:** 1918–1925.
42. Ashraf, J. & E. B. Thompson. 1993. Mol. Endocrinol. **7:** 631–642.
43. Werner, S., M. Thoreu, J. A. Gustafsson & M. Bronnegard. 1992. **75:** 1005–1009.
44. Gehring, H. 1986. Mol. Cell. Endocrinol. **48:** 89–96.
45. Norbiato, G., M. Bevilaqua, T. Vago *et al.* 1992. J. Clin. Endocrinol. Metab. **74:** 608–613.
46. Monig, H., I. U. Ali, E. H. Oldfield & H. M. Schulte. 1993. Exp. Clin. Endocrinol. **101:** 36–38.
47. Chikanza, I. C., P. Petrou, G. P. Chrousos, G. Kingsley & G. S. Panayi. 1992. Arthritis Rheum. **35:** 1281–1288.
48. Lane, S. J. & T. H. Lee. 1991. Am. Rev. Respir. Dis. **143:** 1020–1024.
49. Sher, E. R., Y. M. Leung, W. Surs *et al.* 1994. J. Clin. Invest. **93:** 33–39.

DISCUSSION OF THE PAPER

Question: I am glad you brought up this last possibility, the glucocorticoid hypersensitivity state, because as you were getting into the last part of your presentation I was thinking of a patient that I was recently asked to consult on who looked like that and the endocrine work-up was completely normal. All her levels were normal, though she had a very distinct Cushingoid appearance. She looked like she had Cushing's Syndrome, but all hormone levels were totally normal. She had the Buffalo fat hump and so forth. I do not know how well this is verified, but this change, she said, came on quite suddenly, quite acutely after she had sustained a fall and perhaps a closed head injury, without losing consciousness. But up until that time in the history she gives, she was quite normal. So I wanted to raise this question, since this patient might have been an example of this type of hypersensitivity to glucocorticoids.

Chrousos: It could be. She could have hypersensitivity of the fat tissue to glucocorticoids. There is a theory that obesity may be glucocorticoid hypersensitivity of the lipid tissue. Dr. Bronogan from Sweden introduced a theory a few years ago, and it may be true.

Question: Is the form of the receptor which lacks steroid binding (that you mentioned earlier) a constitutive repressor, or was that what your *in vitro* data suggested, because it seemed to block the normal receptor?

CHROUSOS: The *in vitro* data suggest that it antagonizes the effects of the alpha receptor.

QUESTION: Is it binding then to the GREs?

CHROUSOS: Yes, we have shown that.

QUESTION: So presumably the mutation or the lack of the binding domain also means it is not associated with HSP 90?

CHROUSOS: Yes. Otherwise it would not get into the nucleus.

QUESTION: So it is constitutively activated?

CHROUSOS: Right.

E. R. DE KLOET (*Sylvius Laboratories*): The data you showed intrigued me very much because in most of your examples of glucocorticoid resistance you see a change in affinity of the receptors. This discussion has been dwelling throughout this conference on the implications of changed numbers and changed affinities of the corticosteroid receptors. In the cases I saw in my own laboratory the affinity changes and it was reduced affinity, correlated with effects of antagonists. So the HPA system was less sensitive to glucocorticoid feedback, while changes in numbers were much harder to interpret. Now you showed me the obvious Dutch patients with the congenital change in affinity which was due to a receptor mutation. You said that in the aged patient as well as during fatigue syndrome, there is a changed receptor affinity of the glucocorticoid receptor and that this receptor change could be acquired. You also said that interleukins had the ability to change receptor affinity. Dr. Munck will agree that there are many possibilities to change affinity, for instance, by an altered phosphorylation state, but you also said that there was polymorphism in the glucocorticoid receptor gene. Are you implying that in certain pathological conditions the expression can be changed so receptor mutants show up and that acquired glucocorticoid deficiency is something which you acquire indeed during an endotoxin challenge, an emotional challenge or whatever challenge? This would be a very important concept for stress research.

CHROUSOS: So far it is only an association that has been made and I believe the *in vitro* data that were presented at JCI show that cytokines can change the affinity. But there are many potential mechanisms for that. For example, we know that the glucocorticoid receptor interacts with C2 and that interaction neutralizes both. Also we may have expression of the beta form, which may form heterodimers and alter the affinity of the ligand binding monomer. Recently there was a paper in *Nature* on colriticulum and how colriticulum binds to the DNA binding domain of the glucocorticoid receptor and neutralizes it. So it may be some other protein, histoprotein, who knows, that binds and alters what you see when you measure the affinity. Now as for the polymorphism and the hypertension, that was a purely epidemiological association. We studied about 800 people—normals versus patients with familial hypertension—and they came up with about 27 percent incidence of this particular polymorphism in hypertension. Who knows what it means. In osteoporosis we know that bone density is defined by polymorphism of the vitamin D receptor gene. Something very similar may be happening with the glucocorticoid receptor. We are going to look at that because I think it is very important.

DE KLOET: In a study in rats with Professor Holsboer and Dr. Schöbitz from the Max Planck Institute in Munich, we found that if rats are given interleukin 1 in the brain, a long-term activation of adrenocortical activity is the consequence. Much research has been devoted to turning on the HPA system, but not so much to the question why cytokines also lead to prolonged activation of the pituitary adrenal system. We observed that the mineralocorticoid receptor selectively was changed in its affinity state in the hippocampus. MR had a lower affinity, which would imply that feedback control via this receptor is impaired, leading to disinhibition and thus prolonged activation of the HPA axis where cytokines could induce impaired glucocorticoid action by reducing receptor affinity, so I was very happy to hear your line of reasoning regarding polymorphism of the corticosteroid receptors.

Concluding Remarks

We have now come to the end of three days which Dr. Azmitia, Dr. Landfield, and I feel have been very exciting. It has been an excellent conference. There were group discussions about sometimes sensitive subjects, and we were very happy that there were actually no physical assaults between the discussants! Most of all, we hope that this meeting has led to a new set of experiments and new hypotheses, and that we will have a basis for the next three years to get together again on a topic like this. There are so many people now interested in the function of brain corticosteroid receptors that the field has reached a critical mass in which you can really have a multidisciplinary and integrated type of meeting as we have had during these three days. Hypotheses are being formed and also need to be disproved. We have seen a number of prime examples here.

Finally, we sincerely hope that in December of this year you can give your colleagues the beautiful *Annal* as a Christmas gift, which will contain everything about brain corticosteroid receptors you would like to know. We trust that you will have a pleasant recollection of this meeting, and that we shall see you again soon.

Thank you very much.

E. Ronald de Kloet

Circadian Regulation of Basal Cortisol Levels in Posttraumatic Stress Disorder

RACHEL YEHUDA, MARTIN H. TEICHER, ROBERT A. LEVENGOOD, ROBERT L. TRESTMAN, AND LARRY J. SIEVER

The Mount Sinai Medical Center
One Gustave Levy Place
New York, New York 10029-6574

Posttraumatic stress disorder (PTSD) is a psychiatric condition that can occur following exposure to trauma. Because the hypothalamic-pituitary-adrenal (HPA) axis is a major hormonal system mediating the stress response, and has been found to be altered in response to a variety of acute and chronic stressors, it has been hypothesized that fundamental changes in the HPA axis would be relevant to the pathophysiology of PTSD. There is now converging evidence suggesting that PTSD patients show HPA axis alterations that are substantially different from those that have been described in stress and major depression. For example, in some,[1–3] but not all studies[4] basal 24-hr urinary excretion of cortisol was lower in PTSD compared with normal and depressed patients. Furthermore, in contrast to the "nonsuppression" of cortisol observed in major depression, cortisol levels following dexamethasone (DEX) were lower (*i.e.*, more suppressed) in combat veterans with PTSD than in normals.[5–6]

It has also been observed that combat veterans with PTSD exhibit a larger number of cytosolic glucocorticoid receptors (GR) on circulating lymphocytes compared to other psychiatric groups such as bipolar mania, panic disorder, schizophrenia, and major depression.[3] Combat veterans with PTSD also appear to have a larger number of GR than combat veterans who do not meet criteria PTSD[6] and normal controls.[6,7] Veterans who do not meet criteria for PTSD were found to have a larger number of GR than normals.[6] Although the larger number of cytosolic GR in PTSD may simply reflect the lower circulating levels of cortisol, it is also possible that some long-lasting alterations in GR binding parameters result from exposure to traumatic events, and/or the subsequent development of PTSD.

Based on the above data, we have hypothesized that HPA axis alterations in chronic PTSD appear to reflect an enhanced negative feedback at one or more sites along the axis.[6] In this model, chronic increases in the release of CRF would lead to an altered responsivity of the pituitary, as evidenced by the blunted ACTH response to CRF previously observed.[8] However, because of a primary alteration in GR responsivity, there would be a stronger negative feedback resulting in attenuated baseline ACTH and cortisol levels as well as an enhanced responsivity to DEX. According to this model it would be expected that some target tissue of the HPA axis might be unusually responsive or sensitized.

Recently, in order to further explore the enhanced sensitivity of the HPA axis in PTSD, we conducted a chronobiological analysis to elucidate the intrinsic regulatory

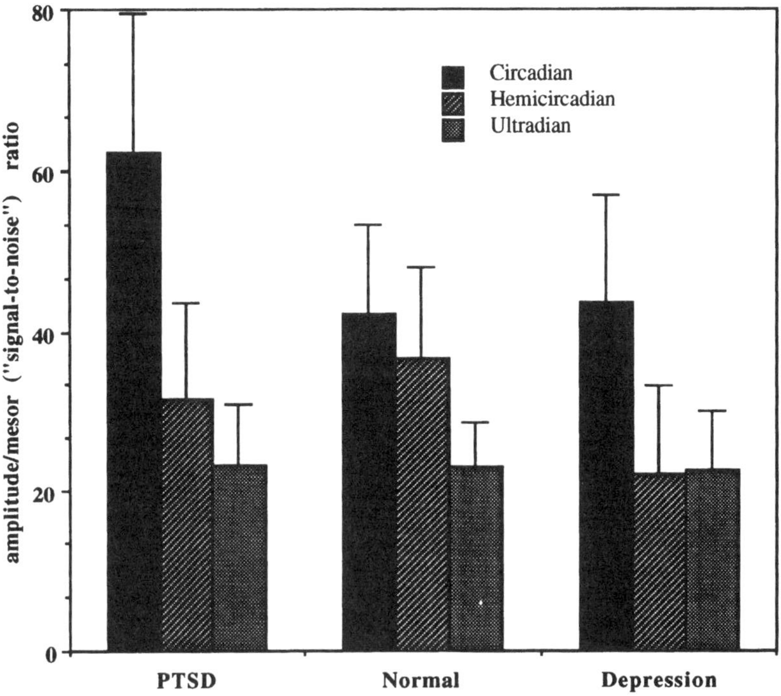

FIGURE 1. "Signal to noise" ratio of basal cortisol secretion over the 24-hr diurnal cycle.

dynamics controlling cortisol levels under baseline conditions. Theoretically, a sensitized system would be expected to show a strengthening of the fundamental regulatory dynamic that would be reflected by a stronger circadian pattern, a greater "signal-to-noise" ratio (*i.e.*, reflecting less random fluctuations) and a resultant decrease in basal cortisol levels secondary to the enhanced regulation. Plasma for cortisol determination was obtained from 15 combat veterans with PTSD, 14 subjects with major depression and 15 normal males every 30 min during a 24-hr period of bed rest. Raw cortisol data were modeled using standard and multioscillator cosinor analyses to determine the best fitting functions for circadian, hemicircadian (*i.e.*, 12-hr rhythm) and ultradian (*i.e.*, greater than three cycles per day) components of cortisol release. The results showed that PTSD subjects had substantially lower cortisol levels and a greater circadian signal-to-noise ratio than the other groups (FIG. 1). The data may reflect a system with a maximally low background and a greater ability to respond to the environment. Thus, alterations of the HPA axis may parallel aspects of the clinical profile observed in PTSD wherein patients appear primed to hyperrespond to stimuli. This type of hypersensitivity appears to be present in other biological systems as well. For example, combat veterans with PTSD appear to show an enhanced MHPG response to yohimbine challenge.[9] It can be concluded from the present results that the pattern of cortisol secretion and regulation observed in the PTSD group under baseline conditions appears to reflect an exaggerated sensitization of the HPA axis.

REFERENCES

1. MASON, J. W., E. L. GILLER, T. R. KOSTEN, R. B. OSTROFF & L. PODD. 1986. Urinary free-cortisol levels in post-traumatic stress disorder patients. J. Nerv. Ment. Dis. **174:** 145–159.
2. YEHUDA, R., S. M. SOUTHWICK, G. NUSSBAUM, E. L. GILLER & J. W. MASON. 1991. Low urinary cortisol excretion in PTSD. J. Nerv. Ment. Dis. **178:** 366–369.
3. YEHUDA, R., D. BOISONEAU, J. W. MASON & E. L. GILLER. 1993. Glucocorticoid receptor number and cortisol excretion in mood, anxiety, and psychotic disorders. Biol. Psychiatry **34**(1–2): 18–25.
4. PITMAN, R. & S. ORR. 1990. Twenty-four hour urinary cortisol and catecholamine excretion in combat-related posttraumatic stress disorder. Biol. Psychiatry **27:** 245–247.
5. YEHUDA, R., S. M. SOUTHWICK, J. H. KRYSTAL, D. S. CHARNEY & J. W. MASON. 1993. Enhanced suppression of cortisol following a low dose of dexamethasone in combat veterans with posttraumatic stress disorder. Am. J. Psychiatry **150:** 83–96.
6. YEHUDA, R., D. BOISONEAU, M. T. LOWY & E. L. GILLER. Prov. accepted. Dose-response changes in plasma cortisol and lymphocyte glucocorticoid receptors following dexamethasone administration in combat veterans with and with posttraumatic stress disorder. Arch. Gen. Psychiatry.
7. YEHUDA, R., M. T. LOWY, S. M. SOUTHWICK, S. SHAFFER & E. L. GILLER. 1991. Increased number of glucocorticoid receptor number in post-traumatic stress disorder. Am. J. Psychiatry **149:** 499–504.
8. SMITH, M. A., J. DAVIDSON, J. C. RITCHIE, H. KUDLER, S. LIPPER, P. CHAPPELL & C. B. NEMEROFF. 1989. The corticotropin releasing hormone test in patients with post traumatic stress disorder. Biol. Psychiatry **26:** 349–355.
9. SOUTHWICK, S. M., J. H. KRYSTAL, C. A. MORGAN, D. JOHNSON, L. M. NAGY, A. NICOLAOU, G. R. HENINGER & D. S. CHARNEY. 1993. Abnormal noradrenergic function in posttraumatic stress disorder. Arch. Gen. Psychiatry **50:** 266–274.

Altered Responses of Cortisol, GH, TSH and Testosterone to Acute Stress after Four Months' Practice of Transcendental Meditation (TM)[a]

C. R. K. MacLEAN, K. G. WALTON, S. R. WENNEBERG, D. K. LEVITSKY, J. V. MANDARINO, R. WAZIRI,[b] AND R. H. SCHNEIDER

Departments of Physiology and Chemistry
Maharishi International University
Faculty Box 1005
1000 North 4th Street
Fairfield, Iowa 52557
and
[b]Department of Psychiatry
University of Iowa
Iowa City, Iowa 52242

Evidence is growing for a role of chronic stress in the etiology of several psychiatric and somatic disorders. Thus, stress reduction techniques are beginning to receive attention as preventative or treatment alternatives. Previous studies suggesting that TM might produce benefits by strengthening adaptive mechanisms implied its effects would be opposite to those resulting from chronic stress.[1–3] The present study investigates changes in neuroendocrine responses to acute laboratory stressors before and after four months of twice-daily practice of TM or of a cognitive approach to stress reduction.

The effects of chronic stress on the hormones investigated in this study have been documented in previous studies on humans and/or nonhuman primates. In the wild baboon, chronic stress appears to elevate baseline cortisol and to reduce the cortisol response to acute stressors.[4] This may be due in part to alterations in hippocampal neurons caused by frequent stress-induced elevations of cortisol. Since the above cortisol profile is associated with poorer health than its opposite, these and other results may support a stress-induced loss of hippocampal regulation of the hypothalamic-pituitary-adrenocortical axis as an important factor in disease and aging.[5] In contrast to its tendency to elevate baseline cortisol, chronic stress appears to decrease testosterone titers, both in animals and in humans.[6–8] Furthermore, a study of the relationship between social status and testosterone in the wild baboon reported an increase of testosterone in response to acute stress in successful males, while subordinate males (who appear chronically stressed) showed a decrease.[8] In humans, thyroid stimulating hormone (TSH) has been reported to increase during strenuous exercise[9] and in a stressful interview.[10] Short-term stressors may increase GH release,

[a]Supported by NIH Grant #1 R15HL 40495 01A1 (to Schneider and Walton), and NIH Grant #RR59, Clinical Research Centers Branch (to U. of Iowa).

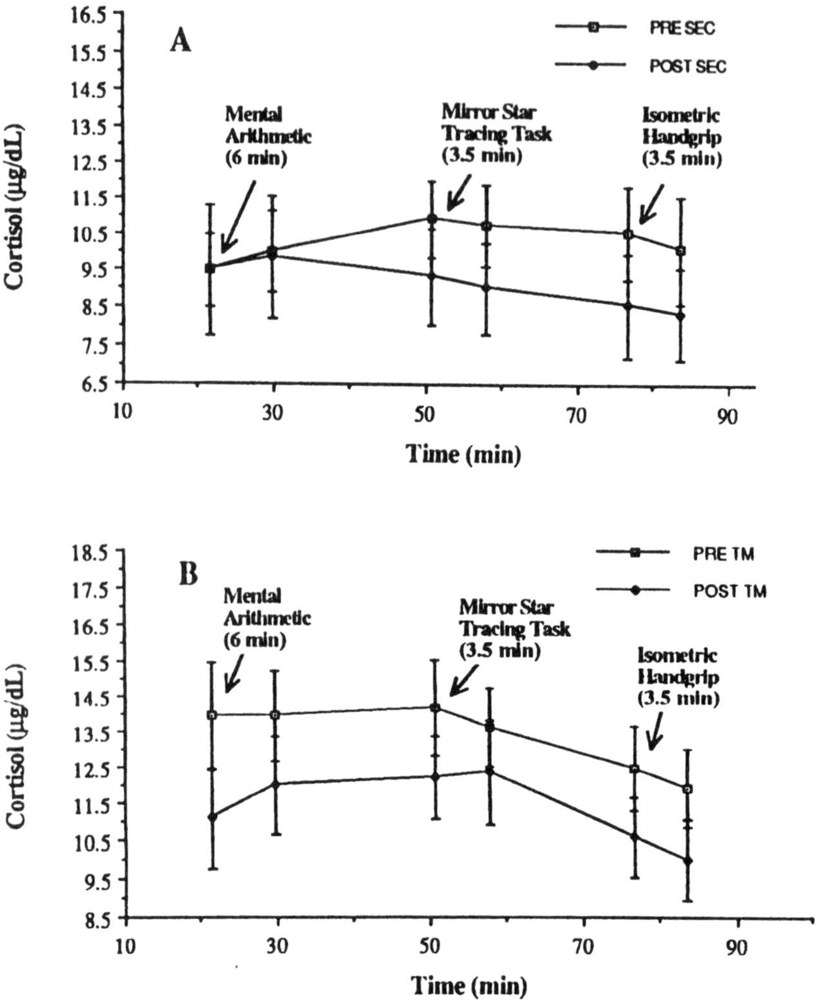

FIGURE 1. Comparison of pre and post changes in cortisol secretion during laboratory stressors for (**A**) the group assigned to SEC (n = 13) and (**B**) the group assigned to TM (n = 16). From pretest to posttest, cortisol decreased significantly at the baseline (the first point) for the TM group (t (15) = 2.21, $p = 0.04$) but not for the SEC group (t (12) = –0.02, $p = 0.98$). In addition, the decrease in average cortisol across the stress session from pre to post was significant for the TM group (t (15) = 2.18, $p = 0.04$) but not for the SEC group (t (12) = 0.775, $p = 0.45$). The two groups differed from pre- to posttest in a mathematically derived, area-under-the-curve variable representing the degree of cortisol response to stressors across the session (F (1, 28) = 5.97, $p = 0.02$).

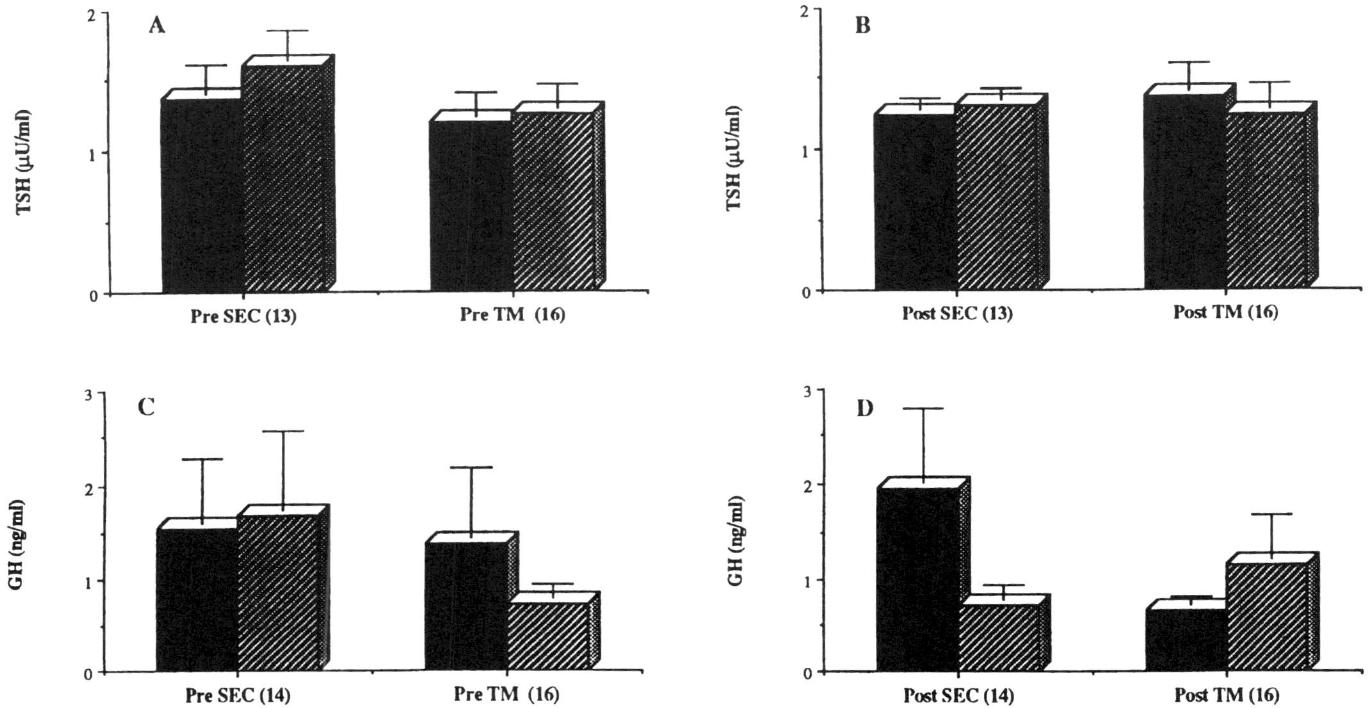

FIGURE 2. Pre and post changes in TSH (A & B) and GH (C & D) from the beginning *(solid bar)* to the end *(diagonal bar)* of the stress session for SEC and TM groups. Posttest TSH response to the stressors was significantly different by ANCOVA between the TM and SEC groups ($F(1, 27) = 6.07$, $p = 0.02$). The posttest response of GH to stressors also differed significantly by ANCOVA between the TM and SEC groups ($F(1, 29) = 4.05$, $p = 0.05$).

but sustained or chronic stress may inhibit.[11] In the case of physical exercise, surgery, stressful interviews or starvation in man, elevated GH secretion has been reported,[10,12,13] whereas severe emotional stress and hypercortisolism appear to inhibit secretion.[12,13]

The present, prospective, randomized, blinded study in healthy 18–32 year old men examined the acute effects of laboratory stressors on plasma hormones before and after four months' participation in either TM or a stress education class (SEC). The stress session began at about 10 AM and included three stressors; mental arithmetic (6 min), a mirror star-tracing task (3.5 min), and isometric hand grip (3.5 min) separated by 20-min rest periods. Samples were assayed by RIA and statistically analyzed by *t* test and ANCOVA.

The TM group decreased significantly from pretest to posttest in baseline cortisol and in overall average of cortisol (FIG. 1). In addition, cortisol response of the TM group increased during the posttest compared to the pretest (FIG. 1). The change in TSH and GH responses to stress from pre- to posttest also differed between the TM and SEC groups (FIG. 2). Furthermore, the two groups differed in testosterone baseline changes from pre- to posttest ($t\ (1, 25) = -2.05$, $p = 0.05$).

These results suggest that practice of the TM technique alters the acute responses of four prominent hormones to laboratory stressors in the direction of more optimal function. These changes in stress responsiveness of multiple hormones may support the conclusion that alterations in neural regulation are involved.

REFERENCES

1. ALEXANDER, C. N., P. ROBINSON, D. W. ORME-JOHNSON, R. H. SCHNEIDER & K. G. WALTON. Homeostasis. In press.
2. JEVNING, R., I. WELLS, A. F. WILSON & S. GUICH. 1987. Physiol. Behav. **40:** 603–606.
3. WERNER, O. R., R. K. WALLACE, B. CHARLES, G. JANSSEN, T. STRYKER & R. A. CHALMERS. 1986. Psychosom. Med. **48**(1/2): 59–66.
4. SAPOLSKY, R. M. 1990. Biol. Psychiatry **28:** 862–878.
5. SAPOLSKY, R. M., L. C. KREY & B. S. MCEWEN. 1986. Endocr. Rev. **7:** 284–301.
6. LEEDY, G. M. & M. S. WILSON. 1985. Psychosom. Med. **47:** 333–337.
7. ALLEN, P. I. M., K. A. BATTY, C. A. S. DODD, J. HERBERT, C. J. HUGH, G. F. MOORE, M. J. SEYMOUR, H. M. SHIERS, P. M. STACEY & S. K. YOUNG. 1985. J. Endocrinol. **107:** 163–170.
8. SAPOLSKY, R. M. 1982. Horm. Behav. **16:** 279–292.
9. ROLANDI, E., E. REGGIANI, R. FRANCESCHINI, G. BAVASTRO, V. MESSINA, G. ODAGLIA & T. BARRECA. 1989. Horm. Res. **21:** 209–221.
10. MOUGEY, E. H., M. A. OLESHANSKY & J. L. MEYEROFF. 1991. Soc. Neurosci. Abstr. **17**(Part 1): 145.
11. SAPOLSKY, R. M. 1992. *In* Behavioral Neuroendocrinology. J. B. Becker, S. M. Breedlove & D. Crews, Eds. 287–324. MIT Press. Cambridge, MA.
12. MALARKEY, W. B., J. C. HALL, D. K. PEARL, J. K. KIECOLT-GLASER & R. GLASER. 1991. J. Clin. Endocrinol. Metab. **75:** 1089–1092.
13. VON ZERSSEN, D., M. BERGER, M. DOSE, P. DOERR, C. KREIG, S. BOSSERT, D. RIEMANN, K.-M. PIRKE, R. DOLHOFER & O.-A. MULLER. 1986. Psychiatr. Dev. **3:** 237–256.

Effects of Hydrocortisone Infusion on rCBF in Schizophrenic Patients during a Memory Task

An O^{15} PET Study[a]

R. GANGULI,[b] M. A. MINTUN, J. T. BECKER, J. S. BRAR, D. DIEHL, M. DELEO, D. MADOFF, AND A. MARTIDIS

Schizophrenia Treatment and Research Center
and
Functional Brain Imaging Center
Department of Psychiatry
Western Psychiatric Institute & Clinic
University of Pittsburgh Medical School
Pittsburgh, Pennsylvania 15213

INTRODUCTION

Regional cerebral blood flow (rCBF) studies have revealed that hippocampal dysfunction may be associated with symptoms of acute schizophrenia, and may be involved in the process that leads to relapse.[1] Neuropathological and neuroanatomical imaging studies in schizophrenia have also reported generalized as well as regional atrophy in temporal lobe structures. Of all limbic structures, the hippocampus is the brain area containing the highest density of cortisol receptors.[2] Any dysregulation of adrenal steroid hormones can significantly affect behavior and memory by interacting hippocampal receptors. It is thus hypothesized that a dysfunction (in the form of insensitivity or failure to downregulate) of cortisol receptors may interfere in the regulation of psychosocial stress in schizophrenic patients, which in turn may precipitate a psychotic relapse.

METHOD

Images of rCBF were obtained on 7 DSM-IIIR schizophrenic patients and 7 healthy controls using a Siemens 95lr/31 PET scanner and intravenous injections of 50 mCi $H_2{}^{15}O$ in 5–7 ml of saline. The scanning session consisted of a transmission scan, followed by eight sequential CBF measures at 12-minute intervals. Each CBF study was obtained either during a rest state (eyes on fixation point) or while performing a memory task (12 word free recall task). Following the first block of four

[a]This work was supported by grants from the National Institute of Mental Health to Rohan Ganguli (RO1-41883 and RSDA MH-00710) and to the Functional Brain Imaging Center at the University of Pittsburgh Medical Center (MH-49815).

[b]Author to whom correspondence should be addressed: Rohan Ganguli, M.D., Associate Professor of Psychiatry and Pathology, Medical and Research Director, Schizophrenia Treatment and Research Center, Western Psychiatric Institute and Clinic, 3811 O'Hara Street, Pittsburgh, PA 15213.

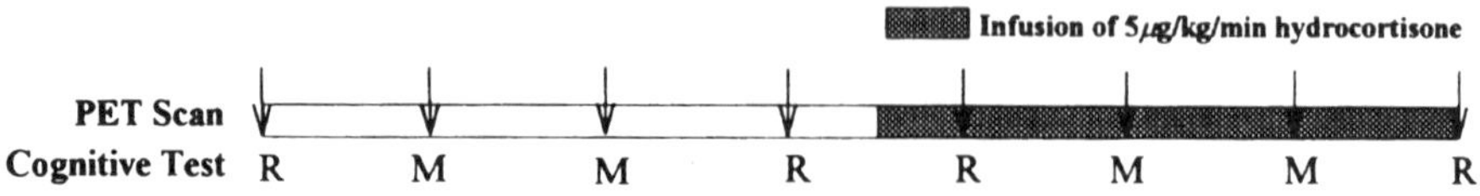

FIGURE 1. Cerebral blood flow measurements at 12-minute intervals. R: rest scan; M: memory test scan.

CBF measures, hydrocortisone (HC) was infused (5 μg/kg^{-1}/min^{-1} for one hour[3]) and a second block using the same task and design obtained (FIG. 1).

Using this design, the following effects were examined: 1. effect of memory task on rCBF in the hippocampus, 2. effect of HC on hippocampal rCBF, and, 3. interaction of memory task and HC infusion on hippocampal rCBF.

RESULTS

Cognitive task performance (12-word recall) was measured before and after infusion of HC. The response from the 12-word recall test was used to characterize "primary" and "secondary" memory. If the recalled word was within 6 words of it being spoken, it was characterized as "primary" or working memory. If the recalled word was beyond the 6 word range it was considered to be "secondary" or short-term memory. Primary memory was enhanced following the HC infusion in both groups. In the control group, secondary or short-term memory deteriorated slightly following the infusion. This component of memory was unaffected by HC in schizophrenic patients (TABLE 1).

Before HC infusion, patients showed less activation in the dorsolateral prefrontal cortex (DLPFC) and anterior cingulate especially on the left side, and less deactivation in the superior parietal and precuneous regions bilaterally. Following the infusion there was a marked increase in the amount of activation and deactivation in both patients and controls. Even though HC produced greater DLPFC activation, and parietal deactivation in patients, it was smaller in magnitude than that produced in controls. HC by itself produced marked right hemispheric deactivation in the parietal and middle temporal regions in controls with very little effect in patients.

DISCUSSION

This study allows the extent of memory impairment and potential drug effects (HC infusion) during the experimental conditions to be correlated with changes in

TABLE 1. Mean (± SD) Numbers of Words Recalled before and after HC Infusion

	Healthy Controls		Schizophrenic Patients	
	Pre-HC	Post-HC	Pre-HC	Post-HC
Primary memory	2.5 ± 0.6	2.8 ± 0.4	2.1 ± 0.6	2.3 ± 0.4
Secondary memory	1.9 ± 0.6	1.6 ± 0.7	1.3 ± 0.7	1.3 ± 0.9

regional blood flow in order to assess their extent and specificity. Memory deficits relating to disturbances in cortisol regulation have significant implications. Hippocampal-mediated cortisol regulation may be helpful in understanding the pathophysiology of stress regulation in schizophrenia, and may help in designing strategies for preventing psychotic relapses in these patients.

REFERENCES

1. Cleghorn, J. M., R. B. Zipursky & S. J. List. 1991. Structural and functional brain imaging in schizophrenia. J. Psychiatry Neurosci. **16**(2): 53–74.
2. Jacobson, L. & R. Sapolsky. 1991. The role of the hippocampus in feedback regulation of the hypothalamic-pituitary-adrenocortical axis. Endocr. Rev. **12**(2): 118–134.
3. Young, E. A., R. F. Haskett, V. Murphy-Weinberg, S. J. Watson & H. Akil. 1991. Loss of glucocorticoid fast feedback in depression. Arch. Gen. Psychiatry **48:** 693–699.

Behavioral and Endocrine Impact of Impaired Type II Glucocorticoid Receptor Function in a Transgenic Mouse Model[a]

S. BEAULIEU,[b] I. ROUSSE,[c] A. GRATTON,[c] N. BARDEN,[d] AND J. ROCHFORD[c]

Biological Psychiatry Branch[b]
National Institute of Mental Health
Building 10, Room 3N212
9000 Rockville Pike
Bethesda, Maryland 20892

[c]Department of Psychiatry
McGill University
Montreal, Quebec, Canada H4H 1R3
and
[d]Department of Physiology
Laval University
Quebec City, Quebec, Canada G1V 4G2

INTRODUCTION

It is well established that a subgroup of depressed patients have an associated hypersecretion of cortisol and loss of normal circadian secretion pattern. Moreover, a negative response to the dexamethasone suppression test has been established as a good state-related marker for depression in nonsuppressors, and this nonsuppressive state usually normalizes with the resolution of the depressive symptoms.[1] The cause of this abnormal glucocorticoid secretion pattern is not clear. Previous studies have demonstrated a higher concentration of corticotropin-releasing factor (CRF) in the cerebrospinal fluid of depressed patients, and results of the CRF stimulation test point to the possibility of a limbic-hypothalamic overdrive. Some studies reported a decrease of the cytoplasmic concentration of glucocorticoid receptors in lymphocytes of depressed patients while others reported a normal concentration but a decreased function of the glucocorticoid receptor. Whether an increased CRF-ergic drive or the recently demonstrated deficient glucocorticoid fast feedback or both concur to produce the abnormal glucocorticoid secretion in this group of depressed patients is not known. A transgenic (TG) mouse model of impaired glucocorticoid type II receptor function which is normalized by antidepressant treatment was previously developed.[2] We now report preliminary results on the behavioral characterization of these animals, as part of a more global endeavor

[a]Supported in part by a Clinician-Scientist Award from the Medical Research Council of Canada to S.B.

to study the role of the limbic-hypothalamo-hypophysio-adrenal axis and its modulation in the symptomatology and etiopathogeny of affective disorders.

MATERIALS AND METHODS

Three sets of experiments were done. First, the locomotor activity of TG mice (1st generation: C57B × C3H = B6C3F1, subsequent generations: B6C3F1 × B6C3F1) and control non-TG mice (B6C3F1) were compared. Locomotor activity was recorded in chambers equipped with photoelectric cells during both dark and light phases of the circadian cycle in different groups of TG and control mice. We also measured the ACTH and corticosterone plasma levels at three time points of the circadian cycle: 8, 14 and 20 h by radioimmunoassay. In the second set of experiments, we measured the locomotor activity during the dark phase of both control and TG animals in response to 1 or 5 mg/kg of d-amphetamine i.p. dissolved in saline 0.9%. The activity was recorded over a period of 2 hours. Finally, we also measured the response of TG and control animals in the Porsolt's forced swim test. Results are reported in terms of immobility time during the last 4 min of the test of 6 min duration.

RESULTS

TG mice exhibited higher basal corticosterone and ACTH plasmatic levels than controls. The amplitude of the circadian variations was also reduced. However, the cortisol and ACTH plasma levels of the controls reached values similar to the TG animals during the dark phase period of the circadian cycle (TABLE 1). TG animals were more active than the controls during the first 30 min of observation of their baseline horizontal locomotor activity, expressed in mean photocell counts,during the light phase (TG = 1133 ± 72, CT = 911 ± 72, p <0.05). However, no difference could be observed during the dark phase of the circadian cycle (TG = 1059 ± 110, CT = 1094 ± 58). These results parallel the endocrine data.

TG animals treated with 1 mg/kg i.p. of d-amphetamine during the dark phase period showed a significantly greater level of locomotor activity than the controls during the first 60 min of recording, although the difference was more impressive during the first 30 min. When receiving a higher dose of 5 mg/kg i.p., the TG animals still demonstrated a higher locomotor activity during the first 30 min (FIG. 1).

In the Porsolt's forced swim test, TG animals had a decreased immobility time compared to controls when tested during the light phase (TG = 167 ± 23 sec, CT = 217 ± 7 sec, p <0.05; second trial: TG = 103 ± 26 sec, CT = 208 ± 17 sec, p <0.01).

TABLE 1. Plasmatic Corticosterone and ACTH Levels at 8, 14 and 20 h in Control (CT) and Transgenic (TG) Animals.

	Corticosterone (ng/ml ± SEM)		ACTH (pg/ml ± SEM)	
TIME	CT	TG	CT	TG
0800 h	41 ± 7	113 ± 22	101 ± 14	178 ± 37
1400 h	85 ± 16	84 ± 11	116 ± 13	173 ± 23
2000 h	155 ± 23	115 ± 21	170 ± 27	169 ± 47

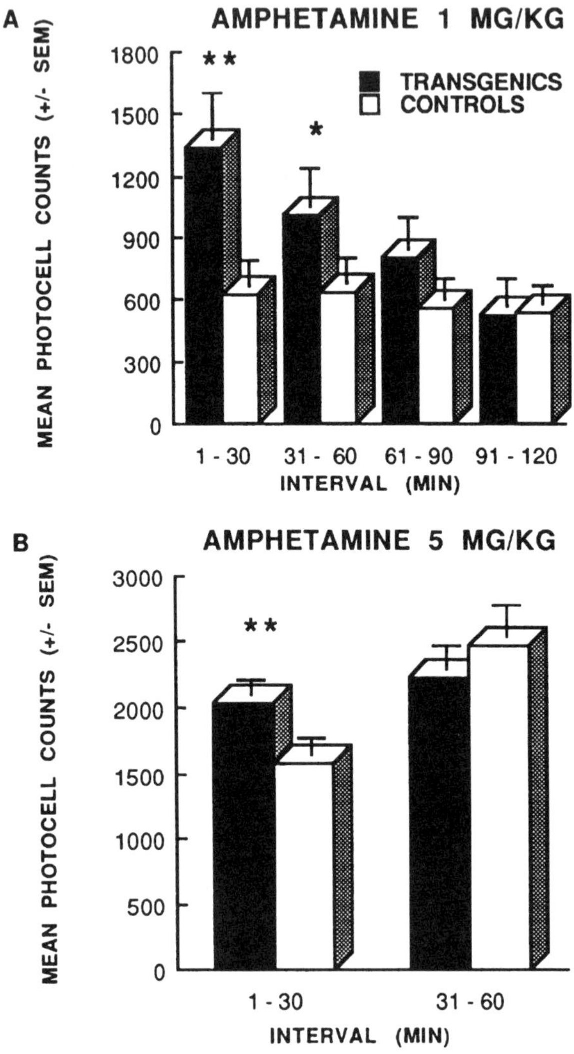

FIGURE 1. **(A)** TG animals treated with 1 mg/kg i.p. of d-amphetamine during the dark phase period, show a significantly greater level of locomotor activity than the controls during the first 60 min of recording (*: p <0.05), although the difference is more impressive during the first 30 min (**: p <0.01). **(B)** When receiving a higher dose of 5 mg/kg i.p., the TG animals still demonstrate a higher locomotor activity during the first 30 min, but not during the second interval of 30 min.

DISCUSSION

The significantly higher plasma levels of ACTH and corticosterone in TG animals in comparison to controls during the light phase of their diurnal cycle might be partly responsible for the higher locomotor activity observed in TG mice during the light phase. The increased sensitivity of the TG animals to d-amphetamine may reflect a higher dopaminergic activity or sensitivity in the central nervous system of these animals, particularly in the nucleus accumbens. Whether these differences of sensitivity to d-amphetamine observed during the dark phase period, *i.e.,* when the ACTH and cortisterone are similar in both control and TG animals, are directly related to the partial deficit in glucocorticoid receptors or to the dampened circadian cycle of TG animals, remains to be studied more precisely. Although there is no clear relationship between motor activity and immobility time in the Porsolt forced swim test,[3] our results obtained using that paradigm parallel those of the horizontal activity test. Recent studies have demonstrated the usefulness of corticosteroid synthesis inhibitors in the treatment of depression.[4] We think that the animal model discussed here might be useful for the study of the role of the glucocorticoids and the type II glucocorticoid receptor in the symptomatology and possibly the etiopathogeny of affective disorders.

REFERENCES

1. RIBEIRO, S. C. M., R. TANDON, L. GRUNHAUS & J. GREDEN. 1993. Am. J. Psychiatry **150:** 1618–1629.
2. PEPIN, M. C., F. POTHIER & N. BARDEN. 1992. Nature **355:** 725–728.
3. BORSINI, F. A. MELI. 1988. Psychopharmacology **94:** 147–160.
4. MURPHY, B. E., V. DHAR, A. M. GHADIRIAN, G. CHOUINARD & R. KELLER. 1991. J. Clin. Psychopharmacol. **11:** 121–126.

Corticosterone Potentiation of Conditioned Fear in Rats[a]

KEITH P. CORODIMAS, JOSEPH E. LeDOUX,
PHILIP W. GOLD,[b] AND JAY SCHULKIN[b]

Center for Neural Science
New York University
6 Washington Place
New York, New York 10003
and
[b]Clinical Neuroendocrinology Branch
National Institute of Mental Health
National Institutes of Health
Bethesda, Maryland 20892

Many people suffer from various types of anxiety disorders, which are characterized by intense fear. Others who suffer from affective disorders such as melancholic depression also display intense fear and is associated with centrally mediated pituitary-adrenal activation.[1] These disorders may represent dysfunctional fear learning and/or fear regulation. Kalin[2] has also shown in young monkeys that a correlation exists between the glucocorticoid corticosterone and fear reactions (freezing behavior). Although many studies have demonstrated an important link between glucocorticoids and stress and CRF and fear, very little is known about the role of glucocorticoids in mediating fear.

The glucocorticoid, corticosterone (CORT), selectively upregulates CRF mRNA in the central nucleus of the amygdala (Ce) in rats.[3] This region plays a crucial role in the learning and expression of conditioned fear reactions.[4] In the present study, we investigated whether CORT would potentiate fear responses (freezing) conditioned to acoustic stimuli.

Twenty adult male Sprague-Dawley rats (250–275 g) were acclimated to the vivarium for approximately 7–10 d before conditioning began. All animals received 10 conditioning trials on one day during which the unconditioned stimulus (US; 1.0 mA footshock) was presented concurrently with the last 500 ms of the 20 s conditioned stimulus (CS; 70 dB/10,000 Hz tone). The intertrial interval was 120 s on the average. Two days after conditioning, subcutaneous hormone pellets (Innovative Research of America, Toledo, OH) were implanted under ketamine (100 mg/kg) and rompun (5 mg/kg) anesthesia and distributed in the following manner: 5 animals received a low dose of CORT (5.0 mg/day/rat); 5 received a high dose of CORT (10.0 mg/day/rat); 5 received a low dose of CORT placebo (5.0 mg/day/rat); and 5 animals received a high dose of CORT placebo (10.0 mg/day/rat). Hormone pellets were left in animals for the entire duration of the experiment.

Freezing was first tested on the morning of the fifth day of hormone exposure (0 h test) in a novel context. While ambulating, animals were exposed to the CS continuously for 600 s and the amount of time they spent freezing was recorded. Freezing was measured at 0 h and 4 h on the fifth day of hormone exposure, then daily until extinction was complete.

A two-way analysis of variance (ANOVA) with one grouping variable (hormone

[a]Supported by MH46516, MH38774, and NSF9209646 awarded to Joseph E. LeDoux.

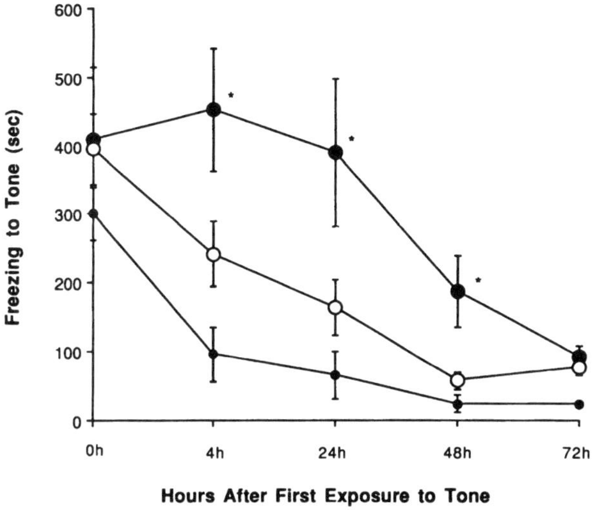

FIGURE 1. Amount of time animals froze in a novel context to a 600 s CS. *Significantly different (p <0.05; Fisher PLSD test) from the low CORT group and the placebo group. Because the freezing responses of both the low and high placebo groups were not significantly different from one another at any timepoint, the data from both groups were collapsed into one group. ● High CORT group; ● low CORT group; ○ placebo group.

treatment) and one repeated measure (test time) was performed on the freezing response data (FIG. 1). The main effect of hormone treatment, $F(2, 17) = 5.1$, $p < 0.025$, and test time, $F(3, 17) = 39.1$ $p < 0.0002$, were both significant, as was the two-way interaction: hormone treatment X test time, $F(6, 17) = 3.1$, $p < 0.02$. Post hoc analysis with the Fisher PLSD test indicated that the high CORT group froze significantly more to the CS compared with the low CORT group and the placebo group at 4 h (p <0.05), 24 h (p <0.05), and 48 h (p <0.05) timepoints (FIG. 1). No significant differences existed between the placebo group and the low CORT group.

High doses of CORT thus potentiated freezing responses to an explicit auditory cue that had previously been paired with footshock. These findings are consistent with our general hypothesis that CORT potentiates fear by altering CRF systems in the Ce. It is also consistent with the known effects of CORT on the Ce,[3] a brain region that is critical for the acquisition and expression of conditioned fear responses.[4]

REFERENCES

1. GOLD, P. W., F. K. GOODWIN & G. P. CHOUSOS. 1988. Clinical and biochemical manifestations of depression. N. Engl. J. Med. **319:** 348–353; 423–420.
2. KALIN, N. H. 1993. The neurobiology of fear. Sci. Am. May: 94–101.
3. MAKINO, S., P. W. GOLD & J. SCHULKIN. 1994. Corticosterone effects on corticotropin-releasing hormone mRNA in the central nucleus of the amygdala and the parvocellular region of the paraventricular nucleus of the hypothalamus. Brain Res. **640:** 105–112.
4. LEDOUX, J. E. 1992. Emotion and the amygdala. *In* The Amygdala. J. Aggleton, Ed. 339–351. Academic Press. New York.

Modulation of the Locomotor Response to Amphetamine by Corticosterone

PIERRE MORMÈDE, JOSETTE DULLUC, AND MARTINE CADOR

Psychobiology Laboratory
INSERM U259–INRA
33077 Bordeaux, France

Repeated exposure to d-amphetamine (AMPH) or to stressors enhances the behavioral response elicited by subsequent exposure to AMPH. This phenomenon, which has been shown to be long-lasting, has been called sensitization. We and others[1–4] have presented experimental evidence that adrenocortical hormones could play a critical role in this process. Indeed, removal of endogenous corticosterone (CS) by adrenalectomy (ADX) reduced the development of sensitization to AMPH, which could be restored by type II (dexamethasone) by not by type I (DOCA or corticosterone) corticosteroid receptor agonists.[1,4] These experiments were designed to investigate further the type of steroid receptors involved and the site of CS action.

Male Wistar rats were adrenalectomized (ADREX) 10 days before studying their response to AMPH sulfate (1.5 mg/kg ip) in a circular corridor with photocell beams to monitor their locomotor activity (FIG. 1). This long-term removal of adrenal steroids reduced the locomotor response to AMPH by approximately one-third as compared to sham-ADREX rats. The response was restored with subcutaneous CS pellets (50 mg) inserted at the time of adrenalectomy. The dose-response study showed that the locomotor response to AMPH was a mirror-image of the thymus weight, a good indicator of glucocorticoid receptor activation. Low CS levels (12.5-mg pellet) could not restore the effect of ADREX on any target studied, and high CS levels (200-mg CS pellet), as observed in chronic social stress,[5] could increase the locomotor response to AMPH over control levels, and also reduced thymus weight. These results bring another argument suggesting that type II CS receptors are involved in modulating the response to AMPH.[1,4]

The response to AMPH was somewhat larger at night, when endogenous CS levels are elevated, than during the day, when CS levels are low. However, this difference was not influenced by chronic changes of CS levels as induced by ADREX and CS replacement at a constant level by subcutaneous hormone pellets (FIG. 1), suggesting (i) that the day/night cycle of the response to AMPH has a neural rather than an adrenal origin, and (ii) that chronic rather than acute changes of CS levels mediate their action on the neural systems involved in the response to AMPH. This finding may be of importance for the search of the biochemical basis of CS action.

To exclude peripheral effects of CS, such as a change in the metabolism of AMPH, the psychostimulant drug was injected directly into the nucleus accumbens,[6] considered as the primary site of its action on locomotor activity (FIG. 2). The same results were obtained as with peripheral administration: the response to locally injected AMPH was strongly attenuated by ADX and restored with CS pellets. This experiment clearly shows that CS acts to change the reactivity of the neuronal systems involved in amphetamine action, as previously suggested.[4] Indeed, type II

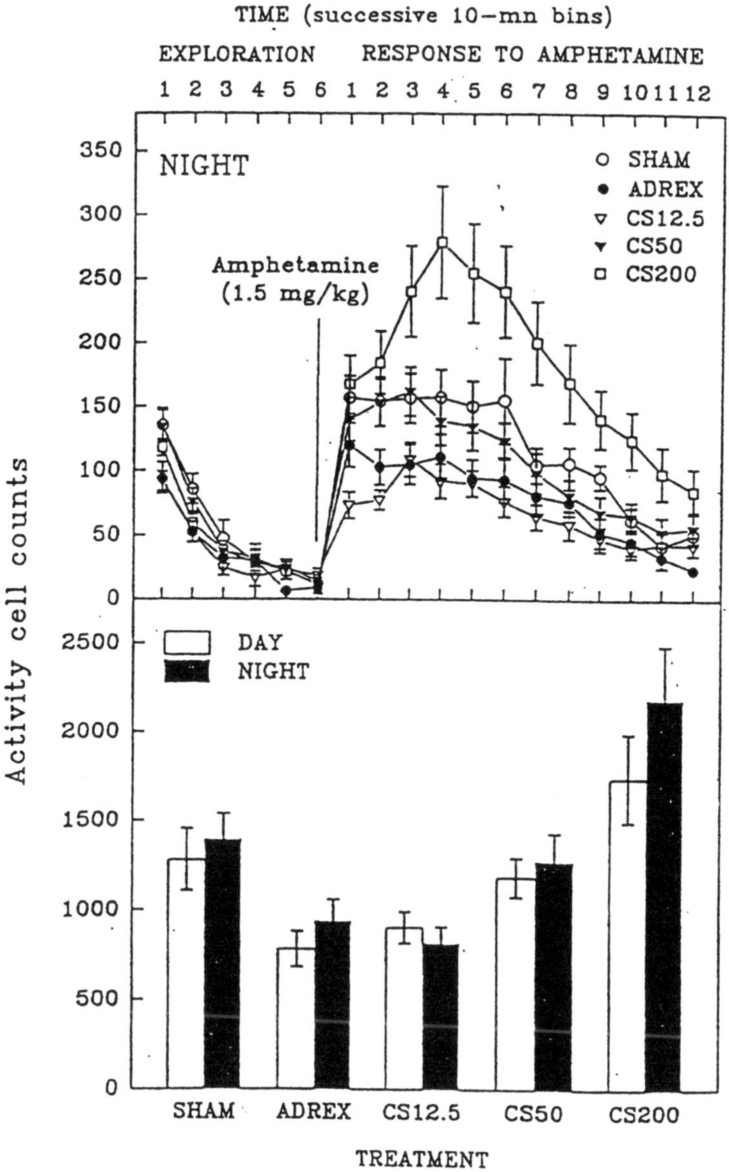

FIGURE 1. *Upper panel.* Locomotor activity during exploration of the circular corridors and in response to amphetamine injected at night, according to the adrenocortical status, 11 animals per group. *Lower panel.* Locomotor activity in response to amphetamine, cumulated over the 2 hrs following injection, according to the period of the day and the adrenocortical status, 11 animals per group. SHAM = SHAM-operated, control animals; ADREX = adrenalectomized animals; CS = subcutaneous replacement theory with pellets containing increasing amount of corticosterone, in mg; mean ± SEM.

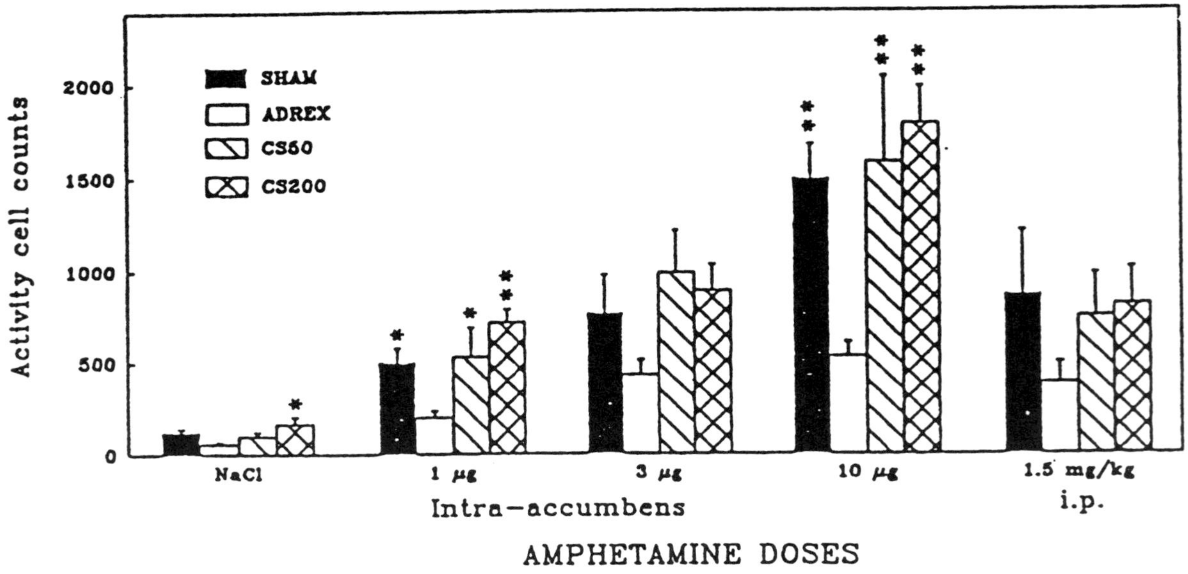

FIGURE 2. Total locomotor activity developed during the 1-hr session in the SHAM (n = 7), ADREX (n = 8), CS50 (n = 6) and CS200 (n = 7) groups following either saline and amphetamine (1, 3 or 10 μg/μl) in the nucleus accumbens and during the 2-hr session following amphetamine 1.5 mg/kg i.p. * $p < 0.05$, ** $p < 0.01$, *** $p < 0.001$ compared with the ADREX group using the Newman Keuls' test.

receptors have been shown to be present in the dopaminergic cell bodies of the ventral tegmental area, where the mesoaccumbens dopaminergic neurons originate.[7] These neurons are therefore primary candidates for the modulation of the behavioral response to psychostimulants by adrenocortical hormones. This mechanism could play a major role in the process of sensitization, when CS secretion is repeatedly activated by stress or drug administrations.

REFERENCES

1. Rivet, J. M., L. Stinus, M. Le Moal & P. Mormède. 1989. Brain Res. **498:** 149–153.
2. Cador, M., B. J. Cole, G. F. Koob, L. Stinus & M. Le Moal. 1993. Brain Res. **606:** 181–186.
3. Deroche, V., P. V. Piazza, S. Maccari, M. Le Moal & H. Simon. 1992. Brain Res. **584:** 309–313.
4. Cools, A. R. 1991. Neuroscience **43:** 419–428.
5. Klein, F., V. Lemaire, C. Sandi, S. Vitiello, J. Van Der Logt, P. E. Laurent, P. Neveu, M. Le Moal & P. Mormède. 1992. Life Sci. **50:** 723–731.
6. Cador, M., J. Dulluc & P. Mormède. 1993. Neuroscience **56:** 981–988.
7. Härfstrand, A., K. Fuxe, A. Cintra, L. F. Agnati, I. Zini, A. C. Wikstrom, S. Okret, Z. Y. Yu, M. Goldstein, H. Steinbuch, A. Verhofstad & J. A. Gustafsson. 1986. Proc. Natl. Acad. Sci. USA **83:** 9779–9783.

Evidence for Nitric Oxide-Mediated Rapid Locomotor Effects of Corticosterone in a Novel Environment

CARMEN SANDI[a] AND CARMEN GUAZA

Neural Plasticity Research Group
Cajal Institute, CSIC
Doctor Arce 37
28002 Madrid, Spain

Glucocorticoids are known to modulate different types of behavior and neural plastic events.[1] They are released during novelty and learning situations and can trigger neural actions through binding to brain receptors. Some of the behavioral actions modulated by glucocorticoids have a rapid onset, suggesting a nongenomic action. Recent evidence in amphibians has shown the presence of neuronal membrane receptors.[2] However, the cellular processes mediating these rapid behavioral effects of corticosteroids are not well understood.

Nitric oxide (NO), a gas molecule which was first identified as the endothelium-derived relaxing factor, is now known to play a wide array of roles in different physiological systems. It has also been identified as an intercellular messenger in the central and peripheral nervous systems.[3] NO is produced by an enzymatic reaction from L-arginine by NO synthase that has been identified in discrete neuronal populations, in particular in areas such as cerebellum, hippocampus, striatum, cortex, hypothalamus, midbrain, and medulla of the rat. In addition, N^G-nitro-L-arginine methyl ester (L-NAME), an inhibitor of NO synthesis, has been shown to reduce morphine-induced changes in locomotion.[4]

This work investigates the possible involvement of NO in certain rapid behavioral changes induced by corticosterone, such as those induced in a novel environment and the startle response.

Male Wistar rats, weighing 200–250 g, were used. Animals were injected i.p. with either saline or N^G-nitro-L-arginine methyl ester (L-NAME, 30 mg/kg; Sigma) 60 min before testing. Half of the animals in each group received a second injection of saline and the other half of corticosterone (Sigma; 5 mg/kg) 15 min before testing. The exploratory activity in a novel environment was evaluated using a digiscan animal activity monitor (activity cage), model RXYZCM TAO (Omnitech Electronics, Inc., Columbus, OH), as previously described.[5] Startle responses were measured in a stabilimeter cage (Responder Economy, Columbus, OH) using acoustic stimulation. The results were analyzed using one-way analysis of variance (ANOVA).

As shown in TABLE 1, the behavioral changes, such as increased locomotion (horizontal activity) and stereotypy counts, induced by corticosterone in a novel environment were prevented by L-NAME, a specific inhibitor of nitric oxide synthase. However, L-NAME did not affect these parameters when injected alone. The NO precursor L-arginine (Sigma; 200 mg/kg) reverted the inhibitory effect of L-

[a]Corresponding author: Dr. Carmen Sandi, Psychobiology Research Group, Cajal Institute, CSIC, Dr. Arce 37, 28002 Madrid, Spain.

TABLE 1. Behavioral Responses (Percentage of SAL-SAL Group) in the Activity Cage of Rats Previously Treated with a First Injection of either Saline or NAME (30 mg/kg) and a Second Injection of either Saline or Corticosterone (5 mg/kg)[a]

Behavioral Parameters	SAL-CORT	NAME-SAL	NAME-CORT
Horizontal activity	127 ± 7.6*	87 ± 5.5	90 ± 9.3+
Stereotypy counts	133 ± 8.8*	86 ± 7.5	84 ± 8.8+
Vertical activity	109 ± 8.5	53 ± 9.1*	54 ± 11.4*
Margin time(s)	104 ± 2.3	110 ± 2*	110 ± 2.1*

[a]Results are the mean ± SEM. * p <0.01 vs SAL-SAL group; + p <0.01 vs SAL-CORT group.

NAME on corticosterone action (data not shown). In addition, administration of L-NAME markedly reduced vertical activity parameters, as well as the time the animals spent on the center of the cage. These effects of L-NAME were not modified by corticosterone treatment. In the startle response, either corticosterone or L-NAME (which failed to modify the corticosterone effect) incudes a marked reduction of the impulse of the acoustic startle response (data not shown).

These results indicate that corticosterone modulates different aspects of the exploratory and acoustic startle responses. Administration of an inhibitor of nitric oxide synthesis prevented the increase of the ambulatory and stereotypical components of exploratory behavior modulated by corticosterone, suggesting that endogenous nitric oxide could play a role with these corticosteroid-induced components. A blood pressure explanation for such antagonism has been discarded (data not shown). However, the inhibitory action of corticosterone on the acoustic startle reflex appears not to be dependent upon endogenous nitric oxide. Since different neural systems have been implicated in locomotor activity and startle responses, this study suggests that endogenous NO could play a role in the neural mechanisms involved in the rapid locomotor enhancement induced by corticosterone in a novel environment. In addition, this study reveals that nitric oxide can mediate certain aspects of the exploratory activity displayed by rats in a novel environment as well as the acoustic startle reaction.

REFERENCES

1. McEwen, B. S., E. R. de Kloet & W. Rostene. 1986. Adrenal steroid receptors and actions in the nervous system. Physiol. Rev. **66**(4): 1121–1188.
2. Orchinik, M., T. F. Murray & F. L. Moore. 1991. A corticosteroid receptor in neuronal membranes. Science **252:** 1848–1851.
3. Garthwaite, J. 1991. Glutamate, nitric oxide and cell-cell signalling in the nervous system. Trends Neurosci. **14**(2): 60–67.
4. Calignano, A., P. Persico, F. Mancuso & L. Sorrentino. 1993. Endogenous nitric oxide modulates morphine-induced changes in locomotion. Eur. J. Pharmacol. **231**(3): 415–420.
5. Sandi, C., J. Borrell & C. Guaza. 1992. Behavioral factors in stress-induced immunomodulation. Behav. Brain Res. **48:** 95–98.

Glucocorticoid Receptor (GR)-Mediated Effects on Cocaine Kindling in Rats

M. A. KLING,[a] D. PLUZNIK, J. R. GLOWA,[c] P. W. GOLD AND M. A. SMITH[b]

Clinical Neuroendocrinology Branch
[b]Biological Psychiatry Branch
National Institute of Mental Health
and
[c]Laboratory of Medicinal Chemistry
NIDDK
Bethesda, Maryland 20892

INTRODUCTION

Glucocorticoids have been shown to have differing effects on neural excitability. Thus, under some conditions they can be shown to decrease brain excitability and elevate the seizure threshold, while other observations indicate that they can increase brain excitability and induce seizures (for reviews see Refs. 1, 2). Subsequent studies have shown that, in addition to their effects at classical intracellular steroid receptors, which result in the modulation of transcriptional events, glucocorticoids and their derivatives have multiple sites of action at the GABA-A macromolecular complex, which mediates inhibitory effects on neuronal firing. These sites of action have different structure-activity requirements and dose-response relationships for naturally occurring and synthetic steroid compounds, perhaps accounting in part for the diversity of effects reported.

We recently showed that repeated administration of dexamethasone (DEX) enhances the development of kindled seizures induced by repeated cocaine administration.[3] In the present study we examined whether this effect is blocked by the synthetic steroid derivative RU 486, which antagonizes glucocorticoid effect at the classical (type II) intracellular glucocorticoid receptor (GR). In addition, because we had previously shown that RU 486 itself enhanced the development of lidocaine kindling,[4] we examined the effect of this agent alone on cocaine kindling.

METHODS

Male Sprague-Dawley rats (350–400 g) were housed in groups of 6 and maintained on a 12:12 h light/dark cycle (lights on from 0600–1800). Water and Purina Rat Chow were provided *ad lib.* Rats were given DEX (25 μg/rat/d), RU 486 (2.5 mg/rat/d), DEX + RU 486, or vehicle i.p. daily beginning 3 days prior to and then 2 h before daily i.p. injections of cocaine (given at 0900 h) for an additional 16 days.

[a]Address correspondence to: Mitchel A. Kling, M.D., Clinical Neuroendocrinology Branch, NIMH, NIH, Building 10, Room 3S-231, 900 Rockville Pike, Bethesda, MD 20892.

Cocaine was given at an initial dose of 35 mg/kg/d; this was increased to 40 mg/kg/d on day 11. Steroid vehicle consisted of normal saline with 2 drops of Tween 80; steroid and vehicle preparations were homogenized for 20–25 strokes with a Potter-Elvejm mortar-and-Teflon pestle apparatus at 400 rpm, then sonicated for 10 minutes.

Rats were observed for 30 minutes following cocaine injection by raters blind to treatment group. The study was continued until 50% of control rats were kindled. Statistical analysis of the cumulative frequency of seizures in each group was done using contingency tables. These studies were conducted under protocols approved by the NIMH Animal Care and Use Committee. Animals were handled in accordance with NIH guidelines for the ethical treatment of experimental animals.

RESULTS

DEX caused a rapid and significant enhancement of kindling (chi-sq = 8.22, d.f. = 1, $p < 0.005$), which was maximal by day 5 of cocaine injections. RU 486 alone resulted in a trend towards enhanced kindling, but the effect was more gradual, peaking on the last day of the study (chi-sq = 2.74, d.f. = 1, $p < 0.1$). RU 486 significantly (chi-sq = 6.17, d.f. = 1, $p < 0.05$) attenuated the DEX-induced enhancement of kindling; rates of kindling in these animals were intermediate between those of RU 486 alone and controls.

DISCUSSION

These results suggest that DEX-induced enhancement of kindling is GR-mediated, since DEX-induced enhancement of kindling was significantly attenuated by concomitant administration of RU 486. An alternative explanation is that RU 486 blocks DEX effects at other sites, possibly at GABA or glutamate receptors. Little data are available to date as to whether DEX or RU 486 interacts with such sites. The trend towards enhancement of cocaine kindling by RU 486 is compatible with our previous observation that RU 486 enhances lidocaine kindling.[4] The effect of RU in the present study may be less robust than previously observed because of the lower dose of RU 486 used here than in the previous studies.

Taken together, these findings suggest that both glucocorticoids and their antagonists may both enhance kindling, but possibly by different mechanisms. In particular, the enhancement of cocaine kindling by DEX can be interpreted in the light of data that glucocorticoids increase corticotropin-releasing hormone (CRH) expression in the central nucleus of the amygdala,[5,6] which may play a role in cocaine kindling.[3] On the other hand, RU 486 given alone may increase CRH functional activity at other sites via blockade of glucocorticoid negative feedback.[4]

REFERENCES

1. MAJEWSKA, M. D. 1985. Steroids and brain activity: essential dialogue between body and mind. Biochem. Pharmacol. **36:** 3781–3788.
2. PAUL, S. M. & R. H. PURDY. 1992. Neuroactive steroids. FASEB J. **6:** 2311–2322.
3. KLING, M. A., M. A. SMITH & J. R. GLOWA *et al.* 1993. Facilitation of cocaine kindling by glucocorticoids in rats. Brain Res. **629:** 163–166.
4. KLING, M. A., M. A. DEMITRACK & M. DE BELLIS *et al.* 1988. Glucocorticoid antagonism

with RU 486 enhances lidocaine kindling. *In* Abstracts of 18th Annual Meeting, Society for Neuroscience, Toronto, Canada. Vol. 2: 411.13:1033.

5. SWANSON, L. W. & D. M. SIMMONS. 1989. Differential steroid hormone and neural influences on peptide mRNA levels in CRH cells of the paraventricular nucleus: a hybridization histochemical study. J. Comp. Neurol. **285:** 413–435.
6. MAKINO, S., P. W. GOLD & J. SCHULKIN. 1994. Corticosterone effects on corticotropin-releasing hormone mRNA in the central nucleus of the amygdala and the parvocellular region of the paraventricular nucleus of the hypothalamus. Brain Res. In press.

Stress-Dependent Impairments of Spatial Memory

Role of 5-HT

V. LUINE, M. VILLEGAS, C. MARTINEZ,
AND B. S. McEWEN[a]

Department of Psychology
Hunter College
City University of New York
695 Park Avenue
and
[a]The Rockefeller University
New York, New York 10021

Exposure to excess corticosterone, either through injection or from repeated restraint stress, damages brain cells, especially hippocampal neurons. Watanabe *et al.*[1] showed that daily restraint stress for 21 days caused atrophy of apical dendrites in CA3 pyramidal neurons of the hippocampus. Since the hippocampus is important for spatial memory function, we investigated whether repeated restraint stress would compromise performance on the eight arm radial maze, a test of spatial memory. In addition, the possible involvement of serotonin (5-HT)-containing neurons in these effects was investigated.

Rats were restrained in Plexiglas containers for 6 h/day for 21 days.[1] Controls were weighed only. Following their last restraint session they were food deprived and received 9 training trials on the maze. Four maze trials were given on days 9–11 following the last restraint session. Stressed rats were impaired compared to control rats. They made their first mistake earlier than controls and also made fewer correct choices in the first eight visits (FIG. 1). These results show that repeated stress is deleterious to spatial memory.

A possible role of serotonin (5-HT)-containing neurons in stress-dependent impairments of spatial memory was investigated by administering tianeptine, an atypical, tricyclic antidepressant which enhances 5-HT uptake. Tianeptine blocks stress-dependent dendritic atrophy in CA3.[2] It was injected to control and restrained rats daily just prior to the restraint session. Tianeptine alone did not alter behavior, and it blocked the stress-dependent impairment in performance (data not show). These

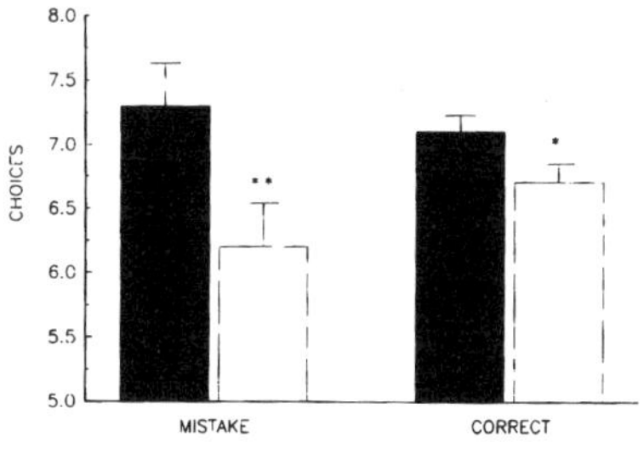

FIGURE 1. Effect of restraint stress on radial arm maze performance. The choice where the first mistake occurred (Mistake) and the number of correct choices in the first eight visits (Correct) are plotted for control *(solid bars)* and stressed *(open bars)* rats. Entries are the average ± SEM for four trials in 13 control and 14 stressed rats. Data analyzed by two-way repeated measures ANOVA (group × trials). For Mistake, $F_{1,25} = 5.31$, $p < 0.03$. For Correct, $F_{1,25} = 3.90$, $p < 0.05$.

TABLE 1. Effect of Stress on Acquisition-Dependent Changes in 5-HT[a]

Group	CA3	CA1	DG	Frontal Cortex
Control (5–7)	2.6 ± 0.2	1.18 ± 0.08	0.71 ± 0.08	0.86 ± 0.12
Acquisition (4)	5.6* ± 0.8	1.36 ± 0.23	0.89 ± 0.11	1.30* ± 0.31
Stress (4)	2.6 ± 0.2	1.12 ± 0.15	0.60 ± 0.11	0.51 ± 0.04
Stress and acquisition (4)	4.3 ± 0.9	0.80 ± 0.13	0.97 ± 0.14	0.90 ± 0.20

[a]Entries are the average ± SEM. Data analyzed by one-way ANOVA and differences between groups tested by Newman-Keuls analysis where * group is different from other groups, p <0.05. See text for further details.

results suggest that 5-HT terminals may contribute to stress-induced, impaired spatial memory.

The role of 5-HT in stress-dependent impairments of spatial memory performance was further investigated by measuring 5-HT levels in the brain of stressed rats. Four groups were utilized: control (food deprivation), acquisition (food deprivation and 6 training sessions), stress (restraint stress for 21 days, food deprivation), and stress and acquisition (restraint stress for 21 days, food deprivation and 6 training sessions). 5-HT was measured by high resolution liquid chromatography with electrochemical detection in CA3, CA1 and dentate gyrus (DG) of the hippocampus and in frontal cortex (TABLE 1). In the acquisition group, levels of 5-HT were increased in CA3 and frontal cortex. No changes in 5-HT levels were found in the stress group. Stress blocked the acquisition-dependent increases in CA3 and cortical 5-HT levels. These results suggest that learning the radial arm maze task increases 5-HT levels in areas important for memory function; more importantly, stress blocks the acquisition-dependent increase.

Results suggest that serotonin-containing neurons are involved in stress-dependent impairments in spatial memory. First, stress-dependent behavioral impairments are blocked by tianeptine. Further, acquisition dependent increases in hippocampal and cortical 5-HT are blocked by stress.

REFERENCES

1. WATANABE, Y., E. GOULD & B. S. MCEWEN. 1992. Stress induces atrophy of apical dendrites of hippocampal CA3 pyramidal neurons, Brain Res. **588:** 341–345.
2. WATANABE, Y., E. GOULD, H. CAMERON, D. DANIELS & B. S. MCEWEN. 1992. Phenytoin prevents stress- and corticosteroid-induced atrophy of CA3 pyramidal neurons. Hippocampus **4:** 431–436.

Adrenalectomy Impairs Spatial Memory in Rats

P. VAHER, V. LUINE, E. GOULD,[a] AND B. S. McEWEN[a]

Department of Psychology
Hunter College
City University of New York
695 Park Avenue
New York, New York 10021
and
[a]The Rockefeller University
New York, New York 10021

Adrenalectomy (ADX) causes death in the granule cells of the hippocampal dentate gyrus (DG) 3 to 4 months later in rats.[1] Neurodegenerative changes are found three days[2] after surgery. Granule cell degeneration caused by adrenal insufficiency is prevented by replacement with corticosterone (CORT). Lesions of the DG by intercranial injection of colchicine produce deficits in radial arm maze performance, a test of spatial memory. Thus, granule cell loss that results from adrenal insufficiency may also produce spatial memory deficits.

Initial experiments determined if there are deficits on the 8 arm radial arm maze shortly after ADX. After being shaped to the maze, sham and ADX groups were tested with all of the arms baited. ADX rats made more errors in completing the maze than shams (TABLE 1). Following these trials, rats were tested with a time delay between the 4th and 5th choices. The ADX group continued to make more errors in completing the task when tested with delays (TABLE 1). ADX-dependent impairments were present for up to 71 days.

Physiological indicators of adrenal function, serum CORT and saline intake, were assessed in order to determine the presence of accessory adrenal tissue in ADX rats. ADX rats had significantly lower levels of serum CORT (%μg) than shams (0.6 ± 0.4 vs 15.0 ± 2.3). As expected, ADX rats consumed amounts of saline (mls/day) well above those of the control group (16.9 ± 1.6 vs 1.3 ± 0.3). The correlation between saline intake and serum CORT level was significant ($r = -0.85, p < 0.0018$); *i.e.*, higher saline intake is associated with lower serum levels of CORT. Saline intake provides a simple means for evaluating the completeness of ADX.

Morphological analysis of the dentate gyrus showed that there were no differences between groups in cross sectional dentate area (10^6 μm^2) in creysl violet sections (0.083 in shams vs 0.087 in ADX). While some ADX rats showed an increase in pyknotic cell density, average pyknotic cell count per 10^6 μm^2 dentate area was also not significantly different. Many ADX rats apparently had some accessory adrenal tissue capable of producing CORT which may have been sufficient to suppress granule neuron pyknosis but not enough to suppress salt intake.

Physiological indicators of ADX and morphological measures of the dentate gyrus cell loss were tested for correlations with spatial memory performance. Serum CORT levels at the time of sacrifice were negatively correlated to the number of errors in completing the task ($p < 0.008$) suggesting that lower levels of serum CORT are associated with impaired radial arm maze performance. Intake of 3% saline was also significantly correlated to errors ($p < 0.04$). Thus, higher saline consumption was associated with poorer spatial memory performance overall. Cross sectional dentate

TABLE 1. Effect of Adrenalectomy (ADX) on Radial Arm Maze Performance[a]

	Number of Errors in Completing the Task						
	Trials			Trials with a Delay			
Group	Block 1	Block 2	Block 3	10 Min	1 Hr	2 Hr	3 Hr
Sham	1.62 ± 0.29	1.60 ± 0.36	1.62 ± 0.18	2.27 ± 0.54	2.36 ± 0.58	2.40 ± 0.43	2.82 ± 0.50
ADX	3.11 ± 0.60	2.85 ± 0.42	2.90 ± 0.49	3.36 ± 0.66	3.82 ± 0.86	3.82 ± 0.60	5.09 ± 1.01

[a]Entries are means ± SEM. Thirteen ADX and eleven sham operated rats started maze testing, but three ADX rats died prior to testing with delay intervals. Since interaction effects in this experiment were absent, main effects were tested by blocks design ANOVA with repeated measures. Significance was tested at $p < 0.05$. Data for the initial 15 maze trials were averaged into blocks of 5 trials for analysis. The ADX effect was significant for the number of errors in completing the maze ($F_{1,67} = 18.03$, $p = 0.0001$) but block effects were not significant. For trials with a delay, the ADX effect was also significant ($F_{1,82} = 10.83$, $p = 0.0015$). Increasing the delay interval did not significantly alter maze performance.

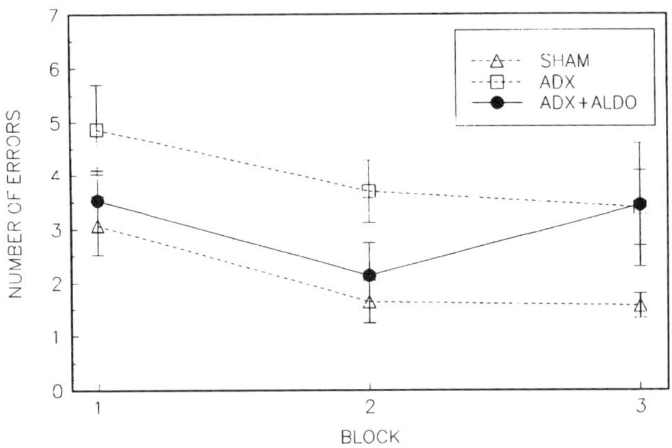

FIGURE 1. Effects of ALDO replacement on radial arm maze performance in ADX rats. ALDO was administered to ADX + ALDO rats by Alzet minipumps implanted one day before testing. Pumps lasted for seven days and secreted ALDO at a rate of 10 μg/hr. Eight rats started in the ADX + ALDO group but two were excluded from the second block of trials (n = 6) and two more from the third (n = 4) because of a decrease in saline consumption. Sham and ADX groups consisted of seven rats each. Data from two cohorts were pooled for analysis. Group effects were tested by blocks design ANOVA with repeated measures and were significant for the number of errors in completing the task ($F(2, 55) = 7.63$, $p = 0.0012$). Pairwise analysis by least significant means procedure showed that the ALDO replacement group is significantly different from ADX controls but not statistically different from shams.

area and pyknotic cell count were not correlated to measures of behavioral performance. Since ADX rats showed deficits in radial arm maze performance in the absence of morphological differences, an endocrine effect on behavior is suggested.

The influence of adrenal hormones on spatial memory function in ADX rats was next examined. Aldosterone (ALDO), a selective type 1 adrenal steroid receptor agonist, enhances long-term potentiation in ADX rats.[3] Replacement of ALDO in ADX rats reversed deficits in radial arm maze performance (FIG. 1). The effect is dependent on circulating ALDO since performance declines when hormone replacement is discontinued.

In conclusion, ADX rats showed deficits in radial arm maze performance in the absence of any significant granule cell loss. Since the effect can be reversed by ALDO administration, ADX-dependent spatial memory impairments may be more a reflection of adrenocortical insufficiency than an instability of the dentate gyrus neuronal population or the destruction of the dentate gyrus.

REFERENCES

1. SLOVITER, R. S., G. VALIQUETTE, G. M. ABRAMS, E. C. RONK, A. L. SOLLAS, L. A. PAUL & S. NEUBORT. 1989. Selective loss of hippocampal granule cells in the mature rat brain after adrenalectomy. Science **243:** 535–538.
2. GOULD, E., C. S. WOOLLY & B. S. MCEWEN. 1990. Short-term glucocorticoid manipulations affect neuronal morphology and survival in the adult dentate gyrus. Neuroscience **37:** 367–375.
3. PAVLIDES, C., Y. WATANABE, A. M. MARGARINOS & B. S. MCEWEN. 1992. Effects of glucocorticoid type I and type II agonists on hippocampal long-term potentiation. Soc. Neurosci. Abstr. 148.2.

Adrenalectomy Leads to a Different Responsiveness to 8-OH-DPAT in a Water-Maze Learning Task

O. C. MEIJER, M. S. OITZL AND E. R. de KLOET

Division of Medical Pharmacology
Leiden/Amsterdam Center for Drug Research
P.O. Box 9503
2300 RA Leiden, The Netherlands

Short-term adrenalectomy (ADX) increases hippocampal 5-HT_{1A} receptor mRNA in the rat[1,2] and also influences other aspects of the brain serotonergic system. We have shown a 40% increase in 5-HT_{1A} receptor mRNA one week after ADX that was specific for the dentate gyrus.[2] In order to see whether this change in mRNA can have functional consequences, we tested the effect of a single injection of 8-OH-DPAT, a specific 5-HT_{1A} receptor agonist, in the free swim trial of the Morris water maze test in ADX and sham-operated animals. This spatial memory test is supposedly sensitive to changes in hippocampal function. It was previously reported that 8-OH-DPAT impairs the performance of intact rats in this paradigm.[3]

Male rats (180 g) were ADX-ed or sham-operated. One week later they were trained for four days to find a submerged platform in a pool. On the fifth day the platform was removed from the pool and the animals were put in the pool for a two-minute period. This free swim trial was taped with a video camera and was analyzed for: i) time spent in the quadrant of the platform (%tp), ii) actual crossings of the (previous) location of the platform (crossings), and iii) distance swum during the trial. 8-OH-DPAT was injected 30 minutes before the trial in two doses (100 μg/kg and 300 μg/kg, s.c.). Saline injection was used as a control. Results were analyzed using ANOVA and post-hoc tukey test. Significance was accepted at $p < 0.05$.

In the sham-operated animals a dose of 100 μg/kg caused a significant decrease in the time spent in the platform quadrant and in the number of crossings of the previous location of the platform, as compared to saline-treated animals. ADX animals showed an inverted-U shaped response to 8-OH-DPAT. Saline-treated ADX animals performed worse than sham/saline-treated animals; they spent significantly less time in the quadrant and had fewer crossings. 100 μg/kg 8-OH-DPAT reversed this bad performance; these animals actually did better than the sham/saline animals both in terms of %tp and in crossings. 300 μg/kg caused the animals again to have lower %tp and crossings. In both groups 8-OH-DPAT increased the track distance of the animals, *i.e.,* stimulated locomotor behavior.

We conclude the following: i) 8-OH-DPAT impairs the performance of sham-operated animals in the free swim trial (cf. Ref. 3). ii) ADX impairs the performance in the free swim trial. iii) A dose of 100 μg/kg 8-OH-DPAT reverses the ADX-induced impairment. iv) A higher dose (300 μg/kg) gives a similar effect in ADX animals as does the low dose of 8-OH-DPAT in sham animals, *i.e.,* a right-shift in sensitivity to the drug seems to occur after removal of the adrenals. v) The effects are specific for the memory-related effect of 8-OH-DPAT as there was no difference between sham and ADX for the stimulatory effect on locomotor activity. vi) These findings may be explained by an inverted-U shaped effect of 5-HT_{1A} receptor

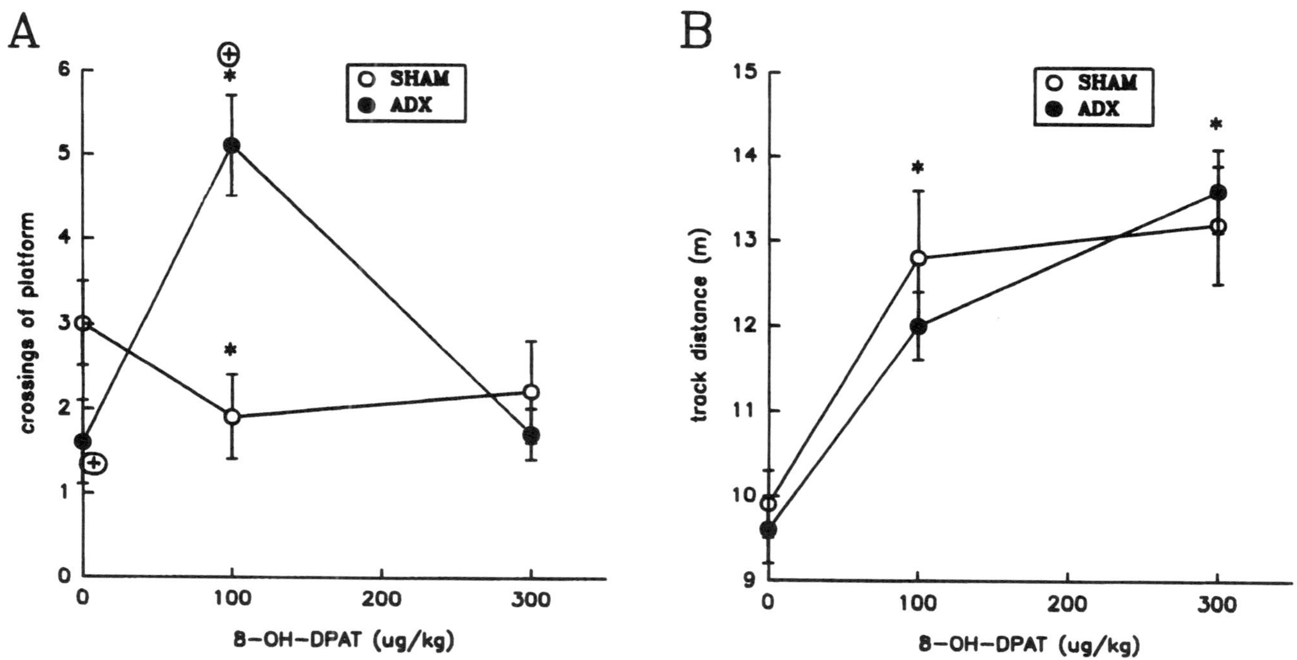

FIGURE 1. The number of times the exact location of the platform was crossed by the animals (**A**) and the distance swum (**B**) in the free swim trial for sham and ADX animals at different doses of 8-OH-DPAT. * indicates a significant difference from saline-treated animals in the corresponding group; + indicates a significant difference from sham animals at the same dose (p <0.05).

activation, when it is assumed that ADX leads to a very low turnover of 5-HT in the brain, which would be normalized by the low dose of 8-OH-DPAT. However, this does not explain the specificity of the effect. vii) It is not clear how these findings relate to the increased amount of 5-HT_{1A} mRNA in these animals.

ACKNOWLEDGMENTS

We would like to thank Mr. M. Fluttert and Ms. J. de Jong for their technical assistance.

REFERENCES

1. Chalmers, D. T. *et al.* 1993. J. Neurosci. **13:** 914–923.
2. Meijer, O. C. & E. R. de Kloet. 1994. Eur. J. Pharmacol. Mol. Pharmacol. Section **266:** 255–261.
3. Carli & Samanin. 1992. Br. J. Pharmacol. **105:** 720–726.

Stress Impairs LTP and Hippocampal-Dependent Memory[a]

DAVID M. DIAMOND[b,d,e] AND GREGORY M. ROSE[b,c,d,]

[b]*Department of Pharmacology*
[c]*Neurosciences Training Program*
University of Colorado Health Sciences Center
Denver, Colorado
and
[d]*Medical Research Service*
Veterans Affairs Medical Center
Denver, Colorado

We studied the effects of stress on hippocampal function from electrophysiological and cognitive approaches. In one approach we used high frequency electrical stimulation to induce a lasting enhancement in synaptic strength. This change in response, which is referred to as long-term potentiation (LTP), is the most widely studied physiological model of memory.[1] In the other approach we studied how stress affects spatial memory.

Exposure of rats to the novelty of a recording chamber, which is a potent stressor, blocked the induction of LTP.[2] When the rats acclimated to the environment LTP occurred (FIG. 1). The hormonal basis of the stress effects may involve corticosterone.[3] There is an inverted-U relationship between the magnitude of PB potentiation and levels of corticosterone, *i.e.* maximal PB potentiation occurred in animals with an intermediate level of corticosterone (10–20 μg/dl) and reduced magnitudes of PB potentiation occurred with low (<10 μg/dl) and stress (21–90 μg/dl) levels of the steroid.[4]

Recently we studied the effects of stress on spatial memory. The hypothesis guiding this work was that since stress impaired the induction of hippocampal LTP, it should also impair hippocampal-dependent memory. We tested this hypothesis by training food-deprived rats on the radial arm maze. In this task there are 14 arms radiating from a central platform. Seven of the arms are never baited and the other 7 are baited with a single food pellet. Over the course of 3–4 weeks of daily training the rats developed an optimal foraging strategy in which they entered each of the 7 baited arms a single time and avoided entry into the 7 nonbaited arms. This training procedure imposes two distinct demands on the rat's memory systems. In one system, which is referred to as reference memory, the information is static and can be used across all trials. The reference memory component of this task is that the rats learn the locations of the 7 arms that are never baited. Once reference memory is well learned it remains intact even after there is damage to the hippocampus.[5,6] Reference memory is therefore a hippocampal-independent form of memory. In the second memory system, which is referred to as working memory, the information is dynamic and is used by the animal for only one training trial at a time. The working memory component of this task is that once a rat eats the food in a baited arm it needs to remember not to return to that same arm during that trial. The ability to remember

[a]Supported by the Veterans Affairs Medical Research Service, NSF and ONR.
[e]Correspondence: David Diamond, Ph.D., Department of Pharmacology, Box C236, University of Colorado Health Sciences Center, Denver, CO 80262.

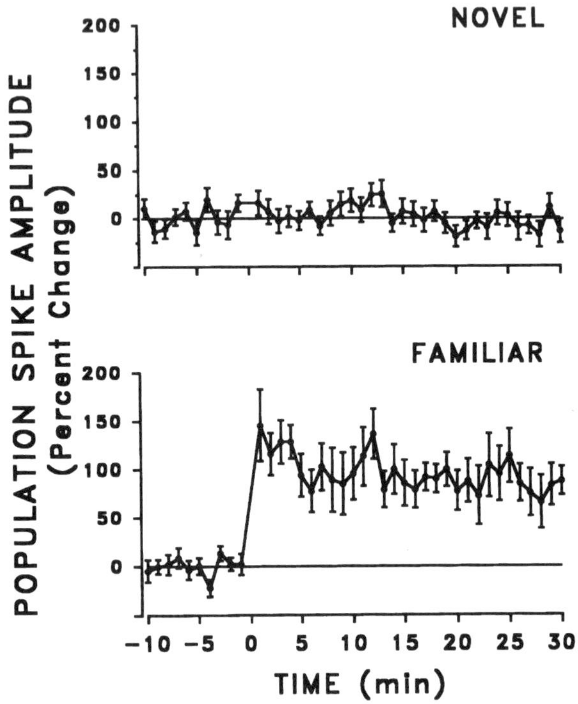

FIGURE 1. The induction of LTP of the CA1 population spike was influenced by environmental manipulations. LTP did not occur when the subjects were placed in the novel environment *(top)*. In contrast, after these same animals were familiarized with the environment (2 weeks of exposure to the recording chamber) LTP occurred *(bottom)*. LTP stimulation, composed of a single pulse followed by a high frequency (200 Hz) bust of 4 pulses to the hippocampal commissure, was delivered at time 0.

which of the baited arms the rats had entered on that trial is dependent on the integrity of the hippocampus.[6–8] Working memory is therefore a hippocampal-dependent form of memory.

We tested the effects of stress on working and reference memory by incorporating the conditions that blocked LTP into the maze training paradigm. Since placement of a rat in a novel environment (the electrophysiological recording chamber) blocked LTP, we used this same stress procedure in the spatial memory testing. Rats were removed from the maze after they had eaten the food in the 4th baited arm for a 4-hr delay period. During the delay period they were placed either in their home cage (familiar environment) or in a recording chamber (novel environment). After the delay the rats were returned to the maze and were given the opportunity to locate the 3 remaining pellets. Placement in the home cage had no effect on either working or reference memory. By contrast, stressed rats made a significant increase in working memory errors, *i.e.*, they were unable to remember which of the baited arms they had entered prior to the delay (FIG. 2). Stress had no effect on reference memory. These findings indicate that the same stress that blocked PB potentiation also impaired hippocampal-dependent memory. Overall, the findings indicate that the stress

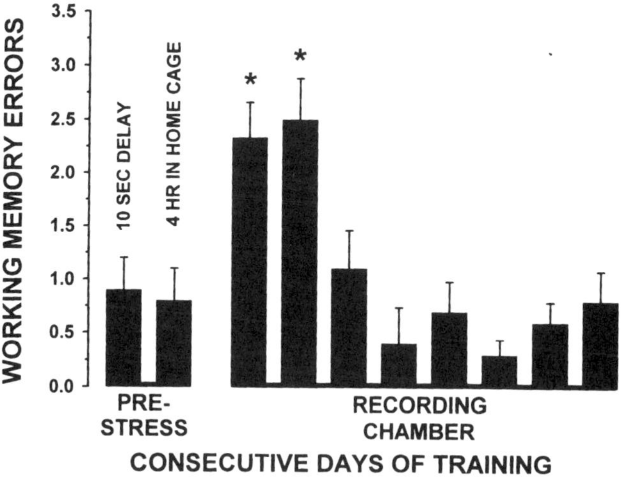

FIGURE 2. This figure illustrates the number of working memory errors the subjects committed, each day, after they were returned to the maze at the end of the delay period. When the delay period was 10 sec the subjects made a mean of 0.9 working memory errors as they attempted to locate the 3 remaining baited arms (Pre-Stress 10 Sec Delay). On the next day of training the duration of the delay was increased to 4 hours which the subjects spent in their home cage. This manipulation did not result in a significant change in the number of working memory errors (Pre-Stress 4 Hr in Home Cage). On the next 8 days of training the subjects were placed in the recording chamber for 4 h during the delay. During the first 2 days of exposure of the rats to the recording chamber the subjects made a significant increase in working memory errors (Recording Chamber, $^{*}p < 0.01$). Reference memory was not affected by the stress manipulation (data not shown).

of exposure to a novel environment interferes with the efficiency of hippocampal functioning.

REFERENCES

1. Bliss, T. V. P. & G. L. Collingridge. 1993. A synaptic model of memory: long-term potentiation in the hippocampus. Nature **361:** 31–39.
2. Diamond, D. M., M. C. Bennett, K. E. Stevens, R. L. Wilson & G. M. Rose. 1990. Exposure to a novel environment interferes with the induction of hippocampal primed burst potentiation. Psychobiology **18:** 273–281.
3. Diamond, D. M., M. C. Bennett, D. A. Engstrom, M. Fleshner & G. M. Rose. 1989. Adrenalectomy reduces the threshold for hippocampal primed burst potentiation in the anesthetized rat. Brain Res. **492:** 356–360.
4. Diamond, D. M., M. C. Bennett, M. Fleshner & G. M. Rose. 1992. Inverted-U relationship between the level of peripheral corticosterone and the magnitude of hippocampal primed burst potentiation. Hippocampus **2:** 421–430.
5. Olton, D. S., J. T. Becker & G. E. Handelmann. 1979. Hippocampus, space, and memory. Behav. Brain Sci. **2:** 313–365.
6. Davis, H. P., J. R. Baranowski, W. A. Pulsinelli & B. T. Volpe. 1987. Retention of reference memory following ischemic hippocampal damage. Physiol. Behav. **39:** 783–786.

7. OLTON, D. S. & W. A. FEUSTLE. 1981. Hippocampal function required for nonspatial working memory. Exp. Brain Res. **41:** 380–389.
8. MCLAMB, R. L., W. R. MUNDY & H. A. TILSON. 1988.Intradentate colchicine disrupts the acquisition and performance of a working memory task in the radial arm maze. Neurotoxicology **9:** 521–528.

Hypertension and the Mineralocorticoid Receptor in the Brain[a]

ELISE P. GÓMEZ SÁNCHEZ

Department of Internal Medicine and Research Service
Harry S. Truman Memorial Veterans Hospital
and
University of Missouri-Columbia
800 Hospital Drive
Columbia, Missouri 65201-5297

It has been proposed that a center within the brain, probably in the AV3V area, is responsible for the orchestration of the multiple homeostatic mechanisms controlling blood pressure and that the pressure "set point" in this center is dependent upon intracellular ion content or transport which is sensitive to mineralocorticoid action.[1] We shall summarize evidence from our laboratory strengthening autoradiographic and ablation studies supporting this hypothesis.

The continuous intracerebroventricular (icv) infusion of aldosterone (ALDO) at a dose which has no systemic effects and does not elevate circulating ALDO values above control, produces an elevation of blood pressure similar both in amplitude and time of onset to that produced by the continuous sc infusion of a 100-fold dose in rats. The response is dose dependent and reversible. The pressor response is accentuated by renal mass reduction and salt loading.[1] Icv ALDO hypertension also occurs in dogs.[2] The concomitant icv infusion of a mineralocorticoid antagonist blocks the hypertension, while the icv infusion of a selective glucocorticoid receptor (GR) agonist has no effect on the blood pressure alone or with icv ALDO.[1]

The hypertension of systemic mineralocorticoid excess also has a central component (FIG. 1A). The icv infusion of a selective mineralocorticoid receptor (MR) antagonist prevents the hypertension produced by the sc administration of ALDO or DOCA.[3,4] As in mineralocorticoid hypertension, adrenalectomy, AV3V lesions and the icv infusion of the MR antagonist prevent the development of hypertension in the Dahl Salt Sensitive rat.[5]

The affinity of the MR *in vitro* for ALDO, corticosterone (CORT), cortisol, or deoxycorticosterone is essentially the same.[6] Glucocorticoids circulate at 100–1000 times the levels of mineralocorticoids, favoring occupancy of the MR by glucocorticoids. Extrinsic specificity for ALDO is conferred upon the MR in ALDO target tissues by CORT/cortisol binding globulin (CBG), which limits the circulating free glucocorticoid, and by 11-β-hydroxysteroid dehydrogenase (11-HSD) which inactivates cortisol and CORT before it can bind the receptor.[7] CBG does not cross the blood brain barrier and the extrinsic specificity imparted by 11β-HSD does not occur in the hippocampus. In intact animals, hippocampal receptors are primarily occupied by CORT, while ALDO binds the hypothalamic MR more competitively.[8,9] The consumption of the 11β-HSD inhibitors glycyrrhetinic acid and carbenoxolone

[a]Supported by Medical Research Funds from the Department of Veterans Affairs and by Grant HL27737 from the National Institutes of Health.

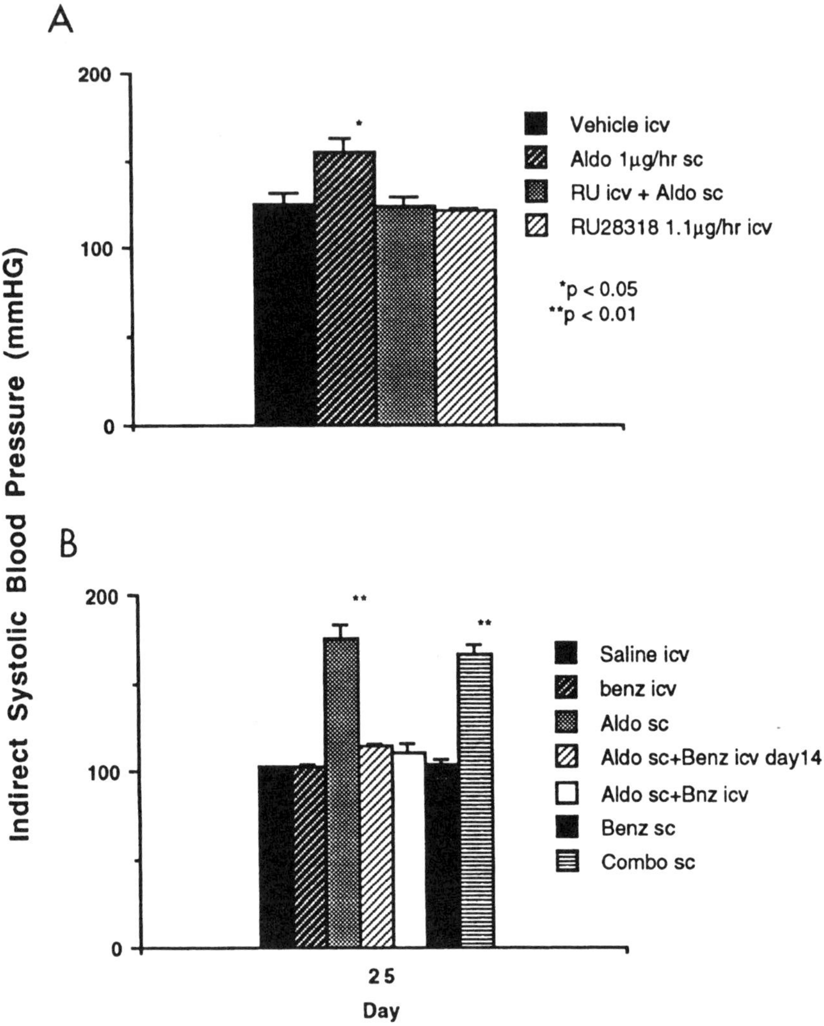

FIGURE 1. Effect of **(A)** RU28318 and **(B)** benzamil on systemic aldosterone excess.

produces hypertension which is effectively antagonized in rats by the icv infusion of a MR antagonist.[10]

Bilateral adrenalectomy prevents the development of hypertension in rats receiving ALDO icv and systemic replacement with CORT restores the response.[1] However, the continuous icv infusion of CORT alone has no effect on blood pressure and, when infused icv with ALDO, CORT antagonizes the pressor response in a dose-dependent manner.[11] Conversely, CORT appears to be the functional ligand for the hippocampal MR. Adrenalectomy-associated behavioral deficits and biochemical changes within the hippocampus are reversed by CORT, but not GR agonists or

ALDO, and ALDO blocks the normalizing effects of CORT.[8] The mechanisms for the different outcomes of MR binding depending upon the ligand are not known.

Mineralocorticoids modulate ion and water movement across transport epithelia, in part by promoting the recruitment or activation of amiloride-sensitive Na^+ channels. Changes in Na^+ transport and intracellular concentration affect that of other ions, particularly Ca^{++}, and alter membrane excitability. The icv infusion of the Na^+ channel-selective amiloride benzamil HCl prevents or attenuates the increase in blood pressure in four models of hypertension in which icv mineralocorticoid receptor antagonist studies indicate a role for central MR: the chronic icv or sc infusion of ALDO,[12] SS/jr rats consuming a high salt diet, and the ingestion of carbenoxolone (unpublished data). It is interesting that the central blood pressure control system appears to have safeguards; the blood pressure of control rats receiving either the mineralocorticoid receptor or Na^+ channel antagonists at the very low doses used in our experiments was not lowered (FIG. 1B).

The MR in the central nervous system clearly is important in the genesis of hypertension in mineralocorticoid excess, 11β-HSD inhibition and in the SS/jr rat. The importance of these data resides in the possibility that the central mechanisms involved may also be operant in the pathogenesis of other forms of hypertension, as well as in the normal control of blood pressure.

REFERENCES

1. GÓMEZ-SÁNCHEZ, E. P. 1991. What is the role of the central nervous system in mineralocorticoid hypertension? Am. J. Hypertens. **4:** 374–381.
2. KAGEYAMA, Y. & E. L. BRAVO. 1988. Hypertensive mechanisms associated with centrally administered aldosterone in dogs. Hypertension **11:** 750–753.
3. GÓMEZ-SÁNCHEZ, E. P., C. M. FORT & C. E. GÓMEZ-SÁNCHEZ. 1990. Intracerebroventricular infusions of RU28318 blocks aldosterone-salt hypertension. Am. J. Physiol. **258:** E482–E484.
4. JANIAK, P. C., S. J. LEWIS & M. J. BRODY. 1990. Role of central mineralocorticoid binding sites in development of hypertension. Am. J. Physiol. **259:** R1025–R1034.
5. GÓMEZ-SÁNCHEZ, E. P., C. FORT & D. THWAITES. 1992. Central mineralocorticoid receptor antagonism blocks hypertension in Dahl S/JR rats. Am. J. Physiol. **262:** E96–E99.
6. KROZOWSKI, Z. S. & J. W. FUNDER. 1983. Renal mineralocorticoid receptors and hippocampal corticosterone binding species have identical intrinsic steroid specificity. Proc. Natl. Acad. Sci. USA **80:** 6056–6060.
7. EDWARDS, C. R. W. & P. M. STEWART. 1991. The cortisol-cortisone shuttle and the apparent specificity of glucocorticoid and mineralocorticoid receptors. J. Steroid Biochem. Mol. Biol. **39:** 859–865.
8. DE KLOET, R. E. 1991. Brain corticosteroid receptor balance and homeostatic control. Front. Neuroendocrinol. **12:** 95–164.
9. YONGUE, B. G. & E. J. ROY. 1987. Endogenous aldosterone and corticosterone in brain cell nuclei of adrenal-intact rats: regional distribution and effects of physiological variations in serum steroids. Brain Res. **436:** 49–61.
10. GÓMEZ-SÁNCHEZ, E. P. & C. E. GÓMEZ-SÁNCHEZ. 1992. Central hypertensinogenic effects of glycyrrhizic acid and carbenoxolone. Am. J. Physiol. **263:** E1125–E1130.
11. GÓMEZ-SÁNCHEZ, E. P., M. T. VENKATARAMAN & D. THWAITES. 1990. ICV infusion of corticosterone antagonizes icv-aldosterone hypertension. Am. J. Physiol. **258:** E649–E653.
12. GÓMEZ-SÁNCHEZ, E. P. & C. E. GÓMEZ-SÁNCHEZ. 1994. Effect of central amiloride infusion on mineralocorticoid hypertension. Am. J. Physiol. Submitted.

The Effects of RU 38486 and RU 28318 on Energy Balance of Ovariectomized Rats

ANNE DAGNAULT, YVES DESHAIES, AND DENIS RICHARD[a]

Department of Physiology
Faculty of Medicine
Laval University
Quebec, Canada G1K 7P4

INTRODUCTION

The understanding of the mechanisms governing the increased rate of energy deposition induced by ovariectomy (OVX) has progressed over the years, in particular from the demonstration that adrenalectomy (ADX) attenuates the stimulating effect of OVX on body weight[1] and energy balance.[2] The observation that ADX attenuates the accelerated rate of growth of castrated females may suggest that the adrenal secretion of corticosterone is increased in castrated rats. However, the measurement of the plasma corticosterone levels in OVX females tends to refute this hypothesis. In fact, castrated animals have been found to have lower corticosterone levels than those of rats with intact gonads.[3]

Assuming that corticosteroids (CS) play a role in the accelerated growth of castrated animals, the measurement of below normal levels of corticosterone in ovariectomized rats may provide evidence for the involvement of the mineralocorticoid receptor (MR) in mediating the effects of CS on energy balance. The MR is one of the two existing types of CS receptors,[4] the other being the glucocorticoid receptor (GR), and both types of CS receptors have been reported to be involved in mediating the influence of CS on energy balance.[5–7] Since the MR mediates the effects of glucocorticoids when their levels are low,[8] it seems reasonable to predict that the MR may be the receptor type involved in the influence of OVX on energy balance. In this respect, the present study was carried out with the specific aim of assessing the involvement of the MR antagonist RU 28318 in the effects of ovariectomy on energy balance. Since the implication of the GR in the effects of OVX cannot be *a priori* ruled out, the effects of OVX were also investigated in female rats with the GR antagonist RU 38486.

MATERIALS AND METHODS

Female Sprague-Dawley rats fed a high carbohydrate purified diet (17.72 kJ/g wet weight) were divided into six groups; surgically ovariectomized or sham-operated rats were treated with a placebo, the MR antagonist RU 23818, or the GR antagonist RU 38486. The antagonists (as well as the placebo) were administered by means of subcutaneously implanted pellets (Innovative Research of America, Toledo, OH), which delivered RU 28318 and RU 38486 at daily rates of 2.75 and 6 mg over a period of 21 days. RU 23818 and RU 38486 were kindly provided by Roussel-UCLAF (Romainville, France).

At the end of the treatment period, the rats were killed and their carcasses were processed to determine the contents in protein, fat, and energy. The carcass content of energy was determined by adiabatic bomb calorimetry (Parr Instruments, Moline, Illinois, USA), that of nitrogen by Kjeldahl analysis, and that of fat by subtracting from the total carcass energy the energy from protein assuming that the energy from nonprotein matter is essentially that of fat. The initial energy, fat and protein contents of the rats were estimated from their body weight by reference to the baseline group of 10 rats that were killed at the beginning of the experimental period and on which carcass analyses were carried out. Such estimations allow the determination of the gains in energy, fat and protein for the treatment period. Rats in the initial group were in every aspect (strain, age, gender), identical to the four experimental groups.

The analysis of variance (ANOVA) was used to determine the main and interaction effects of the factors "surgery," which comprised two levels (OVX and sham-operation), and "CS receptor antagonist," which comprise 3 levels (placebo, RU 28318, RU 38486) on fat and energy gains. The significant "surgery" by "CS receptor antagonist" interactions were analyzed by examining the simple main effects. The analysis of the simple main effects, which consists in determining the effects of a factor at every level of another factor, was used to analyze the effects of "surgery" at every level of "CS receptor antagonist," and vice versa, the effects of "CS receptor antagonist" at every level of "surgery."

RESULTS

The effects of the MR and GR antagonists on energy and fat balances are summarized in TABLE 1. ANOVA revealed significant interactions of "surgery" and "CS receptor antagonist" on both energy and fat gains. From the inspection of the TABLE 1 data and the analysis of the simple main effects, it appeared that the significant interactions were largely attributable to the fact that OVX, which promoted energy and fat gains in rats treated with either the placebo or RU 38486, was without effect in the rats receiving RU 28318. In addition, as also revealed by the analysis of the simple main effects, is the fact that "CS receptor antagonist" significantly affected energy and fat gains in OVX animals whereas this factor was without significant effect in gonadally-intact rats.

CONCLUSION

The present results by emphasizing the ability of the MR receptor antagonist RU 28318 to prevent the accelerated growth induced by OVX provide evidence for the

TABLE 1. Energy and Fat Balance of Ovariectomized Rats Treated with Corticosteroid Receptor Antagonists

	Placebo		RU 38486		RU 28318	
	Intact	OVX	Intact	OVX	Intact	OVX
Energy gain (MJ)	1.2 ± 1.9	2.3 ± 2.2	0.6 ± 0.7	1.9 ± 2.3	1.1 ± 1.4	1.5 ± 1.6
Fat gain (g)	26 ± 5	48 ± 5	12 ± 2	44 ± 5	24 ± 4	30 ± 4

involvement of MR receptors in mediating the influence of CS on energy balance. These results are in agreement with the proposal of Devenport *et al.*,[7] which suggests that the MR is the receptor type mediating the anabolic effect of the corticosteroids.

REFERENCES

1. MOOK, D. G., N. J. KENNEDY, S. ROBERTS, A. I. NUSSBAUM & W. I. RODIER III. 1972. Ovarian-adrenal interactions in regulation of body weight by female rats. J. Comp. Physiol. Psychol. **81:** 198–211.
2. OUERGHI, D., S. RIVEST & D. RICHARD. 1992. Adrenalectomy attenuates the effect of chemical castration on energy balance in rats. J. Nutr. **122:** 369–373.
3. DAGNAULT, A. & D. RICHARD. 1993. The lesion of the paraventricular nucleus of the hypothalamus does not prevent the effect of estradiol on energy balance. Am. J. Physiol. In press.
4. DE KLOET, E. R. 1991. Brain corticosteroid receptor balance and homeostatic control. Front. Neuroendocrinol. **12:** 95–164.
5. LANGLEY, S. C. & D. A. YORK. 1992. Glucocorticoid receptor numbers in the brain and liver of the obese Zucker rat. Int. J. Obes. **16:** 135–143.
7. DEVENPORT, L., A. KNEHANS, A. SUNDSTROM & T. THOMAS. 1989. Corticosterone's dual metabolic actions. Life Sci. **45:** 1389–1396.
8. BRADBURY, M. J., S. F. AKANA, C. S. CASCIO, N. LEVIN, L. JACOBSON & M. F. DALLMAN. 1991. Regulation of basal ACTH secretion by corticosterone is mediated by both type I (MR) and type II (GR) receptors in rat brain. J. Steroid Biochem. Mol. Biol. **40:** 133–142.

Steroid Receptor-Mediated Modulation of CD4+CD62L+ Cell Homing

Implications for Drug Abusers[a]

FRANCESCO CHIAPPELLI,[b] ERCOLANO MANFRINI,
HARRY GWIRTSMAN, CELSA GARCIA, LANCHI PHAM,
PATTY LEE, AND PATRICK FROST

[b]*Laboratory of Human Oral and Molecular Immunology*
Diagnostic Sciences, School of Dentistry
Department of Neurobiology, School of Medicine
Brain & Dental Research Institutes
University of California, Los Angeles
Los Angeles, California 90024-1668
and
Laboratory of Human Immunology & Psychoeuroimmunology
West Los Angeles Veterans Affairs Medical Center

INTRODUCTION

We previously described the role of homing receptors and adhesion molecules in the migration of lymphocytes.[1–3] We and others have shown that glucocorticoids (GC) and other steroids modulate the migration of human and animal lymphocytes.[1–5] Subjects who abuse alcohol or other drugs of abuse characteristically exhibit a deregulated pituitary-adrenocortical axis and elevated GC levels. Associated is a multifaceted deregulation in the distribution of CD4+ T subpopulations. Upon withdrawal, as cortisol levels normalize, so do these cellular immune parameters, as shown in FIG. 1.[6,7]

To begin to test the hypothesis that these outcomes are dependent upon steroid-receptor mediated modulation of lymphocyte homing and adhesion moieties, we designed and characterized human *in vivo* and *in vitro* research systems. We are presently focusing on CD4+ T cell populations, because this subpopulation of lymphocytes is responsible for initiating, regulating and propagating cellular immune response.

RESEARCH PROTOCOLS AND RESULTS

In Vivo *Modulation of CD4+ T Cell Homing by Steroids*

One mg of dexamethasone (DEX) was administered PO or IV to normal healthy young adults at 10:00 P.M. On the morning before and the morning after DEX administration (8:00 A.M.), peripheral blood mononuclear cells (PBMC) were obtained by standard buoyant density centrifugation over Lymphoprep, lymphocyte subpopulations were characterized by FacScan (Becton Dickinson) flow cytometry,

[a]We thank NIH (AI07126; DA07683) and the UCLA Program in Psychoneuroimmunology for support.

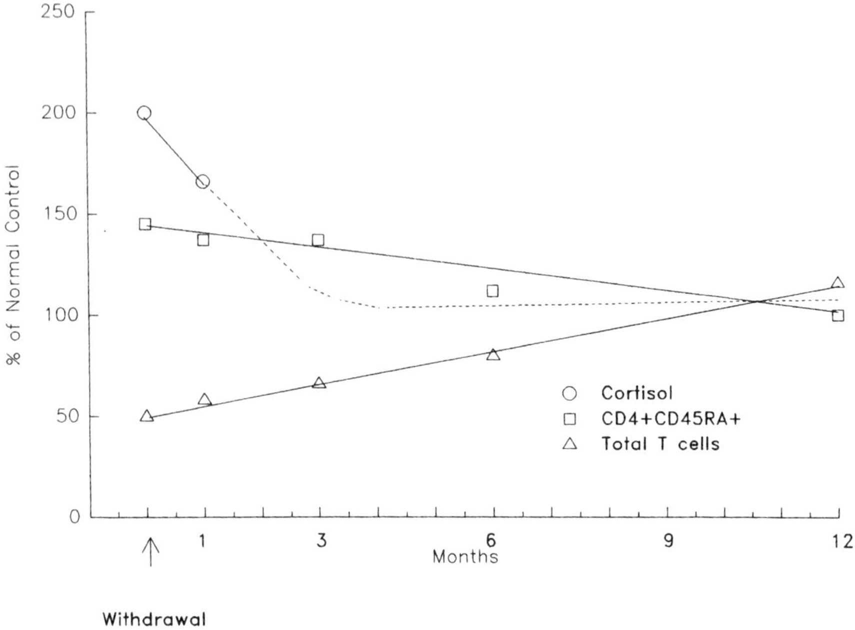

FIGURE 1. Recovery of plasma cortisol, CD4+ lymphocytes and CD4+ subpopulations following alcohol withdrawal. (Data adapted from Wilkins *et al.*[6] and Spinozzi *et al.*[7])

the cytosolic form of the GC receptor (GCCR) was assessed following binding with tritiated triamcinolone, and plasma cortisol, DEX and soluble β2-microglobulin (β2M) levels were determined by radioimmune assay.[1]

We previously reported that the number of circulating CD4+CD62L+ lymphocytes is inversely correlated with basal plasma cortisol levels ($r = -0.69$, $p < 0.01$). CD4+CD62L+ cell number drops by an average of 40% 10 h following DEX administration, in a fashion that appears to be correlated with the decrease in GCCR number ($r = -0.40$, $p < 0.15$).[1,2] Recent analyses show a positive correlation between the percent difference of pre- and post-DEX GCCR number and pre- and post-DEX levels of the T cell-derived soluble marker of immune activation, β2M ($r = 0.68$, $p < 0.05$), when DEX is given IV.

In Vitro *Modulation of CD4+ T Cell Homing by Steroids*

Pure CD4+CD62L+ cells were obtained by sequential incubation over antibody-coated flasks.[3,10] Purity was monitored by flow cytometry. Cultures were treated with DEX (10^{-10}–10^{-6} M) for 12–48 h, and examined phenotypically by flow cytometry and for their ability to express the α-chain of the IL-2 receptor, CD25, upon stimulation with the mitogen concanavalin A.[8]

FIGURE 2 shows a representative flow cytometric profile of purified CD4+ CD62L+ cells (quadrant II = 94%). These cells appropriately respond to mitogen

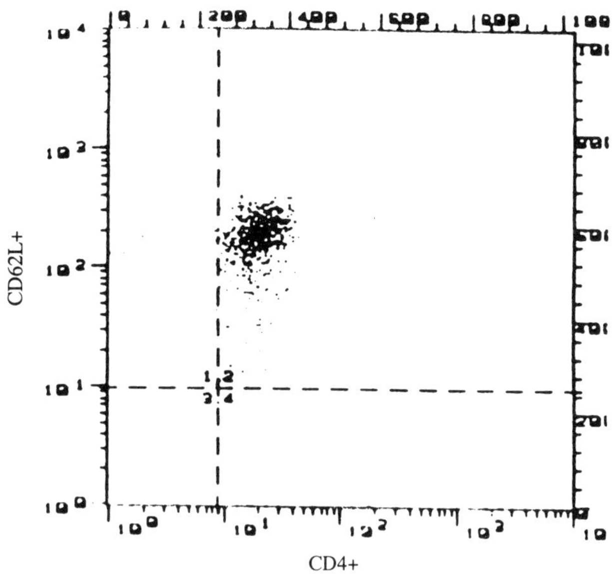

FIGURE 2. Flow cytometric analysis from a representative sample of purified CD4+ CD62L+ cells.

stimulation, as determined by a 4.0 ± 0.5-fold increase in CD25 expression at 24 h. Typical recoveries of >90% pure populations are about 30%. We previously reported that overnight incubation of these cultures with 10^{-9} M DEX leads to a 15–20% drop in cell number.[3] Here, we extend our observations in the micromolar range of DEX, and confirm a dose-dependent toxicity of DEX upon pure CD4+CD62L+ cultures. We also observe a decrease in the mean peak fluorescence value for CD62L expression by these cells following exposure to increasing concentrations of DEX (% drop mean peak channel at 10^{-6} M DEX: −18%, 10^{-7} M DEX: -7%, 10^{-8} M DEX: -7% 10^{-9} M DEX: -2%, 10^{-10} M DEX: +1%).

DISCUSSION AND CONCLUSION

In summary, our results indicate that the number of circulating lymphocytes, CD4+ T cells and CD4+ T cells that express the homing receptor CD62L is inversely correlated with plasma levels of cortisol. We show that the administration of DEX produces significant alterations upon both the hormonal and the cellular immune arms of the neuroendocrine-immune loop. Specifically, we find that the percent difference in GC receptor number in lymphocytes is positively correlated with the percent difference in CD4+CD62L+ cell number and with the cellular immune marker β2M, which is an estimate of all circulating T cells because this peptide is the anchoring protein for the MHC Class I molecule.[9] In a parallel set of studies, we show that pure CD4+CD62L+ cells are sensitive to cytotoxic effects of DEX; this suggests that GC can induce apoptotic-related events in certain T cell subsets.[10] Our data also suggest that increasing concentrations of DEX blunt the expression of

CD62L. This supports the recent report that GC can blunt the expression of intercellular adhesion molecule-1 (ICAM-1, CD54), another adhesion molecule expressed on lymphocyte membranes, and of its mRNA by means of a GC receptor-mediated process.[11]

SUMMARY

Naive human T cells home to peripheral lymph nodes via the leukocyte endothelial cell adhesion molecule-1 (LECAM-1, l-selectin, CD62L, Leu8 antigen) they express. We enriched populations of CD4+CD62L+ cells (attachment of Leu8+ T cells to flasks coated with anti-mouse IgG (AIS); Leu8+ T cells, 82.3% pure (± 2.3%), enriched for CD4+ cells by incubation over flasks coated with anti-CD4 antibody—this 3–4-day procedure yields an 88 ± 1.4% recovery. Cells were treated with dexamethasone *in vitro* for 24–48 h, and monitored by flow cytometry. We found severe toxicity by this steroid at high concentration (10^{-6} M: 35% decrease in CD62L+ T cells, 22% drop specifically in CD4+CD62L+ cells), suggesting the onset of receptor-mediated apoptotic events. The toxicity was dose dependent (5% and 7% drop in CD62L+ T and CD4+CD62L+ cells, respectively, at 10^{-9} M, the concentration found in plasma 10 h following the administration of 1 mg dexamethasone). One mg of dexamethasone given to normal subjects leads to a 15–20% decrease in circulating CD4+CD62L+ cells at 10 h. This tends to be correlated with a drop in the number of glucocorticoid cytosolic receptors. Thus, steroids seem to modulate CD4+CD62L+ cell homing by means of receptor-mediated mechanisms.

ACKNOWLEDGMENTS

We thank the many collaborators who have contributed to these studies, and who are acknowledged in the papers cited.

REFERENCES

1. CHIAPPELLI, F., H. GWIRTSMAN, M. LOWY *et al.* 1991. Pituitary-adrenal-immune system in normal subjects and in patients with anorexia nervosa: the number of circulating helper T lymphocytes (CD4) expressing the homing receptor Leu8 is regulated in part by pituitary-adrenal products. Psychoneuroendocrinology **16:** 423–432.
2. CHIAPPELLI, F. & S. TRIGNANI. 1993. Neuroendocrine-immune interactions: implications for clinical research. *In* Primary and Secondary Eating Disorders: a Psychoneuro-Endocrine and Metabolic Approach. E. Ferrari, F. Brambilla & S. B. Solerte, Eds. 185–198. Pergamon Press. London.
3. CHIAPPELLI, F., C. GARCIA, E. MANFRINI *et al.* 1993. Migration of human CD4+ lymphocytes: implications for cocaine and alcohol abusers. *In* Proc. Int. Lymphol. Congr. M. H. Witte & C. L. Witte, Eds. 206–211. Washington, DC.
4. CHIAPPELLI, F. 1991. Neuroimmunological perspectives on lymphocyte homing and adhesion. Int. J. Immunopathol. Pharmacol. **4:** 115–121.
5. CUPPS, T. R. & A. S. FAUCI. 1982. Corticosteroid-mediated immunoregulation in man. Immunol. Rev. **65:** 133–155.
6. WILKINS, J. N., D. A. GORELICK, K. NADEMANEE *et al.* 1992. Hypothalamic-pituitary function during alcohol exposure and withdrawal and cocaine exposure. Rec. Dev. Alcohol. **10:** 57–71.
7. SPINOZZI, F., E. AGEA, G. BASSOTTI *et al.* 1993. Ethanol-specific impairment of T

lymphocyte activation is caused by a transitory block in signal transduction pathways. Gastroenterology **105:** 1490–1501.
8. Chiappelli, F., P. Frost & T. Wylie. 1993. Alcohol and alcohol/cocaine mixtures blunt the activation of human CD4+ T cells. Alcologia **5:** 221–229.
9. Chiappelli, F. 1991. Immunophysiological role and clinical implications of non-immunoglobulin soluble products of immune effector cells. Adv. Neuroimmunol. **1:** 235–241.
10. Zubiaga, A. M., E. Munoz & B. T. Huber. 1992. IL-4 and IL-2 selectively rescue Th cell subsets from glucocorticoid-induced apoptosis. J. Immunol. **149:** 107–112.
11. Cronstein, B. N., S. C. Kimmel, R. I. Levin *et al.* 1992. A mechanism for the antiinflammatory effects of corticosteroids: the glucocorticoid receptor regulated leukocyte adhesion to endothelial cells and expression of endothelial-leukocyte adhesion molecule 1 and intercellular adhesion molecule 1. Proc. Natl. Acad. Sci. USA **89:** 9991–9995.

Social Defeat Impairs Plasma Corticosterone Response to the 5-HT$_{1A}$ Agonist 8-OH-DPAT in the Rat[a]

S. M. KORTE,[b] B. BUWALDA,[b] E. R. de KLOET,[c]
AND B. BOHUS[b]

[b]*Department of Animal Physiology*
University of Groningen
P.O. Box 14
9750 AA Haren, The Netherlands
and
[c]*Division of Medical Pharmacology*
Leiden/Amsterdam Center for Drug Research
Sylvius Laboratories
University of Leiden, The Netherlands

INTRODUCTION AND METHODS

The purpose of the present study was to analyze the influence of stress of social defeat on postsynaptic 5-HT$_{1A}$ receptor function in male Wistar rats. It is well known that social defeat increases plasma corticosterone levels for several hours[1] and that glucocorticoid treatment impairs 5-HT$_{1A}$ receptor function.[2] In the present challenge study we used the 5-HT$_{1A}$ agonist 8-OH-DPAT to examine the effect of social defeat on postsynaptic 5-HT$_{1A}$ receptor-mediated neuroendocrine activation. Male Wistar rats were threatened and attacked by Tryon Maze Dull S3 rats for 10 min. The experimental animals displayed submissive postures. Defeated and control rats were treated with 8-OH-DPAT (0.05 mg/kg and 0.15 mg/kg i.v.) in their home cages 18–20 h after social defeat. The challenge study was performed under resting conditions. Blood samples were withdrawn for determination of plasma corticosterone.

RESULTS AND DISCUSSION

The plasma corticosterone response to the lower dose of 8-OH-DPAT (0.05 mg/kg, i.v.) was significantly decreased in the defeated rats compared to the controls (FIG. 1). The higher dose of 8-OH-DPAT (0.15 mg/kg, i.v.) produced a higher but similar neuroendocrine response in both groups. The present data indicate that stress of social defeat alters 5-HT$_{1A}$ receptor-mediated neuroendocrine function. Postsynaptic stimulation accounts for the release of CRF and ACTH,[3,4] and consequently the elevation of plasma corticosterone.[5–7] It is suggested that the diminished cortic-

[a]This work was supported by a grant from The Netherlands Organisation For Scientific Research (Grant No. 900-551-057).

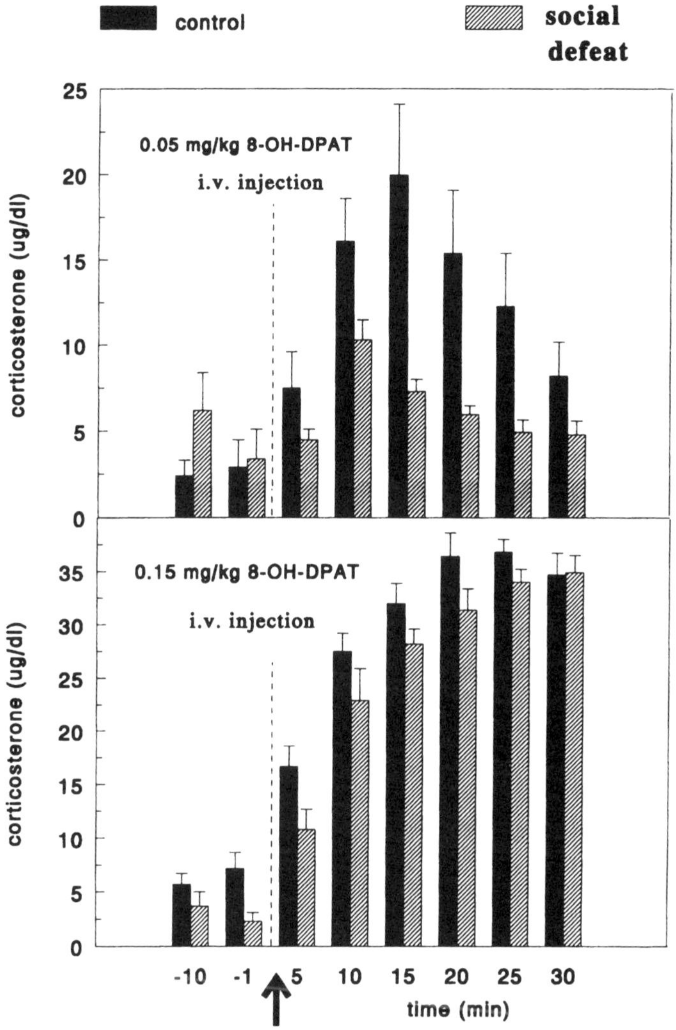

FIGURE 1. Effects of social defeat on the concentration of plasma corticosterone under resting conditions before and after 5-HT_{1A} agonist 8-OH-DPAT treatment (0.05 mg/kg/ 0.15 mg/kg intravenously (i.v.)).

osterone response may be the consequence of hyposensitivity in the postsynaptic 5-HT_{1A} receptor system. It is hypothesized that these changes are induced by a temporal increase in plasma corticosterone levels directly after social defeat. This shift in the sensitivity of the 5-HT neuronal system may be of particular interest for the development of anxiety and/or depression, in which abnormal 5-HT function may have a causal or predisposing role.

REFERENCES

1. SCHUURMAN, T. 1981. Endocrine processes underlying victory and defeat in the male rat. Thesis. University of Groningen. The Netherlands.
2. BAGDY, G., A. E. CALOGERO, C. S. AULAKH, K. SZEMEREDI & D. L. MURPHY. 1989. Neuroendocrinology **50:** 241–247.
3. CALOGERO, A. E., R. BERNARDINI, A. N. MARGIORIS, G. BAGDY, W. T. GALLUCCI, P. J. MUNSON, L. TAMARKIN, T. P. TOMAI, L. BRADY & P. W. GOLD. 1989. Peptides **10:** 189–200.
4. DI SCIULLO, A., M. T. BLUET-PJOT, F. MOUNIER, C. OLIVER & B. SCHMIDT. 1990. Endocrinology **127:** 567–572.
5. KOENIG, J. I., H. Y. MELTZER & G. A. GUDELSKY. 1988. Pharmacol. Biochem. Behav. **29:** 711–715.
6. HALEEM, D. J., G. A. KENNETT, P. S. WHITTON & G. CURZON. 1989. Eur. J. Pharmacol. **164:** 435–443.
7. KORTE, S. M., S. VAN DUIN, G. A. H. BOUWS, J. M. KOOLHAAS & B. BOHUS. 1991. Eur. J. Pharmacol. **197:** 225–228.

Regulation of Hippocampal Glucocorticoid Receptor Gene Expression by Psychosocial Conflict

OLAF JÖHREN, GABRIELE FLÜGGE,
AND EBERHARD FUCHS[a]

German Primate Center
37077 Göttingen, Germany

The activity of the limbic-hypothalamo-pituitary-adrenal (LHPA) axis is regulated by adrenal glucocorticoids exerting a negative feedback control via hippocampal neurons. Glucocorticoid receptors (GRs) in the hippocampus are therefore thought to play an important role in this circuit and their number may be an essential component of the stress response.[1] While much is known about the regulation of GR expression induced by various forms of experimentally induced stress, only little is known about whether GR expression is altered during prolonged psychosocial stress.[2] We therefore addressed the question, whether chronic social conflict in the tree shrew *(Tupaia belangeri)* may alter GR mRNA expression in the hippocampus. In this stress paradigm, the coexistence of two male tree shrews in one cage led to a stable dominant/subordinate relationship resulting in profoundly disturbed patterns of feeding, circadian rhythm and sleep, and in a clear behavioral depression in the subordinate tree shrew which might be related to alterations in the brain adrenoceptor system.[3] Analysis of endocrine parameters in subordinates revealed constantly elevated urinary glucocorticoids indicating a sustained activation of the LHPA axis.[4,5]

The expression of GR mRNA was measured by *in situ* hybridization in the hippocampal formation of subordinate tree shrews and controls. To generate specific probes for the experiments, a 487 base pairs fragment of the tree shrew GR gene encoding a part of the N-terminal domain was amplified by polymerase chain reaction, cloned in a bacterial vector and sequenced. There was a 92% sequence homology between this cloned tree shrew GR DNA fragment and the corresponding human GR gene sequence. As shown by autoradiographic *in situ* hybridization, ^{35}S-labeled riboprobes which had been transcribed from the three shrew-specific GR DNA fragment, hybridized to pyramidal cells in all subregions of the tree shrew hippocampal formation and to granule cells in the dentate gyrus. Expression of GR mRNA was semiquantitatively determined by counting silver grains over single cells of the hippocampal formation.

After 13 days of social conflict, the number of silver grains in CA1 and CA3 pyramidal neurons was significantly lower in subordinates compared to controls. In pyramidal neurons of the CA4 subfield and the subiculum, and in the granule cells of the dentate gyrus, the number of silver grains did not differ from controls (FIG. 1). Since the reduction in the number of silver grains indicates reduced amounts of GR mRNA, the present results suggest that in subordinate tree shrews sustained psychosocial conflict leads to a downregulation of hippocampal GR via modification of mRNA expression by a naturally occurring form of stress and may reflect distinct

[a]Corresponding author: Dr. Eberhard Fuchs, German Primate Center, Kellnerweg 4, 37077 Göttingen, Germany.

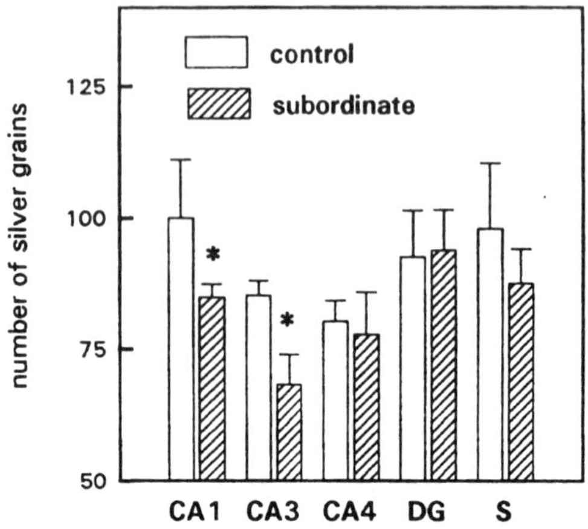

FIGURE 1. Quantification of GR mRNA as determined by autoradiographic *in situ* hybridization with a ^{35}S-labeled riboprobe in hippocampal cells of subordinate male tree shrews after 13 days of social conflict and of controls. The number of silver grains is expressed as percent of number in CA1 pyramidal cells of controls. Note that the silver grain number in CA1 and CA3 pyramidal neurons of subordinates is significantly lower that in controls. *Asterisks* indicate significant differences between ($p \leq 0.05$). *Abbreviations:* CA: pyramidal cells in the subfield of the cornu amonis; DG: granule cells of the dentate gyrus; S: pyramidal cells of the subiculum.

neuroendocrine processes which are presumed to coordinate behavioral and physiological responses to perpetual psychosocial conflict.

REFERENCES

1. DE KLOET, E. R. 1991. Front. Neuroendocrinol. **12:** 95–164.
2. CHAO, H. M., D. C. BLANCHARD, R. J. BLANCHARD, B. S. MCEWEN & R. S. SAKAI. 1993. Mol. Cell. Neurosci. **4:** 543–548.
3. FLÜGGE, G., O. JÖHREN & E. FUCHS. 1992. Brain Res. **597:** 131–137.
4. FUCHS, E., O. JÖHREN & M. GOLDBERG. 1992. Naturwissenschaften **79:** 379–381.
5. FUCHS, E. & G. FLÜGGE. 1994. Psychoneuroendocrinology. In press.

Sex Differences in Type I Corticosteroid Receptor Binding in Selective Brain Areas of Rats

C. M. McCORMICK[a] J. W. SMYTHE,[b] AND D. BEERS[a]

[a]*Department of Psychology*
Bates College
Lewiston, Maine 04240
and
[b]*Douglas Hospital Research Centre*
McGill University
Montreal, Canada H4H 1R3

Despite sex differences in many parameters of hypothalamic-pituitary-adrenal (HPA) function,[1,2] reports of sex differences in type I and type II corticosteroid receptors involved in the control of the HPA axis are few and have varied methodologies and inconsistent results.[3–5] We[6] have found sex differences in type II receptor levels to be contingent upon the developmental history of the animals. Here, we report sex differences in type I binding in specific brain areas based on three separate studies of rats with different developmental histories: (1) prenatally-stressed vs nonstressed, (2) neonatally-handled vs nonhandled, and (3) rats obtained as adults. Receptor densities were obtained from single-point assays of soluble fractions incubated for 44 hours with a saturating concentration (6 nM) of [^{3}H] corticosterone (CORT) and nonlabeled RU28362 (25-fold excess to block type II receptor) and aldosterone (250-fold excess to assess nonspecific binding). Bound and unbound steroid were separated using Sephadex LH-20 columns. Results were expressed as femtomoles binding/mg protein.

Within studies 1 and 2, developmental history was not associated with type I receptor densities, and the data are presented collapsed across developmental condition. In both studies, females had higher levels of [^{3}H] CORT binding than males, most notably in the septum (SEPT) and frontal cortex (FCTX) (Tables 1 and 2). Saturation analyses of SEPT and HYPO (study 1) indicated no difference in receptor affinity between males (x k_D = 0.75 nM) and females (x k_d = 0.5 nM). In the third study (Table 3), we found examined hippocampal type I receptor binding in animals

TABLE 1. Mean (SEM) [^{3}H] Corticosterone Binding (fmol/mg Protein) in Various Areas of the Brain in the Prenatal Stress Study (Study 1)[a]

	HPC	FCTX	AMYG	HYPO	SEPT
Males	65.5 (9.3) n = 5	13.8 (2.8) n = 8	17.8 (4.4) n = 12	23.5 (4.6) n = 4	32.3 (3.2) n = 3
Females	76.1 (7.8) n = 7	47.1 (6.9) n = 10	18.5 (4.4) n = 14	38.3 (6.1) n = 4	62.0 (6.3) n = 4
p value	ns	0.0009	ns	0.099	0.018

[a]HPC, hippocampus; FCTX, frontal cortex; AMYG, amygdala; HYPO, hypothalamus; SEPT, septum. All animals were adrenalectomized 20 hours prior to sacrifice.

TABLE 2. Mean (SEM) [^{3}H] Corticosterone Binding (fmol/mg Protein) in Various Areas of the Brain in the Neonatal Handling Study (Study 2)[a]

	HPC	FCTX	AMYG	HYPO	SEPT
Males	53.9 (6.0) n = 12	26.0 (2.7) n = 14	24.9 (3.7) n = 10	28.9 (4.0) n = 11	26.7 (5.5) n = 10
Females	74.3 (8.2) n = 14	51.3 (3.9) n = 14	34.5 (4.2) n = 14	46.0 (4.8) n = 14	52.1 (6.4) n = 14
p value	0.061	0.0001	0.097	0.015	0.009

[a]HPC, hippocampus; FCTX, frontal cortex; AMYG, amygdala; HYPO, hypothalamus; SEPT, septum. All animals were adrenalectomized 20 hours prior to sacrifice.

TABLE 3 Mean (SEM) [^{3}H] Corticosterone Binding (fmol/mg Protein) in Hippocampus in Rats Obtained as Adults (Study 3)[a]

	HPC-Stress	HPC-Intact	HPC-ADX
Males	38.9 (4.4) n = 5	48.8 (3.1) n = 7	56.5 (6.0) n = 6
Females	49.1 (8.6) n = 7	77.5 (5.0) n = 6	75.8 (4.9) n = 6

[a]Animals were either sacrificed intact following 20 minutes of restraint stress (HPC-Stress), sacrificed intact (HPC-Intact), or adrenalectomized 20 hours prior to sacrifice (HPC-ADX). *p* value for factor of sex = 0.0003; *p* value for factor of condition = 0.0013.

TABLE 4. Mean (SEM) Plasma Corticosterone Values and Type I Corticosteroid Receptor Translocation in Hippocampus of Adrenalectomized (ADX) Rats and Rats That Were Sacrificed Intact or Following a 20-Minute Period of Restraint Stress

	Corticosterone (μg/dl)	Translocation %
Males		
ADX	<1.0	0
Intact	3.3 (1.4)	13
Stress	48.0 (6.6)	31
Females		
ADX	<1.0	0
Intact	18.0 (3.7)	0
Stress	67.8 (3.7)	35

that were adrenalectomized (ADX), intact, or sacrificed following 20 min of restraint stress. RIA of plasma levels of corticosterone from trunk blood confirmed experimental condition: females had higher levels of plasma corticosterone under both intact and stressed conditions (TABLE 4). Females had higher type I binding than males in all three conditions (TABLE 3). Translocation was estimated in intact and stressed groups by calculating [^{3}H] CORT binding capacity as a percentage of that of ADX animals. Males and females had a similar percentage of receptor translocation under stressed conditions, but males had higher percentage translocation than females under intact conditions (TABLE 4).

Thus, our data indicate increased type I binding in females over males in various brain areas and no sex difference in k_D. The results from study 3 indicate there may be functional differences between the sexes in type I receptors. Type I and type II receptors are involved in the regulation of HPA function. Thus, sex differences in these receptor systems are likely related to sex differences in HPA function. Recent research indicates different roles of type I and type II receptors in spatial memory,[7] a sexually dimorphic behavior under septal-hippocampal control. That sex differences in type I receptors in various brain areas may be related to sex differences in behavior is an intriguing possibility.

REFERENCES

1. KITAY, J. I. 1961. Endocrinology **68:** 818–824.
2. CRITCHLOW, V., R. A. LIEBELT, M. BAR-SELA, W. MOUNTCASTLE & H. S. LIPSCOMB. 1963. Am. J. Physiol.**205:** 807–815.
3. FERRINI, M., A. M. MAGARINOS & A. F. DE NICOLA. 1990. J. Steroid Biochem. **35:** 671–677.
4. TURNER, B. B. 1992. Brain Res. **581:** 229–236.
5. AHIMA, R. S., A. N. L. LAWSON, S. Y. S. OSEI & R. E. HARLAN. 1992. Endocrinology **131:** 1409–1416.
6. MCCORMICK, C. M., J. W. SMYTHE, S. SHARMA & M. J. MEANEY. 1992. Soc. Neurosci. Abstr. **18:** 1008.
7. OITZL, M. S. & E. R. DE KLOET. 1992. Behav. Neurosci. **106:** 62–71.

Interaction of Progesterone with the Hippocampal Mineralocorticoid Receptor

M. P. CAREY AND E. R. de KLOET

Division of Medical Pharmacology
Center for Bio-Pharmaceutical Sciences
Sylvius Laboratories
P.O. Box 9503
Wassenaarseweg 72
2300 RA Leiden, The Netherlands

INTRODUCTION

Sex differences in the activity of the hypothalamic-pituitary-adrenal (HPA) axis have been widely reported with an apparent enhanced activity in females. Studies on HPA activity in the female rat have indicated a facilitatory role for oestradiol (E_2) while the effects of progesterone (P) appear inhibitory.[1] However, our investigations in the intact female rat have revealed enhanced HPA activity under both basal and stressful conditions in late pro-oestrus, a time of elevated circulating concentrations of both E_2 and P. To delineate the individual and combined actions of E_2 and P on stress responsiveness, we exposed E_2 and/or P-treated ovariectomized (OVX) rats to the stress of a novel environment and monitored the ACTH response over a two-hour time period (FIG. 1). Clear facilitatory effects of E_2 were observed which were not inhibited by additional P treatment. In contrast, treatment with P alone resulted in an enhanced negative feedback compared to OVX controls. Although the mechanism underlying the facilitatory effects of E_2 on HPA activity have been explored at the level of hypothalamus, little data is available concerning the actions of E_2 and P in the hippocampus. This brain region exerts a negative influence on HPA activity via actions mediated by the adrenocorticoid receptors, the mineralocorticoid (MR) and glucocorticoid (GR) receptors. Recent studies have revealed a pivotal role for the MR in maintaining stability of hippocampal excitability and consequent control of HPA activity. The present report describes our findings on the influence of E_2 and P on binding properties of the hippocampal MR in the female rat which might account for some of the observed effects of these steroids on HPA activity.

METHODS

Female Wistar rats (160–200 gm) were OVX and 14 days later adrenalectomized (ADX). Surgery was performed under ether anaesthesia. Following ADX, animals received subcutaneous (s.c.) injections of E_2 (40 μg/kg), P (500 μg/rat) or the oil vehicle alone, to generate the following treatment groups: (i) OVX/ADX, (ii) OVX/ADX + E_2, (iii) OVX/ADX + E_2 + P, (iv) OVX/ADX + P. P was administered 44 hours following E_2 injection. Animals were decapitated at either 1 or 4 hours following P treatment and hippocampi removed. Saturation MR binding assays were conducted by incubation of hippocampal cytosol with [^{3}H]-aldosterone. The binding

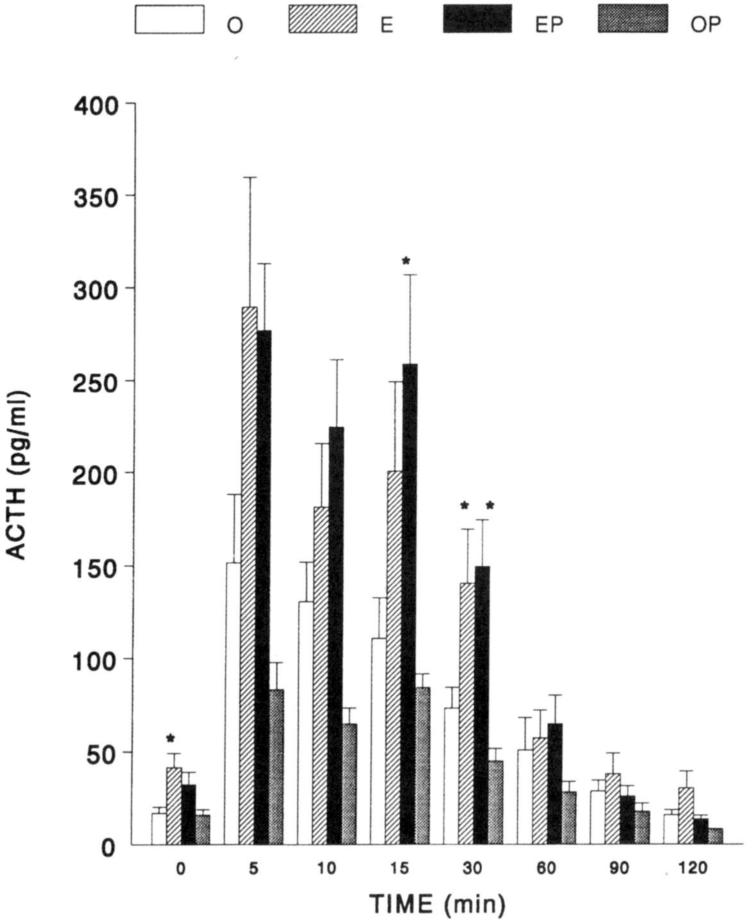

FIGURE 1. Influence of oestradiol and/or progesterone on ACTH concentrations measured from plasma samples obtained before and after exposure of animals to the stress of a novel environment. Animals were fitted with indwelling jugular cannulae to allow sequential blood sampling. Time 0 refers to the pre-stress sample point. The treatment groups were O, OVX; E, OVX + oestradiol (40 μg/kg); EP, OVX + oestradiol (40 μg/kg) + progesterone (500 μg/rat); OP, OVX + progesterone (500 μg/rat). $^{*}p < 0.05$ vs OVX control, Tukey test.

parameters dissociation constant (K_d) and binding maximum (B_{max}) were determined by Scatchard analysis.

The effect of P on MR binding was also determined by *in vivo* autoradiography. OVX/ADX female rats were treated with P or the oil vehicle alone and 3 hours later received an s.c. injection of [^{3}H]-corticosterone (CORT). Animals were decapitated 1 hour following tracer injection and whole brains removed for cryostat sectioning. Sections were exposed to [^{3}H]-ultrafilm and specific finding quantitated by image analysis.

RESULTS AND DISCUSSION

The data demonstrating the *in vitro* MR binding parameters in hippocampal tissue from the 4 steroid-treated groups are shown in TABLE 1.

(1) P treatment of OVX/ADX animals with or without E-priming induced a 5 to 10-fold increase, respectively, in the K_d. This result may reflect the presence of competing steroid in the cytosol rather than an intrinsic change in receptor affinity. Estimation of the binding affinity of P for the MR *in vitro* revealed a K_i of 0.13 nM. Furthermore, addition of 1 nM P to hippocampal cytosol of OVX/ADX animals induced a 4-fold increase in the affinity of the MR for [^{3}H]-AlDO without affecting B_{max}. These data indicate both the competitive nature of P binding to the MR and its very high affinity for this receptor. The differences in the effect of P on K_d depending on E_2 pretreatment may reflect the presence in E_2 + P animals of additional P binding sites, such as the E^2-induced specific P receptor (PR).

(2) Administration of E_2 alone resulted in a 30–40% decrease in the B_{max}. Such an effect of E_2 may be due to an action of the steroid on MR expression in accordance with a previous report.[2]

(3) Treatment of E_2-primed animals with P appears to reverse the E_2-induced decrease in B_{max}. Estimation of B_{max}, 1 hour after P administration revealed a partial reversal to the OVX/ADX level while complete reversal is seen after 4 hours. Enhancing effects of P on MR expression have not been reported but P through its interaction with the PR is well known to antagonize certain gene-mediated actions of the E_2-liganded receptor (E_2R). In the absence of E-priming, a tendency for a P-induced increase in MR B_{max} was also observed. This effect could also be explained by effects of P mediated via the PR, in this case the non-inducible PR.

Recent findings from cell transfection experiments revealed that P is a potent antagonist of the MR.[3] Other investigations have indicated that P binding to the MR promotes receptor transformation to a DNA binding state.[4] Such transformation of the P-liganded MR might be expected to result in a decrease in the available number of binding sites. However, our data from cytosol binding assays did not reveal such a decrease. The effect of P on MR specific binding was further explored by *in vivo* autoradiography. Quantitative analysis of specific [^{3}H]-CORT binding to hippocampal MR revealed no difference between OVX/ADX P-treated animals compared to OVX/ADX controls. Thus, although P appears as a potent competitor of the MR *in vitro,* our evidence to date does not demonstrate competition *in vivo.* However, as indicated above, such *in vivo* competition could be masked by an additional action

TABLE 1. Influence of Oestradiol and Progesterone on MR Binding Parameters[a]

Time after P Treatment	OVX/ADX	OVX/ADX + E_2	OVX/ADX + E_2 + P	OVX/ADX + P
K_d (nM)				
1 hour	0.3	0.3	1.5	3
4 hour	0.2	0.3	1.5	3
B_{max} (fmoles/mg protein)				
1 hour	170	123	147	172
4 hour	194	129	200	219

[a]Values represent the means of duplicate determinations.

of P-induced upregulation of the MR. Further work is necessary to delineate such actions of P.

CONCLUSIONS

Examination of *in vitro* MR binding parameters has revealed the potential for both E_2 and P to affect hippocampal excitability through their modulatory effects on this receptor. Such modulatory effects may be important determinants of observed sex differences in certain aspects of hippocampal function, such as HPA activity and mood.

REFERENCES

1. Viau, V. & M. J. Meaney. 1991. Endocrinology **129:** 2503–2511.
2. Burgess, L. H. & R. J. Handa. 1993. Mol. Cell. Neurosci. **4:** 191–198.
3. Rupprecht, R., J. Reul, B. van Steensel, D. Sprengler, M. Soder, B. Berning, F. Holsboer & K. Damm. 1993. Eur. J. Pharmacol. **247:** 145–154.
4. Couette, B., M. Lombes, E.-E. Baulieu & M. E. Rafestin-Oblin. 1992. Biochem. J. **282:** 697–702.

Lesion of the Central Nucleus of the Amygdala Decreases Basal CRH mRNA Expression and Stress-Induced ACTH Release

CHANTAL M. PREWITT AND JAMES P. HERMAN

Department of Anatomy and Neurobiology
University of Kentucky Medical Center
800 Rose Street
Lexington, Kentucky 40536-0084

Stress integrative neurons localized in the medial parvocellular paraventricular nucleus (PVN) of the hypothalamus are the prime mediators of the hypothalamic-pituitary-adrenocortical (HPA) stress response.[1] In response to stress, these neurons secrete corticotropin releasing hormone (CRH) and arginine vasopressin (AVP), which act at the anterior pituitary to trigger the release of ACTH.[1–5] Ultimate control of the PVN CRH/AVP neurons is accomplished by a complex CNS circuitry. Among such circuitry, the amygdala has been identified as a focal point for initiation of the HPA stress response.[6] Recent data further implicated the central nucleus of the amygdala (Ace) in the mediation of ACTH release, in that lesion of Ace significantly reduces the secretion of ACTH and CORT in response to immobilization stress.[6,7]

These studies were designed to determine whether the Ace is involved in circuit-mediated regulation of hypophysiotrophic PVN neurons. Male Sprague-Dawley rats (250–325 gm) received bilateral injection of ibotenic acid (3.0 μg/side) (IBO) or 0.9% saline (SHAM) into the Ace. After a recovery period of 7 days, groups of IBO and SHAM rats were exposed to one hour of restraint stress (in the AM). Following stress animals were sacrificed by rapid decapitation. Control animals were sacrificed without exposure to stress. Trunk blood was collected and ACTH and CORT levels assayed to confirm the stress response. *In situ* hybridization histochemistry was used to assess localization and regulation of CRH and AVP mRNA in PVN. Adjacent brain sections (15 μm) were postfixed in 4% paraformaldehyde fixative and hybridized with ^{35}S UTP-labeled cRNA probes complementary to AVPexon C (229 bp sequence encoding the C-terminal glycoprotein of preproAVP) and CRH exon 2 (760 bp sequence encoding the coding region and proximal 3′ untranslated region of preproCRH). Images from emulsion-dipped slides were analyzed densitometrically using NIH Image software (CRH) or processed for manual cell counts (AVP). Identification and quantitation of AVP parvocellular neurons was assessed in Nissl counterstained sections. Positive neurons had grains associated with darkly stained nuclei and were localized in the medial parvocellular PVN. Scattered magnocellular neurons identified on the basis of size and grain density were not counted.

A significant increase in ACTH and CORT was seen 60 min following stress in the SHAM rats, indicating activation of the HPA axis by acute stress (TABLE 1). However, IBO rats showed decreased stress-induced ACTH secretion at the 60-min time point relative to SHAM stressed rats. SHAM and IBO animals did not differ in baseline ACTH or CORT (TABLE 1). In the PVN, CRH mRNA significantly decreased (58%; $p = 0.018$) in unstressed IBO rats relative to unstressed SHAM rats. A trend toward significance was seen in the stressed IBO rats relative to unstressed

TABLE 1. Effects of Acute Stress on the HPA Axis

	SHAM		IBO	
	0 m	60 m	0 m	60 m
Plasma ACTH (pg/ml)	125.0 ± 22.1	557.7 ± 108.4[a]	144.9 ± 40.0	231.7 ± 63.3[b]
Plasma CORT (ng/ml)	21.8 ± 5.6	602.6 ± 57.1[a]	33.0 ± 8.9	566.1 ± 67.7[c]

[a]Significantly different from SHAM control (p <0.05).
[b]Significantly different from SHAM stress (p <0.05).
[c]Significantly different from IBO control (p <0.05).

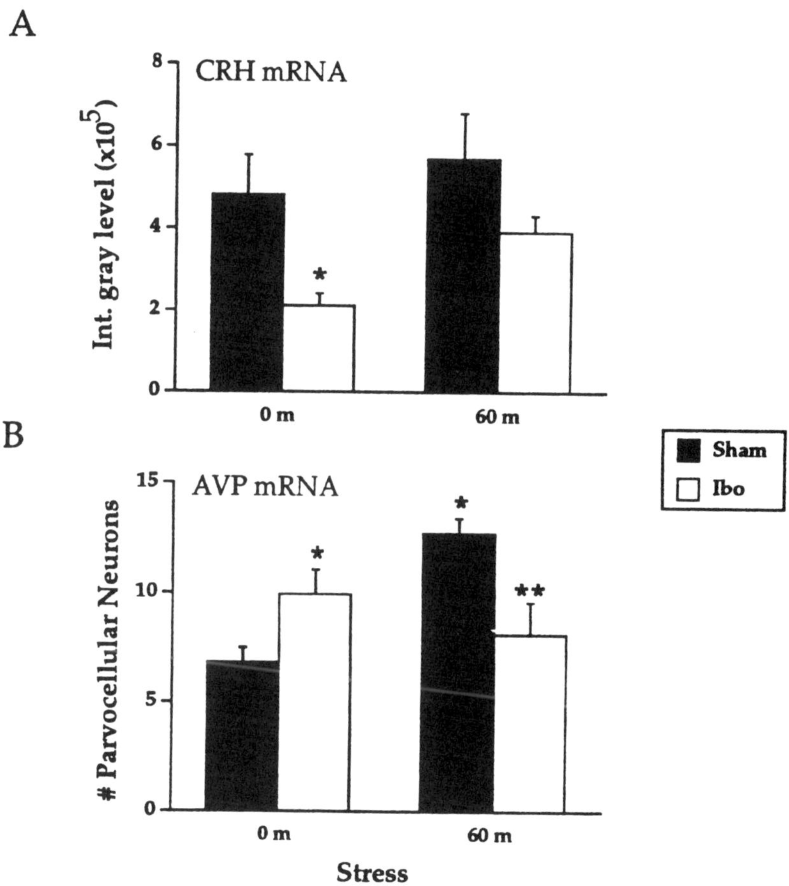

FIGURE 1. (A) Analysis of CRH mRNA in the PVN of control and stress rats. Note significant decrease (58%, p = 0.01) in unstressed IBO group relative to unstressed SHAM group (*). A trend toward significance (p = 0.09) was observed in stressed IBO group relative to stressed SHAM group. **(B)** Analysis of AVP mRNA in PVN control and stress rats. A significant increase (83%, p = 0.002) in the stressed SHAM group is seen relative to the unstressed SHAM group (*). A significant decrease (36%, p = 0.014) is observed in the stressed IBO group relative to the stressed SHAM animals (**). A significant increase was observed in the unstressed IBO group relative to the unstressed SHAM group (*).

IBO rats ($p = 0.06$) (FIG. 1A). The number of parvocellular neurons expressing AVP mRNA increased significantly at 60 min following stress in SHAM rats (83%; $p = 0.002$) relative to unstressed SHAM animals. However, no change was seen between stressed and unstressed IBO animals. AVP mRNA was significantly decreased in the stressed IBO group relative to stressed SHAM animals (36%, $p = 0.014$) (FIG. 1B).

The ACTH response to stress is attenuated in the IBO group, as documented previously.[6] Basal CRH mRNA is decreased in the IBO group, in general agreement with the notion that the Ace is excitatory to PVN neurons (with the observed decrease due to removal of excitatory input). CRH mRNA is increased in the PVN of the IBO group following stress, indicating that the CRH gene is still stress-responsive in the absence of the Ace. Lack of significant change in the stressed SHAM group may be due to selection of the one-hour stress time point, as previous studies note peak CRH induction 2–4 hrs following stress.[8] AVP mRNA increased in the PVN of the SHAM group following stress, indicating induction of AVP neurons by stress and generally consistent with an episodic involvement of the parvocellular AVP system in stress processes. However, AVP mRNA upregulation does not occur in Ace lesion rats, suggesting interaction between the Ace and the parvocellular AVP system.

In conclusion, these observations show excitatory input of the Ace on ACTH secretion, basal CRH tone, and stress induction of AVP. These data suggest that the Ace may form an important link in central pathways regulating the stress response.

REFERENCES

1. ANTONI, F. A. 1986. Endocr. Rev. **7:** 351–378.
2. VALE, W., J. SPIESS, C. RIVIER & J. RIVIER. 1981. Science **213:** 1394–1397.
3. SAWCHENKO, P. E., L. W. SWANSON & W. W. VALE. 1984. Proc. Natl. Acad. Sci. USA **81:** 1883–1887.
4. MURAKAMI, K., S. AKANA, M. F. DALLMAN & W. F. GANONG. 1989. Neuroendocrinology **49:** 233–241.
5. RIVIER, C. L. & W. W. VALE. 1985. Fed. Proc. **44:** 189–195.
6. BEAULIEU, S., T. DIPAOLO & N. BARDEN. 1986. Neuroendocrinology **44:** 247–254.
7. BEAULIEU, S., T. DIPAOLO, J. CÔTÉ & N. BARDEN. 1987. Neuroendocrinology **45:** 37–46.
8. IMAKI, T., T. SHIBASAKI, M. HOTTA & H. DEMURA. 1992. Endocrinology **131:** 240–246.

Lipocortin 1 and Hypothalamo-Pituitary Function in the Rat[a]

A. D. TAYLOR, H. D. LOXLEY, R. J. FLOWER,[b]
AND J. C. BUCKINGHAM[c]

Department of Pharmacology
Charing Cross and Westminster Medical School
St. Dunstan's Road
London W6 8RF, United Kingdom
and
[b]Department of Biochemical Pharmacology
The William Harvey Research Institute
London EC1M 6BQ, United Kingdom

The activity of the hypothalamo-pituitary-adrenocortical (HPA) axis is negatively regulated by the glucocorticoids which act at the levels of the hypothalamus, the anterior pituitary gland and elsewhere in the central nervous system to inhibit the release of the corticotrophin releasing factors (CRF-41 and arginine vasopressin, AVP) and corticotrophin (ACTH). Their actions are mediated by at least three temporally distinct mechanisms which are termed the "rapid," "early delayed" and "late delayed" phases of feedback inhibition respectively.[1] The early delayed phase, unlike the other phases, is dependent on the *de novo* generation of new protein, but the nature of the protein produced is unknown. In the present study we have used *in vitro* models to examine the potential role in this regard of lipocortin 1 (LC1), a Ca^{++} and phospholipid binding protein which has been strongly implicated as a mediator of certain aspects of the anti-inflammatory, anti-mitotic and anti-fever actions of glucocorticoids.[2]

Briefly, anterior pituitary and hypothalamic tissue was incubated according to methods described elsewhere[3,4] in the presence and absence of dexamethasone, a neutralizing monoclonal anti-LC1 antibody (LC1mAb), an N-terminal LC1 fragment ($LC1_{1-188}$) and/or various secretagogues. CRF-41, AVP and ACTH released in the medium were determined by radioimmunoassay and LC1 by Western blotting.[3,4]

IL-1β (500 pg/ml) and IL-1α (200 pg/ml) both caused significant ($p < 0.05$) increases in CRF-41 and AVP release from hypothalamic tissue *in vitro* while ACTH release from the pituitary gland was readily provoked by submaximal concentrations of CRF-41 (1 nM), forskolin (1 mM) and an L-Ca^{++} channel opener, BAY K8644 (1 M). The CRF, AVP and ACTH responses to each of the secretagogues were

[a]We are grateful to the Wellcome Trust and Ono Pharmaceutical for generous financial support. R. J. Flower is supported by an endowment from Lilley Pharmaceuticals.

[c]Corresponding author.

overcome by pre-incubation of the tissues with dexamethasone (0.1/μM) which also caused LC1 to be translocated from within the cells to the outer surface of the cells where it remained attached by a Ca^{++} dependent mechanism. In all instances, the inhibitory actions of the steroid on peptide release were overcome by inclusion of LC1mAb (diluted 1:15000) in the medium but not by an isotype matched control mAb (anti-spectrin α & β, diluted 1:15000, FIG. 1). In addition, like dexamethasone, $LC1_{1-188}$ (1–10 pg/ml) suppressed the secretagogue-induced release of CRF-41 and ACTH (FIG. 2) but, surprisingly, failed to modify the release of AVP.

These results suggest that LC1 acts at the levels of both the hypothalamus and the anterior pituitary gland to mediate certain aspects of the regulatory actions of glucocorticoids on HPA function in the rat. We hypothesize that LC1 is released from target cells in response to a steroid challenge and it acts locally within the hypothalamus and pituitary gland to inhibit peptide release, thereby acting as an autocrine or paracrine agent.

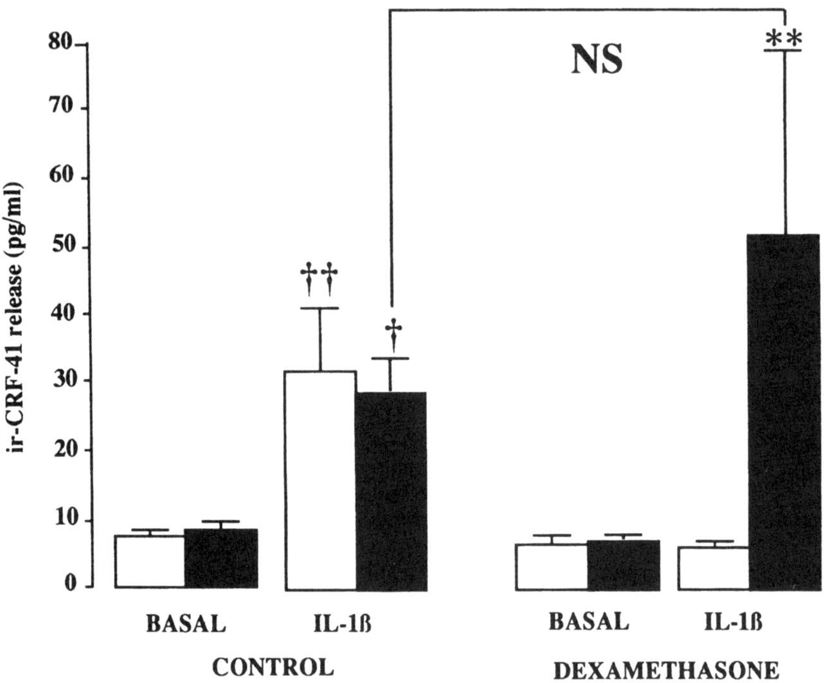

FIGURE 1. Effects of an anti-lipocortin1 monoclonal antibody (LC1mAb) on the capacity of dexamethasone (0.1 μM) to inhibit the increases in CRF-41 release from isolated hypothalami *in vitro* provoked by interleukin 1β (IL-1β, 500 pg/ml). ■ = LC1mAb (1:15000); ❑ = isotype-matched control Ab (anti-spectrin α + β diluted 1:15000). Values represent the mean ± SEM (n = 6), $^{**}p < 0.01$ vs corresponding control Ab-treated group. ††$p < 0.01$, †$p < 0.05$ vs IL-1β free control, Mann Whitney U test.

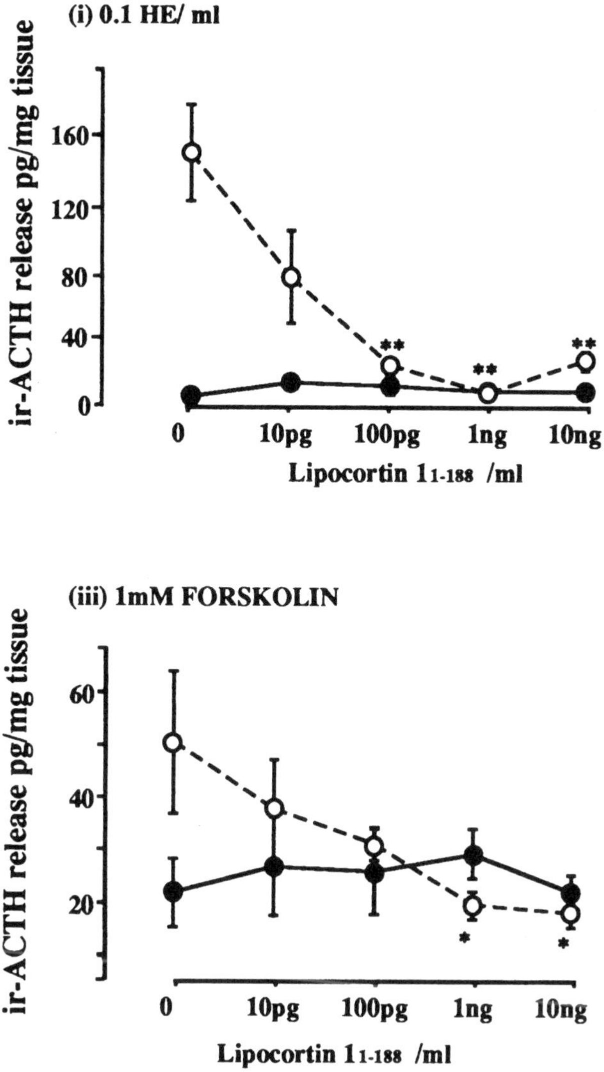

FIGURE 2. The effects of graded concentrations of lipocortin 1_{1-188} on the spontaneous release of ir-ACTH by rat anterior pituitary segments *in vitro* and the secretion evoked by **(i)** 0.1 hypothalamic extracts/ml (HE/mL), **(ii)** corticotrophin releasing factor-41 (1 nM CRF-41), **(iii)** forskolin (1 mM), and **(iv)** BAY K8644 (1 nM). In all four graphs the evoked release is represented by *open circles* (–O–) and basal release by *closed circles* (–●–). Each point represents the mean ± SEM (n = 5–6); p <0.05, **p <0.01 vs corresponding $LC1_{1-188}$ = free control, Duncan's multiple range test. The data are typical of three experiments. (From Taylor *et al.*[4] Reprinted with permission from *Neuroendocrinology*.) (Figure continued on the next page.)

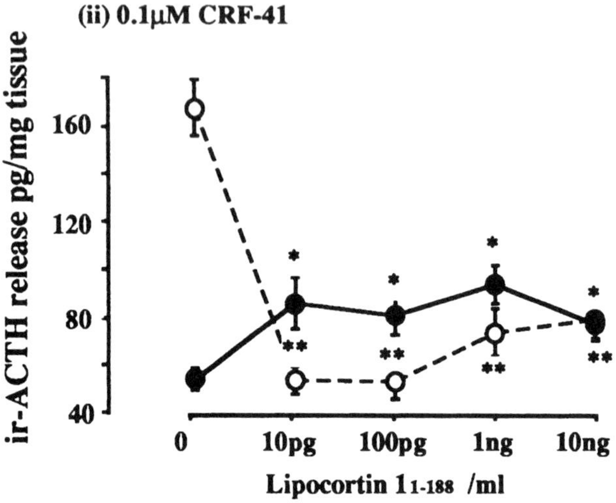

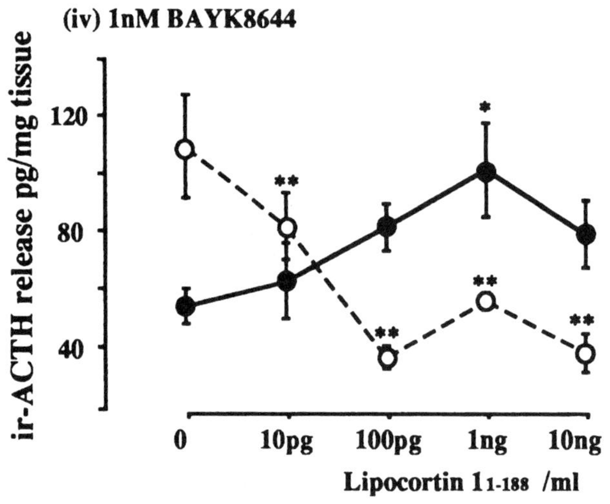

FIGURE 2. Continued.

REFERENCES

1. Buckingham, J. C., T. Smith & H. D. Loxley. 1992. The control of ACTH secretion. *In* The Adrenal Gland. 2nd edit. V. H. T. James, Ed. 131–158. Raven Press. New York.
2. Flower, R. J. 1988. Lipocortin and the mechanism of action of the glucocorticoids. Br. J. Pharmacol. **94:** 987–1015.
3. Loxley, H. D., A-M. Cowell, R. J. Flower & J. C. Buckingham. 1993. Modulation

of the hypothalamo-pituitary-adrenocortical (HPA) responses to cytokines in the rat by lipocortin 1 and glucocorticoids: a role for lipocortin 1 in the feedback inhibition of CRF-41 release? Neuroendocrinology **57:** 801–814.

4. TAYLOR, A. D., A-M. COWELL, R. J. FLOWER & J. C. BUCKINGHAM. 1993. Lipocortin 1 mediates an early delayed action of glucocorticoids on the secretion of ACTH by the rat anterior pituitary gland *in vitro*. Neuroendocrinology **58:** 430–439.

The Role of Lipocortin 1 (LC1) in the Steroid Feedback Control of Hypothalamo-Pituitary-Adrenocortical Function

In Vivo Studies[a]

A. D. TAYLOR, H. D. LOXLEY, R. J. FLOWER,[b]
AND J. C. BUCKINGHAM[c]

Department of Pharmacology
Charing Cross and Westminster Medical School
St. Dunstan's Road
London W6 8RF, United Kingdom
and
[b]*Department of Biochemical Pharmacology*
The William Harvey Research Institute
London EC1M 6BQ, United Kingdom

Lipocortin 1 (LC1), a Ca^{++} and phospholipid binding protein, was first identified as a glucocorticoid-inducible protein in macrophages and has since been shown to be an important mediator of aspects of the anti-inflammatory, anti-mitotic and anti-fever actions of glucocorticoids.[1] Our *in vitro* studies[2,3] strongly suggest that LC1 also contributes to the powerful inhibitory effect of these steroids on the hypothalamo-pituitary-adrenocortical (HPA) axis and that it is released from cells within the hypothalamus and pituitary gland where it acts via cell surface receptors to inhibit the release of the corticotrophin releasing factors (CRF-41 and arginine vasopressin, AVP) and corticotrophin (ACTH) respectively.[2,3] In order to relate our findings *in vitro* to the situation *in vivo* we have now examined the influence of human recombinant LC1 (hu-r-LC1) and a neutralizing polyclonal anti-LC1 antibody (LC1pAb) on the pituitary-adrenocortical responses to IL-1β given centrally or peripherally.

Injection of submaximal doses of IL-1β either centrally (10 ng in a volume of 3 μl via an indwelling intracerebroventricular cannula) or peripherally (3 μg/kg, i.p.) into adult male CFY rats (≈200 g body weight) caused within 1 h significant (p <0.01, vs saline vehicle, Mann Whitney U test) increases in the release of ACTH and corticosterone. These responses were prevented by administration 1 h earlier of dexamethasone (120 μg/kg, i.p.) but not by its saline vehicle (1 ml/kg, i.p.).

[a]We are grateful to the Wellcome Trust and Ono Pharmaceuticals for generous financial support. R. J. F. is funded by an endowment from Lilley Pharmaceuticals.

[c]Corresponding author.

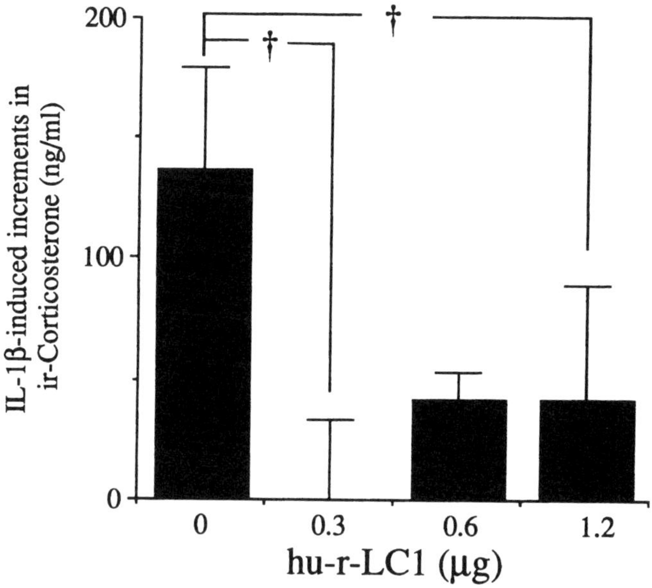

FIGURE 1. Effects of human recombinant LC1 (hu-r-LC1 0.3–1.2 μg/rat, i.c.v. in a volume of 3 μl) on the increments in serum corticosterone induced by the injection 5 min subsequently of IL-1β (10 ng/rat, i.c.v. in a volume of 3 μl). Values represent the mean ± SEM (n = 5–6). †p <0.05 vs cytokine alone, Mann Whitney U test. (From Taylor *et al.*[3] Reprinted by permission from *Neuroendocrinology*.)

Pretreatment of the rats with hu-r-LC1 (0.3–1.2 μg i.c.v. in a volume of 3 μl) 5 min prior to injection of IL-1β (10 ng, i.c.v.) reduced significantly (p <0.05 vs saline vehicle, Mann Whitney U test) the pituitary-adrenocortical responses to the cytokine (FIG. 1). Passive immunization of the rats against LC1 (1 ml/kg/day LC1pAb for two successive days) had no significant effect on the resting concentrations of ACTH or corticosterone in the plasma/serum. However, the anti-LC1 pAb attenuated markedly (p <0.01, Mann Whitney U test) the capacity of dexamethasone (~120 μg/kg i.p.) to inhibit the significant increases (p <0.001 vs control) in ACTH and corticosterone release provoked by IL-1β (3 μg/kg, i.p.) (FIG. 2). By contrast, administration of an equivalent volume of pre-immune sheep serum was without effect on either the basal ACTH and corticosterone concentrations or on the responses to IL-1β in the presence or absence of dexamethasone. Interestingly, the anti-LC1 pAb also specifically attenuated the inhibitory actions of dexamethasone on growth hormone (GH) release evident in this study, but serum luteinizing hormone (LH) concentrations were unaffected by the treatment.

These results are consistent with the data obtained from our *in vitro* experiments and thus support our hypothesis that LC1 plays an important role in effecting the acute inhibitory actions of glucocorticoids on HPA function. These data also suggest that LC1 contributes to the regulatory actions of the glucocorticoids on GH secretion and may therefore fulfil a more widespread role as a mediator of glucocorticoid action within the brain-neuroendocrine system than we envisaged originally.

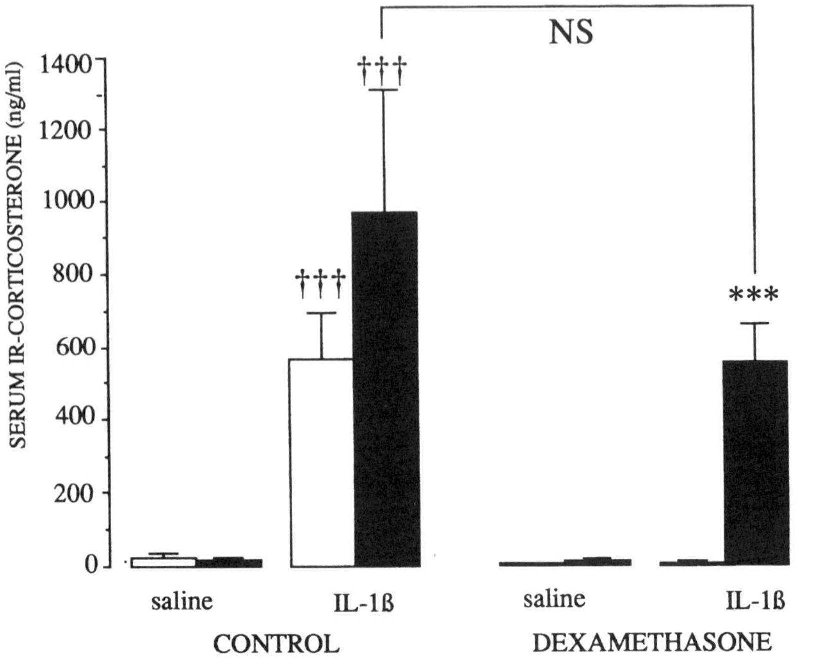

FIGURE 2. Effects of treatment of rats with a neutralizing polyclonal anti-lipocortin1 antibody (anti-LC1pAb, 1 ml/kg/day, s.c. for two days) or a corresponding volume of non-immune control sheep serum on the capacity of dexamethasone (100 µg/kg, 1 ml/kg, i.p.) to inhibit the increases in serum corticosterone concentration provoked by a submaximal dose of rat IL-1β (3 µg/kg, 1 ml/kg, i.p.) ■ = anti-LC1pAb ❑ = control serum. Values given represent the mean ± SEM (n = 6). †††*p* <0.001 vs corresponding IL-1β-free group. ****p* <0.001 vs corresponding control serum-treated group, Mann Whitney U test.

REFERENCES

1. Flower, R. J. 1988. Lipocortin and the mechanism of action of the glucocorticoids. Br. J. Pharmacol. **94:** 987–1015.
2. Loxley, H. D., A-M. Cowell, R. J. Flower & J. C. Buckingham. 1993. Modulation of the hypothalamo-pituitary-adrenocortical responses to cytokines in the rat by lipocortin 1 and glucocorticoids: a role for lipocortin 1 in the feedback inhibition of CRF-41 release. Neuroendocrinology **57:** 801–814.
3. Taylor, A. D., A-M. Cowell, R. J. Flower & J. C. Buckingham. 1993. Lipocortin 1 mediates an early inhibitory action of glucocorticoids on the secretion of ACTH by the rat anterior pituitary gland *in vitro*. Neuroendocrinology **58:** 430–439.

Effects of Chronic Antidepressant Treatment on the Hypothalamic-Pituitary-Adrenal Axis of Wistar Rats

B. W. M. M. PEETERS, R. VAN DER HEIJDEN,
D. G. GUBBELS, AND P. M. L. VANDERHEYDEN

Department of Neuropharmacology
n.v. Organon
P.O. Box 20
5340 BH Oss
The Netherlands

In a subpopulation of endogenously depressed patients, disturbances of the hypothalamic-pituitary-adrenal (HPA) axis are observed. Increased cortisol and CRH levels but normal ACTH concentrations have been reported; often combined with dexamethasone nonsuppression.[1] Successful antidepressant treatment of the patients is associated with a normalization of the HPA function,[2,3] which has been reported to precede clinical amelioration.[4] Pepin *et al.*[5] hypothesized that this could be brought about by an antidepressant-induced increase in corticosteroid receptor numbers, rendering the HPA axis more susceptible to feedback inhibition by cortisol. A number of tricyclic antidepressants were shown to increase glucocorticoid receptor (GR) mRNA concentrations in rat primary neuronal cultures. Follow-up *in vivo* studies investigated the chronic effects of a number of antidepressant treatments on different parameters characterizing HPA status. Repeatedly, increases in corticosteroid receptor (GR and mineralocorticoid (MR)) numbers combined with a diminished HPA activity were found (TABLE 1).

In the present investigations we compared the effects on the HPA axis of the tricyclic antidepressants imipramine (Egis Pharmaceuticals) and amitriptyline (Secifarma spa) with those of the new tetracyclic antidepressant Org 3770[6] (Organon). Male Wistar rats (Harlan, Zeist, the Netherlands) were intraperitoneally (i.p.) injected for 10 days with either placebo (n = 20), imipramine (n = 10; 20 mg/kg), amitriptyline (n = 10; 20 mg/kg) or Org 3770 (n = 10; 20 mg/kg). Two hours before the beginning of the dark period on day 11, animals were killed, blood samples were obtained and the body- and adrenal-weight were determined. Corticosteroid levels were measured using a procedure modified after Rodbard & Lewald.[7] Corticosterone was extracted from the plasma by shaking 25 μl plasma with 475 μl methanol for 3 seconds. After centrifugation at 8700 * g for 3 minutes a 400-μl aliquot of the methanol fraction was evaporated, and the resulting residue dissolved in 400 μl of 0.2% (v/v) ethyleneglycol. Corticosterone was assayed in 50 μl samples at an antibody dilution which yielded 30% binding of [^{3}H]corticosterone. A standard curve of corticosterone (serial dilution at a 1 : 1 ratio) was made in 0.2% (v/v) ethylene glycol.

GR receptor characteristics were determined using routine procedures. In short, hippocampal cytosol was prepared by homogenization in 0.6 ml/hippocampus of 10 mM Tris-HCl pH 7.4 containing 1 mM EDTA, 10 mM $Na_2MoO_4{\cdot}2H_2O$ and 1 mM DTT followed by a centrifugation at 100000 * g for 45 min. 50 μl of the supernatant

TABLE 1. Effects of Antidepressant Treatments on HPA Characteristics[a]

Antidepressant	Treatment	Adrenal Weight	Cort Level	ACTH Level	GR Receptor	MR Receptor	Ref.
Desipramine	10 days 20 mg/kg/day (i.p.)		↓		mRNA↑		Peiffer 1991[9]
	10 days 20 mg/kg/day (i.p.)		0	0	mRNA↑ binding↑		Pepin 1992[10]
	14 days 10 mg/kg/day (i.p.)				mRNA↑	mRNA↑	Seckl 1992[11]
Imipramine	10 days 20 mg/kg/day (i.p.)		↓		mRNA↑		Peiffer 1991[9]
	21 days 5 mg/kg/day (s.c.)		↓	0			Aulakh 1993[12]
	21 days 5 mg/kg/day (s.c.)		0	0			Aulakh 1988[13]
	28 days 20 mg/kg/day (p.o.)				binding↑		Przegalinski 1993[14]
	56 days 5 mg/kg/day (i.p.)	↑	0	↓	mRNA↓	mRNA↑	Brady 1991[15]
Amitriptyline	42 days 4.5 mg/kg/day (p.o.)	↓	↓	↓	binding↓↑	binding↓↑	Reul 1993[8]
	28 days 20 mg/kg/day (p.o.)				binding↑		Przegalinski 1993[14]
	14 days 20 mg/kg/day (i.p.)				mRNA0↑	mRNA↑↑	Seckl 1992[11]
Clomipramine	21 days 5 mg/kg/day (s.c.)		↓	↑			Aulakh 1993[12]
	14 days 15 mg/kg/day (i.p.)	0	0				Armario 1988[16]
Clorgyline	21 days 1 mg/kg/day (s.c.)		↓				Aulakh 1993[12]
Citalopram	14 days 20 mg/kg/day (i.p.)					mRNA↑	Seckl 1992[11]
Chlorimipramine	13 days 15 mg/kg/day (i.p.)	0	0				Armario 1988[16]
ECS	10 days 1/day				binding↑	binding↑	Przegalinski 1993[14]
LiCl	10 days 6 mM/kg/day (i.p.)		↓		mRNA↑		Peiffer 1991[9]
Tianeptine	15 days 20 mg/kg/day (i.p.)		0	↓			Delbende 1994[17]

[a] ↑ = increase; ↓ = decrease; 0 = no effect; ↓↑ = an initial decrease (↓) was followed by an increase (↑) after prolonged administration.

TABLE 2. Effects of Placebo, Imipramine (20 mg/kg), Amitriptyline (20 mg/kg) and Org 3770 (20 mg/kg) on Body Weight, Adrenal Weight, Corticosterone Levels and Hippocampal Glucocorticoid Receptor Characteristics (Bmax, Kd) (Mean ± SEM)

Treatment	Body Weight (g)	Adrenal Weight (mg)	CORT Level (μg/100 ml Plasma)	GR Bmax (fmol/mg Protein)	GR Kd (nM)
Placebo	280.5 ± 3.0	31.5 ± 0.9	9.6 ± 1.7	93.8 ± 17.5	5.0 ± 1.4
Imipramine	265.3 ± 3.5**	30.1 ± 0.8	1.3 ± 0.5**	32.8 ± 4.6**	1.7 ± 0.3
Amitriptyline	257.9 ± 4.4**	32.1 ± 1.1	7.8 ± 1.6	143.0 ± 4.0*	4.1 ± 0.8
Org 3770	253.3 ± 5.4**	32.9 ± 1.3	10.4 ± 1.3	138.5 ± 3.5*	3.7 ± 0.7

*p <0.05 according to Mann-Whitney U test.
**p <0.01 according to Mann-Whitney U test.

was incubated with 30 μl [^{3}H]dexamethasone (s.a. 89 μCi/mmol) at concentrations between 0.05 and 10 nM; and 20 μl with either buffer (for the total binding) or 2 μM unlabeled dexamethasone for measuring the nonspecific binding. The mixture was incubated for 2 hours at room temperature. Free and bound ligand were separated using 100 μl charcoal. After 5 minutes, the mixture was centrifuged for 5 min at 2500 rpm and the bound radioactivity in 100 μl of the supernatant was measured after addition of 2.5 ml Ultima gold MV. The corresponding Kd and Bmax values were calculated by nonlinear regression analysis.

Statistical comparisons were performed by means of Mann-Whitney U tests and compared treatment and placebo scores.

The obtained results are shown in TABLE 2. All three antidepressants decreased the body weight gain after 10 days of treatment. The adrenal weight did not change and the corticosterone level was only decreased after imipramine application. GR densities (Bmax) were modified. While imipramine induced a decrease in receptor numbers, both amitriptyline and Org 3770 induced increases.

The results obtained with amitriptyline and Org 3770 are similar to those reported by others for a number of other antidepressant treatments (TABLE 1). The increase in GR numbers might indicate an improved HPA feedback. The unchanged corticosteroid levels and adrenal weights suggest that the feedback system in Wistar rats is functioning optimally; an improvement of the feedback does not result in a down regulation of the HPA activity.

Imipramine has been reported (TABLE 1) to both up- and downregulate GR numbers. Also corticosteroid levels can be either decreased or unchanged. Our observation that imipramine caused a decrease in GR numbers, combined with a decreased corticosterone level, is surprising. A downregulated HPA feedback would normally result in increased corticosteroid levels. Reul *et al.*[8] observed a decreased GR Bmax after 7 days of treatment with amitriptyline. They explained this by an increased HPA activity at that time, acting suppressive on the hippocampal GR levels. However, they did not measure corticosteroid concentrations at this time point and it is still possible that antidepressants influence the HPA axis at different levels.

In conclusion, our results partly confirm earlier observations which suggest that chronic antidepressant treatment results in a potentiation of the HPA feedback. Whether the treatments act solely at the corticosteroid receptor level can be questioned. This has to be investigated in future experiments.

REFERENCES

1. Murphy, B. E. 1991. J. Steroid Biochem. Mol. Biol. **38:** 537–559.
2. Greden, J. F., R. Gardner, D. King, L. Grunhaus, B. J. Carroll & Z. Kronfol. 1983. Arch. Gen. Psychiatry **40:** 493–500.
3. Holsboer, F., A. Gerken, G. K. Stalla & O. A. Muller. 1985. Biol. Psychiatry **20:** 276–286.
4. Holsboer, F., R. Liebl & R. Hofschuster. 1982. J. Affective Disorders **4:** 93–101.
5. Pepin, M., M. V. Govindan & N. Barden. 1992. Mol. Pharmacol. **41:** 1016–1022.
6. Smith, W. T., V. Glaudin, J. Panagides & E. Gilvary. 1990. Psychopharmacol. Bull. **26:** 191–196.
7. Rodbard, D. & J. E. Lewald. 1970. Acta Endocrinol. **14:** 79–103.
8. Reul, J. M. H. M., I Stec, M. Soder & F. Holsboer. 1993. Endocrinology **133:** 312–320.
9. Peiffer, A., S. Veilleux & N. Barden. 1991. Psychoneuroendocrinology **16:** 505–515.
10. Pepin, M., F. Rothier & N. Barden. 1992. Mol. Pharmacol. **42:** 991–995.
11. Seckl, J. R. & G. Fink. 1992. Neuroendocrinology **55:** 621–626.
12. Aulakh, C. S., J. L. Hill & D. L. Murphy. 1993. Eur. J. Pharmacol. **240:** 85–88.
13. Aulakh, C. S., K. M. Wozniak, J. L. Hill & D. L. Murphy. 1988. Eur. J. Pharmacol. **156:** 395.
14. Przegalinski, E., B. Budziszewska, J. Siwanowicz & L. Jaworska. 1993. Neuropharmacology **32:** 1397–1400.
15. Brady, L. S., H. J. Whitfield, R. J. Fox, P. W. Gold & M. Herkenham. 1991. J. Clin. Invest. **87:** 831–837.
16. Armario, A., C. Garcia-Marquez & M. Giralt. 1988. Psychopharmacology **94:** 217–220.
17. Delbende, C., D. T. Brunel, G. Tarozzo, M. Grino, C. Oliver, E. Mocaer & H. Vaudry. 1994. Eur. J. Pharmacol. **251:** 245–251.

Glucocorticoid Negative Feedback in Pituitary Corticotropes

Pivotal Role for Calcineurin Inhibition of Adenylyl Cyclase

MICHAEL J. SHIPSTON, FERNANDO HERNANDO, RICHARD J. O. BARNARD, AND FERENC A. ANTONI

MRC Brain Metabolism Unit
Department of Pharmacology
University of Edinburgh
Edinburgh, Scotland, UK EH8 9JZ

INTRODUCTION

Early (<2 h) glucocorticoid inhibition of adrenocorticotropin (ACTH) secretion involves induction of the intracellular Ca^{2+}-binding protein, calmodulin, in mouse pituitary corticotrope tumor (AtT20) cells.[1] A potential target of glucocorticoid-induced calmodulin is the Ca^{2+}/calmodulin-dependent protein phosphatase (calcineurin). Previous work demonstrated that the immunosuppressant drugs, FK506 and cyclosporin A (CsA), specifically block calcineurin activity in AtT20 cells.[2] In this study we used FK506 and CsA to examine the potential role of calcineurin in early glucocorticoid inhibition of corticotropin-releasing factor (CRF)-stimulated ACTH release in AtT20 cells.

METHODS

Mouse pituitary corticotrope tumor (AtT20 D16:16) cells were maintained as previously described.[3] FK506, CsA and L685,818 were generous gifts from Fujisawa Ltd., Sandoz Pharma and Merck & Co. respectively. For ACTH release and cAMP accumulation experiments cells were treated as described previously;[3] modifications are described in the respective figure legends. FK506 and CsA were applied at 37°C for 30 min prior to CRF stimulation to allow cellular uptake; fresh drugs were applied during the respective stimulus period. ACTH and cAMP levels were determined by radioimmunoassay.[3]

RESULTS AND DISCUSSION

FK506, which blocks the activity of calcineurin in AtT20 cells by up to 80%,[2] reduced the potency of dexamethasone inhibition of CRF-stimulated ACTH release by 10-fold (FIG. 1A). FK506 produced a small but significant enhancement of CRF-stimulated ACTH release (FIG. 1B) and cAMP accumulation (not shown) under these conditions. Greater enhancement of CRF-stimulated ACTH release and cAMP accumulation (2-fold increase using 1 μM FK506) was observed in experiments performed at 22°C. All effects of FK506 were dose-dependently blocked by the

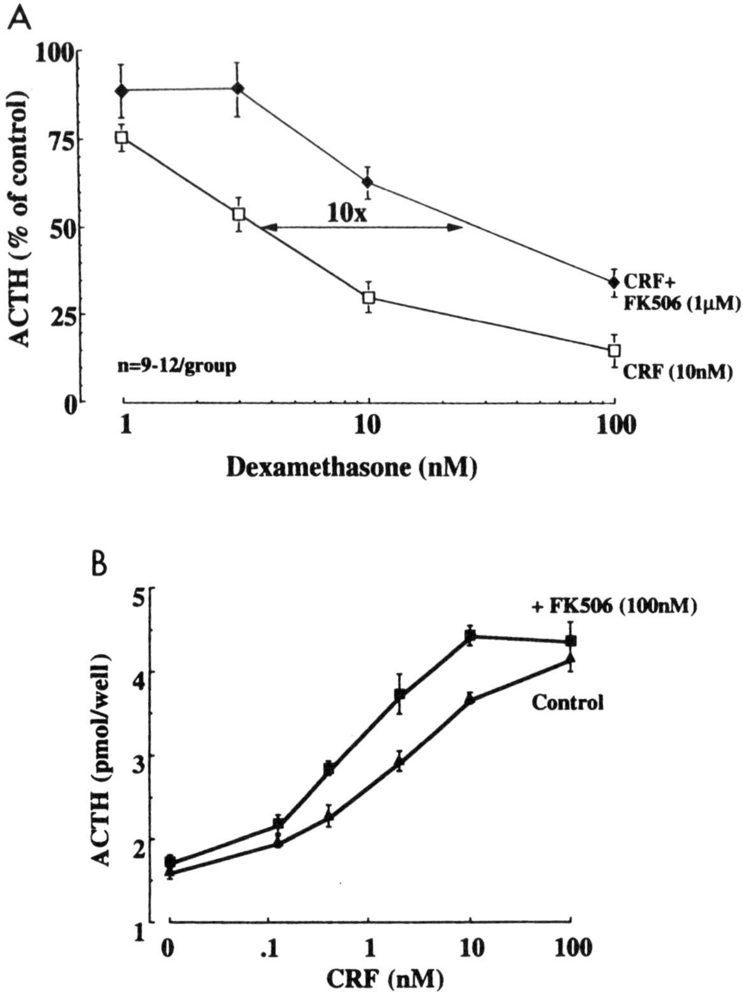

FIGURE 1. (**A**) FK506 reduces the potency of dexamethasone on CRF-stimulated ACTH release in AtT20 cells. Cells were pretreated for 2 h at 37°C with the respective concentration of the synthetic glucocorticoid, dexamethasone. FK506 was added in the last 30 min of dexamethasone application to prevent potential interference of glucocorticoid-induced gene transcription[1,3] on inhibition of calcineurin activity. Fresh FK506 was applied during the 30 min CRF (10 nM) stimulus period at 37°C. Data are expressed as a percentage of the respective CRF stimulus after basal subtraction. (**B**) FK506 enhances CRF-stimulated ACTH release at 37°C. Data are means ± SEM.

FK506 antagonist drug L685,818 and qualitatively reproduced using CsA to block calcineurin activity.

FIGURE 2 demonstrates that CRF-induced cAMP accumulation in Ca^{2+}-depleted AtT20 cells is inhibited by the addition of Ca^{2+} and that this effect is suppressed by FK506. The greatest effect of FK506 was observed at the normal extracellular Ca^{2+}

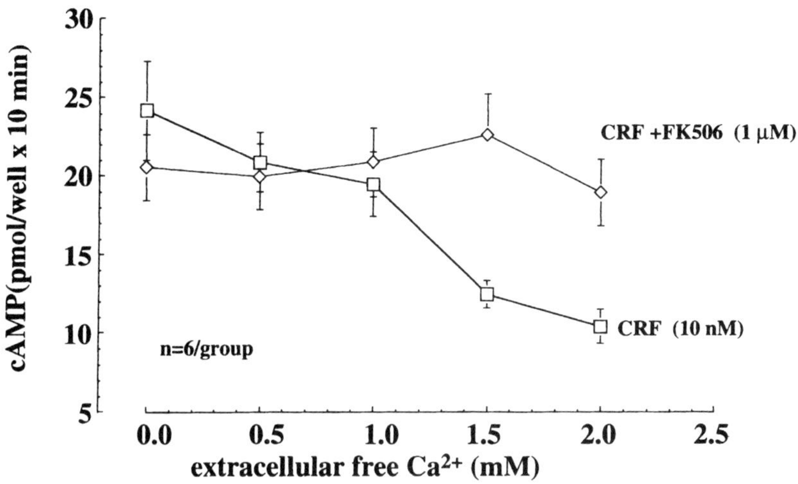

FIGURE 2. Intracellular Ca^{2+} inhibits CRF-stimulated cAMP accumulation in AtT20 cells. Cells were preincubated for 30 min at 37°C in Hepes (25 mM) buffered Hank's balanced salt solution, pH 7.4 (HBSS) containing: 0.25% BSA; 2 mM EGTA; no added Ca^{2+}; 5 μM calcium ionophore, A23187; and 5 μM nimodopine (L-type Ca^{2+} channel blocker) to deplete intracellular free Ca^{2+}. FK506 and 0.5 mM isobutylmethylxantine (phosphodiesterase inhibitor) were applied to treated cells during this period. Subsequently cells were cooled to 22°C and 10 nM CRF was added for 10 min with increasing amounts of added Ca^{2+} in HBSS containing 2 mM EGTA. cAMP content was determined in an acidified cell/medium extract.[3] The *ordinate* shows the calculated concentration of free Ca^{2+} in the extracellular medium, which corresponds to submicromolar intracellular free Ca^{2+} under these conditions. Data are means ± SEM.

concentration of 1.5 mM. This would suggest that calcineurin forms a Ca^{2+} sensor negative feedback loop on CRF-stimulated cAMP accumulation. The potency of dexamethasone inhibition of ACTH release evoked by a cell permeant analog of cAMP (8-CPT-cAMP, 0.1 mM), which would by-pass the negative feedback effect of calcineurin, was reduced 10-fold with respect to dexamethasone inhibition of CRF-stimulated ACTH release (not shown).

These data show that calcineurin forms a Ca^{2+} sensor negative feedback loop on cAMP production evoked by CRF. The feedback action of calcineurin appears to play a pivotal role in determining the efficacy of glucocorticoid inhibition in adenohypophysial corticotrope cells.

REFERENCES

1. SHIPSTON, M. J. & F. A. ANTONI. 1992. Biochem. Biophys. Res. Commun. **189:** 1382–1388.
2. ANTONI, F. A., M. J. SHIPSTON & S. M. SMITH. 1993. Biochem. Biophys. Res. Commun. **194:** 226–233.
3. WOODS, M. D., M. J. SHIPSTON, E. L. MULLENS & F. A. ANTONI. 1992. Endocrinology **131:** 2873–2880.

Early Glucocorticoid Inhibition of Hormone Release in Pituitary Corticotrope Cells Is Voltage Dependent

MARY D. WOODS, MICHAEL J. SHIPSTON, BRIAN McFERRAN,[a] SIMON B. GUILD,[a] AND FERENC A. ANTONI

MRC Brain Metabolism Unit
Department of Pharmacology
University of Edinburgh
Edinburgh, Scotland, UK EH8 9JZ
and
[a]*Department of Biological and Medical Sciences*
University of St. Andrews
St. Andrews, Scotland UK

INTRODUCTION

The suppression of adrenocorticotropic hormone (ACTH) secretion by adrenal corticosteroids is a functionally significant element of homeostatic regulation in the hypothalamic-pituitary-adrenocortical axis.[1] Previous work suggests that a pivotal early effect of corticosteroids in anterior pituitary tumor (AtT20) cells is the reduction of corticotropin releasing factor (CRF)-induced intracellular free calcium transients.[2] Electrophysiological studies in hippocampal neurons[3,4] and pituitary tumor cell lines[5,6] have indicated that glucocorticoids may "stabilize," *i.e.*, hyperpolarize, the membrane potential primarily through the enhancement of potassium currents, which would reduce the activation of voltage-operated calcium currents. As further tests of this hypothesis the present study examined whether 1) the ACTH-release inhibiting action of glucocorticoids could be influenced by manipulations that alter the activity of ion channels in the plasma membrane and bring about depolarization; 2) glucocorticoid inhibition persists in electrically permeabilized AtT20 cells.

RESULTS AND DISCUSSION

The synthetic glucocorticoid dexamethasone caused a dose-dependent inhibition of ACTH release evoked by CRF. When the potassium channel blocker tetraethylammonium (TEA, 25 mM) which blocks about 80% of the total potassium current in AtT20 cells[6] was the stimulus, ACTH release was enhanced to a similar extent as by CRF, but the apparent potency of dexamethasone was reduced by about 30-fold (FIG. 1A). Application of 25 mM choline chloride had no effect on ACTH release (not shown). Depolarization with 50 mM KCl caused greater release of ACTH than 10 nM CRF and a >100-fold reduction of the inhibitory potency of dexamethasone (FIG. 1A). This latter finding is in agreement with a previous report.[7]

These data show that depolarization of the plasma membrane opposes the early effect of glucocorticoids on ACTH release. However, a significant portion of the

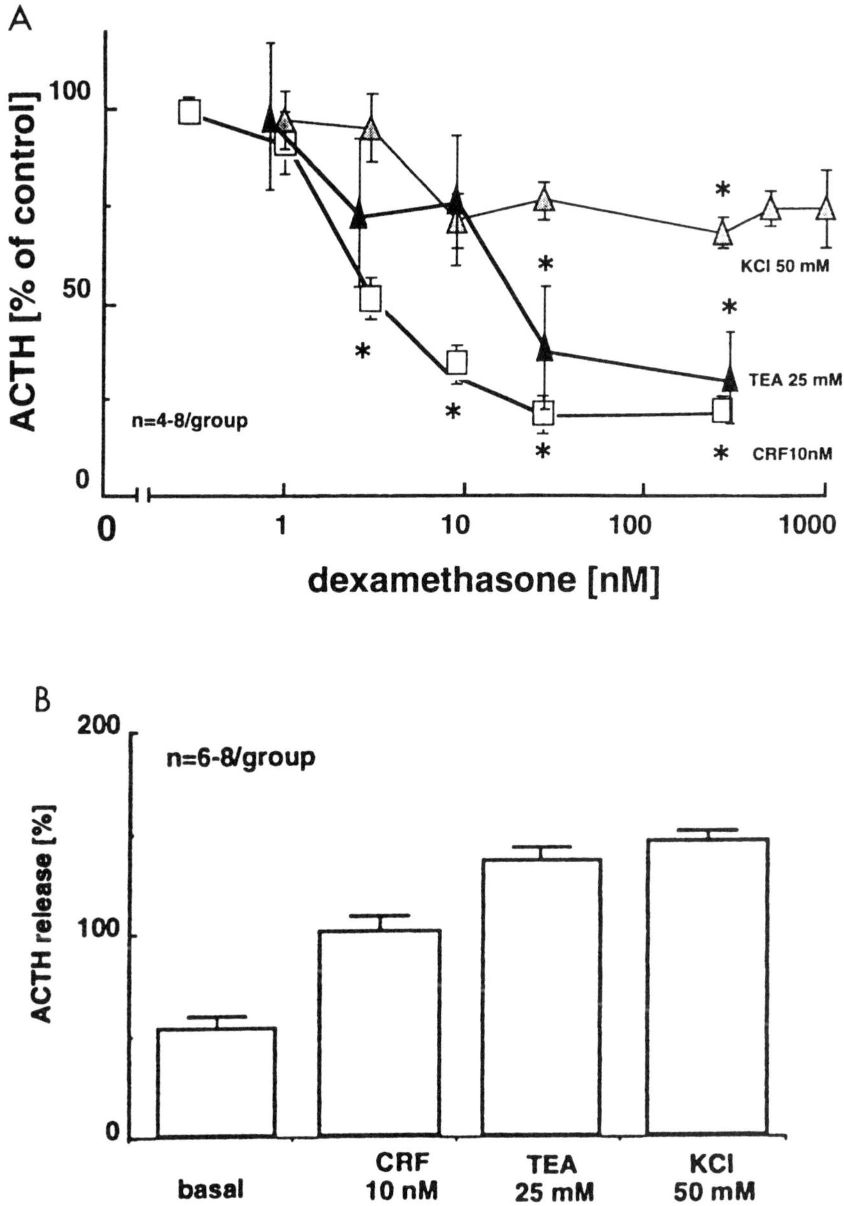

FIGURE 1. Effect of (**A**) the reduction of the transmembrane K^+ gradient and (**C**) the combination of K^+ channel blockade with the dihydropyridine L-channel agonist (±)BayK8644 or the calcineurin blocking immunosuppressant FK 506[9] on dexamethasone inhibition of stimulated ACTH release in AtT20 cells. Cells were preincubated with dexamethasone for 2 h and subsequently challenged with the secretagogues for 30 min.[13] (**B**) shows the

inhibitory effect was still evident even after a drastic reduction of the transmembrane potassium gradient, suggesting that glucocorticoids may act directly on voltage-operated calcium channels. When TEA was combined with the calcium channel agonist (±)Bay K8644 (FIG. 1C) or (+)202-791 (not shown), the effect of dexamethasone was reduced further as seen with 50 mM KCl. Application of the calcineurin antagonist FK506[8,9] with TEA produced a similar effect (FIG. 1C). Thus, early glucocorticoid inhibition is largely overcome by depolarization and calcium channel agonist drugs, suggesting that regulation of the membrane potential is pivotal for early glucocorticoid inhibition of ACTH release.

This conclusion is further supported by experiments in electrically permeabilized[10] AtT20 cells where pretreatment with dexamethasone failed to alter Ca^{2+}-induced ACTH release (FIG. 2). In contrast, enhancement of intracellular free Ca^{2+} by calcium ionophores in cells with otherwise intact plasma membrane has no effect on dexamethasone inhibition of ACTH release.[2]

In summary, early glucocorticoid inhibition in anterior pituitary corticotroph cells apparently involves hyperpolarization of the membrane potential through K^+ channels as well as inhibition of calcium channels plausibly by dephosphorylation *via* calcineurin.[11,12]

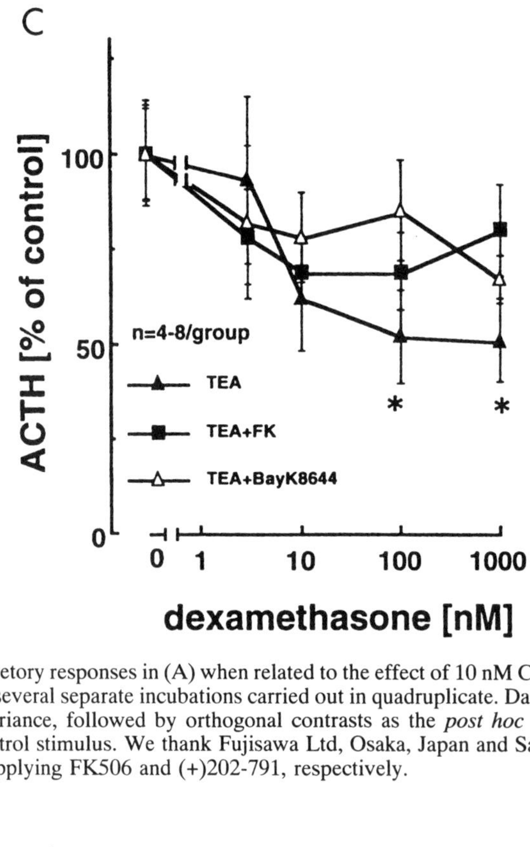

size of the secretory responses in (A) when related to the effect of 10 nM CRF. Data are means ± SEM, from several separate incubations carried out in quadruplicate. Data were analyzed by analysis of variance, followed by orthogonal contrasts as the *post hoc* test. *<0.05 vs the respective control stimulus. We thank Fujisawa Ltd, Osaka, Japan and Sandoz U.K. Ltd, for generously supplying FK506 and (+)202-791, respectively.

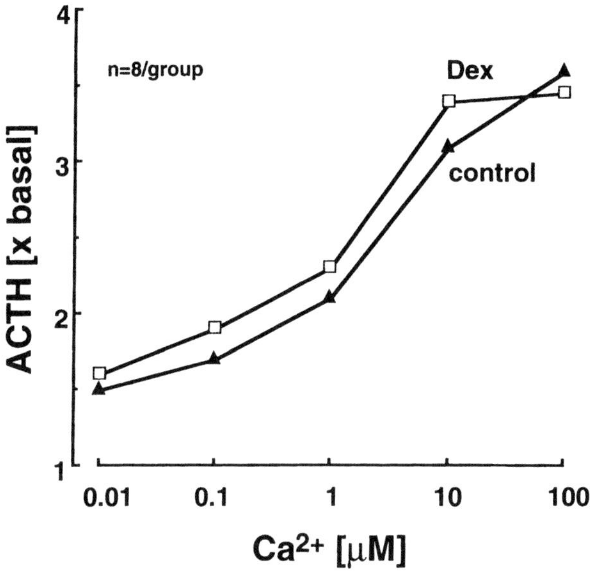

FIGURE 2. Effect of dexamethasone (100 nM) on ACTH release induced by Ca^{2+} in electrically permeabilized[10] AtT20 cells. Cells were preincubated with dexamethasone for 120 min, permeabilized and challenged with Ca^{2+}-buffers for 30 min. The ACTH content of the medium was determined by RIA. Data are means from two separate experiments performed in quadruplicate. The SEMs are less than 10% of the respective means.

REFERENCES

1. Jones, M. T. & B. Gillham. 1988. Physiol. Rev. **68:** 743–818.
2. Antoni, F. A., J. Hoyland, M. D. Woods & W. T. Mason. 1992. J. Endocrinol. **133:** R13–R16.
3. Joëls, M. & E. R. de Kloet. 1989. Science **245:** 1502–1505.
4. Kerr, D. S., L. W. Campbell, S.-Y. Hao & P. W. Landfield. 1989. Science **245:** 1505–1509.
5. Levitan, E. S., L. M. Hemmick, N. C. Birnberg & L. K. Kaczmarek. 1991. Mol. Endocrinol. **5:** 1903–1908.
6. Pennington, A., J. S. Kelly & F. A. Antoni. 1994. J. Neuroendocrinol. **6:** 305–315.
7. Phillips, M. A. & A. H. Tashjian, Jr. 1982. Endocrinology **110:** 892–900.
8. Sigal, N. H. & F. J. Dumont. 1992. Annu. Rev. Immunol. **10:** 519–560.
9. Antoni, F. A., M. J. Shipston & S. M. Smith. 1993. Biochem. Biophys. Res. Commun. **194:** 226–233.
10. Guild, S. 1991. Br. J. Pharmacol. **109:** 117–122.
11. Armstrong, D. L. 1989. Trends Neurosci. **12:** 117–122.
12. Lai, Y., B. Peterson & W. Catterall. 1993. J. Neurochem. **61:** 1333–1339.
13. Woods, M. D., M. J. Shipston, E. L. Mullens & F. A. Antoni. 1992. Endocrinology **131:** 2873–2880.

Modulation of Carbachol Responsiveness by Corticosteroid Hormones in Rat CA1 Pyramidal Neurons

WOUTER HESEN AND MARIAN JOËLS

Graduate School for the Neurosciences
Institute of Neurobiology
Department of Experimental Zoology
Faculty of Biology
University of Amsterdam
Kruislaan 320
1098 SM Amsterdam, The Netherlands

INTRODUCTION

In the CNS two different corticosteroid receptor types have been described: the mineralocorticoid (MR, K_d ~0.3 nM) and the glucocorticoid receptor (GR, K_d ~3–5 nM). Both are present in hippocampal CA1 pyramidal neurons. Their differential affinity for corticosterone, the natural ligand, will have consequences for the activation of relative amounts of MR/GR. At low levels of circulating corticosterone (just before the inactive period) the high-affinity MR will be occupied predominantly, whereas the lower-affinity GR becomes relatively more occupied at increasing levels of corticosterone (circadian peak levels just before the active period or due to stress).[1] In this study we investigated the effect of differential MR- and GR action on cholinergic responsiveness in CA1 pyramidal neurons through differential activation of MR/GR.

METHODS

Adrenalectomized (ADX, 2–6 days before sacrifice) or adrenally intact (control) male Wistar rats (120–180 g) were used. Dorsal hippocampal slices were prepared from and kept in a perfusion system with circulating, carbogenated standard ACSF (32°C). In these ADX-slices, MR and GR will be largely unoccupied due to lack of circulating corticosterone. Selective activation of MR was accomplished by application of the MR ligand aldosterone (3 nM), whereas simultaneous activation of MR and GR was achieved by application of the mixed agonist corticosterone (30 nM). Because of our interest in long-term effects exerted by steroid receptor activation, these hormones were applied to the slices for 20 min and subsequently steroids were washed out for at least 30 min after which cholinergic responsiveness was tested. With microelectrodes, intracellular recordings were made from CA1 pyramidal neurons and effects of bath application of carbachol (Sigma), a stable cholinergic analogue, were studied under current- or voltage clamp (single electrode voltage clamp).

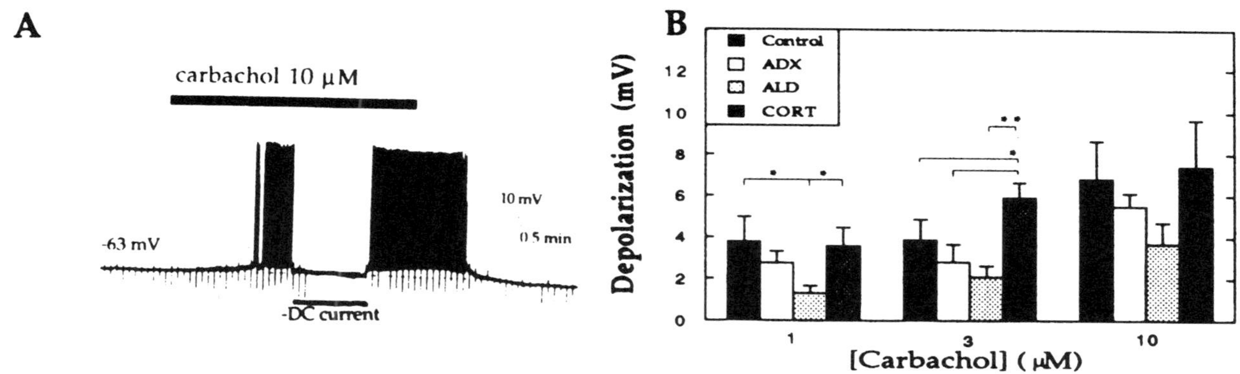

FIGURE 1. **(A)** Typical example of a depolarizing response of a CA1 pyramidal neuron to bath application of carbachol (10 mM). **(B)** Comparison of the carbachol-induced depolarization in the four different experimental groups. (*: $p < 0.05$, **: $p < 0.01$).

RESULTS

Application of carbachol to the perfusion medium induced multiple responses in CA1 pyramidal neurons. Under current clamp, a depolarization (FIG. 1A) was induced, the afterhyperpolarization and accommodation properties were reduced and synaptic responses evoked by stimulation (150 ms, 100–300 μA) of the Schaffer collaterals became attenuated.[2] It appeared that only the depolarization (1 and 3 μM) was modulated by steroids: selective activation of MR (ALD) resulted in a suppression of the depolarization compared to slices in which both MR and GR (CORT) were activated by corticosterone (FIG. 1B).[3] This steroid-mediated effect was also found when carbachol responsiveness was studied under voltage clamp. The carbachol-induced depolarization is then reflected as an inward current shift, which was probably evoked through a reduction of a voltage-independent potassium leak conductance, $I_{k,leak}$.[2] This inward current was clearly reduced after selective activation of MR by aldosterone. No clear steroid-modulation was found for the voltage-activated M-current or the carbachol-mediated attenuation of this current.

DISCUSSION

These experiments show that (apart from other effects), the carbachol-induced depolarization is decreased after selective activation of the MR compared to simultaneous activation of MR and GR. The observations that the inward current induced by carbachol (V_h = −65 mV) is similarly affected and the lack of steroid effects on I_m or the carbachol-induced reduction of I_m point to a selective modulatory effect on the carbachol-induced reduction of $I_{k,leak}$. Previous experiments have shown that under similar experimental conditions activation of MR leads to a reduction of the serotonin ($5HT_{1a}$)-induced hyperpolarization. These and other corticosteroid-mediated effects became apparent only when the membrane potential is hyper- or depolarized since no steroid-effects were observed on resting membrane potential or input resistance.[4] These findings imply that corticosteroids may influence postsynaptic effects of acetylcholine and 5HT on membrane potential through differential activation of MR/GR ratios.

REFERENCES

1. DE KLOET, E. R. 1991. Brain corticosteroid receptor balance and homeostatic control. Front. Neuroendocrinol. **12:** 95–164.
2. MADISON, D. V., B. LANCASTER & R. A. NICOLL. 1987. Voltage clamp analysis of cholinergic action in the hippocampus. J. Neurosci. **7:** 733–741.
3. HESEN, W. & M. JOËLS. 1993. Modulation of carbachol responsiveness in rat CA1 pyramidal neurons by corticosteroid hormones. Brain Res. **627:** 159–167.
4. JOËLS, M. & E. R. DE KLOET. 1992. Control of neuronal excitability by corticosteroid hormones. Trends Neurosci. **15:** 25–30.

Calbindin-D28K and Parvalbumin Immunoreactivity in the Rat Hippocampus following Adrenalectomy and Corticosterone Treatment

HARM J. KRUGERS, RIKJE M. MEDEMA,
FOLKERT POSTEMA, AND JAKOB KORF

Department of Biological Psychiatry
and
Center for Behavioral, Cognitive and Neuro-Sciences
University of Groningen
P.O. Box 30 001
Groningen 9700, The Netherlands

Several lines of evidence suggest effects of corticosteroids on calcium homeostasis in hippocampal neurons.[1,4,6] Intracellular [Ca^{2+}] are maintained by plasma membrane extrusion systems (Ca^{2+}-ATPase and Na^{+}-Ca^{2+} exchanger), organellar sequestration, and calcium binding proteins (CaBP) like calbindin-D28K (CB) and parvalbumin (PV).[7,8,10]

An increase in hippocampal CBmRNA and CB-immunoreactivity in the dentate gyrus following CORT treatment has been reported, whereas adrenalectomy decreased hippocampal CBmRNA and CB-immunoreactivity in the dentate gyrus suggesting a role for corticosteroids in the regulation of CB.[2] The aim of this study was: i) compare the immunocytochemical distribution of CB and PV in the dentate gyrus following ADX, with hippocampal granule cell degeneration following ADX (using silver impregnation,[3,5,11] ii) describe the anatomical distribution of CB and PV immunoreactivity following ADX and CORT treatment in the hippocampus, and iii) investigate whether replacement with low dosages of CORT in ADX animals alter CB and PV immunoreactivity in the rat hippocampus. Male Wistar rats (200–250 g) were bilaterally adrenalectomized (ADX) and maintained on 0.9% NaCl, or sham adrenalectomized (sADX). Four ADX animals were allowed to survive three days, five were allowed to survive three weeks. Four intact animals were daily injected subcutaneously with 10 mg corticosterone, whereas four animals were only injected with vehicle (sesam oil). Ten ADX animals were allowed to survive three weeks from which five received low dosages of corticosterone replacement in their drinking solution (10 mg/L). Four intact and five ADX animals were allowed to survive 24 hours. Animals were perfused, sections were cut out and prepared for silver impregnation[5] and immunostaining. Optical density of CB-immunoreactivity was quantified using an automized imaging analysis system (IBAS). The number of PV-stained neurons and dendrites in the DG and CA1 area were counted. Four hippocampi per animal were analyzed (I.A. 5.7[9]). Statistical analysis was performed using the Mann-Whitney U test, significance level was $p < 0.05$.

Argyrophilic somata appeared in the dentate granule cell layer three days following ADX. Neuritic argyrophilia appeared in the molecular layer of the DG and mossy fiber layer three days following ADX. Degenerating cells outside the DG

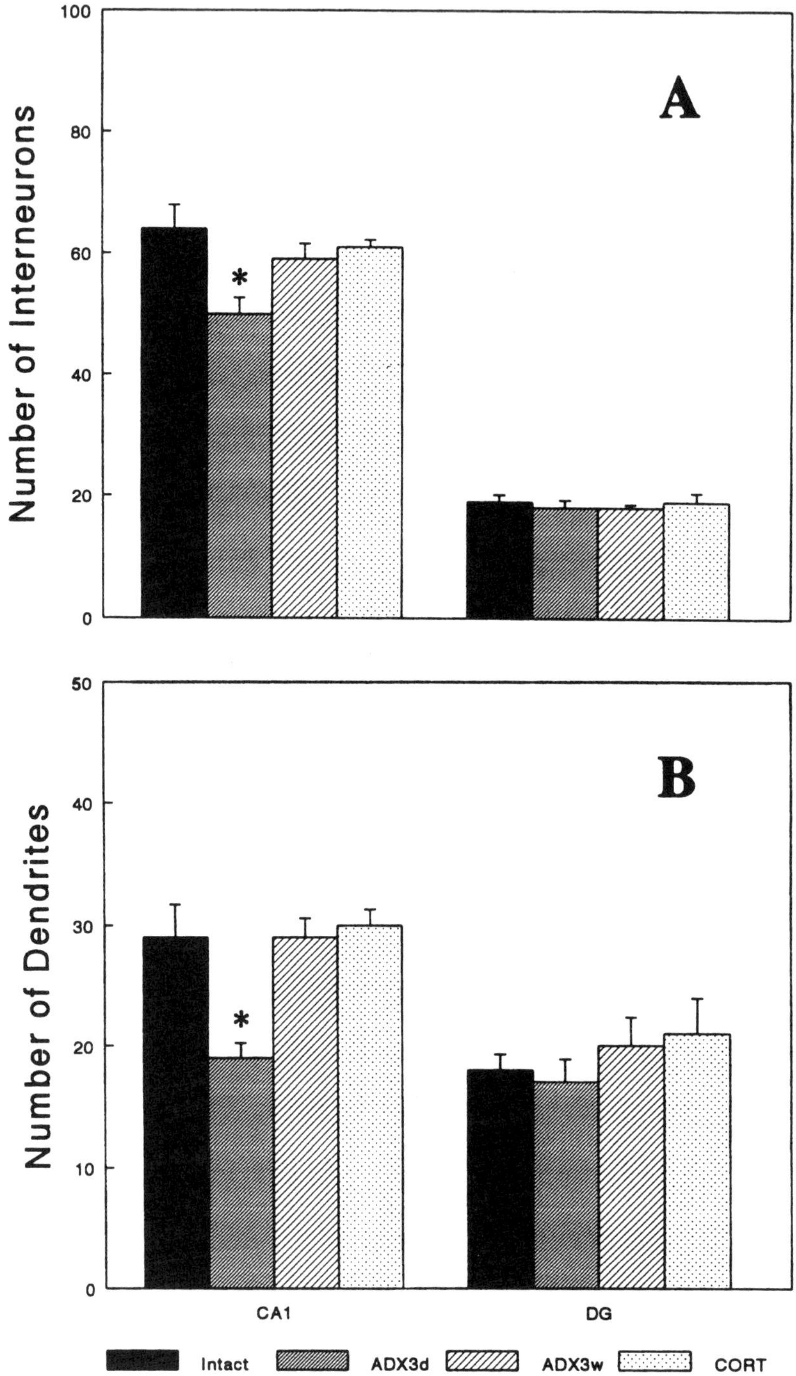
A
Number of Interneurons
100
80
60
40
20
0
*
B
Number of Dendrites
50
40
30
20
10
0
*
CA1
DG
Intact
ADX3d
ADX3w
CORT

FIGURE 1. Parvalbumin (PV) immunoreactive neurons (**A**) and dendrites (**B**) in the CA1 and dentate gyrus (DG) of the hippocampus (I.A. 5.7[9]) in intact rats, three days adrenalectomized rats (ADX3d), three weeks adrenalectomized rats (ADX3w) and rats treated three weeks with corticosterone (CORT, 10 mg/day). Note the reduction of PV immunoreactive neurons and dendrites 3 days following ADX, which is restored three weeks following ADX.

were only found in the CA3c area and were few in number. These changes were absent at 24 hours following ADX and clearly more abundant three weeks following ADX. Degeneration following ADX showed a crest-to-tip but also a medial-to-dorsal gradient. The number of degenerating cells and dendrites at a given time point varied between animals. CORT (10 mg/L) replacement prevented the degenerative changes. No degenerative changes in the hippocampus were found three weeks following daily CORT treatment (10 mg/animal, s.c.). PV immunoreactivity was present in interneurons and their dendrites throughout the different hippocampal subfields. No changes in PV immunoreactivity were found in the dentate gyrus 24 hours, 3 days or three weeks following ADX. A significant reduction of PV immunostained interneurons and their dendrites in the CA1 area was found three days following ADX, which was not present 24 hours or three weeks following ADX (FIG. 1). Neither three weeks CORT treatment in intact nor CORT replacement in ADX animals affected PV immunoreactivity in the CA1 area or DG. A reduction of

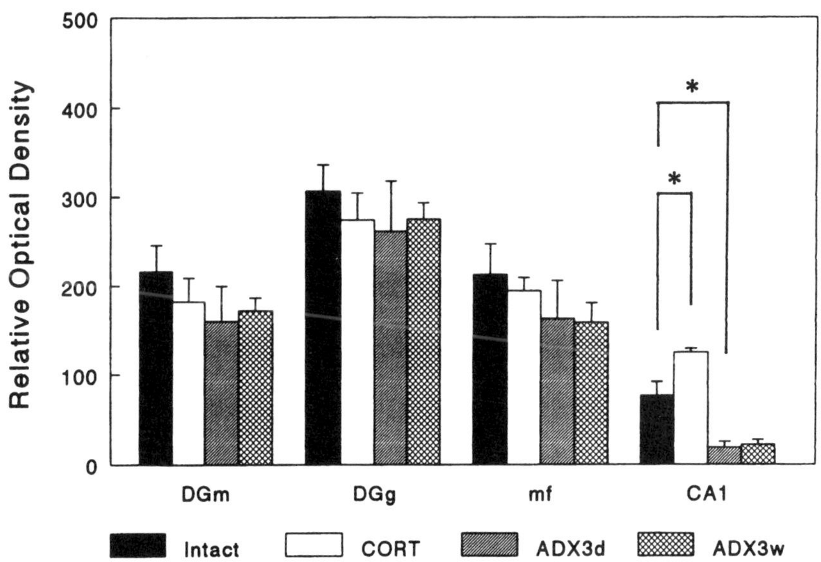

FIGURE 2. Calbindin-D28K immunoreactivity (CB-ir) in the dentate gyrus molecular layer (DGm), granular layer (DGg), mossy fiber layer (mf) and CA1 area. Rats were intact, three weeks treated with corticosterone (CORT, 10 mg/day), three days adrenalectomized (ADX3d) or three weeks adrenalectomized (ADX3w). Three weeks treatment with CORT increased CB-ir in the CA1 area, whereas both 3 days and 3 weeks following ADX a reduction of CB-ir in the CA1 area was observed.

CB-immunoreactive cells was seen in the DG granular layer three days following ADX. This reduction was not seen 24 hours following ADX, but was more pronounced three weeks following ADX. The reduction of CB-ir cells in the DG showed a crest-to-tip and a dorsal-to-medial gradient. Three days and three weeks following ADX a decrease of CB-ir in the pyramidal cell layer of CB-stained dendrites in the CA1 area was found (FIG. 2). This reduction was not present 24 h following ADX. CORT replacement prevented the loss of CB-stained DG granule cells in the DG. Three weeks treatment with high dosages of CORT (10 mg/day) did not affect CB-ir in the DG molecular layer, DG granular layer and mossy fiber layer. Three weeks treatment with CORT (10 mg/day) increased CB-ir in the CA1 area significantly. CORT replacement with low dosages of CORT (10 mg/L) in ADX animals increased CB-ir in the CA1 area. We conclude that i) there is a close relationship between the loss of CB-immunoreactive cells in the DG and the neuronal degeneration in the dentate gyrus following ADX, both in time and space, ii) corticosterone seems to be involved in the regulation of calbindin-D28K in the CA1 area of the hippocampus, and iii) a transient effect on parvalbumin immunoreactive (inter)neurons and their dendrites in the CA1 area is seen following ADX.

REFERENCES

1. ELLIOT, E. M. & R. M. SAPOLSKY. 1993. Brain Res. **602:** 84–90.
2. IACOPINO, A. M. & S. CHRISTAKOS. 1990. J. Biol. Chem. **265:** 10177–10180.
3. JAARSMA, D., F. POSTEMA & J. KORF. 1992. Hippocampus **2:** 143–150.
4. KERR, D. S., L. W. CAMPBELL, S. HAO & P. W. LANDFIELD. 1989. Science **245:** 1505–1509.
5. KRUGERS, H. J., R. M. MEDEMA, F. POSTEMA & J. KORF. Hippocampus. In press.
6. LANDFIELD, P. W., O. THIBAULT, M. L. MAZZANTI, N. M. PORTER & D. S. KERR. 1992. J. Neurobiol. **23:** 1247–1260.
7. LLEDO, P. M., B. SOMASUNDARAM, A. J. MORTON, P. C. EMSON & W. T. MASON. 1992. Neuron **9:** 943–954.
8. MILLER, R. J. 1991. Prog. Neurobiol. **37:** 255–285.
9. PAXINOS, G. & C. WATSON. 1982. The Rat Brain in Stereotaxic Coordinates. Academic Press. New York.
10. SCHARFMAN, H. E. & P. A. SCHWARTZKROIN. 1989. Science **246:** 257–260.
11. SLOVITER, R. S., G. VALIQUETTE, G. M. ABRAMS, E. RONK, A. SOLLAS, L. A. PAUL & S. NEUBORT. 1989. Science **243:** 535–538.

Central and Systemic Corticosterone Differentially Affect Dopamine and Norepinephrine in the Frontal Cortex of the Awake Freely Moving Rat

D. N. THOMAS, R. M. POST, AND A. PERT

Biological Psychiatry Branch
National Institute of Mental Health
Building 10, Room 3N 212
Bethesda, Maryland 20892

There is evidence to suggest important modulatory control of brain amine function by corticosteroids.[1] The perikaryal regions of all three major brain amine systems contain surprisingly high concentrations of glucocorticoid receptors.[2] Furthermore, corticosterone (CORT) and dexamethasone have both been found to alter brain amine metabolism and turnover.[3,4] The purpose of this study was to evaluate the effects of corticosterone on brain amine overflow in the frontal cortex (FCx) of the awake freely moving rat using *in vivo* microdialysis.

Sprague-Dawley rats (250–350 g) were stereotaxically implanted with dialysis probe guide cannulae aimed for the FCx (AP: +12.7 mm, DV: +7.3 mm, Lat: +0.7 mm relative to IAL). In subsequent groups, rats were additionally implanted with bilateral microinjection cannulae aimed for the ventral tegmental area (VTA) or locus coeruleus (LC). Seven days following recovery dialysis probes were introduced into the FCx. The animals were tethered to a liquid swivel and allowed to move freely in a Plexiglas chamber whilst simultaneously being dialyzed. Basal sampling commenced 1–2 hrs post implantation. Dialysis samples were collected every 15 min and analyzed for dopamine (DA) and norepinephrine (NE) content. Drugs were administered when 3 consecutive samples did not differ by more than 10%. CORT (5 mg/kg) or vehicle (Tween 40) were administered s.c. In the rats with microinjection cannulae guides, corticosterone 21-hemisuccinate (CORT-21) bound to BSA (70 μg) was administered into the VTA or LC in a volume of 1 μl of saline. Sampling was continued for 2 hrs following drug administration.

The systemic application of CORT produced transient increases in FCx NE (FIG. 1) and prolonged decreases in DA (FIG. 2). Focal application of CORT-21 into the perikaryal regions was without significant effect, suggesting that the effects of systemically administered CORT on NE and DA function are mediated through actions at nerve terminals or through indirect mechanisms. The relatively immediate effect on NE might suggest a direct action of CORT at the level of the membrane, whilst the DA effect could be mediated through a genomic or some other more persistent mechanism.[1]

It is likely that CORT-induced alterations in frontal cortex catecholamine function may be relevant to normal and pathological states associated with stressors known to enhance glucocorticoid output. The relationship of the CORT-induced increases in NE to anxiety and pathological versus adaptive aspects of the "stress response" deserve further study. Although stress has consistently been found to induce immediate elevations in frontal cortex DA function,[5] the progressive decrease in DA output seen in this study following CORT may reflect the more persistent

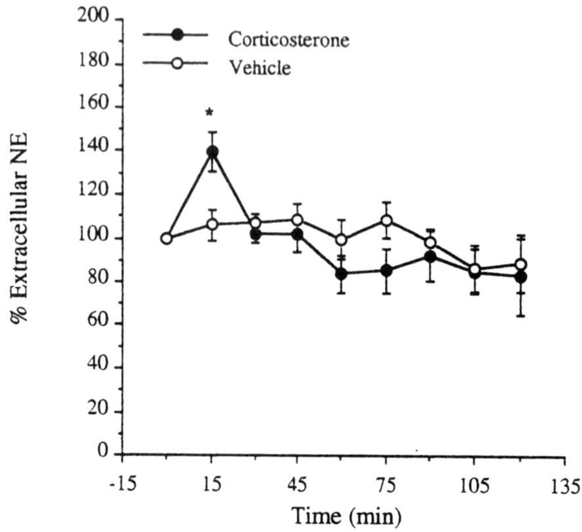

FIGURE 1. Effects of systemic corticosterone (5 mg/kg) or vehicle (Tween 40) on extracellular norepinephrine in the frontal cortex of the awake freely moving rat. Corticosterone or vehicle were administered at time zero. All values are mean + SEM (n = 6). Basal extracellular norepinephrine levels were 3.74 ± 0.52 pg/sample. Statistical analysis using two-way ANOVA with repeated measures showed a significant time × drug interaction (p = 0.04). Post hoc comparisons showed a significant difference (* p = 0.013) at t = 15 min.

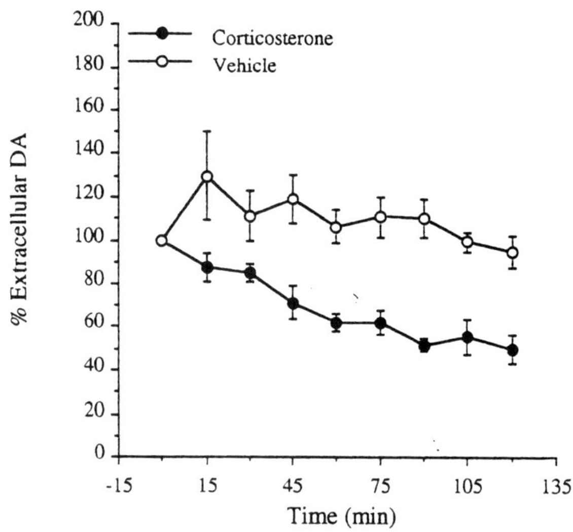

FIGURE 2. Effects of systemic corticosterone (5 mg/kg) or vehicle (Tween 40) on extracellular dopamine in the frontal cortex of the awake freely moving rat. Corticosterone or vehicle were administered at time zero. All values are mean ± SEM (n = 6). Basal extracellular dopamine levels were 0.88 ± 0.1 pg/sample. Statistical analysis using two-way ANOVA with repeated measures showed a significant drug effect (p = 0.011) and a significant time × drug interaction (p = 0.04).

consequences of such input; the importance of these changes for behavior remain to be determined.

REFERENCES

1. McEwen, B. S. 1987. Biochem. Pharmacol. **36:** 1755–1763.
2. Härfstrand, A., K. Fuxe, A. Cintra, L. F. Agnati *et al.* 1986. Proc. Natl. Acad. Sci. USA **83:** 9779–9783.
3. Iuvone, P. M., J. Morasco & A. J. Dunn. 1977. Brain Res. **120:** 571–576.
4. Rastogi, R. B. & R. L. Singhal. 1978. Br. J. Pharmacol. **62:** 131–136.
5. Dunn, A. J. 1988. Ann. N.Y. Acad. Sci. **537:** 188–205.

Dexamethasone Does Not Prevent Apoptosis in the Hippocampus of the Rat[a]

C. D. HORNSBY AND E. R. de KLOET

Division of Medical Pharmacology
Leiden/Amsterdam Center for Drug Research (LACDR)
University of Leiden
P.O. Box 9503
2300 RA Leiden, The Netherlands

INTRODUCTION

It has been known for some time that adrenalectomy (ADX) of young adult rats results in degeneration of the hippocampus.[1] This degeneration is most obvious in the dentate gyrus, and has morphological features, consistent with those seen in apoptosis.[2] It would appear that the mineralocorticoid receptor (MR) plays a pivotal role in this process, as administration of aldosterone, a MR agonist,[3] or corticosterone (CORT) at low doses to ADX animals, can prevent the degeneration.[1] It has additionally been reported, that the GR may also be involved in regulation of dentate gyrus cell survival, as ADX animals which received RU28362, a GR agonist, displayed a lower number of apoptotic cells than vehicle-treated animals.[3] Additionally, unilateral, intrahippocampal infusion of dexamethasone (DEX; GR agonist) has been shown to result in reduced numbers of apoptotic cells only on the infusion side.[4] The aim of the following experiments was to further investigate the mechanism of ADX-induced apoptosis.

METHODS

Male Wistar rats (160 g, Leiden) were subject to ADX and simultaneously implanted subcutaneously with a 100 ± 5-mg cholesterol pellet, containing either 20% or 100% CORT; or, 2% or 25% DEX. Vehicle control (VEH) animals received pellets containing only cholesterol. All operations were completed before 10.00 a.m. on day 1. Animals were subsequently housed in rooms maintained on a 12/12 hours light/dark cycle (lights on at 7.00 a.m.) and allowed free access to food, water and 0.9% saline. On day 7, the animals were deeply anesthetized with pentobarbitone sodium (30 mg/100 g body weight, i.p.) and perfused intracardially with 0.9% saline, followed by ice cold buffered 4% paraformaldehyde. The brains were left *in situ* for 1 hour and then dissected and post-fixed in buffered 4% paraformaldehyde for at least 24 hours. Thirty micron coronal sections were taken and systematically sampled. The sections were then stained either with the Nissl stain, thionine, or the DNA stain, Hoechst 33342. As the phenomenon of apoptosis is difficult to quantify, each section was analyzed purely for the presence or absence of apoptotic cells or nuclei. Plasma CORT concentrations were determined by radioimmunoassay (RIA)

[a]C. D. Hornsby is supported by the Royal Society of Great Britain.

of blood collected immediately before perfusion. Additionally, the thymus gland from each animal was dissected and weighed, and the percentage change in body weight between days 1 and 7 was noted.

RESULTS AND DISCUSSION

The results of the present study show that administration of glucocorticoids to ADX animals resulted in atrophy of the thymus gland and an altered body weight gain over the 7-day period compared to VEH-treated ADX animals (TABLE 1). It has been reported that the thymus gland undergoes apoptosis in response to glucocorticoids.[5] Accordingly, the effects were most pronounced in the DEX-treated animals, the mean thymus weight and % body weight change between days 1 and 7 being 100 ± 14 mg and −19 ± 1.3%, respectively, in 2% DEX animals and 157 ± 45 mg and −22 ± 2.4%, respectively, in 25% DEX animals, whilst in animals receiving 100% CORT pellets, the mean thymus weight was 233 ± 40 mg, and the change in body weight was −2.2 ± 6.7%. Treatment of ADX animals with VEH or 20% CORT had little effect on these parameters, mean thymus weights being, 752 ± 31 mg and 766 ± 64 mg, respectively, and mean body weight gains being +9.1 ± 3.4% and +20 ± 2.1%, respectively.

Light microscopic examination of the brain sections revealed the presence of cells with apoptotic morphology, *i.e.,* shrunken nuclei with condensed chromatin (FIG. 1) in the dentate gyrus of animals in the VEH group (plasma CORT <0.15 μg/dl, TABLE 1). No such cells could be seen in any other hippocampal subfield. ADX animals implanted with 100% CORT pellets had plasma CORT concentrations which would enable occupation of both MR and GR[6] (18 ± 0.7 μg/dl), and no degenerating cells could be detected in the hippocampus of animals in this group. The fact that degeneration was also not apparent in the hippocampus of animals implanted with 20% CORT pellets (resulting in plasma CORT concentrations expected to bind predominantly to MR[6], 0.95 ± 0.08 μg/dl) would suggest that occupation of MR is an important protective factor. In further support of this theory, ADX animals implanted with either 2% or 25% DEX pellets were not able to withstand the ADX-induced effect, and degenerating cells were seen in the dentate gyrus of all animals in these two groups.

TABLE 1. The Presence (+) or Absence (−) of ADX-Induced Apoptosis in the Dentate Gyrus of the Rat, after Replacement Treatment with Cholesterol Pellets Containing 20% or 100% Corticosterone (CORT) or 2% or 25% Dexamethasone (DEX)[a]

Experimental Group	n	Apoptosis	Plasma CORT Concentration (μg/dl)	Thymus Weight (mg)	% Change in Body Weight
ADX + VEH	4	+	<0.15	752 ± 31	+9.1 ± 3.4
ADX + 20% CORT	6	−	0.95 ± 0.08	766 ± 64	+20 ± 2.1
ADX + 100% CORT	5	−	18 ± 0.7	233 ± 40	−2.2 ± 6.7
ADX + 2% DEX	4	+	<0.15	100 ± 14	−19 ± 1.3
ADX + 25% DEX	6	+	<0.15	157 ± 45	−22 ± 2.4

[a]Vehicle-treated animals (VEH) were implanted with pellets containing only cholesterol. The effects of these treatments on thymus gland and change in body weight over the 7-day period are also shown. n = number of individual animals.

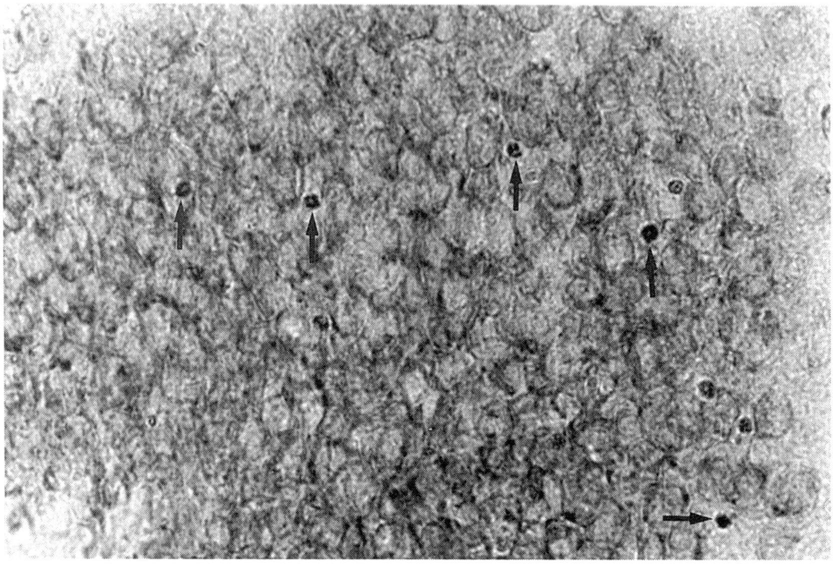

FIGURE 1. Photograph showing thionine-stained dentate gyrus cells in a representative brain section taken from an adrenalectomized animal, implanted subcutaneously with a 25% dexamethasone pellet. Cells with apoptotic morphology are indicated by *arrows* (× 80).

CONCLUSION

From the results of the present study, we conclude that occupation of GR alone, a nonphysiological condition, is not sufficient to prevent ADX-induced degeneration. Instead, it would appear that occupation of MR, either alone or in combination with GR, is the critical factor in the prevention of dentate gyrus cell death.

REFERENCES

1. Sloviter, R. S., G. Valiquette, G. M. Abrams, E. Ronk, A. Sollas, L. A. Paul & S. Neubort. 1989. Science **243:** 535–538.
2. Sloviter, R. S., E. Dean & S. Neubort. 1993. J. Comp. Neurol. **330:** 337–351.
3. Woolley, C. S., E. Gould, R. R. Sakai, R. L. Spencer & B. S. McEwen. 1991. Brain Res. **554:** 312–315.
4. Jaarsma, D., F. Postema & J. Korf. 1992. Hippocampus **2**(2): 143–150.
5. Compton, M. M. & J. A. Cidlowski. 1986. Endocrinology **118:** 38–45.
6. Reul, J. M. H. M. & E. R. de Kloet. 1985. Endocrinology **117**(6): 2505–2511.

Glucocorticoid Receptor-Mediated Repression of GnRH Gene Expression in a Hypothalamic GnRH-Secreting Neuronal Cell Line

D. B. DeFRANCO,[a] B. ATTARDI[b] AND U. R. CHANDRAN[c]

Departments of [a]Biological Sciences and [b]Medicine
University of Pittsburgh
5th and Ruskin Streets
Pittsburgh, Pennsylvania 15260

The synthesis and release of GnRH within a specific subset of neurons in the hypothalamus, which serves as the primary drive to the hypothalamic-pituitary-gonadal (HPG) axis, is subject to various levels of control. Although a number of direct synaptic connections to GnRH-containing neurons have been identified, which presumably provide some regulatory inputs, the mechanisms responsible for hormonal regulation of GnRH synthesis and release mediated by either cell surface or intracellular receptors remain controversial. The recent demonstration that a subset of GnRH-containing neurons in the rat hypothalamus possesses immunoreactive glucocorticoid receptors (GR)[1] implies that this class of steroid hormones could exert a direct effect to regulate the functioning of these neurons and perhaps the HPG axis. We used the GT1-3 and GT1-7 cell lines of immortalized GnRH-secreting hypothalamic neurons[2] as a model to study the direct effects of glucocorticoids on GnRH gene expression.

Using Western blot analysis, we demonstrated that the GT1-3 and GT1-7 cell lines contain GR protein.[3] *In vitro* hormone binding assays using [^{3}H]dexamethasone, a synthetic glucocorticoid that binds with high affinity and selectivity to GR, show that the Kd of binding is ~2–3 nm. GR functions as a ligand-inducible transcription factor in the GT1 cell lines, since it is able to activate transcription from various glucocorticoid-responsive promoter-linked CAT (chloramphenicol acetyltransferase) plasmids transiently transfected into the GT1 cells. In the presence of dexamethasone, Grs in GT1 cells are also able to *repress* transcription from the rat GnRH promoter-linked to the CAT gene (FIG. 1) and this repression is overcome by treatment with the glucocorticoid antagonist, RU486. Northern blot analysis reveals that the Grs are also able to repress the endogenous GnRH messenger RNA level. To localize sequence elements within the GnRH promoter that are responsible for GR-mediated transcriptional repression, the effects of dexamethasone were assessed on a series of reporter plasmids containing 5′-flanking deletions of the GnRH promoter. FIGURE 2 shows the effects of dexamethasone on promoter activity of a series of GnRH-luciferase fusion genes. Luciferase activity was repressed about 40% by dexamethasone with the longest promoter fragments (-3001 and -2281 bp), 60% with the -1580, -1014, and -459 bp deletions and 25% with -200 deletion. It was not possible to conclude whether the shortest promoter fragment tested (−154) retained any significant sensitivity to dexamethasone due to the low basal activity and large standard error. Thus, dexamethasone repression of GnRH promoter activity appears to be mediated by sequence contained with the promoter proximal 459 bp, possibly

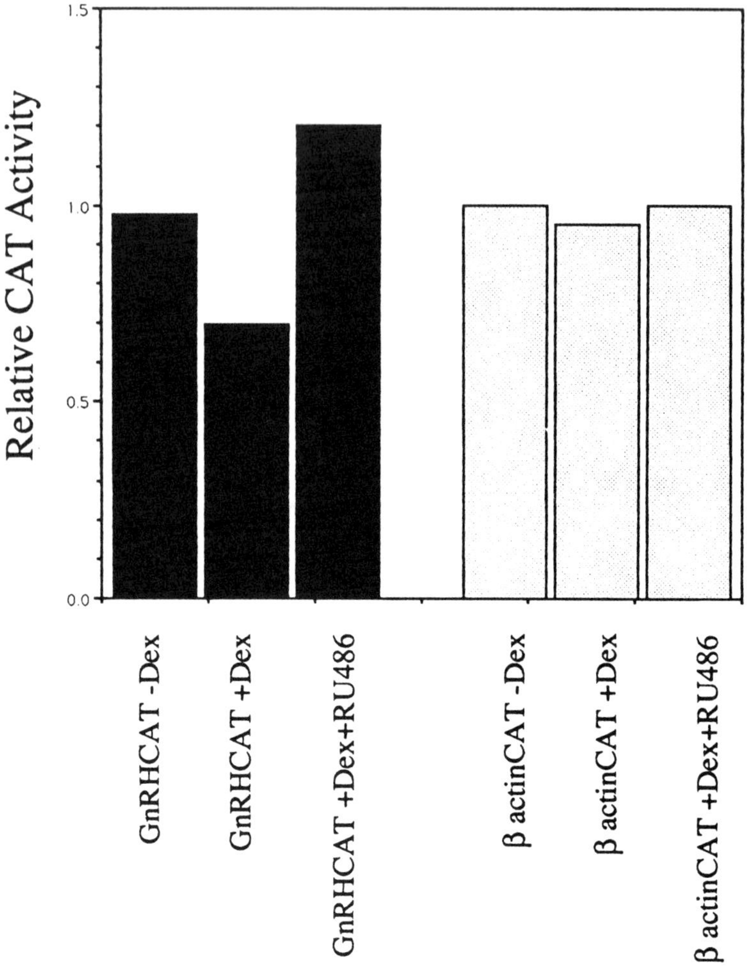

FIGURE 1. Effects of dexamethasone and RU486 on rat GnRH and human β-actin promoters in transiently transfected GT1-3 cells. 10 μg of reporter plasmids that consist of the CAT gene fused to either 3.0 kb of the rat GnRH promoter (GnRHCAT) or human β-actin promoter (β-actinCAT) were transiently transfected into GT1-3 cells. Cells were washed 12 h after transfection and incubated in medium alone, medium containing 10^{-7} M dexamethasone (Dex), or 10^{-7} M Dex and 10^{-6} M RU486. Cells were harvested 24 h later and whole cell extracts prepared and assayed for CAT activity. The effect of hormone treatment on both promoters is expressed relative to the activity from cells that were not treated with hormone. Data shown are the mean of nine experiments for GnRHCAT and three experiments for β-actinCAT ± SEM. 31% repression of GnRH promoter activity by Dex was judged to be statistically significant ($p < 0.005$, n = 9), while β-actin promoter activity was not significantly affected by Dex or Dex and RU486 ($p > 0.5$). The reversal of Dex repression by RU486 was judged to be statistically significant ($p < 0.001$, n = 9).

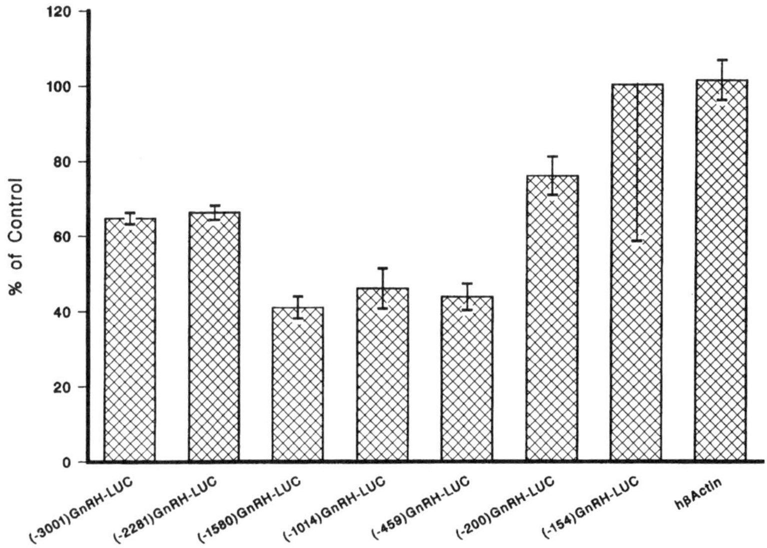

FIGURE 2. Effects of dexamethasone on various 5′-deletions of the GnRH promoter or human β-actin promoter. The indicated 5′-deletions of the GnRH promoter or human β-actin promoter were subcloned into a promoterless luciferase vector. GT1-7 cells were transfected with 10 μg of reporter plasmids, washed after 4 h, and incubated in medium alone or medium containing 10^{-6} M dexamethasone (Dex). Cells were harvested after 20 h and assayed for protein content. Luciferase activity was measured in cell lysates containing 40–160 μg protein. The results shown are the mean ± SEM of triplicate dishes in 2–4 independent experiments and represent the % luciferase activity in dexamethasone-treated cultures compared to the corresponding cultures.

with additional elements between -459 and -200, and not to be influenced by the relative basal activity of the GnRH promoter.

REFERENCES

1. Ahima, R. S. & R. E. Harlan. 1992.Clucocorticoid receptors in LHRH neurons. Neuroendocrinology **56:** 845–850.
2. Mellon, P. L., J. J. Windle, P. C. Goldsmith, C. A. Padula, J. L. Roberts & R. I. Weiner. 1990. Immortalization of hypothalamic GnRH neurons by genetically targeted tumorigenesis. Neuron **5:** 1–10.
3. Chandran, U. R., B. Attardi, R. Friedman, K-W. Dong, J. L. Roberts & D. B. DeFranco. 1994. Endocrinology **134:** 1467–1474.

Effect of Dexamethasone and Forskolin on Neurotensin Expression in Hypothalamic Cells in Culture

V. SCARCÉRIAUX, D. PÉLAPRAT, P. FORGEZ,
A. M. LHIAUBET, AND W. ROSTÈNE

INSERM U339
Hôpital Saint Antoine
184 rue du faubourg Saint Antoine
75571 Paris Cédex 12, France

INTRODUCTION

Neurotensin (NT), a tridecapeptide isolated from bovine hypothalamus,[1] is present in many brain regions with the highest concentrations found in the hypothalamus.[2] In this region, NT was shown to influence the secretion of several pituitary hormones such as thyrotropin, growth hormone and prolactin.[3] The presence on NT gene of inducible *cis*-regulatory sequences highly homologous to glucocorticoid response and cAMP response elements (GRE and CRE, respectively) was recently reported.[4] These data suggested that NT expression could be regulated by glucocorticoids and adenylate cyclase activators. However, whereas such regulations had already been shown in transformed cell lines such as PC12 cells[5] or C cells,[6] they were not yet demonstrated in untransformed neurons.

The aim of the present work was to study the regulation of NT by dexamethasone (dex) and forskolin (fsk) in hypothalamic primary cultures.

METHODS

Neuronal cells were prepared from hypothalami of 17-day-old Wistar rat embryos and grown in serum-free medium as previously described.[7] After 10 days in culture, different agents such as dex, fsk or RU 38486 (glucocorticoid antagonist) were added in the serum-free medium at a concentration of 1 μM, for 48 hours.

NT content was determined by radioimmunoassay and protein content was measured by the Bradford Coomassie blue method. Results were expressed in pmoles of NT per mg of protein.

Total RNA was isolated from hypothalamic cultures by using RNAzol (Bioprobe) for Northern hybridization using a ^{32}P-labeled NT cDNA probe and a β actin cDNA probe.

Statistical analysis (mean ± SEM) was performed by Dunnett *t* test.

RESULTS AND DISCUSSION

In a first step, the study of NT regulation was based upon the measurement of the peptide (FIG. 1). After 10 days in culture, a 48-hour treatment with 1 μM dex induced a 100% increase of hypothalamic NT, which was abolished by RU38486, used at the

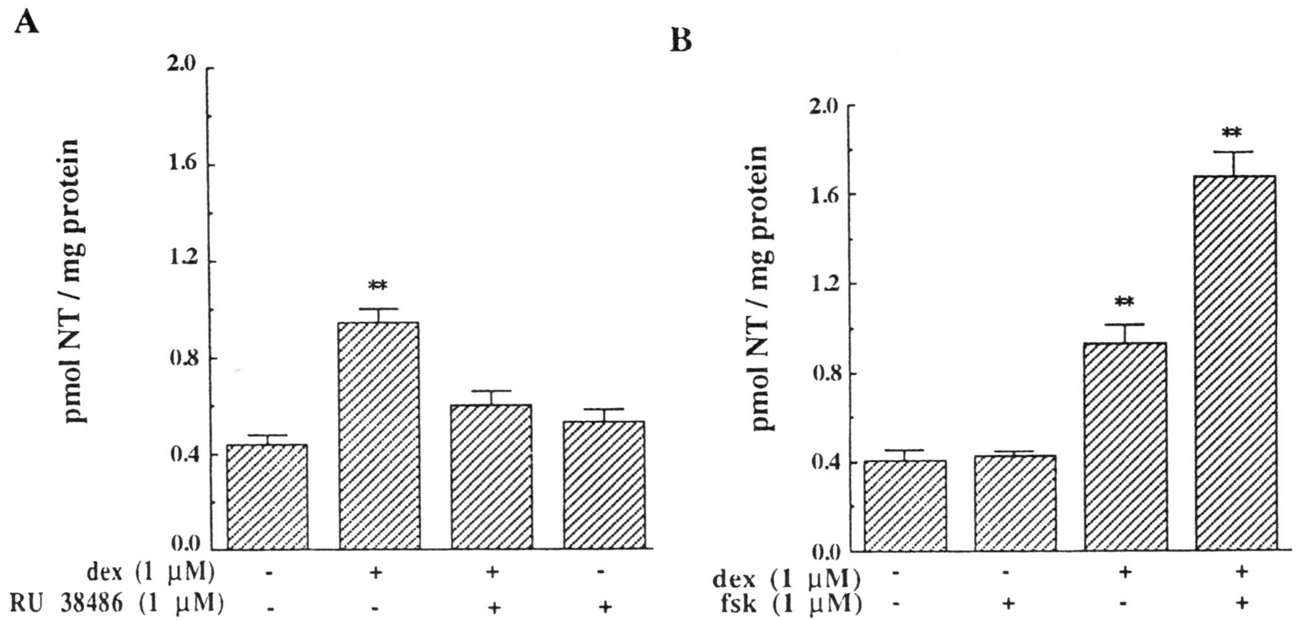

FIGURE 1. Determination of NT content in hypothalamic cultures after 48 hours of treatment with different agents. **(A)** Inhibition of the effect of 1 μM dex by 1 μM RU 38486. **(B)** Potentiation of the effect of 1 μM dex by 1 μM fsk. Values are expressed as the mean ± SEM of 6 different cultures. **: p <0.01.

same concentration as dex, suggesting that the NT increase induced by dex was mediated by glucocorticoid receptors (FIG. 1A). Surprisingly, fsk alone had no effect, whereas it induced an increase in NT levels in PC12 cells.[5] However, in presence of fsk and dex, a 300% increase was obtained, suggesting a potentiation of the dex effect by fsk (FIG. 1B). A similar result was already found in PC12 cells, and explained by a cooperativity between the different response elements in the NT promoter.

In a second step, the study was based upon Northern blot experiments. FIGURE 2 shows that in hypothalamic cultures NT gene was transcribed in two mRNAs of 1.0- and 1.5-kb sizes, similarly to what was observed in hippocampal tissue. Moreover, an increase of NT mRNA levels was shown when hypothalamic cultures were treated with fsk and dex simultaneously.

These findings show that the regulation of NT can be studied in hypothalamic cultures. Our results indicate that dex increased NT production and that fsk, which had no effect alone, could potentiate the action of dex. The lack of effect of fsk alone, observed in hypothalamic cultures, compared with the activity of this agent previously demonstrated in cell lines, shows that the NT levels may be differently regulated in transformed cell lines and in primary neuronal cultures.

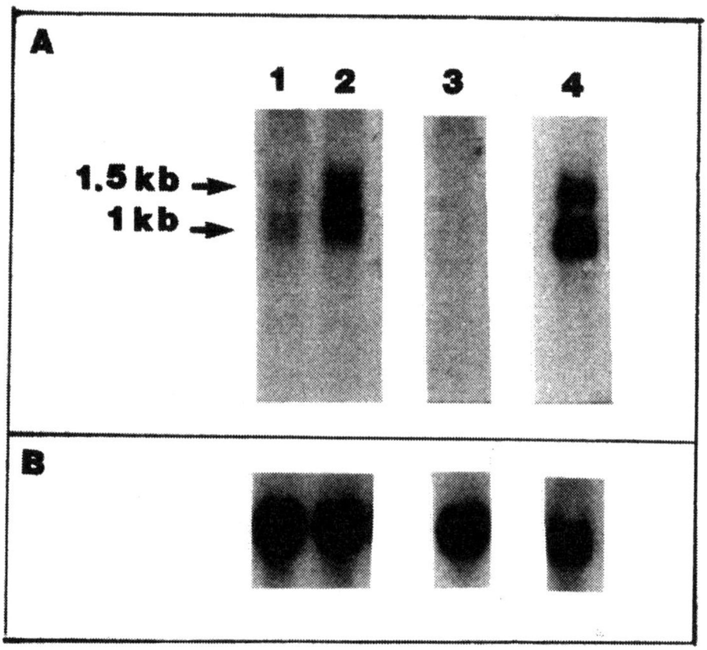

FIGURE 2. Northern blot hybridization of the NT probe **(A)** to total RNA extracted from hypothalamic cultures; control cultures (*lane* 1, 30 μg) and cultures treated with 1 μM dex and 1 μM fsk for 48 hours (*lane* 2, 30 μg). To test the selectivity of the probe, total RNA was extracted from cerebellum tissue as a negative control (*lane* 3, 30 μg), or from hippocampal tissue as a positive control (*lane* 4, 15 μg). After washes, blots were used again for hybridization with the β actin probe **(B).**

REFERENCES

1. CARRAWAY, R. E. & S. E. LEEMAN. 1973. J. Biol Chem. **248:** 6854–6861.
2. UHL, G. R., M. J. KUHAR & S. H. SNYDER. 1977. Proc. Natl. Acad. Sci. USA **74:** 4059–4063.
3. NEMEROFF, C. B. & S. T. CAIN. 1985. Trends Pharmacol. Sci. **6:** 201–205.
4. KISLAUKIS, E. & P. R. DOBNER. 1990. Neuron **4:** 783–795.
5. TISCHLER, A. S., Y. C. LEE, D. COSTOPOULOS, V. W. SLAYTON, W. J. JASON & S. R. BLOOM. 1986. J. Neurosci. **6:** 1719–1725.
6. ZEYTIN, F. N., S. RUSK & S. E. LEFF. 1987. Endocrinology **121:** 361–370.
7. FAIVRE-BAUMAN, A., J. PUYMIRAT, C. LOUDES & A. TIXIER-VIDAL. 1984. Methods for Serum-Free Culture of Neuronal and Lymphoid Cells. 36–57. Alan R. Liss. New York.

Glucocorticoid Stimulation of Sodium Channel Expression in Cultured Astrocytes

P. J. YAROWSKY,[a] D. S. BROUGHER,[b]
AND B. K. KRUEGER[b]

*Departments of [a]Pharmacology and [b]Physiology
University of Maryland School of Medicine
655 West Baltimore Street
Room 4-023 BRB
Baltimore, Maryland 21201*

Glucocorticoid hormones affect brain development and modulate the expression of a variety of genes, including genes for growth factors, growth factor receptors, enzymes (glutamine synthetase and glycerol-3-phosphate dehydrogenase), and ion channels.[1,2] It is now apparent that during development, voltage-gated Na channels are expressed in a number of cell types in addition to neurons and muscle cells. In particular astrocytes from neonatal optic nerve, hippocampus, and cerebral cortex[3,4] express several Na channel types. The role of these Na channel types in development and the role of glucocorticoid hormones in their expression has not been determined. Previously, we found that there are two types of saxitoxin (STX)-sensitive Na channels in astrocytes, high STX affinity and low STX affinity.[5] We now report that hydrocortisone increases high STX affinity Na channels via stimulation of the type II glucocorticoid receptor (GR). This glucocorticoid stimulation of Na channel expression occurs in a time-dependent and dose-dependent fashion. This effect is mimicked by the GR agonist dexamethasone and inhibited by the GR antagonist RU 38486. This regulation of Na channel expression may be important to many cellular processes including the differentiation of astrocytes during development.

Astrocyte cultures were prepared from the brains of one-day-old rats as described previously.[5] After 4 DIV, the culture medium was changed to either cortisol-free or cortisol-containing basal medium which includes DMEM/F12 (1:1) supplemented with 5 ng/ml biotin, 30 nM selenium, 50 μg/ml transferrin, 1.5 μg/ml fibronectin, and 5 μg/ml insulin. RU38486 was generously supplied by Roussel Uclaf, (Romainville, France). Corticosteroids were purchased from Sigma (St. Louis, MO). The primary antibodies utilized were: polyclonal 7493, generated against purified rat brain sodium channel protein[6] (a kind gift of K. Angelides, Baylor College of Medicine) and polyclonal guinea pig GFAP (Advanced Immunological Services, Los Angeles, CA).

We previously found that switching to serum-free, chemically defined medium containing hydrocortisone, insulin, epidermal growth factor and (acidic and basic) fibroblast growth factor (G5[7]), greatly accelerated the expression of the high STX-affinity channels. We screened each of the components of G5 for its ability to affect expression of high STX-affinity Na channels in cultured astrocytes. Hydrocortisone was the only component of G5 that stimulated expression of high STX-affinity Na channels.

Cortisol (1 μM; FIG. 1), when added at 4 DIV, increased ^{3}H-STX binding by about 150% by 8 DIV; cortisol acted without an significant delay. The effects of cortisol were inhibited by actinomycin D (16 nM), a blocker of gene transcription, indicating that cortisol receptor-mediated gene expression is required for high STX-affinity Na channel expression. Three lines of evidence suggest that cortisol is acting via type II glucocorticoid receptors rather than type I mineralocorticoid receptors. First, the effect of cortisol is quantitatively mimicked by the glucocorticoid receptor agonist, dexamethasone (1 μM). Second, the type II glucocorticoid receptor antagonist, RU-38486 (2 μM) inhibited the stimulation of Na channel expression by cortisol (FIG. 1). Third, the stimulation of cortisol was half-maximal at about 10 nM as expected for type II glucorticoid receptors.

Na channel expression was also studied by immunohistochemistry. More than 95% of the cells expressed the astrocyte-specific intermediate filament protein, glial fibrilary acidic protein (GFAP; FIG. 2). In both the absence and presence of cortisol, virtually all of the GFAP-immunoreactive cells were labeled by antibody 7493 (FIG. 2), which is specific for the α subunit of the rat brain Na channel. Thus, cortisol appears to increase the level of Na channel expression in each cell rather than increasing the fraction of cells that express the channels.

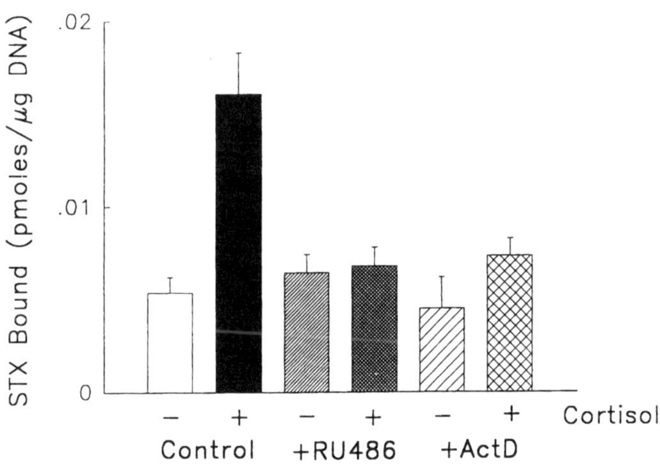

FIGURE 1. Cortisol effects depend on gene transcription and are mediated by type II glucocorticoid receptors. ^{3}H-STX binding to cultured astrocytes was determined by 8 DIV after exposure to CDM in the absence or presence of 10 nM cortisol for 4 days. RU38485 (2 μM), a specific blocker of type II glucocorticoid receptors completely blocked the cortisol-induced increase in high STX-sensitivity Na channel expression. Similar results were obtained with 200 nM of RU38486. Actinomycin D (16 nM), an inhibitor of transcription also completely blocked the cortisol-induced increase in STX binding.

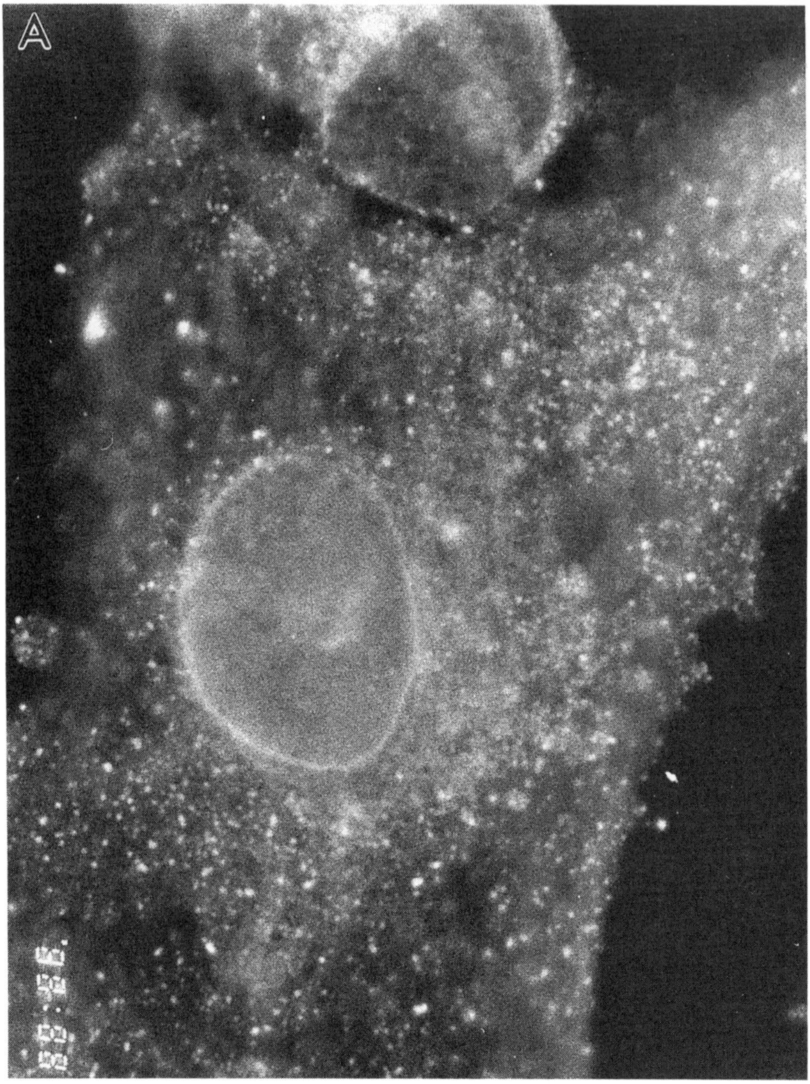

FIGURE 2. Na channel expression in astrocytes detected with antibody 7943. Rat astrocyte cultures on polylysine-coated coverslips were incubated with 1 μM cortisol from 4 to 8 DIV. Cultures were fixed with 4% paraformaldehyde, blocked with 10% normal goat serum in 0.1% Triton X-100/PBS and incubated with either **(A)** antibody 7493 or **(B)** anti-GFAP, which is directed toward an internal epitope of the high STX-affinity rat brain Na channel. Immunoreactivity was labeled with AMCA (GFAP) or fluorescein (Na channel) coupled secondary

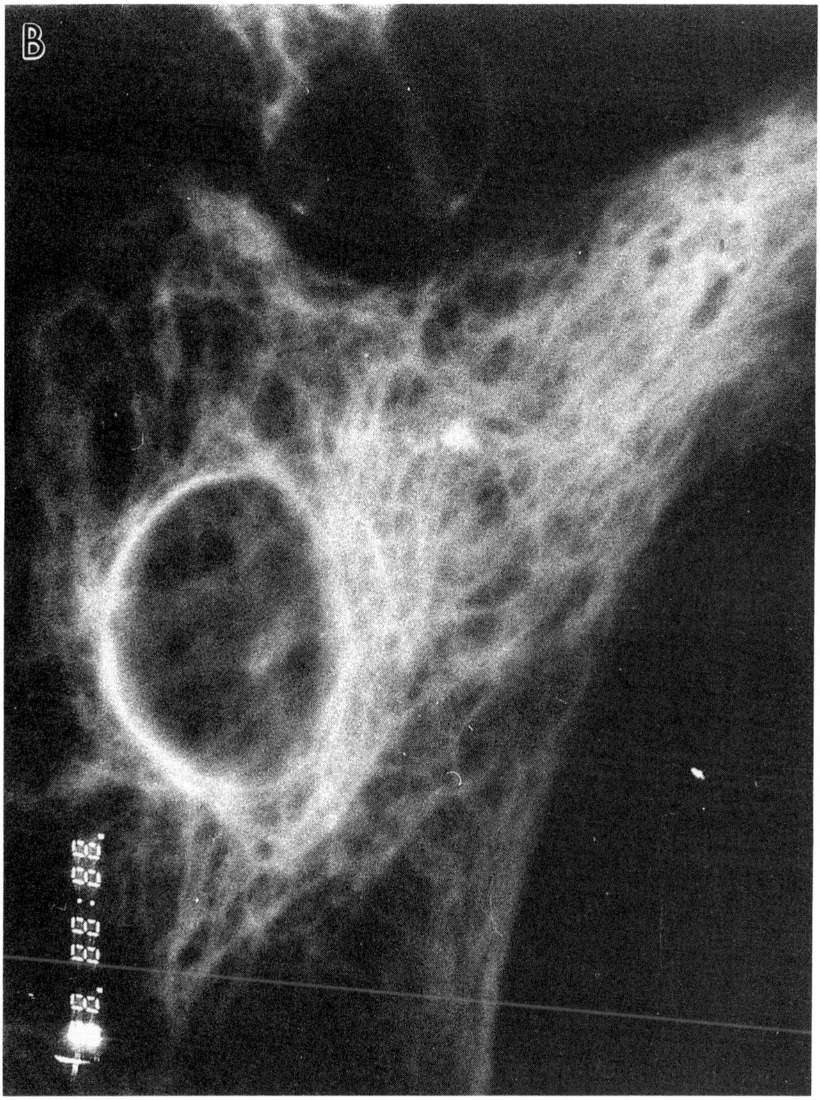

antibodies and visualized by fluorescence microscopy. At least 95% of the cells displayed GFAP-immunoreactive cytoskeletal filaments (B). Virtually all of the GFAP-immunoreactive cells also displayed Na channel immunoreactivity (A), which was distributed primarily in small patches on the cell surface. Some Na channel immunoreactivity appeared to be coincident with GFAP-containing structures suggesting that the channels may preferentially associate with underlying cytoskeletal elements in the astrocytes.

REFERENCES

1. GARBER, S. S., T. HOSHI & R. W. ALDRICH. 1989. J. Neurosci. **9:** 3976–3987.
2. POLLOCK, J. D., M. KREMPIN & B. RUDY. 1990. J. Neurosci. **10:** 2626–2637.
3. BARRES, B. A. 1991. Curr. Opinion Neurobiol. **1:** 354–359.
4. SONTHEIMER, H., B. R. RANSOM, CORNELL-BELL *et al.* 1991. J. Neurophysiol. **65:** 3–19.
5. YAROWSKY, P. J. & B. K. KRUEGER. 1989. J. Neurosci. **9:** 1055–1061.
6. ELMER, L. W., J. A. BLACK, S. G. WAXMAN & K. J. ANGELIDES. 1990. Brain Res. **532:** 222–231.
7. MICHLER-STUKE, A., J. R. WOLFF & J. E. BOTTENSTEIN. 1984. Int. J. Dev. Neurosci. **2:** 575–584.
8. CHOU, Y.-C., W. G. LUTTGE & C. SUMNERS. 1991. Dev. Brain Res. **61:** 55–61.
9. VIELKIND, U., A. WALENCEWICZ, J. M. LEVINE & M. C. BOHN. 1990. J. Neurosci. Res. **27:** 360–373.

Glucocorticoid Regulation of Stress-Induced Mineralocorticoid Receptor Gene Transcription *in Vivo*

JAMES P. HERMAN[a] AND STANLEY J. WATSON[b]

[a]*Department of Anatomy and Neurobiology*
University of Kentucky Medical Center
800 Rose Street
Lexington, Kentucky 40536-0084
and
[b]*Mental Health Research Institute*
University of Michigan
205 Zina Pitcher Place
Ann Arbor, Michigan 48109-0720

The neuronal mineralocorticoid receptor (MR) binds endogenous corticosterone (B) with high affinity.[1] The ligand-activated MR is transported to the cell nucleus, where it is capable of affecting transcription of a wide variety of genes.[2] Within the central nervous system, the MR is of highest abundance within the hippocampus, thus leading to its hypothetical association with hippocampally mediated negative feedback control of the hypothalamo-pituitary-adrenocortical (HPA) stress axis.[3,4]

Understanding regulation of the neuronal MR is critical to elucidating its true function within the nervous system. In the present experiments, we employed a semiquantitative *in situ* hybridization strategy to assess *in vivo* stress regulation of neuronal MR mRNA and gene expression. For characterization of the MR intron probe, brains of normal Sprague-Dawley rats were sectioned at 20 μm and sections through the hippocampus processed for *in situ* hybridization. Adjacent series were hybridized with probes complementary to 1) the 3′ coding and proximal 3′ untranslated region of the MR mRNA (exon probe) and 2) the 5′ portion of a coding region intron lying between exons 2 and 3 (intron probe). The latter probe was designed to recognize unspliced or partially-spliced gene transcripts present in the cell nucleus as heteronuclear (hn) RNA. An additional series was preincubated with RNAse A to verify probe specificity. Sections were dipped in NTB2 liquid emulsion and counterstained with cresyl violet. For analysis of stress and glucocorticoid effects on MR gene and mRNA regulation, rats received either adrenalectomy (ADX) or sham operation (SHAM), and were subsequently implanted with pellets containing either cholesterol (ADX-C and SHAM-C rats) or 30% B/70% cholesterol (ADX-B rats). Acute restraint stress (plastic restraint cages, 60 m) was imposed 5 days following pellet implantation, and groups of animals were killed 60 and 120 m post-stress, with unstressed rats serving as controls. Blood samples from all rats were assayed for B.

Localization of signal generated by MR exon and intron probes are shown in FIGURE 1. MR exon signal was located diffusely throughout the cell (FIG. 1A,B) and showed a characteristic distribution across hippocampal subfields and dentate gyrus (FIG. 1C).[5–7] Cytoplasmic localization was confirmed on Nissl-counterstained sections (data not shown). In contrast, the MR intron probe showed a punctate distribution consistent with nuclear localization (FIG. 1D,E). MR intron signal (FIG. 1F) could be localized to all hippocampal subfields, and was significantly less intense than exon signal at equal exposure times. The localization data indicate 1) compart-

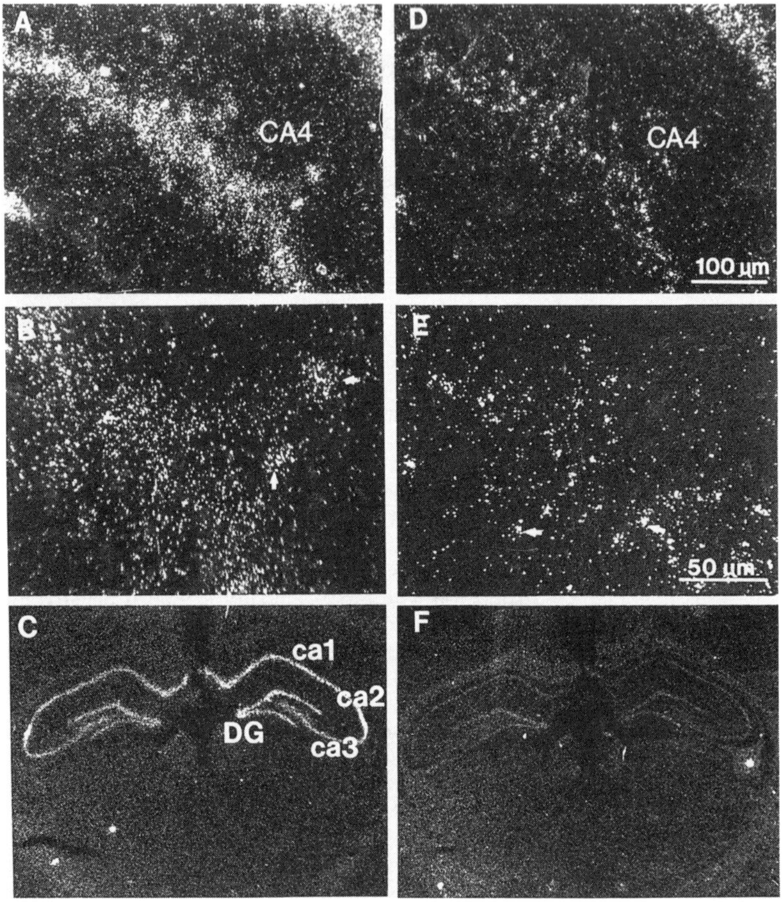

FIGURE 1. Localization of MR mRNA (**A–C**) and hnRNA (**D–F**) in the hippocampal formation of the rat. MR mRNA, as detected by a probe complementary to sequences encoding the protein coding region and 3′-untranslated domain, can be seen as a scattering of grains over hippocampal neurons (**A**). Individual neurons show diffuse labeling over hippocampal cells (**B**). MR mRNA can be localized to all hippocampal subfields and the dentate gyrus (DG) (**C**). MR hnRNA, as detected using a coding region intron probe, can be appreciated as punctate aggregations of grains consistent with nuclear localization (**D,E**). Like MR mRNA, MR hnRNA is localized to all hippocampal subfields, albeit at significantly lower abundance (**F** vs **C**). Note that (**C**) and (**F**) are adjacent sections through the hippocampus of a normal rat, and have been exposed for the same period of time (9 days on Kodak XAR5 film).

mentalization of intron signal within the nucleus, consistent with detection of unspliced or partially spliced transcripts and 2) low resting levels of the intronic sequence, consistent with a high ratio of mature message to primary or partially processed gene transcripts.

The B data verify the efficacy of the stress/replacement regimens. SHAM-C rats showed a predictable stress response (time 0, 40.9 ± 16.7 ng/ml; time 60 m, 284.7 ± 67.2 ng/ml; time 120 min 21.3 ± 11.4 ng/ml). In contrast, ADX-CORT rats showed slightly high basal B levels which did not change in response to stress (time 0, 49.6 ± 18.5 ng/ml; time 60 m, 47.7 ± 7.5 ng/ml; time 120 min 59.2 ± 27.6 ng/ml).

In response to acute stress, MR hnRNA decreased to some 50% of unstressed baseline in SHAM-C rats in subfield CA1, indicative of a rapid inhibition of MR gene transcription following stress in these hippocampal regions (FIG. 2). ADX-B rats, on the other hand, showed no significant change in MR intron hybridization

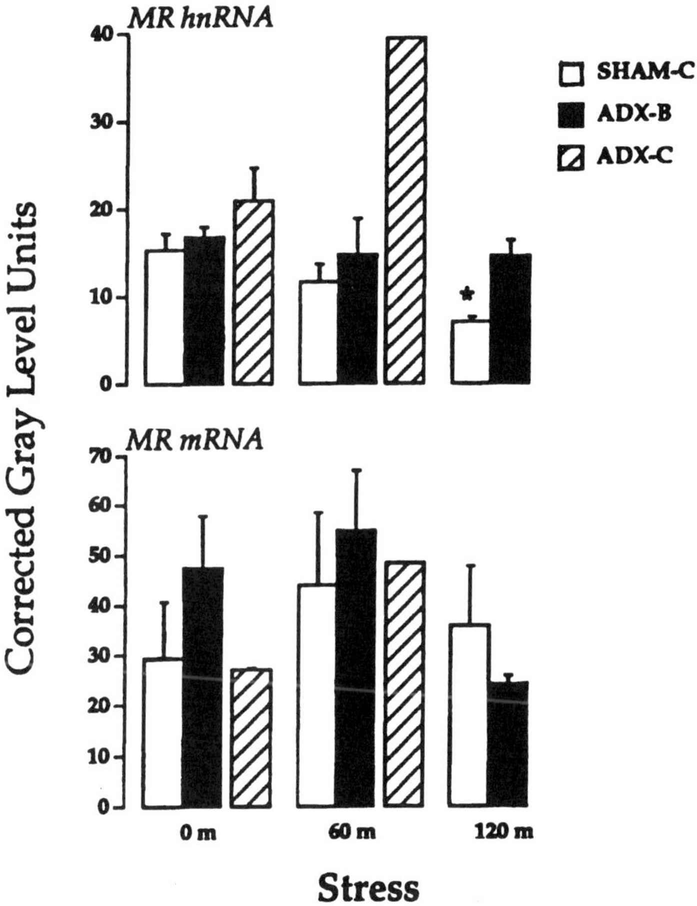

FIGURE 2. Regulation of MR hnRNA and MR mRNA in subfield CA1 of the hippocampus. Two-way ANOVA revealed a significant effect of group on MR hnRNA expression ($p < 0.05$) and a significant interaction effect ($p < 0.05$). Subsequent comparison among means revealed a significant difference between SHAM-C time 0 and SHAM-C time 120, and ADX-B time 120 and SHAM-C time 120. Results indicate that the observed stress-induced decrease in MR hnRNA expression is likely to stem from adrenal glucocorticoid release.

following stress. These results indicate that in the absence of the glucocorticoid flux normally seen following stress, the MR gene does not downregulate. The adrenal stress response is therefore required for normal MR downregulation to occur.

Hybridization with the MR exon probe revealed no significant downregulation by stress in SHAM-C rats or ADX-B rats over the 2-hour post-stress period. These data raise the possibility that steady-state levels of MR mRNA are stable over this time period, resulting in no decrement in message despite decreased transcription, or alternatively that the signal generated by the stress episode acts to stabilize cytoplasmic mRNA. The latter possibility is highlighted by the trend toward a decrease in MR mRNA seen in ADX-B rats at the 2-hour post-stress time point (FIG. 2).

The rapid downregulation of hippocampal MR gene transcription following stress indicates a keen sensitivity of the MR to glucocorticoid fluxes. This observation implies a sensitivity of MR-mediated transcriptional events to stressful information, and further implies a role for the MR in hippocampal functions, including general information processing and possibly HPA regulation. Further analysis is required to determine how transcriptional changes affect receptor protein function and/or selection of MR alternative splicing products.[8]

REFERENCES

1. REUL, J. M. & E. R. DE KLOET. 1985. Endocrinology **117:** 2505–2511.
2. HERMAN, J. P. 1993. Cell. Mol. Neurobiol. **13:** 349–372.
3. DE KLOET, E. R. 1992. Ann. N.Y. Acad. Sci. **663:** 357–371.
4. JACOBSON, L. & R. M. SAPOLSKY. 1990. Endocr. Rev. **12:** 118–134.
5. HERMAN, J. P., P. D. PATEL, H. AKIL & S. J. WATSON. 1989. Mol. Endocrinol. **3:** 1886–1894.
6. ARRIZA, J. L., R. B. SIMERLY, L. W. SWANSON & R. M. EVANS. 1988. Neuron **1:** 887–900.
7. VANEEKELEN, J. A. M., W. JIANG, E. R. DE KLOET & M. C. BOHN. 1988. J. Neurosci. Res. **21:** 88–94.
8. KWAK, S. P., P. D. PATEL, R. C. THOMPSON, H. AKIL & S. J. WATSON. 1993. Endocrinology **133:** 2344–2350.

Subject Index

Index of Contributors

(*Page numbers in italics represent discussion.*)